Inverse Problem Theory

Methods for Data Fitting and Model Parameter Estimation

Inverse Problem Theory

Methods for Data Fitting and Model Parameter Estimation

Albert Tarantola

Institut de Physique du Globe, 4-Place Jussieu, Paris, France

ELSEVIER

Amsterdam — Oxford — New York — Tokyo 1987

ELSEVIER SCIENCE PUBLISHERS B.V.
Sara Burgerhartstraat 25
P.O. Box 211, 1000 AE Amsterdam, The Netherlands

Distributors for the United States and Canada:

ELSEVIER SCIENCE PUBLISHING COMPANY INC.
52, Vanderbilt Avenue
New York, NY 10017, U.S.A.

Library of Congress Cataloging-in-Publication Data

Tarantola, Albert.
Inverse problem theory.

Bibliography: p.
Includes index.
1. Inverse problems (Differential equations)
I. Title.
QA371.T36 1987 515.3'5 87-500
ISBN 0-444-42765-1 (U.S.)

ISBN 0-444-42765-1

Printed in The Netherlands

TABLE OF CONTENTS

PART TWO: GENERAL INVERSE PROBLEMS

PREFACE

Humans were naked worms; yet they had an *internal model* of the world. In the course of time up to the present, this model has been updated many times, following the development of new experimental possibilities (i.e., the developments of their senses) or the development of their intellect. Sometimes the updating has been only quantitative, sometimes it has been qualitative. Inverse problem theory tries to describe the rules human beings should use for quantitative updatings.

Let S represent a physical system (for instance the whole Universe, or a planet, or a quantum particle). Assume that we are able to define a set of *model parameters* which completely describes S . These parameters may not all be directly measurable (for instance, the radius of the Earth's metallic core is not directly measurable). We can operationally define some *observable parameters* whose actual values hopefully depend on the values of the model parameters. To solve the *forward problem* is to predict the values of the observable parameters, given arbitrary values of the model parameters. To solve the *inverse problem* is to infer the values of the model parameters from given observed values of the observable parameters.

The set of observed values usually overdetermines some model parameters while leaving others underdetermined. Schematically, there are two reasons for underdetermination: intrinsic lack of data, and experimental uncertainties. To illustrate the first, consider for instance the problem of estimating the density distribution of matter inside a planet from knowledge of the gravitational field at its surface. It is well known that infinitely many different distributions of matter density give rise to identical exterior gravitational fields (Gauss' theorem), so there is no hope of obtaining a unique solution to the inverse problem using only gravitational data. Additional information has then to be used, such as, for instance, some a priori assumptions on density distribution, or an additional data set, such as seismic observations.

The second reason for underdetermination is uncertainty of knowledge: observed values always have experimental uncertainties, and physical theories allowing the resolution of the forward problem are always approximations of a more complex reality.

Data redundancy can, in general, easily be handled, and present-day methods do not differ essentially from those used, for instance, by Laplace in 1799, who introduced the "least-absolute-values" and the "minimax" criterion for obtaining the "best" solution, or by Legendre in 1801 and Gauss in 1809, who introduced the "least-squares" criterion.

Underdetermination is handled differently by differently thinking schools. Pure mathematicians like to refer to Hadamard's (1902, 1932) definition of "ill-posed problems": a problem is ill-posed if the solution is not unique or if it is not a continuous function of the data (i.e., if to a "small" perturbation of data there corresponds an arbitrarily "large" perturbation of the solution). Examples of ill-posed problems are, for instance: i) the "analytic prolongement" of stationary fields (if a magnetic field of internal origin is given at a height h_1 over the surface of a planet, the problem of calculating the field at h_2 is well-posed if $h_2 \leq h_1$, and is ill-posed if $h_2 > h_1$; ii) the resolution of a diffusion equation (like the heat-transport equation) when the final conditions are given, instead of the initial conditions; iii) the inversion of integral operators (for instance, the typical problem of instrument deconvolution); iv) the resolution of discrete linear systems with a square matrix, if the latter is not regular.

In Hadamard's opinion, ill-posed problems do not have physical sense. General agreement exists today that ill-posed problems have "well-posed extensions" which are very meaningful. These well-posed extensions introduce a priori assumptions as to the unknowns. For example, Tikhonov (1963) assumes some "regularity" properties of the solution, while Franklin (1970) assumes given a priori statistics on the model space.

Some methods of inversion are known as "exact". They concern problems where the data set and the unknown set can be related by an inversible (generally nonlinear) application. Given the application solving the forward problem, the inverse problem consists in discovering the inverse application (in the usual mathematical sense of the word inverse). The whole field of "exact inversion" is neglected in this book, because these methods cannot deal with data uncertainty and data redundancy in a natural manner. They are interesting for solving mathematical inverse problems, not for data interpretation.

Inverse problem theory in the wide sense has been developed by people working with geophysical data. The reason is that geophysicists try to understand the Earth's interior but are doomed to use only data collected at the Earth's surface. Geophysical problems are always underdetermined in some sense, but as geophysical data contain a lot of information, it is worth-while to try to develop methods for extracting it. Since long, such methods have been only empirical. Backus (1970a, 1970b, 1970c) made the first systematic exploration of the mathematical structure of inverse problems. Backus and Gilbert (1967, 1968, 1970) introduced interesting concepts, such as, for instance, that of "model resolution". Their work was at the origin of a very fruitful development of quantitative methods of data interpretation in geophysics.

This book resolutely takes the viewpoint that the most general formulation of Inverse Problems is obtained when using the language of probability calculus, and when using a Bayesian interpretation of probability (Bayes, 1763). Inverse Problem Theory has to be developed from the consideration of uncertainties (either experimental, or in physical laws), and the right (well-posed) question to set is: given a certain amount of (a priori) information on some model parameters, and given an uncertain physical law relating some observable parameters to the model parameters, in which sense should I modify the a priori information, given the uncertain results of some experiments? In my opinion, this in the only approach allowing us to analyze "error and resolution" in the "solution" with a convenient degree of generality, even for nonlinear forward problems.

The techniques used today for solving inverse problems are as multivariate as the problems themselves. One of the purposes of this book is to show that many of the methods used (linear programming, least-squares, maximum likelihood,...) can coherently be described from a few principles, i.e., that it is possible to build a *theory* for inverse problems.

The first part of this book deals exclusively with *discrete* inverse problems with a *finite* number of parameters. Some real problems are naturally discrete, others contain functions of a continuous variable, and can be discretized if the functions under consideration are smooth enough compared to the sampling length, or if the functions can conveniently be described by their development on a truncated basis.

The advantage of a discretized point of view for problems involving functions is that the mathematics are easier. The disadvantage is that some simplifications arising in a general approach can be hidden when using a discrete formulation (discretizing the forward problem and setting a discrete inverse problem is not always equivalent to setting a general inverse problem and discretizing for the practical computations).

The second part of the book deals with general inverse problems, which may contain such functions as data or unknowns. As this general approach contains the discrete case in particular, the separation into two parts corresponds only to a didactical purpose.

Although this book contains a lot of mathematics, it is not a mathematical book. It tries to explain how a method of acquisition of information can be applied to the actual world. Many intuitive arguments are discussed extensively, but not all have yet been justified mathematically. I hope that researchers in the physical sciences will find the compromise acceptable, and that researchers in applied mathematics will find some of the unsolved problems interesting.

Considerable effort has been made so that this book can serve either as a reference manual for researchers neeeding to refresh their memories on a given algorithm, or as a textbook for a course in Inverse Problem Theory.

Albert Tarantola
Paris, October 1986

ACKNOWLEDGEMENTS

I acknowledge all my colleagues from the Institut de Physique du Globe de Paris, for creating the oxygen-rich atmosphere favorable to the growth of this book. In particular, the very long discussions with Georges Jobert and Bernard Valette were of invaluable help.

CHAPTER 1

THE GENERAL DISCRETE INVERSE PROBLEM

Far better an approximate answer to the *right* question,
which is often vague,
than an exact answer to the *wrong* question,
which can always be made precise.

John W. Tukey, 1962.

Central in this chapter is the concept of "state of information" over a parameter set. It is postulated that the most general way of describing a state of information over a parameter set is by defining a probability density over the corresponding parameter space. It follows that the results of the measurements of the observable parameters (data), the a priori information on model parameters, and the information on the physical correlations between observable parameters and model parameters can, all of them, be described using probability densities. The general Inverse Problem can then be set as a problem of "combination" of all this information. Using the point of view developed here, the solution of inverse problems, and the analysis of error and resolution, can be performed in a fully nonlinear way (but perhaps with a prohibitively large amount of computing time). In all usual cases, the results obtained with this method reduce to those obtained from more conventional approaches. All the results of the subsequent chapters are justified by the arguments developed here.

1.1: Model space and Data space

Let S be the *physical system* under study. For instance, S can be a galaxy for an astrophysicist, the Earth for a geophysicist, or a quantum particle for a quantum physicist.

The scientific procedure for the study of a physical system can be (rather arbitrarily) divided into the following three steps.

i) *Parametrization of the system* : discovery of a minimal set of *model parameters* whose values completely characterize the system (from a given point of view).

ii) *Forward modeling* : discovery of the *physical laws* allowing, for given values of the model parameters, of making predictions as to the results of measurements on some *observable parameters.*

iii) *Inverse modeling* : use of the actual results of some measurements of the observable parameters to infer the actual values of the model parameters.

Strong feed-backs exist between these steps, and a dramatic advance in one of them is usually followed by advances in the other two.

While the first two steps are mainly inductive, the third step is mainly deductive. This means that the postulates and rules of thinking that we follow in the two first steps are difficult to make explicit. On the contrary, the mathematical theory of logic (completed with the probability theory) seems to apply quite well to the third step, to which this book is devoted.

1.1.1: The model space

The choice of the model parameters to be used to describe a system is generally not unique.

Example 1: To describe the elastic properties of a solid, it is possible to use the tensor $c^{ijkl}(\mathbf{x})$ of elastic *stiffnesses* relating stress, $\sigma^{ij}(\mathbf{x})$, to strain, $\epsilon^{ij}(\mathbf{x})$, at each point $\mathbf{x}$ of the solid:

$$\sigma^{ij}(\mathbf{x}) = c^{ijkl}(\mathbf{x}) \ \epsilon^{kl}(\mathbf{x}) \ .$$

Alternatively, it is possible to use the tensor $s^{ijkl}(\mathbf{x})$ of elastic *compliances* relating strain to stress:

$$\epsilon^{ij}(x) = s^{ijkl}(x)\ \sigma^{kl}(x) ,$$

where the tensor **s** is the inverse of **c** :

$$c^{ijkl}\ s^{klmn} = \delta^{im}\ \delta^{jn} .$$

The use of stiffnesses or of compliances is completely equivalent, and there is no "natural" choice. ■

A particular choice of model parameters is a *parametrization* of the system. Two different parametrizations are *equivalent* if they are related by a bijection.

Independently of any particular parametrization, it is possible to introduce an abstract space (set) of points, each representing a conceivable "model" of the system. This space is named the *model space* and is denoted by $\aleph$.

For quantitative discussions on the system, a particular parametrization has to be chosen. To define a parametrization means to define a set of experimental procedures allowing, at least in principle, to measure different characteristic of the system. Once a particular parametrization has been chosen, to each point of the model space a set of numerical values is associated, which can be represented by a point in space $\boldsymbol{M}$, isomorphic to a part of $\boldsymbol{R}^n$ ($\boldsymbol{R}$ denotes the real line, and n the number of parameters). The model space $\aleph$ is defined intrinsically, the space $\boldsymbol{M}$ depends on the particular parametrization chosen.

From a mathematical point of view, $\aleph$ is a (nonlinear) *manifold*, and $\boldsymbol{M}$ is a *chart* of $\aleph$.

Example 2: When a nuclear explosion takes place at the Earth's surface, it produces a seismic wave which propagates through the Earth. It is then possible to use the arrival times of the wave at some seismological observatories to estimate the location of the explosion. A "model" is then a particular location for the explosion, i.e., a geometrical point on the surface of the Earth. The model space $\aleph$ can be represented as the surface of a unit sphere (left of Figure 1.1). The model space is defined *intrinsically*, i.e., without reference to any particular system of coordinates. Nevertheless, for numerical computations, it is necessary to represent each point of the model space by its coordinates in a given coordinate system. To define a coordinate system over the model space is to give a *chart* $\boldsymbol{M}$ of the model space. Each model then corresponds to a point (in a part) of $\boldsymbol{R}^2$ (right of Figure 1.1). In this example, the word "chart" can be taken in its etymological sense. By extension, mathematicians call a "chart" any "mapping" between a n-dimensioned nonlinear manifold and a part of $\boldsymbol{R}^n$. It is an abuse of language that a map of the model space is often also named a "model space". The only inconvenience of such abuse of language is that it can lead to the impression that a model space is necessarily a linear space, which is not true in general. For this particular problem, the model space may be furnished with a concept of *distance* between two arbitrary points (which can either be the euclidean length of the straight segment joining the two points through the sphere, or the length of the arc of great circle joining the points on the surface of the sphere). This is an exception rather than the rule, and usual model spaces cannot be furnished with a distance in such a natural way. ■

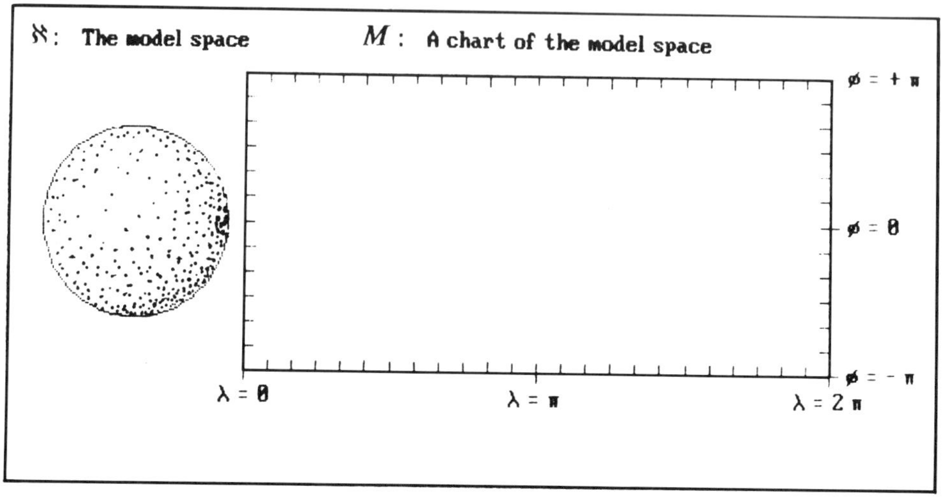

Figure 1.1: The model space ℵ is generally a nonlinear manifold (left of the figure). For numerical computations we introduce a chart of the manifold (right of the figure) which, by language abuse, is also named "model space" (see text for discussion).

The reader interested in the Theory of Differentiable Manifolds may refer, for instance, to Lang (1962), Narasimhan (1968), or Boothby (1975).

We have seen that a chart $\boldsymbol{M}$ of the model space ℵ is "isomorphic" to $\boldsymbol{R}^n$. The difference between a chart (of dimension n) and $\boldsymbol{R}^n$ is that each point of $\boldsymbol{R}^n$ is a set of pure real numbers, i.e., of *dimensionless* quantities. On the contrary, each point of a chart $\boldsymbol{M}$ is a set of n values with given *physical dimensions* (pressure, temperature, electric charge, ...).

Each point of ℵ , or the corresponding point of $\boldsymbol{M}$, is named a *model*, and is represented by m .

Given a model space $\aleph$, when no confusion is possible, and by linguistic abuse, the particular "chart" under consideration will also be named "model space". This corresponds to traditional terminology in inverse problem literature.

The number of model parameters needed for completely describing a system may be either finite or infinite.

Example 3: If the elastic properties of the solid in example 1 effectively vary from point to point, we need an infinite number of values to describe the system completely. If the solid is assumed homogeneous (i.e., if the values of the elastic parameters are independent of $\mathbf{x}$), then 21 parameters suffice for its complete description. ■

Let m^α represent a particular parameter (from a set which can either be finite or infinite). The parameter m^α may take its values in a discrete or in a continuous set. For instance, if m^α represents the mass of the Sun, we can assume a priori that it can take any value from zero to infinity; if m^α represents the spin of a quantum particle, we can assume a priori that it can only take discrete values. We will see later that the use of "delta functions" allows us to consider parameters taking discrete values as a special case of parameters taking continuous values. To simplify the discussions, the terminology used in this book will correspond to the assumption that all the parameters under consideration take their values in a continuous set. If this is not the case in a particular problem, the reader will easily make the corresponding corrections (see in particular problems 1.3 and 1.4).

The theory of infinite dimensional spaces needs a greater technical vocabulary than the theory of finite dimensional spaces. In what follows, and in all the first part of this book, I assume that the model space is *finite-dimensional*. The limitation to systems with finite number of parameters is severe from a mathematical point of view, because the kingdom of functions (even continuous) is infinitely much richer than the kingdom of finite-dimensioned spaces. But, as far as we can accept in some problems that the functions under consideration are *bandlimited*, it is always possible to consider a sampled version of these functions, and in this case there is no difference between the numerical results given by functional and by discrete approaches to Inverse Problem Theory (although the numerical algorithms may differ considerably, as it can be seen by comparing problem 1.2 with problem 7.1).

When a particular parametrization of the system has been chosen, each model can be represented by a particular set of values for the model parameters:

$$\mathbf{m} = \{ m^\alpha \} \qquad (\alpha \in \mathbf{I}_M) \tag{1.1}$$

where I_M represents a discrete finite *index set*. In as far as a particular parametrization of S is interpreted as a choice of coordinate lines over $\aleph$, the variables m^α can be named the *coordinate values* of $\mathbf{m}$ (with respect to the given coordinate lines).

By definition of our terminology, the following are synonymous:

i) to parametrize the physical system S ,

ii) to define a coordinate system over the model space $\aleph$,

iii) to define a chart, M , of $\aleph$.

A chart of $\aleph$ is, by definition, isomorphic to R^n . In particular, a chart is a linear (vector) space. Given a particular chart M , it is then possible to define the *sum of two models*, $\mathbf{m}_1$ and $\mathbf{m}_2$, by the sum of its "components":

$$(\mathbf{m}_1 + \mathbf{m}_2)^\alpha = m_1{}^\alpha + m_2{}^\alpha \qquad (\alpha \in I_M) \tag{1.2}$$

and the *multiplication of a model by a real number* by the multiplication of all its "components":

$$(r\ \mathbf{m})^\alpha = r\ m^\alpha \qquad (\alpha \in I_M)\ (r \in R). \tag{1.3}$$

This justifies the name of "components" given to the coordinates m^α . Nevertheless, it should be emphasized that the previous definitions are *not intrinsic*, in the sense that the sum or the multiplication thus defined *depends* on the particular parametrization chosen for S . It is generally not possible to give an intrinsic definition of $\mathbf{m}_1 + \mathbf{m}_2$ or of $r\ \mathbf{m}$: the model space $\aleph$ is *not* a linear space in general.

Example 4: The homogeneous solid in example 1 can be described by the 21 elastic stiffnesses c^{ijkl} or, alternatively, by the 21 elastic compliances s^{ijkl} . Let $\mathbf{m}_1$ and $\mathbf{m}_2$ represent two different elastic solids. They can be represented either by $c_1{}^{ijkl}$ and $c_2{}^{ijkl}$, or by $s_1{}^{ijkl}$ and $s_2{}^{ijkl}$. The elastic solid defined by $c_1{}^{ijkl} + c_2{}^{ijkl}$ is different from the elastic solid defined by $s_1{}^{ijkl} + s_2{}^{ijkl}$: the sum of two models can not be defined intrinsically. ■

In fact, the theory of inversion can be developed completely without reference to any particular parametrization. We will see below that the only mathematical objects to be defined in order to deal with the most general formulation of inverse problems are *measures* over the model space (in the sense of the mathematical theory of integration). A measure over $\aleph$ is a mapping that, to any subset A of $\aleph$, associates a positive real number:

$$A \rightarrow P(A) \in \boldsymbol{R}^+$$

named the *measure of* A . In general, $P(\aleph) = 1$, and the measure is named a *probability* over $\aleph$. Such measures can, in principle, be defined irrespectively of any particular parametrization of $\aleph$, i.e., independently of any particular chart. But once a particular chart $\boldsymbol{M}$ has been chosen, then it is very easy to describe a probability using a probability density.

In practical applications, we are always faced with a particular parametrization, and so long as it has been astutely chosen, we can forget all the subtleties distinguishing the abstract model space from the corresponding chart, name the last the model space, forget that the linear vector structure is not intrinsic, and go through the computations.

In defining the parametrization of a physical system, it has been said that the parameters have to be defined so as to be measurable "at least in principle". The following example illustrates this notion.

Example 5: The radius of the Earth's metallic core can be defined "experimentally" as the radius of the spherical region of the Earth where the mineral composition is predominantly metallic. Given a sample, we know how to determine its mineralogical composition. To measure the radius of the Earth's core, it "suffices" to make a (quite) deep hole, and analyze the obtained samples. The parameter "radius of the Earth's core" is perfectly defined, although only measurable "in principle". The use of physical laws allows us to predict the behaviour of seismic waves arriving at the surface of the Earth's core. The diffracted-reflected waves are directly observable at the Earth's surface. Thus, inverse problem theory allows us to obtain information on the radius of the core. ■

1.1.2: The data space

To obtain information on model parameters, we have to perform some observations during a physical experiment, i.e., we have to perform a measurement of some observable parameters.

Example 6: As discussed in example 5, for a geophysicist interested in understanding the Earth's deep structure, observations may consist, for instance, in recording a set of seismograms at the Earth's surface. ■

Example 7: For a particle physicist, observations may consist in a mesurement of the flux of particles diffused by a given target at different angles for a given incident particle flux. ■

The task of the experimenter is difficult, not only because he has to perform measurements as accurately as possible, but more essentially because he has to *imagine* new experimental procedures allowing him to measure observable parameters carrying a maximum of information on model parameters. For instance, it may be easy to determine the captain's age (for instance by stealing his passport), but there is little chance that this measurement will carry much information on the number of masts on the ship.

As was the case with model parameters, given experimental equipment, a certain freedom exists in choosing the observable parameters.

Example 8: Given a seismometer, we can choose as "output" a voltage proportional to the displacement of the mass, or to its velocity, or to its acceleration. ■

We thus arrive at the abstract idea of a *data space*, which can be defined as the space of all conceivable instrumental responses. Each particular realization is denoted by $\mathbf{d}$. When a particular choice has been made of the "observable parameters" (in the sense of example 8), a *chart* of the data space has been defined, which will be denoted $\boldsymbol{D}$, and which, by languistic abuse, will also be named "data space". Any conceivable result of the measurements can be written by the "components"

$$\mathbf{d} = \{ d^i \} \qquad (i \in I_D) , \qquad (1.4)$$

where I_D represents a discrete (and finite) index set. $\boldsymbol{D}$ is a linear space with the definitions

$$(\mathbf{d}_1 + \mathbf{d}_2)^i = d_1{}^i + d_2{}^i \qquad (i \in I_D) \qquad (1.5)$$

$$(r\, \mathbf{d})^i = r\, d^i \qquad (i \in I_D) \ (r \in \boldsymbol{R}) . \qquad (1.6)$$

Each vector $\mathbf{d}$ is then named a *data vector*, or *data set.*

1.1.3: The joint space $\boldsymbol{D} \times \boldsymbol{M}$

It is sometimes useful to introduce the product space $\boldsymbol{X} = \boldsymbol{D} \times \boldsymbol{M}$. Its elements are the couples $\mathbf{x} = (\mathbf{d} , \mathbf{m})$. As the elements of $\mathbf{d}$ are termed observable parameters and the elements of $\mathbf{m}$ are termed model parameters, the elements of $\mathbf{x}$ may be called *physical parameters*, or, for short, *parameters.* The space $\boldsymbol{X}$ is then named the *parameter space* .

$\boldsymbol{X}$ can be intuitively interpreted as representing a physical system $\boldsymbol{S}$ *extended* to contain also the measure instruments themselves. This space is much more fundamental than $\boldsymbol{D}$ and $\boldsymbol{M}$ and, in fact, for many problems the separation of $\mathcal{X}$ into a data and a model space may be rather arbitrary.

The components of $\mathbf{x}$ can be viewed as coordinates in the parameter space $\mathcal{X}$. We can arbitrarily define another system of coordinates through a bijection

$$\mathbf{x}^* = \mathbf{x}^*(\mathbf{x}) \qquad\qquad \mathbf{x} = \mathbf{x}(\mathbf{x}^*) \ . \tag{1.7}$$

Such systems of coordinates are named *equivalent* . In section 1.2.3 I introduce the hypothesis that the system of coordinates is *minimal.*

The components of $\mathbf{m}$ are represented by the Greek indexes $\alpha,\beta,...$; the components of $\mathbf{d}$ by the lower-case Latin indexes $i,j,...$. When necessary, the components of $\mathbf{x}$ are represented by upper-case Latin indexes:

$$\mathbf{x} = \{ x^A \} \qquad\qquad (A \in I_X) \ . \tag{1.8}$$

1.1.4: Notations

Authors dealing with discrete inverse problems usually consider a *column-matrix* notation for representing the elements of the model or the data spaces:

$$\mathbf{m} = \begin{bmatrix} m^1 \\ m^2 \\ \cdots \end{bmatrix}$$

$$\mathbf{d} = \begin{bmatrix} d^1 \\ d^2 \\ \cdots \end{bmatrix} .$$

In this book, the word "vector" always means "element of a linear vector space", and it will never be assumed that the natural representation of a vector is by using a column matrix.

Example 9: If the discrete model corresponds to a discretization of a function of *two* variables, say $f(x,y)$, the elements of the model space are naturally represented by

$$\mathbf{m} = \begin{bmatrix} m^{11} & m^{12} & \cdots \\ m^{21} & m^{22} & \cdots \\ \cdots & \cdots & \cdots \end{bmatrix} = \begin{bmatrix} f(x^1,y^1) & f(x^1,y^2) & \cdots \\ f(x^2,y^1) & f(x^2,y^2) & \cdots \\ \cdots & \cdots & \cdots \end{bmatrix}$$

so that the index set $\mathbf{I}_M$ is $N \times N$ (i.e., a set of couples of integers). In that case, a bidimensional array (i.e., a *matrix*) conveniently represents a model, and ranging the components of $\mathbf{m}$ into a column matrix is of no intuitive help (and is never needed in numerical computations). ■

If the components of a vector have to be explicited, the following abstract notation will be used:

$$\mathbf{m} = \{ m^\alpha \} \qquad (\alpha \in \mathbf{I}_M)$$

$$\mathbf{d} = \{ d^i \} \qquad (i \in \mathbf{I}_D) ,$$

which do not assume any particular arrangement.

Let, for instance, $\mathbf{G}$ represent a linear operator from M into D. We write

$$\mathbf{d} = \mathbf{G}\,\mathbf{m} . \tag{1.9a}$$

As $\mathbf{G}$ relates two discrete spaces, it can be shown that there exist constants $G^{i\alpha}$ $(i \in \mathbf{I}_D)$ $(\alpha \in \mathbf{I}_M)$ such that the linear equation (1.9a) can be written

$$d^i = \sum_{\alpha \in \mathbf{I}_M} G^{i\alpha}\, m^\alpha \qquad (i \in \mathbf{I}_D) . \tag{1.9b}$$

In general, the components of a linear operator like $\mathbf{G}$ may be represented as a multidimensional array (not necessarily a "matrix").

Example 10: For the model space of the previous example,

$$d^i = G^{i11}\, m^{11} + G^{i12}\, m^{12} + \dots ,$$

where, in its turn, the index i may be multidimensional. ■

Of course, in as far as the number of components of $\mathbf{m}$ and $\mathbf{d}$ is finite, we can reclass them into column matrices, and linear operators like $\mathbf{G}$ can be represented by ordinary two-dimensional matrices. But the only effect of this is generally to destroy all the symmetries of the problem, and to suggest the use of matricial operations in problems where a slightly more abstract algebra may simplify notations and computations (see, for instance, problem 1.2). For a proper terminology, the array of numerical constants representing a linear operator should be named the *kernel* of the operator, and should not be identified with the operator itself, which is a more fundamental concept (for a given linear operator there are as many different kernels as bases that we may choose in the corresponding linear spaces).

Let ϕ^α (resp. ψ^i) be constants such that $\Sigma\, m^\alpha\, \phi^\alpha$ (resp. $\Sigma\, d^i\, \psi^i$) makes sense (in particular in regard to the homogeneity of physical dimensions), and gives an (adimensional) real number. I define

$$\mathbf{m}^t\, \boldsymbol{\phi} = \sum_{\alpha\in I_M} m^\alpha\, \phi^\alpha \tag{1.10a}$$

$$\mathbf{d}^t\, \boldsymbol{\psi} = \sum_{i\in I_D} d^i\, \psi^i\ . \tag{1.10b}$$

In chapters 4 and 5, $\boldsymbol{\phi}$ (resp. $\boldsymbol{\psi}$) are identified as elements of the *dual* of ***M*** (resp. ***D***).

Given the linear operator $\mathbf{G}$ of equations (1.9), mapping the model space into the data space, the *transpose* of $\mathbf{G}$ is denoted $\mathbf{G}^t$, and is a linear operator defined by the identity

$$\mathbf{m}^t\ (\,\mathbf{G}^t\, \boldsymbol{\psi}\,) = (\,\mathbf{G}\,\mathbf{m}\,)^t\ \boldsymbol{\psi}\ , \tag{1.11a}$$

valid for any $\mathbf{m}$ and $\boldsymbol{\psi}$. In chapters 4 and 5, $\mathbf{G}^t$ is identified as an operator mapping the dual of the data space into the dual of the model space. As $\mathbf{G}^t$ maps two discrete spaces, there exist constants $(\mathbf{G}^t)^{\alpha i}$ such that

$$(\mathbf{G}^t\, \boldsymbol{\psi})^\alpha = \sum_{i\in I_D} (\mathbf{G}^t)^{\alpha i}\ \psi^i\ ,$$

and it can then easily be shown from the general definition (1.11a) that

$$(\mathbf{G}^t)^{\alpha i} = (\mathbf{G})^{i\alpha}\ . \tag{1.11b}$$

In the particular case of matricial kernels, this last formula corresponds to the usual definition of matrix transposition.

The *inverse* of a linear operator, if it exists, is introduced by the usual definition. For instance, the operator $\mathbf{S}$ with components $S^{\alpha\beta}$ $(\alpha\in I_M)$ $(\beta\in I_M)$ is the inverse of $\mathbf{Q}$, with components $Q^{\alpha\beta}$ if

$$\sum_{\beta\in I_M} S^{\alpha\beta}\ Q^{\beta\gamma} = \sum_{\beta\in I_M} Q^{\alpha\beta}\ S^{\beta\gamma} = \delta^{\alpha\gamma}\ , \tag{1.12a}$$

where $\delta^{\alpha\gamma}$ represents the Kronecker's symbol (1 if $\alpha=\gamma$, 0 otherwise). We then write

$$S = Q^{-1} \qquad\qquad Q = S^{-1} . \tag{1.12b}$$

With the notations introduced by equations (1.10), (1.11), and (1.12), general linear equations look like ordinary matricial equations, but keep a much more general sense.

1.2: States of Information

1.2.1: The mathematical concept of probability

Let X represent an arbitrary set. By definition, a measure over X is a rule that to any subset A of X a real positive number $P(A)$ is associated, named the *measure of* A and satisfying the two properties

i) If $\emptyset$ represents the empty set, then

$$P(\emptyset) = 0 . \tag{1.13}$$

ii) If $A_1, A_2, \ldots$ represents a disjoint sequence of sets of X, then

$$P\left(\sum_i A_i \right) = \sum_i P(A_i) . \tag{1.14}$$

$P(X)$ is not necessarily finite. If it is, then P is termed a *probability* (or *probability measure*) over X . In that case, P is usually normalized to unity: $P(X) = 1$.

Example 11: Let X be the set {HEAD, TAIL}. Setting

$$P(\emptyset) = P(\text{neither HEAD nor TAIL}) = 0 ,$$

$$P(\text{HEAD}) = r ,$$

$$P(\text{TAIL}) = 1 - r ,$$

and

$$P(\text{HEAD} \cup \text{TAIL}) = P(\text{HEAD or TAIL}) = 1 ,$$

where r is a real number $0 \le r \le 1$, defines a probability over X . ■

Example 12: Let X be the surface of a sphere, π_0 a particular point on the surface, and H_0 the hemisphere centered at π_0. To any subset A of points on the surface of the sphere, a number $P(A)$ proportional to the surface of the set of points lying on H_0 is associated. This defines a probability over X. This probability has been defined *independently* of any choice of coordinates over the surface. ■

Let P be a measure over a nonlinear manifold X. Assume that a particular coordinate system has been chosen over X. Let $\mathbf{x} = \{x^1, x^2, ...\}$ denote the coordinates of a point. If a function $f(\mathbf{x})$ exists such that for any $A \subset X$,

$$P(A) = \int_A d\mathbf{x}\ f(\mathbf{x}) , \tag{1.15}$$

where, for short, the following notation has been used

$$\int_A d\mathbf{x} \equiv \int dx^1 \int dx^2 \ ... \qquad \text{(over } A\text{)} ,$$

then $f(\mathbf{x})$ is termed the *measure density function* representing P (with respect to the given coordinate system). If P is a probability (i.e., if $P(X)$ is finite), then $f(\mathbf{x})$ is termed a *probability density function.*

Example 13: Let X, π_0, and H_0 be as defined in the previous example. Let us consider a particular choice of coordinates over the surface of the sphere, such as for instance spherical coordinates (θ,ϕ). Letting $f(\theta,\phi)$ be a function defined by

$$f(\theta,\phi) = \sin\theta \qquad \text{on } H_0$$
$$f(\theta,\phi) = 0 \qquad \text{outside } H_0 ,$$

and as the surface element over the surface on a sphere is $dS = \sin\theta\ d\theta\ d\phi$, the probability defined in the previous example can be written

$$P(A) = \int d\theta \int d\phi\ \frac{f(\theta,\phi)}{4\pi} \qquad \text{(over } A\text{)} .$$

The function $f(\theta,\phi)/4\pi$ defines a *probability density* over the surface of the sphere. Let (u,v) be a new choice of coordinates over the sphere. To the *same* probability P there corresponds a new probability density $g(u,v)$ which is, in general, *different* from the old one. ■

Example 14: Let X be the positive part of the real line, $X = R^+ = (0,+\infty)$, and let $f(x)$ be the function $1/x$. The integral (1.15) then defines a measure over X , but not a probability (because $P(R^+) = \infty$). The function $f(x)$ is then a *measure density* but not a probability density. ■

To develop our theory, we will effectively need to consider non-normalizable measures (i.e., measures which are not a probability). These measures cannot describe the probability of a given subset A of the parameter space under consideration: they can only describe the *relative probability* of two subsets A_1 and A_2 . We will see that this is sufficient for our needs. To simplify the discussion, I will use the languistic abuse of naming "probability" an arbitrary measure.

If a measure is normalizable (i.e., if it is a probability), it is immaterial whether it has effectively been normalized or not. In this book, two probabilities P_1 and P_2 which are proportional,

$$P_1(A) \equiv r\ P_2(A) \quad (\text{for any } A\ ) \quad (\text{r given a positive real number}), \tag{1.16}$$

are identified. This gives much lighter notations, because the probability densities under consideration do not need to be systematically normalized. Note that the constant r in (1.16) has to be adimensional.

To allow more generality to the notations, I assume that the "functions" representing probability densitites are, in fact, distributions, i.e., generalized functions containing in particular Dirac's "delta function".

It should be noticed that, as a probability is a real number, and as the components $x^1,x^2,...$ in general have physical dimensions, the physical dimension of a probability density is a *density* of the considered space, i.e., it has as physical dimensions the inverse of the physical dimensions of the volume element of the considered space.

Example 15: Let v be a velocity and m a mass. The respective physical dimensions are $L\ T^{-1}$ and M. Let $f(v,m)$ be a probability density on (v,m). For the probability

$$P(\ v_1 \le v \le v_2 \text{ and } m_1 \le m \le m_2\ ) = \int_{v_1}^{v_2} dv \int_{m_1}^{m_2} dm\ f(v,m)$$

to be a real number, the physical dimensions of f have to be $M^{-1}\ L^{-1}\ T$. ■

Let P be a probability over X , and $f(x)$ be the probability density representing P in a given coordinate system. Let

$$\mathbf{x}^* = \mathbf{x}^*(\mathbf{x}) \tag{1.17}$$

represent a *change of coordinates* over X, and let $f^*(\mathbf{x}^*)$ be the probability density representing P in the new coordinates:

$$P(A) = \int_A d\mathbf{x}^* \; f^*(\mathbf{x}^*) \; .$$

By definition of $f(\mathbf{x})$ and $f^*(\mathbf{x}^*)$, for any $A \subset X$

$$\int_A d\mathbf{x} \; f(\mathbf{x}) = \int_A d\mathbf{x}^* \; f^*(\mathbf{x}^*) \; ,$$

using the elementary properties of the integral, the following important property can be deduced

$$f^*(\mathbf{x}^*) = f(\mathbf{x}) \left| \frac{\partial \mathbf{x}}{\partial \mathbf{x}^*} \right| \; , \tag{1.18}$$

where $|\partial \mathbf{x}/\partial \mathbf{x}^*|$ represents the absolute value of the Jacobian of the transformation.

Let $\mathbf{y}$ and $\mathbf{z}$ be two vector parameter sets, and let $f(\mathbf{y},\mathbf{z})$ be a normalized probability density. Two definitions are important: the *marginal probability density* for $\mathbf{y}$,

$$f_Y(\mathbf{y}) = \int_X d\mathbf{x} \; f(\mathbf{x},\mathbf{y}) \; , \tag{1.19a}$$

and the *conditional probability density* for $\mathbf{x}$ given $\mathbf{y} = \mathbf{y}_0$,

$$f_{X|Y}(\mathbf{x}|\mathbf{y}_0) = \frac{f(\mathbf{x},\mathbf{y}_0)}{\displaystyle\int_X d\mathbf{x} \; f(\mathbf{x},\mathbf{y}_0)} \; . \tag{1.19b}$$

From these definitions it follows that the joint probability density equals the conditional probability density times the marginal probability density:

$$f(\mathbf{x},\mathbf{y}) = f_{X|Y}(\mathbf{x}|\mathbf{y}) \; f_Y(\mathbf{y}) \; , \tag{1.20a}$$

and the *Bayes theorem*:

$$f_{Y|X}(\mathbf{y}|\mathbf{x}) = \frac{f_{X|Y}(\mathbf{x}|\mathbf{y})\, f_Y(\mathbf{y})}{\int_{\mathbf{Y}} d\mathbf{y}\, f_{X|Y}(\mathbf{x}|\mathbf{y})\, f_Y(\mathbf{y})} . \tag{1.20b}$$

The Bayes theorem is a mathematical tautology, which cannot be applied to solve real world problems, unless physical postulates attach a particular interpretation to the probability densities and associated marginal and conditional probabilities.

1.2.2: The interpretation of a probability

It is possible to associate more than one intuitive meaning to any mathematical theory. For instance, the axioms and theorems of a three-dimensional vector space can be interpreted as describing the physical properties of the sum of forces acting on a point material particle, as well as describing the physiological sensations produced in our brain when our retina is excited by a light composed by a mixing of the three fundamental colors (e.g., Feynmann, 1963). Hofstadter (1979) gives a lot of examples of different valid intuitive meanings that can be associated with a given formal system.

There are two different usual intuitive interpretations of the axioms of probability as introduced in the previous section.

The first interpretation is purely statistical: when some physical "random" process takes place it leads to a given "realization". If a great number of realizations have been observed, these can be described in terms of "probabilities", which follow the axioms of the previous section. The physical parameter allowing of describing the different realizations is termed a *random variable.* The mathematical theory of statistics is the natural tool for analyzing the outputs of a random process.

Example 16: After one million throws of a biased coin, I have observed $6.2\ 10^5$ HEADS and $3.8\ 10^5$ TAILS. This gives the probability

$$P(\emptyset) = P(\text{neither HEAD nor TAIL}) = 0 ,$$

$$P(\text{HEAD}) = 0.62 ,$$

$$P(\text{TAIL}) = 0.38 ,$$

and

$$P(\text{HEAD} \cup \text{TAIL}) = P(\text{HEAD or TAIL}) = 1 . \blacksquare$$

The second interpretation is in terms of *subjective degree of knowledge* of the "true" value of a given physical parameter. By "subjective" is meant that it represents the knowledge of a given individual, obtained using rigorous scientific (objective) methods, but that this knowledge may vary from individual to individual because each may possess different data sets.

Example 17: What is the radius of the Earth's metallic core? Nobody knows exactly. But with the increasing accuracy of seismic measurements, the *information* we have on this parameter continuously improves. The opinion maintained in this book is that the more general (and scientifically rigorous) answer it is possible to give at any moment to that question is found by defining a rule giving the probability of the true value of the radius of the Earth's core being within r_1 and r_2 for any couple of values r_1 and r_2 . That is to say, the most general answer consists in the definition of a *probability* over the physical parameter representing the radius of the core. ■

This subjective interpretation of the postulates of the probability theory is usually named *Bayesian*, in honor of Bayes (1763). It is not in contradiction with the statistical interpretation. It simply applies to different situations.

One of the difficulties of the approach is that, given a state of information on a set of physical parameters, it is not always easy to decide which probability "models" it best. I hope that the examples in this book will help to show that it is possible to use some common sense rules to give an adequate solution to this problem.

I set explicitly the following postulate:

> Let $\mathcal{X}$ be a discrete parameter space
> with a finite number of parameters.
> The most general way we have
> for describing any ***state of information*** on $\mathcal{X}$
> is by defining a **probability** (in general, a measure) over $\mathcal{X}$.

Let P denote the probability corresponding to a given state of information on $\mathcal{X}$, and $f(\mathbf{x})$ the associaled probability density:

$$P(A) = \int_A d\mathbf{x} \; f(\mathbf{x}) \qquad \text{for any } A \subset \mathcal{X}.$$

The probability $P(\cdot)$ or the probability density $f(\cdot)$ are said to *represent* the corresponding state of information.

For practical applications of the probability theory, it is also necessary to give an intuitive content to the definition of marginal probability. Let $f(\mathbf{y},\mathbf{z})$ be the probability density representing a certain state of information on the parameters $(\mathbf{y},\mathbf{z})$. From the definition in the preceding section and the previous discussion, we can see that all the information on the parameters $\mathbf{y}$ is contained in the marginal probability density $f_Y(\mathbf{y}) = \int dz\ f(\mathbf{y},\mathbf{z})$. The probability density $f(\mathbf{y},\mathbf{z})$ does not contain more information on $\mathbf{y}$; it contains only information about the "correlations" between the parameters $\mathbf{y}$ and $\mathbf{z}$.

Box 1.1: Central estimators and estimators of dispersion

a) One-dimensional case. Given a normalized one-dimensional probability density function $f(x)$, consider the expression

$$s_p(m) = \left\{ \int_{-\infty}^{+\infty} dx \ |x - m|^p \ f(x) \right\}^{1/p} . \tag{1}$$

For given p , the value of m which makes s_p minimum is termed the *center of* $f(x)$ *in the* ℓ_p *norm sense*, and is denoted by m_p. The value m_1 is termed the *median*, m_2 the *mean* (or *mathematical expectation*), and m_∞ the *mid-range*. The following properties hold:

median (minimum ℓ_1 norm):

$$\int_{-\infty}^{+\infty} dx\ |x-m_1|\ f(x) \ \text{minimum} \ \Leftrightarrow \ \int_{-\infty}^{m_1} dx\ f(x) = \int_{m_1}^{+\infty} dx\ f(x) = \frac{1}{2}$$

mean (minimum ℓ_2 norm):

$$\int_{-\infty}^{+\infty} dx\ (x-m_2)^2\ f(x) \ \text{minimum} \ \Leftrightarrow \ m_2 = \int_{-\infty}^{+\infty} dx\ x\ f(x)$$

mid-range (minimum ℓ_∞ norm):

$$\lim(p\to\infty) \int_{-\infty}^{+\infty} dx\ |x-m_\infty|^p\ f(x) \ \text{minimum} \ \Leftrightarrow \ m_\infty = \frac{x_{sup} + x_{inf}}{2} ,$$

where x_{sup} (resp. x_{inf}) is the maximum (resp. minimum) value of x for which $f(x) \neq 0$.

(...)

The value of $s_p(m)$ at the minimum is termed the *dispersion of* $f(x)$ *in the* ℓ_p *norm sense*, and is denoted by σ_p :

$$\sigma_p = s_p(m_p) . \tag{2}$$

The value σ_1 is termed the *mean deviation* , σ_2 the *standard deviation*, and σ_∞ the *half-range* . The following properties hold:

mean deviation (minimum ℓ_1 norm):

$$\sigma_1 = \int_{-\infty}^{+\infty} dx\ |x-m_1|\ f(x) \quad \Leftrightarrow \quad \sigma_1 = \int_{m_1}^{+\infty} dx\ x\ f(x) - \int_{-\infty}^{m_1} dx\ x\ f(x)$$

standard deviation (minimum ℓ_2 norm):

$$\sigma_2{}^2 = \int_{-\infty}^{+\infty} dx\ (x-m_2)^2\ f(x) \quad \Leftrightarrow \quad \sigma_2{}^2 = \int_{-\infty}^{+\infty} dx\ x^2\ f(x) - m_2{}^2$$

half-range (minimum ℓ_∞ norm):

$$\sigma_\infty = \lim(p\to\infty) \left[\int dx\ |x-m_\infty|^p\ f(x) \right]^{1/p} \quad \Leftrightarrow \quad \sigma_\infty = \frac{x_{sup} - x_{inf}}{2}$$

b): Multidimensional case. Given a probability density function $f(\mathbf{x})$ defined for the vector variable $\mathbf{x} = \{x^i \ ; \ i \in I_X\}$, consider the operator $\mathbf{C}_2(\mathbf{m})$ defined by its components

$$C_2{}^{ij}(\mathbf{m}) = \int_{-\infty}^{+\infty} d\mathbf{x}\ (x^i-m^i)\ (x^j-m^j)\ f(\mathbf{x}) . \tag{3}$$

The vector $\mathbf{m}_2$ which minimizes the diagonal elements of $\mathbf{C}_2(\mathbf{m})$ is termed the *mean* (or mathematical expectation) of $\mathbf{x}$ *in the* ℓ_2 *norm sense.* It is given by

$$\mathbf{m}_2 = \int_{-\infty}^{+\infty} d\mathbf{x}\ \mathbf{x}\ f(\mathbf{x}) .$$

The value at $\mathbf{m} = \mathbf{m}_2$ of the operator (3) is termed the *covariance* of $\mathbf{x}$ *in the* ℓ_2 *norm sense*, and is simply denoted by $\mathbf{C}_2$:

(...)

$$\mathbf{C}_2 = \mathbf{C}_2(\mathbf{m}_2) . \tag{4}$$

The diagonal elements of $\mathbf{C}_2$ clearly equal the variances (square of standard deviations) previously defined:

$$C_2^{ii} = (\sigma_2^{i})^2 .$$

The covariance operator in the ℓ_2 norm sense (or *ordinary* covariance operator) has the following properties (see, for instance, Pugachev 1965):

i) $\mathbf{C}_2$ is symmetric:

$$C_2^{ij} = C_2^{ji} .$$

ii) $\mathbf{C}_2$ is definite nonnegative: for any vector $\mathbf{x}$,

$$\mathbf{x}^t \; \mathbf{C}_2^{-1} \; \mathbf{x} \geq \mathbf{0} .$$

iii) if $\mathbf{C}_2$ is positive definite, then, for any vector $\mathbf{x}$, the quantity

$$\| \mathbf{x} \|_2 = (\mathbf{x}^t \; \mathbf{C}_2^{-1} \; \mathbf{x})^{1/2}$$

has the properties of a norm. It is termed the *weighted* ℓ_2 *norm* of the vector $\mathbf{x}$.

iv) The *correlation coefficients* ρ_2^{ij} defined by

$$\rho_2^{ij} = \frac{C_2^{ij}}{\sigma_2^{i} \, \sigma_2^{j}}$$

have the property

$$-1 \leq \rho_2^{ij} \leq +1 .$$

v) The probability density

$$f(\mathbf{x}) = \left((2\pi)^N \; \det \mathbf{C}_2 \right)^{-1/2} \; \exp\left[-\frac{1}{2} \; (\mathbf{x} - \mathbf{x}_0)^t \; \mathbf{C}_2^{-1} \; (\mathbf{x} - \mathbf{x}_0) \right] ,$$

where N is the dimension of the vector $\mathbf{x}$, is normalized, with a mean value $\mathbf{x}_0$, and covariance operator $\mathbf{C}_2$ (e.g., Dubes, 1968). From the results of problem 1.15 it follows that among all the probability densities with given ℓ_2 norm covariance operator, the Gaussian function has (...)

minimum information content (i.e., it has maximum "spreading").

The discussion of the multidimensional spaces has been limited to the ℓ_2 norm case. It is not clear at present which is the right generalization of these concepts to the general ℓ_p norm case.

1.2.3: The state of Perfect Knowledge

If we definitely know that the true value of $\mathbf{x}$ is $\mathbf{x} = \mathbf{x}_0$, it is clear that the corresponding probability density is

$$f(\mathbf{x}) = \delta(\mathbf{x} - \mathbf{x}_0) , \tag{1.21}$$

where $\delta(.)$ represents Dirac's delta function. The probability density (1.21) gives null probability to $\mathbf{x} \neq \mathbf{x}_0$, and probability 1 to $\mathbf{x} = \mathbf{x}_0$. The use of such a state of knowledge does not make sense in itself, because all our knowledge of the real world is subjected to uncertainties, but it is often justified when a certain type of error is negligible *compared* to another type of error (see for instance section 1.5.3).

1.2.4: The state of Total Ignorance (or the reference state of information)

Given a parameter set, it is useful to define a certain state of information which represents a *reference* state of information: in some sense the state of "lowest" information. The probability representing this particular state of information is termed *non-informative*, and is represented by $M(\cdot)$. The associated probability density is denoted $\mu(\cdot)$:

$$M(A) = \int_A d\mathbf{x} \;\; \mu(\mathbf{x}) ,$$

and is termed the *non-informative probability density*.

Example 18: Assume that our problem is the estimation of the spatial location of some event. We can intuitively accept that the non-informative probability gives equal probabilitites of containing the event to all regions of the space with equal volume. Using, for instance, cartesian coordinates (x,y,z) , the volume element of the space is

$$dV = dx\, dy\, dz .$$

The requirement

$$P(V) = \iiint_V dx\, dy\, dz \;\; \mu(x,y,z) \qquad \text{proportional to } V \tag{1.22}$$

shows that the non-informative probability density for a location is, in cartesian coordinates,

$$\mu(x,y,z) \;=\; \text{const} \; . \tag{1.23}$$

It is of course possible to choose other systems of coordinates to represent a spatial location. Using, for instance, spherical coordinates (r,θ,ϕ) , equation (1.22) becomes

$$P(V) = \iiint_V dr\, d\theta\, d\phi \;\; \mu^*(r,\theta,\phi) \qquad \text{proportional to } V \tag{1.24}$$

and, as the volume element of the space is, in spherical coordinates,

$$dV \;=\; r^2 \sin\theta \; dr\, d\theta\, d\phi \; ,$$

we deduce that the non-informative probability density for a location is, in spherical coordinates,

$$\mu^*(r,\theta,\phi) \;=\; \text{const}\; r^2 \sin\theta \tag{1.25}$$

(this last result can also be directly obtained from (1.23) using (1.18)). This example shows that there is no intuitive reason for assuming that a non-informative density function has to be uniform. It also shows that there may exist a particular choice of parameters for which it is uniform. ■

All situations are not so obvious, and explicit notions of invariance may have to be invoked in order to define the non-informative probability density.

Example 19: Let $\mathbf{v}$ be the velocity of a non-relativistic particle:

$$\mathbf{v} \;=\; \frac{d\mathbf{r}}{dt} \; ,$$

where $\mathbf{r}$ denotes the spatial position of the particle, and t a Newtonian time. Let v denote the euclidean norm of $\mathbf{v}$:

$$v \;=\; \frac{\| d\mathbf{r} \|}{| dt |} \; .$$

Using cartesian coordinates, we have

$$v = \frac{\sqrt{dx^2+dy^2+dz^2}}{|dt|} .$$

Let $\mu(v)$ denote the non-informative probability density for v .

Assume that a second observer uses another Galilean coordinate system (x^*,y^*,z^*,t^*) . It is then related with the previous one by a change of origin and of scale,

$$x^* = x_0 + a\,x \ , \qquad y^* = y_0 + a\,y \ , \qquad z^* = z_0 + a\,z \ ,$$

and

$$t^* = t_0 + b\,t \ ,$$

where a and b are some constants. Let $\mu^*(v^*)$ be the non-informative probability density for the second observer. The postulate of space-time homogeneity implies that the two coordinate systems have to be equivalent. In particular, $\mu(\cdot)$ and $\mu^*(\cdot)$ have to be the same function, i.e., for any w,

$$\mu(w) = \mu^*(w) . \qquad (1.26)$$

We have

$$v^* = \frac{\sqrt{dx^{*2}+dy^{*2}+dz^{*2}}}{|dt^*|} = c\,v \ ,$$

where $c = |a/b|$. Using (1.18),

$$\mu(v) = \mu^*(v^*) \left| \frac{dv^*}{dv} \right| = c\ \mu^*(c\,v) .$$

Condition (1.26) then gives

$$\mu(v) = c\ \mu(c\,v) \ ,$$

i.e.,

$$\mu(v) = \frac{\text{const}}{v} . \qquad (1.27)$$

It should be noticed that (1.27) defines a measure density which is not a probability density (it is not normalizable). ■

Example 20: Assume that an observer prefers to use slowness $n = 1/v$ instead of velocity. Which is the function $\mu^*(n)$ representing the non-informative probability density for n ?

It is possible here to follow exactly the same argument as in the previous example. More simply, using the properties of the change of variables (equation 1.18), we directly obtain

$$\mu^*(n) = \mu(v) \left| \frac{dv}{dn} \right| = \frac{const}{n} . \qquad (1.28)$$

It is interesting to note that in addition to the invariance of form of μ with respect to a change of space-time coordinates, we also have invariance of form with respect to the choice of *reciprocal parameters* (the choice $\mu(v) =$ const would not be consistent with the choice $\mu^*(n) =$ const) . ■

There is a lot of controversy in the literature as to the possibility of effectively defining non-informative probability densities. Jaynes (1968), for instance, suggests that for a given definition of the physical parameters $\mathbf{x}$ it is possible to find a *unique* density function $\mu(\mathbf{x})$ which has the strong property of being *form invariant* under the transform groups which leave the fundamental equations of physics invariant. He then suggests taking such a density function as the non-informative one. Additional discussion can be found in Box and Tiao (1973), Rietsch (1977), or Savage (1954, 1962).

In the rest of this book I assume that, for any parameter set, it is possible to exhibit some probability density (or measure density) which, by consensus, will be termed the *reference probability density*, and which some (as myself) will call the *non-informative probability density* .

The parameterizations for which the probability density is uniform ($\mu(\mathbf{x})$ = const) play an important practical role. This justifies the following definition: *a parameter set* $\mathbf{x}$ *is termed* ***cartesian*** *if the corresponding non-informative probability is represented by a uniform probability density.*

Example 21: The cartesian (in the ususal sense) coordinates of example 18. ■

Example 22: The non-informative probability density for a velocity has been obtained in example 19:

$$\mu(v) = \frac{const}{v} .$$

Introducing the "log-velocity" v^* by

$$v^* = \alpha \ \mathrm{Log}\left(\frac{v}{v_0} \right) ,$$

where v_0 is an arbitrary fixed velocity, and α an arbitrary constant, gives the non-informative probability density

$$\mu^*(v^*) = \mu(v) \left| \frac{dv}{dv^*} \right| = \frac{const}{v} \frac{v}{\alpha} = \frac{const}{\alpha} = const ,$$

thus showing that the log-velocity is a cartesian parameter, while the velocity is not. ■

Example 23: The slowness $n = 1/v$ of example 20 is not a cartesian parameter. But the log-slowness

$$n^* = \alpha \operatorname{Log}\left(\frac{n}{n_0} \right)$$

is. ■

Examples 19 and 20 suggest that the probability density

$$f(x) = 1/x$$

plays an important role in practical applications. As taking the logarithm of the parameter

$$x^* = \alpha \operatorname{Log}\left(\frac{x}{x_0} \right)$$

transforms the probability density into a uniform one,

$$f^*(x^*) = const ,$$

the function $1/x$ will be termed the *log-uniform probability density* . It is shown in box 1.3 to be a particular case of the log-normal probability density. As discussed by Jeffreys (1939, 1957), parameters with a log-uniform non-informative probability density are characterized by being *positive* by definition (a temperature T , a density of matter ρ , a wave-celerity v ,...). Their reciprocal parameters ($\beta = 1/kT$, lightness of matter $\ell = 1/\rho$, slowness $n = 1/v$,...) can naturally be introduced, and also have a log-uniform non-informative probability density.

It should be mentioned that, although no coherent inverse theory can be set without the introduction of the non-informative probability, generally it does not play an important role, and, except in highly degenerated problems, numerical inverse results do not critically depend on the particular form of $\mu(\mathbf{x})$.

There are two ways of defining a physical parameter. It can be defined *operationally*, or it can be defined *mathematically* as a function of other parameters already defined. In the latter case, the equation relating the parameters is *not* a physical law. I postulate that if a parameter set contains only *independently defined* (i.e., operationally defined) parameters, the non-informative joint probability density $\mu(\mathbf{x})$ is given by

$$\mu(\mathbf{x}) = \prod_A \mu_A(x^A) \qquad (A \in \mathbf{I}_X) \tag{1.29}$$

where $\mu_A(x^A)$ is the non-informative probability density for the parameter x^A. Such a parameter set is termed *minimal*. Unless otherwise stated, I assume that this is always the case (this avoids, for instance, having, say, a velocity, *and* the corresponding slowness *defined* by $n = 1/v$ in a parameters set).

Box 1.2: Generalized Gaussian

As shown in problem 1.15, among all the normalized probability densities $f(x)$ with fixed ℓ_p norm estimator of dispersion,

$$\int_{-\infty}^{+\infty} dx \; | \, x - x_0 \, |^p \; f(x) = (\sigma_p)^p ,$$

the one with *minimum information content* (i.e., with maximum "spreading") is given by

$$f_p(x) = \frac{p^{1-1/p}}{2 \, \sigma_p \, \Gamma(1/p)} \exp\left(- \frac{1}{p} \frac{| \, x - x_0 \, |^p}{(\sigma_p)^p} \right) , \tag{1}$$

where $\Gamma(.)$ denotes the Gamma function.

Figure 1.2 shows some examples with p respectively equal to 1.0 , 1.5 , 2.0 , 3.0 , and 100.0 . For $p = 1$,

$$f_1(x) = \frac{1}{2 \, \sigma_1} \exp\left(- \frac{| \, x - x_0 \, |}{\sigma_1} \right) ,$$

and $f_1(x)$ is a symmetric exponential, centered at $x = x_0$ with *mean deviation* equal to σ_1 . For $p = 2$,

$$f_2(x) = \frac{1}{\sqrt{2\pi} \, \sigma_2} \exp\left(- \frac{1}{2} \frac{(\, x - x_0 \,)^2}{\sigma_2{}^2} \right) ,$$

and $f_2(x)$ is a Gaussian function, centered at $x = x_0$ with *standard deviation* equal to σ_2 . For $p \rightarrow \infty$,

(...)

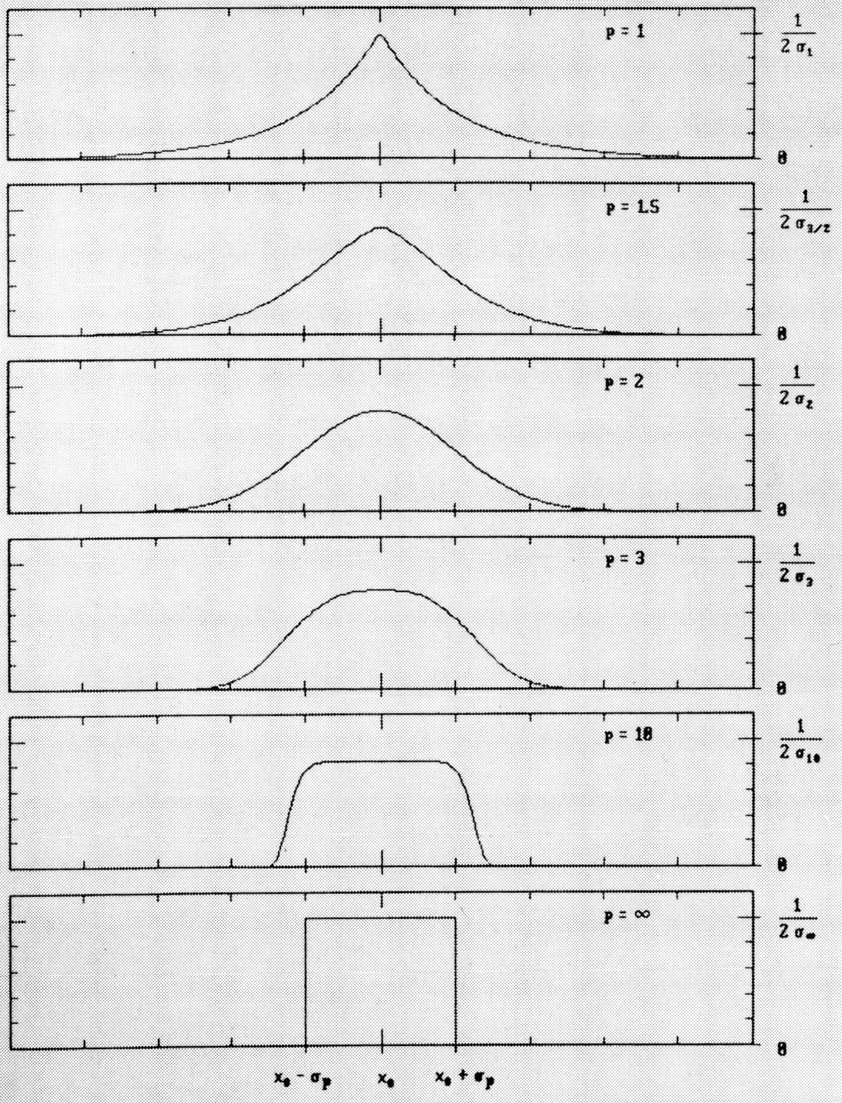

Figure 1.2: Generalized Gaussian of order p . The value p = 1 gives a double exponential, p = 2 gives an ordinary Gaussian, and p = ∞ gives a box-car.

(...)

$$f_\infty(x) = \begin{cases} 1/(2\,\sigma_\infty) & \text{for } x_0 - \sigma_\infty \le x \le x_0 + \sigma_\infty \\ 0 & \text{otherwise} \,, \end{cases}$$

and $f_\infty(x)$ is a box function, centered at $x = x_0$ with *mid-range* equal to σ_∞ . Problem 1.21 shows that $f_p(x)$ is normalized to unity.

The function $f_p(x)$ defined in (1) can be termed *generalized Gaussian*, because it generates a family of well-behaved functions containing the Gaussian function as a particular case. Symmetric exponentials, Gaussian functions, and box-car functions are often used to model error distribution. The definition of generalized Gaussian slightly widens the possibility of choice.

1.2.5: Shannon's measure of Information Content

Given two normalized probability density functions $f_1(\mathbf{x})$ and $f_2(\mathbf{x})$, the *relative information content* of $\mathbf{f}_1$ with respect to $\mathbf{f}_2$ is defined by

$$I(f_1;f_2) = \int d\mathbf{x}\ f_1(\mathbf{x})\ \mathrm{Log}\ \frac{f_1(\mathbf{x})}{f_2(\mathbf{x})} \ . \qquad (1.30a)$$

When the logarithm base is 2, the unit of information is termed a *bit*; if the base is e=2.71828..., the unit is the *nep*; if the base is 10, the unit is the *digit*.

The relative information content of a probability density $f(\mathbf{x})$ with respect to the (normalized) non-informative probability density $\mu(\mathbf{x})$,

$$I(f;\mu) = \int d\mathbf{x}\ f(\mathbf{x})\ \mathrm{Log}\ \frac{f(\mathbf{x})}{\mu(\mathbf{x})} \ , \qquad (1.30b)$$

is simply called the *information content* of $f(\mathbf{x})$.

Equation (1.30b) generalizes Shannon's (1948) original definition for discrete probabilities

$$I(P) = \sum_i P_i\ \mathrm{Log}\ P_i$$

to density functions. It should be noticed that the expression $\int d\mathbf{x}\ f(\mathbf{x})\ \mathrm{Log}\ f(\mathbf{x})$ cannot be used as a definition because it is not consistent (because, besides the fact that the logarithm of a dimensional quantity is not defined, a bijective change of variables $\mathbf{x}^* = \mathbf{x}^*(\mathbf{x})$ would alter the information content, which is not the case with the right definition (1.30b) (see problem

1.13).

It can be shown (problem 1.14) that the information content is always positive

$$I(f;\mu) \geq 0 \ ,$$

and that it is null only if $f(x) \equiv \mu(x)$: the non-informative probability density function $\mu(x)$ represents the *state of null information.*

1.2.6: The Combination of States of Information

The classical theory of logic gives the rules human beings use to handle information. As an example, let r denote the radius of the Earth's metallic core, and let A_1 and A_2 be the propositions

A_1 : " 3300 km < r < 3500 km "

A_2 : " 3400 km < r < 3600 km " .

These two propositions can be combined to obtain new propositions. For instance, the *disjunction,* (A_1 *or* A_2) , and the *conjunction,* (A_1 *and* A_2) , are

(A_1 *or* A_2) : " 3300 km < r < 3600 km "

(A_1 *and* A_2) : " 3400 km < r < 3500 km " .

Let $P(A)$ be the "value of truth" of the proposition A , i.e., $P(A) = 1$ if A is "true", and $P(A) = 0$ if A is "false". The combination of logical propositions can be defined by their "table of truth". For instance, the propositions (A_1 *or* A_2) and (A_1 *and* A_2) are characterized in Table 1.1 .

$P(A_1)$	$P(A_2)$	$P(A_1$ *or* $A_2)$	$P(A_1$ *and* $A_2)$
0	0	0	0
0	1	1	0
1	0	1	0
1	1	1	1

Table 1.1: Values of truth of the logical propositions *and* and *or* .

More formally,

$$P(A_1 \textit{ or } A_2) = 0 \quad \Leftrightarrow \quad \begin{cases} P(A_1) = 0 \\ \text{and} \\ P(A_2) = 0 \end{cases} \tag{1.31}$$

$$P(A_1 \textit{ and } A_2) = 0 \quad \Leftrightarrow \quad \begin{cases} P(A_1) = 0 \\ \text{or} \\ P(A_2) = 0 . \end{cases} \tag{1.32}$$

The value of truth of a proposition can also be referred to its "probability": it is equal to 1 for a (certainly) true proposition, and equal to 0 for a (certainly) false proposition.

In section 1.2.2 a state of information on a parameter set $\boldsymbol{X}$ has been defined as a probability over X. Let P_1 and P_2 be two probabilities over X representing two states of information. For the development of our theory, we need to define the conjunction of two states of information; this will be a generalization of the conjunction of logical propositions. The probability representing the new state of information will be denoted $(P_1 \textit{ and } P_2)$.

The equivalent of (1.32) for states of information is as follows: for any $A \subset X$,

$$(P_1 \textit{ and } P_2)(A) = 0 \quad \Leftrightarrow \quad \begin{cases} P_1(A) = 0 \\ \text{or} \\ P_2(A) = 0 , \end{cases}$$

or, equivalently,

$$\left.\begin{matrix} P_1(A) = 0 \\ \text{or} \\ P_2(A) = 0 \end{matrix}\right\} \Rightarrow \quad (P_1 \textit{ and } P_2)(A) = 0 \tag{1.33a}$$

$$\left.\begin{matrix} P_1(A) \neq 0 \\ \text{and} \\ P_2(A) \neq 0 \end{matrix}\right\} \Rightarrow \quad (P_1 \textit{ and } P_2)(A) \neq 0. \tag{1.33b}$$

These conditions are not strong enough to define the conjunction $(P_1 \textit{ and } P_2)$ uniquely. It seems reasonable to impose that the conjunction of any state of information P with the state of null information, M , does not modify the information:

$$(P \textit{ and } M)(A) = P(A) \qquad \text{for any } A \subset X \tag{1.34}.$$

Equations (1.33a) and (1.34) now define a probability uniquely. For greater compactness, I write the definition of the conjunction $(P_1 \textit{ and } P_2)$ as follows:

$$\text{for any } P_1 \text{ and } P_2 : \qquad (P_1 \textit{ and } P_2) = (P_2 \textit{ and } P_1) \tag{1.35a}$$

$$\text{for any } P_1 , P_2 , \text{ and any } A \subset X : P_1(A) = 0 \Rightarrow (P_1 \textit{ and } P_2)(A) = 0 \tag{1.35b}$$

$$\text{for any } P : \qquad (P \textit{ and } M) = P , \tag{1.35c}$$

where M represents the state of null information. In problem 1.17 it is shown that if $f_1(\mathbf{x})$, $f_2(\mathbf{x})$,and $\mu(\mathbf{x})$ are the probability densities representing P_1 , P_2 , and M respectively:

$$P_1(A) = \int_A d\mathbf{x} \quad f_1(\mathbf{x}) ,$$

$$P_2(A) = \int_A d\mathbf{x} \quad f_2(\mathbf{x}) ,$$

$$M(A) = \int_A d\mathbf{x} \quad \mu(\mathbf{x}) ,$$

and $\sigma(\mathbf{x})$ is the probability density representing $(P_1 \textit{ and } P_2)$,

$$(P_1 \textit{ and } P_2)(A) = \int_A d\mathbf{x} \quad \sigma(\mathbf{x}) ,$$

then we have

$$\boxed{\sigma(\mathbf{x}) = \frac{f_1(\mathbf{x}) \, f_2(\mathbf{x})}{\mu(\mathbf{x})} .} \tag{1.36}$$

The conjunction of states of information was first defined by Tarantola and Valette (1982a).

Example 24: Let x and y be cartesian coordinates on a cathodic screen. A random device projects electrons on the screen with a known probability density $f_1(x,y)$. We are interested in the coordinates (x,y) at which a particular electron will hit the screen, and we build an experimental device for measuring them. The measuring instrument is not perfect, and in performing the experiment we can only get the information that the true coordinates of the impact point had the probability density $f_2(x,y)$. We wish to combine this experimental information with the previous knowledge of the random device, and obtain a better estimate of the impact point. This example is developed in problem 1.12. ■

Let P_1 be a probability measure on the space $X = Y \times Z$, with probability density $f_1(y,z)$, and let P_2 be the probability measure with probability density $f_2(y,z)$, giving a probability of 1 to the event that $z = z_0$:

$$f_2(y,z) = \mu_Y(y)\,\delta(z-z_0) ,$$

where $\mu_Y(y)$ denotes the null information probability density on y . Using (1.36), the conjunction of these two states of information gives the probability density (1.36)

$$\sigma(y,z) = \frac{f_1(y,z)\, f_2(y,z)}{\mu(y,z)} = \frac{f_1(y,z)\,\delta(z-z_0)}{\mu_Z(z)} ,$$

where $\mu_Z(z)$ is the null information probability density for z (equation (1.29) has been used).

The a posteriori marginal probability density for z is

$$\sigma_Z(z) = \int_Y dy\ \sigma(y,z) = \delta(z-z_0) ,$$

in accordance with the information that the true value of z is z_0 . The a posteriori marginal probability density for y is

$$\sigma_Y(y) = \int_Z dz\ \sigma(y,z) = \frac{f_1(y,z_0)}{\mu_Z(z_0)} ,$$

or, if it can be normalized,

$$\sigma_Y(y) = \frac{f_1(y,z_0)}{\int_Y dy\ f_1(y,z_0)} . \tag{1.37}$$

This expression corresponds to the usual definition of *conditional probability density* for y , given $f_1(y,z)$ and $z = z_0$, which is usually denoted $f_1(y|z_0)$. We see that this concept is here contained in the more general concept of combination of states of information (see also box 1.4).

In section 1.5.1, the conjunction of states of information is used to combine information obtained from measurements with information obtained from a physical theory, and is shown to be the basis of the Inverse Problem Theory.

1.3: Information obtained from Physical Theories (Solving the Forward Problem)

Strictly speaking, to solve the forward problem means to predict the error-free values of data, **d** , that would correspond to a given model, **m** . I denote this theoretical prediction by

$$\mathbf{d}_{cal} = \mathbf{g}(\mathbf{m}) \ , \tag{1.38}$$

which is a short notation for the set of equations

$$d^i_{cal} = g^i(\mathbf{m}) \qquad (i \in I_D) \ .$$

Example 25: Some physical quantity, d , depends on time, t , through the equation

$$d = m^1 \, t + m^2 \ .$$

If the parameters m^1 and m^2 are known, for any value t^i we can estimate the corresponding value d^i by

$$d^i_{cal}(\mathbf{m}) = g^i(\mathbf{m}) = g^i(m^1,m^2) = m^1 \, t^i + m^2 \ . \ \blacksquare$$

Example 26: The model parameters may represent a discretization of the Potential V(r) describing the spherically symmetric electric field created by an atomic nucleus. The observable parameters d^i may represent the flux of electrons diffused at different directions for a given incident flux. To solve the forward problem corresponds to the resolution of the Schrödinger equation. ■

The predicted values cannot, in general, be identical to the true "observed" values, for two reasons: experimental uncertainties and modelization errors. It is important to note that these two very different sources of error generally produce errors which are of the *same order of magnitude*, because, due to the continuous progress of scientific research, as soon as new experimental methods are capable of decreasing the experimental uncertainty, new theories and new models arise which allow of explaining the observations more accurately. For this reason, it is generally not possible to set inverse problems properly without a careful analysis of modelization errors.

The way of describing experimental uncertainties will be studied in section 1.4. Let us see here how to describe uncertainties due to modelization errors. Following our postulate that the more general way of describing any state of information is to define a probability density function, we have to accept here that the most general way of describing uncertainties due to rough modelization is by defining, for given values of the model parameters $\mathbf{m}$, a probability density over $\mathbf{d}$, i.e., a *conditional probability density* which will be denoted by $\Theta(\mathbf{d}|\mathbf{m})$.

Example 27: For an exact theory

$$\Theta(\mathbf{d}|\mathbf{m}) = \delta(\mathbf{d}-\mathbf{g}(\mathbf{m})) , \tag{1.39}$$

where $\mathbf{g}(\mathbf{m})$ represents the vector function introduced in equation (1.38). ■

Example 28: Theory with independent error bars. An example can be

$$\Theta(\mathbf{d}|\mathbf{m}) = \left(\prod_{i\in I_D} \frac{1}{2\,\sigma^i(\mathbf{m})} \right) \exp\left(- \sum_{i\in I_D} \frac{| d^i - g^i(\mathbf{m}) |}{\sigma^i(\mathbf{m})} \right) , \tag{1.40}$$

where it is assumed that a double exponential function conveniently models the error distribution, and where the "error bar" of the i-th predicted data value is $\sigma^i(\mathbf{m})$ and is independent of the error bar of the j-th predicted data (null covariances), but depending on the current value of $\mathbf{m}$. ■

Example 29: Gaussian errors. Letting $\mathbf{C}_T(\mathbf{m})$ be a covariance operator (see box 1.2) describing the estimated modeling errors for a model $\mathbf{m}$, it is possible to take

$$\Theta(\mathbf{d}|\mathbf{m}) = \left((2\pi)^{ND} \det \mathbf{C}_T(\mathbf{m})\right)^{-\frac{1}{2}} \exp\left(- \frac{1}{2}(\mathbf{d}-\mathbf{g}(\mathbf{m}))^t \; \mathbf{C}_T^{-1}(\mathbf{m}) \; (\mathbf{d}-\mathbf{g}(\mathbf{m})) \right), \tag{1.41}$$

where ND represents the dimension of the data space (number of data parameters). ■

Example 30: Errors due to linearization. Sometimes, the resolution of the forward problem (i.e., the computation of $\mathbf{d}_{cal} = \mathbf{g}(\mathbf{m})$) is too difficult or too expensive, and a *linearization* around a reference model is used:

$$\mathbf{g}(\mathbf{m}) \simeq \mathbf{g}(\mathbf{m}_{ref}) + \mathbf{G}_{ref}\,(\mathbf{m}-\mathbf{m}_{ref}) , \tag{1.42}$$

where

$$G^{i\alpha}_{ref} = \left(\frac{\partial g^i}{\partial m^\alpha} \right)_{m_{ref}} .$$

In this case, modelization errors are usually strongly correlated, and neglecting them may severely alter the solution of the problem. It is often difficult to give a realistic estimation of the linearization errors. ■

Example 31: Errors independent of the model value $\mathbf{m}$. A useful approximation is sometimes that the true value $\mathbf{d}$ differs from the "computed" value $\mathbf{g(m)}$ by an "error" ϵ_T :

$$\mathbf{d} = \mathbf{g(m)} + \epsilon_T , \tag{1.43}$$

where the error ϵ_T is assumed to have known statistics described by the probability density function $f_T(\epsilon_T)$. This gives

$$\Theta(\mathbf{d}|\mathbf{m}) = f_T(\epsilon_T) = f_T(\mathbf{d}-\mathbf{g(m)}) . \ ■ \tag{1.44}$$

In section 1.2.4, the density function $\mu_M(\mathbf{m})$ describing the state of null information on model parameters was introduced. It is clear that the function defined by

$$\Theta(\mathbf{d},\mathbf{m}) = \Theta(\mathbf{d}|\mathbf{m}) \, \mu_M(\mathbf{m}) \tag{1.45}$$

does not contain any information on $\mathbf{m}$ (the marginal probability density for $\mathbf{m}$ is the null information probability density), and it still describes the physical correlations between $\mathbf{d}$ and $\mathbf{m}$ that the physical theory is able to predict. So, for greater generality, I will assume that the description of the information concerning the resolution of the forward problem is not given by a conditional density function $\Theta(\mathbf{d}|\mathbf{m})$ but by a joint density function $\Theta(\mathbf{d},\mathbf{m})$ over the space $\boldsymbol{D} \times \boldsymbol{M}$. Equation (1.45) will only represent a particular (very current) case.

In fact, there exists a class of problems in which the correlations between $\mathbf{d}$ and $\mathbf{m}$ are not predicted by a formal theory, but result from an accumulation of observations. In this case, the joint density function $\Theta(\mathbf{d},\mathbf{m})$ is the natural description of the information.

Example 32: The data parameters d^i may represent the current state of a volcano (intensity of seismicity, rate of accumulation of strain, ...). The model parameters m^α may represent, for instance, the time interval to the next volcanic eruption, the magnitude of this eruption, etc. Our present knowledge of volcanoes does not allow of relating these parameters realistically using physical laws, so that, at present, the only scientific description is statistical. Provided that in the past we were able to observe a significative number of eruptions of this volcano, we can construct a histogram in the space *DxM* which describes all our information correlating the parameters (see Tarantola, 1983, for an example). This histogram can directly be identified with $\Theta(\mathbf{d},\mathbf{m})$. ■

Briefly synthesizing the conclusions of this section: the expected physical correlations between model and observable parameters can be described using a joint density function $\Theta(\mathbf{d},\mathbf{m})$. When these correlations are predicted by a (necessarily inexact) physical theory, $\Theta(\mathbf{d},\mathbf{m})$ is given by $\Theta(\mathbf{d},\mathbf{m}) = \Theta(\mathbf{d}|\mathbf{m})\, \mu_M(\mathbf{m})$, where $\Theta(\mathbf{d}|\mathbf{m})$ represents the probability density of $\mathbf{d}$ for any given value $\mathbf{m}$. This usually corresponds to put some "error bars" around a "predicted value" $\mathbf{g}(\mathbf{m})$ (see figure 1.3).

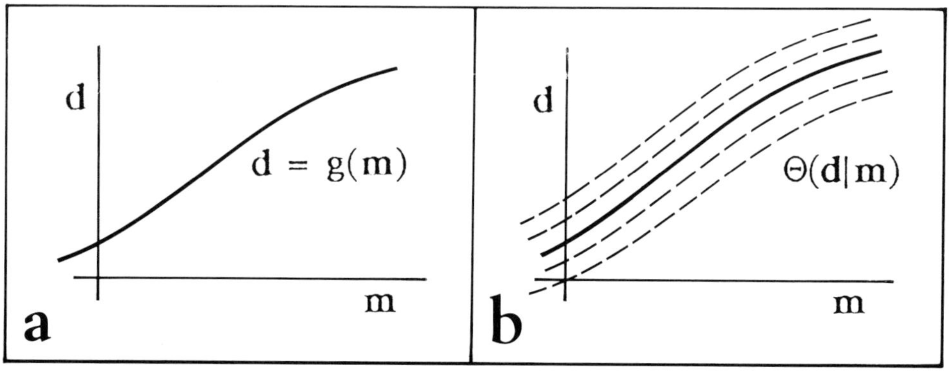

Figure 1.3: a) If uncertainties in the forward modelization can be neglected, a functional relationship $\mathbf{d} = \mathbf{g}(\mathbf{m})$ can be introduced which gives, for each model $\mathbf{m}$, the predicted (or "calculated") data values, $\mathbf{d}$. b) If forward-modeling uncertainties cannot be neglected, they can be described using a conditional probability density, $\Theta(\mathbf{d}|\mathbf{m})$, giving, for each model $\mathbf{m}$, a probability density for $\mathbf{d}$. Roughly speaking, this corresponds to putting "error bars" on the theoretical relation $\mathbf{d} = \mathbf{g}(\mathbf{m})$.

1.4: Information obtained from measurements, and a priori information

1.4.1: Results of the measurements

The measurement experiment will give a certain amount of information on the true values of the observable parameters. Let $\rho_D(\mathbf{d})$ be the probability density function describing this information.

Example 33: Observations are the output of an instrument with known statistics. To simplify the discussion, I will refer to "the instrument" as if all the measurements could result from a single reading on a large apparatus although, more realistically, we generally have some readings from several apparatuses. Assume that at each measurement the instrument delivers a given value of $\mathbf{d}$, denoted $\mathbf{d}_{out}$. Ideally, the supplier of the apparatus should provide a statistical analysis of the errors of the instrument (if he does not, we should not pay for it!). The most useful and general way of giving the results of the statistical analysis is to define the probability density for the value of the output, $\mathbf{d}_{out}$, when the actual input is $\mathbf{d}$. Let $\nu(\mathbf{d}_{out}|\mathbf{d})$ be this conditional probability density. If $f(\mathbf{d}_{out},\mathbf{d})$ denotes the joint probability density for $\mathbf{d}_{out}$ and $\mathbf{d}$, and if we don't use any information on the input, we have

$$f(\mathbf{d}_{out},\mathbf{d}) = \nu(\mathbf{d}_{out}|\mathbf{d})\ \mu_D(\mathbf{d}) \ .$$

If the actual result of a measurement is $\mathbf{d}_{out} = \mathbf{d}_{obs}$, then we can assimilate $\rho_D(\mathbf{d})$ to the conditional probability density for $\mathbf{d}$ given $\mathbf{d}_{out} = \mathbf{d}_{obs}$:

$$\rho_D(\mathbf{d}) = f_{D|D_{out}}(\mathbf{d}|\mathbf{d}_{out}=\mathbf{d}_{obs}) = \frac{f(\mathbf{d}_{obs},\mathbf{d})}{\int_D d\mathbf{d}\ f(\mathbf{d}_{obs},\mathbf{d})} ,$$

i.e.,

$$\rho_D(\mathbf{d}) = \frac{\nu(\mathbf{d}_{obs}|\mathbf{d})\ \mu_D(\mathbf{d})}{\int_D d\mathbf{d}\ \nu(\mathbf{d}_{obs}|\mathbf{d})\ \mu_D(\mathbf{d})} . \ \blacksquare \tag{1.46}$$

Example 34: Perfect instrument. In that case,

$$\nu(\mathbf{d}_{obs}|\mathbf{d}) = \delta(\mathbf{d}-\mathbf{d}_{obs}) , \tag{1.47}$$

thus giving

$$\rho_D(\mathbf{d}) = \delta(\mathbf{d}-\mathbf{d}_{obs}) . \tag{1.48}$$

The assumption of a perfect instrument may be made when measuring errors are negligible *compared* to modelization errors (see section 1.5.3). ∎

Example 35: Gaussian uncertainties. Taking

$$\nu(\mathbf{d}_{obs}|\mathbf{d}) = ((2\pi)^{ND}\det \mathbf{C}(\mathbf{d}_{obs}))^{-1/2}\exp\left[-\frac{1}{2}(\mathbf{d}-\mathbf{d}_{obs})^t\,\mathbf{C}(\mathbf{d}_{obs})^{-1}(\mathbf{d}-\mathbf{d}_{obs})\right], \tag{1.49a}$$

where ND represents the number of data, corresponds to the assumption that estimated experimental errors can be described by the covariance operator $\mathbf{C}$, which may depend on the observed values $\mathbf{d}_{obs}$. Using (1.46) this gives

$$\frac{\rho_D(\mathbf{d})}{\mu_D(\mathbf{d})} = ((2\pi)^{ND}\det \mathbf{C}(\mathbf{d}_{obs}))^{-1/2}\exp\left[-\frac{1}{2}(\mathbf{d}-\mathbf{d}_{obs})^t\,\mathbf{C}(\mathbf{d}_{obs})^{-1}(\mathbf{d}-\mathbf{d}_{obs})\right]. \;\blacksquare \tag{1.49b}$$

Example 36: Errors of the measuring instrument are independent of the input. Assume that the output $\mathbf{d}_{obs}$ is related to the input $\mathbf{d}$ through the simple relation

$$\mathbf{d}_{obs} = \mathbf{d} + \epsilon_D , \tag{1.50}$$

where ϵ_D is an unknown error with known statistics described by the probability density function $f_D(\epsilon_D)$. In that case,

$$\nu(\mathbf{d}_{obs}|\mathbf{d}) = f_D(\epsilon_D) = f_D(\mathbf{d}_{obs}-\mathbf{d}) . \;\blacksquare \tag{1.51}$$

Example 37: Outliers in a data set. Some data sets contain outliers which are difficult to eliminate, in particular when the data space is highly dimensioned, because it is difficult to visualize such data sets. Problem (1.9) shows that a single outlier in a data set can lead to unacceptable inverse results if the Gaussian assumption is used. This problem suggests using "long-tailed" density functions to represent uncertainties on this kind of data sets. Examples of long-tailed density functions are the symmetric exponential (Laplace) function

$$\rho_D(\mathbf{d}) = \prod_{i \in I_D} \left(\frac{1}{2\sigma^i} \exp\left(- \frac{\left| d^i - d^i_{obs} \right|}{\sigma^i} \right) \right) = \prod_i \frac{1}{2\sigma^i} \exp\left(- \sum_i \frac{\left| d^i - d^i_{obs} \right|}{\sigma^i} \right) ,$$

or the Cauchy function

$$\rho_D(\mathbf{d}) = \prod_{i \in I_D} \left(\frac{1}{\pi \, \sigma^i} \frac{1}{1 + \left(\frac{d^i - d^i_{obs}}{\sigma^i} \right)^2} \right) ,$$

which has the nice particularity of having infinite standard deviations. ■

Example 38: Consider a measurement made to obtain the arrival time of a given seismic wave recorded by a seismograph. Sometimes, the seismogram is simple enough to give a simple result (Figure 1.4a). But sometimes, due to strong noise (with unknown statistics), the measurement is not trivial. Figure 1.4b shows a particular example where it is difficult to obtain a numerical value, say t_{obs} , for the arrival time. The use of a probability density function allows of describing information on the arrival time with a sufficient degree of generality (Figure 1.4c). With such kinds of data, it is clear that the subjectivity of the scientist plays a major role. It is indeed the case, *whichtever inverse method is used*, that results obtained by different scientists (for instance, for the location of a hypocenter) from such data sets are different. Objectivity can only be attained if the data redundancy is great enough that differences in data interpretation among different observers do not significantly alter the models obtained. ■

Example 39: Assume that the only instrument we have for measuring a given observable is a buzzer that responds when the true value d is in the range $\mathbf{d}_{inf} \leq \mathbf{d} \leq \mathbf{d}_{sup}$. We make the measurement, and the buzzer does *not* respond. The corresponding probability density is then

$$\rho_D(\mathbf{d}) = \begin{cases} 0 & \text{for } \mathbf{d}_{inf} \leq \mathbf{d} \leq \mathbf{d}_{sup} \\ \mu_D(\mathbf{d}) & \text{otherwise ,} \end{cases} \tag{1.52}$$

where $\mu_D(\mathbf{d})$ is the non-informative probability density for observable parameters. ■

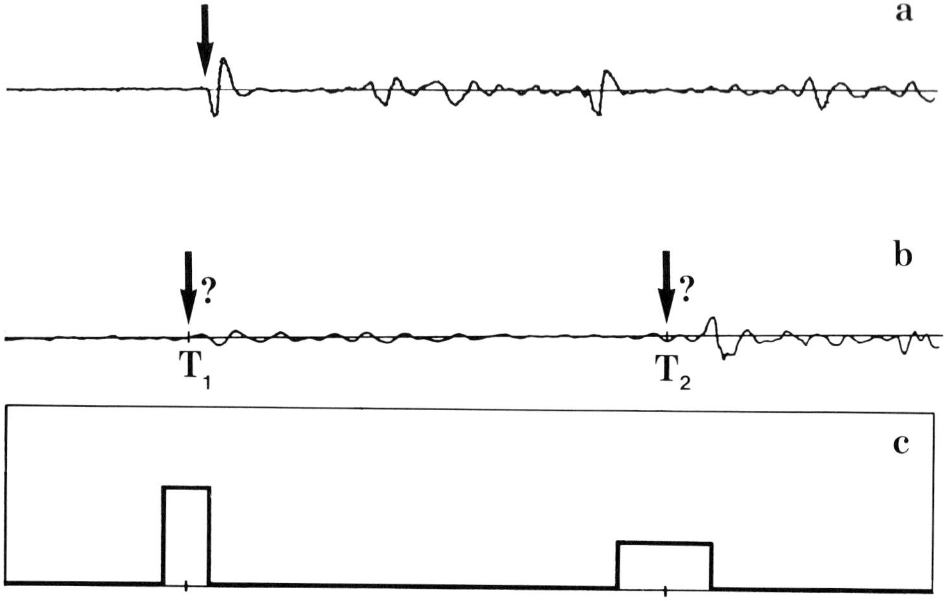

Figure 1.4: a) Seismogram corresponding to an earthquake that occurred in the south of Honshu (Japan) on April 24, 1984, recorded in Paris. The time of first arrival of the seismic wave is clearly visible. b) Seismogram corresponding to an earthquake that occurred east of New Guinea on June 27, 1974, recorded in the south of France. Due to the presence of "ambient noise", it is difficult to pick the first arrival time of the waves. In particular, one may hesitate between times T_1 and T_2 . If an expert gives, say, a 50% probability of the first arrival time being in the vicinity of T_1 and a 50% probability of its being in the vicinity of T_2 , it is possible to represent this information by the probability density shown in c). The width of each "peak" represents the uncertainty of the reading of each of the possible arrivals, while the separation of the peaks, represents the overall uncertainty.

1.4.2: A priori information on model parameters

By a priori information (or prior information) I mean information which is obtained independently of the results of measurements. The probability density function representing this a priori information will be denoted by $\rho_M(\mathbf{m})$.

Example 40: We have no a priori information. In that case,

$$\rho_M(\mathbf{m}) = \mu_M(\mathbf{m}) \ , \tag{1.53}$$

where $\mu_M(\mathbf{m})$ is the noninformative probability density for model parameters. ■

Example 41: For a given parameter m^α we have only the information that it is strictly bounded by the two values m^α_{inf} and m^α_{sup}. We can take

$$\rho_M(\mathbf{m}) = \prod_{\alpha \in \mathbf{I}_M} \rho_\alpha(m^\alpha)$$

where

$$\rho_\alpha(m^\alpha) = \begin{cases} \mu_\alpha(m^\alpha) & \text{for} \quad m^\alpha_{inf} \le m^\alpha \le m^\alpha_{sup}. \\ 0 & \text{otherwise} \ , \end{cases}$$

and where $\mu_\alpha(m^\alpha)$ represents the non-informative probability density for m^α . ■

Example 42: The parameters m^α represent a discretization of an unknown continuous function $\Psi(t)$. Assume, for instance, that the a priori information we have on the true value of $\Psi(t)$ is that it belongs to the class of functions represented by the members shown in Figure 1.5. This means that we know that the unknown function is smooth, with a given smoothness length, and that, for each value of t , it takes a value $\Psi_{Mean}(t) \pm \sigma(t)$. In order for the discretization $m^\alpha = \Psi(t_\alpha)$ to be acceptable, we have first to assume that the discretization interval is significantly smaller than the smoothness length of the function. The family of functions in Figure 1.5 seems to be reasonably well represented by a gaussian process with given mean value $\Psi_{Mean}(t)$, and given covariance function $C(t,t')$, both of them grossly estimated from the Figure. Taking

$$m^{\alpha}_{prior} = \Psi_{Mean}(t_\alpha)$$

$$C_M{}^{\alpha\beta} = C(t_\alpha, t_\beta) ,$$

the density function representing the a priori information in the model space is

$$\rho_M(\mathbf{m}) = ((2\pi)^{NM} \det C_M)^{-1/2} \exp\left[-\frac{1}{2}(\mathbf{m}-\mathbf{m}_{prior})^t \mathbf{C}_M^{-1} (\mathbf{m}-\mathbf{m}_{prior}) \right]. \tag{1.54}$$

This density function gives a high probability density to models $\mathbf{m}$ which are close to $\mathbf{m}_{prior}$ in the sense of the covariance operator $\mathbf{C}_M$, i.e., models in which the difference $\mathbf{m}-\mathbf{m}_{prior}$ is small at each point (with respect to standard deviations in C_M) and smooth (with respect to correlations in $\mathbf{C}_M$). Covariance operators are defined in box 1.2. ■

Example 43: Let us consider a particular atomic nucleus, and let T denote its *half-life*. By definition of half-life, T is necessarily positive. A Gaussian probability density cannot be used to represent any a priori information on T , because a Gaussian function gives a non-vanishing probability for T being negative. Figure 1.6 shows a histogram of the half-lives of the first 580 atomic nuclei quoted in the 1984 CRC Handbook of Chemistry and Physics (I got tired before arriving at the end of the list!). The abscissa of the Figure is

$$T^* = \mathrm{Log}_{10}\left(\frac{T}{1 \text{ second}} \right) . \tag{1.55a}$$

The logarithmic scale has been chosen for the time axis because, as the half-lives span many orders of magnitude, it is difficult to show the histogram in a linear time axis. With this logarithmic scale, the histogram may conveniently be approximated by a Gaussian function:

$$f^*(T^*) = \frac{1}{(2\pi)^{1/2}\,\sigma^*} \exp\left(-\frac{1}{2} \frac{(T^*-T_0{}^*)^2}{(\sigma^*)^2} \right) , \tag{1.56a}$$

with $T_0{}^* \simeq 3$ and $\sigma^* \simeq 3$. For greater generality, let me write (1.55a) as

$$T^* = \beta \,\mathrm{Log}\left(\frac{T}{\tau_0} \right) \qquad T = \tau_0 \exp\left(\frac{T^*}{\beta} \right) , \tag{1.55b}$$

where, in our example, $\tau_0 = 1$ second , $\beta = \mathrm{Log}\ 10$. Using (1.18), we can then obtain the probability density for the time variable T :

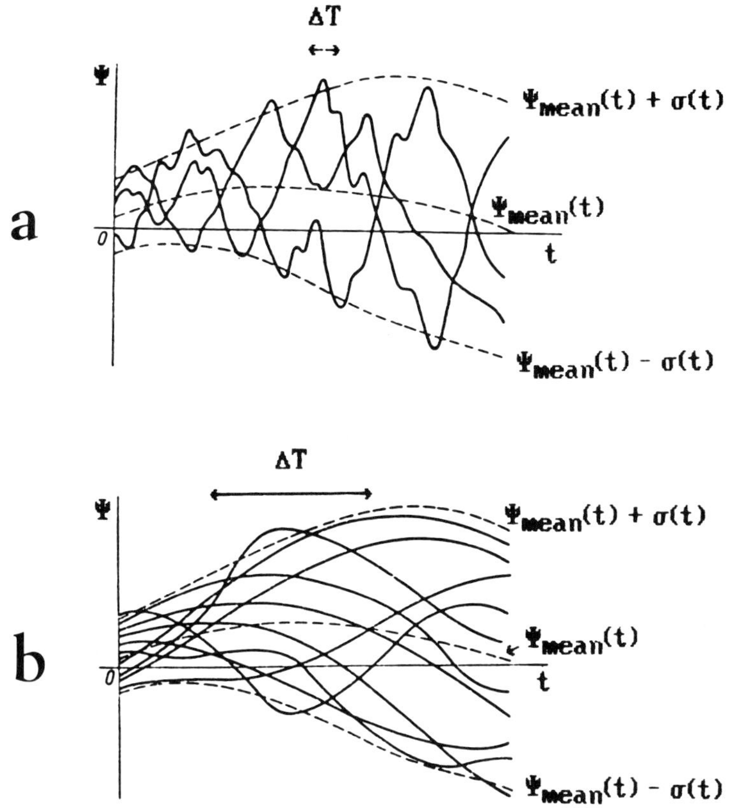

Figure 1.5: Each family of functions (a) and (b) represents some realizations of two different random functions (from Pugachev, 1965). In both cases the mean value, $\Psi_{Mean}(t)$, of the random function is the same. At a given value t , the variance $\sigma^2(t) = C(t,t)$ is also the same. But the covariance $C(t,t')$ between the value at t and the value at t' is different in the two examples. The "correlation length", ΔT , of the family (a) is shorter than the correlation length of the family (b). These two random families can be (quite simplistically) modeled by a Gaussian process (equation (1.54)) with mean $\Psi_{Mean}(t)$ and with the covariance function

$$C(t,t') \quad = \quad \sigma^2\left(\frac{t+t'}{2}\right) \quad \exp\left(-\frac{1}{2}\ \frac{(t-t')^2}{\Delta T^2}\right) ,$$

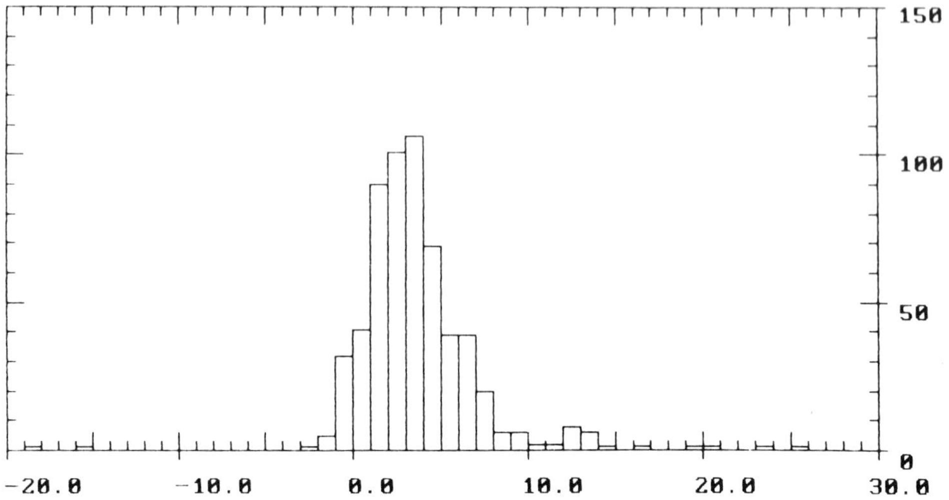

Figure 1.6: Histogram of disintegration periods (half-lives) of the first 580 atomic nuclei in the CRC Handbook of Chemistry and Physics (1984). The horizontal axis represents the logarithm of the half life: $T^* = \mathrm{Log}_{10}(\ T\ /\ 1\ \mathrm{second}\ )$. It is very difficult to show the histogram in a linear time axis, because observed disintegration periods span 45 orders of magnitude in time. Note that the use of a logarithmic time axis allows the histogram to be approximated by a Gaussian probability density. This implies a lognormal probability density in a linear time axis.

$$f\ (T) = \frac{1}{(2\pi)^{1/2}\ s}\ \frac{1}{T}\ \exp\left(-\frac{1}{2\ s^2} \left(\mathrm{Log}\ \frac{T}{T_0} \right)^2 \right) , \tag{1.56b}$$

with

$$T_0^* = \beta\ \mathrm{Log}\left(\frac{T_0}{\tau_0} \right) \qquad\qquad T_0 = \tau_0\ \exp\left(\frac{T_0^*}{\beta} \right)$$

and

$$s = \frac{\sigma^*}{\beta} \qquad\qquad \sigma^* = \beta\ s\ .$$

The density function (1.56b) is well known and is termed the *log-normal* probability density (because the logarithm of the variable has a normal [Gaussian] probability density). The log-normal density function is studied in box 1.3.

This example suggests that the use of a log-normal probability density is well adapted to modelling a priori information of the type $T \simeq T_0 \pm \Delta T$ for a positive parameter.

Nevertheless, as shown in box 1.3, if the "dispersion" s in (1.56b) is very small, the log-normal function tends to the normal function

$$f(T) \rightarrow \frac{1}{(2\pi)^{1/2}\,\sigma} \exp\left[-\frac{1}{2}\frac{(T-T_0)^2}{\sigma^2}\right] \qquad (\sigma = s\,T_0 << 1)\ , \qquad (1.57)$$

and the subtleties between normal and log-normal probabilities can be neglected. This corresponds to the case where the a probability density gives negligible probabilities to the negative values of the parameter.

The opposite limit (σ very large) is also interesting. As seen in box 1.3, we then have

$$f(T) \quad \rightarrow \quad \frac{1}{(2\,\pi)^{1/2}\ s}\ \frac{1}{T} \qquad (s >> 1)\ , \qquad (1.58)$$

which is the log-uniform probability density introduced in section 1.2.4. In his 1968 paper, Jaynes uses invariance arguments to obtain the non-informative probability density for the half-life of an atomic nucleus, and he obtains (1.58). It is remarkable that that the experimental histogram suggests the same conclusion. ■

Example 44: Figure 1.7a shows the histogram of densities of different known materials in the Earth's crust (independently of their relative abundance). In 1.7b, the same histogram is shown on a logarithmic scale. If one should stumble over a stone, one may wonder (whilst falling) what the density of the stone may be. If you bear in mind Figure 1.7b, you can take this log-normal function as representing your prior state of information as to its density. If you do not have this Figure in mind, the log-uniform density function will represent your ignorance well. (If, in going to measure the actual density, you stumble over the stone again, better go to another example...). ■

The examples in this section show how it is possible to use probability densities to describe prior information. I have never found a state of information (in the intuitive sense) which cannot be very precisely stated using a probability density. On the other hand, it may seem that probability densities have too many degreees of freedom to allow a definite choice that represents a given state of information.

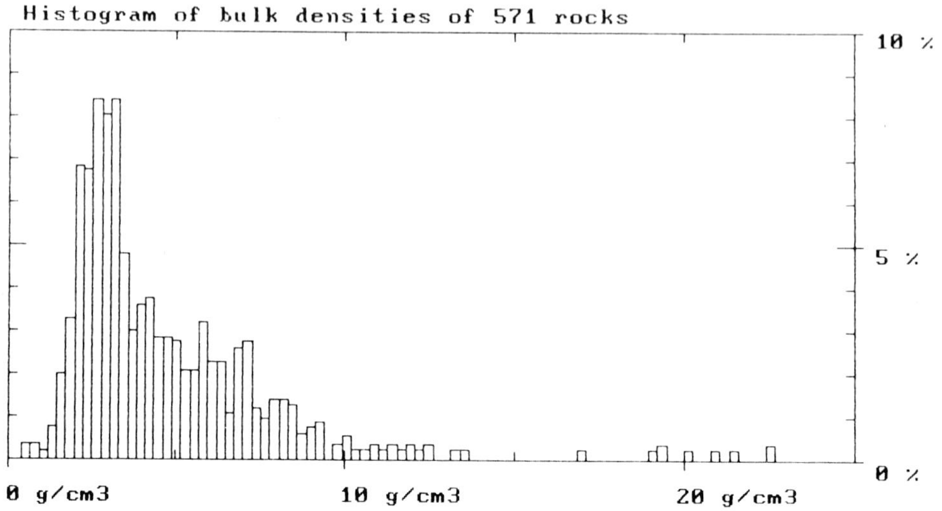
Histogram of bulk densities of 571 rocks
10 %
5 %
0 %
0 g/cm3
10 g/cm3
20 g/cm3

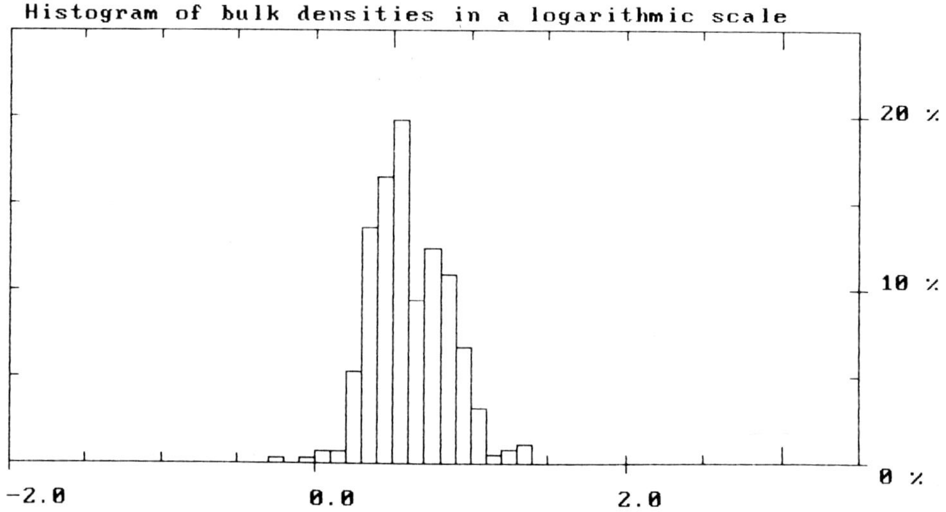
Histogram of bulk densities in a logarithmic scale
20 %
10 %
0 %
-2.0
0.0
2.0

← *Figure 1.7:* Top: histogram of bulk densities of the 571 different rock types quoted by Johnson & Olhoeft in the CRC Handbook of Physical Properties of Rocks, Vol. III, 1984. Bottom: the same histogram in a logarithmic horizontal scale $\rho^* = \mathrm{Log}_{10}(\rho/\mathrm{g\ cm}^{-3})$. The top histogram is very asymmetric, due to the existence of very heavy minerals ($\rho \simeq 20$ g cm^{-3}) . In a logarithmic scale, it is much more symmetric (bottom). A Gaussian probability density is a reasonably good approximation of the histogram in the ρ^* variable, which means that the corresponding probability density in the ρ variable is log-normal.

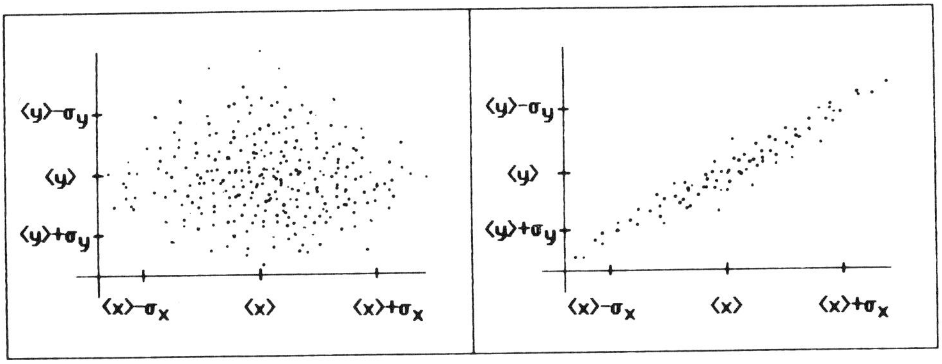

↑ *Figure 1.8:* The random points (x_i,y_i) of these diagrams have been generated using 2-dimensional probability densities $f_1(x,y)$ (left) and $f_2(x,y)$ (right). The two probability densities have identical mean values and standard deviations. Only the covariance C_{xy} is different. On the left, the covariance is small, on the right it is large. The probability density $f_2(x,y)$ is more "informative" than $f_1(x,y)$, because it demarcates a smaller region in the space $X \times Y$. This example suggests that if off-diagonal elements of a covariance operator are difficult to estimate, setting them to zero corresponds to neglecting information.

In fact, only a few characteristics of a density function are usually relevant, such as for instance, the position of the "center", the degree of asymmetry, the size of the "error bounds", the "correlations" between different parameters, and the behaviour of the density function "far from the center".

If hesitation exists in choosing the a priori error bars, it is of course best to be overconservative and to choose them very large. A conservative choice for correlations is to neglect them (see Figure 1.8 for an example). The behaviour of the density functions far from the center is only crucial if outliers may exist: the choice of functions tending too rapidly to zero (box-car functions or even Gaussian functions) may lead to inconsistencies; the solution to the problem (as defined in the next section) may not exist, or may senseless.

Usually the a priori states of information have the form of "soft bounds"; the normal or log-normal density functions generally apply well to that case. If the normal function is thought to vanish too rapidly when the parameter's value tends to infinity, longer tailed functions may be used, such as for instance, the symmetric-exponential function (see box 1.2).

Box 1.3: The Log-normal probability density

It is defined by

$$f(x) = \frac{1}{(2\pi)^{1/2}\, s} \; \frac{1}{x} \exp\left(-\frac{1}{2\, s^2} \left(\text{Log} \frac{x}{x_0} \right)^2 \right) . \tag{1}$$

Figure 1.9 shows some examples for $x_0 = 1$ and s respectively equal to 0.1 , 0.2 , 0.4 , 0.8 , 1.6 , and 3.2 .

The mode, median, and mean of $f(x)$ are respectively

$$m_0 = x_0 \exp(-s^2) \qquad (f(x) \text{ maximum for } x = m_0)$$

$$m_1 = x_0 \tag{2}$$

$$m_2 = x_0 \exp(s^2/2) \; .$$

The mean deviation and standard deviation are respectively

$$\sigma_1 = x_0 \exp(s^2/2) \; (\text{Erf}(1/s) - \text{Erf}(-1/s))$$

$$\tag{3}$$

$$\sigma_2 = x_0 \exp(s^2/2) \; (\exp(s^2) - 1)^{1/2} ,$$

(...)

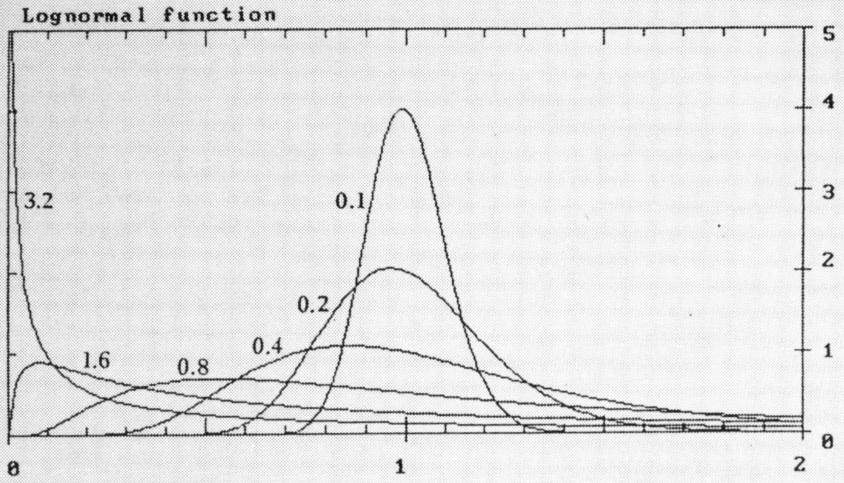

Figure 1.9: The Log-normal probability density.

where Erf(·) denotes the error function

$$\mathrm{Erf}(u) \;=\; \frac{1}{(2\pi)^{1/2}} \int_{-\infty}^{u} dt \;\exp\left(-\frac{t^2}{2}\right)$$

The log-normal probability density is so called because the *logarithm* of the variable has a normal (Gaussian) probability density. For the change of variables

$$x^* \;=\; \beta \,\mathrm{Log}\left(\frac{x}{\gamma}\right) \qquad\qquad x \;=\; \gamma \exp\left(\frac{x^*}{\beta}\right) \tag{4}$$

transforms f(x) into

$$f^*(x^*) \;=\; \frac{1}{(2\pi)^{1/2}\,\sigma} \exp\left(-\frac{1}{2}\,\frac{(x^*-x_0^*)^2}{\sigma^2}\right), \tag{5}$$

with

(...)

$$\sigma = s\,\beta \qquad\qquad s = \frac{\sigma}{\beta} \tag{6a}$$

and

$$x_0^* = \beta \,\mathrm{Log}\left(\frac{x_0}{\gamma}\right) \qquad\qquad x_0 = \gamma \exp\left(\frac{x_0^*}{\beta}\right) . \tag{6b}$$

In (4), the constant β is often $\mathrm{Log}_e 10$, which corresponds to defining x^* by $x^* = \mathrm{Log}_{10}(x/\gamma)$. The constant γ often corresponds to the physical unit used for x (see example 43). Alternatively, the particular choice

$$x^* = \frac{1}{s} \mathrm{Log}\left(\frac{x}{x_0}\right) \qquad\qquad x = x_0 \exp(s\, x^*)$$

leads to a Gaussian density with zero mean and unit standard deviation:

$$f^*(x^*) = \frac{1}{(2\pi)^{1/2}} \exp\left(-\frac{(x^*)^2}{2}\right) .$$

Figure 1.9 suggests that, for given x_0 , when the "dispersion" s is very small, the log-normal probability density tends to a Gaussian function. This is indeed the case. For, when $s \to 0$, $f(x)$ takes significant values only in the vicinity of x_0 , and

$$f(x) = \frac{1}{(2\pi)^{1/2}\, s}\, \frac{1}{x_0} \exp\left[-\frac{1}{2\, s^2}\left(-1 + \frac{x}{x_0} - \dots\right)^2\right]$$

$$\simeq \frac{1}{(2\pi)^{1/2}\, (s\, x_0)} \exp\left[-\frac{1}{2}\,\frac{(x-x_0)^2}{(s\, x_0)^2}\right] . \tag{7}$$

If, for given x_0 , the dispersion s is very large, the log-normal probability density tends to a log-uniform probability density (i.e., proportional to $1/x$; see section 1.2.4). For any x not too close to the origin, the argument of the exponential in (1) can be taken as null, thus showing that, at the limit $s \to \infty$,

$$f(x) \simeq \frac{1}{(2\pi)^{1/2}\, s}\, \frac{1}{x} . \tag{8}$$

The convergence of (1) into (8) is not a uniform convergence, in the sense that while the function (8) tends to infinity when x tends to 0 , the log-normal (1) takes the value 0 at the origin. But for values of x of the same order of magnitude as x_0 , the approximation (8) is adequate (for instance, for the values $x_0 = 1$, $s = 10.$, the log-normal function

(...)

and the log-uniform function are indistinguishable in Figure 1.9).

As suggested in section 1.2.4, the log-normal probability density is often adequate to represent probability distributions for variables which by definition are constrained to be positive.

The reader will easily verify that if a variable x has a log-normal distribution, the variable $y = 1/x$ has the same distribution.

The function

$$f(x) = \frac{p^{1-1/p}}{2\, s\, \Gamma(1/p)} \frac{1}{x} \exp\left(- \frac{1}{p\, s^p} \left| \text{Log} \frac{x}{x_0} \right|^p \right) , \qquad (9)$$

transforms, under the change of variables (4), into the generalized Gaussian

$$f^*(x^*) = \frac{p^{1-1/p}}{2\, \sigma\, \Gamma(1/p)} \exp\left(- \frac{1}{p} \frac{\left| x^* - x_0^* \right|^p}{\sigma^p} \right) , \qquad (10)$$

where σ and x_0^* are given by (6). This suggests that (9) can be referred to as the "generalized log-normal in the ℓ_p norm sense".

1.4.3: Joint prior information in the $\boldsymbol{D} \times \boldsymbol{M}$ *space*

By definition, the a priori information on model parameters is independent of observations. The information we have in both model parameters and observable parameters can then be described in the $\boldsymbol{D} \times \boldsymbol{M}$ space by the joint density function

$$\rho(\mathbf{d},\mathbf{m}) = \rho_D(\mathbf{d})\, \rho_M(\mathbf{m}) \ . \qquad (1.59)$$

It may happen that part of the "a priori" information has been obtained from a first, rough analysis of the data set. Rigorously then, there exist correlations between $\mathbf{d}$ and $\mathbf{m}$ in $\rho(\mathbf{d},\mathbf{m})$, and equation (1.59) no longer holds. For a maximum of generality we thus have to assume the existence of a general probability density $\rho(\mathbf{d},\mathbf{m})$, not necessarily satisfying (1.59), and representing all the information we have in data and model parameters *independently* of the use of any theoretical information (which is described by the probability density $\Theta(\mathbf{d},\mathbf{m})$ introduced in section 1.3).

1.5: Defining the solution of the Inverse Problem

1.5.1: Combination of experimental, a priori, and theoretical information

We have seen in the previous section that the *prior probability density* $\rho(\mathbf{d},\mathbf{m})$, defined in the space $D \times M$, represents both information obtained on the observable parameters (data) $\mathbf{d}$, and a priori information on model parameters $\mathbf{m}$. We have also seen that the *theoretical probability density* $\Theta(\mathbf{d},\mathbf{m})$ represents the information on the physical correlations between $\mathbf{d}$ and $\mathbf{m}$, as obtained from a physical law, for instance.

These two states of information combine to produce the *a posteriori state of information*. I postulate here that the way used in the previous sections to introduce the a priori and the theoretical states of information is such that the a posteriori state of information is given by the *conjunction* of these two states of information.

From (1.36), the probability density $\sigma(\mathbf{d},\mathbf{m})$ representing the a posteriori information is then

$$\boxed{\sigma(\mathbf{d},\mathbf{m}) = \frac{\rho(\mathbf{d},\mathbf{m})\ \Theta(\mathbf{d},\mathbf{m})}{\mu(\mathbf{d},\mathbf{m})} ,} \tag{1.60}$$

where $\mu(\mathbf{d},\mathbf{m})$ represents the state of null information.

Like all postulates, this one is justified by the correctness of its consequences. All the rest of this book is based on (1.60). It will be seen that the conclusions obtained from this equation, although more general than those obtained from more traditional approaches, reduce to them in all particular cases. Equation (1.60) first appeared in Tarantola and Valette (1982a).

Once the a posteriori information in the $D\mathrm{x}M$ space has been defined, the a posteriori information in the model space is given by the marginal density function

$$\sigma_M(\mathbf{m}) = \int_D d\mathbf{d}\ \ \sigma(\mathbf{d},\mathbf{m}) , \tag{1.61}$$

while the a posteriori information in the data space is given by

$$\sigma_D(\mathbf{d}) = \int_M d\mathbf{m}\ \ \sigma(\mathbf{d},\mathbf{m}) . \tag{1.62}$$

Figure 1.10 illustrates geometrically the determination of $\sigma_M(\mathbf{m})$ and $\sigma_D(\mathbf{d})$ from $\rho(\mathbf{d},\mathbf{m})$ and $\Theta(\mathbf{d},\mathbf{m})$.

1.5.2: Resolution of Inverse Problems

Equation (1.60) solves a very general problem. Inverse Problems correspond to the particular case where the spaces $\boldsymbol{D}$ and $\boldsymbol{M}$ have fundamentally different physical meaning, and where we are interested in "translating" information from the data space $\boldsymbol{D}$ into the model space $\boldsymbol{M}$. Let us make the assumptions usual in this sort of problems.

First, the theoretical information takes the form:

$$\Theta(\mathbf{d},\mathbf{m}) = \Theta(\mathbf{d}|\mathbf{m}) \; \mu_M(\mathbf{m}) \; , \tag{1.63}$$

i.e., is a conditional probability density in $\mathbf{d}$ for given $\mathbf{m}$ (see section 1.3). Second, the prior information $\rho(\mathbf{d},\mathbf{m})$ takes the form:

$$\rho(\mathbf{d},\mathbf{m}) = \rho_D(\mathbf{d}) \quad \rho_M(\mathbf{m}) \; , \tag{1.64}$$

which means that information in the data space has been obtained (from measurements) independently of the prior information in the model space (see section 1.4). This gives, for the posterior information in the model space,

$$\sigma_M(\mathbf{m}) = \int_D d\mathbf{d} \; \sigma(\mathbf{d},\mathbf{m}) = \rho_M(\mathbf{m}) \int_D d\mathbf{d} \; \frac{\rho_D(\mathbf{d}) \;\; \Theta(\mathbf{d}|\mathbf{m})}{\mu_D(\mathbf{d})} \; , \tag{1.65}$$

where it has been assumed that $\mu(\mathbf{d},\mathbf{m}) = \mu_D(\mathbf{d}) \; \mu_M(\mathbf{m})$ (see section 1.2.4).

Equation (1.65) gives ***the*** solution of the general inverse problem. From $\sigma_M(\mathbf{m})$ is is possible to obtain any sort of information we wish on model parameters: mean values, median values, maximum likelihood values, error bars,... Section 1.6 gives a discussion.

The "existence" of the solution simply means that $\sigma_M(\mathbf{m})$, as defined by (1.65), is not identically null. If this were the case, it would indicate the incompatibility of the experimental results, the a priori hypothesis on model parameters, and the theoretical information, thus showing that some "error bars" have beeen underestimated. I have not been able to define, in the general case, a "test" that would measure the degree of compatibility of the a posteriori information with respect to the a priori one (like the χ^2 test introduced in chapter 4 for least squares problems).

The "uniqueness" of the solution is evident when by solution we mean the probability density $\sigma_M(\mathbf{m})$ itself, and is simply a consequence of the uniqueness of the conjunction of states of information. Of course, $\sigma_M(\mathbf{m})$ may be very pathological (non-normalizable, multimodal,...) but that would

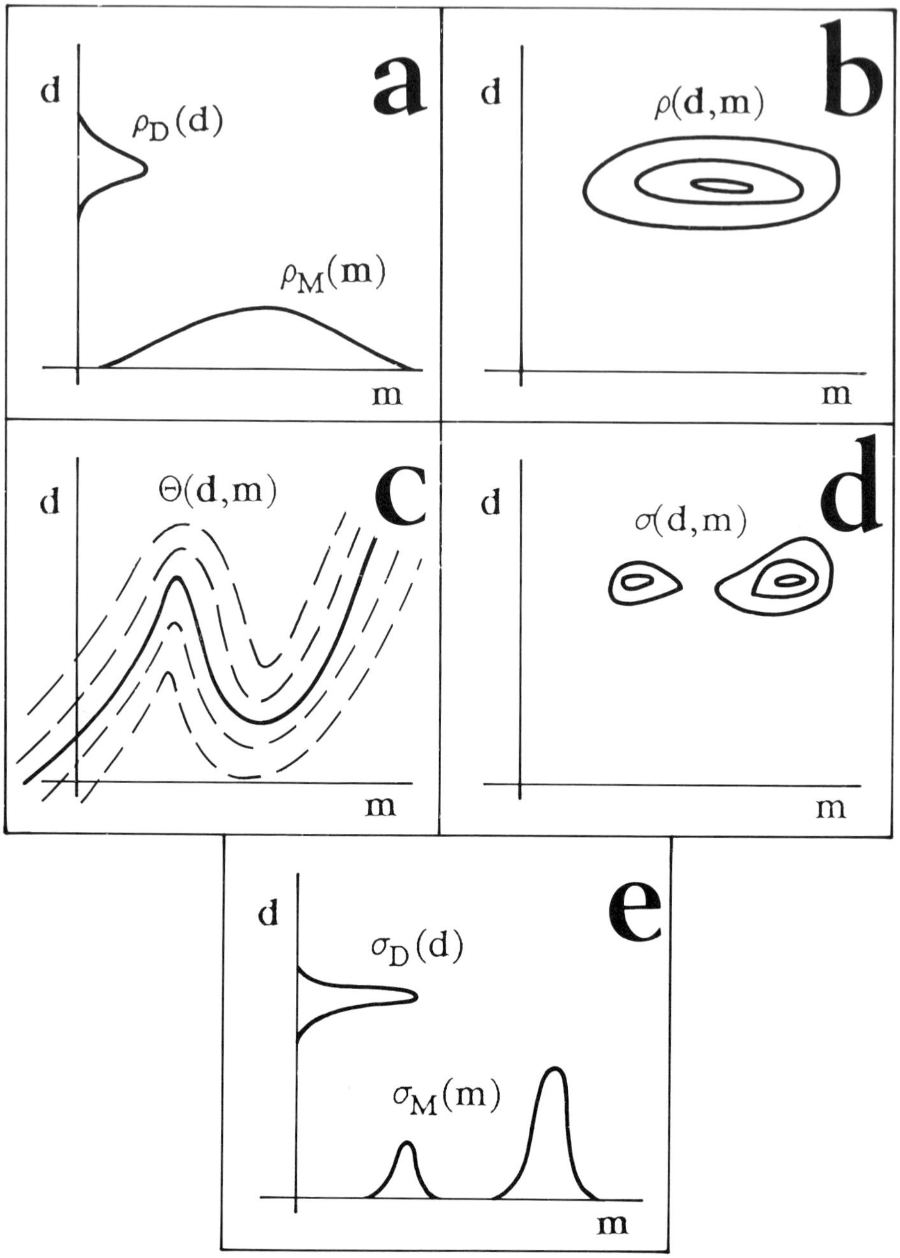
a
d
$\rho_D(d)$
$\rho_M(m)$
m
b
d
$\rho(d,m)$
m
c
d
$\Theta(d,m)$
m
d
d
$\sigma(d,m)$
m
e
d
$\sigma_D(d)$
$\sigma_M(m)$
m

← *Figure 1.10:* a) The probability densities $\rho_D(\mathbf{d})$ and $\rho_M(\mathbf{m})$ respectively represent the information on observable parameters (data) and the prior information on model parameters. b) As the prior information on model parameters is, by definition, independent of the information on observable parameters (measurements), the joint probability density in the space $\boldsymbol{D} \times \boldsymbol{M}$ representing both informations is $\rho(\mathbf{d},\mathbf{m}) = \rho_D(\mathbf{d})\, \rho_M(\mathbf{m})$. c) $\Theta(\mathbf{d},\mathbf{m})$ represents the information on the physical correlations between $\mathbf{d}$ and $\mathbf{m}$, as predicted by a physical theory (usually, $\Theta(\mathbf{d},\mathbf{m}) = \Theta(\mathrm{d}|\mathbf{m})\ \mu\_M(\mathbf{m})$) . d) Given the two states of information represented by $\rho(\mathbf{d},\mathbf{m})$ and $\Theta(\mathbf{d},\mathbf{m})$, their conjuction is $\sigma(\mathbf{d},\mathrm{m}) = \dfrac{\rho(\mathbf{d},\mathrm{m})\ \Theta(\mathbf{d},\mathrm{m})}{\mu(\mathbf{d},\mathbf{m})}$, and represents the "combination" of the two states of information. e) From $\sigma(\mathrm{d},\mathbf{m})$ it is possible to obtain the marginal probability densities $\sigma_M(\mathbf{m}) = \int \mathrm{dd}\ \sigma(\mathbf{d},\mathbf{m})$ and $\sigma_D(\mathrm{d}) = \int \mathrm{d}\mathbf{m}\ \sigma(\mathbf{d},\mathbf{m})$. By comparison of the posterior probability density $\sigma_M(\mathbf{m})$ with the prior one, $\rho_M(\mathbf{m})$, we see that some information has been gained on the model parameters, thanks to the data $\rho_D(\mathrm{d})$ and to the theoretical information $\Theta(\mathbf{d},\mathrm{m})$.

simply mean that such is the information we possess on model parameters. The information itself is uniquely defined.

Using (1.63) and (1.64), the posterior probability density in the data space is

$$\sigma_D(\mathbf{d}) = \int_{\boldsymbol{M}} \mathrm{d}\mathbf{m}\ \sigma(\mathbf{d},\mathbf{m}) = \frac{\rho_D(\mathbf{d})}{\mu_D(\mathbf{d})} \int_{\boldsymbol{M}} \mathrm{d}\mathbf{m}\ \Theta(\mathrm{d}|\mathbf{m})\ \rho_M(\mathbf{m})\ . \tag{1.66}$$

While the probability density (1.65) allows us to estimate the posterior values of the model parameters, the probability density (1.66) allows of estimating the posterior values of data parameters ("recalculated data").

1.5.3: Some special cases

a) Results of the measurements are the output of an instrument with known statistics: Example 33 has shown that if a measuring instrument delivers the value d_{obs} , then

$$\rho_D(\mathbf{d}) = \frac{\nu(\mathbf{d}_{obs}|\mathbf{d})\ \mu_D(\mathbf{d})}{\int_D d\mathbf{d}\ \nu(\mathbf{d}_{obs}|\mathbf{d})\ \mu_D(\mathbf{d})} ,$$

where $\nu(\mathbf{d}_{obs}|\mathbf{d})$ describes the statistics of the instrument, and represents the probability density of the output being $\mathbf{d}_{obs}$ when the input is $\mathbf{d}$. Equation (1.65) then gives

$$\sigma_M(\mathbf{m}) = \text{const.}\ \rho_M(\mathbf{m}) \int_D d\mathbf{d}\ \ \nu(\mathbf{d}_{obs}|\mathbf{d})\ \theta(\mathbf{d}|\mathbf{m}) ,$$

or, in normalized form,

$$\sigma_M(\mathbf{m}) = \frac{\rho_M(\mathbf{m}) \int_D d\mathbf{d}\ \nu(\mathbf{d}_{obs}|\mathbf{d})\ \theta(\mathbf{d}|\mathbf{m})}{\int_M d\mathbf{m}\ \rho_M(\mathbf{m}) \int_D d\mathbf{d}\ \nu(\mathbf{d}_{obs}|\mathbf{d})\ \theta(\mathbf{d}|\mathbf{m})} . \tag{1.67}$$

This equation is identical to equation (6) of box 1.2 obtained using a strict Bayesian approach. ■

b) Errors of the measure instrument are independent of the input, and errors in the theory are independent of the model value. This case corresponds to examples 31 and 36. The output of the measure instrument is related to the input though

$$\mathbf{d}_{obs} = \mathbf{d} + \epsilon_D ,$$

where ϵ_D is un unknown error with known statistics described by the probability density function $f_D(\epsilon_D)$. Then

$$\nu(\mathbf{d}_{obs}|\mathbf{d}) = f_D(\mathbf{d}_{obs}-\mathbf{d}) .$$

If the true value $\mathbf{d}$ differs from the computed value $\mathbf{g}(\mathbf{m})$ by an error ϵ_T,

$$\mathbf{d} = \mathbf{g}(\mathbf{m}) + \epsilon_T ,$$

independent of $\mathbf{m}$, with known statistics described by the probability density function $f_T(\epsilon_T)$, then

$$\theta(\mathbf{d}|\mathbf{m}) = f_T(\mathbf{d}-\mathbf{g}(\mathbf{m})) .$$

Equation (1.67) then gives

$$\sigma_M(\mathbf{m}) = \frac{f(\mathbf{d}_{obs}-\mathbf{g}(\mathbf{m}))\ \rho_M(\mathbf{m})}{\int_{M} d\mathbf{m}\ f(\mathbf{d}_{obs}-\mathbf{g}(\mathbf{m}))\ \rho_M(\mathbf{m})} , \tag{1.68a}$$

where $f(\epsilon)$ is the convolution of $f_D(\epsilon)$ and $f_T(\epsilon)$,

$$f(\epsilon) = f_D(\epsilon) * f_T(\epsilon) , \tag{1.68b}$$

and represents the probability density of the sum of observational and theoretical errors:

$$\epsilon = \epsilon_D + \epsilon_T .$$

Equations (1.68) were used by Duijndam (1987) to show, in this particular example, the equivalence between the strict Bayesian approach and the approach introduced by Tarantola and Valette (1982a) (and developed in this book). ■

c) Negligible modelisation errors: If modelisation errors are negligible compared to observational errors, we can take

$$\Theta(\mathbf{d}|\mathbf{m}) = \delta(\mathbf{d}-\mathbf{g}(\mathbf{m})) ,$$

where $\mathbf{d} = \mathbf{g}(\mathbf{m})$ denotes the (exact) resolution of the forward problem. Equation (1.65) then gives

$$\sigma_M(\mathbf{m}) = \rho_M(\mathbf{m}) \left(\frac{\rho_D(\mathbf{d})}{\mu_D(\mathbf{d})} \right)_{\mathbf{d}=\mathbf{g}(\mathbf{m})} . \blacksquare \tag{1.69}$$

d) Negligible observational errors: Letting $\mathbf{d}_{obs}$ denote the observed data values, the hypothesis of negligible observational errors (with respect to modelization errors) is written

$$\rho_D(\mathbf{d}) = \delta(\mathbf{d}-\mathbf{d}_{obs}) . \tag{1.70}$$

Equation (1.65) then gives

$$\sigma_M(\mathbf{m}) = \rho_M(\mathbf{m}) \frac{\Theta(\mathbf{d}_{obs}|\mathbf{m})}{\mu_D(\mathbf{d}_{obs})} , \tag{1.71a}$$

or, more simply, if we can normalize to unity,

$$\sigma_M(\mathbf{m}) = \frac{\rho_M(\mathbf{m}) \ \Theta(\mathbf{d}_{obs}|\mathbf{m})}{\int_{\mathcal{M}} d\mathbf{m} \ \rho_M(\mathbf{m}) \ \Theta(\mathbf{d}_{obs}|\mathbf{m})} . \ \blacksquare \tag{1.71b}$$

e) Gaussian modelisation and observational errors: This corresponds respectively to (equation (1.41) of example 29)

$$\Theta(\mathbf{d}|\mathbf{m}) = ((2\pi)^{ND} \det \mathbf{C}_T)^{-\frac{1}{2}} \exp\left[-\frac{1}{2}(\mathbf{d}-\mathbf{g}(\mathbf{m}))^t \ \mathbf{C}_T^{-1} \ (\mathbf{d}-\mathbf{g}(\mathbf{m}))\right], \tag{1.72}$$

and (equation (1.49b) of example 35)

$$\frac{\rho_D(\mathbf{d})}{\mu_D(\mathbf{d})} = ((2\pi)^{ND} \det \mathbf{C}_d)^{-\frac{1}{2}} \exp\left[-\frac{1}{2}(\mathbf{d}-\mathbf{d}_{obs})^t \ \mathbf{C}_d^{-1} \ (\mathbf{d}-\mathbf{d}_{obs})\right]. \tag{1.73}$$

As demonstrated in problem 1.20, equation (1.65) then gives

$$\sigma_M(\mathbf{m}) = \rho_M(\mathbf{m}) \exp\left[-\frac{1}{2} (\mathbf{g}(\mathbf{m})-\mathbf{d}_{obs})^t \ \mathbf{C}_D^{-1} \ (\mathbf{g}(\mathbf{m})-\mathbf{d}_{obs})\right] , \tag{1.74}$$

where

$$\mathbf{C}_D = \mathbf{C}_d + \mathbf{C}_T . \tag{1.75}$$

Result (1.74)-(1.75) is important because it shows that, *in the Gaussian assumption, observational errors and modelization errors simply combine by addition of the respective covariance operators*, even when the forward problem is nonlinear. ■

Box 1.4: *Solving Inverse Problems using the Bayesian paradigm.*

Let $f(\mathbf{d},\mathbf{m})$ be a joint probability density on the parameters $(\mathbf{d},\mathbf{m})$. From $f(\mathbf{d},\mathbf{m})$ one can ***define*** the *marginal* probability densities

(...)

$$f_M(\mathbf{m}) = \int_D d\mathbf{d}\ f(\mathbf{d},\mathbf{m}) \tag{1a}$$

and

$$f_D(\mathbf{d}) = \int_M d\mathbf{m}\ f(\mathbf{d},\mathbf{m}) \tag{1b}$$

and the *conditional* probability densities

$$f_{M|D}(\mathbf{m}|\mathbf{d}_0) = \frac{f(\mathbf{d}_0,\mathbf{m})}{\int_M d\mathbf{m}\ f(\mathbf{d}_0,\mathbf{m})} , \tag{2a}$$

and

$$f_{D|M}(\mathbf{d}|\mathbf{m}_0) = \frac{f(\mathbf{d},\mathbf{m}_0)}{\int_D d\mathbf{d}\ f(\mathbf{d},\mathbf{m}_0)} . \tag{2b}$$

It is easy to see that, for a given $\mathbf{d}_0$, $f_{M|D}(\mathbf{m}|\mathbf{d}_0)$ is a probability density in $\mathbf{m}$. Respectively, for a given $\mathbf{m}_0$, $f_{D|M}(\mathbf{d}|\mathbf{m}_0)$ is a probability density for $\mathbf{d}$. See Figure 1.11.

One can easily obtain the identities

$$f_{M|D}(\mathbf{m}|\mathbf{d}_0) = \frac{f_{D|M}(\mathbf{d}_0|\mathbf{m})\ f_M(\mathbf{m})}{\int d\mathbf{m}\ f_{D|M}(\mathbf{d}_0|\mathbf{m})\ f_M(\mathbf{m})} , \tag{3a}$$

and

$$f_{D|M}(\mathbf{d}|\mathbf{m}_0) = \frac{f_{M|D}(\mathbf{m}_0|\mathbf{d})\ f_D(\mathbf{d})}{\int d\mathbf{d}\ f_{M|D}(\mathbf{m}_0|\mathbf{d})\ f_D(\mathbf{d})} , \tag{3b}$$

which are known as the *theorem of Bayes*. Replacing $\mathbf{d}_0$ by $\mathbf{d}_{obs}$ in (4a) gives

(...)

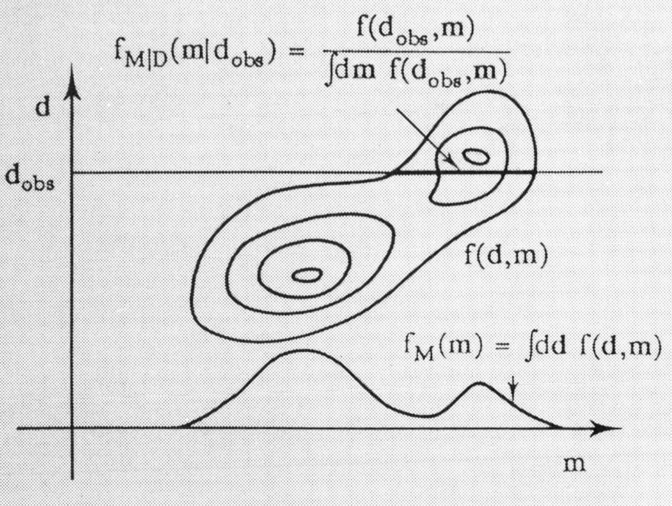

Figure 1.11: A joint probability density $f(\mathbf{d},\mathbf{m})$, the marginal (prior) probability density for $\mathbf{m}$, $f_M(\mathbf{m})$, and the conditional (posterior) probability density for $\mathbf{m}$, $f_{M|D}(\mathbf{m}|\mathbf{d}_{obs})$, given the observation $\mathbf{d} = \mathbf{d}_{obs}$.

$$f_{M|D}(\mathbf{m}|\mathbf{d}_{obs}) = \frac{f_{D|M}(\mathbf{d}_{obs}|\mathbf{m})\ f_M(\mathbf{m})}{\int d\mathbf{m}\ f_{D|M}(\mathbf{d}_{obs}|\mathbf{m})\ f_M(\mathbf{m})} . \tag{4}$$

The interpretation of a probability density as representing an state of information over a parameter set is coherent with the interpretation of the conditional probability density (4) as the *posterior* probability density for $\mathbf{m}$ if an actual measurement of $\mathbf{d}$ gives unambiguosly the result $\mathbf{d} = \mathbf{d}_{obs}$. Then, $f_M(\mathbf{m})$ corresponds to the prior probability density for $\mathbf{m}$. The conditional probability density $f_{D|M}(\mathbf{d}|\mathbf{m})$ is interpreted as the probability density for the observed data value to be $\mathbf{d}$ when the true model value is $\mathbf{m}$. This describes the forward modeling of data, taking into account all sources of uncertainties in the prediction of $\mathbf{d}$ (in particular, observational uncertainties and uncertainties in the physical theory used to predict the data values). This interpretation justifies the use of (4) for the resolution of inverse problems.

(...)

Example: Let $\theta(\mathbf{d}|\mathbf{m})$ be the conditional probability density describing the theoretical relationship between $\mathbf{d}$ and $\mathbf{m}$ (as described in section 1.3), and let $\nu(\mathbf{d}_{obs}|\mathbf{d})$ be the probability density for the output of a measuring instrument to be $\mathbf{d}_{obs}$ when the input is $\mathbf{d}$ (example 33). Then, the conditional probability density $f_{D|M}(\mathbf{d}_{obs}|\mathbf{m})$ defined above is given by

$$f_{D|M}(\mathbf{d}_{obs}|\mathbf{m}) = \int_{D} d\mathbf{d}\ \nu(\mathbf{d}_{obs}|\mathbf{d})\ \theta(\mathbf{d}|\mathbf{m}) . \tag{5}$$

Then, equation (4) gives

$$f_{M|D}(\mathbf{m}|\mathbf{d}_{obs}) = \frac{f_M(\mathbf{m}) \int_{D} d\mathbf{d}\ \nu(\mathbf{d}_{obs}|\mathbf{d})\ \theta(\mathbf{d}|\mathbf{m})}{\int_{M} d\mathbf{m}\ f_M(\mathbf{m}) \int_{D} d\mathbf{d}\ \nu(\mathbf{d}_{obs}|\mathbf{d})\ \theta(\mathbf{d}|\mathbf{m})} . \tag{6}$$

Although with different notations, this equation is identical to equation (1.67), obtained from the general concept of combination of information developed in this book. Of course, equation (1.60), at the basis of this book, applies to much more general situations than (6) does. For instance, Figure (1.4) shows that situations exist such that the result of a measurement does not give a thing which can be named $\mathbf{d}_{obs}$, which is central in the concept of conditional probability, and thus, in the strict Bayesian approach.

1.6: Using the solution of the Inverse Problem

1.6.1: Describing the a posteriori information in the model space

What does it mean to "solve" an inverse problem? This depends on the sort of practical application we have in view.

Very often we are interested in the model parameters *per se*. The most general way of studying the information obtained on the parameter values is by a direct study of the probability density $\sigma_M(\mathbf{m})$. As a probability density may be quite complicated (multimodal, infinite variances,...) there is no general procedure for obtaining simple pieces of information. The most comprehensive understanding is obtained by directly discussing the *probability* that the true value of the model parameters lies in a given range (i.e., it belongs to a given subset):

$$P(m \in A) = \frac{1}{N} \int_A dm \ \sigma_M(\mathbf{m}) \ , \tag{1.76}$$

where N is the norm of σ_M :

$$N = \int_M dm \ \sigma_M(\mathbf{m}) \tag{1.77}$$

(if σ_M is not normalizable, only *relative* probabilities can be computed).

Choosing different subsets $\boldsymbol{A} \subset M$ it is possible to get quite a good idea of the actual information we possess on the true values of the model parameters. If $\sigma_M(\mathbf{m})$ does not have a very complicated shape, it is possible to describe it adequately by its central estimator and estimators of dispersion (box 1.2 recalls the general definitions in norm ℓ_p for $1 \le p \le \infty$). Among the central estimators, the easiest to obtain are generally the maximum likelihood value $\mathbf{m}_{ML}$

$$\mathbf{m}_{ML} \qquad : \qquad \sigma_M(\mathbf{m}_{ML}) \ \ MAX \ ; \tag{1.78}$$

and the mean value (or mathematical expectation) $\langle \mathbf{m} \rangle$

$$\langle \mathbf{m} \rangle = \frac{1}{N} \int_M dm \ \mathbf{m} \ \sigma_M(\mathbf{m}) \ . \tag{1.79}$$

Among the estimators of dispersion, the easiest to obtain is generally the posterior covariance operator $\mathbf{C}_{M'}$

$$\mathbf{C}_{M'} = \frac{1}{N} \int_M dm \ (\mathbf{m} - \langle \mathbf{m} \rangle) (\mathbf{m} - \langle \mathbf{m} \rangle)^t \ \sigma_M(\mathbf{m})$$

$$= \frac{1}{N} \int_M dm \ \mathbf{m} \ \mathbf{m}^t \ \sigma_M(\mathbf{m}) - \langle \mathbf{m} \rangle \ \langle \mathbf{m} \rangle^t \ , \tag{1.80}$$

but it has to be emphasized that the covariance operator gives understandable information only in the case when the probability density $\sigma_M(\mathbf{m})$ can be fitted reasonably well by a Gaussian function.

Sometimes the inverse problem is solved as an intermediate one in a more general decision problem in which the decision maker has to combine information obtained from the inverse problem with economic considerations such as, for instance, in operational research. As an example, consider the oil company which, in the light of the results obtained after a 1 million dollar seismic exploration experiment has to decide on the eventual drilling of a 5 million dollar exploratory well. Unfortunately, although the present state of computer technology allows a reasonably general resolution of the inverse problem in seismic exploration, it does not yet allow a general resolution of the coupled inverse problem / decision problem in realistically complex cases. This field will certainly undergo a rapid growth in the coming years. Readers interested in Bayesian decision theory can refer to Box and Tiao (1973), Morgan (1968), Schmitt (1969), or Winkler (1972).

1.6.2: Analysis of error and resolution

When the solution of the inverse problem is given as a central estimator (such as mean, median, or maximum likelihood), it is necessary to discuss error and resolution.

By discussion of errors is usually meant the obtainment of significant "error bars". What the meaning of an error bar may be is not always evident. For instance, while the probability of the parameter being at more than 3.5 standard deviations from the mean is null for a box-car function, it is approximately 10% for a symmetric-exponential function. For multimodal density functions, the standard deviation can be completely meaningless.

In the traditional "analysis of resolution", two different concepts are involved. First, a parameter is *well resolved* by the data set if its posterior error bar is much smaller than the prior one. More generally, if its posterior marginal probability density is significantly different from the prior one. If, for instance, the prior and posterior probability densities are identical, the parameter is completely unresolved.

The second concept involved in the analysis of resolution arises when the parameters m^α represent a discretization of a function $\Psi(t)$ of a continuous variable t (representing for instance a location in time or space). Assume that the posterior covariance operator adequately represents the dispersion around the mean. Usual covariance operators (prior and posterior) are not diagonal, but are "band diagonal" (covariances between neighbouring parameters are not null). This means that neighbouring parameters have errors which are correlated. The greater the correlation length of errors, the worse is the (temporal, spatial,...) resolution attained with the data set. In Chapter 9 the different (although equivalent) point of view of Backus and Gilbert (1968) is discussed.

Stability is defined as the property of a central estimator of being insensitive to small random errors in the data values. If the a priori information in the model space has been properly introduced, stability is generally warranted.

Robustness is the property of insensitivity with respect to a small number of big errors (outliers) in the data set. For instance, the hypothesis that errors are described by a symmetric exponential is robust; the hypothesis that errors are distributed following a Gaussian function is *not* robust (see Chapter 6).

Another important concept is that of the *importance* of a particular datum for a particular model parameter. If, for instance, the conditional probability density of the parameter m^α with respect to d^i , $\sigma(m^\alpha|d^i)$, is, in fact, independent of d^i, the datum d^i has a null importance for m^α . A possible definition of the importance of the datum d^i for the model parameter m^α is the change of information content of the posterior marginal probability density for m^α when suppressing the datum d^i from the data set. See, for instance, Chapter 4, or Minster et al. (1974), for a discussion of the importance of data in least-squares problems. The analysis of data importance may be helpful for optimizing experimental configurations.

It may happen that some of the hypotheses made are inconsistent. For instance, in least-squares problems, the posterior data residuals may be much greater than experimental or modelization uncertainties (Chapter 4), and a χ^2 test can easily detect such an inconsistency. In the general case, inconsistency is detected if the product $\rho(\mathbf{d},\mathbf{m})\ \Theta(\mathbf{d},\mathbf{m})$ is very small everywhere in the space $D \times M$, but I have not yet been able to obtain a quantitative test. Possibly, Shannon's concept of information content (section 1.2.4) should be applied for the general study of resolution, data importance, or consistency, but no general result yet exists.

Let me now review some of the numerical techniques that may be used to solve the inverse problem.

1.6.3: Analytic solutions

In some cases, it is possible to obtain a simple analytical expression for the posterior probability density. For instance, if probability densities used to describe observational uncertainties, forward modelization uncertainties, and prior uncertainties on model parameters are *Gaussian*, and if the forward equation $\mathbf{d}_{cal} = \mathbf{g}(\mathbf{m})$ is linear in $\mathbf{m}$, then it can be shown that the posterior probability density $\sigma_M(\mathbf{m})$ is also Gaussian. It can then be completely described by its central value and its covariance operator, for which it is possible to obtain explicit expressions (see section 1.7).

Sometimes, the forward equation $\mathbf{d}_{cal} = \mathbf{g}(\mathbf{m})$ is not linear, but it can be linearized around some *reference* model $\mathbf{m}_0$:

$$\mathbf{d}_{cal} = \mathbf{g}(\mathbf{m}) \simeq \mathbf{g}(\mathbf{m}_0) + \mathbf{G}_0\,(\mathbf{m}-\mathbf{m}_0) \ ,$$

where the linear operator $\mathbf{G}_0$ represents the derivative of $\mathbf{g}$ at $\mathbf{m} = \mathbf{m}_0$. Then, the analytic advantages of linear problems are preserved (see also section 1.7 for an example).

For nonlinear forward problems, there is generally no explicit expression of the solution.

1.6.4: Systematic exploration of the model space

If the number of model parameters is very small (≤ 4), and if the computation of the numerical value of $\sigma_M(\mathbf{m})$ for an arbitrary $\mathbf{m}$ is inexpensive (i.e., not consuming too much computer-time), we can define a grid over the model space, compute $\sigma_M(\mathbf{m})$ everywhere in the grid, and directly use these results to discuss the information obtained on model parameters. This is certainly the most general way of solving the inverse problem. Problem 1.1 gives an illustration of the method.

1.6.5: Monte Carlo methods

If the number of parameters is not small, and if the computation of $\sigma_M(\mathbf{m})$ at any point $\mathbf{m}$ is not expensive, the systematic exploration of the model space can advantageously be replaced by a random (Monte Carlo) exploration. Monte Carlo methods are discussed in Chapter 3. For instance, the method of *simulated annealing* allows of obtaining the maximum likelihood point, even when the probability density $\sigma_M(\mathbf{m})$ is multimodal. Also, the computation of the mathematical expectation and of the posterior covariance operator can be made by evaluating the sums (1.79) and (1.80) by a Monte Carlo method of numerical integration.

Monte Carlo methods have the advantage of not using any linear approximation.

1.6.6: Computation of the maximum likelihood point

Usual problems do not have a small number of parameters and the computation of $\sigma_M(\mathbf{m})$ at any point $\mathbf{m}$ is expensive. The only practical approach is often to try to define an astute strategy which, in a small number of moves, will give the point $\mathbf{m}_{ML}$ maximizing $\sigma_M(\mathbf{m})$, i.e., the maximum likelihood point.

If $\sigma_M(\mathbf{m})$ is a differentiable function of $\mathbf{m}$, the maximum likelihood point can be obtained using gradient methods. The gradient of $\sigma_M(\mathbf{m})$ has components $\partial\sigma_M/\partial m^\alpha$. As discussed in Chapters 4 and 5, "gradient" is *not* synonymous with "direction of steepest ascent": to define the latter it is necessary to define a *distance* over the model space (or, more precisely, over the chosen *chart* of the model space). Except for very special cases (e.g. example 2), there is no "natural" definition of distance. The simplest choice (not necessarily very good) is the ℓ_2 distance

$$D(\mathbf{m}_1,\mathbf{m}_2) = \left[\sum_{\alpha\in \mathbf{I}_M} \frac{(m_1{}^\alpha - m_2{}^\alpha)^2}{(\sigma^\alpha)^2} \right]^{1/2} ,$$

where σ^α is any characteristic length for m^α, such as for instance, the corresponding estimator of dispersion in the prior probability density, $\rho_M(\mathbf{m})$. Although any definition of distance is acceptable, astute choices may speed the convergence of a gradient method. For instance, in example 2, the parameters $(m^1,m^2) = (\lambda,\theta)$ were geographical coordinates; the best definition of distance in the map (λ,θ) between two points $\mathbf{m}_1 = (\lambda_1,\theta_1)$ and $\mathbf{m}_2 = (\lambda_2,\theta_2)$ is the length of the corresponding great circle in the sphere:

$$D(\mathbf{m}_1,\mathbf{m}_2) = \text{Arc cos}(\sin\theta_1 \sin\theta_2 + \cos\theta_1 \cos\theta_2 \cos(\lambda_2-\lambda_1)) . \tag{1.81}$$

To take this definition of distance to compute the direction of steepest descent simply correspond to the computation of the gradient "in spherical coordinates". Figure 1.12 illustrates the dependence of the direction of steepest descent on the particular choice of distance. For more details on gradient methods, see Chapters 4 and 5.

As $\sigma_M(\mathbf{m})$ is, in general, an arbitrarily complicated function of $\mathbf{m}$, there is no warranty that the maximum likelihood point is unique, or that a given point which is locally maximum, is the absolute maximum. Only a full exploration of the space would give the proof, but this is generally too expensive to make. Only a skilful blend of mathematical discussion and physical arguments usually gives some insight into the uniqueness of the maximum likelihood point.

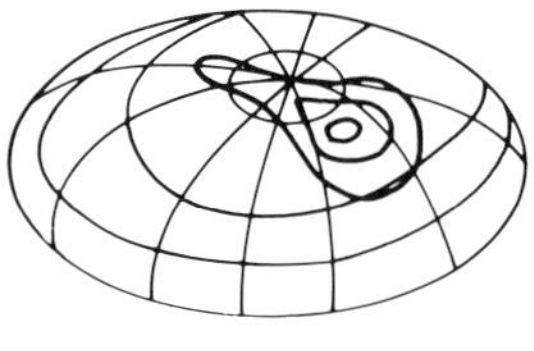

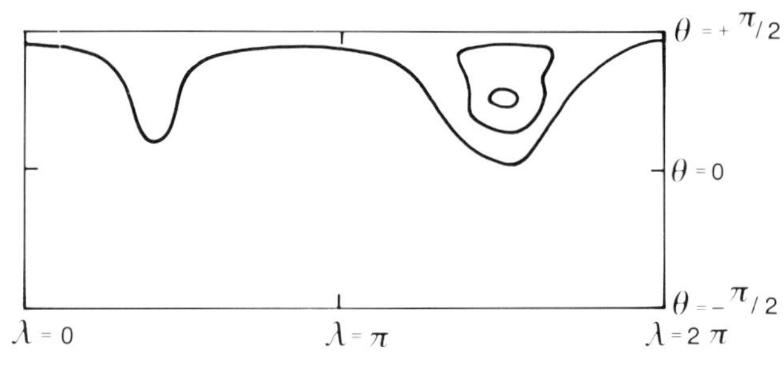

1.7: Special cases

1.7.1: The Gaussian Hypothesis (least-squares criterion). Case $\mathbf{d} = \mathbf{g(m)}$.

As discussed in section 1.5, the general solution of inverse problems is

$$\sigma_M(\mathbf{m}) = \rho_M(\mathbf{m}) \int_D d\mathbf{d} \, \frac{\rho_D(\mathbf{d}) \, \Theta(\mathbf{d}|\mathbf{m})}{\mu_D(\mathbf{d})} , \qquad (1.65 \text{ again})$$

where $\rho_D(\mathbf{d})$ represents the available information on the true values of observable parameters (data), $\mu_D(\mathbf{d})$ is the non-informative probability density on observable parameters, $\Theta(\mathbf{d}|\mathbf{m})$ is the conditional probability density representing the forward modelization, $\rho_M(\mathbf{m})$ represents the a priori information on model parameters, and $\sigma_M(\mathbf{m})$ represents the a posteriori information.

If the observations consist of the observed output, $\mathbf{d}_{obs}$, of a measuring instrument with known statistics, and if $\nu(\mathbf{d}_{obs}|\mathbf{d})$ represents the conditional probability density for the observed value to be $\mathbf{d}_{obs}$ when the true value is $\mathbf{d}$, then, as discussed in example 33,

$$\frac{\rho_D(\mathbf{d})}{\mu_D(\mathbf{d})} = \frac{\nu(\mathbf{d}_{obs}|\mathbf{d})}{\int_D d\mathbf{d} \, \nu(\mathbf{d}_{obs}|\mathbf{d}) \, \mu_D(\mathbf{d})} . \qquad (1.46 \text{ again})$$

If the statistic of the instrument is Gaussian, then (equation (1.49a))

$$\nu(\mathbf{d}_{obs}|\mathbf{d}) = ((2\pi)^{ND} \det \mathbf{C}(\mathbf{d}_{obs}))^{-1/2} \exp\left[-\frac{1}{2} \, (\mathbf{d}-\mathbf{d}_{obs})^t \, \mathbf{C}_d^{-1} \, (\mathbf{d}-\mathbf{d}_{obs}) \right],$$

where $\mathbf{C}_d$ is the covariance operator describing "experimental uncertainties".

If the forward modelization can be written

$$\mathbf{d}_{cal} = \mathbf{g}(\mathbf{m}) , \qquad (1.82)$$

$\mathbf{g}$ represents a nonlinear operator from the model space into the data space, and if estimated modelization errors are assumed Gaussian, then

---

← *Figure 1.12:* A function $\sigma(\lambda,\theta)$ defined over the sphere has to be minimized. A naïve use of gradient methods would correspond to the choice of an euclidean distance *over the chart*. A right use of gradient methods implies the definition of a distance over the chart (equation (1.81)) which corresponds to the (geodetic) distance over the sphere.

$$\theta(\mathbf{d}|\mathbf{m}) = ((2\pi)^{ND} \det \mathbf{C}_T)^{-1/2} \exp\left[-\frac{1}{2}(\mathbf{d}-\mathbf{g}(\mathbf{m}))^t\, \mathbf{C}_T^{-1}\, (\mathbf{d}-\mathbf{g}(\mathbf{m}))\right], \qquad (1.41 \text{ again})$$

where $\mathbf{C}_T$ is the covariance operator describing "forward modelization uncertainties".

With these assumptions, the sum over the data space in (1.65) can be performed analytically (problem 1.20), and we obtain

$$\sigma_M(\mathbf{m}) = \rho_M(\mathbf{m}) \exp\left[-\frac{1}{2}\,(\,\mathbf{g}(\mathbf{m})-\mathbf{d}_{obs}\,)^t\, \mathbf{C}_D^{-1}\,(\,\mathbf{g}(\mathbf{m})-\mathbf{d}_{obs}\,)\right], \qquad (1.75 \text{ again})$$

where the covariance operator $\mathbf{C}_D$ combines experimental and theoretical uncertainties:

$$\mathbf{C}_D = \mathbf{C}_d + \mathbf{C}_T\,. \qquad (1.83)$$

Thanks to the simplicity of (1.83), when using Gaussian models for uncertainties, we can forget that there are two different sources of uncertainties in the data space. All happens as if the forward modelization were exact, and $\mathbf{C}_D$ represented only experimental uncertainties (or, conversely, as if observations were exact, and $\mathbf{C}_D$ represented only forward modelization errors).

In this section we examine the case where the probability density representing the a priori information on model parameters is also Gaussian:

$$\rho_M(\mathbf{m}) = ((2\pi)^{NM} \det \mathbf{C}_M)^{-1/2} \exp\left[-\frac{1}{2}\,(\mathbf{m}-\mathbf{m}_{prior})^t\, \mathbf{C}_M^{-1}\,(\mathbf{m}-\mathbf{m}_{prior})\right]. \qquad (1.84)$$

$\mathbf{m}_{prior}$ is the "a priori model", and $\mathbf{C}_M$ is the covariance operator describing estimated uncertainties in $\mathbf{m}_{prior}$.

We then have, up to a real (adimensional) normalizing constant,

$$\sigma_M(\mathbf{m}) = \text{const} \qquad (1.85)$$
$$\exp\left[-\frac{1}{2}\left((\mathbf{g}(\mathbf{m})-\mathbf{d}_{obs})^t\, \mathbf{C}_D^{-1}\,(\mathbf{g}(\mathbf{m})-\mathbf{d}_{obs}) + (\mathbf{m}-\mathbf{m}_{prior})^t\, \mathbf{C}_M^{-1}\,(\mathbf{m}-\mathbf{m}_{prior})\right)\right].$$

Let me discuss linear and nonlinear problems separately.

a) The forward problem is linear. Instead of writing $\mathbf{d} = \mathbf{g}(\mathbf{m})$, we then write

$$\mathbf{d} = \mathbf{G}\,\mathbf{m}\,, \qquad (1.86)$$

where $\mathbf{G}$ represents a *linear* operator acting from the model space into the data space.

This gives

$$\sigma_M(\mathbf{m}) = \text{const. } \exp(- S(\mathbf{m})) , \tag{1.87}$$

where $S(\mathbf{m})$ is the quadratic function

$$S(\mathbf{m}) = \frac{1}{2}\left[(\mathbf{G}\,\mathbf{m} - \mathbf{d}_{obs})^t\, \mathbf{C}_D^{-1}\, (\mathbf{G}\,\mathbf{m} - \mathbf{d}_{obs}) + (\mathbf{m}-\mathbf{m}_{prior})^t\, \mathbf{C}_M^{-1}\, (\mathbf{m}-\mathbf{m}_{prior}) \right]. \tag{1.88}$$

Defining

$$\boxed{\langle\, \mathbf{m}\, \rangle = \left(\mathbf{G}^t\, \mathbf{C}_D^{-1}\, \mathbf{G} + \mathbf{C}_M^{-1} \right)^{-1} \left(\mathbf{G}^t\, \mathbf{C}_D^{-1}\, \mathbf{d}_{obs} + \mathbf{C}_M^{-1}\, \mathbf{m}_{prior} \right) ,} \tag{1.89}$$

and

$$\boxed{\mathbf{C}_{M'} = \left(\mathbf{G}^t\, \mathbf{C}_D^{-1}\, \mathbf{G} + \mathbf{C}_M^{-1} \right)^{-1} ,} \tag{1.90}$$

we obtain

$$2\, S(\mathbf{m}) = (\, \mathbf{m} - \langle\, \mathbf{m}\, \rangle\,)^t\, \mathbf{C}_{M'}^{-1}\, (\, \mathbf{m} - \langle\, \mathbf{m}\, \rangle\,)$$

$$- \langle\, \mathbf{m}\, \rangle^{t}\, \mathbf{C}_M^{-1}\, \langle\, \mathbf{m}\, \rangle + \mathbf{d}_{obs}{}^t\, \mathbf{C}_D^{-1}\, \mathbf{d}_{obs} + \mathbf{m}_{prior}{}^t\, \mathbf{C}_M^{-1}\, \mathbf{m}_{prior} \,.$$

All the right-hand terms but the first are constant (independent of $\mathbf{m}$), and can be absorbed in the constant factor of (1.87). This gives

$$\sigma_M(\mathbf{m}) = \text{const. } \exp\left[-\frac{1}{2}\, (\, \mathbf{m} - \langle\, \mathbf{m}\, \rangle\,)^t\, \mathbf{C}_{M'}^{-1}\, (\, \mathbf{m} - \langle\, \mathbf{m}\, \rangle\,) \right] ,$$

or in normalized form,

$$\sigma_M(\mathbf{m}) = ((2\pi)^{NM} \det \mathbf{C}_{M'})^{-1/2}$$

$$\exp\left[-\frac{1}{2}\, (\, \mathbf{m} - \langle\, \mathbf{m}\, \rangle\,)^t\, \mathbf{C}_{M'}^{-1}\, (\, \mathbf{m} - \langle\, \mathbf{m}\, \rangle\,) \right] . \tag{1.91}$$

Equation (1.91) shows the important result that, when the forward problem is linear, the a posteriori probability density in the model space is Gaussian. The center of this Gaussian is given by (1.89), while its covariance operator is given by (1.90).

We successively have

$$\langle \mathbf{m} \rangle = \mathbf{C}_{M'} \left(\mathbf{G}^t \mathbf{C}_D^{-1} \mathbf{d}_{obs} + \mathbf{C}_M^{-1} \mathbf{m}_{prior} \right)$$

$$= \mathbf{C}_{M'} \mathbf{G}^t \mathbf{C}_D^{-1} \mathbf{d}_{obs} + \mathbf{C}_{M'} \mathbf{C}_M^{-1} \mathbf{m}_{prior}$$

$$= \mathbf{C}_{M'} \mathbf{G}^t \mathbf{C}_D^{-1} \mathbf{d}_{obs} + \mathbf{C}_{M'} \left(\mathbf{G}^t \mathbf{C}_D^{-1} \mathbf{G} + \mathbf{C}_M^{-1} - \mathbf{G}^t \mathbf{C}_D^{-1} \mathbf{G} \right) \mathbf{m}_{prior}$$

$$= \mathbf{C}_{M'} \mathbf{G}^t \mathbf{C}_D^{-1} \mathbf{d}_{obs} + \mathbf{C}_{M'} \left(\mathbf{C}_{M'}^{-1} - \mathbf{G}^t \mathbf{C}_D^{-1} \mathbf{G} \right) \mathbf{m}_{prior}$$

$$= \mathbf{C}_{M'} \mathbf{G}^t \mathbf{C}_D^{-1} \mathbf{d}_{obs} + \left(\mathbf{I} - \mathbf{C}_{M'} \mathbf{G}^t \mathbf{C}_D^{-1} \mathbf{G} \right) \mathbf{m}_{prior}$$

$$= \mathbf{m}_{prior} + \mathbf{C}_{M'} \mathbf{G}^t \mathbf{C}_D^{-1} (\mathbf{d}_{obs} - \mathbf{G}\, \mathbf{m}_{prior}) ,$$

i.e.,

$$\boxed{\langle \mathbf{m} \rangle = \mathbf{m}_{prior} + \left(\mathbf{G}^t \mathbf{C}_D^{-1} \mathbf{G} + \mathbf{C}_M^{-1} \right)^{-1} \mathbf{G}^t \mathbf{C}_D^{-1} (\mathbf{d}_{obs} - \mathbf{G}\, \mathbf{m}_{prior}) .} \tag{1.92}$$

In problem 1.19 it is shown that expressions (1.90) and (1.92) can also be written:

$$\boxed{\langle \mathbf{m} \rangle = \mathbf{m}_{prior} + \mathbf{C}_M \mathbf{G}^t (\mathbf{G}\, \mathbf{C}_M \mathbf{G}^t + \mathbf{C}_D)^{-1} (\mathbf{d}_{obs} - \mathbf{G}\, \mathbf{m}_{prior}) ,} \tag{1.93}$$

and

$$\boxed{\mathbf{C}_{M'} = \mathbf{C}_M - \mathbf{C}_M \mathbf{G}^t (\mathbf{G}\, \mathbf{C}_M \mathbf{G}^t + \mathbf{C}_D)^{-1} \mathbf{G}\, \mathbf{C}_M .} \tag{1.94}$$

As $\langle \mathbf{m} \rangle$ clearly maximizes $\sigma_M(\mathbf{m})$, it minimizes the quadratic expression

$$S(\mathbf{m}) = \frac{1}{2}\left((\mathbf{G}\, \mathbf{m} - \mathbf{d}_{obs})^t \mathbf{C}_D^{-1} (\mathbf{G}\, \mathbf{m} - \mathbf{d}_{obs}) + (\mathbf{m} - \mathbf{m}_{prior})^t \mathbf{C}_M^{-1} (\mathbf{m} - \mathbf{m}_{prior}) \right), \tag{1.95}$$

thus justifying the usual terminology of a *least-squares estimator*.

From equations (1.92) and (1.93) it is clear that $\langle \mathbf{m} \rangle - \mathbf{m}_{prior}$ depends linearly on $\mathbf{d}_{obs} - \mathbf{G}\, \mathbf{m}_{prior}$. It can be shown (see for instance Rao

(1973)) that among all such linear estimators, the least squares estimator has the property of "minimum variance", whatever the a priori statistics in the data and model spaces may be (Gaussian or not). This result is often used as a justification of the least-squares criterion. This is a mistake, because the minimization of the variances is a *bad* criterion when dealing with non-Gaussian functions. We have thus to emphasize here that, even if data errors have $\mathbf{C}_D$ as covariance operator, if the a priori model has $\mathbf{C}_M$ as covariance operator, we should *not* accept the minimization of $S(\mathbf{m})$ defined in equation (1.95) as a good criterion *unless* these error distributions can reasonably be modeled by Gaussian functions. Problem 1.9 shows an example where this is not the case.

Numerical aspects of the use of the previous equations are given in Chapter 4.

So far we have only been interested in the posterior probability density for model parameters. For data parameters this density is:

$$\sigma_D(\mathbf{d}) = \int_M d\mathbf{m} \ \sigma(\mathbf{d},\mathbf{m}) = \frac{\rho_D(\mathbf{d})}{\mu_D(\mathbf{d})} \int_M d\mathbf{m} \ \Theta(\mathbf{d}|\mathbf{m}) \ \rho_M(\mathbf{m}) \ . \qquad \text{(1.66 again)}$$

Using the assumptions in this paragraph, it can be shown (see section 1.7.2 for a demontration) that $\sigma_D(\mathbf{d})$ is also Gaussian:

$$\sigma_D(\mathbf{d}) = (2\pi)^{ND} \det \mathbf{C}_{D'})^{1/2} \exp\left[-\frac{1}{2} (\mathbf{d} - \langle \mathbf{d} \rangle)^t \ \mathbf{C}_{D'}^{-1} \ (\mathbf{d} - \langle \mathbf{d} \rangle) \right] . \qquad (1.96)$$

For instance, for negligible modeling errors ($\mathbf{C}_T = \mathbf{0}$), we obtain the following simple results (see section 1.7.2):

$$\boxed{\langle \mathbf{d} \rangle = \mathbf{G} \langle \mathbf{m} \rangle \ ,} \qquad (1.97a)$$

and

$$\boxed{\mathbf{C}_{D'} = \mathbf{G} \ \mathbf{C}_{M'} \ \mathbf{G}^t \ .} \qquad (1.97b)$$

b) The forward problem is nonlinear In this case, equation (1.85) cannot be further simplified:

$$\sigma_M(\mathbf{m}) = \text{const.} \qquad \text{(1.85 again)}$$

$$\exp\left[-\frac{1}{2} \left[(\mathbf{g}(\mathbf{m})-\mathbf{d}_{obs})^t \ \mathbf{C}_D^{-1} \ (\mathbf{g}(\mathbf{m})-\mathbf{d}_{obs}) + (\mathbf{m}-\mathbf{m}_{prior})^t \ \mathbf{C}_M^{-1} \ (\mathbf{m}-\mathbf{m}_{prior}) \right] \right] .$$

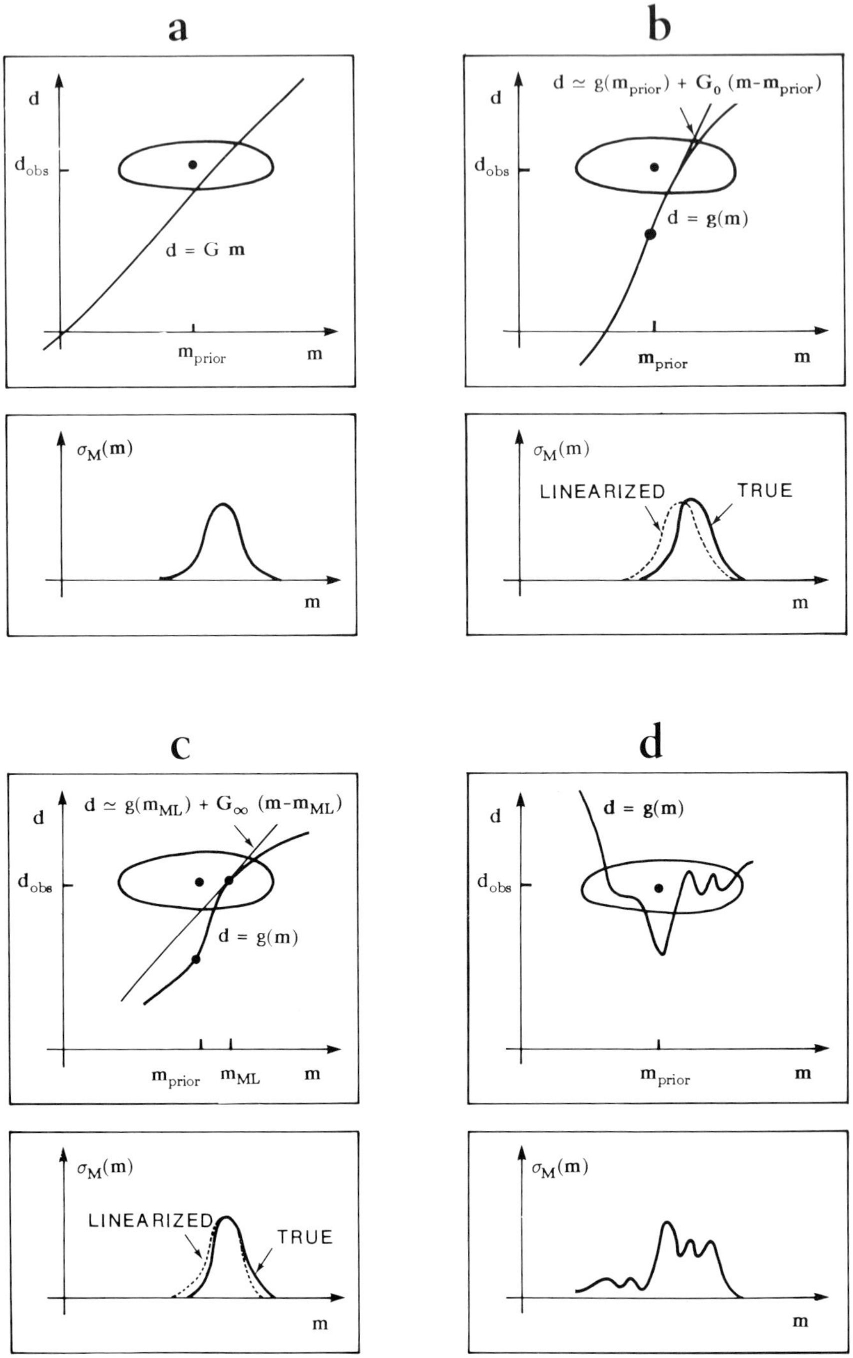
a
d
d_{obs}
d = G m
m_{prior}
m
$\sigma_M(\mathbf{m})$
m
b
$d \simeq g(m_{prior}) + G_0 (m - m_{prior})$
d
d_{obs}
d = g(m)
m_{prior}
m
$\sigma_M(m)$
LINEARIZED
TRUE
m
c
$d \simeq g(m_{ML}) + G_\infty (m - m_{ML})$
d
d_{obs}
d = g(m)
m_{prior}
m_{ML}
m
$\sigma_M(m)$
LINEARIZED
TRUE
m
d
d = g(m)
d
d_{obs}
m_{prior}
m
$\sigma_M(m)$
m

← *Figure 1.13:* In sketches (a) to (d), the top is a representation of the probability density $\rho(\mathbf{d},\mathbf{m})$ and of the theoretical relationship $\mathbf{d} = \mathbf{g}(\mathbf{m})$ (resolution of the forward problem), with increasing nonlinearity from (a) to (d). The bottom is a representation of the corresponding a posteriori probability density $\sigma_M(\mathbf{m})$.

a) The forward equation $\mathbf{d} = \mathbf{G}\ \mathbf{m}$ is linear. The posterior probability density $\sigma_M(\mathbf{m})$ is Gaussian:

$$\sigma_M(\mathbf{m}) = ((2\pi)^N \det \mathbf{C}_{M'})^{-1/2} \exp\left[-\frac{1}{2}(\mathbf{m}-\langle\ \mathbf{m}\ \rangle\ )^t\ \mathbf{C}_{M'}{}^{-1}\ (\mathbf{m}-\langle\ \mathbf{m}\ \rangle\ )\right] ,$$

with

$$\langle\ \mathbf{m}\ \rangle\ = \mathbf{m}_{prior} + \left[\mathbf{G}^t\ \mathbf{C}_D{}^{-1}\ \mathbf{G} + \mathbf{C}_M{}^{-1}\right]^{-1}\ \mathbf{G}^t\ \mathbf{C}_D{}^{-1}\ (\mathbf{d}_{obs} - \mathbf{G}\ \mathbf{m}_{prior})$$

$$\mathbf{C}_{M'} = \left[\mathbf{G}^t\ \mathbf{C}_D{}^{-1}\ \mathbf{G} + \mathbf{C}_M{}^{-1}\right]^{-1} .$$

b) The forward equation $\mathbf{d} = \mathbf{g}(\mathbf{m})$ can be linearized around $\mathbf{m}_{prior}$:

$$\mathbf{g}(\mathbf{m}) \simeq \mathbf{g}(\mathbf{m}_{prior}) + \mathbf{G}_0\ (\mathbf{m} - \mathbf{m}_{prior}) ,$$

where $\mathbf{G}_0$ represents the derivative operator with elements

$$G_0{}^{i\alpha} = \left[\frac{\partial g^i}{\partial m^\alpha}\right]_{\mathbf{m}_{prior}} .$$

The posterior probability density $\sigma_M(\mathbf{m})$ is approximately Gaussian. Its maximum likelihood point is given approximately by

$$\mathbf{m}_{ML} \simeq \mathbf{m}_{prior} + \left[\mathbf{G}_0{}^t\ \mathbf{C}_D{}^{-1}\ \mathbf{G} + \mathbf{C}_M{}^{-1}\right]^{-1}\ \mathbf{G}_0{}^t\ \mathbf{C}_D{}^{-1}\ (\mathbf{d}_{obs} - \mathbf{g}(\mathbf{m}_{prior})) ,$$

and the a posteriori covariance operator is approximately given by

$$\mathbf{C}_{M'} \simeq \left[\mathbf{G}_0{}^t\ \mathbf{C}_D{}^{-1}\ \mathbf{G}_0 + \mathbf{C}_M{}^{-1}\right]^{-1} .$$

c) The forward equation $\mathbf{d} = \mathbf{g}(\mathbf{m})$ can be linearized around the true maximum likelihood point, $\mathbf{m}_{ML}$:

$$\mathbf{g}(\mathbf{m}) \simeq \mathbf{g}(\mathbf{m}_{ML}) + \mathbf{G}_\infty\ (\mathbf{m} - \mathbf{m}_{ML}) ,$$

where $\mathbf{G}_\infty$ represents the derivative operator with elements

$$G_\infty{}^{i\alpha} = \left[\frac{\partial g^i}{\partial m^\alpha}\right]_{\mathbf{m}_{ML}} .$$

The point $\mathbf{m}_{ML}$ has to be obtained by the non-quadratic minimization of

$$S(\mathbf{m}) = \frac{1}{2}\left[(\mathbf{g}(\mathbf{m})-\mathbf{d}_{obs})^t \, \mathbf{C}_D^{-1} \, (\mathbf{g}(\mathbf{m})-\mathbf{d}_{obs}) + (\mathbf{m}-\mathbf{m}_{prior})^t \, \mathbf{C}_M^{-1} \, (\mathbf{m}-\mathbf{m}_{prior})\right] .$$

This can be achieved using an iterative algorithm (see Chapter 4):

$$\mathbf{m}_{n+1} = \mathbf{m}_n + \delta\mathbf{m}_n .$$

Denote by $\mathbf{m}_\infty$ the point where we decide to stop the iterations $(\mathbf{m}_{ML}) \simeq \mathbf{m}_\infty)$. The posterior covariance operator can then be estimated by

$$\mathbf{C}_{M'} \simeq \left[\mathbf{G}_\infty{}^t \, \mathbf{C}_D^{-1} \, \mathbf{G}_\infty + \mathbf{C}_M^{-1}\right]^{-1} .$$

d) The forward equation $\mathbf{d} = \mathbf{g}(\mathbf{m})$ cannot be linearized. The a posteriori probability density is far from Gaussian and special methods should be used (see text).

It is clear that if $\mathbf{g}(\mathbf{m})$ is not a linear function of $\mathbf{m}$, $\sigma_M(\mathbf{m})$ is not Gaussian. The more nonlinear $\mathbf{g}(\mathbf{m})$ is, the more remote is $\sigma_M(\mathbf{m})$ from a Gaussian function.

Figure 1.13 schematically represents different degrees of nonlinearity. In 1.13a, the problem is linear. In 1.13b it can be linearized around the a priori model, $\mathbf{m}_{prior}$. In 1.13c the linearization around $\mathbf{m}_{prior}$ is no longer acceptable, but the problem is still quasi-linear. In figure 1.13d the problem is strongly nonlinear. Let us examine these cases separately.

The weakest case of nonlinearity arises when the function $\mathbf{g}(\mathbf{m})$ can be linearized around $\mathbf{m}_{prior}$:

$$\mathbf{g}(\mathbf{m}) \simeq \mathbf{g}(\mathbf{m}_{prior}) + \mathbf{G}_0 \, (\mathbf{m} - \mathbf{m}_{prior}) , \tag{1.98}$$

where

$$G_0{}^{i\alpha} = \left(\frac{\partial g^i}{\partial m^\alpha}\right)_{\mathbf{m}_{prior}} . \tag{1.99}$$

The symbol $\simeq$ in equation (1.98) means precisely that second-order terms can be neglected compared to observational and modelization errors (i.e., compared with standard deviations and correlations in $\mathbf{C}_D$). This is the sense of figure 1.13b.

Replacing (1.98) in (1.85), we see that the a posteriori probability density is then approximately Gaussian, the center being given by

$$\langle \, \mathbf{m} \, \rangle \simeq \mathbf{m}_{prior} + \left[\mathbf{G}_0{}^t \, \mathbf{C}_D^{-1} \, \mathbf{G} + \mathbf{C}_M^{-1}\right]^{-1} \mathbf{G}_0{}^t \, \mathbf{C}_D^{-1} \, (\mathbf{d}_{obs}-\mathbf{g}(\mathbf{m}_{prior})) \tag{1.100}$$

$$= \mathbf{m}_{prior} + \mathbf{C}_M \, \mathbf{G}_0{}^t \, (\mathbf{G}_0 \, \mathbf{C}_M \, \mathbf{G}_0{}^t + \mathbf{C}_D)^{-1} \, (\mathbf{d}_{obs} - \mathbf{g}(\mathbf{m}_{prior})) \, , \qquad (1.101)$$

and the a posteriori covariance operator being given by

$$\mathbf{C}_{M'} \simeq \left(\mathbf{G}_0{}^t \, \mathbf{C}_D{}^{-1} \, \mathbf{G}_0 + \mathbf{C}_M{}^{-1} \right)^{-1} \qquad (1.102)$$

$$= \mathbf{C}_M - \mathbf{C}_M \, \mathbf{G}_0{}^t \, (\mathbf{G}_0 \, \mathbf{C}_M \, \mathbf{G}_0{}^t + \mathbf{C}_D)^{-1} \, \mathbf{G}_0 \, \mathbf{C}_M \, . \qquad (1.103)$$

We see thus that solving linearizable problems is not more difficult than solving strictly linear problems.

In the case shown in figure 1.13c the linearization (1.98) is no longer acceptable, but the function $\mathbf{g}(\mathbf{m})$ is still quasi-linear inside the region of the $D \times M$ space of significant posterior probability density. The right strategy for these problems is to obtain (using some iterative algorithm) the maximum likelihood of $\sigma_M(\mathbf{m})$, say $\mathbf{m}_{ML}$, and to use a linearization of $\mathbf{g}(\mathbf{m})$ around $\mathbf{m}_{ML}$ for estimating the a posteriori covariance operator.

Let us see how this can work. The point maximizing $\sigma_M(\mathbf{m})$ clearly minimizes

$$S(\mathbf{m}) =$$

$$\frac{1}{2}\left((\mathbf{g}(\mathbf{m})-\mathbf{d}_{obs})^t \, \mathbf{C}_D{}^{-1} \, (\mathbf{g}(\mathbf{m})-\mathbf{d}_{obs}) + (\mathbf{m}-\mathbf{m}_{prior})^t \, \mathbf{C}_M{}^{-1} \, (\mathbf{m}-\mathbf{m}_{prior}) \right) , \qquad (1.104)$$

(the factor 1/2 is left for subsequent simplifications). As (1.104) is quadratic in data and model residuals, it justifies the name of *least-squares estimator* for $\mathbf{m}_{ML}$. The obtainment of the minimum of $S(\mathbf{m})$ corresponds to a classical problem of nonlinear optimization. Using, for instance, a quasi-Newton method, an iterative algorithm is obtained which corresponds to the three equivalent equations (see Chapter 4 for more details) :

see p. 194 ?

$$\mathbf{m}_{n+1} = \mathbf{m}_n - \left(\mathbf{G}_n{}^t \, \mathbf{C}_D{}^{-1} \, \mathbf{G}_n + \mathbf{C}_M{}^{-1} \right)^{-1}$$

$$\left(\mathbf{G}_n{}^t \, \mathbf{C}_D{}^{-1} \, (\mathbf{g}(\mathbf{m}_n)-\mathbf{d}_{obs}) + \mathbf{C}_M{}^{-1} \, (\mathbf{m}_n-\mathbf{m}_{prior}) \right) \qquad (1.105)$$

$$= \mathbf{m}_{prior} - \left(\mathbf{G}_n{}^t \, \mathbf{C}_D{}^{-1} \, \mathbf{G}_n + \mathbf{C}_M{}^{-1} \right)^{-1}$$

$$\mathbf{G}_n{}^t \, \mathbf{C}_D{}^{-1} \, ((\mathbf{g}(\mathbf{m}_n)-\mathbf{d}_{obs}) - \mathbf{G}_n \, (\mathbf{m}_n-\mathbf{m}_{prior})) \qquad (1.106)$$

$$= \mathbf{m}_{prior} - \mathbf{C}_M \, \mathbf{G}_n{}^t \, (\mathbf{C}_D + \mathbf{G}_n \, \mathbf{C}_M \, \mathbf{G}_n{}^t)^{-1}$$

$$((\mathbf{g}(\mathbf{m}_n)-\mathbf{d}_{obs}) - \mathbf{G}_n \, (\mathbf{m}_n-\mathbf{m}_{prior})). \qquad (1.107)$$

In these equations,

$$\mathbf{G}_n = \left(\frac{\partial \mathbf{g}}{\partial \mathbf{m}} \right)_{\mathbf{m}_n} , \tag{1.108}$$

or, explicitly,

$$G_n{}^{i\alpha} = \left(\frac{\partial g^i}{\partial m^\alpha} \right)_{m_n} .$$

If the problem is quasi-linear (see figure 1.13c) such a method will not present convergence troubles. For more details of the numerical aspects, the reader is referred to Chapter 4.

Once the maximum likelihood point $\mathbf{m}_{ML} = \mathbf{m}_\infty$ has been conveniently approached, the a posteriori covariance operator can be estimated by

$$\mathbf{C}_{M'} \simeq \left(\mathbf{G}_\infty{}^t \, \mathbf{C}_D{}^{-1} \, \mathbf{G}_\infty + \mathbf{C}_M{}^{-1} \right)^{-1} \tag{1.109}$$

$$= \mathbf{C}_M - \mathbf{C}_M \, \mathbf{G}_\infty{}^t \, (\mathbf{G}_\infty \, \mathbf{C}_M \, \mathbf{G}_\infty{}^t + \mathbf{C}_D)^{-1} \, \mathbf{G}_\infty \, \mathbf{C}_M \, . \tag{1.110}$$

The main computational difference between this "nonlinear" solution and the linearized solution is that here, $\mathbf{g}(\mathbf{m})$, the predicted data for the current model, has to be computed at each iteration without using any linear approximation. In usual problems it is more difficult to compute $\mathbf{g}(\mathbf{m})$ than $\mathbf{g}(\mathbf{m}_0) + \mathbf{G}_0 \, (\mathbf{m}-\mathbf{m}_0)$: nonlinear problems are in general more expensive to solve than linearizable problems.

Of course, even if a problem is linearizable, it can be solved nonlinearly, but the gain in accuracy usually does not justify the computational effort.

Last, we have to discuss the strongly nonlinear problems represented by Figure 1.13d. The a posteriori probability density is then clearly non-Gaussian, and no general discussion of the solution can be made without a rather exhaustive exploration of the model space. If the number of model parameters is small, then we can use the general methods described in section 1.6.3. As, in that case, no advantage is taken of the Gaussian hypothesis, we do better to drop it and use a more realistic error modelization. If the number of parameters is great, we can always use equations (1.108)-(1.110) to obtain a local maximum likelihood point, to fix all its components except a few, and to explore the subspace thus defined.

Sometimes, the number of secondary maxima of $\sigma_M(\mathbf{m})$ (i.e., of secondary minima of $S(\mathbf{m})$) is not too large. Starting the iterative algorithms (1.108)-(1.110) at different points $\mathbf{m}_0$, we can check the existence of secondary minima in the region of interest. At the least, this strategy will allow of making a semi-quantitative discussion of the a posteriori state of information in the model space.

1.7.2: The Gaussian Hypothesis (least-squares criterion). Case $f(\mathbf{d},\mathbf{m}) = 0$.

In the previous section I have assumed that the equations solving the forward problem were written under the form $\mathbf{d} = \mathbf{g}(\mathbf{m})$. In fact, the distinctions between data vector and model vector do not necessarily correspond to observable parameters and model parameters, and it is the equation $\mathbf{d} = \mathbf{g}(\mathbf{m})$, expressing the theoretical correlations between all the parameters, that allows of *defining* the data and model vectors as those appearing in the left and right-hand sides of the equation $\mathbf{d} = \mathbf{g}(\mathbf{m})$, respectively.

Sometimes, it may be easy to consider the more general equation

$$\mathbf{f}(\mathbf{d},\mathbf{m}) = \mathbf{0} , \tag{1.111}$$

and this section gives the corresponding formulas. It is not more difficult to examine the most general problem that can be solved, within the Gaussian hypothesis, using the approach developed here.

Let $\mathbf{x} = (\mathbf{d},\mathbf{m})$ be a generic vector of the joint space $D \times M$ defined in section 1.1.3, containing observable and model parameters. The a priori information on $\mathbf{x}$ is assumed Gaussian,

$$\frac{\rho(\mathbf{x})}{\mu(\mathbf{x})} = \text{const.} \exp\left[-\frac{1}{2} (\mathbf{x}-\mathbf{x}_{\text{prior}})^t \, \mathbf{C}_X^{-1} \, (\mathbf{x}-\mathbf{x}_{\text{prior}}) \right] , \tag{1.112}$$

with center $\mathbf{x}_{\text{prior}}$ and covariance operator $\mathbf{C}_X$. Instead of writting the theoretical information on $\mathbf{x}$ as

$$\mathbf{f}(\mathbf{x}) = \mathbf{0} , \tag{1.113}$$

we allow theoretical uncertainties which, if they are assumed Gaussian, can be described using the probability density

$$\Theta(\mathbf{x}) = \text{const.} \exp\left[-\frac{1}{2} \mathbf{f}(\mathbf{x})^t \, \mathbf{C}_T^{-1} \, \mathbf{f}(\mathbf{x}) \right] . \tag{1.114}$$

The combination of $\rho(\mathbf{x})$ with $\Theta(\mathbf{x})$ leads to (see section 1.5)

$$\sigma(\mathbf{x}) = \frac{\rho(\mathbf{x}) \, \Theta(\mathbf{x})}{\mu(\mathbf{x})} , \tag{1.115}$$

i.e.,

$$\boxed{\sigma(\mathbf{x}) = \text{const.} \exp\left[-\frac{1}{2} \left[\mathbf{f}(\mathbf{x})^t \, \mathbf{C}_T^{-1} \, \mathbf{f}(\mathbf{x}) + (\mathbf{x}-\mathbf{x}_{\text{prior}})^t \, \mathbf{C}_X^{-1} \, (\mathbf{x}-\mathbf{x}_{\text{prior}}) \right] \right] .} \tag{1.116}$$

As for the explicit case, let me discuss linear and nonlinear problems separately,

a) The theoretical equation is linear. Instead of writing $\mathbf{f}(\mathbf{x}) = \mathbf{0}$, we then write

$$\mathbf{F}\ \mathbf{x} = 0 \ , \tag{1.117}$$

where $\mathbf{F}$ represents a linear operator acting from the total space of parameters into a space of "residuals".

This gives

$$\sigma(\mathbf{x}) \ = \ \text{const} \ \exp(\ - \ S(\mathbf{x}) \ ) \ , \tag{1.118}$$

where $S(\mathbf{x})$ is the quadratic function

$$S(\mathbf{x}) \ = \ \frac{1}{2} \left[(\mathbf{F}\ \mathbf{x})^t \ \mathbf{C}_T^{\ -1} \ (\mathbf{F}\ \mathbf{x}) + (\mathbf{x}-\mathbf{x}_{prior})^t \ \mathbf{C}_X^{\ -1} \ (\mathbf{x}-\mathbf{x}_{prior}) \right] . \tag{1.119}$$

Defining

$$\langle \ \mathbf{x} \ \rangle \ = \ \left[\mathbf{F}^t \ \mathbf{C}_T^{\ -1} \ \mathbf{F} + \mathbf{C}_X^{\ -1} \right]^{-1} \ \mathbf{C}_X^{\ -1} \ \mathbf{x}_{prior} \ , \tag{1.120}$$

and

$$\mathbf{C}_{X'} \ = \ \left[\mathbf{F}^t \ \mathbf{C}_T^{\ -1} \ \mathbf{F} + \mathbf{C}_X^{\ -1} \right]^{-1} \ , \tag{1.121}$$

we obtain

$$S(\mathbf{x}) \ = \ (\ \mathbf{x} \ - \ \langle \ \mathbf{x} \ \rangle \ )^t \ \mathbf{C}_{X'}^{\ -1} \ (\ \mathbf{x} \ - \ \langle \ \mathbf{x} \ \rangle \ ) \ - \ \langle \ \mathbf{x} \ \rangle^{\ t} \ \mathbf{C}_X^{\ -1} \ \langle \ \mathbf{x} \ \rangle$$
$$+ \ \ \mathbf{x}_{prior}^{\ \ t} \ \mathbf{C}_X^{\ -1} \ \mathbf{x}_{prior} \ .$$

All the right-hand terms but the first are constant (independent of $\mathbf{x}$), and can be absorbed in the constant facto of (1.118). This gives

$$\sigma(\mathbf{x}) \ = \ \text{const.} \ \exp\left[\ -\frac{1}{2} \ (\ \mathbf{x} \ - \ \langle \ \mathbf{x} \ \rangle \ )^t \ \ \mathbf{C}_{X'}^{\ -1} \ \ (\ \mathbf{x} \ - \ \langle \ \mathbf{x} \ \rangle \ ) \ \right] ,$$

or, in normalized form,

$$\sigma(\mathbf{x}) = ((2\pi)^N \ \det \mathbf{C}_{X'})^{-1/2} \ \exp\left[-\frac{1}{2}(\mathbf{x}-\langle \ \mathbf{x} \ \rangle \ )^t \ \mathbf{C}_{X'}^{\ -1} \ (\mathbf{x}-\langle \ \mathbf{x} \ \rangle \ ) \right] . \tag{1.122}$$

Equation (1.122) shows the important result that, when the theoretical equation is linear, the a posteriori probability density in the model space is Gaussian. The center of this Gaussian is given by (1.120), while its covariance operator is given by (1.121).

Successively we have

$$\begin{aligned}\langle \mathbf{x} \rangle &= \mathbf{C}_{X'}\, \mathbf{C}_X^{-1}\, \mathbf{m}_{prior} \\ &= \mathbf{C}_{X'} \left(\mathbf{F}^t\, \mathbf{C}_T^{-1}\, \mathbf{F} + \mathbf{C}_X^{-1} - \mathbf{F}^t\, \mathbf{C}_T^{-1}\, \mathbf{F} \right) \mathbf{x}_{prior} \\ &= \mathbf{C}_{X'} \left(\mathbf{C}_{X'}^{-1} - \mathbf{F}^t\, \mathbf{C}_T^{-1}\, \mathbf{F} \right) \mathbf{x}_{prior} \\ &= \left(\mathbf{I} - \mathbf{C}_{X'}\, \mathbf{F}^t\, \mathbf{C}_T^{-1}\, \mathbf{F} \right) \mathbf{x}_{prior} \\ &= \mathbf{x}_{prior} - \mathbf{C}_{X'}\, \mathbf{F}^t\, \mathbf{C}_T^{-1}\, \mathbf{F}\, \mathbf{x}_{prior} ,\end{aligned}$$

i.e.

$$\langle \mathbf{x} \rangle = \mathbf{x}_{prior} - \left(\mathbf{F}^t\, \mathbf{C}_T^{-1}\, \mathbf{F} + \mathbf{C}_X^{-1} \right)^{-1} \mathbf{F}^t\, \mathbf{C}_T^{-1}\, \mathbf{F}\, \mathbf{x}_{prior} . \tag{1.123}$$

In problem 1.19 it is shown that expressions (1.120) and (1.123) can also be written:

$$\langle \mathbf{x} \rangle = \mathbf{x}_{prior} - \mathbf{C}_X\, \mathbf{F}^t\, (\mathbf{F}\, \mathbf{C}_X\, \mathbf{F}^t + \mathbf{C}_T)^{-1}\, \mathbf{F}\, \mathbf{x}_{prior} , \tag{1.124}$$

and

$$\mathbf{C}_{X'} = \mathbf{C}_X - \mathbf{C}_X\, \mathbf{F}^t\, (\mathbf{F}\, \mathbf{C}_X\, \mathbf{F}^t + \mathbf{C}_T)^{-1}\, \mathbf{F}\, \mathbf{C}_X . \tag{1.125}$$

As $\langle \mathbf{x} \rangle$ clearly maximizes $\sigma(\mathbf{x})$, it minimizes the quadratic expression

$$S(\mathbf{x}) = \frac{1}{2} \left((\mathbf{F}\,\mathbf{x})^t\, \mathbf{C}_T^{-1}\, (\mathbf{F}\,\mathbf{x}) + (\mathbf{x} - \mathbf{x}_{prior})^t\, \mathbf{C}_X^{-1}\, (\mathbf{x} - \mathbf{x}_{prior}) \right) , \tag{1.126}$$

thus justifying the usual terminology of *least-squares estimator*.

Let us now direct our interest to the special case where we can divide the parameter set X into two subsets D and M such that the theoretical equation $\mathbf{F}\,\mathbf{x} = \mathbf{0}$ simplifies to

$$\mathbf{F}\,\mathbf{x} = \mathbf{G}\,\mathbf{m} - \mathbf{d} = \mathbf{0} , \tag{1.127}$$

i.e.,

$$\mathbf{d} = \mathbf{G}\,\mathbf{m} , \tag{1.128}$$

where $\mathbf{G}$ is a linear operator from M into D. Formally,

$$\mathbf{F}\,\mathbf{x} = [\,-\mathbf{I} \quad \mathbf{G}\,] \begin{bmatrix} \mathbf{d} \\ \mathbf{m} \end{bmatrix} = \mathbf{0} ,$$

i.e.,

$$\mathbf{F} = [\ -\mathbf{I} \quad \mathbf{G}\ ] . \tag{1.129}$$

Using the notations

$$\mathbf{x} = \begin{bmatrix} \mathbf{d} \\ \mathbf{m} \end{bmatrix} \qquad \mathbf{x}_{prior} = \begin{bmatrix} \mathbf{d}_{prior} \\ \mathbf{m}_{prior} \end{bmatrix} \qquad \langle\ \mathbf{x}\ \rangle = \begin{bmatrix} \langle\ \mathbf{d}\ \rangle \\ \langle\ \mathbf{m}\ \rangle \end{bmatrix} , \tag{1.130a}$$

and

$$\mathbf{C_X} = \begin{bmatrix} \mathbf{C_{DD}} & \mathbf{C_{MD}} \\ \mathbf{C_{DM}} & \mathbf{C_{MM}} \end{bmatrix} \qquad \mathbf{C_{X'}} = \begin{bmatrix} \mathbf{C_{DD'}} & \mathbf{C_{MD'}} \\ \mathbf{C_{DM'}} & \mathbf{C_{MM'}} \end{bmatrix} , \tag{1.130b}$$

and using (1.124) and (1.125) we easily obtain

$$\langle\ \mathbf{m}\ \rangle = \mathbf{m}_{prior} - (\mathbf{C_{MM}}\ \mathbf{G}^t - \mathbf{C_{MD}}) \tag{1.131}$$
$$(\mathbf{C_T} + \mathbf{C_{DD}} + \mathbf{G}\ \mathbf{C_{MM}}\ \mathbf{G}^t - \mathbf{C_{DM}}\ \mathbf{G}^t - \mathbf{G}\ \mathbf{C_{MD}})^{-1}\ (\mathbf{G}\ \mathbf{m}_{prior} - \mathbf{d}_{obs}) ,$$

$$\langle\ \mathbf{d}\ \rangle = \mathbf{G}\ \mathbf{m}_{prior} - (\mathbf{C_T} + \mathbf{G}\ (\mathbf{C_{MM}}\ \mathbf{G}^t - \mathbf{C_{MD}})) \tag{1.132}$$
$$(\mathbf{C_T} + \mathbf{C_{DD}} + \mathbf{G}\ \mathbf{C_{MM}}\ \mathbf{G}^t - \mathbf{C_{DM}}\ \mathbf{G}^t - \mathbf{G}\ \mathbf{C_{MD}})^{-1}\ (\mathbf{G}\ \mathbf{m}_{prior} - \mathbf{d}_{obs}) ,$$

$$\mathbf{C_{MM'}} = \mathbf{C_{MM}} - (\mathbf{C_{MM}}\ \mathbf{G}^t - \mathbf{C_{MD}}) \tag{1.133}$$
$$(\mathbf{C_T} + \mathbf{C_{DD}} + \mathbf{G}\ \mathbf{C_{MM}}\ \mathbf{G}^t - \mathbf{C_{DM}}\ \mathbf{G}^t - \mathbf{G}\ \mathbf{C_{MD}})^{-1}\ (\mathbf{G}\ \mathbf{C_{MM}} - \mathbf{C_{DM}}) ,$$

$$\mathbf{C_{DD'}} = \mathbf{C_{DD}} - (\mathbf{C_{DM}}\ \mathbf{G}^t - \mathbf{C_{DD}}) \tag{1.134a}$$
$$(\mathbf{C_T} + \mathbf{C_{DD}} + \mathbf{G}\ \mathbf{C_{MM}}\ \mathbf{G}^t - \mathbf{C_{DM}}\ \mathbf{G}^t - \mathbf{G}\ \mathbf{C_{MD}})^{-1}\ (\mathbf{G}\ \mathbf{C_{MD}} - \mathbf{C_{DD}}) ,$$

$$\mathbf{C_{DM'}} = \mathbf{C_{DM}} - (\mathbf{C_{DM}}\ \mathbf{G}^t - \mathbf{C_{DD}}) \tag{1.135a}$$
$$(\mathbf{C_T} + \mathbf{C_{DD}} + \mathbf{G}\ \mathbf{C_{MM}}\ \mathbf{G}^t - \mathbf{C_{DM}}\ \mathbf{G}^t - \mathbf{G}\ \mathbf{C_{MD}})^{-1}\ (\mathbf{G}\ \mathbf{C_{MM}} - \mathbf{C_{DM}}) ,$$

and

$$\mathbf{C_{MD'}} = \mathbf{C_{MD}} - (\mathbf{C_{MM}}\ \mathbf{G}^t - \mathbf{C_{MD}})\ (\mathbf{C_T} + \mathbf{C_{DD}} + \mathbf{G}\ \mathbf{C_{MM}}\ \mathbf{G}^t - \mathbf{C_{DM}}\ \mathbf{G}^t - \mathbf{G}\ \mathbf{C_{MD}})^{-1}\ (\mathbf{G}\ \mathbf{C_{MD}} - \mathbf{C_{DD}}) = \mathbf{C_{DM'}}^t .$$

Using some algebra, many equivalent expressions may be obtained. Among them, we will need the following:

$$\mathbf{C_{DD'}} = \mathbf{C_T} + \mathbf{G}\ \mathbf{C_{MM}}\ \mathbf{G}^t - (\mathbf{C_T} + \mathbf{G}\ (\mathbf{C_{MM}}\ \mathbf{G}^t - \mathbf{C_{MD}}))$$

$$(\mathbf{C_T} + \mathbf{C_{DD}} + \mathbf{G}\ \mathbf{C_{MM}}\ \mathbf{G^t} - \mathbf{C_{DM}}\ \mathbf{G^t} - \mathbf{G}\ \mathbf{C_{MD}})^{-1}$$

$$(\mathbf{C_T} + (\mathbf{G}\ \mathbf{C_{MM}} - \mathbf{C_{DM}})\ \mathbf{G^t}) , \qquad (1.134b)$$

and

$$\mathbf{C_{DM'}} = \mathbf{G}\ \mathbf{C_{MM}} - (\mathbf{C_T} + \mathbf{G}\ (\mathbf{C_{MM}}\ \mathbf{G^t} - \mathbf{C_{MD}}))$$
$$(\mathbf{C_T} + \mathbf{C_{DD}} + \mathbf{G}\ \mathbf{C_{MM}}\ \mathbf{G^t} - \mathbf{C_{DM}}\ \mathbf{G^t} - \mathbf{G}\ \mathbf{C_{MD}})^{-1}$$
$$(\mathbf{G}\ \mathbf{C_{MM}} - \mathbf{C_{DM}}) . \qquad (1.135b)$$

In the particular case where errors in the theoretical relationship $\mathbf{d} = \mathbf{G}\ \mathbf{m}$ can be neglected,

$$\mathbf{C_T} = 0 ,$$

and equations (1.131)-(1.135) reduce to the simple expressions

$$\langle\ \mathbf{m}\ \rangle = \mathbf{m}_{prior} - (\mathbf{C_{MM}}\ \mathbf{G^t} - \mathbf{C_{MD}}) \qquad (1.136)$$
$$(\mathbf{C_{DD}} + \mathbf{G}\ \mathbf{C_{MM}}\ \mathbf{G^t} - \mathbf{C_{DM}}\ \mathbf{G^t} - \mathbf{G}\ \mathbf{C_{MD}})^{-1}\ (\mathbf{G}\ \mathbf{m}_{prior} - \mathbf{d}_{obs}) ,$$

$$\langle\ \mathbf{d}\ \rangle = \mathbf{G}\ \langle\ \mathbf{m}\ \rangle , \qquad (1.137)$$

$$\mathbf{C_{MM'}} = \mathbf{C_{MM}} - (\mathbf{C_{MM}}\ \mathbf{G^t} - \mathbf{C_{MD}}) \qquad (1.138)$$
$$(\mathbf{C_{DD}} + \mathbf{G}\ \mathbf{C_{MM}}\ \mathbf{G^t} - \mathbf{C_{DM}}\ \mathbf{G^t} - \mathbf{G}\ \mathbf{C_{MD}})^{-1}\ (\mathbf{G}\ \mathbf{C_{MM}} - \mathbf{C_{DM}}) ,$$

$$\mathbf{C_{DD'}} = \mathbf{G}\ \mathbf{C_{MM'}}\ \mathbf{G^t} , \qquad (1.139)$$

$$\mathbf{C_{DM'}} = \mathbf{G}\ \mathbf{C_{MM'}} , \qquad (1.140)$$

and

$$\mathbf{C_{MD'}} = \mathbf{C_{MM'}}\ \mathbf{G^t} . \qquad (1.141)$$

It should be noticed that, in the particular case where a priori information on $\mathbf{m}$ is uncorrelated with the a priori information on $\mathbf{d}$,

$$\mathbf{C_{DM}} = \mathbf{C_{MD}}^t = 0 ,$$

and all the above equations reduce to those shown in the previous section.

b) The theoretical equation is nonlinear. The point maximizing $\sigma(\mathbf{x})$ (the maximum likelihood point) will minimize the exponent in (1.116), i.e.,

$$S(\mathbf{x}) = \frac{1}{2} \left[\mathbf{f}(\mathbf{x})^t \, C_T^{-1} \, \mathbf{f}(\mathbf{x}) + (\mathbf{x}-\mathbf{x}_{prior})^t \, \mathbf{C}_X^{-1} \, (\mathbf{x}-\mathbf{x}_{prior}) \right] , \tag{1.142}$$

which intuitively means that the maximum likelihood point will approximately satisfy the theoretical equations ($\mathbf{f}(x) \simeq 0$) and will be close to the a priori point $\mathbf{x}_{prior}$, the appropriate trade-off being imposed by the relative values of $\mathbf{C}_T$ and $\mathbf{C}_X$.

Using a quasi-Newton method (see Chapter 4 for more details) to obtain the maximum likelihood point leads to the algorithm

$$\mathbf{x}_{n+1} = \mathbf{x}_n - \left(\mathbf{F}_n^t \, C_T^{-1} \, \mathbf{F}_n + C_X^{-1} \right)^{-1} \left(\mathbf{F}_n^t \, C_T^{-1} \, f(x_n) + C_X^{-1} \, (x_n - x_{prior}) \right) \tag{1.143}$$

$$= x_{prior} - \left(\mathbf{F}_n^t \, C_T^{-1} \, \mathbf{F}_n + C_X^{-1} \right)^{-1} \mathbf{F}_n^t \, C_T^{-1} \, (f(\mathbf{x}_n) - \mathbf{F}_n \, (\mathbf{x}_n - \mathbf{x}_{prior})) \tag{1.144}$$

$$= x_{prior} - C_X \, \mathbf{F}_n^t \, (C_T + \mathbf{F}_n \, C_X \, \mathbf{F}_n^t)^{-1} \, (f(\mathbf{x}_n) - \mathbf{F}_n \, (\mathbf{x}_n - \mathbf{x}_{prior})) \, . \tag{1.145}$$

In these equations,

$$\mathbf{F}_n = \left(\frac{\partial f}{\partial x} \right)_{\mathbf{x}_n} , \tag{1.146}$$

or, explicitly,

$$F_n^{iA} = \left(\frac{\partial f^i}{\partial x^A} \right)_{\mathbf{x}_n} .$$

Once the maximum likelihood point $\mathbf{x}_{ML} = \mathbf{x}_\infty$ has been conveniently approached, the a posteriori covariance operator can be estimated by

$$C_{X'} \simeq \left(\mathbf{F}_\infty^t \, \mathbf{C}_T^{-1} \, \mathbf{F}_\infty + \mathbf{C}_X^{-1} \right)^{-1} \tag{1.147}$$

$$= C_X - C_X \, \mathbf{F}_\infty^t \, (\mathbf{F}_\infty \, C_X \, \mathbf{F}_\infty^t + C_T)^{-1} \, \mathbf{F}_\infty \, C_X \, . \tag{1.148}$$

The solution of the formulas (1.143)-(1.145) for the case $\mathbf{f}(\mathbf{x}) = \mathbf{f}(\mathbf{d},\mathbf{m}) = \mathbf{g}(\mathrm{m}) - \mathbf{d} = \mathbf{0}$ is left as an exercise for the reader.

1.7.3: Generalized Gaussian (least-absolute-values criterion, minimax criterion)

In this section we explore the implications of assuming that experimental uncertainties on observations, or uncertainties in the a priori model, can be described using the Generalized Gaussian functions introduced in box 1.2.

Let d^i_{obs} represent the observed data values. We assume that errors are uncorrelated, that they can be estimated by the values σ^i_D , and that the density function describing uncertainties can be modeled by the Generalized Gaussian of order p

$$\frac{\rho_D(\mathbf{d})}{\mu_D(\mathbf{d})} = \text{const.} \prod_{i \in I_D} \exp\left(-\frac{1}{p} \frac{\left| d^i - d^i_{obs} \right|^p}{\left(\sigma^i_D\right)^p} \right) . \tag{1.149}$$

Similarly, let m^α_{prior} represent the a priori model values, and σ^α_M the estimated errors. We represent the a priori density function in the model space by

$$\rho_M(\mathbf{m}) = \text{const.} \prod_{i \in I_M} \exp\left(-\frac{1}{p} \frac{\left| m^\alpha - m^\alpha_{prior} \right|^p}{\left(\sigma^\alpha_M\right)^p} \right) . \tag{1.150}$$

Simple results are obtained only when modelization errors are neglected. Thus, let us assume here that the forward modelization can be written

$$\mathbf{d} = \mathbf{g}(\mathbf{m}) \ ,$$

and that observational errors are predominant with respect to errors in the modelization. From equation (1.72) we obtain the a posteriori density function in the model space:

$$\sigma_M(\mathbf{m}) = \exp\left[-\frac{1}{p} \left(\sum_{i \in \mathbf{I}_D} \frac{\left| g^i(\mathbf{m}) - d^i_{obs} \right|^p}{\left(\sigma^i_D\right)^p} + \sum_{i \in \mathbf{I}_M} \frac{\left| m^\alpha - m^\alpha_{prior} \right|^p}{\left(\sigma^\alpha_M\right)^p} \right) \right] . \tag{1.151}$$

The maximum likelihood $\mathbf{m}_{ML}$ maximizes $\sigma_M(\mathbf{m})$, i.e., minimizes

$$S(\mathbf{m}) = \frac{1}{p} \left(\sum_{i \in \mathbf{I}_D} \frac{\left| g^i(\mathbf{m}) - d^i_{obs} \right|^p}{\left(\sigma^i_D\right)^p} + \sum_{i \in \mathbf{I}_M} \frac{\left| m^\alpha - m^\alpha_{prior} \right|^p}{\left(\sigma^\alpha_M\right)^p} \right) , \tag{1.152}$$

which clearly corresponds to the minimization of an ℓ_p norm . We see thus how ℓ_p norm minimization problems arise in Inverse Problem Theory.

Two important particular cases are usually considered, namely, the ℓ_1 norm and the ℓ_∞ norm problems (in addition, of course, to the ℓ_2 norm already considered). For $p=1$ we have to minimize the quantity

$$S(\mathbf{m}) = \sum_{i \in \mathbf{I}_D} \frac{\left| g^i(\mathbf{m}) - d^i_{obs} \right|}{\sigma^i_D} + \sum_{i \in \mathbf{I}_M} \frac{\left| m^\alpha - m^\alpha_{prior} \right|}{\sigma^\alpha_M} , \tag{1.153}$$

which corresponds to a *minimum-absolute-values criterion.* See Chapter 5 for more details.

For $p \to \infty$ it can be shown (see Chapter 5) that the minimization of $S(\mathbf{m})$ is equivalent to the *minimization of the maximum* of

$$\left(\frac{\left| g^i(\mathbf{m}) - d^i_{obs} \right|}{\sigma^i_D} \quad (i \in \mathbf{I}_D) \quad ; \quad \frac{\left| m^\alpha - m^\alpha_{prior} \right|}{\sigma^\alpha_M} \quad (\alpha \in \mathbf{I}_M) \right) . \tag{1.154}$$

This is the reason why the case $p \to \infty$ is known as the *minimax* criterion.

Techniques for solving intermediate cases ($1 < p < \infty$) are also described in Chapter 5.

Sometimes the following question arises: Which is the best criterion to use for the resolution of inverse problems, the least-absolute-values criterion, the least-squares criterion, or the minimax criterion? We have seen in this section that each of these criteria can be considered as implied by a particular assumption about the probability densities under consideration. It is thus clear that, depending on the particular form of these probability densities, any one of the criteria can be better than the other two. As an illustration see problems 1.5, 1.9, and 1.11.

Problems for chapter 1:

Problem 1.1: Estimation of the epicentral coordinates of a seismic event: A nuclear explosion took place at time $T=0$ in an unknown location at the surface of the Earth. The seismic waves produced by the explosion have been recorded at a network of six seismic stations whose coordinates in a rectangular system are

$$\begin{aligned} (x^1,y^1) &= (3\text{ km}\ ,\ 15\text{ km}) \\ (x^2,y^2) &= (3\text{ km}\ ,\ 16\text{ km}) \\ (x^3,y^3) &= (4\text{ km}\ ,\ 15\text{ km}) \\ (x^4,y^4) &= (4\text{ km}\ ,\ 16\text{ km}) \\ (x^5,y^5) &= (5\text{ km}\ ,\ 15\text{ km}) \\ (x^6,y^6) &= (5\text{ km}\ ,\ 16\text{ km})\ . \end{aligned} \tag{1}$$

The observed arrival times of the seismic waves at these stations are

$$\begin{aligned} d^1{}_{obs} &= 3.12\text{ s} \ \pm\ \sigma \\ d^2{}_{obs} &= 3.26\text{ s} \ \pm\ \sigma \\ d^3{}_{obs} &= 2.98\text{ s} \ \pm\ \sigma \\ d^4{}_{obs} &= 3.12\text{ s} \ \pm\ \sigma \\ d^5{}_{obs} &= 2.84\text{ s} \ \pm\ \sigma \\ d^6{}_{obs} &= 2.98\text{ s} \ \pm\ \sigma\ , \end{aligned} \tag{2}$$

where

$$\sigma = 0.10\text{ s}\ , \tag{3}$$

and where $\pm\,\sigma$ is a short notation indicating that experimental uncertainties are independent and can be modeled using a Gaussian probability density with a standard deviation equal to σ .

Estimate the epicentral coordinates (X,Y) of the explosion, assuming a velocity of

$$v = 5\text{ km/s} \tag{4}$$

for the seismic waves. Use the approximation of a flat Earth's surface, and consider that the coordinates in (1) are cartesian.

Discuss the generalization of the problem to the case where the time of the explosion, the locations of the seismic observatories, or the velocity of the seismic waves is not perfectly known, and to the case of a realistic Earth.

Solution: The model parameters are the coordinates of the epicenter of the explosion:

$$\mathbf{m} = (X,Y) , \tag{5}$$

and the data parameters are the arrival times at the seismic network:

$$\mathbf{d} = (d^1,d^2,d^3,d^4,d^5,d^6) , \tag{6}$$

while the coordinates of the seismic stations and the velocity of seismic waves are assumed perfectly known (i.e., known with errors which are negligible with respect to errors in the observed arrival times).

The probability density representing the state of null information on the epicentral coordinates is

$$\mu_M(X,Y) = \text{const} , \tag{7}$$

because, as we use cartesian coordinates, (7) assigns equal probabilities to identical volumes (see example 18). Let $\rho_M(X,Y)$ be the probability density representing the a priori information on the epicentral location. As the statement of the problem does not give any a priori information,

$$\rho_M(X,Y) = \mu_M(X,Y) = \text{const} . \tag{8}$$

As data uncertainties are Gaussian and independent, the probability density representing the information we have on the true values of the arrival times is

$$\rho_D(d^1,d^2,d^3,d^4,d^5,d^6) = \text{const} \exp\left(-\frac{1}{2} \sum_{i=1}^{6} \frac{\left(d^i - d^i_{obs}\right)^2}{\sigma^2} \right) . \tag{9}$$

As d^i are Newtonian times, the null information probability density in the data space is (see section 1.2.4)

$$\mu_D(d^1,d^2,d^3,d^4,d^5,d^6) = \text{const} . \tag{10}$$

For a given (X,Y) , the arrival times of the seismic wave at the seismic stations can be computed using the (exact) equation

$$d^i = g^i(X,Y) = \frac{1}{v} \sqrt{(x^i-X)^2 + (y^i-Y)^2} \qquad (i=1,...,6) , \tag{11}$$

which solves the forward problem. As the theoretical relationship between data and model parameters is assumed to be error free, the probability density representing this theoretical information is the conditional probability density

$$\theta(\mathbf{d}|(X,Y)) = \delta(\mathbf{d}-\mathbf{g}(X,Y)) , \tag{12}$$

where $\delta(\mathbf{d})$ is the delta function in the data space, and where $g(X,Y)$ denotes the (vector) function (11).

The posterior information resulting from the combination of $\rho_D(\mathbf{d})$, $\rho_M(X,Y)$, and $\theta(\mathbf{d}|(X,Y))$ is (equations (1.60), (1.63), and (1.64)):

$$\sigma(\mathbf{d},X,Y) = \frac{\rho(\mathbf{d},X,Y)\ \theta(\mathbf{d},X,Y)}{\mu(\mathbf{d},X,Y)} = \frac{\rho_D(\mathbf{d})}{\mu_D(\mathbf{d})}\ \rho_M(X,Y)\ \theta(\mathbf{d}|(X,Y)) . \tag{13}$$

The marginal a posteriori information for the model parameters is

$$\sigma_M(X,Y) = \int_D d\mathbf{d}\ \ \sigma(\mathbf{d},X,Y) , \tag{14}$$

which, using (12) and (13), gives

$$\sigma_M(X,Y) = \rho_M(X,Y) \left(\frac{\rho_D(\mathbf{d})}{\mu_D(\mathbf{d})} \right)_{\mathbf{d}=\mathbf{g}(X,Y)} . \tag{15}$$

Using equations (8) to (11) gives

$$\sigma_M(X,Y) = \text{const}\ \exp\left(-\frac{1}{2\sigma^2} \sum_{i=1}^{6} \left(d^i_{cal} - d^i_{obs} \right)^2 \right) , \tag{16}$$

where

$$d^i_{cal} = \frac{1}{v} \sqrt{(x^i-X)^2 + (y^i-Y)^2} . \tag{17}$$

The probability density $\sigma_M(X,Y)$ describes all the a posteriori information we have on the epicentral coordinates. As we only have two parameters, the simplest (and more general) way of studying this information is to plot the values of $\sigma_M(X,Y)$ directly in the region of the plane where it takes significant values. Figure 1.14 shows the corresponding result.

We see that the zone of non-vanishing probability density is crescent-shaped. This can be interpreted as follows. The arrival times of the seismic wave at the seismic network (top left of the figure) is of the order of 3 s , and as we know that the explosion took place at time T = 0. , and the velocity of the seismic wave is 5 km/s , this gives the reliable information that the explosion took place at a distance of approximately 15 km from the seismic network. But as the observational errors (±0.1 s) in the arrival times are of the order of the travel times of the seismic wave between the stations, the azimuth of the epicenter is not well resolved. As the distance is well determined but not the azimuth, it is natural to obtain a probability density with a crescent shape.

From the values shown in Figure 1 it is possible to obtain any estimator of the epicentral coordinates we wish, such as, for instance, the median values, the mean values, the maximum likelihood values, and so on. But the general solution of the inverse problem is the probability density itself. Notice in particular that a computation of the covariance between X and Y will miss the circular aspect of the correlation.

If the time of the explosion was not known, or the coordinates of the seismic stations were not perfectly known, or if the velocity of the seismic waves was only imperfectly known, the model vector would contain all these parameters:

$$\mathbf{m} = (X,Y,T,x^1,y^1,...,x^6,y^6,v) \; . \tag{18}$$

After properly introducing the a priori information on T (if any), on (x^i,y^i) , and on v , the posterior probability density $\sigma_M(X,Y,T,x^1,y^1,...,x^6,y^6,v)$ should be defined as before, from where the marginal probability density on the epicentral coordinates (X,Y) could be obtained through

$$\sigma_{X,Y}(X,Y) = \int_{-\infty}^{\infty} dT \int_{-\infty}^{\infty} dx^1 ... \int_{-\infty}^{\infty} dy^6 \int_{0}^{\infty} dv \; \sigma_M(X,Y,T,x^1,y^1,...,x^6,y^6,v), \tag{19}$$

while the posterior probability density on the time T of the explosion is

$$\sigma_T(T) = \int_{-\infty}^{\infty} dX \int_{-\infty}^{\infty} dY \int_{-\infty}^{\infty} dx^1 \; ... \int_{-\infty}^{\infty} dy^6 \int_{0}^{\infty} dv \; \sigma_M(X,Y,T,x^1,y^1,...,x^6,y^6,v). \tag{20}$$

As computations rapidly become expensive, it may be necessary to make some simplifying assumptions. The most drastic one is to neglect uncertainties on (x^i,y^i) and v , and to increase the estimated error in the observed arrival times artificially, to compensate approximately for the simplification.

A realistic Earth is tridimensional and heterogeneous. It is generally simpler to use spherical coordinates (R,Θ,Φ) . Then the null information probability density is no longer uniform (see example 18). Also, for a realistic Earth, errors made in computing the travel times of seismic waves are not negligible compared to errors in the observation of arrival times at the seismic stations. Equation (12) should then be replaced, for instance, by the

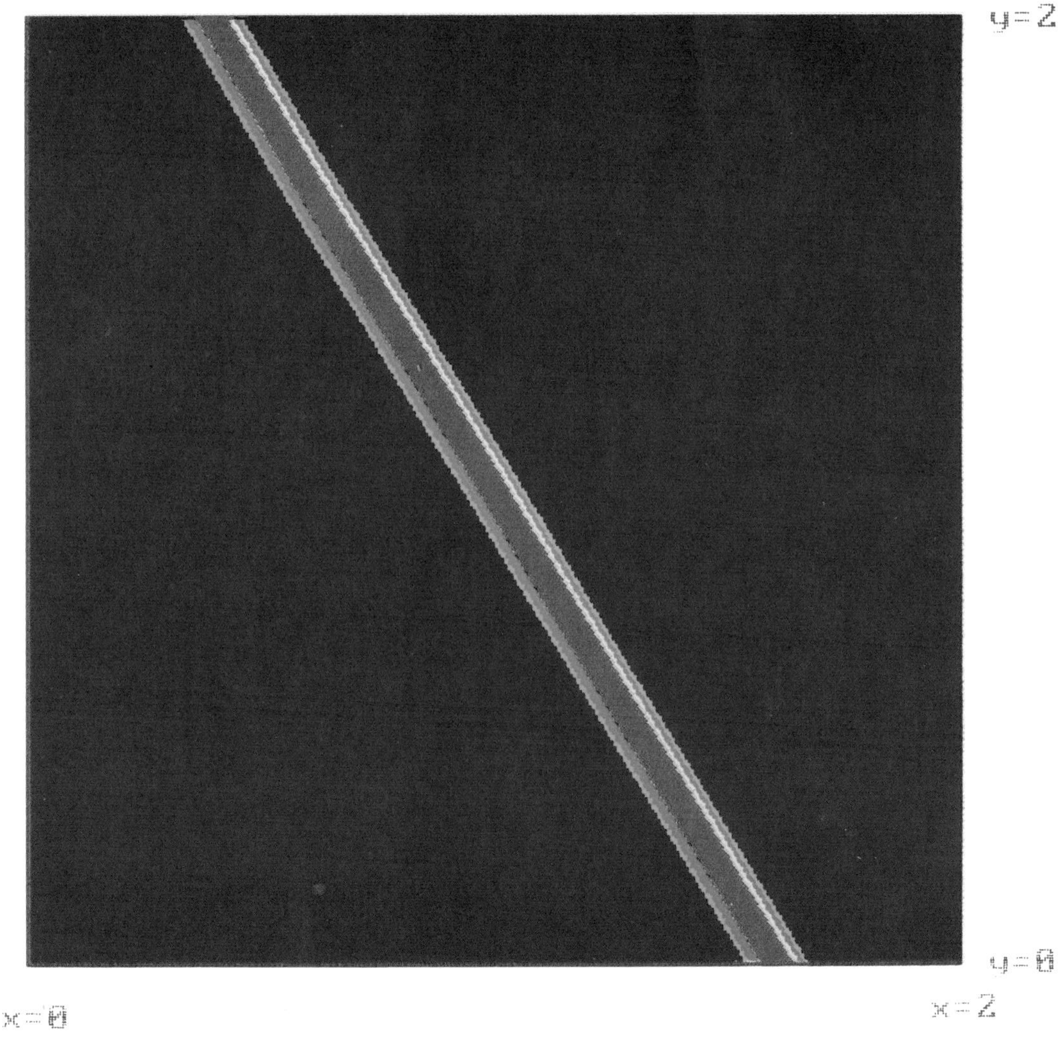

Figure 1.36 (from problem (1.10): Same as the previous figure, with finer detail.

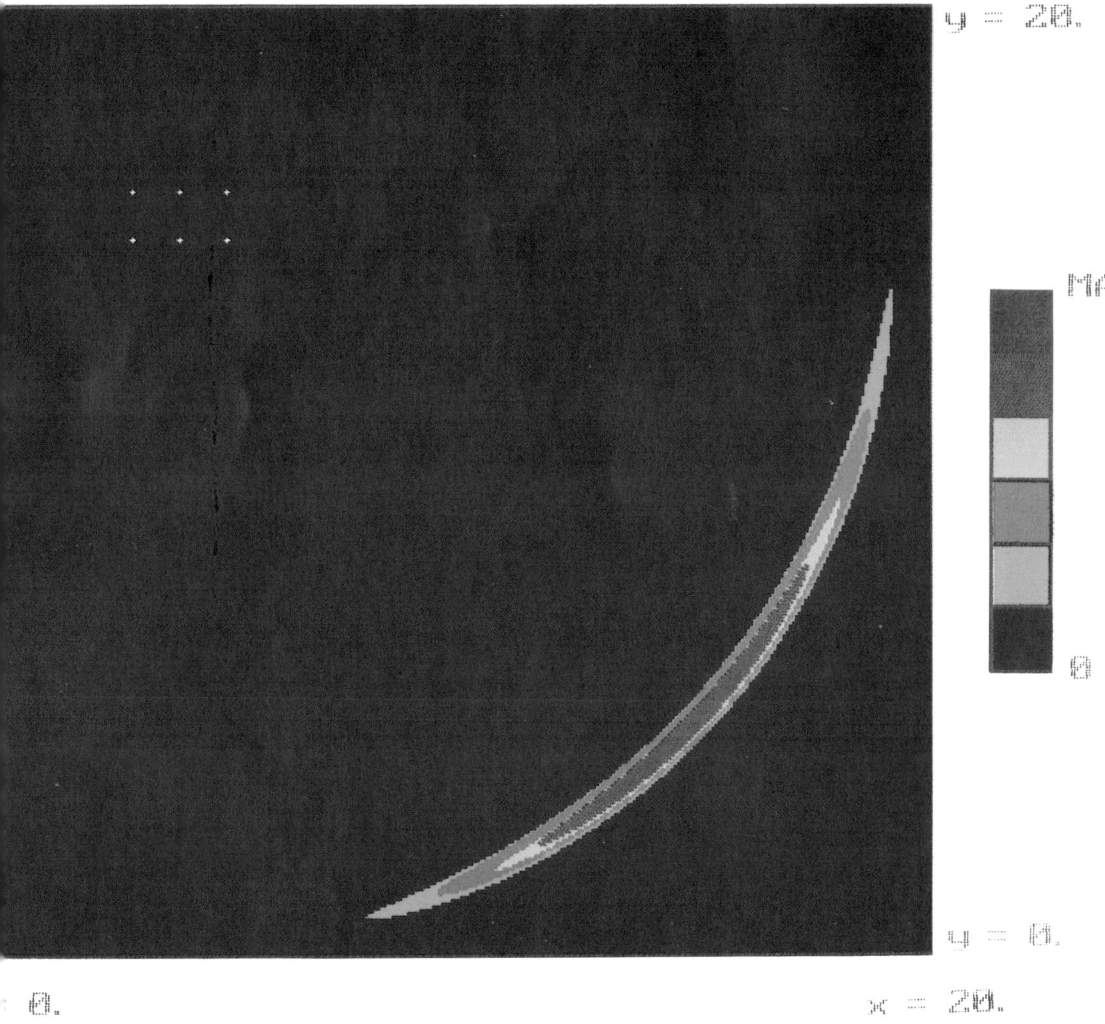

Figure 1.14: Probability density for the true epicentral coordinates of the seismic event, obtained using as data the arrival times of the seismic wave at six seismic stations (points in the top of the figure). The color scale is linear, between zero and the maximum value of the probability density. The crescent-shape of the region of significant probability density cannot be described using a few numbers (mathematical expectation, variances, covariances,...) as in usual solutions to inverse problems.

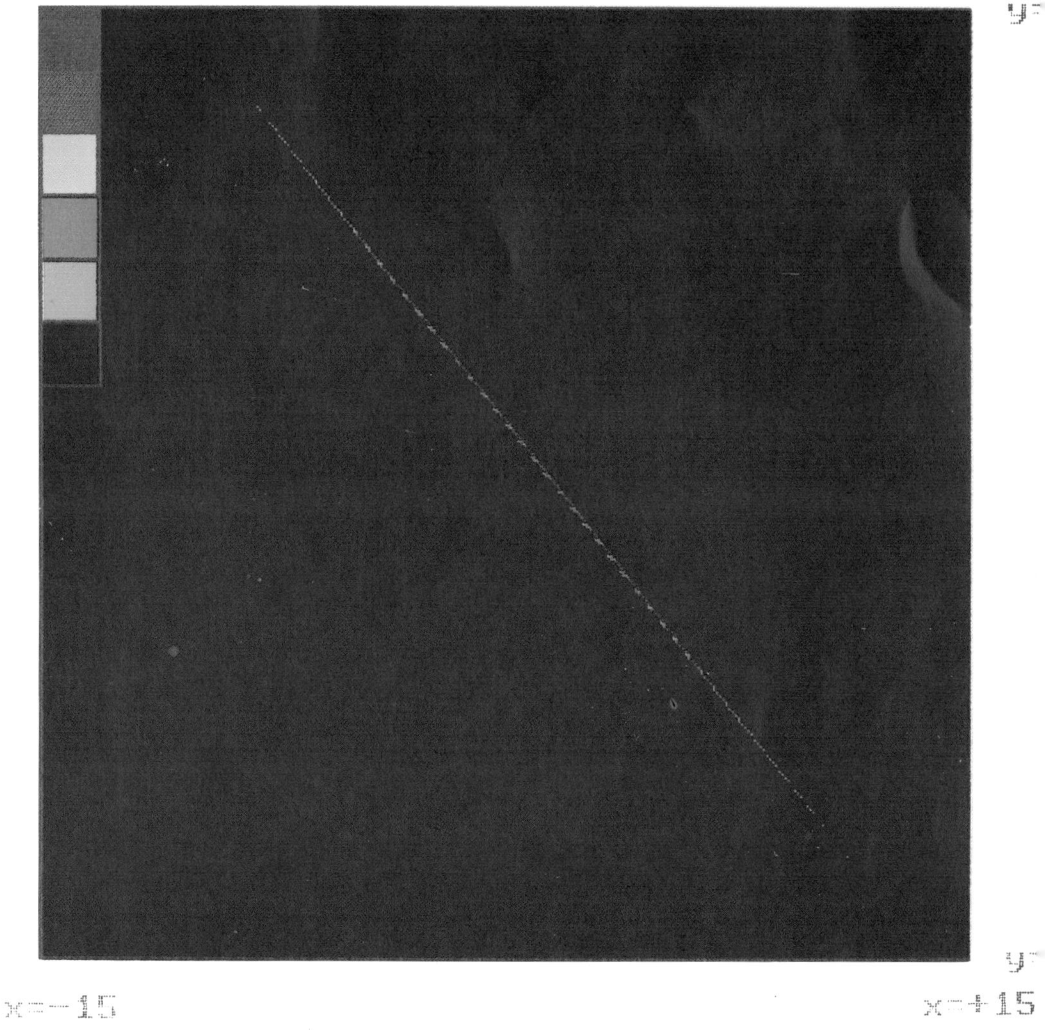

Figure 1.35 (from problem 1.10): Marginal probability density for the parameters m^1 and m^2 . Uncertainties are so strongly correlated that it is difficult to distinguish the ellipsoid of errors from a segment. Although the standard deviations for each of the parameters are large, we have much information on these parameters, because their true values must lie on the "line". The resolution of the plotting device (300 pixel x 300 pixel) is not fine enough for a good representation, so that the colors obtained for the ellipsoid are aliased. The next figure shows a zoom of the central region.

equation

$$\theta(\mathbf{d}|r,\theta,\phi,T) = \exp\left\{ -\frac{1}{2}(\mathbf{d}-\mathbf{g}(r,\theta,\phi,T))^t \; C_T^{-1} \; (\mathbf{d}-\mathbf{g}(r,\theta,\phi,T)) \right\}, \tag{21}$$

where $\mathbf{C}_T$ is an ad-hoc covariance matrix approximately describing the errors made in estimating arrival times theoretically. For more details, the reader may refer to Tarantola and Valette (1982a).

Problem 1.2: First (elementary) approach to tomography. Figure 1.15 shows an object composed of 9 homogeneous portions. The values indicated correspond to the *linear attenuation coefficients* (relative to some reference medium, for instance, water) for X-rays (in given units). An X-ray experiment using the geometry shown in Figure 1.16 allows of measuring the *transmittance* ρ^{ij} along each ray, which is given by

$$\rho^{ij} = \exp\left[- \int_{R^{ij}} ds^{ij} \; m(\, x(s^{ij})\,) \right], \tag{1}$$

where $m(\mathbf{x})$ represents the linear attenuation coefficient at point $\mathbf{x}$, R^{ij} represents the ray between source i and receiver j, and ds^{ij} is the element of length along the ray R^{ij}. Assume that instead of measuring ρ^{i} we measure

$$d^{ij} = - \text{Log}\, \rho^{ij} = \int_{R^{ij}} ds^{ij} \; m(\, \mathbf{x}(s^{ij})\,)\, , \tag{2}$$

which is termed the *integrated attenuation.*

If the medium is a priori assumed to be composed of the 9 homogeneous portions of Figure 1, any model of the medium may be represented using the notation

$$\mathbf{m} = \begin{bmatrix} m^{11} & m^{12} & m^{13} \\ m^{21} & m^{22} & m^{23} \\ m^{31} & m^{32} & m^{33} \end{bmatrix}, \tag{3}$$

where the first index represents the column and the second index represents the row. Any possible set of numerical values in (3) is a *model vector*. For instance, the true model is (Figure 1)

$$\begin{bmatrix} m^{11} & m^{12} & m^{13} \\ m^{21} & m^{22} & m^{23} \\ m^{31} & m^{32} & m^{33} \end{bmatrix} = \begin{bmatrix} 50 & 60 & 50 \\ 60 & 58 & 60 \\ 50 & 60 & 50 \end{bmatrix}. \tag{4}$$

A *data vector* is represented by

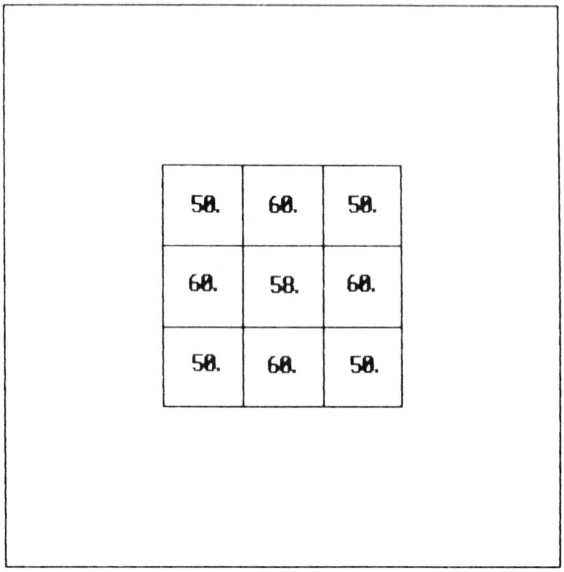

Figure 1.15: A bidimensional medium is composed of 3×3 homogeneous blocks. Indicated are the "true values" of the linear attenuation coefficient for X-rays (with respect to the surrounding medium).

$$\mathbf{d} = \begin{bmatrix} d^{11} & d^{12} & d^{13} & d^{14} & d^{15} & d^{16} \\ d^{21} & d^{22} & d^{23} & d^{24} & d^{25} & d^{26} \end{bmatrix} , \tag{5}$$

where the first index denotes the source number, and the second index denotes the receiver number. Equation (2) then simplifies to the discrete equation

$$d^{ij} = \sum_{\alpha=1}^{3} \sum_{\beta=1}^{3} G^{ij\alpha\beta} \; m^{\alpha\beta} \qquad \text{for } i=1,2 \qquad j=1,2,3,4,5,6 \ , \tag{6}$$

where $G^{ij\alpha\beta}$ represents the length of the ray ij inside the block $\alpha\beta$. An actual measurement of the integrated attenuation gives the values

$$\begin{bmatrix} d^{11} & d^{12} & d^{13} & d^{14} & d^{15} & d^{16} \\ d^{21} & d^{22} & d^{23} & d^{24} & d^{25} & d^{26} \end{bmatrix} =$$

$$\begin{bmatrix} 341.9\pm0.1 & 353.1\pm0.1 & 356.2\pm0.1 & 356.2\pm0.1 & 353.1\pm0.1 & 341.9\pm0.1 \\ 341.9\pm0.1 & 353.1\pm0.1 & 356.2\pm0.1 & 356.2\pm0.1 & 353.1\pm0.1 & 341.9\pm0.1 \end{bmatrix} , \tag{7}$$

where ±0.1 indicates the standard deviation of the estimated (Gaussian) error.

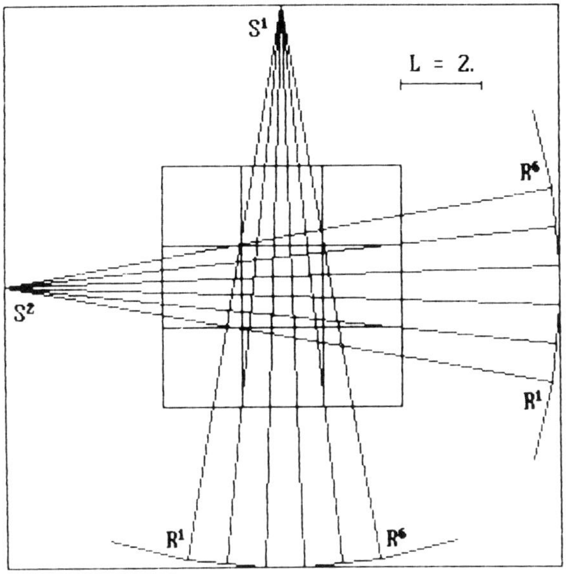

Figure 1.16: In order to infer the true (unknown) values of the linear attenuation coefficient, an X-ray transmission tomographic experiment is performed. Each block measures $L = 2$ units of length, and the figure is to scale (the angular separation between rays is 4 degrees). S^1 and S^2 represent the two source locations, and $R^1,...,R^6$ the six receivers used. Let $m(\mathbf{x})$ represent the linear attenuation coefficient at point $\mathbf{x}$ of the medium under study, R^{ij} the ray between source 1 and receiver j , s^{ij} the position along ray R^{ij} ,and d^{ij} the integrated attenuation along ray R^{ij} : $d^{ij} = \int ds^{ij}\ m(\mathbf{x}(s^{ij}))$ (along R^{ij}) . The measured values of the integrated attenuation along each ray are, in order for each receiver, 341.9±0.1, 353.1±0.1, 356.2±0.1 356.2±0.1 , 353.1±0.1, and 341.9±9±0.1 for source 1 , and 341.9±0.1, 353.1±0.1, 356.2±0.1 356.2±0.1 , 353.1±0.1, and 341.9±9±0.1 for source 2 (these values correspond in fact to the actual values as they can be computed from the true linear attenuation values of Figure 1, plus a Gaussian noise with standard deviation 0.1, and are rounded to the first decimal). These values are assumed to be corrected for the effect of the propagation outside the 3×3 model, so that the linear attenuation coefficient outside the model can be taken as null. The inverse problem consists in using these "observed" values of integrated attenuation to infer the actual model values. Remark that the upper-left block is explored with very short ray lengths, and owing to the relatively high noise in the data, the actual value of this block will probably be poorly resolved.

Assume that you have the a priori information that the model values of the linear attenuation coefficients equal 55 ± 15 (Figure 1.17). Give a better estimation of them using the data (7) and the least squares theory. Discuss.

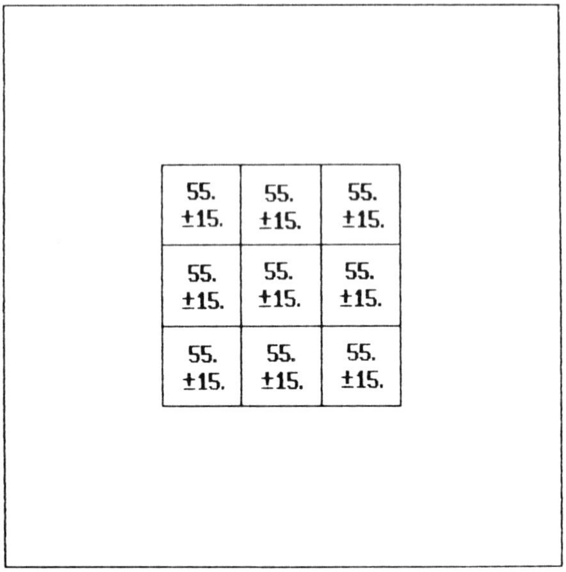

Figure 1.17: We have the a priori information that the true linear attenuation coeffients are 55 ± 15 . It is assumed that a Gaussian probability density well represents this a priori information (in particular, ± 15 represent "soft limits", which can be outpassed with a probability corresponding to the Gaussian density function).

Solution: We wish here to obtain the model $\mathbf{m}$ minimizing

$$S(\mathbf{m}) = \tag{8}$$

$$\frac{1}{2}\Big((\mathbf{Gm}-\mathbf{d}_{obs})^t \, \mathbf{C}_D{}^{-1} \, (\mathbf{Gm}-\mathbf{d}_{obs}) + (\mathbf{m}-\mathbf{m}_{prior})^t \, \mathbf{C}_M{}^{-1} \, (\mathbf{m}-\mathbf{m}_{prior}) \Big) ,$$

where

$$\mathbf{d}_{obs} = \begin{bmatrix} 341.9 & 353.1 & 356.2 & 356.2 & 353.1 & 341.9 \\ 341.9 & 353.1 & 356.2 & 356.2 & 353.1 & 341.9 \end{bmatrix} , \tag{9}$$

$$(C_D)^{ijkl} = 0.1^2 \, \delta^{ik} \, \delta^{jl} , \tag{10}$$

$$\mathbf{m}_{\text{prior}} = \begin{bmatrix} 55 & 55 & 55 \\ 55 & 55 & 55 \\ 55 & 55 & 55 \end{bmatrix} , \tag{11}$$

$$(C_M)^{\alpha\beta\gamma\delta} = 15^2 \; \delta^{\alpha\gamma} \; \delta^{\beta\delta} , \tag{12}$$

and where the elements of the kernel of the linear operator **G** can be obtained from Figure 1.16 using a simple geometrical computation:

$$\begin{bmatrix} G^{1111} & G^{1112} & G^{1113} \\ G^{1121} & G^{1122} & G^{1123} \\ G^{1131} & G^{1132} & G^{1133} \end{bmatrix} = \begin{bmatrix} 0.3338 & 1.6971 & 0.0000 \\ 2.0309 & 0.0000 & 0.0000 \\ 2.0309 & 0.0000 & 0.0000 \end{bmatrix} , \tag{13a}$$

$$\begin{bmatrix} G^{1211} & G^{1212} & G^{1213} \\ G^{1221} & G^{1222} & G^{1223} \\ G^{1231} & G^{1232} & G^{1233} \end{bmatrix} = \begin{bmatrix} 0.0000 & 2.0110 & 0.0000 \\ 0.0000 & 2.0110 & 0.0000 \\ 0.4883 & 1.5227 & 0.0000 \end{bmatrix} , \tag{13b}$$

$$\begin{bmatrix} G^{1311} & G^{1312} & G^{1313} \\ G^{1321} & G^{1322} & G^{1323} \\ G^{1331} & G^{1332} & G^{1333} \end{bmatrix} = \begin{bmatrix} 0.0000 & 2.0012 & 0.0000 \\ 0.0000 & 2.0012 & 0.0000 \\ 0.0000 & 2.0012 & 0.0000 \end{bmatrix} , \tag{13c}$$

$$\begin{bmatrix} G^{1411} & G^{1412} & G^{1413} \\ G^{1421} & G^{1422} & G^{1423} \\ G^{1431} & G^{1432} & G^{1433} \end{bmatrix} = \begin{bmatrix} 0.0000 & 2.0012 & 0.0000 \\ 0.0000 & 2.0012 & 0.0000 \\ 0.0000 & 2.0012 & 0.0000 \end{bmatrix} , \tag{13d}$$

$$\begin{bmatrix} G^{1511} & G^{1512} & G^{1513} \\ G^{1521} & G^{1522} & G^{1523} \\ G^{1531} & G^{1532} & G^{1533} \end{bmatrix} = \begin{bmatrix} 0.0000 & 2.0110 & 0.0000 \\ 0.0000 & 2.0110 & 0.0000 \\ 0.0000 & 1.5227 & 0.4883 \end{bmatrix} , \tag{13e}$$

$$\begin{bmatrix} G^{1611} & G^{1612} & G^{1613} \\ G^{1621} & G^{1622} & G^{1623} \\ G^{1631} & G^{1632} & G^{1633} \end{bmatrix} = \begin{bmatrix} 0.0000 & 1.6971 & 0.3338 \\ 0.0000 & 0.0000 & 2.0309 \\ 0.0000 & 0.0000 & 2.0309 \end{bmatrix} , \tag{13f}$$

$$\begin{bmatrix} G^{2111} & G^{2112} & G^{2113} \\ G^{2121} & G^{2122} & G^{2123} \\ G^{2131} & G^{2132} & G^{2133} \end{bmatrix} = \begin{bmatrix} 0.0000 & 0.0000 & 0.0000 \\ 1.6971 & 0.0000 & 0.0000 \\ 0.3338 & 2.0309 & 2.0309 \end{bmatrix} , \tag{13g}$$

$$\begin{bmatrix} G^{2211} & G^{2212} & G^{2213} \\ G^{2221} & G^{2222} & G^{2223} \\ G^{2231} & G^{2232} & G^{2233} \end{bmatrix} = \begin{bmatrix} 0.0000 & 0.0000 & 0.0000 \\ 2.0110 & 2.0110 & 1.5227 \\ 0.0000 & 0.0000 & 0.4883 \end{bmatrix} , \tag{13h}$$

$$\begin{bmatrix} G^{2311} & G^{2312} & G^{2313} \\ G^{2321} & G^{2322} & G^{2323} \\ G^{2331} & G^{2332} & G^{2333} \end{bmatrix} = \begin{bmatrix} 0.0000 & 0.0000 & 0.0000 \\ 2.0012 & 2.0012 & 2.0012 \\ 0.0000 & 0.0000 & 0.0000 \end{bmatrix} , \tag{13i}$$

$$\begin{bmatrix} G^{2411} & G^{2412} & G^{2413} \\ G^{2421} & G^{2422} & G^{2423} \\ G^{2431} & G^{2432} & G^{2433} \end{bmatrix} = \begin{bmatrix} 0.0000 & 0.0000 & 0.0000 \\ 2.0012 & 2.0012 & 2.0012 \\ 0.0000 & 0.0000 & 0.0000 \end{bmatrix} , \tag{13j}$$

$$\begin{bmatrix} G^{2511} & G^{2512} & G^{2513} \\ G^{2521} & G^{2522} & G^{2523} \\ G^{2531} & G^{2532} & G^{2533} \end{bmatrix} = \begin{bmatrix} 0.0000 & 0.0000 & 0.4883 \\ 2.0110 & 2.0110 & 1.5227 \\ 0.0000 & 0.0000 & 0.0000 \end{bmatrix} , \tag{13k}$$

and

$$\begin{bmatrix} G^{2611} & G^{2612} & G^{2613} \\ G^{2621} & G^{2622} & G^{2623} \\ G^{2631} & G^{2632} & G^{2633} \end{bmatrix} = \begin{bmatrix} 0.3338 & 2.0309 & 2.0309 \\ 1.6971 & 0.0000 & 0.0000 \\ 0.0000 & 0.0000 & 0.0000 \end{bmatrix} . \tag{13l}$$

The minimum of expression (8) can, for instance, be obtained using equation (1.92) of the text:

$$\mathbf{m} = \mathbf{m}_{\text{prior}} + \left(\mathbf{G}^t \, \mathbf{C}_D^{-1} \, \mathbf{G} + \mathbf{C}_M \right)^{-1} \mathbf{G}^t \, \mathbf{C}_D^{-1} \left(\mathbf{d}_{\text{obs}} - \mathbf{G} \, \mathbf{m}_{\text{prior}} \right) . \tag{14}$$

This gives

$$\mathbf{m} = \begin{bmatrix} 55.9 & 59.3 & 50.3 \\ 59.3 & 58.4 & 60.2 \\ 50.3 & 60.2 & 50.3 \end{bmatrix} . \tag{15}$$

The covariance operator describing a posteriori uncertainties in the model parameters is (equation 1.90 of the text)

$$\mathbf{C}_{M'} = \left(\mathbf{G}^t \, \mathbf{C}_D^{-1} \, \mathbf{G} + \mathbf{C}_M^{-1} \right)^{-1} . \tag{16}$$

Instead of representing variances and covariances of $\mathbf{C}_{M'}$, it is more useful to represent standard deviations and correlations (see box 1.1). This gives the standard deviations

$$\sigma_{M'} = \begin{bmatrix} 14.7 & 1.7 & 0.7 \\ 1.7 & 1.0 & 0.7 \\ 0.7 & 0.7 & 0.6 \end{bmatrix} , \tag{17}$$

and the coefficients of correlation

$$\begin{bmatrix} R^{1111} & R^{1112} & R^{1113} \\ R^{1121} & R^{1122} & R^{1123} \\ R^{1131} & R^{1132} & R^{1133} \end{bmatrix} = \begin{bmatrix} 1.0000 & -0.9977 & 0.9536 \\ -0.9977 & 0.9958 & 0.9874 \\ 0.9536 & 0.9874 & 0.9901 \end{bmatrix} , \tag{18a}$$

$$\begin{bmatrix} R^{1211} & R^{1212} & R^{1213} \\ R^{1221} & R^{1222} & R^{1223} \\ R^{1231} & R^{1232} & R^{1233} \end{bmatrix} = \begin{bmatrix} -0.9977 & 1.0000 & -0.9710 \\ 0.9997 & -0.9972 & -0.9897 \\ -0.9708 & -0.9902 & -0.9896 \end{bmatrix} , \tag{18b}$$

$$\begin{bmatrix} R^{1311} & R^{1312} & R^{1313} \\ R^{1321} & R^{1322} & R^{1323} \\ R^{1331} & R^{1332} & R^{1333} \end{bmatrix} = \begin{bmatrix} 0.9536 & -0.9710 & 1.0000 \\ -0.9708 & 0.9657 & 0.9624 \\ 0.9948 & 0.9632 & 0.9515 \end{bmatrix} , \qquad (18c)$$

$$\begin{bmatrix} R^{2111} & R^{2112} & R^{2113} \\ R^{2121} & R^{2122} & R^{2123} \\ R^{2131} & R^{2132} & R^{2133} \end{bmatrix} = \begin{bmatrix} -0.9977 & 0.9997 & -0.9708 \\ 1.0000 & -0.9972 & -0.9902 \\ -0.9710 & -0.9897 & -0.9896 \end{bmatrix} , \qquad (18d)$$

$$\begin{bmatrix} R^{2211} & R^{2212} & R^{2213} \\ R^{2221} & R^{2222} & R^{2223} \\ R^{2231} & R^{2232} & R^{2233} \end{bmatrix} = \begin{bmatrix} 0.9958 & -0.9972 & 0.9657 \\ -0.9972 & 1.0000 & 0.9776 \\ 0.9657 & 0.9776 & 0.9963 \end{bmatrix} , \qquad (18e)$$

$$\begin{bmatrix} R^{2311} & R^{2312} & R^{2313} \\ R^{2321} & R^{2322} & R^{2323} \\ R^{2331} & R^{2332} & R^{2333} \end{bmatrix} = \begin{bmatrix} 0.9874 & -0.9897 & 0.9624 \\ -0.9902 & 0.9776 & 1.0000 \\ 0.9632 & 0.9977 & 0.9622 \end{bmatrix} , \qquad (18f)$$

$$\begin{bmatrix} R^{3111} & R^{3112} & R^{3113} \\ R^{3121} & R^{3122} & R^{3123} \\ R^{3131} & R^{3132} & R^{3133} \end{bmatrix} = \begin{bmatrix} 0.9536 & -0.9708 & 0.9948 \\ -0.9710 & 0.9657 & 0.9632 \\ 1.0000 & 0.9624 & 0.9515 \end{bmatrix} , \qquad (18g)$$

$$\begin{bmatrix} R^{3211} & R^{3212} & R^{3213} \\ R^{3221} & R^{3222} & R^{3223} \\ R^{3231} & R^{3232} & R^{3233} \end{bmatrix} = \begin{bmatrix} 0.9874 & -0.9902 & 0.9632 \\ -0.9897 & 0.9776 & 0.9977 \\ 0.9624 & 1.0000 & 0.9622 \end{bmatrix} , \qquad (18h)$$

and

$$\begin{bmatrix} R^{3311} & R^{3312} & R^{3313} \\ R^{3321} & R^{3322} & R^{3323} \\ R^{3331} & R^{3332} & R^{3333} \end{bmatrix} = \begin{bmatrix} 0.9901 & -0.9896 & 0.9515 \\ -0.9896 & 0.9963 & 0.9622 \\ 0.9515 & 0.9622 & 1.0000 \end{bmatrix} . \qquad (18i)$$

The solution (15) with the errors (17) is represented in Figure 1.18 (to be compared with Figure 1.15 and Figure 1.17). The a priori information was that the values in each block were 55 ± 15 . We see that the a posteriori errors are much smaller except in block (1 1) , where the solution, 55.9 ± 14.7 , practically coincides with the a priori solution. As can be seen in Figure 1.16, this block contains very short lengths of rays, so that it has practically not been explored by our data; the value of the attenuation coefficient is practically not resolved by the data set used. If a least squares inversion was performed with the data (7) but without using a priori information, that block would certainly take very arbitrary values, thus polluting the values of the attenuation coefficient in all the other blocks. More dramatically, numerical instabilities could arise (because the the operator $\mathbf{G}^t \mathbf{C}_D^{-1} \mathbf{G}$ could become numerically not positive definite due to computer rounding errors) and the used computer code would clash with a "zero divide" diagnostic

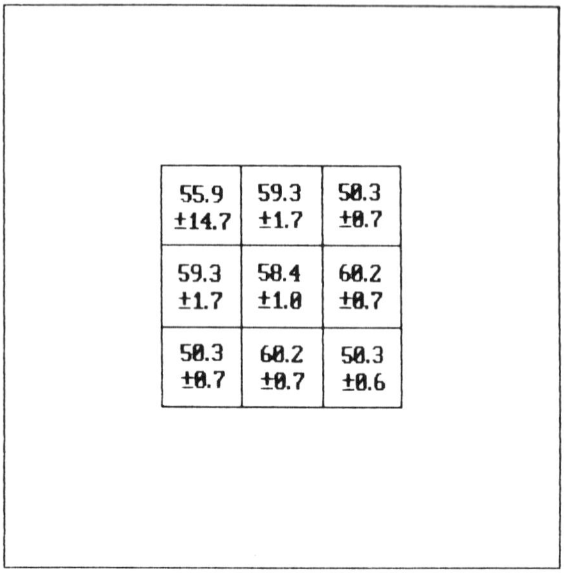

Figure 1.18: The a posteriori solution obtained by inversion of the available data. Remark that the value of the upper-left block has not been resolved (a posteriori value and estimated error almost identical to the a priori values). The values of all other blocks have been estimated with a relative error of less than 3%.

Except for the unresolved block m^{11} , the values obtained are close to the true values, and within the estimated error bar. Of course, as the data used were noise-corrupted, the obtained values cannot be identical to the true values. Using more rays would give a more precise solution.

The data values recalculated from the solution (15) are

$$\mathbf{d}_{obs} = \begin{bmatrix} 341.92 & 353.10 & 356.20 & 356.20 & 353.10 & 341.90 \\ 341.89 & 353.10 & 356.20 & 356.20 & 353.10 & 341.92 \end{bmatrix} , \tag{19}$$

which are almost identical to the observed values (9).

The coefficients of correlation as shown in (18) are all very close to unity. This is due to the fact that there is no independent information (all rays traverse at least three blocks), and there is not much data redundancy.

Remark number 1: Assume that a new experiment produces one new datum, corresponding to a new ray (equal to or different from the previous rays). In order to incorporate this new information, we can either take the a priori model (11)-(12) and perform an inversion using 13 data, or, more simply, we can take the a posteriori solution (15)-(17)-(18) as a priori solution for an inverse problem with a single datum (the new one). As demonstrated in Chapter 4, this gives exactly the same solution (thus showing the coherence of the "a priori information" approach).

Remark number 2: Usual computer codes consider that vectors (i.e., elements of a linear space) are necessarily represented using column matrices, and that the kernels of linear operators are then represented using two-dimensional matrices. It may then be simpler for numerical computations to replace the previous notations by the matricial notations

$$\mathbf{d}_{obs} = \begin{bmatrix} 341.9 \\ 353.1 \\ 356.2 \\ 356.2 \\ 353.1 \\ 341.9 \\ 341.9 \\ 353.1 \\ 356.2 \\ 356.2 \\ 353.1 \\ 341.9 \end{bmatrix} , \qquad (20)$$

$$(C_D)^{ij} = 0.1^2 \, \delta^{ij} , \qquad (21)$$

$$\mathbf{m}_{prior} = \begin{bmatrix} 55 \\ 55 \\ 55 \\ 55 \\ 55 \\ 55 \\ 55 \\ 55 \\ 55 \end{bmatrix} , \qquad (22)$$

$$(C_M)^{\alpha\beta} = 15^2 \, \delta^{\alpha\beta} , \qquad (23)$$

and

$$\mathbf{G} = \begin{bmatrix} 0.3338 & 1.6971 & 0.0000 & 2.0309 & 0.0000 & 0.0000 & 2.0309 & 0.0000 & 0.0000 \\ 0.0000 & 2.0110 & 0.0000 & 0.0000 & 2.0110 & 0.0000 & 0.4883 & 1.5227 & 0.0000 \\ 0.0000 & 2.0012 & 0.0000 & 0.0000 & 2.0012 & 0.0000 & 0.0000 & 2.0012 & 0.0000 \\ 0.0000 & 2.0012 & 0.0000 & 0.0000 & 2.0012 & 0.0000 & 0.0000 & 2.0012 & 0.0000 \\ 0.0000 & 2.0110 & 0.0000 & 0.0000 & 2.0110 & 0.0000 & 0.0000 & 2.0110 & 0.4883 \\ 0.0000 & 1.6971 & 0.3338 & 0.0000 & 0.0000 & 2.0309 & 0.0000 & 0.0000 & 2.0309 \\ 0.0000 & 0.0000 & 0.0000 & 1.6971 & 0.0000 & 0.0000 & 0.3338 & 2.0309 & 2.0309 \\ 0.0000 & 0.0000 & 0.0000 & 2.0110 & 2.0110 & 1.5227 & 0.0000 & 0.0000 & 0.4883 \\ 0.0000 & 0.0000 & 0.0000 & 2.0012 & 2.0012 & 2.0012 & 0.0000 & 0.0000 & 0.0000 \\ 0.0000 & 0.0000 & 0.0000 & 2.0012 & 2.0012 & 2.0012 & 0.0000 & 0.0000 & 0.0000 \\ 0.0000 & 0.0000 & 0.4883 & 2.0110 & 2.0110 & 1.5227 & 0.0000 & 0.0000 & 0.0000 \\ 0.3338 & 2.0309 & 2.0309 & 1.6971 & 0.0000 & 0.0000 & 0.0000 & 0.0000 & 0.0000 \end{bmatrix} . \tag{24}$$

Equations (8) and (14) are then usual matricial equations.

Problem 1.3: Formulas for inverse problems when the (discrete) parameters can take only discrete values. The theory developed in this chapter applies to the case where each of the discrete parameters $d^1, d^2, \ldots$ and $m^1, m^2, \ldots$ can take continuous values. In some problems, the data parameters and/or the model parameters can only take some discrete values. Obtain the corresponding formulas for the resolution of inverse problems.

Solution: Let $\mathbf{x} = \{x^A\} = (x^1, x^2, \ldots)$ denote a discrete (and finite) parameter set. In the text it has been assumed that each of the parameters x^A $(A=1,2,\ldots)$ takes its values on some continuous set. The vector $\mathbf{x}$ then takes its values on some "volume" $\mathbf{V}$. We have seen that the most general way of describing a *state of information* on the true values of the parameters is by defining a ***probability density*** $\rho(\mathbf{x})$ $(\mathbf{x} \in \mathrm{V})$ over the parameter space.

If each of the parameters x^A $(A=1,2,\ldots)$ take its values on some discrete set , $x^A \in (x^A{}_1, x^A{}_2, \ldots)$, then the vector $\mathbf{x}$ can only take some discrete values $\mathbf{x}_1$, $\mathbf{x}_2$,... , and a state of information is then described by a ***probability*** $\tilde{\rho}(\mathbf{x}_u)$ $(u=1,2,\ldots)$: the probability of each of the discrete values $\mathbf{x}_1$, $\mathbf{x}_2$,...

The formulas corresponding to an inverse problem where the (discrete) parameters can only take discrete values should be obtained from a theory more general than the one developed in the text, or from an ad-hoc discrete theory. Instead, although less rigorous, it is simpler to formally consider a probability as a special case of a probability density:

$$\rho(\mathbf{x}) = \sum_{u} \tilde{\rho}(\mathbf{x}_u)\ \delta(\mathbf{x}-\mathbf{x}_u) \ , \tag{1}$$

where, if $\rho(\mathbf{x})$ is a probability density and $\delta(\mathbf{x})$ is Dirac's delta "function", then $\tilde{\rho}(\mathbf{x}_u)$ clearly corresponds to the probability of the point $\mathbf{x}_u$.

I consider here that the model vector $\mathbf{m}$ can only take discrete values $\mathbf{m}_1,\mathbf{m}_2,...$, and for any value $\mathbf{m}_w$, the data vector $\mathbf{d}$ can only take discrete values $\mathbf{d}_1,\mathbf{d}_2,...$ As discussed in the text, the a posteriori probability density in the space DxM is (equations 1.60, 1.63, and 1.64)

$$\sigma(\mathbf{d},\mathbf{m}) = \frac{\rho(\mathbf{d},\mathbf{m})\ \theta(\mathbf{d},\mathbf{m})}{\mu(\mathbf{d},\mathbf{m})} = \frac{\rho_D(\mathbf{d})\ \rho_M(\mathbf{m})\ \mu_M(\mathbf{m})}{\mu_D(\mathbf{d})\ \mu_M(\mathbf{m})}\ \theta(\mathbf{d}|\mathbf{m}) \ , \tag{2}$$

i.e.,

$$\sigma(\mathbf{d},\mathbf{m}) = \frac{\rho_D(\mathbf{d})\ \rho_M(\mathbf{m})}{\mu_D(\mathbf{d})}\ \theta(\mathbf{d}|\mathbf{m}) \ . \tag{3}$$

In equation (3), $\rho_D(\mathbf{d})$ is the probability density representing the knowledge obtained on the true values of $\mathbf{d}$ through our experiments (measurements), $\rho_M(\mathbf{m})$ is the probability density representing our a priori information on the model parameters, $\theta(\mathbf{d}|\mathbf{m})$ is the conditional probability density for $\mathbf{d}$, given $\mathbf{m}$, representing our knowledge on the theoretical correlations existing between $\mathbf{m}$ and $\mathbf{d}$, and where $\mu_D(\mathbf{d})$ is the non-informative probability density in the data space.

If $\mathbf{d}$ and $\mathbf{m}$ can only take discrete values, then

$$\rho_D(\mathbf{d}) = \sum_{v} \tilde{\rho}_D(\mathbf{d}_v)\ \delta(\mathbf{d}-\mathbf{d}_v) \ , \tag{4a}$$

$$\mu_D(\mathbf{d}) = \sum_{v} \tilde{\mu}_D(\mathbf{d}_v)\ \delta(\mathbf{d}-\mathbf{d}_v) \ , \tag{4b}$$

$$\rho_M(\mathbf{m}) = \sum_{w} \tilde{\rho}_M(\mathbf{m}_w)\ \delta(\mathbf{m}-\mathbf{m}_w) \ , \tag{4c}$$

and

$$\theta(\mathbf{d}|\mathbf{m}_w) = \sum_{v} \tilde{\theta}(\mathbf{d}_v|\mathbf{m}_w)\ \delta(\mathbf{d}-\mathbf{d}_v) \ , \tag{4d}$$

where $\tilde{\rho}_D(\mathbf{d}_v)$ is the probability that we assing of $\mathbf{d}_v$ being the true value of $\mathbf{d}$ which has been realized in our measurement (ambiguous results of the measurements), $\tilde{\mu}_D(\mathbf{d}_v)$ is the non-informative probability on $\mathbf{d}$ (usually uniform), $\tilde{\rho}_M(\mathbf{m}_w)$ is the a priori probability of the true value of $\mathbf{m}$ being $\mathbf{m}_w$, and $\theta(\mathbf{d}_v|\mathbf{m}_w)$ is the conditional probability of the true value of $\mathbf{d}$ being $\mathbf{d}_v$, if the true value of $\mathbf{m}$ was $\mathbf{m}_w$. Introducing the a posteriori probability $\tilde{\sigma}(\mathbf{d}_v,\mathbf{m}_w)$ by

$$\sigma(\mathbf{d},\mathbf{m}) = \sum_v \sum_w \tilde{\sigma}(\mathbf{d}_v,\mathbf{m}_w)\ \delta(\mathbf{d}-\mathbf{d}_v)\ \delta(\mathbf{m}-\mathbf{m}_w) , \tag{4e}$$

we obtain, *at each point* $(\mathbf{d}_v,\mathbf{m}_w)$,

$$\boxed{\tilde{\sigma}(\mathbf{d}_v,\mathbf{m}_w) = \frac{\tilde{\rho}_D(\mathbf{d}_v)\ \tilde{\rho}_M(\mathbf{m}_w)\ \tilde{\theta}(\mathbf{d}_v|\mathbf{m}_w)}{\tilde{\rho}_D(\mathbf{d}_v)} .} \tag{5}$$

where the symbol $\delta(\mathbf{0})$ has been formally manipulated as an ordinary (finite) constant.

The (marginal) a posteriori probability in the model space is then

$$\tilde{\sigma}_M(\mathbf{m}_w) = \sum_v \tilde{\sigma}(\mathbf{d}_v,\mathbf{m}_w) , \tag{6}$$

i.e.,

$$\tilde{\sigma}_M(\mathbf{m}_w) = \tilde{\rho}_M(\mathbf{m}_w) \sum_v \frac{\tilde{\rho}_D(\mathbf{d}_v)}{\tilde{\mu}_D(\mathbf{d}_v)}\ \tilde{\theta}(\mathbf{d}_v|\mathbf{m}_w) , \tag{7}$$

or, if the probability is normalized,

$$\boxed{\tilde{\sigma}_M(\mathbf{m}_w) = \frac{\tilde{\rho}_M(\mathbf{m}_w) \displaystyle\sum_v \frac{\tilde{\rho}_D(\mathbf{d}_v)\ \tilde{\theta}(\mathbf{d}_v|\mathbf{m}_w)}{\tilde{\mu}_D(\mathbf{d}_v)}}{\displaystyle\sum_w \tilde{\rho}_M(\mathbf{m}_w) \sum_v \frac{\tilde{\rho}_D(\mathbf{d}_v)\ \tilde{\theta}(\mathbf{d}_v|\mathbf{m}_w)}{\tilde{\mu}_D(\mathbf{d}_v)}} .} \tag{8}$$

The (marginal) a posteriori probability in the data space is

$$\tilde{\sigma}_D(\mathbf{d}_v) = \sum_w \tilde{\sigma}(\mathbf{d}_v,\mathbf{m}_w) , \tag{9}$$

i.e.,

$$\tilde{\sigma}_D(\mathbf{d}_v) = \frac{\tilde{\rho}_D(\mathbf{d}_v)}{\tilde{\mu}_D(\mathbf{d}_v)} \sum_w \tilde{\rho}_M(\mathbf{m}_w)\, \tilde{\theta}(\mathbf{d}_v|\mathbf{m}_w) , \tag{10}$$

or, if the probability is normalized,

$$\boxed{\tilde{\sigma}_D(\mathbf{d}_v) = \frac{\dfrac{\tilde{\rho}_D(\mathbf{d}_v)}{\tilde{\mu}_D(\mathbf{d}_v)} \displaystyle\sum_w \tilde{\rho}_M(\mathbf{m}_w)\, \tilde{\theta}(\mathbf{d}_v|\mathbf{m}_w)}{\displaystyle\sum_v \dfrac{\tilde{\rho}_D(\mathbf{d}_v)}{\tilde{\mu}_D(\mathbf{d}_v)} \sum_w \tilde{\rho}_M(\mathbf{m}_w)\, \tilde{\theta}(\mathbf{d}_v|\mathbf{m}_w)} .} \tag{11}$$

In the particular case where the measurement is perfect and gives an unambiguous result, $\mathbf{d} = \mathbf{d}_{obs}$;

$$\tilde{\rho}_D(\mathbf{d}_v) = \delta_{v v_{obs}} = \begin{cases} 1 & \text{for } v = v_{obs} \\ 0 & \text{for } v \neq v_{obs} . \end{cases} \tag{12}$$

Equation (8) then becomes

$$\boxed{\tilde{\sigma}_M(\mathbf{m}_w) = \frac{\tilde{\theta}(\mathbf{d}_{obs}|\mathbf{m}_w)\, \tilde{\rho}_M(\mathbf{m}_w)}{\displaystyle\sum_w \tilde{\theta}(\mathbf{d}_{obs}|\mathbf{m}_w)\, \tilde{\rho}_M(\mathbf{m}_w)} ,} \tag{13}$$

while equation (11) degenerates into

$$\tilde{\sigma}_D(\mathbf{d}_v) = \tilde{\rho}_D(\mathbf{d}_v) . \tag{14}$$

Formula (13) encounters a large domain of application. Let us recall the sense of each of the terms. $\tilde{\rho}_M(\mathbf{m}_w)$, for w=1,2,..., is the a priori (subjective) probability we assign of each of the $\mathbf{m}_w$ being the true model vector, $\tilde{\theta}(\mathbf{d}_v|\mathbf{m}_w)$, for v=1,2,..., is the probability we assign of each of the $\mathbf{d}_v$ being the true data vector if the true model vector is $\mathbf{m}_v$, and $\tilde{\sigma}_M(\mathbf{m}_w)$ is the a posteriori (subjective) probability we assign to each of the $\mathbf{m}_w$, after a measurement of the true value of the data vector which has given the unambiguous result that the true value is $\mathbf{d}_{v_{obs}}$.

For the sake of completeness, let me give the formulas corresponding to (8), (11), and (13) in the case where only one of the data vector or the model vector is discrete.

If only the model vector $\mathbf{m}$ takes discrete values, equation (8) becomes

$$\tilde{\sigma}_M(\mathbf{m}_w) = \frac{\tilde{\rho}_M(\mathbf{m}_w) \displaystyle\int_D d\mathbf{d} \, \frac{\rho(\mathbf{d}) \, \theta(\mathbf{d}|\mathbf{m}_w)}{\mu_D(\mathbf{d})}}{\displaystyle\sum_w \tilde{\rho}_M(\mathbf{m}_w) \int_D d\mathbf{d} \, \frac{\rho_D(\mathbf{d}) \, \theta(\mathbf{d}|\mathbf{m}_w)}{\mu_D(\mathbf{d}_v)}} , \tag{15}$$

equation (11) becomes

$$\sigma_D(\mathbf{d}) = \frac{\dfrac{\rho_D(\mathbf{d})}{\mu_D(\mathbf{d})} \displaystyle\sum_w \tilde{\rho}_M(\mathbf{m}_w) \, \theta(\mathbf{d}|\mathbf{m}_w)}{\displaystyle\int_D d\mathbf{d} \, \frac{\rho_D(\mathbf{d})}{\mu_D(\mathbf{d})} \sum_w \tilde{\rho}_M(\mathbf{m}_w) \, \theta(\mathbf{d}|\mathbf{m}_w)} , \tag{16}$$

and equation (13) becomes

$$\tilde{\sigma}_M(\mathbf{m}_w) = \frac{\theta(\mathbf{d}_{obs}|\mathbf{m}_w) \, \tilde{\rho}_M(\mathbf{m}_w)}{\displaystyle\sum_w \theta(\mathbf{d}_{obs}|\mathbf{m}_w) \, \tilde{\rho}_M(\mathbf{m}_w)} . \tag{17}$$

If only the data vector $\mathbf{d}$ takes discrete values, equation (8) becomes

$$\sigma_M(\mathbf{m}) = \frac{\rho_M(\mathbf{m}) \displaystyle\sum_v \frac{\tilde{\rho}_D(\mathbf{d}_v)\, \tilde{\theta}(\mathbf{d}_v|\mathbf{m})}{\tilde{\mu}_D(\mathbf{d}_v)}}{\displaystyle\int_{\boldsymbol{M}} d\mathbf{m}\; \rho_M(\mathbf{m}) \sum_v \frac{\tilde{\rho}_D(\mathbf{d}_v)\, \tilde{\theta}(\mathbf{d}_v|\mathbf{m})}{\tilde{\mu}_D(\mathbf{d}_v)}} , \tag{18}$$

equation (11) becomes

$$\tilde{\sigma}_D(\mathbf{d}_v) = \frac{\dfrac{\tilde{\rho}_D(\mathbf{d}_v)}{\tilde{\mu}_D(\mathbf{d}_v)} \displaystyle\int_{\boldsymbol{M}} d\mathbf{m}\;\; \rho_M(\mathbf{m})\, \tilde{\theta}(\mathbf{d}_v|\mathbf{m})}{\displaystyle\sum_v \frac{\tilde{\rho}_D(\mathbf{d}_v)}{\tilde{\mu}_D(\mathbf{d}_v)} \int_{\boldsymbol{M}} d\mathbf{m}\;\; \rho_M(\mathbf{m})\, \tilde{\theta}(\mathbf{d}_v|\mathbf{m})} , \tag{19}$$

and equation (13) becomes

$$\sigma_M(\mathbf{m}) = \frac{\tilde{\theta}(\mathbf{d}_{obs}|\mathbf{m})\; \rho_M(\mathbf{m})}{\displaystyle\int_{\boldsymbol{M}} d\mathbf{m}\; \tilde{\theta}(\mathbf{d}_{obs}|\mathbf{m})\; \rho_M(\mathbf{m})} . \tag{20}$$

Problem 1.4: Approximately one in each 100,000 individuals of a given population is affected by a very dangerous disease, which is only apparent in the final stages. A medical test has been designed to indicate if a given individual is affected by the disease. If that person is affected, the response of the test is always positive, but if not, it gives a positive (erroneous) response in one per cent of the cases. I have been submitted to the test and have obtained a positive response. Am I affected by the disease?

Solution: Let m_1 and m_2 be defined as follows

m_1 = I am affected by the disease , (1a)

$$m_2 \;=\; \text{I am not affected by the disease ,} \tag{1b}$$

and let d_1 and d_2 be defined as follows

$$d_1 \;=\; \text{The test gives a positive response ,} \tag{2a}$$
$$d_2 \;=\; \text{The test gives a negative response .} \tag{2b}$$

As approximately one of every 100,000 individuals is affected by the disease, the a priori information I have on the value of m is described by the probability

$$\tilde{\rho}_M(m_1) \;=\; 0.000\,01 \;, \tag{3a}$$

$$\tilde{\rho}_M(m_2) \;=\; 0.999\,99 \;. \tag{3b}$$

Let $\tilde{\theta}(d_v|m_w)$ be the probability of the result of the test being d_v if the value of m is m_w . As, if an individual is affected, the response of the test is always positive,

$$\tilde{\theta}(d_1|m_1) \;=\; 1. \;\;, \tag{4a}$$

$$\tilde{\theta}(d_2|m_1) \;=\; 0. \;\;, \tag{4b}$$

and as if the individual is not affected, the test gives a positive (erroneous) response in one percent of the cases,

$$\tilde{\theta}(d_1|m_2) \;=\; 0.01 \;, \tag{5a}$$

$$\tilde{\theta}(d_2|m_2) \;=\; 0.99 \;. \tag{5b}$$

The application of the test in my case has unambiguously given the value

$$d_{obs} \;=\; d_1 \;, \tag{6}$$

so that the information I have on the true value of d is represented by the probability

$$\tilde{\rho}_D(d_1) \;=\; 1. \;\;, \tag{7a}$$

$$\tilde{\rho}_D(d_2) \;=\; 0. \;\;. \tag{7b}$$

The a posteriori information on the true value value of m can then be computed using equation (8) of the previous problem:

$$\tilde{\sigma}_M(m_w) = \frac{\tilde{\rho}_M(m_w) \sum_v \frac{\tilde{\rho}_D(d_v)\,\tilde{\theta}(d_v|m_w)}{\tilde{\mu}_D(d_v)}}{\sum_w \tilde{\rho}_M(m_w) \sum_v \frac{\tilde{\rho}_D(d_v)\,\tilde{\theta}(d_v|m_w)}{\tilde{\mu}_D(d_v)}} , \tag{8}$$

which, using (7), becomes

$$\tilde{\sigma}_M(m_w) = \frac{\tilde{\theta}(d_{obs}|m_w)\,\tilde{\rho}_M(m_w)}{\sum_w \tilde{\theta}(d_{obs}|m_w)\,\tilde{\rho}_M(m_w)} . \tag{9}$$

This gives

$$\begin{aligned} \tilde{\sigma}_M(m_1) &= \frac{\tilde{\theta}(d_1|m_1)\,\tilde{\rho}_M(m_1)}{\tilde{\theta}(d_1|m_1)\,\tilde{\rho}_M(m_1) + \tilde{\theta}(d_1|m_2)\,\tilde{\rho}_M(m_2)} \\ &= \frac{1.\cdot 0.000\,01}{1.\cdot 0.000\,01 + 0.01\cdot 0.999\,99} \simeq 0.000\,999\,0 \\ &\simeq 10^{-3} , \end{aligned} \tag{10a}$$

and

$$\begin{aligned} \tilde{\sigma}_M(m_2) &= \frac{\tilde{\theta}(d_1|m_2)\,\tilde{\rho}_M(m_2)}{\tilde{\theta}(d_1|m_1)\,\tilde{\rho}_M(m_1) + \tilde{\theta}(d_1|m_2)\,\tilde{\rho}_M(m_2)} \\ &= \frac{0.01\cdot 0.999\,99}{1.\cdot 0.000\,01 + 0.01\cdot 0.999\,99} \simeq 0.999\,001\,0 \\ &\simeq 1 - 10^{-3} , \end{aligned} \tag{10b}$$

i.e., in spite of the fact that the response of the test (which makes only one per cent of errors) has been positive, I only have one chance per thousand of being affected by the disease. The intuitive interpretation is as follows: although the test makes very few errors, the percentage of diseased individuals in the population is so low that the test may make many errors before it is used with a diseased individual. As the a priori probability of my having the disease was of one in 100,000 , and the a posteriori probability is 100 times greater, the test effectively generates information, although I do not yet have to worry very much about it, because a probability of one per thousand is still very low. However, further health checks are justified.

Assume now that the response to the test had not been unambiguous (Hi, Pr. Tarantola, here is Dr. Jekyll. My assistant just gave me the response of the test, and she says that the result was positive. Nevertheless, I have to inform you that my assistant is not a very reliable woman: each time she has a result, she throws a dice, and if she obtains a six, she lies to me about the result...).

As the probability $\tilde{\rho}_D(d_v)$ is the (a priori) probability that the true value of d is d_v , instead of (7) I should now take

$$\tilde{\rho}_D(d_1) = 5/6 \ , \qquad (11a)$$

$$\tilde{\rho}_D(d_2) = 1/6 \ , \qquad (11b)$$

and representing the null information by the probability

$$\tilde{\mu}_D(d_1) = \tilde{\mu}_D(d_2) = 0.5 \ , \qquad (12)$$

the a posteriori probability for m can be obtained from equation (8), which gives

$$\tilde{\sigma}_M(m_1) \simeq 4.8 \ 10^{-5} \ . \qquad (13)$$

The a posteriori probability for the true result of the test can be computed using equation (11) of the previous problem:

$$\tilde{\sigma}_D(d_v) = \frac{\dfrac{\tilde{\rho}_D(d_v)}{\tilde{\mu}_D(d_v)} \displaystyle\sum_w \tilde{\rho}_M(m_w)\, \tilde{\theta}(d_v|m_w)}{\displaystyle\sum_v \dfrac{\tilde{\rho}_D(d_v)}{\tilde{\mu}_D(d_v)} \sum_w \tilde{\rho}_M(m_w)\, \tilde{\theta}(d_v|m_w)} \ , \qquad (14)$$

which gives

$$\tilde{\sigma}_D(d_1) \simeq 4.8 \ 10^{-2} \ . \qquad (15)$$

Result (13) shows that the a posteriori probability of being affected by the disease is only five times greater than the a priori one, despite the fact that the assistant says the response was positive. Result (15) explains this result by the fact that the a posteriori probability of the assistant of telling the untruth is very high.

Should the assistant tell the truth only if she throws a six, then

$$\tilde{\rho}_D(d_1) = 1/6 \ , \tag{16a}$$

$$\tilde{\rho}_D(d_2) = 5/6 \ , \tag{16b}$$

which gives

$$\tilde{\sigma}_M(m_1) \simeq 2.0 \ 10^{-6} \ . \tag{17}$$

If she always lies, then

$$\tilde{\rho}_D(d_1) = 0. \ , \tag{18a}$$

$$\tilde{\rho}_D(d_2) = 1. \ , \tag{18b}$$

and

$$\tilde{\sigma}_M(m_1) = 0. \ . \tag{19}$$

Problem 1.5: Some physical quantity d is related with the physical quantity x through the equation

$$d = m^1 + m^2 \ x \ , \tag{1}$$

where m^1 and m^2 are unknown parameters. Equation (1) represents a straight line on the plane (d,x) . In order to estimate m^1 and m^2 , the parameter d has been experimentally measured for some selected values of x , and the following results have been obtained (Figure 1.19):

$$\begin{array}{ll} x^1 = 03.500 & d^1_{obs} = 2.0 \ \pm 0.5 \\ x^2 = 05.000 & d^2_{obs} = 2.0 \ \pm 0.5 \\ x^3 = 07.000 & d^3_{obs} = 3.0 \ \pm 0.5 \\ x^4 = 07.500 & d^4_{obs} = 3.0 \ \pm 0.5 \\ x^5 = 10.000 & d^5_{obs} = 4.0 \ \pm 0.5 \ , \end{array} \tag{2}$$

where $\pm$ 0.5 denotes *rounding* errors (to the nearest integer). Estimate m^1 and m^2 . (Note: this problem is nonclasical in the sense that experimental errors are not Gaussian, and the usual least squares regression is not adapted).

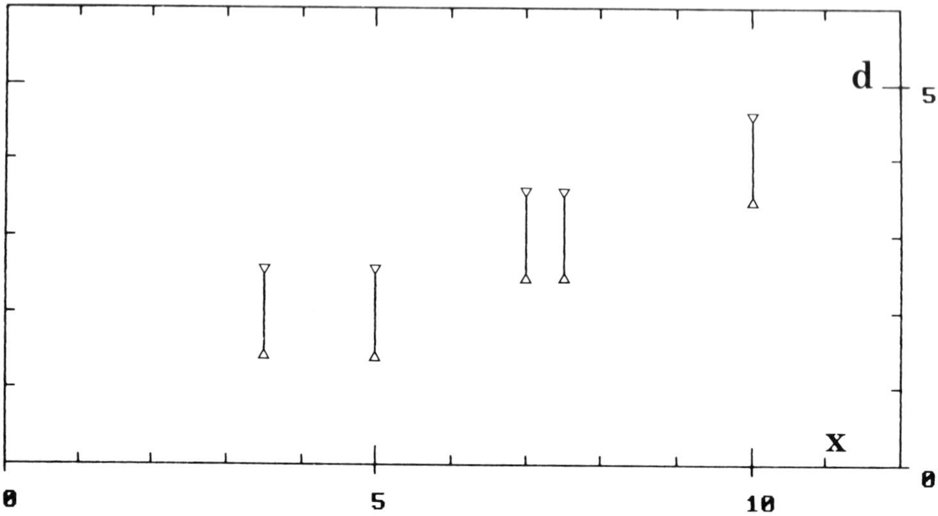

Figure 1.19: Some experimental points. Error "bars" represent *rounding* errors to the nearest integer. Solve the general problem of estimating a regression line.

Solution: Let an arbitrary set $(d^1, d^2, d^3, d^4, d^5)$ be called a data vector and be denoted by **d** , let an arbitrary set (m^1, m^2) be called a parameter vector and be denoted by **m**. Let

$$\mathbf{d} = \mathbf{g}(\mathbf{m}) \tag{3}$$

denote the (linear) relationship

$$\begin{aligned} d^1 &= m^1 + m^2 x^1 \\ d^2 &= m^1 + m^2 x^2 \\ d^3 &= m^1 + m^2 x^3 \\ d^4 &= m^1 + m^2 x^4 \\ d^5 &= m^1 + m^2 x^5 . \end{aligned} \tag{4}$$

Let

$$\rho_M(\mathbf{m}) = \rho_M(m^1,m^2) \tag{5}$$

be the probability density representing the a priori information (if any) on model parameters. Let

$$\rho_D(\mathbf{d}) = \rho_D(d^1,d^2,d^3,d^4,d^5) \tag{6}$$

be the probability density describing the "experimental" uncertainties (see text). As rounding errors are mutually independent,

$$\rho_D(\mathbf{d}) = \rho_D(d^1,d^2,d^3,d^4,d^5) = \rho^1{}_D(d^1)\,\rho^2{}_D(d^2)\,\rho^3{}_D(d^3)\,\rho^4{}_D(d^4)\,\rho^5{}_D(d^5)\ , \tag{7}$$

where $\rho^i_D(d^i)$ denotes the probability density describing the "experimental" uncertainty for the observed data d^i . As the errors are only rounding errors, they can be conveniently modeled using box-car probability density functions:

$$\rho^i_D(d^i) = \begin{cases} \text{const} & \text{if} \quad d^i_{obs} - 0.5 < d^i < d^i_{obs} + 0.5 \\ 0 & \text{otherwise .} \end{cases} \tag{8}$$

This gives

$$\rho_D(\mathbf{d}) = \begin{cases} \text{const} \quad \text{if} & \begin{cases} 1.5 < d^1 < 2.5 \\ \text{and} \\ 1.5 < d^2 < 2.5 \\ \text{and} \\ 2.5 < d^3 < 3.5 \\ \text{and} \\ 2.5 < d^4 < 3.5 \\ \text{and} \\ 3.5 < d^5 < 4.5 \end{cases} \\ 0 \qquad \text{otherwise.} \end{cases} \tag{9}$$

The general solution of an inverse problem is obtained when the posterior probability density in the model space has been defined. It is given by equation (1.65) of the text:

$$\sigma_M(\mathbf{m}) = \rho_M(\mathbf{m}) \int_D d\mathbf{d} \; \frac{\rho_D(\mathbf{d})}{\mu_D(\mathbf{d})} \; \Theta(\mathbf{d}|\mathbf{m}) \; , \tag{10}$$

or more particularly, as the relationship between $\mathbf{d}$ and $\mathbf{m}$ is exact, by equation (1.69) of the text:

$$\sigma_M(\mathbf{m}) = \rho_M(\mathbf{m}) \; \frac{\rho_D(\mathbf{g}(\mathbf{m}))}{\mu_D(\mathbf{g}(\mathbf{m}))} \; , \tag{11}$$

where $\mu_D(\mathbf{d})$ represents the non-informative probability density on data parameters. Equations (9) and (11) solve the problem.

For instance, if we take as non-informative prior in the data space:

$$\mu_D(\mathbf{d}) = \mu_D(d^1,d^2,d^3,d^4,d^5) = \text{const} \; , \tag{12}$$

and if we accept a priori all pairs (m^1,m^2) as equally probable:

$$\rho_M(\mathbf{m}) = \rho_M(m^1,m^2) = \text{const} \; , \tag{13}$$

then we obtain

$$\sigma_M(\mathbf{m}) = \sigma_M(m^1,m^2) = \begin{cases} \text{const} \quad \text{if} & \begin{cases} 1.5 < m^1 + m^2\, x^1 < 2.5 \\ \text{and} \\ 1.5 < m^1 + m^2\, x^2 < 2.5 \\ \text{and} \\ 2.5 < m^1 + m^2\, x^3 < 3.5 \\ \text{and} \\ 2.5 < m^1 + m^2\, x^4 < 3.5 \\ \text{and} \\ 3.5 < m^1 + m^2\, x^5 < 4.5 \end{cases} \\ 0 \quad \text{otherwise.} \end{cases} \tag{14}$$

This result is represented graphically in figure 1.20. The dark region has a positive (constant) probability density. All pairs (m^1,m^2) inside this region have equal probability density, and all pairs (m^1,m^2) outside it are impossible, so that this region represents the "domain of admissible solutions". Which is *the best* "regression line"? There is no such a thing: all lines inside the domain are equally good. Figure 1.21 shows two particular solutions (giving extremal values for m^1 and m^2).

Figure 1.22 shows the computer code effectively used for obtaining the general solution shown in Figure 1.20. For problems with few model parameters (2 in this example), the full exploration of the model space is, in general, the easiest strategy (it takes approximately 2 minutes to go from the statement of the problem to the result in figure 1.20.

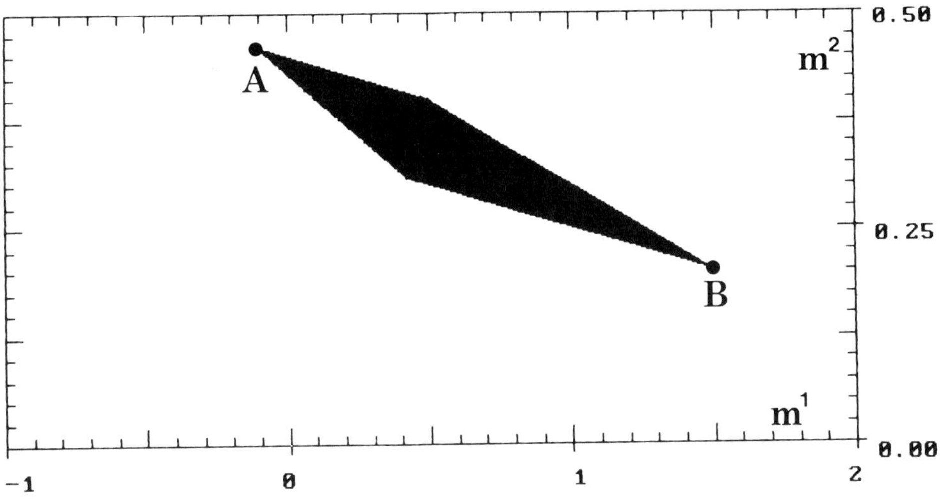

Figure 1.20: The general solution of the problem is given by the probability density $\sigma_M(m^1,m^2)$ for the parameters of the regression line. It is constant inside the dark region and null outside. The dark region represents the domain of admissible solutions. There is not any "best line": all pairs (m^1,m^2) inside the region are equally likely.

Problem 1.6 (Usual least-squares regression): Find the best regression line for the experimental points in figure 1.23, assuming Gaussian uncertainties.

Solution: Figure 1.23 suggests that errors in the t^i are negligible, while errors in the y^i are uncorrelated. Let us introduce

$$\mathbf{m} = \begin{bmatrix} a \\ b \end{bmatrix} \qquad \mathbf{d} = \begin{bmatrix} y^1 \\ y^2 \\ \cdots \\ y^n \end{bmatrix} \qquad \mathbf{G} = \begin{bmatrix} t^1 & 1 \\ t^2 & 1 \\ \cdots & \cdots \\ t^n & 1 \end{bmatrix} \tag{1}$$

($y^1,y^2,\ldots,t^1,t^2,\ldots$ are indexes, not powers). The equations

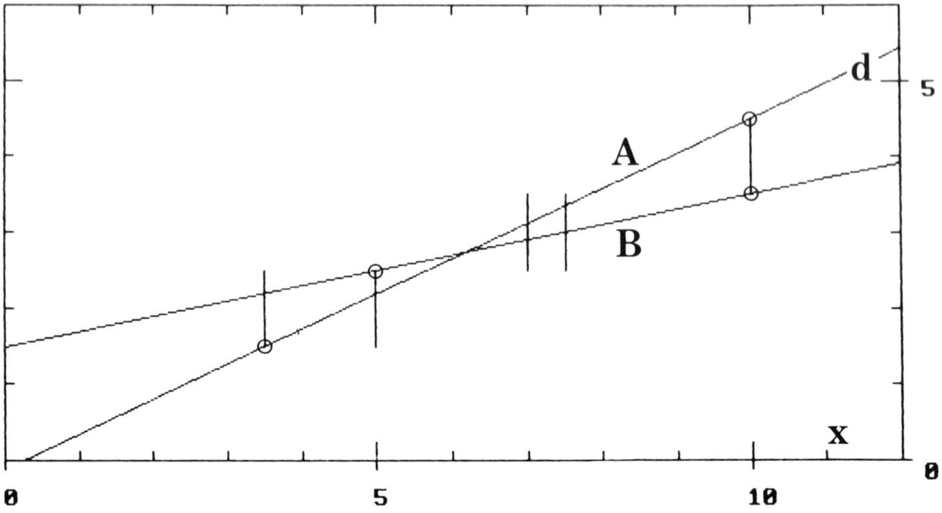

Figure 1.21: Two particular solutions (A and B in figure 1.20), corresponding to extremal values of m^1 and m^2 . Notice that they touch the extremities of the error bars (circles).

$$y^1 = a\ t^i + b \qquad (i=1,2,...,n) \tag{2}$$

can be written

$$\mathbf{d} = \mathbf{G}\ \mathbf{m}\ . \tag{3}$$

The matrix $\mathbf{G}$ is assumed perfectly known. We have some information on the true values of $\mathbf{d}$, and we wish to estimate the true value of $\mathbf{m}$.

As it is assumed that errors in the y^i are uncorrelated Gaussian, the information we have on the true value of $\mathbf{d}$ can be represented using a Gaussian probability density with mathematical expectation

$$\mathbf{d}_{obs} = \begin{bmatrix} y_0^1 \\ y_0^2 \\ \cdots \\ y_0^n \end{bmatrix} , \tag{4}$$

and covariance matrix

```
110 FOR M1 = -1 TO 2 STEP .005
120 FOR M2 = 0 TO .5 STEP .002
130 D1 = M1 + 3.5 * M2
140 IF ( D1 < 1.5 ) OR ( D1 > 2.5 ) THEN GOTO 300
150 D2 = M1 + 5 * M2
160 IF ( D2 < 1.5 ) OR ( D2 > 2.5 ) THEN GOTO 300
170 D3 = M1 + 7 * M2
180 IF ( D3 < 2.5 ) OR ( D3 > 3.5 ) THEN GOTO 300
190 D4 = M1+ 7.5 * M2
200 IF ( D4 < 2.5 ) OR ( D4 > 3.5 ) THEN GOTO 300
210 D5 = M1 + 10 * M2
220 IF ( D5 < 3.5 ) OR ( D5 > 4.5 ) THEN GOTO 300
250 DRAW POINT (M1,M2)
300 NEXT M2
310 NEXT M1
```

Figure 1.22: Computer code in BASIC-like notation effectively used for obtaining the result in figure 1.20. The limits for m^1 and m^2 in lines 110-120 have been chosen after trial and error. The steps 0.005 and 0.002 in lines 110-120 have been chosen small enough not to be visible on the graphic device used to generate figure 1.20. The command DRAW POINT (X,Y) in line 250 simply plots a point on the graphic device at coordinates (X,Y).

$$\mathbf{C}_D = \begin{bmatrix} (\sigma^1)^2 & 0 & 0 & \cdots \\ 0 & (\sigma^2)^2 & 0 & \cdots \\ 0 & 0 & (\sigma^3)^2 & \cdots \\ \cdots & \cdots & \cdots & \cdots \end{bmatrix} . \tag{5}$$

We now need to introduce the a priori information (if any) on the parameters $\mathbf{m}$. The simplest results are obtained when using a Gaussian probability density in the model space with mathematical expectation

$$\mathbf{m}_{prior} = \begin{bmatrix} a_0 \\ b_0 \end{bmatrix} , \tag{6}$$

and covariance matrix

$$\mathbf{C}_M = \begin{bmatrix} \sigma^2{}_a & \rho\, \sigma_a\, \sigma_b \\ \rho\, \sigma_a\, \sigma_b & \sigma^2{}_b \end{bmatrix} . \tag{7}$$

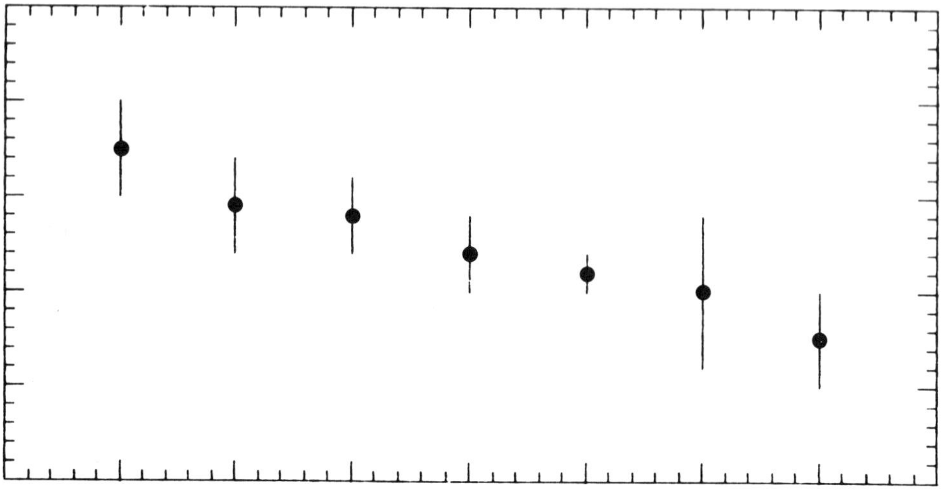

Figure 1.23: The physical parameter y (ordinate) is related with the physical parameter t (abscissa) through the equation $y = a\,t + b$, where the parameters a and b are unknown. The experimental points in the figure have to be used to estimate the best values for a and b , in the least squares sense.

As the information on both data and model parameters is Gaussian, we are in the hypothesis of section (1.7.1). The a posteriori information on the model parameters is then also Gaussian, with mathematical expectation given by

$$\mathbf{m}_{post} = \mathbf{m}_{prior} + \left(\mathbf{G}^t\, \mathbf{C}_D^{-1}\, \mathbf{G} + \mathbf{C}_M^{-1}\right)^{-1} \mathbf{G}^t\, \mathbf{C}_D^{-1}\, (\mathbf{d}_{obs} - \mathbf{G}\, \mathbf{m}_{prior}) \tag{8}$$

$$= \mathbf{m}_{prior} + \mathbf{C}_M\, \mathbf{G}^t\, (\mathbf{G}\, \mathbf{C}_M\, \mathbf{G}^t + \mathbf{C}_D)^{-1}\, (\mathbf{d}_{obs} - \mathbf{G}\, \mathbf{m}_{prior}) \ , \tag{9}$$

and covariance matrix given by

$$\mathbf{C}_{M'} = \left(\mathbf{G}^t\, \mathbf{C}_D^{-1}\, \mathbf{G} + \mathbf{C}_M^{-1}\right)^{-1} \tag{10}$$

$$= \mathbf{C}_M - \mathbf{C}_M\, \mathbf{G}^t\, (\mathbf{G}\, \mathbf{C}_M\, \mathbf{G}^t + \mathbf{C}_D)^{-1}\, \mathbf{G}\, \mathbf{C}_M \ . \tag{11}$$

The a posteriori (i.e., recalculated) data values are then (equation 1.97a)

$$\mathbf{d}_{post} = \mathbf{G}\,\mathbf{m}_{post} \ , \tag{12}$$

and the a posteriori data errors are given by (equation 1.97b)

$$\mathbf{C}_{D'} = \mathbf{G}\,\mathbf{C}_{M'}\,\mathbf{G}^t \ . \tag{13}$$

As we have only two model parameters, expressions (8) and (10) should be preferred to (9) and (11). An easy computation gives the a posteriori values of a and b :

$$a = a_0 + \frac{A\,P - C\,Q}{A\,B - C^2} \tag{14}$$

$$b = b_0 + \frac{B\,Q - C\,P}{A\,B - C^2} \ , \tag{15}$$

and the posteriori standard deviations and correlation:

$$\sigma_{a}' = \frac{1}{\sqrt{B - C^2/A}} \ , \tag{16a}$$

$$\sigma_{b}' = \frac{1}{\sqrt{A - C^2/B}} \ , \tag{16b}$$

$$\rho' = \frac{-1}{\sqrt{AB/C^2}} \ , \tag{16c}$$

where

$$A = \sum_i \frac{1}{(\sigma^i)^2} + \frac{1}{(1-\rho^2)\,\sigma_b^{\,2}} \ , \tag{17a}$$

$$B = \sum_i \frac{(t^i)^2}{(\sigma^i)^2} + \frac{1}{(1-\rho^2)\,\sigma_a^{\,2}} \ , \tag{17b}$$

$$C = \sum_i \frac{t^i}{(\sigma^i)^2} - \frac{\rho}{(1-\rho^2)\,\sigma_a\,\sigma_b} \ , \tag{17c}$$

$$P = \sum_i \frac{t^i}{(\sigma^i)^2}\,[y_0^{\,i} - (a_0\,t^i + b_0)] \ , \tag{17d}$$

and

$$Q = \sum_i \frac{1}{(\sigma^i)^2} [y_0{}^i - (a_0 \, t^i + b_0)] \, . \tag{17e}$$

Usually, a priori errors on model parameters are uncorrelated. Then

$$\rho = 0 \, . \tag{18}$$

This gives

$$A = \sum_i \frac{1}{(\sigma^i)^2} + \frac{1}{\sigma_b{}^2} \, , \tag{19a}$$

$$B = \sum_i \frac{(t^i)^2}{(\sigma^i)^2} + \frac{1}{\sigma_a{}^2} \, , \tag{19b}$$

and

$$C = \sum_i \frac{t^i}{(\sigma^i)^2} \, . \tag{19c}$$

If there is no a priori information on model parameters,

$$\sigma_a \rightarrow \infty \tag{20}$$

and

$$\sigma_b \rightarrow \infty \, . \tag{21}$$

Instead of taking the limits (20)-(21) in the last equations, it is simpler to use

$$\mathbf{C_M}^{-1} = \mathbf{0} \tag{22}$$

in equations (8) and (10). This gives

$$\mathbf{m}_{post} = \left(\mathbf{G}^t \, \mathbf{C_D}^{-1} \, \mathbf{G}\right)^{-1} \mathbf{G}^t \, \mathbf{C_D}^{-1} \, \mathbf{d}_{obs} \tag{23}$$

and

$$\mathbf{C_{M'}} = \left(\mathbf{G}^t \, \mathbf{C_D}^{-1} \, \mathbf{G}\right)^{-1} \, . \tag{24}$$

Equations (14) and (15) then become

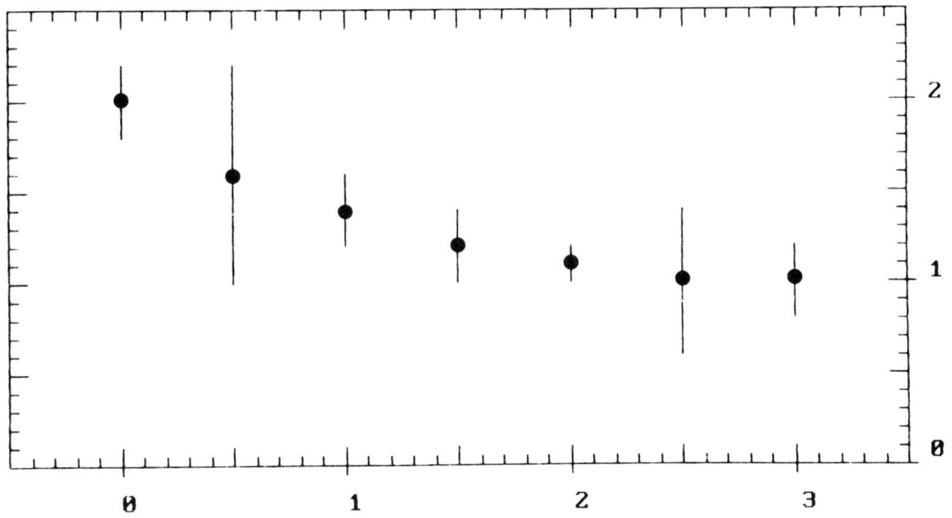

Figure 1.24: See text.

$$a = \frac{A\,P - C\,Q}{A\,B - C^2} \tag{25}$$

and

$$b = \frac{B\,Q - C\,P}{A\,B - C^2}\ , \tag{26}$$

while the constants A , B , C , P , and Q simplify to

$$A = \sum_i \frac{1}{(\sigma^i)^2} \qquad B = \sum_i \frac{(t^i)^2}{(\sigma^i)^2} \qquad C = \sum_i \frac{t^i}{(\sigma^i)^2}$$

$$P = \sum_i \frac{t^i\,y_0{}^i}{(\sigma^i)^2} \qquad Q = \sum_i \frac{y_0{}^i}{(\sigma^i)^2}\ . \tag{27}$$

If all data uncertainties are identical,

$$\sigma^i = \sigma\ , \tag{28}$$

then

$$A = \frac{n}{\sigma^2} \qquad B = \frac{1}{\sigma^2} \sum_i (t^i)^2 \qquad C = \frac{1}{\sigma^2} \sum_i t^i$$

$$P = \frac{1}{\sigma^2} \sum_i t^i \, y_0{}^i \qquad Q = \frac{1}{\sigma^2} \sum_i y_0{}^i \ . \tag{29}$$

Problem 1.7: The two variables y and t are related through the parabolic relationship

$$y = a\,t^2 + b\,t + c \ . \tag{1}$$

The points (y^i, t^i) shown in Figure 1.24 have been obtained experimentally. Error bars denote Gaussian errors. Estimate the parameters a , b , and c , and analyze uncertainties.

Answer the same question if the assumed relationship between y and t is

$$y = a\,e^{-bt} + c \ . \tag{2}$$

Problem 1.8 (Two-axes least-squares regression): Find the best regression line for the experimental points in figure 1.25, assuming Gaussian uncertainties.

Solution: There are some equivalent ways of properly setting this problem. The approach followed here has the advantage of giving a symmetrical treatment to both axes.

As the statement of the problem refers to a regression *line*, a linear relationship has to be assumed between the variables y and t :

$$\alpha\,y + \beta\,t = 1 \ . \tag{1}$$

We have measured some pairs (x^i,y^i) and wish to estimate the true values of α and β .

I introduce a parameter vector $\mathbf{m}$ which contains the y^i , the t^i , α , and β :

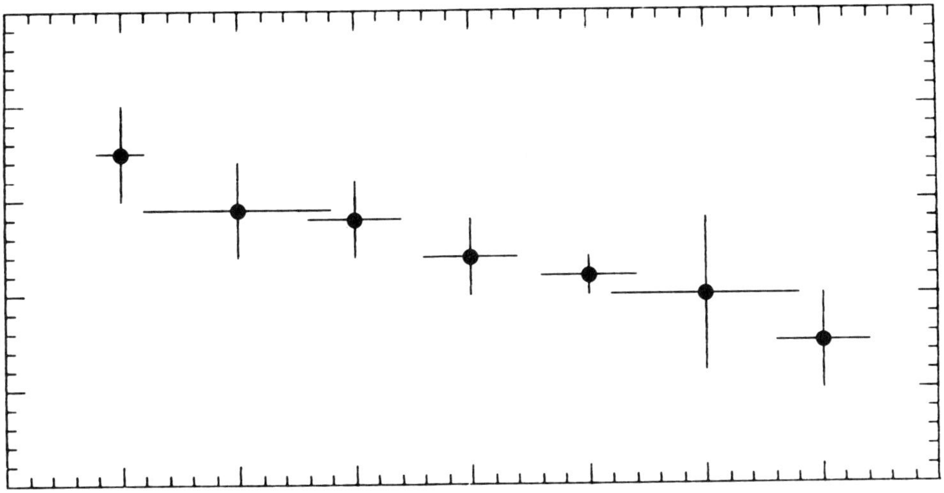

Figure 1.25: The physical parameter y (ordinate) is related to the physical parameter t (abscissa) through the equation $y = a\,t + b$, where the parameters a and b are unknown. The experimental points in the figure have to be used to estimate the best values for a and b, in the least squares sense. This problem is nonclassical in the sense that uncertainties are present in both coordinates.

$$\mathbf{m} = \begin{bmatrix} \mathbf{y} \\ \mathbf{t} \\ \mathbf{u} \end{bmatrix} = \begin{bmatrix} y^1 \\ y^2 \\ \dots \\ t^1 \\ t^2 \\ \dots \\ \alpha \\ \beta \end{bmatrix} , \tag{2}$$

and, for each conceivable value of $\mathbf{m}$, I define a vector

$$\mathbf{d} = \begin{bmatrix} d^1 \\ d^2 \\ \dots \end{bmatrix} \tag{3}$$

by

$$d^i = g^i(\mathbf{m}) = \alpha\, y^i + \beta\, t^i \qquad (i=1,2,...) \ . \tag{4}$$

Defining the "observed value" of $\mathbf{d}$ by

$$\mathbf{d}_{obs} = \begin{bmatrix} 1 \\ 1 \\ 1 \\ ... \end{bmatrix} , \tag{5}$$

and the "a priori" value of $\mathbf{m}$ by

$$\mathbf{m}_{prior} = \begin{bmatrix} y^1{}_0 \\ y^2{}_0 \\ ... \\ t^1{}_0 \\ t^2{}_0 \\ ... \\ \alpha_0 \\ \beta_0 \end{bmatrix} , \tag{6}$$

where $\mathbf{y}_0$ and $\mathbf{t}_0$ are the experimental values, and α_0 and β_0 the a priori values of α and β , the inverse problem can now be set as the problem of obtaining a vector $\mathbf{m}$ such that $\mathbf{g}(\mathbf{m})$ is close (or identical) to $\mathbf{d}_{obs}$, and such that $\mathbf{m}$ is close to $\mathbf{m}_{prior}$. We see thus that this "relabeling" of the variables allows an immediate use of the standard equations. Nevertheless, this problem is less simple than the previous problem of one-axis regression, because here we have twice the number of points + 2 "unknowns" instead of 2 , and the forward equation $\mathbf{d} = \mathbf{g}(\mathbf{m})$ is nonlinear (because it contains the mutual product of parameters).

More precisely, I assume that the a priori information on $\mathbf{m}$ can be described using a Gaussian probability density with mathematical expectation $\mathbf{m}_{prior}$ and covariance matrix

$$\mathbf{C}_M = \begin{bmatrix} \mathbf{C}_y & \mathbf{0} & 0 & 0 \\ \mathbf{0} & \mathbf{C}_t & 0 & 0 \\ 0 & 0 & \sigma^2{}_\alpha & 0 \\ 0 & 0 & 0 & \sigma^2{}_\beta \end{bmatrix} , \tag{7}$$

where independence of errors has been assumed only to simplify the notations. The a priori information on $\mathbf{d}$ is also assumed to be Gaussian, with mathematical expectation $\mathbf{d}_{obs}$ and covariance matrix $\mathbf{C}_D$. Later, we may take $\mathbf{C}_D = \mathbf{0}$, so that the observed values (5) may be fitted exactly by the a posteriori solution. Instead, we may keep $\mathbf{C}_D$ finite to allow for errors in the hypothesis of a strictly linear relationship between y and t .

Now we are exactly in the hypothesis of section 1.7.1. The a posteriori probability density for **m** is (equation 1.85):

$$\sigma_M(m) = \text{const.} \cdot \tag{8}$$
$$\exp\left[-\frac{1}{2}\left((\mathbf{g}(\mathbf{m})-\mathbf{d}_{obs})^t\, \mathbf{C}_D^{-1}\, (\mathbf{g}(\mathbf{m})-\mathbf{d}_{obs}) + (\mathbf{m}-\mathbf{m}_{prior})^t\, \mathbf{C}_M^{-1}\, (\mathbf{m}-\mathbf{m}_{prior})\right)\right] ,$$

which, owing to the nonlinearity of $\mathbf{g}(\mathbf{m})$, is not Gaussian. The maximum likelihood value of **m** can be obtained using, for instance, the iterative algorithm suggested in equation (1.107):

$$\mathbf{m}_{n+1} = \tag{9}$$
$$\mathbf{m}_{prior} - \mathbf{C}_M\, \mathbf{G}_n^{\,t}\, (\mathbf{G}_n\, \mathbf{C}_M\, \mathbf{G}_n^{\,t} + \mathbf{C}_D)^{-1}\, [(\mathbf{g}(\mathbf{m}_n)-\mathbf{d}_{obs}) - \mathbf{G}_n\, (\mathbf{m}_n-\mathbf{m}_{prior})] .$$

We have

$$\mathbf{G}_n = \left[\left(\frac{\partial \mathbf{g}}{\partial \mathbf{y}}\right)_{\mathbf{m}_n} \quad \left(\frac{\partial \mathbf{g}}{\partial \mathbf{t}}\right)_{\mathbf{m}_n} \quad \left(\frac{\partial \mathbf{g}}{\partial \alpha}\right)_{\mathbf{m}_n} \quad \left(\frac{\partial \mathbf{g}}{\partial \beta}\right)_{\mathbf{m}_n} \right] , \tag{10}$$

which gives

$$\mathbf{G}_n = [\, \alpha_n\, \mathbf{I} \quad \beta_n\, \mathbf{I} \quad \mathbf{y}_n \quad \mathbf{t}_n\,] , \tag{11}$$

$$\mathbf{C}_M\, \mathbf{G}_n^{\,t} = \begin{bmatrix} \alpha_n\, \mathbf{C}_y \\ \beta_n\, \mathbf{C}_t \\ \sigma_a^{\,2}\, \mathbf{y}_n^{\,t} \\ \sigma_\beta^{\,2}\, \mathbf{t}_n^{\,t} \end{bmatrix} , \tag{12}$$

$$\mathbf{G}_n\, \mathbf{C}_M\, \mathbf{G}_n^{\,t} + \mathbf{C}_D =$$
$$= \sigma_a^{\,2}\, \mathbf{y}_n\, \mathbf{y}_n^{\,t} + \sigma_b^{\,2}\, \mathbf{t}_n\, \mathbf{t}_n^{\,t} + \alpha_n^{\,2}\, \mathbf{C}_y + \beta_n^{\,2}\, \mathbf{C}_t + \mathbf{C}_D , \tag{13}$$

and

$$\mathbf{g}(\mathbf{m}_n)-\mathbf{d}_{obs} - \mathbf{G}_n\, (\mathbf{m}_n-\mathbf{m}_{prior})$$
$$= (\alpha_0-\alpha_n)\, \mathbf{y}_n + (\beta_0-\beta_n)\, \mathbf{t}_n - \mathbf{d}_{obs} + \alpha_n\, \mathbf{y}_0 + \beta_n\, \mathbf{t}_0 . \tag{14}$$

Denoting

$$\delta\hat{\mathbf{d}}_n = (\, \mathbf{G}_n\, \mathbf{C}_M\, \mathbf{G}_n^{\,t} + \mathbf{C}_D)^{-1}\, [\, (\mathbf{g}(\mathbf{m}_n)-\mathbf{d}_{obs}) - \mathbf{G}_n\, (\mathbf{m}_n-\mathbf{m}_{prior})] , \tag{15}$$

the iterative algorithm (9) can be written

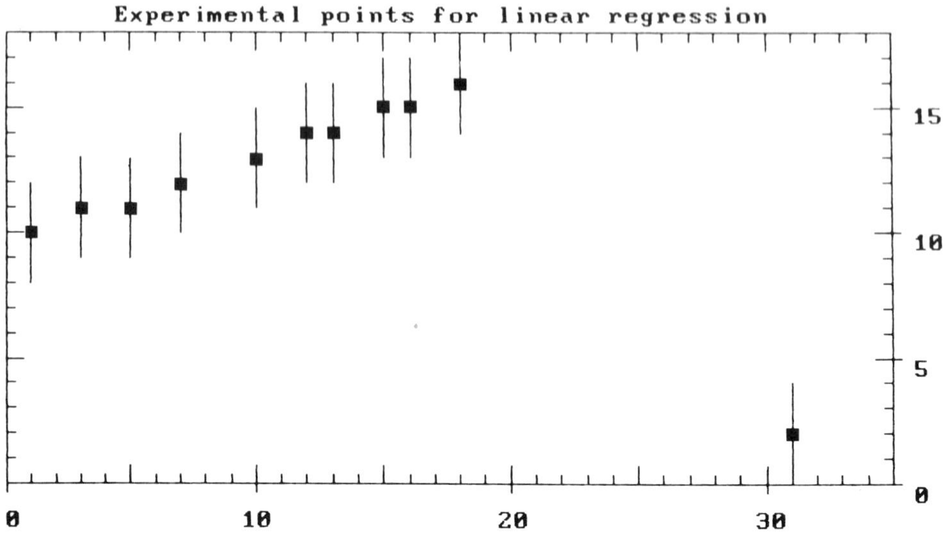

Figure 1.26: Two variables y (ordinate) and t (abscissa) are related by the relationship y = a t + b , where a and b are unknown parameters. In order to estimate a and b , an experiment has been performed which has furnished the 11 experimental points shown in the figure. The exact meaning of the "error bars" is not indicated.

$$\mathbf{y}_{n+1} = \mathbf{y}_0 - \alpha_n \mathbf{C}_y \, \delta\hat{\mathbf{d}}_n \, , \tag{16a}$$

$$\mathbf{t}_{n+1} = \mathbf{t}_0 - \beta_n \mathbf{C}_t \, \delta\hat{\mathbf{d}}_n \, , \tag{16b}$$

$$\alpha_{n+1} = \alpha_0 - \sigma_\alpha^2 \, \mathbf{y}_n^t \, \delta\hat{\mathbf{d}}_n \, , \tag{16c}$$

and

$$\beta_{n+1} = \beta_0 - \sigma_\beta^2 \, \mathbf{t}_n^t \, \delta\hat{\mathbf{d}}_n \, . \tag{16d}$$

The algorithm usually converges in a few iterations ($\simeq$3). The values α_∞ and β_∞ are the estimated values of the parameters defining the regression

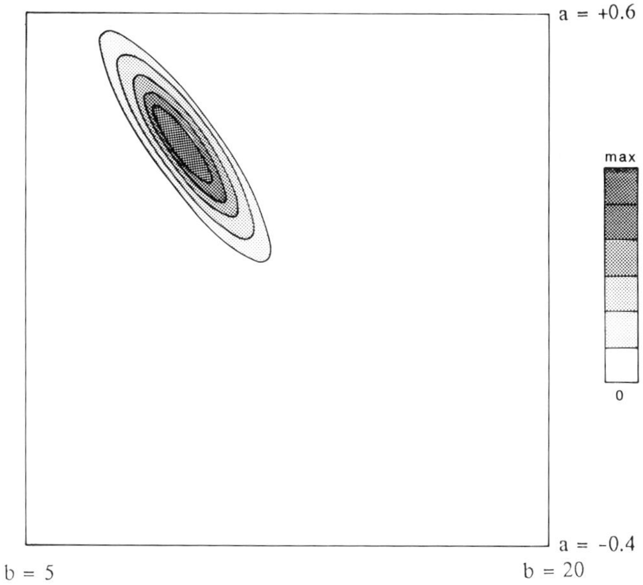

Figure 1.27: The probability density for the parameters (a,b) obtained using the Gaussian hypothesis for experimental uncertainties, and without using the blunder.

line, and the values (t^i_∞, y^i_∞) (i=1,2,...) are the a posteriori values of the experimental points. If $\mathbf{C}_D = \mathbf{0}$, the a posteriori points belong to the straight line.

Problem 1.9: Two variables y and t are related through a linear relationship

$$y = a\,t + b\,. \qquad (1)$$

In order to estimate the parameters a and b , the 11 experimental points (y^i, t^i) shown in Figure 1.26 have been obtained.

It is clear that if the linear relationship (1) applies, then the point indicated with an arrow must be an outlier. Suppress that point and solve the problem of estimating a and b , under the hypothesis of Gaussian errors. Does the solution change very much if the outlier is included?

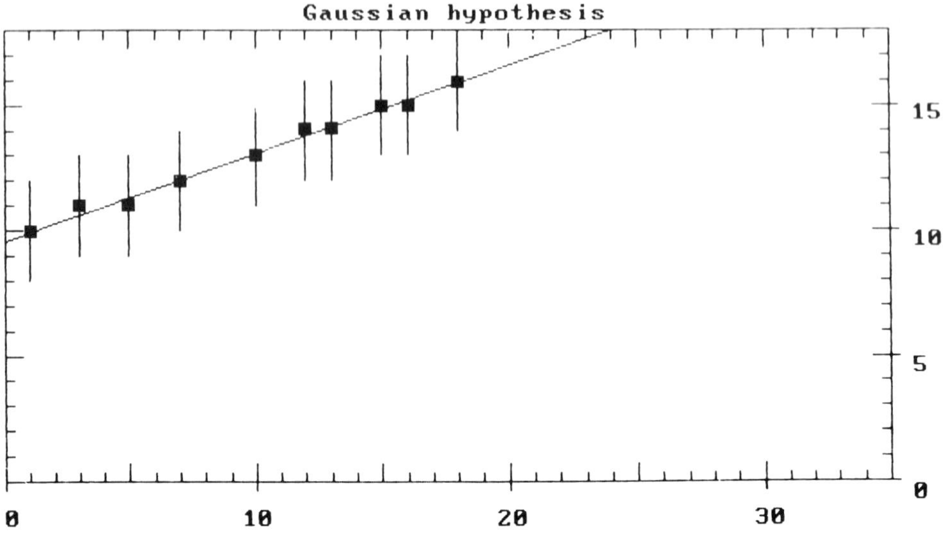

Figure 1.28: The maximum likelihood line for the probability density in figure 1.27.

Assume now that errors can be modeled using an exponential probability density, and solve the problem again. Discuss the relative robustness of the Gaussian and exponential hypotheses with respect to the existence of outliers on a data set.

Solution: Let

$$\mathbf{m} = (a,b) \tag{2}$$

denote a model vector, and

$$\mathbf{d} = (d^1,d^2,...) \tag{3}$$

a data vector. The (exact) theoretical relationship between **d** and **m** is linear:

$$d^i = a\, t^i + b \ , \tag{4a}$$

or, for short,

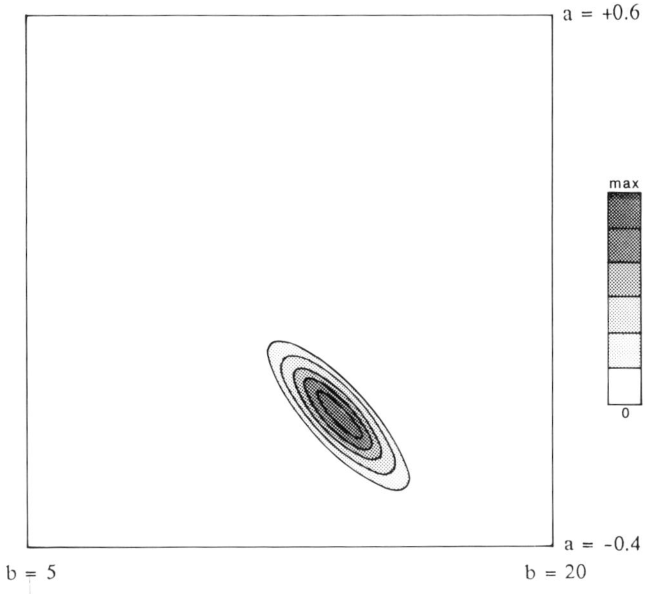

Figure 1.29: Same as Figure 1.27, but the 11 experimental points have been used. The blunder has "translated" the probability density. This shows that the Gaussian hypothesis is not very robust with respect to the existence of a small number of blunders in a data set.

$$\mathbf{d} = \mathbf{G}\,\mathbf{m} , \qquad (4b)$$

where $\mathbf{G}$ is a linear operator.

Assume that the null information probability density on model parameters is

$$\mu_M(a,b) = \text{const} , \qquad (5)$$

and that we do not have a priori information on model parameters:

$$\rho_M(a,b) = \mu_M(a,b) = \text{const.} . \qquad (6)$$

Assume that the null information probability density on data parameters is

$$\mu_D(d^1,d^2,...) = \text{const} . \qquad (7)$$

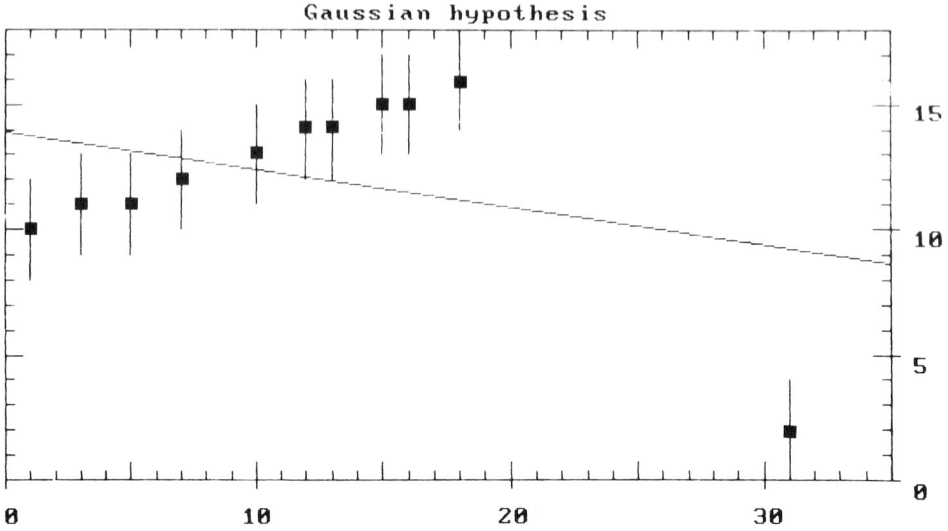

Figure 1.30: The maximum likelihood line for the probability density in figure 1.29.

If $\rho_D(d^1,d^2,...)$ is the probability density representing the information on the true values of $(d^1,d^2,...)$ as obtained through the measurements, then the Gaussian hypothesis gives (for independent uncertainties)

$$\rho_D(d^1,d^2,...) = \exp\left(-\frac{1}{2} \sum_i \frac{\left(d^i - d^i_{obs} \right)^2}{\sigma^2} \right) , \tag{8}$$

where $\mathbf{d}_{obs}$ is the vector of observed values

$$\mathbf{d}_{obs} = (10. , 11. , 11. , 12. , 13. , 14. , 14. , 15. , 15. , 16. , 2.) , \tag{9}$$

and where, if we interpret the error bars in Figure 1 as standard deviations,

$$\sigma = 2. \tag{10}$$

Let $\theta(\mathbf{d},\mathbf{m})$ be the probability density representing the information we have on the theoretical relationship between $\mathbf{d}$ and $\mathbf{m}$. As (4) is an exact relationship

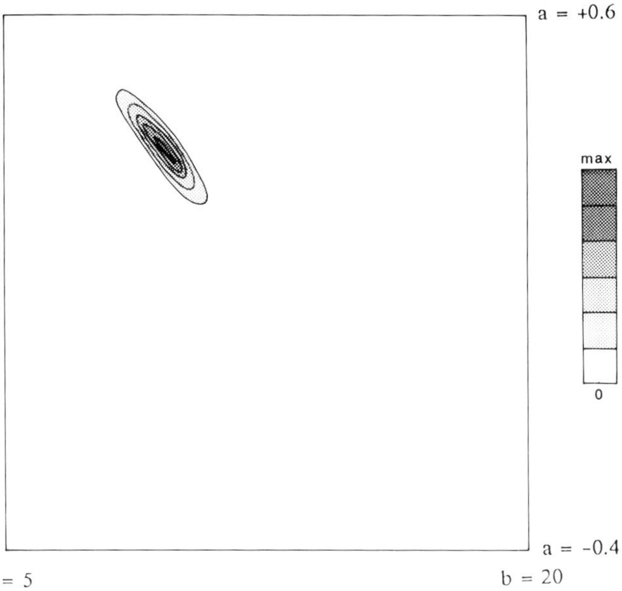

Figure 1.31: The exponential hypothesis for data uncertainties has been used instead of the Gaussian hypothesis. Here the blunder has not been used. The solution looks similar to the solution corresponding to the Gaussian hypothesis in Figure 1.27.

$$\theta(\mathbf{d},\mathbf{m}) = \theta(\mathrm{d}|\mathbf{m})\,\mu_M(\mathbf{m}) = \delta(\mathbf{d}\text{-}\mathbf{G}\ \mathbf{m})\,\mu_M(\mathrm{m}) . \tag{11}$$

The posterior information on the parameters $(\mathbf{d},\mathbf{m})$ is given by (equation 1.60)

$$\sigma(\mathbf{d},\mathbf{m}) = \frac{\rho(\mathbf{d},\mathbf{m})\ \theta(\mathbf{d},\mathbf{m})}{\mu(\mathbf{d},\mathbf{m})} = \frac{\rho_D(\mathbf{d})\ \rho_M(\mathbf{m})}{\mu_D(\mathbf{d})}\ \theta(\mathrm{d}|\mathbf{m}) , \tag{12}$$

and the posterior information on model parameters alone is given by the marginal probability density

$$\sigma_M(\mathrm{m}) = \int_D \mathrm{dd}\ \ \sigma(\mathbf{d},\mathrm{m}) . \tag{13}$$

Using (4), (6), (7), (8), and (11) easily gives

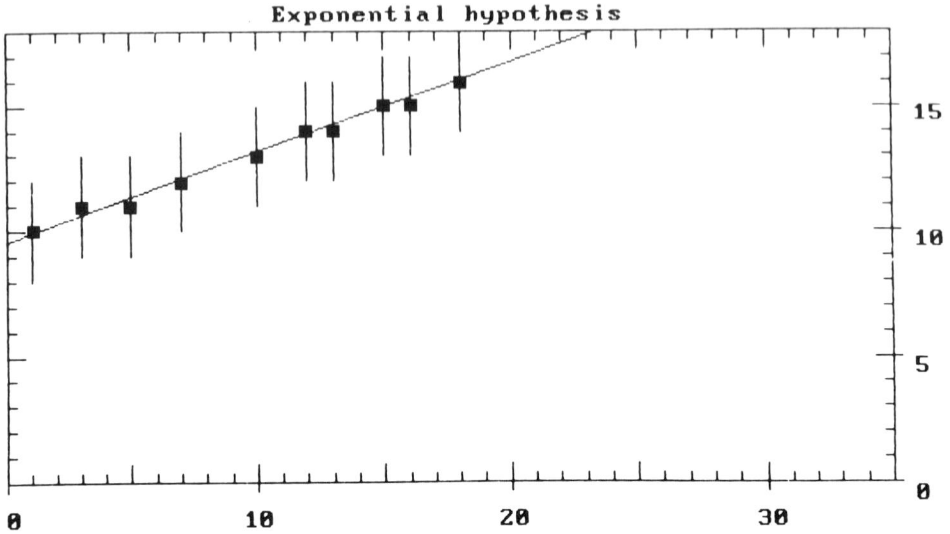

Figure 1.32: The maximum likelihood line for the probability density in Figure 1.31.

$$\sigma_M(a,b) = \exp\left(-\frac{1}{2} \sum_i \frac{\left(d^i_{obs} - d^i_{cal}(a,b) \right)^2}{\sigma^2} \right) , \tag{14}$$

where

$$d^i_{cal}(a,b) = a\, t^i + b \, . \tag{15}$$

As this problem only has two model parameters, the simplest way to analyze the a posteriori information we have on model parameters is to directly compute the values $\sigma_M(a,b)$ in a given grid, and to plot the results. Figure 1.27 shows the corresponding result, if the outlier is suppressed from the data set (only 10 points have been used). This probability density is Gaussian, and the line corresponding to its center is shown in Figure 1.28. If the outlier is not suppressed, so that the 11 points are used, the probability density $\sigma_M(a,b)$ obtained is shown in figure 1.29. The probability density has been essentially "translated" by the outlier. The line corresponding to the

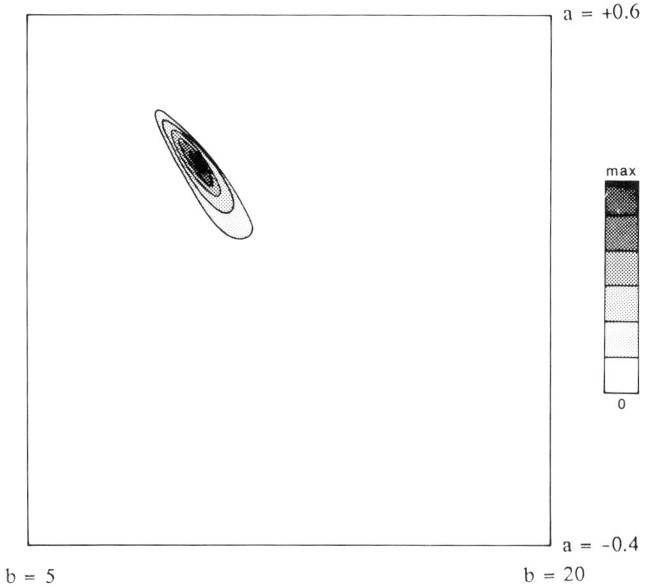

Figure 1.33: The probability density using all 11 experimental points in the exponential hypothesis. By comparison with Figure 1.31, we see that the introduction of the blunder does not completely distort the solution. This shows that the exponential hypothesis is more robust than the Gaussian hypothesis with respect to the existence of a few blunders in a data set.

center of the probability density is shown in figure 1.30. Figures 1.29 and 1.30 show that the Gaussian assumption gives results which are not robust with respect to the existence of outliers in a data set. This may be annoying, because in multidimensional problems it is not always easy to detect outliers.

If instead of assuming uncorrelated Gaussian, we assume uncorrelated exponential uncertainties, equation (8) is replaced by

$$\rho_D(d^1,d^2,...) = \exp\left(- \sum_i \frac{\left| d^i - d^i_{obs} \right|}{\sigma} \right) . \tag{16}$$

The a posteriori probability density is then

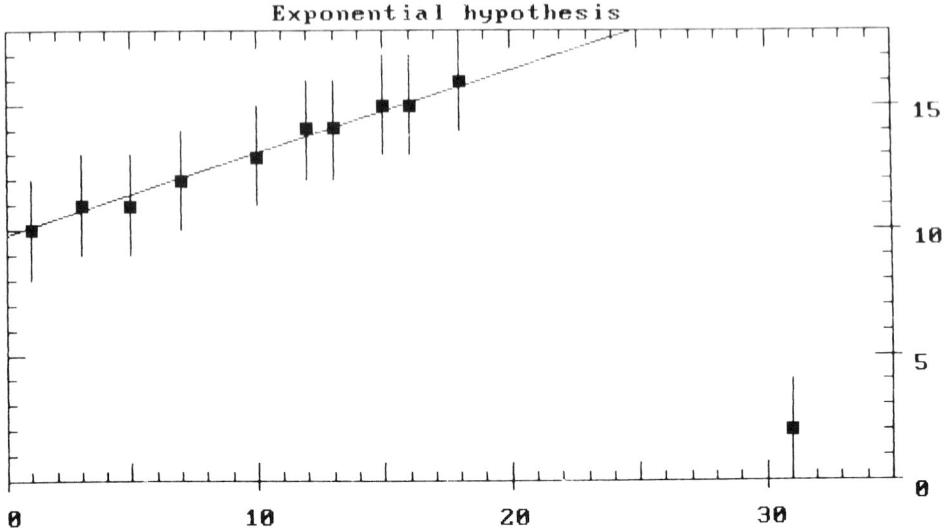

Figure 1.34: The maximum likelihood line for the probability density in Figure 1.33.

$$\sigma_M(a,b) = \exp\left(- \sum_i \frac{\left| d^i_{obs} - d^i_{cal}(a,b) \right|}{\sigma} \right) . \tag{17}$$

This probability density is shown in Figure 1.31 for all points but the outlier, and in Figure 1.33 for all 11 points. The corresponding maximum likelihood lines are shown in Figures 1.32 and 1.34. We see that the introduction of the outlier "deforms" the posterior probability density, but it does not "translate" it. The exponential hypothesis for data uncertainties is more robust than the Gaussian hypothesis.

It should be noticed that the question of which probability density may truly represent the experimental uncertainties for the data in Figure 1.26 has not been adressed. Obviously, it is not Gaussian, because the probability of a outlier like the one present in the figure is extremely low. But the probability of such an outlier is also very low in the exponential hypothesis. A careful examination of the experimental conditions can, in principle, suggest a realistic choice of probability density for representing uncertainties, but this is not always easy. The conclusion of this numerical example is that if a probability density adequately representing experimental uncertainties is unknown, but we suspect a small number of large errors, we should not take the Gaussian probability density, but a more long-tailed one.

Problem 1.10: Condition number and a posteriori errors. The (Cramer's) solution of the system

$$\begin{bmatrix} 10 & 7 & 8 & 7 \\ 7 & 5 & 6 & 5 \\ 8 & 6 & 10 & 9 \\ 7 & 5 & 9 & 10 \end{bmatrix} \begin{bmatrix} m^1 \\ m^2 \\ m^3 \\ m^4 \end{bmatrix} = \begin{bmatrix} 32.0 \\ 23.0 \\ 33.0 \\ 31.0 \end{bmatrix} \tag{1}$$

is

$$\begin{bmatrix} m^1 \\ m^2 \\ m^3 \\ m^4 \end{bmatrix} = \begin{bmatrix} 1.0 \\ 1.0 \\ 1.0 \\ 1.0 \end{bmatrix} , \tag{2}$$

while the solution of the system

$$\begin{bmatrix} 10 & 7 & 8 & 7 \\ 7 & 5 & 6 & 5 \\ 8 & 6 & 10 & 9 \\ 7 & 5 & 9 & 10 \end{bmatrix} \begin{bmatrix} m^1 \\ m^2 \\ m^3 \\ m^4 \end{bmatrix} = \begin{bmatrix} 32.1 \\ 22.9 \\ 33.1 \\ 30.9 \end{bmatrix} , \tag{3}$$

where the right hand member has been slightly modified, is completely different:

$$\begin{bmatrix} m^1 \\ m^2 \\ m^3 \\ m^4 \end{bmatrix} = \begin{bmatrix} 9.2 \\ -12.6 \\ 4.5 \\ -1.1 \end{bmatrix} . \tag{4}$$

This result may be surprising, because the determinant of the matrix of the system is not "small" (it equals one), and the inverse matrix looks as ordinary as the original one:

$$\begin{bmatrix} 10 & 7 & 8 & 7 \\ 7 & 5 & 6 & 5 \\ 8 & 6 & 10 & 9 \\ 7 & 5 & 9 & 10 \end{bmatrix}^{-1} = \begin{bmatrix} 25 & -41 & 10 & -6 \\ -41 & 68 & -17 & 10 \\ 10 & -17 & 5 & -3 \\ -6 & 10 & -3 & 2 \end{bmatrix} . \qquad (5)$$

This nice example is due to R.S. Wilson, and is quoted by Ciarlet (1982). Clearly, the matrix in the example has some special property, which it is important to identify. In classical numerical analysis it is usual to introduce the concept of "condition number" of a matrix. It is defined by

$$\text{cond}(\ \mathbf{A}\ ) = ||\ \mathbf{A}\ ||\ \ ||\ \mathbf{A}^{-1}\ ||\ , \qquad (6)$$

where $||\ \mathbf{A}\ ||$ denotes a given matricial norm. For instance, the ℓ_p matricial norms can be defined by

$$||\ A\ ||_1 = \sup \frac{||\ \mathbf{A\ v}\ ||_1}{||\ \mathbf{v}\ ||_1}\ , \qquad (7a)$$

$$||\ A\ ||_2 = \sup \frac{||\ \mathbf{A\ v}\ ||_2}{||\ \mathbf{v}\ ||_2}\ , \qquad (7b)$$

$$||\ A\ ||_\infty = \sup \frac{||\ \mathbf{A\ v}\ ||_\infty}{||\ \mathbf{v}\ ||_\infty}\ , \qquad (7c)$$

and verify (e.g. Ciarlet, 1982):

$$||\ A\ ||_1 = \max \sum_i |\ A^{ij}\ |\ , \qquad (8a)$$

$$||\ A\ ||_2 = \sqrt{\max\ \lambda_i(\mathbf{A}^*\ \mathbf{A})}\ , \qquad (8b)$$

$$||\ A\ ||_\infty = \max \sum_j |\ A^{ij}\ |\ , \qquad (8c)$$

where $\lambda_i(\mathbf{B})$ denotes the eigenvalues of the matrix **B**, and where $\mathbf{A}^*$ denotes the adjoint of **A** (in chapter 4 the difference between adjoint and transpose will be explained; for the while, let us simply admit that we only consider euclidean scalar products, and adjoint and transpose coincide).

The interpretation of the condition number is obtained as follows. Let **A** and **d** respectively represent a given regular matrix and a given vector, and let **m** represent the solution of

$$\mathbf{A\ m} = \mathbf{d}\ , \qquad (9)$$

i.e.,

$$\mathbf{m} = \mathbf{A}^{-1}\,\mathbf{d} \quad . \tag{10}$$

Let now $\mathbf{m} + \delta\mathbf{m}$ represent the solution of the "perturbed system"

$$\mathbf{A}\,(\,\mathbf{m} + \delta\mathbf{m}\,) = \mathbf{d} + \delta\mathbf{d} \quad . \tag{11}$$

From $\mathbf{d} = \mathbf{A}\,\mathbf{m}$ and $\delta\mathbf{m} = \mathbf{A}^{-1}\,\delta\mathbf{d}$ it can be deduced that

$$\|\,\mathbf{d}\,\| \leq \|\,\mathbf{A}\,\|\;\|\,\mathbf{m}\,\| \tag{12a}$$

$$\|\,\delta\mathbf{m}\,\| \leq \|\,\mathbf{A}^{-1}\,\|\;\|\,\delta\mathbf{d}\,\| \quad , \tag{12b}$$

i.e.,

$$\frac{\|\,\delta\mathbf{m}\,\|}{\|\,\mathbf{m}\,\|} \leq \|\,\mathbf{A}\,\|\;\|\,\mathbf{A}^{-1}\,\|\;\frac{\|\,\delta\mathbf{d}\,\|}{\|\,\mathbf{d}\,\|} \quad , \tag{13}$$

which, using the definition of condition number, can be written

$$\frac{\|\,\delta\mathbf{m}\,\|}{\|\,\mathbf{m}\,\|} \leq \mathrm{cond}(\,\mathbf{A}\,)\;\frac{\|\,\delta\mathbf{d}\,\|}{\|\,\mathbf{d}\,\|} \quad . \tag{14}$$

This shows that for given "relative data error" $\|\,\delta\mathbf{d}\,\| \,/\, \|\,\mathbf{d}\,\|$, the "relative solution error" $\|\,\delta\mathbf{m}\,\| \,/\, \|\,\mathbf{m}\,\|$ may be large if the condition number is large. As it can be shown that

$$1 \leq \mathrm{cond}(\,\mathbf{A}\,) \leq \infty \quad , \tag{15}$$

a linear system for which $\mathrm{cond}(\,\mathbf{A}\,) \simeq 1$ is called *well conditioned*; a linear system for which $\mathrm{cond}(\,\mathbf{A}\,) >> 1$ is called *ill conditioned.*

The following properties which are sometimes useful can be demonstrated (Ciarlet, 1982):

$$\mathrm{cond}(\,\mathbf{A}\,) = \mathrm{cond}(\,\mathbf{A}^{-1}\,) \tag{16}$$

$$\mathrm{cond}_2(\,\mathbf{A}\,) = \frac{\sqrt{\max\ \lambda_i(\,\mathbf{A}^*\,\mathbf{A}\,)}}{\sqrt{\min\ \lambda_i(\,\mathbf{A}^*\,\mathbf{A}\,)}} \tag{17}$$

$$A^* A = M^2 \quad ==> \quad \text{cond}_2(A) = \frac{\max |\lambda_i(M)|}{\min |\lambda_i(M)|} \tag{18}$$

Coming back to the numerical example, the eigenvalues of **A** are

$$\begin{aligned} \lambda_1 &\simeq 0.010 \\ \lambda_2 &\simeq 0.843 \\ \lambda_3 &\simeq 3.858 \\ \lambda_4 &\simeq 30.289 \end{aligned} \tag{19}$$

and using, for instance, (18) gives

$$\text{cond}_2(A) = \frac{\lambda_4}{\lambda_1} \simeq 3\ 10^3 \ , \tag{20}$$

which shows that the system is ill conditioned, and the relative error of the solution may amount to $\simeq 3\ 10^3$ times the relative data error (as is almost the case in the example).

In fact, the introduction of the concept of condition number is only useful when a simplistic approach is used for the resolution of "linear systems". More generally, the reader is asked to solve the following problem:

The observable values $\mathbf{d} = (d^1 , d^2 , d^3 , d^4)$ are known to depend on the model values $\mathbf{m} = (m^1 , m^2 , m^3 , m^4)$ through the (exact) equation

$$\begin{bmatrix} d^1 \\ d^2 \\ d^3 \\ d^4 \end{bmatrix} = \begin{bmatrix} 10 & 7 & 8 & 7 \\ 7 & 5 & 6 & 5 \\ 8 & 6 & 10 & 9 \\ 7 & 5 & 9 & 10 \end{bmatrix} \begin{bmatrix} m^1 \\ m^2 \\ m^3 \\ m^3 \end{bmatrix} , \tag{21}$$

or, for short,

$$\mathbf{d} = \mathbf{G}\, \mathbf{m} \ . \tag{22}$$

A measurement of the observable values gives

$$\begin{bmatrix} d^1 \\ d^2 \\ d^3 \\ d^4 \end{bmatrix} = \begin{bmatrix} 32.0 \pm 0.1 \\ 23.0 \pm 0.1 \\ 33.0 \pm 0.1 \\ 31.0 \pm 0.1 \end{bmatrix} . \tag{23}$$

Use the least squares theory to solve the inverse problem and *discuss error and resolution*

Solution: The best solution (in the least squares sense) for a linear problem is (equations 1.90 and 1.92):

$$\langle \mathbf{m} \rangle = \mathbf{m}_{prior} + \left(\mathbf{G}^t \, \mathbf{C_D}^{-1} \, \mathbf{G} + \mathbf{C_M}^{-1} \right)^{-1} \mathbf{G}^t \, \mathbf{C_D}^{-1} \, [\mathbf{d}_{obs} - \mathbf{G} \, \mathbf{m}_{prior}] \, , \qquad (24)$$

$$\mathbf{C_{M'}} = \left(\mathbf{G}^t \, \mathbf{C_D}^{-1} \, \mathbf{G} + \mathbf{C_M}^{-1} \right)^{-1} \, . \qquad (25)$$

If there is no a priori information, $\mathbf{C_M} \rightarrow \infty \mathbf{I}$, and

$$\langle \mathbf{m} \rangle = \left(\mathbf{G}^t \, \mathbf{C_D}^{-1} \, \mathbf{G} \right)^{-1} \mathbf{G}^t \, \mathbf{C_D}^{-1} \, \mathbf{d}_{obs} \qquad (26)$$

$$\mathbf{C_{M'}} = \left(\mathbf{G}^t \, \mathbf{C_D}^{-1} \, \mathbf{G} \right)^{-1} \, . \qquad (27)$$

In our numerical example,

$$\mathbf{d}_{obs} = \begin{bmatrix} 32.0 \\ 23.0 \\ 33.0 \\ 31.0 \end{bmatrix} \qquad (28)$$

$$\mathbf{C_D} = \sigma^2 \, \mathbf{I} = 0.01 \, \mathbf{I} = \begin{bmatrix} 0.01 & 0 & 0 & 0 \\ 0 & 0.01 & 0 & 0 \\ 0 & 0 & 0.01 & 0 \\ 0 & 0 & 0 & 0.01 \end{bmatrix} \qquad (29)$$

$$\mathbf{G} = \begin{bmatrix} 10 & 7 & 8 & 7 \\ 7 & 5 & 6 & 5 \\ 8 & 6 & 10 & 9 \\ 7 & 5 & 9 & 10 \end{bmatrix} \, . \qquad (30)$$

As, in this particular example, $\mathbf{G}$ is squared and regular, we successively have

$$\langle \mathbf{m} \rangle = \left(\mathbf{G}^t \, \mathbf{C_D}^{-1} \, \mathbf{G} \right)^{-1} \mathbf{G}^t \, \mathbf{C_D}^{-1} \, \mathbf{d}_{obs}$$

$$= \mathbf{G}^{-1} \, \mathbf{C_D} \, (\mathbf{G}^t)^{-1} \, \mathbf{G}^t \, \mathbf{C_D}^{-1} \, \mathbf{d}_{obs} = \mathbf{G}^{-1} \, \mathbf{d}_{obs} \, , \qquad (31)$$

i.e.,

$$\langle \mathbf{m} \rangle = \begin{bmatrix} \langle m^1 \rangle \\ \langle m^2 \rangle \\ \langle m^3 \rangle \\ \langle m^4 \rangle \end{bmatrix} = \begin{bmatrix} 1.0 \\ 1.0 \\ 1.0 \\ 1.0 \end{bmatrix} \, . \qquad (32)$$

The posterior covariance operator is given by

$$C_{M'} = \left[G^t \, C_D^{-1} \, G\right]^{-1} = G^{-1} \, C_D \, (G^t)^{-1} = \sigma^2 \, G^{-1} \, (G^t)^{-1} \quad , \tag{33}$$

and, as **G** is symmetric,

$$C_{M'} = \sigma^2 \, G^{-1} \, G^{-1} \quad , \tag{34}$$

i.e.,

$$C_{M'} = 0.01 \begin{bmatrix} 2442 & -4043 & 1015 & -602 \\ -4043 & 6694 & -1681 & 997 \\ 1015 & -1681 & 423 & -251 \\ -602 & 997 & -251 & 149 \end{bmatrix} . \tag{35}$$

From $C_{M'}$ it is easy to obtain the standard deviations of model parameters

$$\begin{aligned} \sigma^1{}_M &= 4.94 \\ \sigma^2{}_M &= 8.18 \\ \sigma^3{}_M &= 2.06 \\ \sigma^4{}_M &= 1.22 \ , \end{aligned} \tag{36}$$

and the correlation matrix (see box 1.1)

$$R = \begin{bmatrix} 1 & -0.99997 & +0.99867 & -0.99800 \\ -0.99997 & 1 & -0.99898 & +0.99830 \\ +0.99867 & -0.99898 & 1 & -0.99979 \\ -0.99800 & +0.99830 & -0.99979 & 1 \end{bmatrix} . \tag{37}$$

The overall information on the solution can thus be expressed by this correlation matrix and the short notation

$$\langle m \rangle = \begin{bmatrix} \langle m^1 \rangle \\ \langle m^2 \rangle \\ \langle m^3 \rangle \\ \langle m^4 \rangle \end{bmatrix} = \begin{bmatrix} 1.00 \pm 4.94 \\ 1.00 \pm 8.18 \\ 1.00 \pm 2.06 \\ 1.00 \pm 1.22 \end{bmatrix} . \tag{38}$$

The interpretation of these results is as follows.

The least-squares approach is only fully justified if errors (in this example, data errors) are modeled using Gaussian probability densities. For a linear problem, as discussed in section 1.7, the a posteriori errors are then also Gaussian. Taking, for instance, twice the standard deviation, the probability of the true value of the parameter $m^1{}_{true}$ verifying the inequality

$$-8.88 \le m^1{}_{true} \le +10.88 \tag{39a}$$

is about 95% , *independently of the respective values of* $m^2{}_{true}$, $m^3{}_{true}$, *and* $m^4{}_{true}$. Similarly, the probability of the true values of each of the parameters m^2_{true} , m^3_{true} , and m^4_{true} verifying the inequalities

$$
\begin{aligned}
-15.36 &\le m^2_{true} \le +17.36 \\
-3.12 &\le m^3_{true} \le +5.12 \\
-1.44 &\le m^4_{true} \le +3.44
\end{aligned}
\qquad (39b)
$$

is also about 95% .

This gives information on the true value of each parameter, considered independently, but the correlation matrix gives additional information on error correlation. For instance, the correlation of m^1 with m^2 is -0.99997 . This means that if the estimated value for m^1 , $\langle m^1 \rangle$, is in error (with respect to the true unknown value) it is *almost certain* that the the estimated value for m^2 , $\langle m^2 \rangle$, will also be in error (because the absolute value of the correlation is close to 1), and the the sign of the error will be opposite to the error in $\langle m^1 \rangle$ (because the correlation is negative). For instance, if the true value of m^1 was $m^1_{true} = \langle m^1 \rangle + 2\,\sigma^1_M$ it is almost certain that the true value of m^2 will be $m^2_{true} = \langle m^2 \rangle - 2\,\sigma^2_M$.

The easiest way to understand this is to consider the a posteriori probability density in the parameter space (equation 1.91):

$$
\sigma_M(\mathbf{m}) = \left((2\pi)^2 \ \det C_{M'} \right)^{-1/2} \exp\left[-\frac{1}{2} \left(\mathbf{m} - \langle \mathbf{m} \rangle \right)^t \ C_{M'}^{-1} \ \left(\mathbf{m} - \langle \mathbf{m} \rangle \right) \right] . \qquad (40)
$$

To simplify the discussion, let us first analyze the two parameters m^1 and m^2 . Their marginal probability density is

$$
\sigma_{12}(m^1,m^2) = \left(2\pi \ \det C_{12} \right)^{-1/2} \exp\left(-\frac{1}{2} \begin{bmatrix} m^1-1.0 \\ m^2-1.0 \end{bmatrix}^t \begin{bmatrix} 24.42 & -40.43 \\ -40.43 & 66.94 \end{bmatrix}^{-1} \begin{bmatrix} m^1-1.0 \\ m^2-1.0 \end{bmatrix} \right) \qquad (41)
$$

(it is well known [e.g. Dubes, 1968] that marginal probability densities corresponding to a mutidimensional Gaussian are simply obtained by "picking" the corresponding covariances in the joint covariance operator). Figures 1.35 and 1.36 illustrate this probability density. The correlation between m^1 and m^2 is so strong in this numerical example, that the 95% confidence ellipsoid is undistinguishable from a segment. This means that, although the data set used in this example is not able to give an accurate location for the true values of m^1 or m^2 independently, it imposes that these true values must lie on the segment of the figure. As the volume of the allowed region is almost null, this gives, in fact, a lot of information.

See color plate of page 89

Figure 1.35: Marginal probability density for the parameters m^1 and m^2 . Uncertainties are so strongly correlated that it is difficult to distinguish the ellipsoid of errors from a segment. Although the standard deviations for each of the parameters are large, we have much information on these parameters, because their true values must lie on the "line". The resolution of the plotting device (300 pixel x 300 pixel) is not fine enough for a good representation, so that the colors obtained for the ellipsoid are aliased. The next figure shows a zoom of the central region.

See color plate of page 89

Figure 1.36: Same as the previous figure, with finer detail.

Similarly, the four-dimensional probability density $\sigma_M(\mathbf{m})$ defines a 95% confidence "ellipsoid" on the parameter space which corresponds to the extra-long "cigar" joining the point (-8.88 , +17.36 , -3.12 , +3.44) to the point (+10.88 , -15.36 , +5.12 , -1.44) . The reader will easily verify that the two "solutions" of the linear system obtained using Cramer's method for two slightly different data vectors correspond to two points on the cigar.

It should be noticed that if a further experiment gives accurate information on the true value of one of the parameters, the values of the other three parameters can readily be deduced, with very small uncertainties.

This example has shown that:

i) a careful analysis of the a posteriori covariance operator always has to be made when solving least-squares inverse problems,

ii) the information given by the "condition number" is very rough compared with the information given by the covariance operator (it only gives information about the ratio between the largest and shortest diameters of the ellipsoid of errors in the model space).

Problem 1.11: We wish to measure a single quantity x_{true} . To do this we have a digital instrument which delivers 5-digit decimal results. We have performed a great number (101) of measurements, and because the instrument has some intrinsic uncertainty, we obtain a different value at each measurement:

$x_{000} = 21.738$	$x_{001} = 21.273$	$x_{002} = 21.300$	$x_{003} = 21.540$	$x_{004} = 21.878$
$x_{005} = 21.052$	$x_{006} = 21.066$	$x_{007} = 21.894$	$x_{008} = 21.922$	$x_{009} = 21.536$
$x_{010} = 21.575$	$x_{011} = 21.990$	$x_{012} = 21.013$	$x_{013} = 21.488$	$x_{014} = 21.421$
$x_{015} = 21.296$	$x_{016} = 21.951$	$x_{017} = 21.717$	$x_{018} = 21.675$	$x_{019} = 21.136$
$x_{020} = 21.029$	$x_{021} = 21.515$	$x_{022} = 21.104$	$x_{023} = 21.872$	$x_{024} = 21.789$
$x_{025} = 21.191$	$x_{026} = 21.882$	$x_{027} = 21.578$	$x_{028} = 21.658$	$x_{029} = 21.069$
$x_{030} = 21.030$	$x_{031} = 21.217$	$x_{032} = 21.651$	$x_{033} = 21.285$	$x_{034} = 21.659$
$x_{035} = 21.965$	$x_{036} = 21.816$	$x_{037} = 21.535$	$x_{038} = 21.715$	$x_{039} = 21.104$
$x_{040} = 21.044$	$x_{041} = 21.977$	$x_{042} = 21.711$	$x_{043} = 21.758$	$x_{044} = 21.751$
$x_{045} = 21.489$	$x_{046} = 21.087$	$x_{047} = 21.814$	$x_{048} = 21.104$	$x_{049} = 21.971$
$x_{050} = 21.625$	$x_{051} = 21.581$	$x_{052} = 21.076$	$x_{053} = 21.648$	$x_{054} = 21.983$
$x_{055} = 21.888$	$x_{056} = 21.121$	$x_{057} = 21.239$	$x_{058} = 21.474$	$x_{059} = 21.788$
$x_{060} = 21.414$	$x_{061} = 21.930$	$x_{062} = 21.353$	$x_{063} = 21.001$	$x_{064} = 21.863$
$x_{065} = 21.087$	$x_{066} = 21.931$	$x_{067} = 21.776$	$x_{068} = 21.065$	$x_{069} = 21.664$
$x_{070} = 21.421$	$x_{071} = 21.127$	$x_{072} = 21.746$	$x_{073} = 21.110$	$x_{074} = 21.102$
$x_{075} = 21.947$	$x_{076} = 21.128$	$x_{077} = 21.610$	$x_{078} = 21.465$	$x_{079} = 21.822$
$x_{080} = 21.617$	$x_{081} = 21.675$	$x_{082} = 21.898$	$x_{083} = 21.573$	$x_{084} = 21.397$
$x_{085} = 21.309$	$x_{086} = 21.485$	$x_{087} = 21.018$	$x_{088} = 21.132$	$x_{089} = 21.462$
$x_{090} = 21.146$	$x_{091} = 21.391$	$x_{092} = 21.222$	$x_{093} = 21.034$	$x_{094} = 21.977$
$x_{095} = 21.079$	$x_{096} = 21.713$	$x_{097} = 21.028$	$x_{098} = 21.598$	$x_{099} = 21.105$
$x_{100} = 21.333$.				

The uncertainty ϵ of the instrument has an unknown probability density function $f(\epsilon)$ which is known to be symmetric and centered at $\epsilon = 0$. Under these conditions, the median m_1 , the mean m_2 , and the mid-range m_∞ are unbiased central estimators. They are given by

$$
\begin{aligned}
m_1 &= 21.422 \\
m_2 &= 21.482 \\
m_\infty &= 21.492 \ .
\end{aligned}
$$

Which of these is to be preferred for estimating x_{true} ?.

Solution: One response to the question is that it has no response. The median m_1 minimizes the mean deviation of the residuals, the mean m_2 minimizes the standard deviation, and the mid-range m_∞ minimizes the maximum deviation. So the "best" central estimator will depend on which criterion we use to measure "goodness".

That is not the good response, because with the experimental results in the previous table we can do better than computing a central estimator: we can estimate the probability density function $f(\epsilon)$ itself. Figure 1.37 shows a histogram of the results. It has the striking feature of taking strictly null values outside a given range where the values are rather uniform. As $f(\epsilon)$ is known to be symmetric, we can try to fit a Generalized Gaussian to that histogram (see box 1.2). The value $p = \infty$ (box-car function) seems not too bad a candidate (in any case is clearly better than $p = 1$ or $p = 2$). It is clear that to estimate the center of a generalized Gaussian of order p , we should use the central estimator in norm ℓ_p . The value $m_\infty = 21.492$ then is to be preferred for estimating x_{true} .

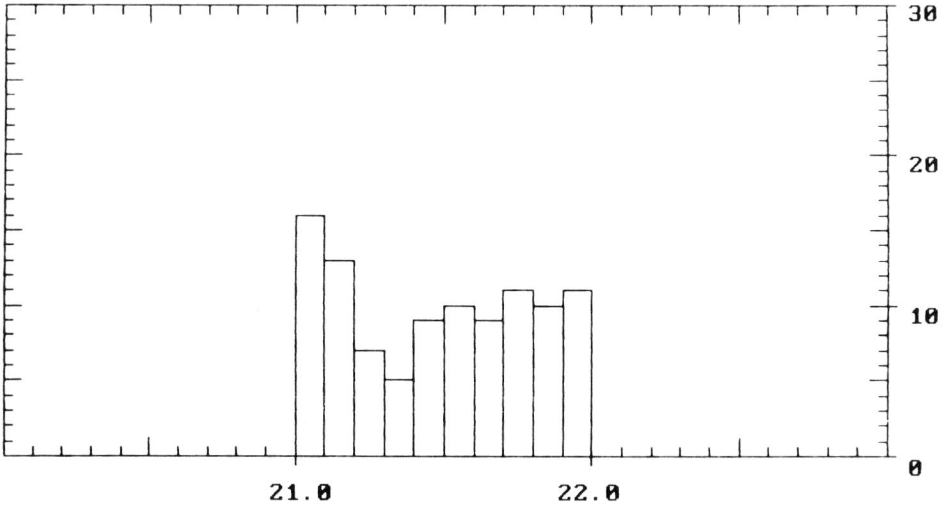

Figure 1.37: Histogram of the 101 values obtained when measuring x_{true} . The statistics of the errors are unknown, but the probability density is known to be unbiased (centered on x_{true}). Which is the best estimator of x_{true} ?

Actually, the 101 numbers used as data have been generated using the pseudo-random computer code

$$x_i = 21. + \mathrm{RND} ,$$

where RND is a routine generating numbers with a theoretically uniform probability density over $(0,1)$. That m_∞ is a better estimator than m_1 or m_2 in some absolute sense can be seen by a simple numerical test. The previous experiment can be repeated a great number of times ($\simeq$100), and we can make histograms of the values thus obtained for m_1 , m_2 , and m_∞ in each experiment. They are shown in Figure 1.38. We clearly see that m_∞ is *less scattered* around the true value x_{true} than m_1 or m_2 *independently of the criterion used to measure the scatter.*

Problem 1.12: Let x and y be cartesian coordinates on a cathodic screen. A random device projects electrons on the screen with a known probability density:

$$\Theta(x,y) = \begin{cases} \text{const. } r\ (2-r) & \text{if } 0 \leq r \leq 2 \\ 0 & \text{if } r > 2 , \end{cases} \qquad (1)$$

where $r = \sqrt{x^2+y^2}$.

We are interested in the coordinates (x,y) at which a particular electron will hit the screen, and we build an experimental device to measure them. The measuring instrument is not perfect, and when we perform the experiment we can only get the information that the true coordinates of the impact point had the probability density

$$\rho(x,y) = \text{const. } \exp\left[-\frac{1}{2} \begin{bmatrix} x-x_0 \\ y-y_0 \end{bmatrix} \begin{bmatrix} \sigma^2 & \rho\sigma^2 \\ \rho\sigma^2 & \sigma^2 \end{bmatrix}^{-1} \begin{bmatrix} x-x_0 \\ y-y_0 \end{bmatrix} \right] \qquad (2)$$

with $(x_0,y_0) = (0,0)$, $\sigma = 2$, and $\rho = 0.99$. Combine this experimental information with the previous knowledge of the random device, and obtain a better estimate of the impact point.

Solve the problem again, using everywhere polar coordinates instead of cartesian coordinates.

Solution: As x and y are cartesian coordinates, the null information probability density for the impact point is

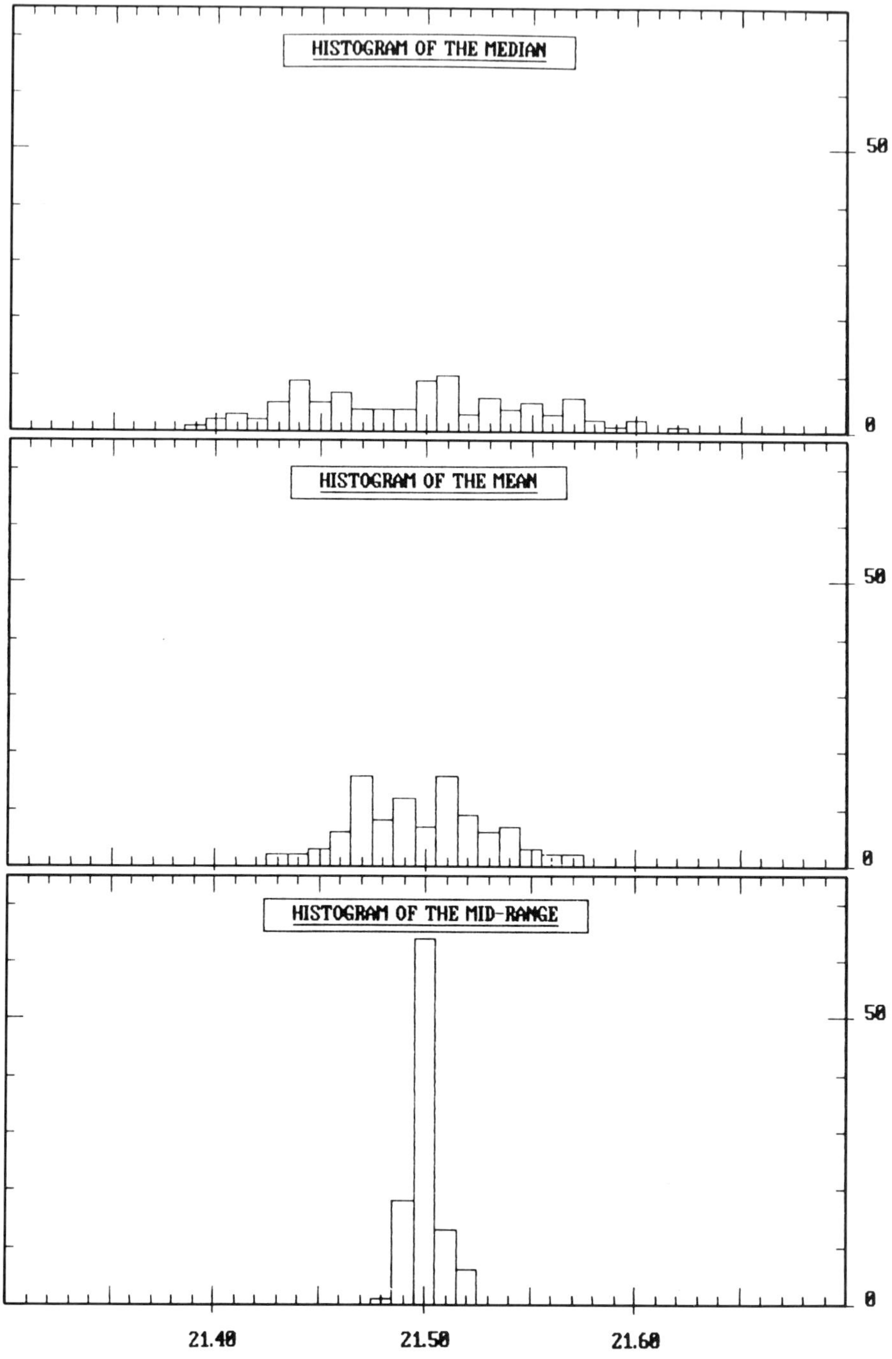
HISTOGRAM OF THE MEDIAN
50
0
HISTOGRAM OF THE MEAN
50
0
HISTOGRAM OF THE MID-RANGE
50
0
21.40
21.50
21.60

← *Figure 1.38:* The histogram in Figure 1.37 suggests that the statistics of errors correspond to a box-car probability density. In that case the best estimator of x_{true} is the mid-range of the 101 values. These values were, in fact, generated using a computer code simulating a box-car probability density. This figure shows the histograms obtained for the median (top), mean (middle), and mid-range (bottom) when repeating the whole experiment (generation of the 101 random points) a large number of times. Undoubtedly, the mid-range is (in this example) the best estimator, whatever criterion of goodness we may use.

$$\mu(x,y) = \text{const.} \tag{3}$$

The information represented by $\Theta(x,y)$ and $\rho(x,y)$ are independent in the sense discussed in section 1.2.6 . Combination of these data then corresponds to the conjunction

$$\sigma(x,y) = \frac{\rho(x,y)\ \Theta(x,y)}{\mu(x,y)} , \tag{4}$$

which is plotted in Figure 1.39.

The polar coordinates verify

$$r = (x^2+y^2)^{1/2} \qquad \text{tg}\ \phi = \frac{x}{y} , \tag{5}$$

so that the Jacobian of the transformation is

$$J(r,\phi) = \begin{vmatrix} \frac{\partial r}{\partial x} & \frac{\partial r}{\partial y} \\ \frac{\partial \phi}{\partial x} & \frac{\partial \phi}{\partial y} \end{vmatrix} = \frac{1}{r} . \tag{6}$$

Let $f(x,y)$ be a probability density in cartesian coordinates. To any surface S of the plane it assigns the probability

$$P(S) = \iint_S dx\ dy\ f(x,y) . \tag{7}$$

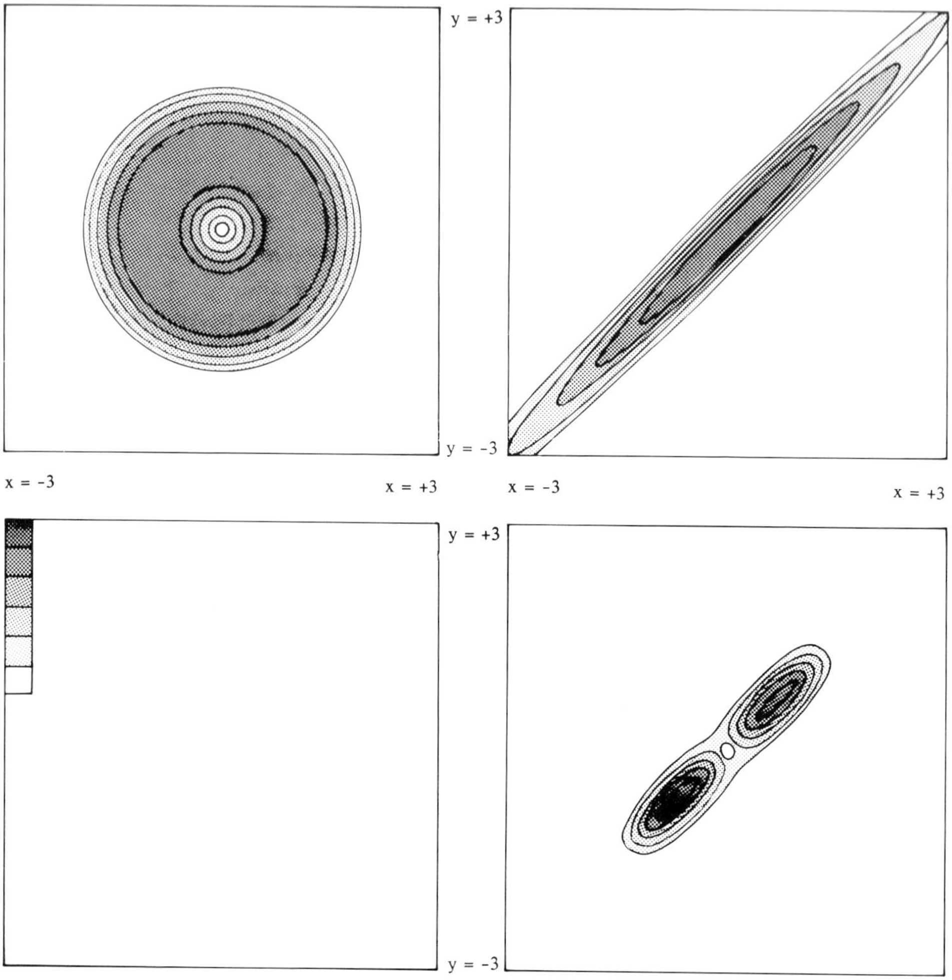

Let $\tilde{f}(r,\phi)$ be a probability density in polar coordinates. If we wish $\tilde{f}(r,\phi)$ to assign to **S** the same probability as $f(x,y)$,

$$P(S) = \iint_S dr\, d\phi\, f(r,\phi) \; , \tag{8}$$

then necessarily

← *Figure 1.39:* A random device has been built which projects electrons on a cathodic screen with the probability density shown in the top-left. Coordinates are cartesian. Independently of this probability, a measurement of the impact point of a particular electron gives the information represented by the probability density shown in the top right. The null information probability density (which is uniform, and has been represented in arbitrary color) is shown in the bottom left. It is then possible to combine all these states of information to obtain the posterior probability density, shown in ther bottom right.

$$\tilde{f}(r,\phi) \;=\; f(\, x(r,\phi)\, ,\, y(r,\phi)\,)\; |J(r,\phi)| \;. \tag{9}$$

This is the usual formula for the change of variables in a probability density. In our case

$$\tilde{f}(r,\phi) \;=\; r\, f(r \sin\phi\, ,\, r\cos\phi)\;. \tag{10}$$

This gives

$$\tilde{\Theta}(x,y) \;=\; \begin{cases} \text{const.}\;\; r^2\;\; (1-r) & \text{if}\;\; 0 \le r \le 2 \\ 0 \quad \text{if}\;\; r > 2\;, & \end{cases} \tag{11}$$

$$\tilde{\rho}(r,\phi) \;=\; \text{const.}\;\; r\;\; \exp\left[\, -\, \frac{r^2\,(\,1-2\,\rho\,\sin\phi\,\cos\phi\,)}{2\,\sigma^2\,(1-\rho^2)}\,\right]\;, \tag{12}$$

and

$$\tilde{\mu}(r,\phi) \;=\; \text{const}\;\; r\;. \tag{13}$$

The combination of $\Theta(r,\phi)$ with $\tilde{\rho}(r,\phi)$ is given by the conjunction

$$\tilde{\sigma}(r,\phi) \;=\; \frac{\tilde{\rho}(r,\phi)\;\tilde{\Theta}(r,\phi)}{\tilde{\mu}(r,\phi)}\;, \tag{14}$$

and is shown in Figure 1.40.

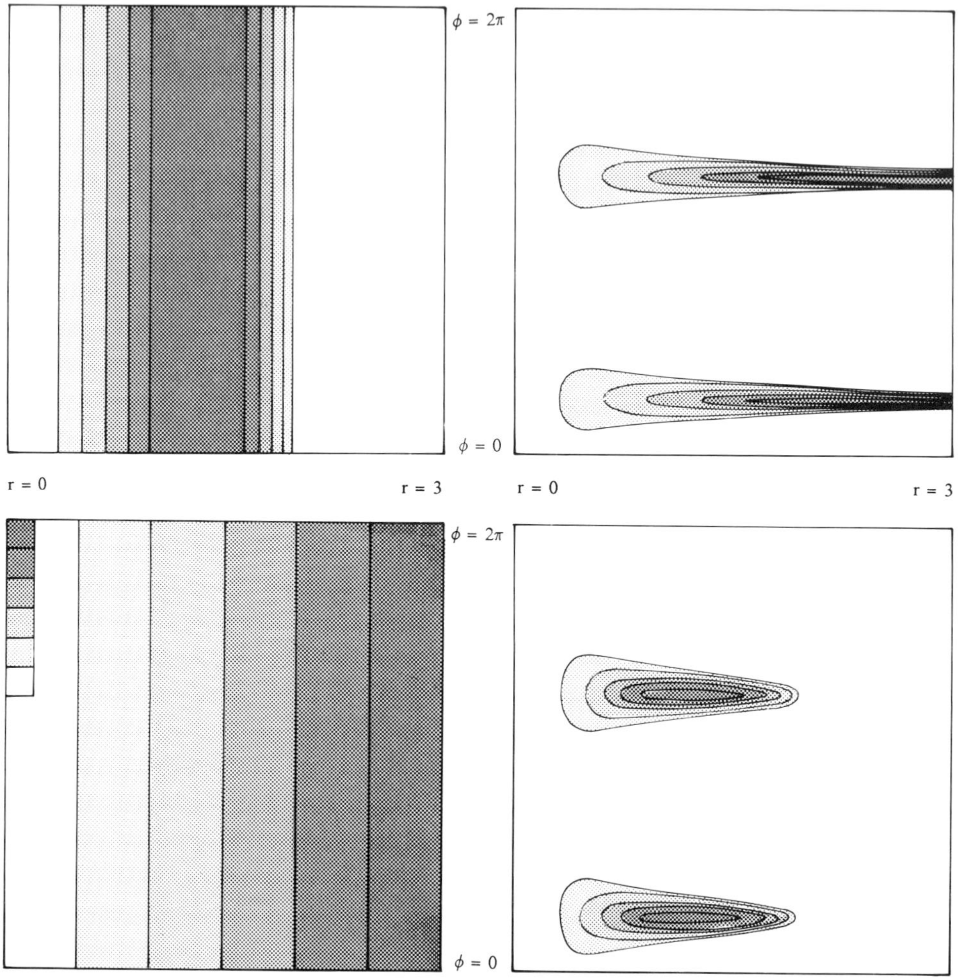

It should be noticed that the probability density representing the null information probability is not uniform in polar coordinates: the probability density (13) assigns equal probabilities to equal "volumes", as it must.

The solution obtained for this problem using cartesian coordinates (Figure 1.39) , and the solution obtained using polar coordinates (figure 1.40) are coherent: Figure 1.39 can be deduced from Figure 1.40 using (9) , and viceversa.

← *Figure 1.40:* It is also possible to solve the problem using polar coordinates throughout. The top-left represents the probability density of an impact on the screen, as imposed by the experimantal device. The probability density is constant for given r . The top-right shows the result of the measurement. In the bottom-left, the noninformative probability density in polar coordinates is shown. It assigns equal probability to equal surfaces of the screen. The combination of these states of information gives the posterior probability density shown in the bottom-right. This probability is completely equivalent to the probability density in the bottom-right of the previous figure, as they can be deduced one from another through the usual formula of change of variables between cartesian and polar coordinates $\tilde{\sigma}(\theta,\phi) = r\ \sigma(x,y)$.

Using more elementary approaches, this problem may present some pathologies. In particular, the result cannot be expressed using a single estimator of the impact point, because the probability density is bimodal. The mean value and median value are meaningless, and only the two maximum likelihood points make clear sense. But we should be aware that the maximum likelihood points obtained using cartesian coordinates and using polar coordinates are not identical.

It should be mentioned that the usual Bayesian approach does not apply directly to this problem.

Problem 1.13: Demonstrate that the relative information of two probabity densisities is invariant under a bijective change of variables.

Solution: Let f(x) be a probability density function representing a given state of information on the parameters $\mathbf{x}$. The information content of $f(\mathbf{x})$ has been defined by

$$I(f;\mu) = \int d\mathbf{x}\ f(\mathbf{x})\ \mathrm{Log}\ \frac{f(\mathbf{x})}{\mu(\mathbf{x})} , \tag{1}$$

where $\mu(\mathbf{x})$ represents the reference state of information. If instead of the parameters $\mathbf{x}$ we decide to use the parameters

$$\mathbf{x}^* = \mathbf{x}^*(\mathbf{x}) , \tag{2}$$

the same state of information is described in the new variables by

$$f^*(\mathbf{x}^*) = f(\mathbf{x}) \left| \frac{\partial \mathbf{x}}{\partial \mathbf{x}^*} \right| , \tag{3}$$

while the reference state of information is described by

$$\mu^*(x^*) = \mu(x) \left| \frac{\partial \mathbf{x}}{\partial \mathbf{x}^*} \right| , \tag{4}$$

where $|\partial x/\partial x^*|$ denotes the absolute value of the Jacobian of the transformation. A computation of the information content in the new variables gives

$$I(f^*;\mu^*) = \int dx^* \; f^*(x^*) \; \mathrm{Log} \frac{f^*(x^*)}{\mu^*(\mathbf{x}^*)}$$

$$= \int dx^* \left| \frac{\partial x}{\partial x^*} \right| f(x) \; \mathrm{Log} \frac{f(x)}{\mu(x)} , \tag{5}$$

and, using

$$dx^* \left| \frac{\partial x}{\partial x^*} \right| = d\mathbf{x} , \tag{6}$$

we directly obtain

$$I(f^*;\mu^*) = I(f;\mu) . \tag{7}$$

Problem 1.14: Demonstrate the equivalence

$$I(f;\mu) = 0 \quad \Leftrightarrow \quad f(\mathbf{x}) \equiv \mu(\mathbf{x}) . \tag{1}$$

Solution: If $f(\mathbf{x}) = \mu(\mathbf{x})$, it is evident that $I(\mathbf{f};\mu) = 0$. To demonstrate the reciprocal, let us, for given $\mu(\mathbf{x})$, search the minimumof $S(\mathbf{f}) = \int d\mathbf{x} \; f(\mathbf{x}) \; \mathrm{Log}\, f(\mathbf{x})/\mu(\mathbf{x})$ under the constraint $\int d\mathbf{x} \; f(x) = 1$. Using the method of Lagrange's multipliers (see Chapter 4), we can introduce the function

$$S(\mathbf{f},\lambda) = \int dx \; f(\mathbf{x}) \; \mathrm{Log} \frac{f(\mathbf{x})}{\mu(\mathbf{x})} - \lambda \left[1 - \int dx \; f(x) \right] , \tag{2}$$

to be minimized with respect to the function $f(\mathbf{x})$ and the parameter λ. We have first

$$\frac{\partial S}{\partial \lambda} = 0 \quad \Leftrightarrow \quad \int dx \;\; f(\mathbf{x}) = 1 \; , \tag{3}$$

which simply is the normalization condition. The definition of functional derivative is written (see, for instance, Chapter 7) , for arbitrary $\delta \mathbf{f}$,

$$S(\mathbf{f}+\delta\mathbf{f};\lambda) - S(\mathbf{f};\lambda) = \frac{\partial S}{\partial \mathbf{f}} \; \delta\mathbf{f} + O(||\delta\mathbf{f}||^2) \; . \tag{4}$$

We have successively

$$S(\mathbf{f}+\delta\mathbf{f};\lambda) - S(\mathbf{f};\lambda) =$$

$$= \int d\mathbf{x} \;\; (\; f(\mathbf{x}) + \delta f(\mathbf{x}) \;) \;\; \mathrm{Log} \; \frac{f(\mathbf{x}) + \delta f(\mathbf{x})}{\mu(\mathbf{x})} \;\; - \lambda \; (\; 1 \; - \int d\mathbf{x} \; (\; f(\mathbf{x}) + \delta f(\mathbf{x}) \;) \;)$$

$$- \int d\mathbf{x} \;\; f(\mathbf{x}) \;\; \mathrm{Log} \; \frac{f(\mathbf{x})}{\mu(\mathbf{x})} \; + \lambda \; (\; 1 \; - \int d\mathbf{x} \; f(\mathbf{x}) \;)$$

$$= \int d\mathbf{x} \left[\; \mathrm{Log} \; \frac{f(\mathbf{x})}{\mu(\mathbf{x})} \; + 1 + \lambda \; \right] \; \delta f(\mathbf{x}) \;\; + \;\; O(||\delta\mathbf{f}||^2). \tag{5}$$

The condition $\partial S/\partial \mathbf{f} = \mathbf{0}$ then gives successively

$$\mathrm{Log} \; \frac{f(\mathbf{x})}{\mu(\mathbf{x})} \; \equiv \; - \lambda - 1 \; , \tag{6}$$

$$f(\mathbf{x}) \; \equiv \; e^{-\lambda-1} \;\; \mu(\mathbf{x}) \; , \tag{7}$$

and, as the information content is only defined if $\mu(\mathbf{x})$ is normalized, this gives $\lambda = -1$ and

$$f(\mathbf{x}) \; \equiv \; \mu(\mathbf{x}) \; . \tag{8}$$

Problem 1.15: Maximum entropy probability density. Let $\mathbf{V}(x)$ be an arbitrary given vector function of $\mathbf{x}$. Demonstrate that among all probability densities $f(\mathbf{x})$ for which the mathematical expectation for $\mathbf{V}(x)$ equals $\mathbf{V}_0$,

$$\int d\mathbf{x} \;\; \mathbf{V}(\mathbf{x}) \;\; f(\mathbf{x}) \; = \mathbf{V}_0 \; , \tag{1}$$

the one which has ***minimum information*** (maximum extropy) with respect to a given probability density $\mu(\mathbf{x})$,

$$\int d\mathbf{x} \;\; f(\mathbf{x}) \;\; \mathrm{Log} \left(\frac{f(x)}{\mu(\mathbf{x})} \right) \qquad \text{MINIMUM} \; , \tag{2}$$

necessarily has the form

$$f(\mathbf{x}) = k\ \mu(\mathbf{x}) \exp(- \mathbf{W}^t\ \mathbf{V}(\mathbf{x})) , \tag{3}$$

where k and $\mathbf{W}$ are constants (independent of $\mathbf{x}$).

Solution: The problem is:

$$\text{Minimize} \qquad S'(f(\cdot)) = \int d\mathbf{x}\ f(\mathbf{x})\ \mathrm{Log}\left(\frac{f(\mathbf{x})}{\mu(\mathbf{x})} \right) \tag{4a}$$

$$\text{Under the constraints} \quad \begin{cases} \int d\mathbf{x}\ f(\mathbf{x}) = 1 & \text{(4b)} \\ \int d\mathbf{x}\ \mathbf{V}(\mathbf{x})\ f(\mathbf{x}) = \mathbf{V}_0 , & \text{(4c)} \end{cases}$$

i.e., a problem of constrained minimization, which is non-typical in the sense that the variable is a function (i.e., a variable in an infinite dimensional space). Nevertheless, the problem can be solved using the classical method of Lagrange's parameters (see Chapter 4). The problem of minimization of S' under the constraints (4b)-(4c) is equivalent to the problem of unconstrained minimization of

$$S(f(\cdot) , U , \mathbf{W}) = \int d\mathbf{x}\ f(\mathbf{x})\ \mathrm{Log}\left(\frac{f(\mathbf{x})}{\mu(\mathbf{x})} \right) - U (1 - \int d\mathbf{x}\ f(\mathbf{x}))$$

$$- \mathbf{W}^t (\mathbf{V}_0 - \int d\mathbf{x}\ \mathbf{V}(\mathbf{x})\ f(\mathbf{x})) \tag{5}$$

because the conditions $\partial S/\partial U = 0$ and $\partial S/\partial \mathbf{W} = 0$ directly impose the constraints (4b)-(4c) . We have

$$S(f(\cdot)+\delta f(\cdot) , U , \mathbf{W}) - S(f(\cdot) , U , \mathbf{W}) =$$

$$= \int d\mathbf{x}\ (f(\mathbf{x}) + \delta f(\mathbf{x}))\ \mathrm{Log}\left(\frac{f(\mathbf{x})+\delta f(\mathbf{x})}{\mu(\mathbf{x})} \right) - \int d\mathbf{x}\ f(\mathbf{x})\ \mathrm{Log}\left(\frac{f(\mathbf{x})}{\mu(\mathbf{x})} \right)$$

$$+ U \int d\mathbf{x}\ \delta f(\mathbf{x}) + \mathbf{W}^t \int d\mathbf{x}\ \mathbf{V}(\mathbf{x})\ \delta f(\mathbf{x}) , \tag{6}$$

and using the first order development

$$\mathrm{Log}(1+u) = u + O(u^2) , \tag{7}$$

gives, everywhere $f(\mathbf{x}) \neq 0$,

$$\text{Log}\left(\frac{f(\mathbf{x})+\delta f(\mathbf{x})}{\mu(\mathbf{x})} \right) = \text{Log}\left(\frac{f(\mathbf{x})}{\mu(\mathbf{x})} \right) + \frac{\delta f(\mathbf{x})}{f(\mathbf{x})} + O(\delta f^2) \ , \tag{8}$$

and then,

$$S(\ f(\cdot)+\delta f(\cdot)\ ,\ U\ ,\ \mathbf{W}\ ) - S(\ f(\cdot)\ ,\ U\ ,\ \mathbf{W}\ ) =$$

$$= \int dx \left[\text{Log}\left(\frac{f(x)}{\mu(\mathbf{x})} \right) + 1 + U + \mathbf{W}^t\ \mathbf{V}(\mathbf{x}) \right] \delta f(\mathbf{x}) + O(\delta \mathbf{f}^2)\ . \tag{9}$$

The condition of minimum of S with respect to $f(\mathbf{x})$ causes the factor of $\delta f(\mathbf{x})$ in the right hand of (9) to vanish, from which result (3) follows.

Problem 1.16: Let $f_1(x)$ and $f_0(\mathbf{x})$ represent two normalized probability density functions. The *relative information* on f_1 with respect to f_0 is defined by

$$I(f_1;f_0) = \int dx\ f_1(\mathbf{x})\ \text{Log}\left(\frac{f_1(\mathbf{x})}{f_0(\mathbf{x})} \right) \ . \tag{1}$$

Demonstrate that if f_1 and f_0 are Gaussian probability densities with mathematical expectations respectively equal to $\mathbf{x}_1$ and $\mathbf{x}_0$ and covariance operators respectively equal to C_1 and C_0, then

$$I(f_1;f_0) = \text{Log}\left(\frac{\det^{1/2}C_0}{\det^{1/2}C_1} \right) + \frac{1}{2}\ (\mathbf{x}_1-\mathbf{x}_0)^t\ C_0^{-1}\ (\mathbf{x}_1-\mathbf{x}_0)$$

$$+ \frac{1}{2}\ \text{Trace}\left(C_1\ C_0^{-1} - \mathbf{I} \right) . \tag{2}$$

Solution: By definition,

$$f_1(x) = \frac{1}{(2\pi)^{n/2}\ \det^{1/2}C_1}\ \exp\left(-\frac{1}{2}\ (\ x-\mathbf{x}_1\ )^t\ C_1^{-1}\ (\ x-\mathbf{x}_1\ ) \right) \tag{3a}$$

and

$$f_0(x) = \frac{1}{(2\pi)^{n/2}\ \det^{1/2}C_0}\ \exp\left(-\frac{1}{2}\ (\ x-\mathbf{x}_0\ )^t\ C_0^{-1}\ (\ x-\mathbf{x}_0\ ) \right) . \tag{3b}$$

Replacing (3) in (1) gives

$$I(f_1;f_0) = \text{Log}\left(\frac{\det^{1/2}C_0}{\det^{1/2}C_1} \right)$$

$$- \frac{1}{2} E_1\Big((\mathbf{x}-\mathbf{x}_1)^t \, C_1^{-1} \, (\mathbf{x}-\mathbf{x}_1) \Big) + \frac{1}{2} E_1\Big((\mathbf{x}-\mathbf{x}_0)^t \, C_0^{-1} \, (\mathbf{x}-\mathbf{x}_0) \Big) , \tag{4}$$

where $E_1(\cdot)$ denotes the mathematical expectation with respect to f_1 :

$$E_1(\ \Psi(\mathbf{x})\ ) \ = \ \int d\mathbf{x} \ \ f_1(\mathbf{x}) \ \ \Psi(\mathbf{x}) \ . \tag{5}$$

From the definition of covariance operator, and using the linearity of the mathematical expection, we obtain

$$C_1 \ = \ E_1(\ (\mathbf{x}-\mathbf{x}_1)\ (\mathbf{x}-\mathbf{x}_1)^t\ ) \ = \ E_1(\ \mathbf{x}\ \mathbf{x}^t \ - \ 2\ \mathbf{x}_1\ \mathbf{x}^t + \mathbf{x}_1\ \mathbf{x}_1^{\ t}\ ) \tag{6}$$

$$= \ E_1(\ \mathbf{x}\ \mathbf{x}^t\ ) - 2\ \mathbf{x}_1\ E_1(\ \mathbf{x}^t\ ) + \mathbf{x}_1\ \mathbf{x}_1^{\ t} \ = \ E_1(\ \mathbf{x}\ \mathbf{x}^t) - \mathbf{x}_1\ \mathbf{x}_1^{\ t} ,$$

whence, using a tensor notation we deduce

$$E_1(\ x^\alpha\ x^\beta\ ) \ = \ C_1^{\alpha\beta} \ + \ x_1^{\ \alpha}\ x_1^{\ \beta} \ . \tag{7}$$

We have

$$E_1\Big(\ (\mathbf{x}-\mathbf{x}_1)^t \ C_1^{-1} \ (\mathbf{x}-\mathbf{x}_1)\ \Big) \ = \ \ldots \ = \ E_1\Big(\ \mathbf{x}^t \ C_1^{-1}\ \mathbf{x}\ \Big) \ - \ \mathbf{x}_1^{\ t}\ C_1^{-1}\ \mathbf{x}_1$$

$$= \ (\mathbf{C}_1^{-1})^{\alpha\beta}\ E_1(\ x^\alpha\ x^\beta\ ) \ - \ (\mathbf{C}_1^{-1})^{\alpha\beta}\ x_1^{\ \alpha}\ x_1^{\ \beta} , \tag{8}$$

whence, using (7), we deduce

$$E_1\Big(\ (\mathbf{x}-\mathbf{x}_1)^t \ C_1^{-1} \ (\mathbf{x}-\mathbf{x}_1)\ \Big) \ = \ (C_1^{-1})^{\alpha\beta}\ C_1^{\alpha\beta} \ = \ \text{Trace } \mathbf{I} \ . \tag{9}$$

We also have

$$E_1\Big(\ (\mathbf{x}-\mathbf{x}_0)^t \ C_0^{-1} \ (\mathbf{x}-\mathbf{x}_0)\ \Big)$$

$$= \ \ldots \ = \ E_1\Big(\ \mathbf{x}^t \ C_0^{-1}\ \mathbf{x}\ \Big) \ - \ 2\ \mathbf{x}_0^{\ t}\ C_0^{-1}\ \mathbf{x}_1 \ + \ \mathbf{x}_0^{\ t}\ C_0^{-1}\ \mathbf{x}_0$$

$$= \ (\mathbf{C}_0^{-1})^{\alpha\beta}\ E_1(\ x^\alpha\ x^\beta\ ) \ - \ 2\ (\mathbf{C}_0^{-1})^{\alpha\beta}\ x_0^{\ \alpha}\ x_1^{\ \beta}$$

$$+ \ (\mathbf{C}_0^{-1})^{\alpha\beta}\ x_0^{\ \alpha}\ x_0^{\ \beta} , \tag{10}$$

whence, using (7), we deduce

$$E_1\Big(\ (\mathbf{x}-\mathbf{x}_0)^t \ C_0^{-1} \ (\mathbf{x}-\mathbf{x}_0)\ \Big)$$

$$= (C_0^{-1})^{\alpha\beta} \, C_1^{\alpha\beta} + (C_0^{-1})^{\alpha\beta} \, (x_1^{\alpha}-x_0^{\alpha}) \, (x_1^{\beta}-x_0^{\beta})$$

$$= \text{Trace}\left(C_0^{-1} \, C_1 \right) + (\mathbf{x}_1-\mathbf{x}_0)^t \, C_0^{-1} \, (\mathbf{x}_1-\mathbf{x}_0) \, . \qquad (11)$$

Inserting (9) and (11) in (4), result (2) follows. Notice that the factor $\det^{1/2}\mathbf{C}$ represents the (hyper) volume of the hyper-ellipsoid representing the covariance operator $\mathbf{C}$.

Problem 1.17: Let X be a parameter space, and P_1 , P_2 ,... probability distributions representing different states of information on X. In section 1.2.6, the conjunction (P_1 *and* P_2) has been defined by

$$(P_1 \textit{ and } P_2) = (P_2 \textit{ and } P_1) \qquad \text{for any } P_1 \text{ and } P_2 \qquad (1)$$

$$P_1(A) = 0 \Rightarrow (P_1 \textit{ and } P_2)(A) = 0 \qquad \text{for any } P_1 , P_2 , \text{ and any } A \subset X \qquad (2)$$

$$(P \textit{ and } M) = P \qquad \text{for any } P , \qquad (3)$$

where M represents the state of null information. If the probability densities representing P_1 , P_2 , and M , are respectively $f_1(\mathbf{x})$, $f_2(\mathbf{x})$, and $\mu(\mathbf{x})$, show that the probability density representing (P_1 *and* P_2) is $\dfrac{f_1(\mathbf{x}) \, f_2(\mathbf{x})}{\mu(\mathbf{x})}$

Solution: In mathematical terminology, the condition (2) means that the measure (P_1 *and* P_2) is "absolutely continuous" with respect to the measure P_1. The Radon-Nikodym theorem (e.g., Taylor, 1966) states that there then exists a unique function $\phi_2(\mathbf{x})$ such that, for any $A \subset X$,

$$(P_1 \textit{ and } P_2)(A) = \int_A d\mathbf{x} \;\; \phi_2(\mathbf{x}) \;\; f_1(\mathbf{x}) \, . \qquad (4)$$

Using conditions (1) and (2) and the Radon-Nikodym theorem again, we see that there also exists a unique function $\phi_1(\mathbf{x})$ such that, for any $A \subset X$,

$$(P_1 \textit{ and } P_2)(A) = \int_A d\mathbf{x} \;\; \phi_1(\mathbf{x}) \;\; f_2(\mathbf{x}) \, . \qquad (5)$$

At any point $\mathbf{x}$ where the product $f_1(\mathbf{x}) \, f_2(\mathbf{x})$ is non-vanishing, I define

$$\omega_1(\mathbf{x}) = \frac{f_1(\mathbf{x})}{\phi_1(\mathbf{x})} \qquad (6a)$$

$$\omega_2(\mathbf{x}) = \frac{f_2(\mathbf{x})}{\phi_2(\mathbf{x})} \ . \tag{6b}$$

For any $A \subset X$ we then have

$$(P_1 \ \mathit{and} \ P_2)(A) = \int_A d\mathbf{x} \ \frac{f_1(\mathbf{x}) \ f_2(\mathbf{x})}{\omega_2(\mathbf{x})} = \int_A d\mathbf{x} \ \frac{f_1(\mathbf{x}) \ f_2(\mathbf{x})}{\omega_1(\mathbf{x})} \ . \tag{7}$$

This gives

$$\omega_1(\mathbf{x}) = \omega_2(\mathbf{x}) = \omega(\mathbf{x}) \tag{8}$$

and

$$(P_1 \ \mathit{and} \ P_2)(A) = \int_A d\mathbf{x} \ \frac{f_1(\mathbf{x}) \ f_2(\mathbf{x})}{\omega(\mathbf{x})} \ . \tag{9}$$

Condition (3) then gives

$$\int_A d\mathbf{x} \ \frac{f(\mathbf{x}) \ \mu(\mathbf{x})}{\omega(\mathbf{x})} = \int_A d\mathbf{x} \ f(\mathbf{x}) \ , \tag{10}$$

i.e.,

$$\omega(\mathbf{x}) = \mu(\mathbf{x}) \ . \tag{11}$$

We finally obtain

$$(P_1 \ \mathit{and} \ P_2)(A) = \int_A d\mathbf{x} \ \frac{f_1(\mathbf{x}) \ f_2(\mathbf{x})}{\mu(\mathbf{x})} \ . \tag{12}$$

Equation (11) has been obtained only for the points $\mathbf{x}$ where the product $f_1(\mathbf{x}) \ f_2(\mathbf{x})$ is not vanishing. But as elsewhere the probability density vanishes, the result (12) is valid for the whole space X.

Problem 1.18: Assume the very particular case where the *exact* relationship $\mathbf{d} = g(\mathbf{m})$ between model parameters and observable parameters is a bijection (i.e., we can also write $\mathbf{m} = \mathbf{g}^{-1}(\mathbf{d})$). In that case, find the relationship between the a posteriori p.d.f. in the model space, $\sigma_M(\mathbf{m})$, and the a posteriori p.d.f. in the data space $\sigma_D(\mathbf{d})$.

Solution: We start from the general solution

$$\sigma(\mathbf{d},\mathbf{m}) = \frac{\rho(\mathbf{d},\mathbf{m})\ \Theta(\mathbf{d},\mathbf{m})}{\mu(\mathbf{d},\mathbf{m})} \ . \tag{1}$$

The assumption of an exact theory is written

$$\Theta(\mathbf{d},\mathbf{m}) = \Theta(\mathbf{d}|\mathbf{m})\ \mu_M(\mathbf{m}) = \delta(\mathbf{d} - \mathbf{g}(\mathbf{m}))\ \mu_M(\mathbf{m}) \ , \tag{2}$$

where $\mu_M(\mathbf{m})$ is the marginal reference p.d.f. for $\mathbf{m}$.

The a posteriori p.d.f. in the model space is

$$\sigma_M(\mathbf{m}) = \int d\mathbf{d}\ \sigma(\mathbf{d},\mathbf{m}) = \frac{\rho(\mathbf{g}(\mathbf{m}),\mathbf{m})}{\mu(\mathbf{g}(\mathbf{m}),\mathbf{m})}\ \mu_M(\mathbf{m}) \ , \tag{3}$$

while the a posteriori p.d.f. in the data space is

$$\sigma_D(\mathbf{d}) = \int d\mathbf{m}\ \sigma(\mathbf{d},\mathbf{m}) = \int d\mathbf{m}\ \frac{\rho(\mathbf{d},\mathbf{m})}{\mu(\mathbf{d},\mathbf{m})}\ \delta(\mathbf{d}-\mathbf{g}(\mathbf{m}))\ \mu_M(\mathbf{m}) \ . \tag{4}$$

Using the bijection, the last sum can be transformed on a sum over a variable $\mathbf{d}' = \mathbf{g}(\mathbf{m})$:

$$\sigma_D(\mathbf{d}) = \int d\mathbf{d}'\ \frac{1}{\left|\frac{\partial \mathbf{g}}{\partial \mathbf{m}}\right|}\ \frac{\rho(\mathbf{d},\mathbf{g}^{-1}(\mathbf{d}'))}{\mu(\mathbf{d},\mathbf{g}^{-1}(\mathbf{d}'))}\ \delta(\mathbf{d}-\mathbf{d}')\ \mu_M(\mathbf{g}^{-1}(\mathbf{d}'))$$

$$= \frac{1}{\left|\frac{\partial \mathbf{g}}{\partial \mathbf{m}}\right|}\ \frac{\rho(\mathbf{d},\mathbf{g}^{-1}(\mathbf{d}))}{\mu(\mathbf{d},\mathbf{g}^{-1}(\mathbf{d}))}\ \mu_M(\mathbf{g}^{-1}(\mathbf{d})) \ . \tag{5}$$

In particular, we have

$$\sigma_D(\mathbf{g}(\mathbf{m})) = \frac{1}{\left|\frac{\partial \mathbf{g}}{\partial \mathbf{m}}\right|}\ \frac{\rho(\mathbf{g}(\mathbf{m}),\mathbf{m})}{\mu(\mathbf{g}(\mathbf{m}),\mathbf{m})}\ \mu_M(\mathbf{m}) \ , \tag{6}$$

and, by comparison with (eq1) we deduce

$$\sigma_M(\mathbf{m}) = \sigma_D(\mathbf{g}(\mathbf{m}))\ \left|\frac{\partial \mathbf{g}}{\partial \mathbf{m}}\right| \ , \tag{7}$$

which is the usual formula relating information in variables related through an exact bijection.

It should be noticed that $\rho_D(\mathbf{g}(\mathbf{m}))$ and $\sigma_M(\mathbf{m}))$ are **not** related by such an equation, as can be expected using more naïve approaches to inverse problem theory.

Problem 1.19: Letting $\mathbf{G}$ be an arbitrary linear operator from a vector space $\boldsymbol{M}$ into a vector space $\boldsymbol{D}$, and $\mathbf{C_M}$ and $\mathbf{C_D}$ two covariance operators acting respectively on $\boldsymbol{M}$ and $\boldsymbol{D}$ (i.e., two linear, symmetric, positive definite operators), demonstrate the following identities:

$$\boxed{(\mathbf{G}^t\ \mathbf{C_D}^{-1}\ \mathbf{G} + \mathbf{C_M}^{-1})^{-1}\ \mathbf{G}^t\ \mathbf{C_D}^{-1} = \mathbf{C_M}\ \mathbf{G}^t\ (\mathbf{C_D} + \mathbf{G}\ \mathbf{C_M}\ \mathbf{G}^t)^{-1}\ ,} \tag{1}$$

$$\boxed{(\mathbf{G}^t\ \mathbf{C_D}^{-1}\ \mathbf{G} + \mathbf{C_M}^{-1})^{-1} = \mathbf{C_M} - \mathbf{C_M}\ \mathbf{G}^t\ (\mathbf{C_D} + \mathbf{G}\ \mathbf{C_M}\ \mathbf{G}^t)^{-1}\ \mathbf{G}\ \mathbf{C_M}\ .} \tag{2}$$

Solution: The first equation follows from the following obvious identities

$$\begin{aligned}\mathbf{G}^t + \mathbf{G}^t\ \mathbf{C_D}^{-1}\ \mathbf{G}\ \mathbf{C_M}\ \mathbf{G}^t &= \mathbf{G}^t\ \mathbf{C_D}^{-1}\ (\mathbf{C_D} + \mathbf{G}\ \mathbf{C_M}\ \mathbf{G}^t) \\ &= (\mathbf{G}^t\ \mathbf{C_D}^{-1}\ \mathbf{G} + \mathbf{C_M}^{-1})\ \mathbf{C_M}\ \mathbf{G}^t \end{aligned} \tag{3}$$

since $\mathbf{G}^t\ \mathbf{C_D}^{-1}\ \mathbf{G} + \mathbf{C_M}^{-1}$ and $\mathbf{C_D} + \mathbf{G}\ \mathbf{C_M}\ \mathbf{G}^t$ are positive definite and, thus, regular matrices. Furthermore,

$$\begin{aligned}&\mathbf{C_M} - \mathbf{C_M}\ \mathbf{G}^t\ (\mathbf{C_D} + \mathbf{G}\ \mathbf{C_M}\ \mathbf{G}^t)^{-1}\ \mathbf{G}\ \mathbf{C_M} \\ &= \mathbf{C_M} - (\mathbf{G}^t\ \mathbf{C_D}^{-1}\ \mathbf{G} + \mathbf{C_M}^{-1})^{-1}\ \mathbf{G}^t\ \mathbf{C_D}^{-1}\ \mathbf{G}\ \mathbf{C_M} \\ &= (\mathbf{G}^t\ \mathbf{C_D}^{-1}\ \mathbf{G} + \mathbf{C_M}^{-1})^{-1}\ ((\mathbf{G}^t\ \mathbf{C_D}^{-1}\ \mathbf{G} + \mathbf{C_M}^{-1})\ \mathbf{C_M} - \mathbf{G}^t\ \mathbf{C_D}^{-1}\ \mathbf{G}\ \mathbf{C_M}) \\ &= (\mathbf{G}^t\ \mathbf{C_D}^{-1}\ \mathbf{G} + \mathbf{C_M}^{-1})^{-1}\ . \end{aligned} \tag{4}$$

Problem 1.20 (the convolution of two Gaussians is Gaussian): Evaluate the sum

$$I = \int d\mathbf{d}\ \exp\left[-\frac{1}{2}\left((\mathbf{d}-\mathbf{d}_0)^t\ C_d^{-1}\ (\mathbf{d}-\mathbf{d}_0) + (\mathbf{d}-\mathbf{g}(\mathbf{m}))^t\ C_T^{-1}\ (\mathbf{d}-\mathbf{g}(\mathbf{m}))\right)\right]\ . \tag{1}$$

Solution: The separation of the quadratic terms from the linear terms leads to:

$$I = \int dd \exp\left[-\frac{1}{2}(\mathbf{d}^t \mathbf{A}\, \mathbf{d} - 2\, \mathbf{b}^t\, \mathbf{d} + c\,)\right] , \tag{2}$$

where

$$\mathbf{A} = \mathbf{C_d}^{-1} + \mathbf{C_T}^{-1} \tag{3a}$$

$$\mathbf{b}^t = \mathbf{d_0}^t\, \mathbf{C_d}^{-1} + \mathbf{g(m)}^t\, \mathbf{C_T}^{-1} \tag{3b}$$

$$c = \mathbf{d_0}^t\, \mathbf{C_T}^{-1}\, \mathbf{d_0} + \mathbf{g(m)}^t\, \mathbf{C_T}^{-1}\, \mathbf{g(m)}\ . \tag{3c}$$

Since $\mathbf{A}$ is positive definite, it follows:

$$I = \int dd \exp\left[-\frac{1}{2}((\mathbf{d}-\mathbf{A}^{-1}\, \mathbf{b})^t\, \mathbf{A}\, (\mathbf{d}-\mathbf{A}^{-1}\, \mathbf{b}) + (c-\mathbf{b}^t\, \mathbf{A}^{-1}\, \mathbf{b}))\right]$$

$$= \exp\left[-\frac{1}{2}(c-\mathbf{b}^t\, \mathbf{A}^{-1}\, \mathbf{b})\right] \int dd \exp\left[-\frac{1}{2}(\mathbf{d}-\mathbf{A}^{-1}\, \mathbf{b})^t\, \mathbf{A}\, (\mathbf{d}-\mathbf{A}^{-1}\, \mathbf{b})\right]$$

$$= (2\pi)^{n/2}\, (\det \mathbf{A})^{-1/2} \exp\left[-\frac{1}{2}\, (\, c - \mathbf{b}^t\, \mathbf{A}^{-1}\, \mathbf{b}\,)\right] . \tag{4}$$

By substitution we obtain

$$\begin{aligned} c - \mathbf{b}^t\, \mathbf{A}^{-1}\, \mathbf{b} = {} & \mathbf{d_0}^t \left[\mathbf{C_d}^{-1} - \mathbf{C_d}^{-1} \left(\mathbf{C_d}^{-1} + \mathbf{C_T}^{-1} \right) \mathbf{C_d}^{-1} \right] \mathbf{d_0} \\ & + \mathbf{g(m)}^t \left[\mathbf{C_T}^{-1} - \mathbf{C_T}^{-1} \left(\mathbf{C_d}^{-1} + \mathbf{C_T}^{-1} \right) \mathbf{C_T}^{-1} \right] \mathbf{g(m)} \\ & - 2\, \mathbf{g(m)}^t\, \mathbf{C_T}^{-1} \left(\mathbf{C_d}^{-1} + \mathbf{C_T}^{-1} \right) \mathbf{C_d}^{-1}\, \mathbf{d_0}\ . \end{aligned} \tag{5}$$

Thus, by using the two identities demonstrated in the previous problem, we get

$$\begin{aligned} c - \mathbf{b}^t\, \mathbf{A}^{-1}\, \mathbf{b} = {} & \mathbf{d_0}\, (\, \mathbf{C_d} + \mathbf{C_T}\,)^{-1}\, \mathbf{d_0} + \mathbf{g(m)}^t\, (\, \mathbf{C_d} + \mathbf{C_T}\,)^{-1}\, \mathbf{g(m)} \\ & - 2\, \mathbf{g(m)}^t\, (\, \mathbf{C_d} + \mathbf{C_T}\,)^{-1}\, \mathbf{d_0} \end{aligned}$$

$$= (\mathbf{d}_0 - \mathbf{g}(\mathbf{m}))^t (\mathbf{C}_d + \mathbf{C}_T)^{-1} (\mathbf{d}_0 - \mathbf{g}(\mathbf{m})) . \qquad (6)$$

Finally, we obtain:

$$I = (2\pi)^{n/2} \det(\mathbf{C}_D{}^{-1}+\mathbf{C}_T{}^{-1})^{-1/2} \exp\left[-\frac{1}{2}(\mathbf{d}_0-\mathbf{g}(\mathbf{m}))^t (\mathbf{C}_D+\mathbf{C}_T)^{-1} (\mathbf{d}_0-\mathbf{g}(\mathbf{m}))\right] . \qquad (7)$$

Problem 1.21: The Generalized Gaussian of order p is defined by

$$f_p(x) = \frac{p^{1-1/p}}{2\,\sigma\,\Gamma(1/p)} \exp\left[- \frac{1}{p} \frac{| x - x_0 |^p}{\sigma^p} \right] . \qquad (1)$$

Demonstrate that it is normalized. Give a direct computation of its ℓ_p norm estimator of dispersion.

Solution: We have

$$I_p = \int_{-\infty}^{+\infty} dx\ f_p(x) = \int_{-\infty}^{+\infty} dx\ f_p(x+x_0) = 2 \int_0^{\infty} dx\ f_p(x+x_0) =$$

$$= \frac{p^{1-1/p}}{\sigma\,\Gamma(1/p)} \int_0^{\infty} dx\ \exp\left(- \frac{1}{p} \frac{x^p}{\sigma^p} \right) . \qquad (2)$$

Introducing the variable

$$u = \frac{x^p}{p\ x^{p-1}} , \qquad (3)$$

we successively have

$$du = \frac{x^{p-1}\ dx}{\sigma^p} , \qquad (4)$$

$$dx = \frac{\sigma^p}{x^{p-1}} du = \frac{\sigma\ u^{1-1/p}}{p^{1-1/p}} du , \qquad cc(5)$$

and

$$I_p = \frac{1}{\Gamma(1/p)} \int_0^{\infty} du\ u^{1-1/p}\ e^{-u} . \qquad (6)$$

Using the definition of the Gamma function

$$\Gamma(t) = \int_0^{\infty} du \; u^{1-t} \; e^{-u} \, , \tag{7}$$

we directly obtain

$$I_p = \int_{-\infty}^{+\infty} dx \; f_p(x) = 1 \, . \tag{8}$$

By definition, the estimator of dispersion in norm ℓ_p is (see Box 1.2)

$$\sigma_p = \left[\int_{-\infty}^{+\infty} dx \; | \, x - x_0 \, |^p \; f_p(x) \right]^{1/p} . \tag{9}$$

We successively have

$$\sigma_p = \left[\int_{-\infty}^{+\infty} dx \; | \, x \, |^p \; f_p(x+x_0) \right]^{1/p} = \left[2 \int_0^{\infty} dx \; | \, x \, |^p \; f_p(x+x_0) \right]^{1/p}$$

$$= \left[\frac{p^{1-1/p}}{\sigma \, \Gamma(1/p)} \int_0^{\infty} dx \; x^p \; \exp\left(- \frac{1}{p} \frac{x^p}{\sigma^p} \right) \right]^{1/p} , \tag{10}$$

and, again using the change of variables previously defined

$$\sigma_p = \left[\frac{p \, \sigma^p}{\Gamma(1/p)} \int_0^{\infty} du \; u^{1-(1+1/p)} \; e^{-u} \right]^{1/p} = \left(\frac{p \, \sigma^p \, \Gamma(1+1/p)}{\Gamma(1/p)} \right)^{1/p} . \tag{11}$$

Finally, using the property $\Gamma(1+t) = t \, \Gamma(t)$, we obtain

$$\sigma_p = \sigma \, . \tag{12}$$

CHAPTER 2

THE TRIAL AND ERROR METHOD

Would you tell me, please, which way I ought to go from here?,
said Alice,
That depends a good deal on where you want to get to,
said the Cat,
I don't much care where,
said Alice,
Then it doesn't matter which way you go,
said the Cat.

Lewis Carroll, 1865.

Let $\mathbf{m}$ represent an arbitrary model, $\mathbf{d}_{obs}$ the observed data values, and

$$\mathbf{d}_{cal} = \mathbf{g}(\mathbf{m})$$

the predicted data values for the model $\mathbf{m}$. *Trial and* (correction of) *error* is a method in which a user starts from some initial model $\mathbf{m}_0$, computes $\mathbf{d}_{cal} = \mathbf{g}(\mathbf{m}_0)$, compares $\mathbf{d}_{cal}$ with $\mathbf{d}_{obs}$, and appeals to his physical intuition to guess a new model $\mathbf{m}_1$ for which $\mathbf{g}(\mathbf{m}_1)$ fits the observed data values better than $\mathbf{g}(\mathbf{m}_0)$. The procedure is iterated until successive updatings of the model do not significantly improve the fit between observed and computed data values.

Usually, the method is worked interactively on a computer terminal which allows a convenient display of the data. The misfit between observed and computed data values may be measured through some "cost function" (such as, for instance, equation (1.104)), or may simply be qualitatively estimated. Only model updatings are considered which are compatible with the user's a priori information on model parameters.

Example 1 (from Hirn et al., 1984): In Figure 2.1 a seismic reflection experiment to probe the Earth's crust is described. An explosion at the Earth's surface produces elastic waves which propagate into the Earth and which are reflected at the main discontinuities of its elastic properties. In this figure, the approximation of a perfectly layered Earth is used. An array of seismometers is deployed at distances adapted to the recording of some par-

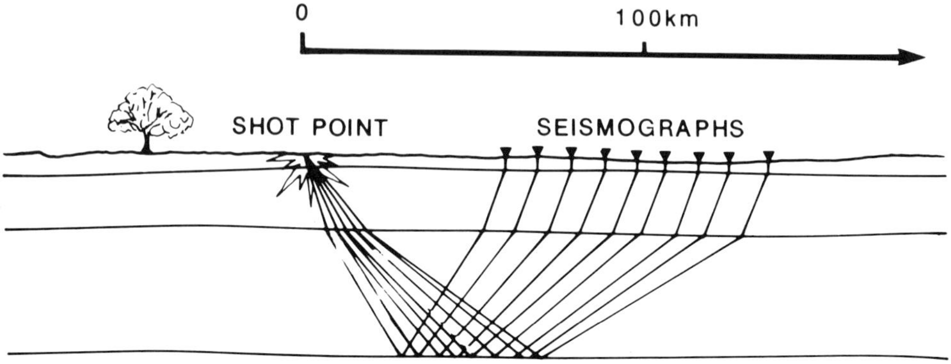

Figure 2.1: Typical seismic reflexion experiment for probing the Earth's crust. The shot is usually a dynamite charge in a drill hole. Seismographs record the displacement of the Earth's surface.

ticular reflected waves. The top of Figure 2.2 shows the seismograms obtained in such an experiment performed in the Tibet area. Each seismogram represents the vertical displacement of the Earth's surface recorded at the seismometer location. As the authors are only interested in a reflector about 80 km deep, only the displacements recorded in a window between 10 s and 18 s after the shot time are shown. These seismograms correspond to the observed data $\mathbf{d}_{obs}$. If the Earth's crust is assumed layered, it is conveniently parameterized by three functions of depth: the density $\rho(z)$, the velocity of longitudinal waves $\alpha(z)$, and the velocity of transverse waves $\beta(z)$. As the data do not contain enough information to estimate these three parameters, the two less significant parameters, $\rho(z)$ and $\beta(z)$, are assumed given. There remains the velocity of longitudinal waves, $\alpha(z)$. Previous experiments in the area suggest that $\alpha(z)$ should belong to the class of models shown in the left of the figure. Here the authors are interested in the nature of the reflector at $\simeq$80 km depth. A lot of different models are tried, and a computer code is used which, for each input model $\alpha(z)$, exhibits the predicted data $\mathbf{d}_{cal}$. This code is used as a black box (the user does not need to understand the underlying mathematics). Three of the trials are shown here. As the simplifying hypotheses used are very strong, the predicted seismograms look much more simplistic than the observed ones. It is then only possible to define a subjective criterion of goodness of fit, which can only be used without gross mistakes by an expert seismologist. The seismograms corresponding to the model at bottom left predict an arrival of reflected energy at the right time (as do the other two models shown) with a maximum of amplitude at about 240 km distant from the shot point (which is better than the prediction of the other two models). This suggests that the reflector at 80 km depth corresponds to a gradual transition, over a ≃20 km thickness, of the elastic parameters of the Tibetan crust. ■

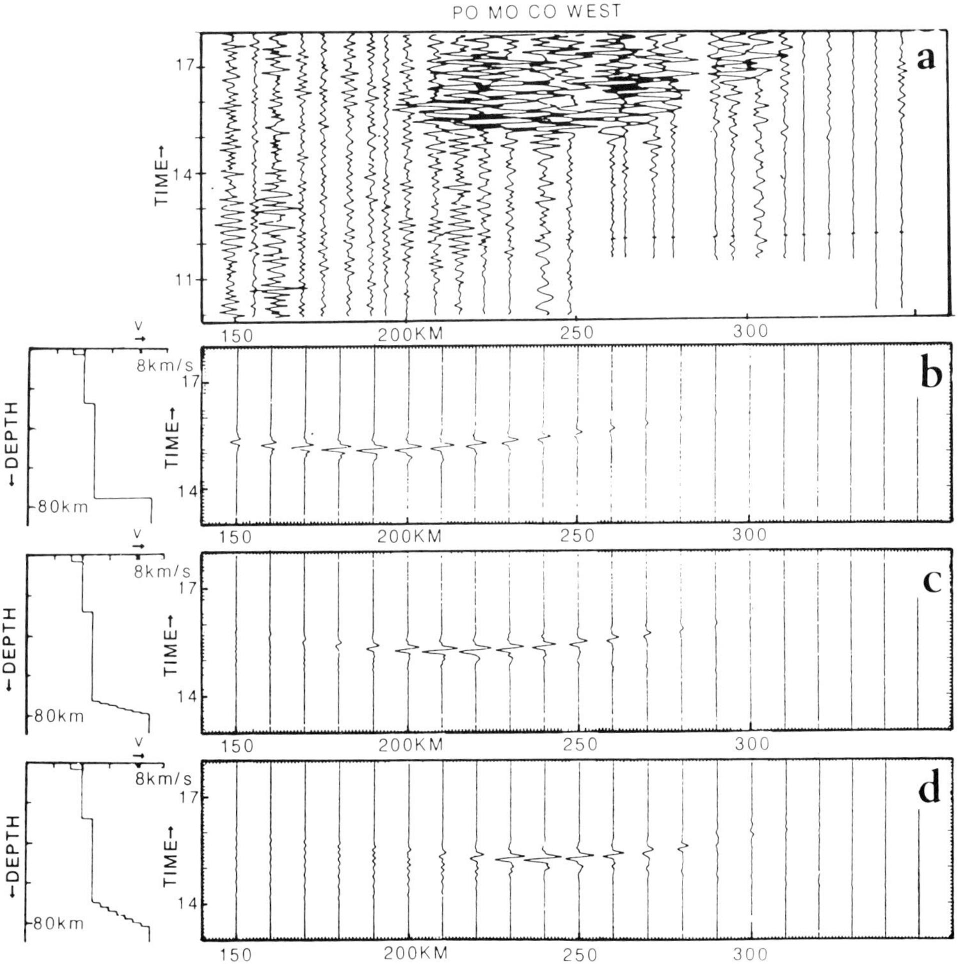

Figure 2.2: a) displacement of the earth's surface recorded during a seismic reflection experiment for probing the Tibetan crust. b to d) different crust model trials. The model (d) is preferred on a subjective basis (see text for discussion). From Hirn et al., 1984.

The major advantages of the trial and error method are:

i) no mathematics are needed other than those used for solving the forward problem;

ii) it allows the effective solution of problems in which the complexity of observations prevents any reasonable definition of a cost function to measure the data misfit, so that the comparison is better made qualitatively by the user (as in the example in Figure 2.2);

iii) it sometimes gives the user a good intuitive feeling as to which kind of information is carried by the observations.

The major drawbacks are:

i) it is usually difficult to assess the unicity of the solution thus obtained (there is no warranty that a different model will not fit the data adequately);

ii) it can be tedious for problems with more than a few parameters (and not usable at all for large-sized problems).

Typically, this method is only to be recommended to persons who have a good intuition of the physics involved in the problem, a good computer code solving the forward problem realistically, the possibility of a synthetic display of observations on a computer terminal, and a problem with few parameters, but who have "no time" for a more detailed, quantitative, analysis of their data set.

Problems for chapter 2:

Problem 2.1: Solve problem 1.1 (estimation of the epicentral coordinates of an earthquake) using trial and error.

Problem 2.2: Solve the tomography problem 1.2 using trial and error.

CHAPTER 3

MONTE CARLO METHODS

Dans l'ordre biologique établi par la sélection,
le hasard vient de temps en temps semer le désordre.
Il secoue périodiquement ces barrières trop contraignantes
et permet à l'évolution de changer de cap.
Le hasard est anti-conservateur.

Jacques Ruffié, 1982

We have seen in Chapter 1 that the most general method for solving nonlinear inverse problems needs a complete exploration of the model space. For problems other than academic, the method is in general too computer intensive to be useful. This is why usual methods limit their scope to obtaining some "best" model, i.e., a model maximizing the probability density $\sigma_M(\mathbf{m})$ or minimizing some misfit function $S(\mathbf{m})$.

If the forward problem is not excessively nonlinear, the functions $\sigma_M(\mathbf{m})$ and/or $S(\mathbf{m})$ are well behaved and usually have a single extremum, which can be obtained, for instance, using gradient methods, i.e., methods that use the local properties of the function at a current point $\mathbf{m}_n$ to decide on a direction of search for the updated model $\mathbf{m}_{n+1}$. For highly nonlinear problems, there is a considerable risk that gradient methods will converge to secondary solutions. It happens that, for model spaces with more than a few parameters, it is dramatically more economical to select points in the model space randomly, than to define a regular grid dense enough to ensure that at least one point will be in the optimal area.

Any method which uses a random (or pseudo-random) generator at any stage is named *Monte Carlo*, in homage to the famous casino. For instance, we can use a Monte Carlo method for computing the number π : on a regular floor, made of strips of equal width w , we throw needles of length $l = w/2$. The probability that a needle will intersect a groove in the floor equals $1/\pi$ (Georges Louis Leclerc, Comte de Buffon (1707-1788)). Deltheil (1926) quotes that a series of 50 observations, each one with 100 trials, made by Wolff in Zurich in 1850 leads to a value for π of 3.1596 ± 0.0524 , which is not so bad. In numerical methods, the throwing of needles is replaced by a pseudo-random generation of numbers by a computer code.

The interest of Monte Carlo methods of inversion is that they can solve problems of relatively large size in a fully nonlinear form (i.e., without any linearization).

Interesting references to the use of Monte Carlo methods of inversion in geophysics are Keilis-Borok and Yanovskaya (1967), Press (1968, 1971), Anderssen and Seneta (1971, 1972), and Rothman (1985, 1986).

Section 3.1 describes the Press (1968, 1971) method of searching for the domain of admissible models. In section 3.2 I suggest a method for nonlinear computation of variances and covariances. Finally, section 3.3 introduces the simulated annealing method (Kirkpatrick et al., 1983; Rothman, 1985, 1986), well suited for obtaining the maximum likelihood point. If the general ideas are understood, the reader will easily derive the particular method adapted to his particular problem.

Throughout this chapter, $\boldsymbol{M}$ represents the model space, and $m = \{ m^\alpha \}$ $(\alpha \in I_M)$ a particular point in $\boldsymbol{M}$. A set of N points in $\boldsymbol{M}$ will be denoted by $m_1, m_2, ..., m_N$. $\boldsymbol{D}$ represents a discrete data space, its elements are denoted by $\mathbf{d}$, and the solution of the forward problem is denoted by $\mathbf{d}_{cal} = \mathbf{g}(\mathbf{m})$.

3.1: Search for the domain of admissible models.

Assume that we have some a priori information on the parameter space, which can be set in the simple form

$$m^\alpha_{inf} \leq m^\alpha \leq m^\alpha_{sup} \qquad (\alpha \in \mathbf{I}_M) \tag{3.1}$$

As defined by Press (1968, 1971), the Monte Carlo method of inversion consists in using a pseudo-random number generator to generate random models inside the region defined by (3.1), in computing for each one of these models, say $\mathbf{m}$, the corresponding predicted data, $\mathbf{d}_{cal} = \mathbf{g}(\mathbf{m})$, and in using some quantitative criterion of comparison between $\mathbf{d}_{cal}$ and $\mathbf{d}_{obs}$ to decide if $\mathbf{m}$ is *acceptable* or not. The computations are stopped when the number of accepted models is sufficient to suggest that the model space has been conveniently explored.

Example 1: In his 1968 paper, Press studied the density of the Earth's mantle, as well as the velocity of seismic waves, as a function of radius r, using as data measured eigenperiods of the Earth's vibration, some measured travel times of seismic waves, the total Earth's mass, and the Earth's moment of inertia.

The parameters to be evaluated were the density $\rho(r)$, the velocity of longitudinal waves $\alpha(r)$, and the velocity of transverse waves $\beta(r)$. These functions were considered at 23 different values of r (the remaining values were defined by interpolation). This makes a total of 23x3=69 parameters:

$$\mathbf{m} = (m^1,...,m^{69}) = (\rho(r^1),...,\rho(r^{23}),\alpha(r^1),...,\alpha(r^{23}),\beta(r^1),...,\beta(r^{23})).$$

Approximately *five million* models were randomly generated and tested. Of these, six were acceptable (they gave predicted data values close enough to the observed values). Figure 3.1 shows three of the Earth models thus obtained. In as far as we can accept that this figure gives a quite good idea of the domain of admissible Earth models, the inverse problem has essentially been solved.

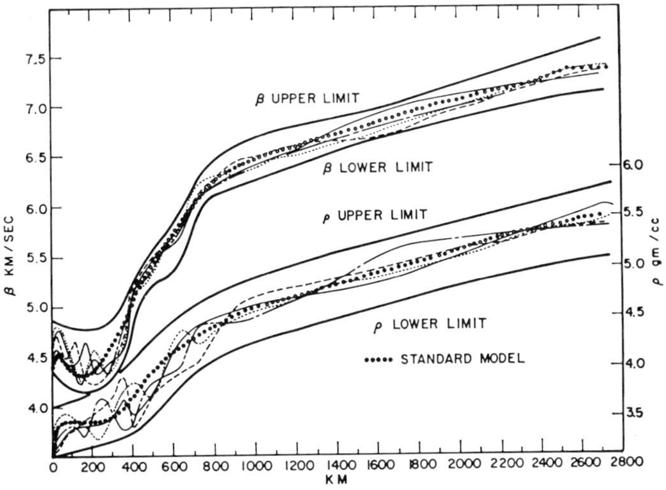

Figure 3.1: The four acceptable Earth models of velocity of transverse waves (β) and of density (ρ) found by Press (1968) out of five million random models tested. Prior permissible regions for β and ρ are shown by heavy curves. The standard model is also shown (dashes).

Figure 3.2 is from Press (1971) and shows many randomly generated models of shear velocity of seismic waves in the Earth's mantle to test for bias in the selection procedure. Figure 3.3 shows the *acceptable* models for the Earth's density. Notice in particular that all acceptable models present a relatively high density at about 100 km depth. ■

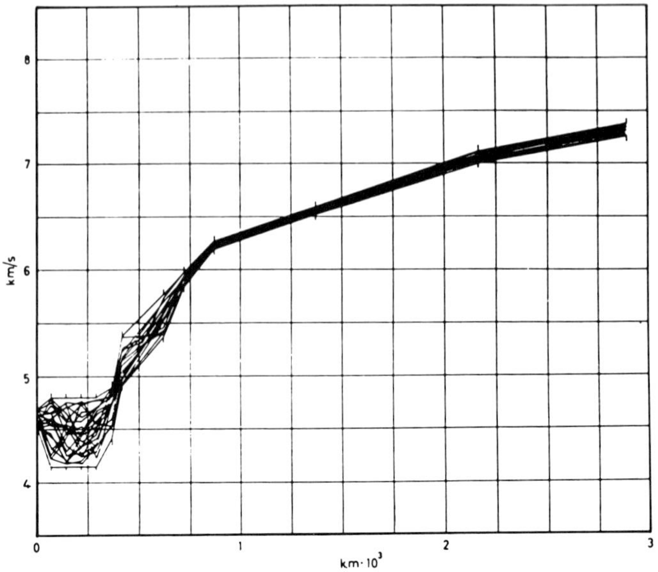

Figure 3.2: Random models of velocity of transverse waves, to test for bias in the selection procedure (Press, 1971).

These examples suggest some of the difficulties of this method. In particular, if the a priori error bars (3.1) are chosen too large, the number of trials may be very large; if they are chosen too small, they will in fact control the results. It is also difficult to ascertain that the number of trials has been sufficient.

3.2: Nonlinear computation of variances and covariances

We assume here that, for any model $\mathbf{m}$, we are able to compute the posterior probability density function in the model space, $\sigma_M(\mathbf{m})$, as defined in Chapter 1. This section is concerned with problems where:

i) the number of components of the vector $\mathbf{m}$, say NM , is large (in fact, $NM \geq 4$).

ii) for a given $\mathbf{m}$, the computation of $\sigma_M(\mathbf{m})$ is inexpensive (so we can compute $\sigma_M(\mathrm{m})$ for a great number of models).

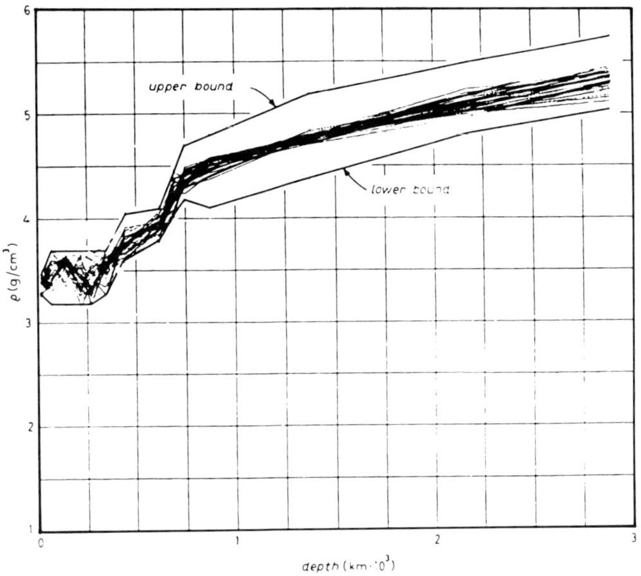

Figure 3.3: Successful density models for the mantle. Notice the relatively high density presented by all acceptable models at about 100 km depth (Press, 1971).

Let us write the posterior probability density in the model space as

$$\sigma_M(m) = \rho_M(m) \ L(\mathbf{m}) , \tag{3.2}$$

where $\rho_M(m)$ represents the prior information in the model space ($L(\mathbf{m})$ may be termed the "likelihood"). For instance, if d^i_{obs} $(i \in I_D)$ represents the observed data values and σ^i_D the estimated mean deviations, assuming double exponentially distributed observational errors gives (equation (1.151):

$$L(\mathbf{m}) = \exp\left(- \sum_{i \in \mathbf{I}_D} \frac{| g^i(\mathbf{m}) - d^i_{obs} |}{\sigma^i_D} \right) .$$

If $\mathbf{C}_D$ represents the covariance operator describing estimated errors and error correlations, assuming a Gaussian error distribution gives (equation (1.75)):

$$L(\mathbf{m}) = \exp\left[-\frac{1}{2} (\mathbf{g}(\mathbf{m})-\mathbf{d}_{obs})^t \, \mathbf{C}_D^{-1} \, (\mathbf{g}(\mathbf{m})-\mathbf{d}_{obs}) \right] .$$

Some other examples are given in Chapter 1.

Our purpose here is to compute the mathematical expectation

$$\langle \mathbf{m} \rangle = \frac{1}{\nu} \int_{\boldsymbol{M}} d\mathbf{m} \;\; \mathbf{m} \;\; \sigma_M(\mathbf{m}) \tag{3.3}$$

and the posterior covariance operator

$$\mathbf{C}_M = \frac{1}{\nu} \int_{\boldsymbol{M}} d\mathbf{m} \;\; \mathbf{m} \;\; \mathbf{m}^t \, \sigma_M(\mathbf{m}) \; - \; \langle \mathbf{m} \rangle \; \langle \mathbf{m} \rangle^t \tag{3.4}$$

where ν is the norm of $\sigma_M(\mathbf{m})$:

$$\nu = \int_{\boldsymbol{M}} d\mathbf{m} \;\; \sigma_M(\mathbf{m}) \, . \tag{3.5}$$

Using (3.2) we can rewrite

$$\nu = \int_{\boldsymbol{M}} d\mathbf{m} \;\; \rho_M(\mathbf{m}) \;\; L(\mathbf{m}) \, . \tag{3.6}$$

$$\langle \mathbf{m} \rangle = \frac{1}{\nu} \int_{\boldsymbol{M}} d\mathbf{m} \;\; \rho_M(\mathbf{m}) \;\; \mathbf{m} \;\; L(\mathbf{m}) \tag{3.7}$$

$$\mathbf{C}_M = \frac{1}{\nu} \int_{\boldsymbol{M}} d\mathbf{m} \;\; \rho_M(\mathbf{m}) \;\; \mathbf{m} \;\; \mathbf{m}^t \;\; L(\mathbf{m}) \; - \; \langle \mathbf{m} \rangle \; \langle \mathbf{m} \rangle^t \, . \tag{3.8}$$

Introducing the components of $\langle \mathbf{m} \rangle$ and $\mathbf{C}_M$, we arrive at the final equations:

$$\nu = \int_{\boldsymbol{M}} d\mathbf{m} \;\; \rho_M(\mathbf{m}) \;\; L(\mathbf{m}) \qquad \text{(1 sum)} \tag{3.9}$$

$$\langle m^\alpha \rangle = \frac{1}{\nu} \int_{\boldsymbol{M}} d\mathbf{m} \;\; \rho_M(\mathbf{m}) \;\; m^\alpha \;\; L(\mathbf{m}) \qquad \text{(NM sums)} \tag{3.10}$$

$$C_M{}^{\alpha\beta} = \frac{1}{\nu} \int_M dm \ \rho_M(\mathbf{m}) \ m^\alpha \ m^\beta \ L(\mathbf{m}) \ - \ \langle \ m^\alpha \ \rangle \ \langle \ m^\beta \ \rangle$$

$$\left(\frac{NM(NM+1)}{2} \text{ sums } \right). \tag{3.11}$$

We see thus that to compute the mean model value $\langle \ \mathbf{m} \ \rangle$ or the components of the covariance operator $\mathbf{C}_M$, we have essentially to perform integrations over the model space. If the dimension of the space is large (say more than 4) it is well known that Monte Carlo methods of numerical integration should be preferred to more elementary methods using regular grids in the space, because the number of points needed with regular grids grows too rapidly with the dimension of the space. Box 3.1 introduces the basic method of Monte Carlo numerical integration.

Comparison of equations (3.9)-(3.11) with equation (2) of Box 3.1 suggests using the function $\rho_M(\mathbf{m})$ as density of probability of generating random points in the model space, i.e. identifying $\rho_M(\mathbf{m})$ with $p(\mathbf{m})$ of equation (2). The effect of this choice will be to sample the regions of the space more densely where a priori we expect significant values of the integrands, so the convergence of the method can be reasonably good. In addition, the identification of $\rho_M(\mathbf{m})$ with $p(\mathbf{m})$ allows a nice simplification of the formulas.

Explicitly, the unbiased estimators of $\langle \ m^\alpha \ \rangle$ and $C_M{}^{\alpha\beta}$, after generation of N points, are

$$\langle \ m^\alpha \ \rangle = \frac{1}{N \ \nu} \sum_{n=1}^{N} m^\alpha \ L(\mathbf{m}_n) \tag{3.12}$$

$$C_M{}^{\alpha\beta} = \frac{1}{N \ \nu} \sum_{i=1}^{N} m^\alpha \ m^\beta \ L(\mathbf{m}_n) - \langle \ m^\alpha \ \rangle \ \langle \ m^\beta \ \rangle \ , \tag{3.13}$$

where

$$\nu = \sum_{i=0}^{N} L(\mathbf{m}_n) \ , \tag{3.14}$$

and where it is assumed that the points $\mathbf{m}_1$, $\mathbf{m}_2$,... have been generated with the probability density $\rho_M(\mathbf{m})$.

As explained in Box 3.1, the generation of random points may be stopped as soon as the criterion of relative precision is verified.

While computing $\langle \ \mathbf{m} \ \rangle$ and $\mathbf{C}_M$, it is desirable to keep in the computer memory the models which have obtained high values for $\sigma_M(\mathbf{m})$. They will approximately represent the domain of acceptable models as defined in the previous section. If these models are reasonably close to $\langle \mathbf{m} \rangle$, we can have some confidence in that the probability density function $\sigma_M(\mathbf{m})$ is not very asymmetric or multimodal.

As a concluding remark, the user of any Monte Carlo method should beware of the fact that pseudo-random number generators are never truly random. If one is not sure of the quality of the generator used, some statistical tests should be made (see, for instance, Knuth, 1981, or Press et al., 1986).

Box 3.1: The Monte Carlo method of numerical integration.

Let $\phi(\mathbf{m})$ be an arbitrary scalar function defined over a discrete, s-dimensional space $\boldsymbol{M}$ $(\mathbf{m} \in \boldsymbol{M})$. Assume that we need to evaluate the sum

$$I = \int d\mathbf{m} \;\; \phi(\mathbf{m}) = \int dm^1 \dots \int dm^s \;\; \phi(m^1,\dots,m^s) \qquad (1)$$

over a given domain $\boldsymbol{M}' \subset \boldsymbol{M}$.

If $\boldsymbol{M}'$ has finite volume, the simplest method of evaluating I numerically is by defining a regular grid of points in $\boldsymbol{M}'$, by computing $\phi(\mathbf{m})$ at each point of the grid, and by approximating the integral in equation (1) by a discrete sum. But as the number of points in a regular grid is a rapidly increasing function of the dimension of the space ($N \propto const^s$), the method becomes impractical for large-dimension spaces (say $s \geq 4$). The Monte Carlo method of numerical integration consists in replacing the regular grid of points by a pseudo-random grid generated by a computer code based on a pseudo-random number generator. Although it is not possible to give any general rule for the number of points needed for an accurate evaluation of the sum (because this number is very much dependent on the form of $\phi(\mathbf{m})$), it turns out *in practical applications* that, for "well behaved" functions $\phi(\mathbf{m})$, this number can be smaller, by some orders of magnitude, than the number of points needed in a regular grid.

Let us note $p(\mathbf{m})$ as an arbitrary probability density function over $\boldsymbol{M}$ that we choose to use for generating pseudo-random points over $\boldsymbol{M}$; ($p(\mathbf{m}) = const$ is the simplest choice, but more astute choices can improve the efficacy of the algorithm).

(...)

Defining $\psi(\mathbf{m}) = \frac{\phi(\mathbf{m})}{p(\mathbf{m})}$, the sum we wish to evaluate can be written

$$I = \int_{M'} d\mathbf{m} \;\; p(\mathbf{m}) \;\; \psi(\mathbf{m}) \; . \tag{2}$$

Let $\mathbf{m}_1,...,\mathbf{m}_N$ be a suite of N points collectively independent and randomly distributed over M' with a density of probability $p(\mathbf{m})$. We define

$$\psi_n \; = \; \psi(\mathbf{m}_n) \tag{3}$$

$$I_N \; = \; \frac{1}{N} \sum_{n=1}^{N} \psi_n \tag{4}$$

$$V_N \; = \; \frac{N}{N-1} \left[\frac{1}{N} \sum_{n=1}^{N} \psi_n{}^2 \; - \; I_N{}^2 \right] \tag{5}$$

It can easily be seen that the mathematical expectation of I_N is

$$\langle \; I_N \; \rangle \; = \; \int_{M'} d\mathbf{m} \;\; p(\mathbf{m}) \;\; \psi(\mathbf{m}) \; = \; I \; , \tag{6}$$

so that I_N *is an unbiased estimate of* I . Using the central limit theorem it can be shown (see, for instance, Bakhvalov, 1977) that, for large N, the probability of the relative error $| \, I_N - I \, | \, / \, | \, I \, |$ being bounded by

$$\frac{| \, I_N - I \, |}{| \, I \, |} \; \leq \; \frac{k \, \sqrt{V}}{| \, I \, | \; \sqrt{N}} \; , \tag{7}$$

where V is the (unknown) variance of $\psi(\mathbf{m})$, is asymptotically equal to

$$P(k) \; = \; 1 - \frac{2}{\sqrt{2\pi}} \int_k^{+\infty} dt \;\; \exp\left(- \; \frac{t^2}{2} \right) . \tag{8}$$

For large N, a useful estimate of the right-hand member of (7) is

(...)

$$\frac{k\sqrt{V}}{|I|\sqrt{N}} \simeq \frac{k\sqrt{V_N}}{|I_N|\sqrt{N}} , \tag{9}$$

where I_N and V_N are defined by equations (4) and (5).

This method of numerical integration is used as follows: first, one selects the value of the confidence level, $P(k)$, at which the bound equation (7) is required to hold (for instance, $P(k) = 0.99$). The corresponding value of k is easily deduced using equation (8) and the error-function tables ($k \simeq 3$ for $P(k) = 0.99$). A pseudo-random number generator is then used to obtain the points $\mathbf{m}_1, \mathbf{m}_2, \ldots$ distributed with the probability $p(\mathbf{m})$, and, for each new point, the right-hand member of equation (7) is estimated using equation (9). The computations are stopped when this number equals the relative accuracy desired (for instance, 10^{-3}). The typical statement that can then be made is as follows: "The value of I can be estimated by I_N, with a probability of $P(k)$ (e.g., 99%) for the relative error being smaller than ϵ (e.g., 10^{-3})".

It should be noticed that the Metropolis algorithm (see Section 3.3) is indeed a Monte Carlo integration technique, for cases in which $\phi(\mathbf{m})$ takes significant values in some very small regions of the space.

For more details, see, for instance, Hammersley and Handscomb (1964).

3.3: Simulated Annealing.

Annealing consists of heating a solid until thermal stresses are released, then in cooling it very slowly to the ambient temperature. Ideally, the substance is heated until it melts then, cooled very slowly until a perfect crystal is formed. The substance then reaches the state of lowest energy. If the cooling is not slow enough, a metastable glass can be formed.

Simulated annealing (Kirkpatrick et al., 1983) is a numerical method, using an analogy between the process of physical annealing and the mathematical problem of obtaining the global minimum of a function (assimilated to an energy) which many have local minima (metastable states). It has been introduced in the framework of inverse theory by Rothman (1985a, 1985b, 1986).

Let $\sigma_M(\mathbf{m})$ be the (not necessarily normalized) posterior probability density in the model space. We wish to obtain the maximum likelihood point $\mathbf{m}_{ML}$:

$$\sigma_M(\mathbf{m}) \qquad \text{MAXIMUM FOR} \qquad \mathbf{m} = \mathbf{m}_{ML} . \tag{3.15}$$

This problem is not trivial if the probability density is multimodal and if the dimension of the model space is large enough to prevent a systematic exploration of the space.

First we define the *energy function*

$$S(\mathbf{m}) \;=\; -\,T_0 \,\mathrm{Log}\left(\frac{\sigma_M(\mathbf{m})}{\rho_M(\mathbf{m})} \right) , \tag{3.16}$$

where T_0 is an arbitrary, but fixed, real (adimensional) positive number, termed the *ambient temperature* (for instance, $T_0 = 1$). $\rho_M(\mathbf{m})$ is the prior probability density in the model space. This gives

$$\sigma_M(\mathbf{m}) \;=\; \rho_M(\mathbf{m}) \;\; e^{-\frac{S(\mathbf{m})}{T_0}} \; . \tag{3.17}$$

Defining, for arbitrary *temperature* T , the new function

$$\sigma_M(\mathbf{m},T) \;=\; \rho_M(\mathbf{m}) \;\; e^{-\frac{S(\mathbf{m})}{T}} \; , \tag{3.18}$$

we have the following properties (see Figures 3.4 and 3.5)

$$\sigma_M(\mathbf{m},T{=}\infty) \;=\; \rho_M(\mathbf{m}) \; , \tag{3.19a}$$

$$\sigma_M(\mathbf{m},T{=}T_0) \;=\; \sigma_M(\mathbf{m}) \; , \tag{3.19b}$$

$$\sigma_M(\mathbf{m},T{=}0) \;=\; \mathrm{const}\; \delta(\mathbf{m}-\mathbf{m}_{ML}) \; . \tag{3.19c}$$

For $\rho_M(\mathbf{m}) = \mathrm{const}$, equation (3.18) clearly resembles the Gibbs distribution (also called the canonical distribution), giving the probability of a "state" $\mathbf{m}$ with energy $S(\mathbf{m})$ of a statistical system at temperature T (the Boltzmann constant k is taken here equal to 1). This justifies both the name of "energy function" for $S(\mathbf{m})$ and of "temperature" for T . The factor $\rho_M(\mathbf{m})$ slightly generalizes the Gibbs distribution: the probability density at infinite temperature is $\rho_M(\mathbf{m})$, and not necessarily uniform (equation (3.19a)).

Example 3.1: In nonlinear least squares, usually $\rho_M(\mathbf{m})$ = const, and

$$\sigma_M(\mathbf{m}) \;=\; \rho_M(\mathbf{m}) \;\exp\left[-\frac{1}{2}\,(\mathbf{g}(\mathbf{m})-\mathbf{d}_{obs})^t\; \mathbf{C}_D^{\;-1}\; (\mathbf{g}(\mathbf{m})-\mathbf{d}_{obs}) \right] .$$

The energy function $S(\mathbf{m})$ is then simply T_0 times the usual data misfit function:

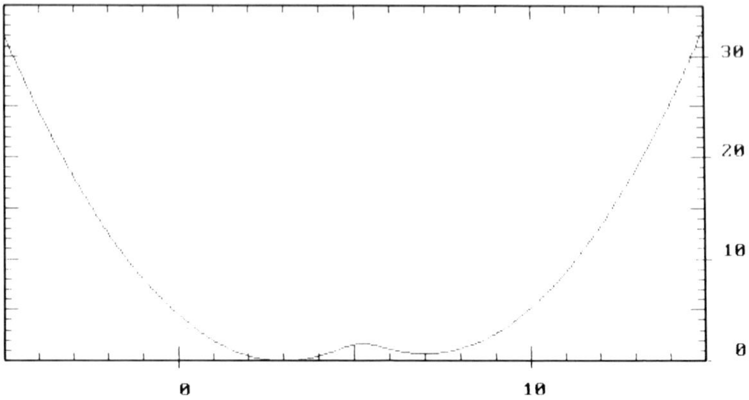

Figure 3.4: Schematic example of a function S(m) with a global minimum at m = 3 and one secondary minimum at m = 7 . We wish to obtain the global minimum using simulated annealing.

$$S(\mathbf{m}) = \frac{T_0}{2} (\mathbf{g}(\mathbf{m})-\mathbf{d}_{obs})^t \, \mathbf{C}_D^{-1} \, (\mathbf{g}(\mathbf{m})-\mathbf{d}_{obs}) \ .$$

Then

$$\sigma_M(\mathbf{m},T) = \frac{1}{2} (\mathbf{g}(\mathbf{m})-\mathbf{d}_{obs})^t \, \tilde{\mathbf{C}}_D^{-1}(T) \, (\mathbf{g}(\mathbf{m})-\mathbf{d}_{obs}) \ , \tag{3.20a}$$

where

$$\tilde{\mathbf{C}}_D(T) = \frac{T}{T_0} \mathbf{C}_D \ . \tag{3.21}$$

This shows that heating the system corresponds to increasing the variances representing data uncertainties (with fixed correlations). ■

Assume we have a method that, for any value of T , can generate random models $\mathbf{m}_1$, $\mathbf{m}_2$, ... with probability density (3.18) (see below). To simulate annealing, first take a high value of T (heat the system), and start generating random models. They will be distributed following the prior probability density $\rho_M(\mathbf{m})$. Now make T tend to zero (cool the system) very slowly while continuing to generate random models. When you arrive at zero temperature, the system is frozen in the state of lower energy $\mathbf{m}_{ML}$. And the problem is solved.

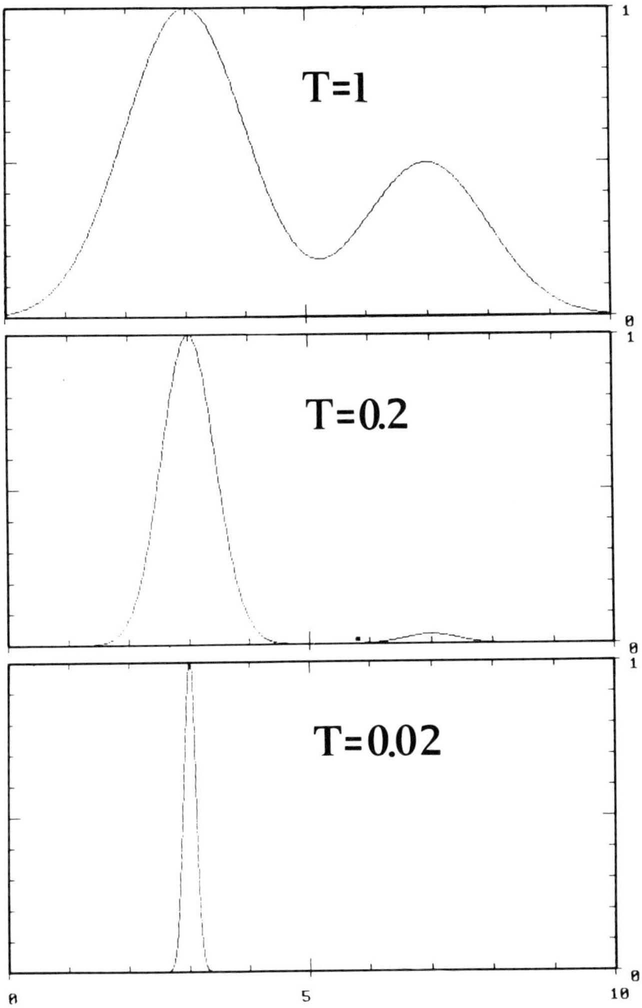

Figure 3.5: From the function S(m) of Figure 3.4, we define, for each value of T , the probability density $\sigma(m,T) = \exp(-S(m)/T)$. In the top of the figure, T = 1 , in the middle, T = 0.2 , and in the bottom, T = 0.01 . If we start with T = 1 , and we slowly decrease T , while generating random values of m according to the probability density corresponding to the current value of T , we will converge to the value m = 3 , which is the global minimum of S(m) . A too rapid decrease of T may trap the algorithm in the secondary solution, m = 7 .

If the cooling is too rapid, the system can be trapped in a metastable state (secondary minima of the energy function $S(\mathbf{m})$). This is why the cooling has to be **very** slow.

Simulated annealing has been used in examples where $S(\mathbf{m})$ is a sum of "local" terms, i.e., where the modification of the value of a single component of $\mathbf{m}$ only imposes the reevaluation of a few terms of the sum. Then, from a starting model $\mathbf{m}_0$, new random models are generated which only differ by the value of one component. The most widely used method for generating random models in simulated annealing is due to Metropolis et al. (1953). Instead, I adapt here the method suggested by Rothman (1985a, 1985b, 1986).

Two successively generated random models $\mathbf{m}_n$ and $\mathbf{m}_{n+1}$ will only differ by the value of one of the components. All the components are updated in turn, and after the last, we retake the first. Assume, for instance, that to generate the random model m_{n+1} it is the turn of the i-th component to be updated. The new value for this i-th component is simply chosen according to the marginal probability density for m^i , given the current values of all other components. Namely, the new value is chosen according the probability density

$$f(m^i) = \sigma_M(m^i \mid (m^1{=}m^1{}_n,...,m^{i-1}{}_n,m^{i+1}{}_n,...;T)) , \qquad (3.22)$$

which is directly computed for some different values of m^i . Using the arguments of Rothman, it can be shown that if we iterate long enough, we reach equilibrium, i.e., the probability density of any state $\mathbf{m}$ equals (asymptotically) $\sigma_M(\mathbf{m},T)$.

Kirkpatrick et al. (1983) show that the computing time needed for simulated annealing depends almost linearly on the number of parameters being estimated. See Geman and Geman (1984) for an analysis of the convergence properties of simulated annealing.

If the cooling rate is too slow, the method is too expensive. If the cooling rate is too rapid, we can converge to a secondary solution. The right cooling rate can be obtained after some experimentation. In the example in Figures 3.6 to 3.13 (due to Rothman, 1986)), a further simplification has been used. Instead of a continuous lowering of the temperature until a sufficiently low value is obtained, Rothman chooses, after some experimentation, a convenient value of T , which is kept constant all through the iterations, which are continued until equilibrium is reached.

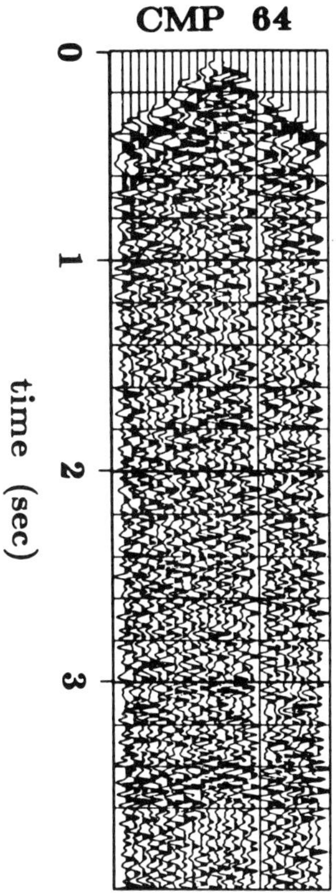

Figure 3.6: This panel displays, as a function of time, 24 "Common Mid-Point" (CMP) seismograms (see Figure 3.7) recorded in a petroleum exploration seismic experiment. They correspond to the CMP number 64, so they are denoted $s^1{}_{64}(t)$, $s^2{}_{64}{}^2(t)$, ... , $s^{24}{}_{64}(t)$. All the seismograms in this panel have to be added together (stacked) to produce a single seismogram, $S_{64}(t)$, but they have to be shifted relatively prior to addition (see Figure 3.7),

$$S_{64}(t) = \sum_{i=1}^{24} s^i{}_{64}(t+\tau^i{}_{64}) ,$$

in order to maximize the coherence

$$E_{64} = \int_0^4 dt \; (S_{64}(t))^2 .$$

For this panel, this gives the 24 unknowns $\tau^i{}_{64}$ (i=1,...,24) which have to be estimated. The single seismogram $S_{64}(t)$ obtained from this panel is numbered 64 in Figures 3.8 to 3.12. The starting data are 170 panels such as the one shown here (so there are 170 stacked seismograms in Figures 3.8 to 3.12).

This would make, in principle, 170x24 unknowns to estimate, but many of the shifts are common for seismograms in different panels (the unknowns are not independent), and the effective number of unknown shifts is 175. The problem is then to obtain the values of these 175 parameters for which the total coherence

$$E = \sum_{\alpha=1}^{170} E_\alpha = \sum_{\alpha=1}^{170} \int_0^4 dt \; (S_\alpha(t))^2$$

is maximized. As the individual seismograms are highly oscillating, shifts of the order of the dominating period produce periodicities, and E has many secondary maxima. The unknown shifts are constrained to fall within ±160 ms , in 8 ms increments. Each shift may then take 41 values. 175 parameters, with 41 possible values for each one, makes a total of 41^{175} ($\simeq 10^{282}$) possibilities. From Rothman (1985, 1986).

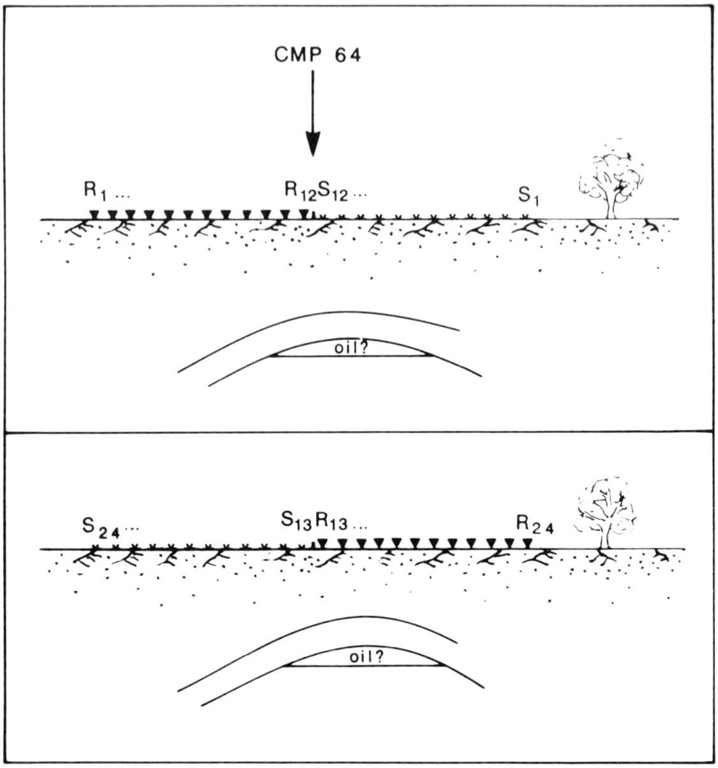

Figure 3.7: For preliminary interpretation of seismic reflection data, all seismograms recorded with sources (S_1 , S_2 , ... , S_{24}) and receivers (R_1 , R_2 , ... , R_{24}) placed symmetrically with respect to a "Common Mid-Point" (CMP) are grouped (Figure 3.6) and added (after correction of geometrical effects) to produce a single seismogram, corresponding approximately to the one that should be recorded by placing source and receiver at the CMP. Near-surface heterogeneities introduce strong noise in a CMP section, consisting essentially in static shifts of the seismograms. For a correct stacking of the seismograms of all CMP seismic sections, the shifts corresponding to all different source and receiver locations have to be estimated.

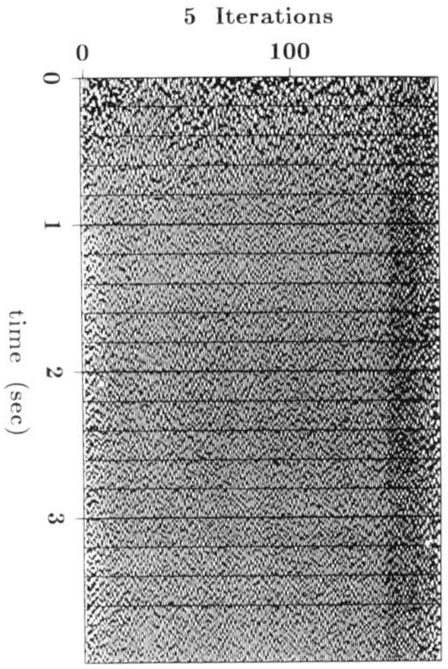

Figure 3.8: The stacked seismograms obtained after 5 iterations. The initial temperature was sufficiently high so that all structure in the stack was destroyed. Mathematically, this is equivalent to beginning the search at one of the highest points on the function to be minimized, obviating any initial need to climb out of a local minimum. From Rothman (1985, 1986).

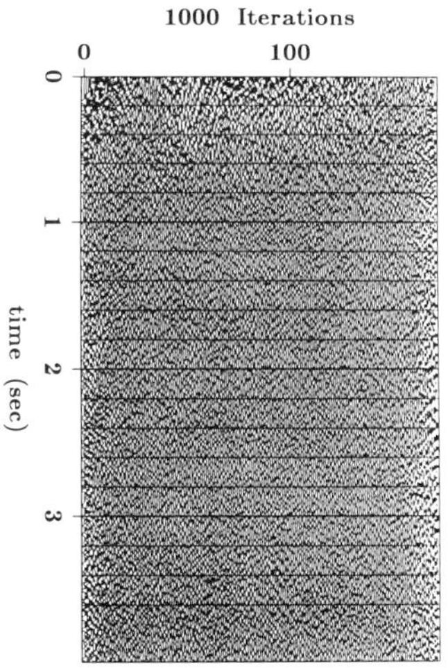

Figure 3.9: The stacked seismograms after 1000 iterations. Temperature is kept constant from iteration 53 onward. No appreciable differences between this stack and the result after 5 iterations (Figure 3.8) are evident. From Rothman (1985, 1986).

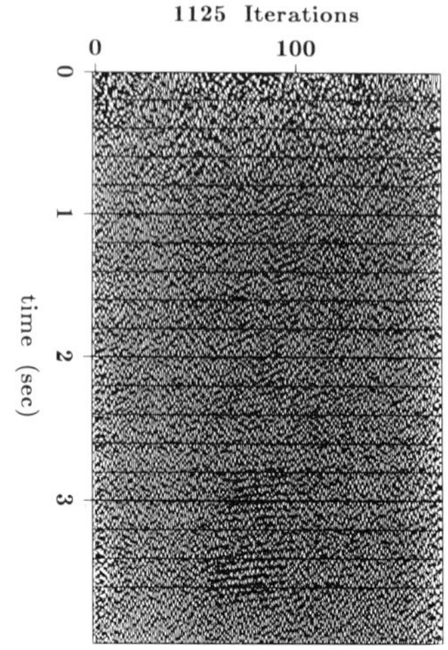

Figure 3.10: The stacked seismograms after 1125 iterations. The faint spatial coherence in the middle of the section shows that convergence is now beginning. From Rothman (1985, 1986).

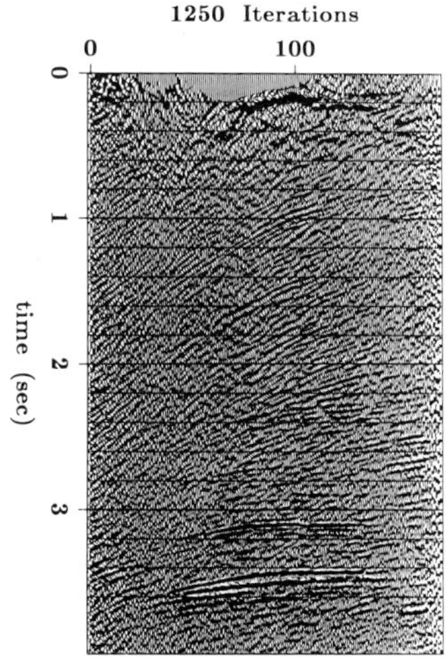

Figure 3.11: The stacked seismograms after 1250 iterations. Convergence is now almost complete; only at the extremities of the panel has the solution not converged. From Rothman (1985, 1986).

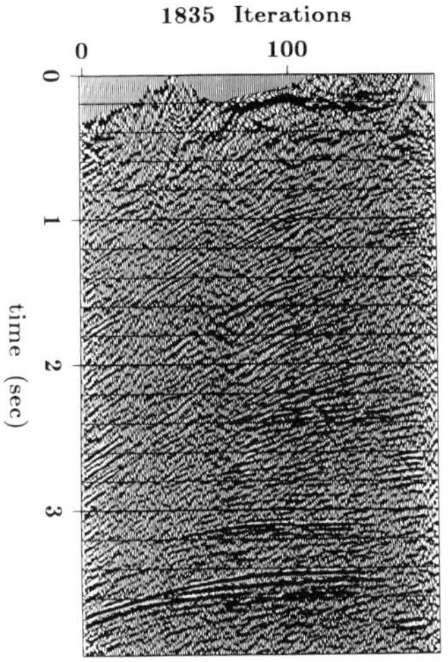

Figure 3.12: The stacked seismograms after 1835 iterations. Of the total of 2000 iterations performed by Rothman, this stack yielded the greatest value for the coherence function. The final 600 iterations yielded essentially similar stacks except for the behaviour at the far right of the seismic section (which did not converge). This section gives an approximate image of the subsurface (time corresponding to depth). From Rothman (1985, 1986).

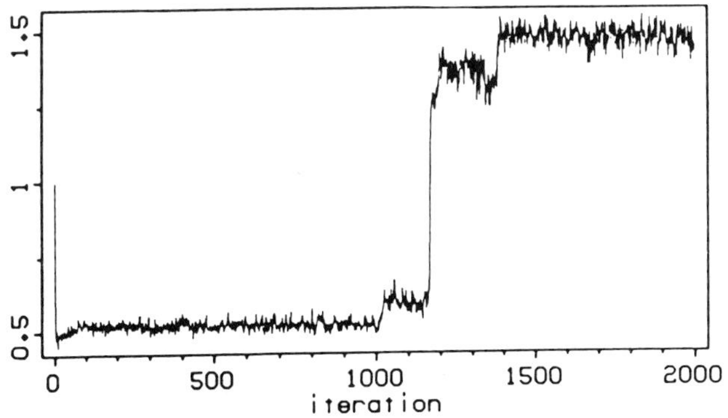

Figure 3.13: Normalized value of the coherency as a function of iteration number, for the example illustrated in Figures 3.8 to 3.12 . Convergence begins at approximately iteration 1000. A sharp increase occurs near iteration 1150; this rapid change is analogous to abrupt changes in a system's energy upon crystallization. Global convergence occurs after about 1400 iterations. From Rothman (1985, 1986).

CHAPTER 4

THE LEAST-SQUARES (ℓ_2-norm) CRITERION

If we know that our individual errors and fluctuations
follow the magic bell-shaped curve *exactly*,
then the resulting estimates are known to have
almost all the nice properties that people have been able to think of.

John W. Tukey, 1965.

Least squares are so popular for solving inverse problems because they lead to the easiest computations. Their only drawback is their lack of robustness, i.e., their strong sensitivity to a small number of large errors (outliers) in a data set.

In this book, the least-squares criterion is justified by the hypothesis that all sources of errors present in the problem can be modeled using Gaussian functions. Covariance operators play a central role in the method. The underlying mathematics are simple and beautiful.

The methods of resolution suggested in this chapter are based on the classical optimization theory. Gradient and Newton methods for the resolution of nonlinear problems are introduced. The approach followed in this chapter is fully nonlinear, and linear or linearized problems are treated as special cases.

4.1: Introducing least squares

Let $\mathbf{d}$ denote a generic data vector, $\mathbf{m}$ a generic model vector, and $\mathbf{d} = \mathbf{g}(\mathbf{m})$ the theoretical relationship between data and model parameters. A measurement of the true value of $\mathbf{d}$ gives $\mathbf{d}_{obs}$, with Gaussian uncertainties described by the covariance operator $\mathbf{C}_d$. The a priori information on $\mathbf{m}$ may be described by the a priori value $\mathbf{m}_{prior}$, with Gaussian uncertainties described by the covariance operator $\mathbf{C}_M$. Finally, the theoretical relationship $\mathbf{d} = \mathbf{g}(\mathbf{m})$ holds only approximately, and the corresponding uncertainties are Gaussian and are described by the covariance operator $\mathbf{C}_T$. As demonstrated in Chapter 1, the probability density representing the posterior information in the model space is then given by (equation 1.85):

$$\sigma_M(\mathbf{m}) = \text{const} \times \tag{4.1}$$

$$\times \exp\left[-\frac{1}{2}\left((\mathbf{g}(\mathbf{m})-\mathbf{d}_{obs})^t \, \mathbf{C}_D^{-1} \, (\mathbf{g}(\mathbf{m})-\mathbf{d}_{obs}) + (\mathbf{m}-\mathbf{m}_{prior})^t \, \mathbf{C}_M^{-1} \, (\mathbf{m}-\mathbf{m}_{prior}) \right) \right] .$$

where

$$\mathbf{C}_D = \mathbf{C}_d + \mathbf{C}_T . \tag{4.2}$$

If the equation solving the forward problem is linear

$$d^i_{cal} = g^i(\mathbf{m}) = \sum_{\alpha \in \mathbf{I}_M} G^{i\alpha} \, m^\alpha \qquad (i \in \mathbf{I}_D) , \tag{4.3a}$$

or, for short,

$$\mathbf{d}_{cal} = \mathbf{G}\,\mathbf{m} , \tag{4.3b}$$

the posterior probability density $\sigma_M(\mathbf{m})$ *is then Gaussian* (equations (1.89) to (1.94)):

$$\sigma_M(\mathbf{m}) = ((2\pi)^{NM} \det \mathbf{C}_{M'})^{1/2} \exp\left[-\frac{1}{2} \, (\mathbf{m}-\langle \, \mathbf{m} \, \rangle \,)^t \, \mathbf{C}_{M'}^{-1} \, (\mathbf{m}-\langle \, \mathbf{m} \, \rangle \,) \right], \tag{4.4}$$

with center

$$\langle \, \mathbf{m} \, \rangle = \left[\mathbf{G}^t \, \mathbf{C}_D^{-1} \, \mathbf{G} + \mathbf{C}_M^{-1} \right]^{-1} \left[\mathbf{G}^t \, \mathbf{C}_D^{-1} \, \mathbf{d}_{obs} + \mathbf{C}_M^{-1} \, \mathbf{m}_{prior} \right] , \tag{4.5a}$$

$$= \mathbf{m}_{prior} + \left[\mathbf{G}^t \, \mathbf{C}_D^{-1} \, \mathbf{G} + \mathbf{C}_M^{-1} \right]^{-1} \mathbf{G}^t \, \mathbf{C}_D^{-1} \, (\mathbf{d}_{obs} - \mathbf{G} \, \mathbf{m}_{prior}) , \tag{4.5b}$$

$$= \mathbf{m}_{prior} + \mathbf{C}_M \, \mathbf{G}^t \left(\mathbf{G} \, \mathbf{C}_M \, \mathbf{G}^t + \mathbf{C}_D \right)^{-1} (\mathbf{d}_{obs} - \mathbf{G} \, \mathbf{m}_{prior}) \, , \tag{4.5c}$$

and covariance operator

$$\mathbf{C}_{M'} = \left(\mathbf{G}^t \, \mathbf{C}_D^{-1} \, \mathbf{G} + \mathbf{C}_M^{-1} \right)^{-1} , \tag{4.6a}$$

$$= \mathbf{C}_M - \mathbf{C}_M \, \mathbf{G}^t \left(\mathbf{G} \, \mathbf{C}_M \, \mathbf{G}^t + \mathbf{C}_D \right)^{-1} \mathbf{G} \, \mathbf{C}_M \, . \tag{4.6b}$$

The value $\langle \mathbf{m} \rangle$, center of the Gaussian, is both the mean value of $\sigma_M(\mathbf{m})$ and its maximum likelihood point. It is abusively referred to as *the* solution of the inverse problem. We see thus that, for linear problems, we have explicit expressions for the solution and for the posterior covariance operator.

If the forward problem is nonlinear, the posterior probability density $\sigma_M(\mathbf{m})$ is not Gaussian, and the analysis of the solution is not so straightforward. For strongly nonlinear problems, $\sigma_M(\mathbf{m})$ may be quite chaotic (multimodal, with infinite dispersions, etc.), and the general methods described in Chapter 1 should be used to analyse the solution. If $\sigma_M(\mathbf{m})$ is reasonably well behaved, the a posteriori information in the model space may be well represented by a central estimator of $\sigma_M(\mathbf{m})$ and a properly defined covariance operator. Among all the central estimators, the easiest to compute is generally the maximum likelihood point $\mathbf{m}_{ML}$:

$$\sigma_M(\mathbf{m}) \qquad MAXIMUM \qquad \text{for} \qquad \mathbf{m} = \mathbf{m}_{ML} \, , \tag{4.7a}$$

because the obtainment of $\mathbf{m}_{ML}$ corresponds to a problem of optimization of a scalar function, and many methods exist allowing an economical resolution of that problem.

Defining the *misfit* function (or *cost* function, or *objective* function, or *least-squares* function, or *chi-squared* function) by

$$\boxed{S(\mathbf{m}) = \frac{1}{2} \left((\mathbf{g}(\mathbf{m}) - \mathbf{d}_{obs})^t \, \mathbf{C}_D^{-1} \, (\mathbf{g}(\mathbf{m}) - \mathbf{d}_{obs}) + (\mathbf{m} - \mathbf{m}_{prior})^t \, \mathbf{C}_M^{-1} \, (\mathbf{m} - \mathbf{m}_{prior}) \right) ,} \tag{4.8}$$

the maximum likelihood point is clearly defined by

$$S(\mathbf{m}) \qquad MINIMUM \qquad \text{for} \qquad \mathbf{m} = \mathbf{m}_{ML} \, . \tag{4.7b}$$

For uncorrelated errors,

$$(C_D)^{ij} = (\sigma_D^i)^2 \, \delta^{ij} ,$$

$$(C_M)^{\alpha\beta} = (\sigma_M^\alpha)^2 \, \delta^{\alpha\beta} ,$$

the misfit function $S(\mathbf{m})$ becomes

$$S(\mathbf{m}) = \frac{1}{2} \left[\sum_{i \in I_D} \frac{\left(g^i(\mathbf{m}) - d^i_{obs} \right)^2}{\left(\sigma^i_D \right)^2} + \sum_{\alpha \in \mathbf{I}_M} \frac{\left(m^\alpha - m^\alpha_{prior} \right)^2}{\left(\sigma^\alpha_M \right)^2} \right] ,$$

which justifies the name of *least squares* for the criterion (4.7b).

As introduced here, the least-squares criterion is intimately related with the Gaussian probability assumption. Many people like to justify least squares from a statistical point of view. They deal almost exclusively with linear problems. For them, $\mathbf{d}$ and $\mathbf{m}$ are random variables with known covariance operators C_D and C_M , and unknown mathematical expectations $\mathbf{d}_{true}$ and $\mathbf{m}_{true}$. $\mathbf{d}_{obs}$ and $\mathbf{m}_{prior}$ are then interpreted as two particular *realizations* of the random variables $\mathbf{d}$ and $\mathbf{m}$. The problem is then to obtain an estimator of $\mathbf{m}_{true}$, $\mathbf{m}_{est}$, which is, in some sense, optimum. The Gauss-Markoff theorem (see, for instance, Plackett, 1949, or Rao, 1973) shows that, *for linear problems*, the least-squares estimator has *minimum variance* among all the estimators which are linear functions of $\mathbf{d}_{obs}$ and $\mathbf{m}_{prior}$, *irrespectively of the particular form of the probability density functions of the random variables* $\mathbf{d}$ *and* $\mathbf{m}$. This is why the least-squares criterion is sometimes used even if the form of the density functions is not Gaussian. The trouble is that the criterion of minimum variance is not magic, and, in fact, it may be a *very bad* criterion in some cases, such as, for instance, when a small number of large, uncontrolled errors are present in a data set (see problem 1.9). As the general approach developed in Chapter 1 justifies the least-squares criterion only when all errors (modelization errors, observational errors, errors in the a priori model) are Gaussian, I urge the reader to limit the use of the techniques described in this chapter to the cases where this assumption is not too strongly violated.

It often happens that some model parameters are, by definition, positive. To take a Gaussian function to model the a priori information is not coherent because a Gaussian function gives a non-nul probability to negative values. Sometimes, a least-squares criterion is used for such parameters, completed with a positivity constraint. This is not the most rigorous nor the easiest way to attack this sort of problem. As suggested in section 1.2.4, these parameters usually accept a log-normal function as a prior probability density. Taking the *logarithm* of the positive parameter defines a new (unbounded) parameter whose a priori probability density is Gaussian, and for which standard least-squares techniques apply.

4.2: Methods of resolution (I)

We have just seen that the central problem in general least squares consists in the minimization of the misfit function

$$S(\mathbf{m}) = \frac{1}{2} \left[(\mathbf{g}(\mathbf{m})-\mathbf{d}_{obs})^t \, \mathbf{C}_D^{-1} \, (\mathbf{g}(\mathbf{m})-\mathbf{d}_{obs}) + (\mathbf{m}-\mathbf{m}_{prior})^t \, \mathbf{C}_M^{-1} \, (\mathbf{m}-\mathbf{m}_{prior}) \right] .$$

This section reviews the simplest (although not necessarily the more economical) methods for the resolution of this minimization problem. After a few mathematics, section 4.5 will deal with more sophisticated methods.

4.2.1: Systematic exploration of the model space

A regular grid is defined over the model space $\mathcal{M}$, and the value $S(\mathbf{m})$ is computed at each point. Taking a dense enough grid, the point minimizing $S(\mathbf{m})$ can be approached with arbitrary accuracy. The method can only be used if the model space has a small number of dimensions.

If this method has to be used, instead of plotting $S(\mathbf{m})$ I recommend dealing directly with

$$\sigma_M(\mathbf{m}) \;=\; \text{const} \;\; \exp(-S(\mathbf{m})) \; ,$$

because, after proper normalization, the results obtained are directly interpretable in terms of probabilities.

4.2.2: Trial and error

If the number of dimensions of the model space is small, it is also possible to work interactively with a computer terminal, modifying the current point on an intuitive basis, until a point $\mathbf{m}$ which gives an acceptably low value for the misfit $S(\mathbf{m})$ is found. See Chapter 2 for more details.

4.2.3: Relaxation

Let $\mathbf{m}_0$ represent the "starting point". To obtain the updated point $\mathbf{m}_1$, all the components of $\mathbf{m}$ but one are fixed, and the value of the free component for which $S(\mathbf{m})$ has a minimum is obtained using an arbitrary method (trial and error, interpolation,...). All the components of **m** are chosen in turn, and the procedure is iterated until subsequent updatings alter the result only negligibly. Figure 4.1 illustrates the method. The convergence rate is, in general, poor.

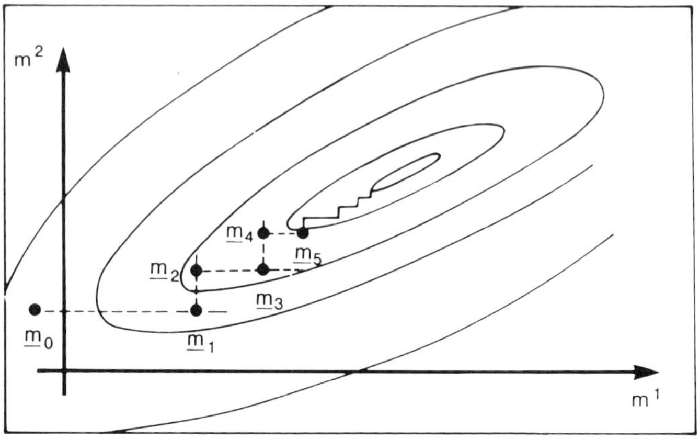

Figure 4.1: The path obtained using the relaxation method of minimization in a schematic problem with two variables.

4.2.4: Monte Carlo methods

As explained in Chapter 3, a Monte Carlo method is any method using a random generator at any stage. These methods are particularly useful for highly nonlinear problems. In its simplest formulation, a threshold value of S is fixed. Any model **m** for which S(**m**) is less than the threshold value is named an *acceptable* model. Random points are generated over the model space ***M*** until a sufficient number of acceptable models has been obtained as being representative of the acceptable domain of the model space (see Chapter 3 for more details).

4.2.5: The Gauss-Newton method

If the functions $g^i(\mathbf{m})$ solving the forward problem are differentiable, i.e., if the derivatives

$$G_n^{i\alpha} = \left(\frac{\partial g^i}{\partial m^\alpha} \right)_{\mathbf{m}_n} \tag{4.9}$$

can be defined at any point $\mathbf{m}_n$ (or at "almost" any point), and if they can easily be computed, then the derivatives of $S(\mathbf{m})$ can also be easily obtained, and the very powerful gradient and Newton methods can be used for minimizing $S(\mathbf{m})$. Although their detailed study is left for section 4.5, the Gauss-Newton method is introduced here in an elementary way.

Equation (4.8) can be written explicitly as

$$S(\mathbf{m}) = \frac{1}{2}\left\{ \sum_{i\in\mathbf{I}_D} \sum_{j\in\mathbf{I}_D} \left(g^i(\mathbf{m})-d^i_{obs}\right) \left(C_D{}^{-1}\right)^{ij} \left(g^j(\mathbf{m})-d^j_{obs}\right) + \sum_{\alpha\in\mathbf{I}_M} \sum_{\beta\in\mathbf{I}_M} \left(m^\alpha-m^\alpha_{prior}\right) \left(C_M{}^{-1}\right)^{\alpha\beta} \left(m^\beta-m^\beta_{prior}\right) \right\} , \tag{4.10}$$

We obtain easily

$$\left(\frac{\partial S}{\partial m^\alpha}\right)_{\mathbf{m}_n} = \sum_{i\in\mathbf{I}_D} \sum_{j\in\mathbf{I}_D} G_n^{i\alpha} \left(C_D{}^{-1}\right)^{ij} \left(g^j(\mathbf{m}_n)-d^j_{obs}\right) + \sum_{\beta\in\mathbf{I}_M} \left(C_M{}^{-1}\right)^{\alpha\beta} \left(m^\beta-m^\beta_{prior}\right) , \tag{4.11}$$

or, in more compact notation,

$$\left(\frac{\partial S}{\partial \mathbf{m}}\right)_n = \mathbf{G}_n{}^t\, \mathbf{C}_D{}^{-1} \left(\mathbf{g}(\mathbf{m}_n)-\mathbf{d}_{obs}\right) + \mathbf{C}_M{}^{-1} \left(\mathbf{m}_n-\mathbf{m}_{prior}\right) . \tag{4.12}$$

The vector defined in (4.11) and (4.12) is named the *gradient* of $S(\mathbf{m})$ at $\mathbf{m} = \mathbf{m}_n$ (see Box 4.6 for more details). At the minimum of S , the gradient will vanish:

$$S(\mathbf{m}) \quad \text{minimum} \quad \text{for} \quad \mathbf{m} = \mathbf{m}_n \quad \Rightarrow \quad \left(\frac{\partial S}{\partial \mathbf{m}}\right)_{\mathbf{m}_n} = 0 ,$$

the reciprocal of course not necessarily being true. In practical applications, once we get a point at which the gradient vanishes, in general it is easy to verify that it is not a maximum of S or a saddle point, because when any iterative method provides a sequence of points $\mathbf{m}_1$, $\mathbf{m}_2$, ..., it is easy to verify that the sequence $S(\mathbf{m}_1)$, $S(\mathbf{m}_2)$,..., is decreasing. Saddle points are in general not obtained with iterative methods because the path leading to them is very unstable (the only possible way for a drop of rain to reach the saddle point of a horse saddle is just to fall on that point).

Another more difficult question is to know whether the point reached is *the* absolute minimum, or if it only is a secondary minimum. As no local test can answer the question, the only possibility is to start iterating at different points and to check if we converge into the same point.

A traditional way of obtaining the minimum of a function (i.e., the zero of its grtadient) is to use the Newton method:

$$\mathbf{m}_{n+1} = \mathbf{m}_n - \left[\frac{\partial^2 S}{\partial \mathbf{m}^2}\right]_{\mathbf{m}_n}^{-1} \left[\frac{\partial S}{\partial \mathbf{m}}\right]_{\mathbf{m}_n} , \tag{4.13}$$

where the operator $\partial S^2/\partial \mathbf{m}^2$ is named the *Hessian* of S, and is here defined by its components:

$$\left[\frac{\partial S^2}{\partial m^2}\right]^{\alpha\beta} = \frac{\partial}{\partial m^\beta} \frac{\partial S}{\partial m^\alpha} = \frac{\partial^2 S}{\partial m^\alpha \partial m^\beta} . \tag{4.14}$$

From equation (4.11) we readily obtain

$$\left[\frac{\partial^2 S}{\partial m^2}\right]_{m_n}^{\alpha\beta} = \sum_{i \in I_D} \sum_{j \in I_D} G_n^{i\alpha} \left[C_D^{-1}\right]^{ij} G_n{}^{j\beta} + \left[C_M^{-1}\right]^{\alpha\beta}$$

$$+ \sum_{i \in I_D} \sum_{j \in I_D} \left[\frac{\partial G^{i\alpha}}{\partial m^\beta}\right]_n \left[C_D^{-1}\right]^{ij} \left[g^j(\mathbf{m}_n) - d^j_{obs}\right] . \tag{4.15}$$

The last term is small if: i) the residuals are small, or ii) the forward equation is quasi linear. As in Newton methods we never need to know the Hessian with great accuracy, and as the last term in equation (4.15) is, in general, difficult to handle, it is generally dropped off, thus giving the approximation

$$\left[\frac{\partial^2 S}{\partial m^2}\right]_{m_n}^{\alpha\beta} \simeq \sum_{i \in I_D} \sum_{j \in I_D} G_n^{i\alpha} (C_D^{-1})^{ij} G_n^{j\beta} + (C_M^{-1})^{\alpha\beta} , \tag{4.16}$$

or, in compact form,

$$\left[\frac{\partial S^2}{\partial m^2}\right]_n \simeq \mathbf{G}_n{}^t \, \mathbf{C}_D{}^{-1} \, \mathbf{G}_n + \mathbf{C}_M{}^{-1} . \tag{4.17}$$

Using equations (4.11), (4.13), and (4.17), we arrive at the following algorithm

? see p. 244

$$\mathbf{m}_{n+1} = \mathbf{m}_n + \left[\mathbf{G}_n{}^t \, \mathbf{C}_D{}^{-1} \, \mathbf{G}_n + \mathbf{C}_M{}^{-1}\right]^{-1} \left[\mathbf{G}_n{}^t \, \mathbf{C}_D{}^{-1} \, (\mathbf{g}(\mathbf{m}_n) - \mathbf{d}_{obs}) + \mathbf{C}_M{}^{-1} \, (\mathbf{m}_n - \mathbf{m}_{prior})\right] . \tag{4.18}$$

If the components of the data and model vectors have been ordered in a column matrix, this equation can then be interpreted as an ordinary matricial equation.

As covariance operators are, by definition, positive definite (if they do not contain null variances or perfect correlations), the matrix $\mathbf{G}_n{}^t\, \mathbf{C}_D{}^{-1}\, \mathbf{G}_n + \mathbf{C}_M{}^{-1}$ is symmetric and positive definite, so that Cholesky's decomposition methods can be used (see Appendix 4.5). It should be noticed that, given a regular matrix $\mathbf{A}$ and a vector $\mathbf{y}$, the most economical way of computing the vector $\mathbf{x} = \mathbf{A}^{-1}\, \mathbf{y}$, is *not* to compute the inverse matrix $\mathbf{A}^{-1}$, and then to multiply $\mathbf{A}^{-1}$ by $\mathbf{y}$, but to use an algorithm allowing the direct resolution of the matricial equation $\mathbf{A}\, \mathbf{x} = \mathbf{y}$, which can be done much more economically (see, for instance, Appendices 4.4 and 4.5).

If the functions $g^i(\mathbf{m})$ are linear, then the function $S(\mathbf{m})$ is a quadratic function of $\mathbf{m}$, in which case the Newton algorithm converges in only one iteration. For nonlinear functionals $g^i(\mathbf{m})$, not only do we need to iterate, but we cannot be sure that the Newton algorithm will converge. Thus, at each iteration we have to compute $S_{n+1} = S(\mathbf{m}_{n+1})$, and check if the condition

$$S_{n+1} \le S_n \ ,$$

is fulfilled. If not, equation (4.18) has to be replaced by

$$\mathbf{m}_{n+1} = \mathbf{m}_n + \epsilon_n \left(\mathbf{G}_n{}^t\, \mathbf{C}_D{}^{-1}\, \mathbf{G}_n + \mathbf{C}_M{}^{-1} \right)^{-1} \left(\mathbf{G}_n{}^t\, \mathbf{C}_D{}^{-1}\, (\mathbf{g}(\mathbf{m}_n) - \mathbf{d}_{obs}) + \mathbf{C}_M{}^{-1}\, (\mathbf{m}_n - \mathbf{m}_{prior}) \right) , \qquad (4.19)$$

where ϵ_n is an ad-hoc constant less than unity. Alternatively, it is also possible to dump the Newton updates by artificially diminishing the a priori variances in $\mathbf{C}_M$. This last method is almost equivalent to the method proposed by Levenberg (1944), or Marquardt (1963). See section 4.5 for more details.

4.2.6: Characterization of the solution.

At the minimum of the misfit function, the gradient must vanish. This gives, using (4.12),

$$\mathbf{C}_M{}^{-1}\, (\mathbf{m} - \mathbf{m}_{prior}) = -\, \mathbf{G}^t\, \mathbf{C}_D{}^{-1}\, (\mathbf{g}(\mathbf{m}) - \mathbf{d}_{obs}) \ .$$

Adding $\mathbf{G}^t\, \mathbf{C}_D{}^{-1}\, \mathbf{G}\, (\mathbf{m} - \mathbf{m}_{prior})$ to both sides we obtain

$$\left(\mathbf{G}^t\ \mathbf{C}_D^{-1}\ \mathbf{G} + \mathbf{C}_M^{-1}\right)\ (\mathbf{m}-\mathbf{m}_{prior}) \ = \ -\ \mathbf{G}^t\ \mathbf{C}_D^{-1}\ (\mathbf{g}(\mathbf{m})-\mathbf{d}_{obs}-\mathbf{G}\ (\mathbf{m}-\mathbf{m}_{prior})\ ),$$

i.e.,

$$\mathbf{m} \ = \ \mathbf{m}_{prior} - \left(\mathbf{G}^t\ \mathbf{C}_D^{-1}\ \mathbf{G} + \mathbf{C}_M^{-1}\right)^{-1}\ \mathbf{G}^t\ \mathbf{C}_D^{-1}\ (\mathbf{g}(\mathbf{m})-\mathbf{d}_{obs}-\mathbf{G}\ (\mathbf{m}-\mathbf{m}_{prior}))\ , \tag{4.20a}$$

or, using the identity demonstrated in problem 1.19,

$$\mathbf{m} \ = \ \mathbf{m}_{prior} - \mathbf{C}_M\ \mathbf{G}^t\ \left(\mathbf{C}_D + \mathbf{G}\ \mathbf{C}_M\ \mathbf{G}\right)^{-1}\ (\mathbf{g}(\mathbf{m})-\mathbf{d}_{obs}-\mathbf{G}\ (\mathbf{m}-\mathbf{m}_{prior}))\ . \tag{4.20b}$$

Equations (4.20) characterize the solution. Notice that these are implicit equations (the unknown $\mathbf{m}$ appears at both sides). In fact it is possible to obtain the solution from (4.20) using a fixed point algorithm:

$$\mathbf{m}_{n+1} \ = \ \mathbf{m}_{prior} - \left(\mathbf{G}_n{}^t\ \mathbf{C}_D^{-1}\ \mathbf{G}_n + \mathbf{C}_M^{-1}\right)^{-1}\ \mathbf{G}_n{}^t\ \mathbf{C}_D^{-1}\ (\mathbf{g}(\mathbf{m}_n)-\mathbf{d}_{obs}-\mathbf{G}_n\ (\mathbf{m}_n-\mathbf{m}_{prior}))\ , \tag{4.21a}$$

$$\mathbf{m}_{n+1} \ = \ \mathbf{m}_{prior} - \mathbf{C}_M\ \mathbf{G}_n{}^t\ \left(\mathbf{C}_D + \mathbf{G}_n\ \mathbf{C}_M\ \mathbf{G}_n\right)^{-1}\ (\mathbf{g}(\mathbf{m}_n)-\mathbf{d}_{obs}-\mathbf{G}_n\ (\mathbf{m}_n-\mathbf{m}_{prior}))\ . \tag{4.21b}$$

4.3: Analysis of error and resolution

4.3.1: Computation of the posterior covariance operator

We have seen in section 4.1 that if the equation solving the forward problem is linear,

$$\mathbf{g}(\mathbf{m}) \ = \ \mathbf{G}\ \mathbf{m}\ ,$$

the posterior probability density is then Gaussian (equation (4.4)), with center given by (4.5), and with covariance operator given by

$$\mathbf{C}_{M'} \ = \ \left(\mathbf{G}^t\ \mathbf{C}_D^{-1}\ \mathbf{G} + \mathbf{C}_M^{-1}\right)^{-1} \tag{4.22a}$$

$$= \mathbf{C}_M - \mathbf{C}_M \mathbf{G}^t \left(\mathbf{G} \mathbf{C}_M \mathbf{G}^t + \mathbf{C}_D \right)^{-1} \mathbf{G} \mathbf{C}_M \ . \tag{4.22b}$$

A problem is "linearizable around $\mathbf{m}_{ref}$ " if, for any $\mathbf{m}$ of interest,

$$\mathbf{g}(\mathbf{m}) \simeq \mathbf{g}(\mathbf{m}_{ref}) + \mathbf{G}_{ref} (\mathbf{m} - \mathbf{m}_{ref}) \tag{4.23}$$

where

$$\mathbf{G}_{ref} = \left(\frac{\partial \mathbf{g}}{\partial \mathbf{m}} \right)_{\mathbf{m}_{ref}} \ . \tag{4.24}$$

$\mathbf{m}_{ref}$ is named the reference model and usually (but not necessarily) corresponds to the a priori model

$$\mathbf{m}_{ref} = \mathbf{m}_{prior} \ . \tag{4.25}$$

It is easy to see that within the approximation (4.23), the a posteriori probability density in the model space is approximately Gaussian, with center

$$\langle \mathbf{m} \rangle = \mathbf{m}_{prior} - \left(\mathbf{G}^t_{ref} \mathbf{C}_D^{-1} \mathbf{G}_{ref} + \mathbf{C}_M^{-1} \right)^{-1} \mathbf{G}^t_{ref} \mathbf{C}_D^{-1} (\mathbf{g}(\mathbf{m}_{prior}) - \mathbf{d}_{obs}) \tag{4.26a}$$

$$= \mathbf{m}_{prior} - \mathbf{C}_M \mathbf{G}^t_{ref} \left(\mathbf{G}_{ref} \mathbf{C}_M \mathbf{G}^t_{ref} + \mathbf{C}_D \right)^{-1} (\mathbf{g}(\mathbf{m}_{prior}) - \mathbf{d}_{obs}) \ , \tag{4.26b}$$

and covariance operator

$$\mathbf{C}_{M'} = \left(\mathbf{G}^t_{ref} \mathbf{C}_D^{-1} \mathbf{G}_{ref} + \mathbf{C}_M^{-1} \right)^{-1} \tag{4.27a}$$

$$= \mathbf{C}_M - \mathbf{C}_M \mathbf{G}^t_{ref} \left(\mathbf{G}_{ref} \mathbf{C}_M \mathbf{G}^t_{ref} + \mathbf{C}_D \right)^{-1} \mathbf{G}_{ref} \mathbf{C}_M \ . \tag{4.28b}$$

For true nonlinear problems, the approximation (4.23) is no longer acceptable. As explained in Figure 1.13, least squares can only be applied to problems where the nonlinearity is not too strong, i.e., in problems where the function $\mathbf{g}(\mathbf{m})$ is linearizable in the region of significant posterior probability density. Let $\mathbf{m}_\infty$ be the maximum likelihood of $\sigma_M(\mathbf{m})$ obtained by any of the methods described in this chapter, *without* linearizing $\mathbf{g}(\mathbf{m})$ (the index "∞" stands because the maximum likelihood point is always obtained as the limit point of an iterative algorithm). That $\mathbf{g}(\mathbf{m})$ is linearizable in the region of significant posterior probability density means that, once the point

$\mathbf{m}_\infty$ has been obtained using a nonlinear algorithm, for evaluating $\sigma_M(\mathbf{m})$ we can use the linear approximation

$$g(\mathbf{m}) \simeq g(\mathbf{m}_\infty) + \mathbf{G}_\infty \, (\mathbf{m} - \mathbf{m}_\infty) \ , \tag{4.29}$$

where

$$\mathbf{G}_\infty = \left(\frac{\partial g}{\partial \mathbf{m}} \right)_{\mathbf{m}_\infty} \tag{4.30}$$

Inserting (4.29) in (4.1) gives

$$\sigma_M(\mathbf{m}) \simeq \text{const} \exp\left[-\frac{1}{2} \, (\mathbf{m}-\mathbf{m}_\infty)^t \, \mathbf{C}_{M'}^{-1} \, (\mathbf{m}-\mathbf{m}_\infty) \right] , \tag{4.31}$$

where $\mathbf{m}_\infty$ has been obtained by a nonlinear computation, and where

$$\mathbf{C}_{M'} = \left(\mathbf{G}_\infty^t \, \mathbf{C}_D^{-1} \, \mathbf{G}_\infty + \mathbf{C}_M^{-1} \right)^{-1} \tag{4.32a}$$

$$= \mathbf{C}_M - \mathbf{C}_M \, \mathbf{G}_\infty^t \left(\mathbf{G}_\infty \, \mathbf{C}_M \, \mathbf{G}_\infty^t + \mathbf{C}_D \right)^{-1} \mathbf{G}_\infty \, \mathbf{C}_M \ . \tag{4.32b}$$

Kennett (1978) studied the first-order modification of the posterior covariance operator due to non linearity.

If you are not sure of the validity of the linearized estimation of posterior errors, you can solve the inverse problem a few times, with different values of $\mathbf{d}_{obs}$ and of $\mathbf{m}_{prior}$, and make a rough statistics of the results. If you have more time, generate Gaussian random data vectors with mean $\mathbf{d}_{obs}$ and covariance operator $\mathbf{C}_D$ and Gaussian random model vectors with mean $\mathbf{m}_{prior}$ and covariance operator $\mathbf{C}_M$, solve the nonlinear inverse problem for each realization, and make a proper computation of the mean value and covariance operator of the results.

4.3.2: Interpretation of the posterior covariance operator

The most trivial use of the posterior covariance operator $\mathbf{C}_{M'}$ is to interpret the square roots of the diagonal elements (variances) as "error bars" describing the uncertainities on the posterior values of the model parameters.

A direct examination of the off-diagonal elements (covariances) of a covariance operator is not easy, and it is much better to introduce the *correlations*

$$\rho^{\alpha\beta} = \frac{C^{\alpha\beta}}{(C^{\alpha\alpha})^{1/2} \, (C^{\beta\beta})^{1/2}} \ ,$$

which have the well known property

$-1 \leq \rho^{\alpha\beta} \leq +1$.

If the posterior correlation between parameters m^α and m^β is close to zero, the posterior uncertainties are uncorrelated (in the intuitive sense). If the correlation is close to $+1$ (resp. -1), the uncertainties are highly correlated (resp. anticorrelated). A strong correlation on uncertainties means that the two parameters have not been independently resolved by the data set, and that only some linear combination of the parameters is resolved.

Sometimes the parameters $m^1, m^2, \ldots$ represent the discretized values of some spatial (or temporal) function. Each row (or column) of the posterior covariance operator can then directly be interpreted in terms of spatial (or temporal) resolution power of the data set. See the color plate in Chapter 7 for a nice example.

It should not be forgotten that a covariance operator over the model space can be represented by its "ellipsoid of error". For instance, if $\mathbf{m}_\infty$ is the maximum likelihood posterior point, and $\mathbf{C}_{M'}$ is the posterior covariance operator, then we know that the posterior probability density in the model space is Gaussian (or approximately Gaussian), as given by (4.31). We can then represent the iso-density values corresponding, for instance, to probabilities of 10% , 20% , etc.

In all usual examples, the prior covariance operators $\mathbf{C}_D$ and $\mathbf{C}_M$ are chosen regular (no null or infinite variances, no perfect correlations), so that their inverse exists. But it is possible to give more general sense to the obtained results. For instance, in equation (4.22b) the prior covariance operator $\mathbf{C}_D$ may well be singular, if $\mathbf{G}\ \mathbf{C}_M\ \mathbf{G}^t$ is not. If in equation (4.22a) the operator $\mathbf{W}_{M'} = (\mathbf{G}^t\ \mathbf{C}_D{}^{-1}\ \mathbf{G} + \mathbf{C}_M{}^{-1})$ was singular, the posterior covariance operator $\mathbf{C}_{M'} = \mathbf{W}_{M'}{}^{-1}$ would not be defined, but all useful information could be extracted from the weighting operator $\mathbf{W}_{M'}$. From a geometrical point of view, a regular covariance operator defines a probability density which takes its maximum at a single point. If the covariance operator is singular, the maximum corresponds to a subspace of dimension ≥ 1 .

4.3.3: The resolution operator

In the approaches to inversion not directly based on probabilistic concepts, it is usual to introduce the "resolution operator". In order to make the link with these methods, let me briefly introduce this concept.

Let $\mathbf{m}_{true}$ represent "the true" model (which is known only by the gods). The observed data values $\mathbf{d}_{obs}$ generally do not equal the computed values $\mathbf{g}(\mathbf{m}_{true})$ because of observational errors and of modelization errors. The concept of resolution operator arises when the relationship is sought between the least-squares solution $\langle\ \mathbf{m}\ \rangle$ and the true model $\mathbf{m}_{true}$,

$$\langle \mathbf{m} \rangle = \mathbf{r}(\mathbf{m}_{true}) \tag{4.33}$$

in the optimum case where, by chance, data are error free:

$$\mathbf{d}_{obs} = \mathbf{g}(\mathbf{m}_{true}) \ .$$

Equation (4.33) defines $\mathbf{r}(\cdot)$, the *nonlinear resolution operator*. Using equation (4.20a) and linearizing around $\langle \mathbf{m} \rangle$ we obtain

$$\langle \mathbf{m} \rangle - \mathbf{m}_{prior} = \mathbf{R} \, (\mathbf{m}_{true} - \mathbf{m}_{prior}) \ , \tag{4.34}$$

where

$$\mathbf{R} = \left(\mathbf{G}^t \, \mathbf{C}_D^{-1} \, \mathbf{G} + \mathbf{C}_M^{-1} \right)^{-1} \mathbf{G}^t \, \mathbf{C}_D^{-1} \, \mathbf{G} \tag{4.35a}$$

$$= \mathbf{C}_M \, \mathbf{G}^t \left(\mathbf{G} \, \mathbf{C}_M \, \mathbf{G}^t + \mathbf{C}_D \right)^{-1} \mathbf{G} \ , \tag{4.35b}$$

thus defining the *linearized resolution operator*, $\mathbf{R}$. From the last expression we obtain

$$\mathbf{R} = \mathbf{I} - \left(\mathbf{C}_M - \mathbf{C}_M \, \mathbf{G}^t \left(\mathbf{G} \, \mathbf{C}_M \, \mathbf{G}^t + \mathbf{C}_D \right)^{-1} \mathbf{G} \, \mathbf{C}_M \right) \mathbf{C}_M^{-1} \ ,$$

and, using the expression (4.6b) for the posterior covariance operator,

$$\mathbf{R} = \mathbf{I} - \mathbf{C}_{M'} \, \mathbf{C}_M^{-1} \ . \tag{4.36}$$

If the resolution operator is the identity operator, equations (4.33) or (4.34) show that $\langle \mathbf{m} \rangle = \mathbf{m}_{true}$, and the model is perfectly resolved. The farther the resolution operator is from the identity, the worse the resolution is. Following Backus and Gilbert (1968), we can consider the resolution operator as a filter: the computed a posteriori model equals the true model filtered by the resolution operator. We cannot see the real world; we can only see a filtered version. For more details (in linearized problems), the reader is referred to Backus and Gilbert (1968). Interesting examples are also given by Aki and Lee (1976) or by Aki, Christofferson and Husebye (1977).

Usually, we do not wish to examine the whole kernel $R^{\alpha\beta}$ of $\mathbf{R}$. We only wish to analyze $R^{\alpha\alpha_0}$ for some selected index α_0 (a "column" of $\mathbf{R}$, if using matrix notations). We have

$$R^{\alpha\alpha_0} = \sum_{\beta} R^{\alpha\beta} \, \delta^{\beta\alpha_0} \ .$$

Introducing the vectors $\rho_0{}^\alpha = R^{\alpha\alpha_0}$ and $u_0{}^\alpha = \delta^{\alpha\alpha_0}$,we have

$$\rho_0 = \mathbf{R}\,\mathbf{u}_0 ,$$

i.e.,

$$\rho_0 = \left(\mathbf{G}^t\,\mathbf{C}_D{}^{-1}\,\mathbf{G} + \mathbf{C}_M{}^{-1}\right)^{-1}\,\mathbf{G}^t\,\mathbf{C}_D{}^{-1}\,\mathbf{G}\,\mathbf{u}_0$$

$$= \mathbf{C}_M\,\mathbf{G}^t\left(\mathbf{G}\,\mathbf{C}_M\,\mathbf{G}^t + \mathbf{C}_D\right)^{-1}\,\mathbf{G}\,\mathbf{u}_0 .$$

As noticed by Hirahara (1986) the computation of ρ_0 will need the same operations as the computation of the solution of a linear (or linearized) problem (see for instance equations (4.5)). It follows that all the gradient methods developed in sections 4.5 to 4.7 for computing $\langle\ \mathbf{m}\ \rangle$, can also be used to compute the resolution vector ρ_0 .

Taking the trace of equation (4.36) gives

$$\mathrm{Trace}(\ \mathbf{I}\ ) = \mathrm{Trace}(\ \mathbf{R}\ ) + \mathrm{Trace}\left(\ \mathbf{C}_{M'}\,\mathbf{C}_M{}^{-1}\ \right) , \tag{4.37a}$$

an equation that that can be interpreted as follows:

$$\begin{bmatrix}\text{TOTAL}\\ \text{NUMBER}\\ \text{OF}\\ \text{MODEL}\\ \text{PARAMETERS}\end{bmatrix} = \begin{bmatrix}\text{NUMBER OF}\\ \text{PARAMETERS}\\ \text{RESOLVED}\\ \text{BY THE DATA}\\ \text{SET}\end{bmatrix} + \begin{bmatrix}\text{NUMBER OF}\\ \text{PARAMETERS}\\ \text{RESOLVED BY}\\ \text{THE A PRIORI}\\ \text{INFORMATION}\end{bmatrix} . \tag{4.37b}$$

4.3.4: Eigenvector analysis

I have to start this section by stating an elementary fact which is generally ignored: *a covariance operator do not has eigenvectors*, i.e., in general the eigenvector-eigenvalue equation

$$\mathbf{C}\,\phi = \lambda\,\phi , \tag{4.38}$$

makes no sense.

Example 1: Let **C** be the covariance operator with kernel

$$C = \begin{bmatrix} (\sigma^1)^2 & 0 \\ 0 & (\sigma^2)^2 \end{bmatrix} .$$

From the equation

$$\begin{bmatrix} (\sigma^1)^2 & 0 \\ 0 & (\sigma^2)^2 \end{bmatrix} \begin{bmatrix} \phi^1 \\ \phi^2 \end{bmatrix} = \lambda \begin{bmatrix} \phi^1 \\ \phi^2 \end{bmatrix}$$

we deduce

$$(\sigma^1)^2 \ \phi^1 = \lambda \ \phi^1$$

and

$$(\sigma^2)^2 \ \phi^2 = \lambda \ \phi^2 .$$

From the first equation, the physical dimension of λ is the square of the physical dimension of σ^1 , while from the second equation it is the square of the physical dimension of σ^2 . As σ^1 and σ^2 may have different physical dimensions, the equation (4.91) does not respect the physical homogeneity of the terms: it is inconsistent. ■

From a more mathematical point of view, the eigenvector-eigenvalue equation is defined only for an *automorphism* (an operator mapping a space into itself), and we have seen that a covariance operator maps dual spaces.

If, in performing eigenvector-eigenvalue analysis, we do not consider the physical dimensions of the quantities involved, the formal results numerically obtained will not be intrinsic, i.e., they will depend on the particular units chosen, so that the conclusions reached by someone working with, say Pascals, will be incompatible with the those reached by someone working with, say bars.

Which is the precise sense of the intuitive meaning of "principal axis" of the ellipsoid representing a covariance operator? The following problem makes sense. Among those vectors $\boldsymbol{\delta m}$ with constrained length measured with respect to the prior covariance,

$$\| \boldsymbol{\delta m} \|_0^{\,2} = \boldsymbol{\delta m}^t \, \mathbf{C_M}^{-1} \, \boldsymbol{\delta m} = 1 , \tag{4.39a}$$

which one has extremal length measured with respect to the posterior covariance:

$$\| \, \delta \mathbf{m} \, \|^2 \ = \ \delta \mathbf{m}^t \, \mathbf{C}_{M'}^{-1} \, \delta \mathbf{m} \qquad \text{EXTREMAL ?} \tag{4.39b}$$

Using the method of Lagrange's multipliers (see Appendix 4.6), the problem is solved by optimizing

$$S(\delta \mathbf{m},\lambda) \ = \ \delta \mathbf{m}^t \, \mathbf{C}_{M'}^{-1} \, \delta \mathbf{m} \ - \ \lambda \left(\delta \mathbf{m}^t \, \mathbf{C}_M^{-1} \, \delta \mathbf{m} \ - \ 1 \right) .$$

The condition $\partial S/\partial \delta \mathbf{m} = \mathbf{0}$ directly gives

$$\mathbf{C}_M \, \mathbf{C}_{M'}^{-1} \, \delta \mathbf{m} \ = \ \lambda \, \delta \mathbf{m} \ , \tag{4.40}$$

which is clearly an eigenvector-eigenvalue equation for the positive definite operator $\mathbf{C}_M \ \mathbf{C}_{M'}^{-1}$ (which is an automorphism). All the eigenvalues of a positive definite operator are positive. The eigenvector associated with the maximum eigenvalue is, among all vectors with a priori length given, the one with maximum a posteriori length. This means that it is directed along the shortest axis of the ellipsoid of errors representing $\mathbf{C}_{M'}$ (with respect to the metric defined by $\mathbf{C}_M$). The eigenvector associated with the smallest eigenvalue is directed along the largest axis of the ellipsoid, and the intermediate eigenvectors correspond to the intermediate axis (they do not correspond to extrema of S , but to saddle points).

In least squares, the posterior covariance operator is $\mathbf{C}_{M'} = (\mathbf{G} \ \mathbf{C}_D^{-1} \ \mathbf{G} + \mathbf{C}_M^{-1})^{-1}$. The principal axes of the posterior covariance operator (with respect to the metric defined by $\mathbf{C}_M$) are obtained then by the eigenvector-eigenvalue analysis of the operator $\mathbf{I} + \mathbf{C}_M \ \mathbf{G}^t \ \mathbf{C}_D^{-1} \ \mathbf{G}$:

$$\left(\mathbf{I} + \mathbf{C}_M \ \mathbf{G}^t \ \mathbf{C}_D^{-1} \ \mathbf{G} \right) \delta \mathbf{m} \ = \ \lambda \ \delta \mathbf{m} \ . \tag{4.41}$$

We then see that the eigenvector $\delta \mathbf{m}_1$ associated with the largest eigenvalue of $\mathbf{I} + \mathbf{C}_M \ \mathbf{G}^t \ \mathbf{C}_D^{-1} \ \mathbf{G}$ represents a linear combination of parameters which is particularly *well resolved* ($\delta \mathbf{m}$ "points" in the direction of the shorter axis of the ellipsoid of posterior errors). Using a different argument, Wiggins (1972) also emphasized the importance of the eigenvector-eigenvalue analysis for the identification of well-resolved parameters. Unfortunately, such an analysis is linear (or linearized), and the most interesting problems concerning the choice of parameters are nonlinear: to describe an elastic solid should we use the tensor of elastic stiffnesses c^{ijkl} , or its inverse, the tensor of elastic compliances s^{ijklm} , or the velocities of the different elastic waves propagating in the solid, or... ? No general treatment of this problem is known.

The operator appearing in (4.41) is self-adjoint but it is not symmetric.

Introducing the square roots of $\mathbf{C}_M$ and $\mathbf{C}_D$ (see Box 4.1)

$$\mathbf{C}_M = \mathbf{C}_M{}^{1/2} \, \mathbf{C}_M{}^{t/2} , \tag{4.42a}$$

$$\mathbf{C}_D = \mathbf{C}_D{}^{1/2} \, \mathbf{C}_D{}^{t/2} , \tag{4.43b}$$

introducing adimensioned spaces through

$$\tilde{\boldsymbol{M}} = \mathbf{C}_M{}^{-1/2} \, \boldsymbol{M} , \tag{4.44a}$$

$$\tilde{\boldsymbol{D}} = \mathbf{C}_D{}^{-1/2} \, \boldsymbol{D} , \tag{4.45b}$$

and defining

$$\tilde{\mathbf{G}} = \mathbf{C}_D{}^{-1/2} \, \mathbf{G} \, \mathbf{C}_M{}^{-1/2} , \tag{4.46}$$

the eigenvalue-eigenvector equation (4.41) becomes

$$\left(\mathbf{I} + \tilde{\mathbf{G}}^t \, \tilde{\mathbf{G}} \right) \delta\tilde{\mathbf{m}} = \lambda \, \delta\tilde{\mathbf{m}} , \tag{4.47}$$

which can be solved using standard computer routines (the eigenvectors of $\mathbf{I} + \tilde{\mathbf{G}}^t \, \tilde{\mathbf{G}}$ and of $\tilde{\mathbf{G}}^t \, \tilde{\mathbf{G}}$ are the same).

Let

$$\left(\mathbf{I} + \tilde{\mathbf{G}}^t \, \tilde{\mathbf{G}} \right) = \mathbf{U} \, \boldsymbol{\Lambda} \, \mathbf{U}^t \tag{4.48}$$

be the Lanczos decomposition of $\mathbf{I} + \tilde{\mathbf{G}}^t \, \tilde{\mathbf{G}}$. Then (see Box 4.2),

$$\left(\mathbf{I} + \tilde{\mathbf{G}}^t \, \tilde{\mathbf{G}} \right)^{-1} = \mathbf{U} \, \boldsymbol{\Lambda}^{-1} \, \mathbf{U}^t . \tag{4.49}$$

The last equation shows that if an eigenvalue-eigenvector analysis of $\mathbf{I} + \tilde{\mathbf{G}}^t \, \tilde{\mathbf{G}}$ has been performed, the inverse Hessian required by the Newton method for the computation of the least-squares solution is known. Unfortunately, complete eigenvector-eigenvalue analysis is only possible for problems with not too many parameters.

Box 4.1: The square root of a covariance matrix.

Let E be a vector space, which may represent either the data or the model space, and let $\mathbf{C}$ be a covariance matrix over E. The *correlation operator* $\mathbf{R}$ is defined by

$$\mathbf{C} = \boldsymbol{\Sigma}\ \mathbf{R}\ \boldsymbol{\Sigma} \tag{1}$$

where $\boldsymbol{\Sigma}$ is the diagonal operator whose elements are the standard deviations $(C^{\alpha\alpha})^{1/2}$. $\mathbf{R}$ is symmetric:

$$\mathbf{R} = \mathbf{R}^t . \tag{2}$$

Let

$$\mathbf{R} = \mathbf{U}\ \boldsymbol{\Lambda}\ \mathbf{U}^t \tag{3}$$

be the Lanczos' decomposition of $\mathbf{R}$ (see Box 4.2). Using $\mathbf{U}^t\ \mathbf{U} = \mathbf{I}$, we can write

$$\mathbf{R} = \mathbf{U}\ \boldsymbol{\Lambda}^{1/2}\ \mathbf{U}^t\ \mathbf{U}\ \boldsymbol{\Lambda}^{1/2}\ \mathbf{U}^t , \tag{4}$$

(as $\boldsymbol{\Lambda}$ is diagonal, the definition of $\boldsymbol{\Lambda}^{1/2}$ is straightforward). The positive definite operator $\mathbf{R}^{1/2}$ defined by

$$\mathbf{R}^{1/2} = \mathbf{U}\ \boldsymbol{\Lambda}^{1/2}\ \mathbf{U}^t , \tag{5}$$

is termed the *square root* of $\mathbf{R}$. It is clearly symmetric

$$\mathbf{R}^{1/2} = \mathbf{R}^{t/2} \tag{6}$$

and verifies

$$\mathbf{R} = (\mathbf{R}^{1/2})^2 = \mathbf{R}^{1/2}\ \mathbf{R}^{1/2} . \tag{7}$$

We have

$$\mathbf{C} = \boldsymbol{\Sigma}\,\mathbf{R}\,\boldsymbol{\Sigma} = \boldsymbol{\Sigma}\,\mathbf{R}^{1/2}\,\mathbf{R}^{1/2}\,\boldsymbol{\Sigma} = (\boldsymbol{\Sigma}\,\mathbf{R}^{1/2})\,(\boldsymbol{\Sigma}\,\mathbf{R}^{1/2})^t , \tag{8}$$

and defining

(...)

$$\mathbf{C}^{1/2} = \mathbf{\Sigma}\,\mathbf{R}^{1/2} = \mathbf{\Sigma}\,\mathbf{U}\,\mathbf{\Lambda}^{1/2}\,\mathbf{U}^t \tag{9}$$

gives

$$\boxed{\mathbf{C} = \mathbf{C}^{1/2}\,\mathbf{C}^{t/2}\,.} \tag{10}$$

$\mathbf{C}^{1/2}$ can be termed the *square root* of $\mathbf{C}$. Using, for instance, dimensionality arguments, it can be seen that in general no operator $\mathbf{\Psi}$ exists such that

$$\mathbf{C} = \mathbf{\Psi}^2 = \mathbf{\Psi}\,\mathbf{\Psi} \quad \text{(in general } \mathbf{\Psi} \text{ does not exist)}\,. \tag{11}$$

For the same reason, the expression $\mathbf{C}^{t/2}\,\mathbf{C}^{1/2}$ is usually undefined.

Defining

$$\mathbf{C}^{-1/2} = (\mathbf{C}^{1/2})^{-1} = \mathbf{R}^{-1/2}\,\mathbf{\Sigma}^{-1} = \mathbf{U}\,\mathbf{\Lambda}^{-1/2}\,\mathbf{U}^t\,\mathbf{\Sigma}^{-1} \tag{12}$$

gives

$$\mathbf{C}^{1/2}\,\mathbf{C}^{-1/2} = \mathbf{C}^{-1/2}\,\mathbf{C}^{1/2} = \mathbf{I}\,, \tag{13}$$

and, using $\mathbf{U}\,\mathbf{U}^t = \mathbf{I}$ gives,

$$\mathbf{C}^{-1/2}\,\mathbf{C}\,\mathbf{C}^{-t/2} = \mathbf{C}^{t/2}\,\mathbf{C}^{-1}\,\mathbf{C}^{1/2} = \mathbf{I}\,. \tag{14}$$

Let $\mathbf{e}$ be an element of E . The element

$$\tilde{\mathbf{e}} = \mathbf{C}^{-1/2}\,\mathbf{e} \tag{15}$$

belongs to an adimensional space denoted $\tilde{E}$. If $\mathbf{C}$ is a covariance operator for $\mathbf{e}$, then, the covariance operator for $\tilde{\mathbf{e}}$ is the identity:

$$\tilde{\mathbf{C}} = \mathbf{C}^{-1/2}\,\mathbf{C}\,\mathbf{C}^{-t/2} = \mathbf{I}\,. \tag{16}$$

Box 4.2: The Lanczos decomposition.

Let $\mathbf{A}$ be a linear operator, and $\mathbf{A}^*$ its adjoint with respect to a certain scalar product $(\cdot , \cdot)$. Although the Lanczos decomposition can be defined for arbitrary operators, we only need to consider self-adjoint

$$\mathbf{A}^* = \mathbf{A} , \tag{1a}$$

positive definite operators (see note at the end of the box). By definition (see Box 4.5), a self-adjoint operator is an endomorphism (i.e., it maps a linear space into itself). The eigenvector-eigenvalue equation

$$\mathbf{A}\, \mathbf{m}_i = \lambda_i\, \mathbf{m}_i \tag{2}$$

then makes always sense (if $\mathbf{A}\,\mathbf{m}$ does not belong to the same space as $\mathbf{m}$, equation (2) makes *no* sense).

It is well known that the eigenvalues of a self-adjoint positive definite operator are all real and positive. In what follows, the eigenvectors $\mathbf{m}_i$ have been normalized with respect to the scalar product under consideration:

$$(\mathbf{m}_i,\mathbf{m}_i) = \| \mathbf{m}_i \|^2 = 1 . \tag{3a}$$

Let $\boldsymbol{\Lambda}$ be a diagonal matrix whose diagonal elements are the eigenvalues of $\mathbf{A}$ (classified in an arbitrary order), and let $\mathbf{U}$ be the matrix whose columns are the respective components of the normalized eigenvectors of $\mathbf{A}$:

$$\boldsymbol{\Lambda} = \begin{bmatrix} \lambda_1 & 0 & 0 & \cdots \\ 0 & \lambda_2 & 0 & \cdots \\ 0 & 0 & \lambda_3 & \cdots \\ \cdots & \cdots & \cdots & \cdots \end{bmatrix} ,$$

$$\mathbf{U} = \begin{bmatrix} m_1{}^1 & m_2{}^1 & m_3{}^1 & \cdots \\ m_1{}^2 & m_2{}^2 & m_3{}^2 & \cdots \\ m_1{}^3 & m_2{}^3 & m_3{}^3 & \cdots \\ \cdots & \cdots & \cdots & \cdots \end{bmatrix} . \tag{4}$$

Using for instance the results of Lanczos (1957), the following properties can easily be shown:

(...)

$$\mathbf{U}^* \mathbf{U} = \mathbf{I} , \tag{5a}$$

$$\mathbf{U} \mathbf{U}^* = \mathbf{I} , \tag{6a}$$

$$\mathbf{U}^{-1} = \mathbf{U}^* , \tag{7a}$$

$$\boxed{\mathbf{A} = \mathbf{U} \mathbf{\Lambda} \mathbf{U}^* ,} \tag{8a}$$

and

$$\boxed{\mathbf{A}^{-1} = \mathbf{U} \mathbf{\Lambda}^{-1} \mathbf{U}^* .} \tag{9a}$$

More generally, letting $f(x)$ be an arbitrary function of a real variable x , the following definition can be introduced

$$f(\mathbf{A}) = \mathbf{U} \; f(\mathbf{\Lambda}) \; \mathbf{U}^* , \tag{10}$$

where $f(\mathbf{\Lambda})$ represents the diagonal matrix whose elements are $f(\lambda_i)$. For instance,

$$\mathbf{A}^{1/2} = \mathbf{U} \; \mathbf{\Lambda}^{1/2} \; \mathbf{U}^* . \tag{11}$$

It should be noticed that if $\mathbf{A}$ maps the space $\boldsymbol{M}$ into itself, then $\mathbf{U}$ maps $\boldsymbol{R}^n$ into $\boldsymbol{M}$.

If the space $\boldsymbol{M}$ can be furnished with the euclidean scalar product, the operator $\mathbf{A}$ is at the same time self-adjoint and symmetric, and the symbol $(\;)^*$ (adjoint) can be replaced everywhere by the symbol $(\;)^t$ (transpose).

For an arbitrary scalar product, introducing the explicit notation

$$(\mathbf{m}_1,\mathbf{m}_2) = \mathbf{m}_1{}^t \; \mathbf{C}^{-1} \; \mathbf{m}_2 , \tag{12}$$

then

$$\mathbf{A}^* = \mathbf{C} \; \mathbf{A}^t \; \mathbf{C}^{-1} , \tag{13}$$

$$\mathbf{U}^* = \mathbf{U}^t \; \mathbf{C}^{-1} , \tag{14}$$

and the previous equations can be written explicitly

(...)

$$\mathbf{C}\,\mathbf{A}^t\,\mathbf{C}^{-1} = \mathbf{A}\,, \tag{1b}$$

$$\mathbf{m}_i{}^t\,\mathbf{C}^{-1}\,\mathbf{m}_i = \|\,\mathbf{m}_i\,\|^2 = 1\,. \tag{3b}$$

$$\mathbf{U}^t\,\mathbf{C}^{-1}\,\mathbf{U} = \mathbf{I}\,, \tag{5b}$$

$$\mathbf{U}\,\mathbf{U}^t = \mathbf{C}\,, \tag{6b}$$

$$\mathbf{U}^{-1} = \mathbf{U}^t\,\mathbf{C}^{-1}\,, \tag{7b}$$

$$\boxed{\mathbf{A} = \mathbf{U}\,\boldsymbol{\Lambda}\,\mathbf{U}^t\,\mathbf{C}^{-1}\,,} \tag{8b}$$

and

$$\boxed{\mathbf{A}^{-1} = \mathbf{U}\,\boldsymbol{\Lambda}^{-1}\,\mathbf{U}^t\,\mathbf{C}^{-1}\,.} \tag{9b}$$

If the matrix is not positive definite, but only definite nonnegative, some of the eigenvalues are null. The subspace of vectors for which

$$\mathbf{A}\,\mathbf{u} = \mathbf{0} \tag{15}$$

is called the *null space* of $\mathbf{A}$. Let $\boldsymbol{\Lambda}'$ be the matrix deduced from $\boldsymbol{\Lambda}$ by suppressing the columns and rows corresponding to the null eigenvalues, and let $\mathbf{U}'$ be the matrix deduced from $\mathbf{U}$ by suppressing the columns of the corresponding eigenvectors. It is easy to see that we still have

$$\mathbf{A} = \mathbf{U}'\,\boldsymbol{\Lambda}'\,\mathbf{U}'^*\,. \tag{16}$$

The *generalized inverse* of $\mathbf{A}$ is denoted $\mathbf{A}^{(-1)}$ and is defined by

$$\boxed{\mathbf{A}^{(-1)} = \mathbf{U}'\,\boldsymbol{\Lambda}'^{-1}\,\mathbf{U}'^*\,.} \tag{17}$$

It corresponds to the inverse of the restriction of $\mathbf{A}$ to the subspace orthogonal to its null space.

In least-squares theory, the self-adjoint operator $\mathbf{I} + \mathbf{G}^*\,\mathbf{G} = \mathbf{I} + \mathbf{C}_M\,\mathbf{G}^t\,\mathbf{C}_D{}^{-1}\,\mathbf{G}$ has to be considered. It is always positive definite, and its inverse always exists. In the approaches to inversion where the

(...)

a priori information on the model space is not explicitly introduced, the operator considered is $G^* G = C_M G^t C_D^{-1} G$, which may be singular. A unique solution is given to the inverse problem by replacing the inverse of $C_M G^t C_D^{-1} G$ by its generalized inverse. Wiggins (1972) suggests suppressing not only the null eigenvalues of $C_M G^t C_D^{-1} G$, but also the "small" eigenvalues. Matsu'ura and Hirata (1982) give a criterion, based on probabilistic arguments, for deciding the optimum "cutoff" value for small eigenvalues. These viewpoints are not considered in this book.

4.3.5: The posterior covariance operator in the data space. Importance of a datum

If $C_{M'}$ is the covariance operator describing a posteriori uncertainties in the solution $\mathbf{m}_\infty$, then the recalculated data

$$\mathbf{d}_\infty = \mathbf{g}(\mathbf{m}_\infty) \tag{4.50}$$

have uncertainties described by the covariance operator

$$C_{D'} \simeq G_\infty C_{M'} G_\infty{}^t . \tag{4.51}$$

Sometimes we may be interested in these posterior values and uncertainities per se (data-fitting problems), but is also possible to use the operator $C_{D'}$ to analyze the mutual independence of different data. Assume, for instance, that a priori data uncertainties were independent. If the posterior correlation between two different data is null, this means that they are resolving different parameters, and are independent.

Minster et al. (1974) introduce the concept of *data importance*. Using (4.5b)-(4.5c), the recalculated data for a linear problem are given by

$$\mathbf{d} = \mathbf{G\,m} = \mathbf{Q\,d}_{obs} + (\mathbf{I} - \mathbf{Q})\,\mathbf{G\,m}_{prior} , \tag{4.52}$$

where

$$Q = G \left[G^t C_D^{-1} G + C_M^{-1} \right]^{-1} G^t C_D^{-1} \tag{4.53a}$$

$$= G C_M G^t \left[G C_M G^t + C_D \right]^{-1} . \tag{4.53b}$$

Using (4.6a) and (4.51) gives

$$Q = C_{D'} \, C_D^{-1} . \tag{4.54}$$

The analysis of the rows of $\mathbf{Q}$ gives information on the relative data independence (Minster et al., 1974) (I prefer the term data independence, instead of data importance, because the importance of the captain's age should not equal one).

4.3.6: Are the residuals too large?

It may happen that some of the assumptions are violated. For instance, observational errors or modeling errors may be underestimated, or too much confidence may be given to the a priori model. Often, blunders exist in the data set. The Gaussian assumption is then not adequate, and other long-tailed distributions should be chosen (see Chapter 5).

It is generally not very easy to check the correctness of the assumptions. The examination of the residuals $\mathbf{d}_{obs}-\mathbf{g}(\mathbf{m}_\infty)$ and $\mathbf{m}_{prior}-\mathbf{m}_\infty$ may be of some help. The most important is, of course, a qualitative examination, after a convenient display (see, for instance, Draper and Smith, 1981), but some easy numerical tests can be performed. The easiest concerns the value of the misfit function at the minimum, $S(\mathbf{m}_\infty)$.

Let me first recall some elementary results concerning the so-called χ^2 (chi-squared) probability density.

i) Definition: Let $x_1, x_2, \dots , x_\nu$ be Gaussian independent random variables, each with zero mean and unit standard deviation, and let χ^2 be the random variable

$$\chi^2 = \sum_{i=1}^{\nu} x_i^2 . \tag{4.55}$$

The probability density of χ^2 is (see, e.g., Rao, 1973, or Afifi and Azen, 1979)

$$f(\chi^2 ; \nu) = \frac{(\chi^2)^{\frac{\nu}{2}-1}}{2^{\nu/2} \; \Gamma(\nu/2)} \exp(-\chi^2/2) \tag{4.56}$$

and is called the "chi-squared probability density with ν degrees of freedom". The reader should notice that χ^2 represents the random variable, not the square of the variable). The mathematical expectation and variance of χ^2 are respectively

$$\langle\ \chi^2\ \rangle\ =\ \nu\ , \tag{4.57}$$

and

$$\sigma^2\ =\ 2\nu\ . \tag{4.58}$$

ii) First property: Let $\mathbf{x} = \{\ x_1,\ x_2,\ ...\ ,\ x_\nu\ \}$ be a vector random variable, multivariate Gaussian, with mean $\boldsymbol{\mu}$ and covariance operator $\mathbf{C}$. The random variable

$$\chi^2\ =\ (\mathbf{x}-\boldsymbol{\mu})^t\ \mathbf{C}^{-1}\ (\mathbf{x}-\boldsymbol{\mu})\ .$$

is distributed following a χ^2 probability density with ν degrees of freedom (see, e.g., Rao, 1973, or Afifi and Azen, 1979).

iii) The standard theorem of least squares (see, e.g., Rao, 1973): Let $\mathbf{x}$ be a multivariate Gaussian random variable with covariance operator $\mathbf{C}$. For each realization $\mathbf{x}_0$ of $\mathbf{x}$, let $\hat{\beta}$ be defined by the minimization of $(x_0-\mathbf{A}\beta)^t\ \mathbf{C}^{-1}\ (x_0-\mathbf{A}\beta)$, and let χ^2 be the random variable defined by

$$\chi^2\ =\ (\mathbf{x}_0-\mathbf{A}\ \hat{\beta})^t\ \mathbf{C}^{-1}(\mathbf{x}_0-\mathbf{A}\ \hat{\beta})\ . \tag{4.59}$$

Then χ^2 is distributed following a χ^2 probability density with

$$\nu\ =\ \text{DIMENSION OF } \mathbf{x}\ \ -\ \text{RANK OF } \mathbf{A} \tag{4.60}$$

degrees of freedom.

iv) Letting

$$\mathbf{x}\ =\ \begin{bmatrix}\mathbf{d}\\ \mathbf{m}\end{bmatrix} \qquad \mathbf{x}_0\ =\ \begin{bmatrix}\mathbf{d}_{obs}\\ \mathbf{m}_{prior}\end{bmatrix}$$

$$\mathbf{C}\ =\ \begin{bmatrix}\mathbf{C}_D & 0\\ 0 & \mathbf{C}_M\end{bmatrix} \qquad \mathbf{A}\ =\ \begin{bmatrix}\mathbf{G}\\ \mathbf{I}\end{bmatrix} ,$$

the previous result shows that the minimum of (twice) the misfit function

$$\chi^2\ =\ 2\ S\ =\ (\mathbf{G}\ \mathbf{m}-\mathbf{d}_{obs})^t\ \mathbf{C}_D^{\ -1}\ (\mathbf{G}\ \mathbf{m}-\mathbf{d}_{obs})\ +\ (\mathbf{m}-\mathbf{m}_{prior})^t\ \mathbf{C}_M^{\ -1}\ (\mathbf{m}-\mathbf{m}_{prior})$$

arising in linear inverse problems, is distributed following a χ^2 distribution with

ν = DIMENSION OF THE DATA SPACE *D*

degrees of freedom (because the rank of the operator **A** in (iii) equals the dimension of the model space *M*). It is easy to show that the value of (twice) the misfit function at the minimum can be obtained by

$$\chi^2 = 2\, S_{Min}$$
$$= (\mathbf{G}\, \mathbf{m}_{prior} - \mathbf{d}_{obs})^t\, (\mathbf{G}\, \mathbf{C}_M\, \mathbf{G}^t + \mathbf{C}_D)^{-1}\, (\mathbf{G}\, \mathbf{m}_{prior} - \mathbf{d}_{obs}) \ . \qquad (4.61)$$

If, in the effective resolution of an inverse problem, a too large value of χ^2 is obtained ("too large" being understood with respect to the probability density defined by equation (4.56)), then some violation of the hypothesis has to be feared. Also, an improbably small value of χ^2 would suggest error overestimation.

The previous result applies only to linear problems. In the quasi-linear problems where least-squares apply (see section 1.7), the result remains approximately true.

4.4: The mathematics of discrete least-squares

In the methods in section 4.2 few mathematics were involved. For a full understanding of least-squares methods, we need a more precise terminology.

4.4.1: The least-squares definition of distance, norm and scalar product.

Let $\mathbf{d}_{obs}$ be the vector of observed data values, and $\mathbf{m}_{prior}$ the prior model vector. Experimental uncertainties in $\mathbf{d}_{obs}$ are described by the covariance operator $\mathbf{C}_D$, while uncertainties in $\mathbf{m}_{prior}$ are described by the covariance operator $\mathbf{C}_M$. Assuming that these covariance operators are positive definite (i.e., that they do not contain null or infinite variances, or perfect correlations), their inverses $\mathbf{C}_D^{-1}$ and $\mathbf{C}_M^{-1}$ are defined. The usual definition of distance between an arbitrary data vector **d** (resp. model vector **m**) and the observed vector $\mathbf{d}_{obs}$ (resp. the a priori vector $\mathbf{m}_{prior}$) is then

$$D(\mathbf{d},\mathbf{d}_{obs})_D = \left[(\mathbf{d}-\mathbf{d}_{obs})^t\, \mathbf{C}_D^{-1}\, (\mathbf{d}-\mathbf{d}_{obs}) \right]^{1/2} \qquad (4.62a)$$

$$D(\mathbf{m},\mathbf{m}_{prior})_M = \left[(\mathbf{m}-\mathbf{m}_{prior})^t \, C_M^{-1} \, (\mathbf{m}-\mathbf{m}_{prior}) \right]^{1/2} . \tag{4.62b}$$

With these distances we can associate the norms

$$\| \mathbf{d}_1 - \mathbf{d}_2 \|_D = D(\mathbf{d}_1,\mathbf{d}_2)_D \tag{4.63a}$$

$$\| \mathbf{m}_1 - \mathbf{m}_2 \|_M = D(\mathbf{m}_1,\mathbf{m}_2)_M \tag{4.63b}$$

which, in turn, can be associated with the scalar products

$$(\mathbf{d}_1,\mathbf{d}_2)_D = \mathbf{d}_1{}^t \, C_D^{-1} \, \mathbf{d}_2 \tag{4.64a}$$

$$(\mathbf{m}_1,\mathbf{m}_2)_M = \mathbf{m}_1{}^t \, C_M^{-1} \, \mathbf{m}_2 \tag{4.64b}$$

through

$$\| \mathbf{d} \|_D = (\mathbf{d},\mathbf{d})_D{}^{1/2} \tag{4.65a}$$

$$\| \mathbf{m} \|_M = (\mathbf{m},\mathbf{m})_M{}^{1/2} . \tag{4.65b}$$

It is well known (see, e.g., Pugachev, 1965) that covariance operators are symmetric and positive definite (if they do not contain null variances or perfect correlations). It is then easy to see that equations (4.64) satisfy the usual properties of a scalar product (in particular that they define real, adimensional, numbers). It follows (see Box 4.3) that equations (4.65) satisfy the properties of a norm, and (4.63) those of a distance.

In all rigor, the notation $\| \delta\mathbf{d} \|$ (resp. $\| \delta\mathbf{m} \|$) should be preferred to $\| \mathbf{d} \|$ (resp. $\| \mathbf{m} \|$) because the norm of a data vector (resp. a model vector) in general makes no physical sense. Only the norm of a difference has physical meaning. Although this subtlety is essential in infinite-dimensional spaces (see Chapter 7), it can be ignored here.

Norms which are defined through a scalar product which, in turn, is defined through (the inverse of) a covariance operator are termed "least-squares norms". Their relationship with usual "ℓ_2 norms" will be examined in the following section.

Box 4.3: Distances, norms and scalar products

Let E be an arbitrary space, with elements denoted $\mathbf{e}_1, \mathbf{e}_2, \ldots$ A *distance* over E associates a real number to any pair of elements $(\mathbf{e}_1,\mathbf{e}_2)$ of E. This distance is denoted $D(\mathbf{e}_1,\mathbf{e}_2)$, and has the properties

$$D(\mathbf{e}_1,\mathbf{e}_2) = 0 \Leftrightarrow \mathbf{e}_1 = \mathbf{e}_2 \qquad \text{for any } \mathbf{e}_1 \text{ and } \mathbf{e}_2 \tag{1a}$$

$$D(\mathbf{e}_1,\mathbf{e}_2) = D(\mathbf{e}_2,\mathbf{e}_1) \qquad \text{for any } \mathbf{e}_1 \text{ and } \mathbf{e}_2 \tag{1b}$$

$$D(\mathbf{e}_1,\mathbf{e}_3) \le D(\mathbf{e}_1,\mathbf{e}_2) + D(\mathbf{e}_2,\mathbf{e}_3) \qquad \text{for any } \mathbf{e}_1 , \mathbf{e}_2 \text{ and } \mathbf{e}_3 . \tag{1c}$$

If a distance over E has been defined, E is termed a *metric space.* Each element of a metric space E is called a *point* of E.

Let E be an arbitrary space, with elements denoted $\mathbf{e}_1, \mathbf{e}_2, \ldots$ If we can define the sum $\mathbf{e}_1 + \mathbf{e}_2$ of two elements of E, and the multiplication $\lambda\,\mathbf{e}$ of a real number by an element of E verifying the following conditions,

$$\mathbf{e}_1 + \mathbf{e}_2 = \mathbf{e}_2 + \mathbf{e}_1 \qquad \text{for any } \mathbf{e}_1 \text{ and } \mathbf{e}_2 \tag{2a}$$

$$(\mathbf{e}_1 + \mathbf{e}_2) + \mathbf{e}_3 = \mathbf{e}_1 + (\mathbf{e}_2 + \mathbf{e}_3) \qquad \text{for any } \mathbf{e}_1 , \mathbf{e}_2 \text{, ans } \mathbf{e}_3 \tag{2b}$$

$$\text{For any } \mathbf{e} \text{ there exists } \mathbf{0} \in M \text{ such that } \mathbf{e} + \mathbf{0} = \mathbf{e} \tag{2c}$$

$$\text{For any } \mathbf{e} \text{ there exists } (-\mathbf{e}) \text{ such that } \mathbf{e} + (-\mathbf{e}) = \mathbf{0} \tag{2d}$$

$$\lambda(\mathbf{e}_1 + \mathbf{e}_2) = \lambda\mathbf{e}_1 + \lambda\mathbf{e}_2 \qquad \text{for any } \mathbf{e}_1 , \mathbf{e}_2 \text{, and any real } \lambda . \tag{2e}$$

$$(\lambda + \mu)\mathbf{e} = \lambda\mathbf{e} + \mu\mathbf{e} \qquad \text{for any } \mathbf{e} \text{ and any real } \lambda \text{ and } \mu \tag{2f}$$

$$(\lambda\mu)\mathbf{e} = \lambda(\mu\mathbf{e}) \qquad \text{for any } \mathbf{e} \text{ and any real } \lambda \text{ and } \mu \tag{2g}$$

$$1\mathbf{e} = \mathbf{e} \qquad \text{for any } \mathbf{e} , \tag{2h}$$

then E is termed a (real) *linear space*, or *vector space*, or *linear vector space*. The elements of E are termed *vectors*.

Let E be a linear vector space. A *norm* over E associates a positive real number to any element $\mathbf{e}$ of E. This norm is denoted $\| \mathbf{e} \|$, and has the properties

(...)

$$\| \mathbf{e} \| = 0 \quad \Leftrightarrow \quad \mathbf{e} = \mathbf{0} \qquad \text{for any } \mathbf{e} \tag{3a}$$

$$\| \lambda \mathbf{e} \| = | \lambda | \; \| \mathbf{e} \| \qquad \text{for any } \mathbf{e} \text{ and any real } \lambda \tag{3b}$$

$$\| \mathbf{e}_1 + \mathbf{e}_2 \| \leq \| \mathbf{e}_1 \| + \| \mathbf{e}_2 \| \qquad \text{for any } \mathbf{e}_1 \text{ and } \mathbf{e}_2 \text{ ,} \tag{3c}$$

the last property being called the *triangular property* (the length of one side of a triangle is less than or equal to the sum of the lengths of the other two sides). A linear vector space furnished with a norm is termed a *normed linear vector space.*

Let E be a real linear vector space. A real *scalar product* over E is an application $(\mathbf{e}_1,\mathbf{e}_2) \rightarrow W(\mathbf{e}_1,\mathbf{e}_2)$ from $E \times E$ into R with the following properties:

$$W(\mathbf{e}_1+\mathbf{e}_2,\mathbf{e}_3) = W(\mathbf{e}_1,\mathbf{e}_3) + W(\mathbf{e}_2,\mathbf{e}_3) \qquad \text{for any } \mathbf{e}_1,\mathbf{e}_2, \text{ and } \mathbf{e}_3 \tag{4a}$$

$$W(\mathbf{e}_1,\mathbf{e}_2) = W(\mathbf{e}_2,\mathbf{e}_1) \qquad \text{for any } \mathbf{e}_1 \text{ and } \mathbf{e}_2 \tag{4b}$$

$$W(\lambda\mathbf{e}_1,\mathbf{e}_2) = W(\mathbf{e}_1,\lambda\mathbf{e}_2) = \lambda\, W(\mathbf{e}_1,\mathbf{e}_2) \qquad \text{for any } \mathbf{e}_1, \mathbf{e}_2, \text{ and any real } \lambda \tag{4c}$$

$$\mathbf{e} \neq \mathbf{0} \quad \Rightarrow \quad W(\mathbf{e},\mathbf{e}) > 0 \qquad \text{for any } \mathbf{e} \tag{4d}$$

Common notations for a scalar product are

$$W(\mathbf{e}_1,\mathbf{e}_2) = (\mathbf{e}_1,\mathbf{e}_2) = \mathbf{e}_1^t \, \mathbf{W} \, \mathbf{e}_2 \text{ ,} \tag{5}$$

the last having the advantage of recalling the scalar product encountered in least squares theory, where $\mathbf{W}$ is a weighting operator, the inverse of a covariance operator.

From

$$0 \leq (\mathbf{e}_1 - \lambda\mathbf{e}_2 \, , \, \mathbf{e}_1 - \lambda\mathbf{e}_2) = (\mathbf{e}_1,\mathbf{e}_1) - 2\lambda(\mathbf{e}_1,\mathbf{e}_2) + \lambda^2(\mathbf{e}_2,\mathbf{e}_2)$$

there follows, taking $\lambda = (\mathbf{e}_1,\mathbf{e}_2)/(\mathbf{e}_2,\mathbf{e}_2)$,

$$|(\mathbf{e}_1,\mathbf{e}_2)| \leq (\mathbf{e}_1,\mathbf{e}_1)^{1/2} \, (\mathbf{e}_2,\mathbf{e}_2)^{1/2} \qquad \text{(Cauchy-Schwarz inequality).} \tag{6}$$

Let $(\mathbf{e}_1,\mathbf{e}_2)$ denote a scalar product. Then

$$\| \mathbf{e} \| = (\mathbf{e},\mathbf{e})^{1/2} \tag{7}$$

(...)

is a norm. Only the triangular property needs to be proved. We have

$$\| e_1+e_2 \|^2 = (e_1+e_2 , e_1+e_2) = (e_1,e_1) + 2 (e_1,e_2) + (e_2,e_2)$$
$$= \| e_1 \|^2 + 2 (e_1,e_2) + \| e_2 \|^2$$
$$\leq \| e_1 \|^2 + 2 |(e_1,e_2)| + \| e_2 \|^2 ,$$

which, using the Schwarz inequality (6), gives

$$\| e_1+e_2 \|^2 \leq \| e_1 \|^2 + 2 \| e_1 \| \; \| e_2 \| + \| e_2 \|^2$$
$$= (\| e_1 \| + \| e_2 \|)^2 ,$$

from whence the triangular inequality follows.

4.4.2: Dual spaces:

The *dual* of a linear vector space is defined as the space of the *linear forms* over the space (i.e., the linear applications from the space into the real line $\boldsymbol{R}$).

The dual of the data space $\boldsymbol{D}$ will be denoted $\hat{\boldsymbol{D}}$, while the dual of the model space $\boldsymbol{M}$ will be denoted $\hat{\boldsymbol{M}}$.

The result of the action of $\hat{\mathbf{d}}_1 \in \hat{\boldsymbol{D}}$ (resp. $\hat{\mathbf{m}}_1 \in \hat{\boldsymbol{M}}$) over an arbitrary $\mathbf{d}_2 \in \boldsymbol{D}$ (resp. $\mathbf{m}_2 \in \boldsymbol{M}$) is denoted by $\langle \hat{\mathbf{d}}_1,\mathbf{d}_2 \rangle_D$ (resp, $\langle \hat{\mathbf{m}}_1,\mathbf{m}_2 \rangle_M$).

It is easy to see that with the definitions

$$\langle (\hat{\mathbf{d}}_1+\hat{\mathbf{d}}_2),\mathbf{d} \rangle_D = \langle \hat{\mathbf{d}}_1,\mathbf{d} \rangle_D + \langle \hat{\mathbf{d}}_2,\mathbf{d} \rangle_D ,$$

$$\langle \alpha\hat{\mathbf{d}}_1,\mathbf{d} \rangle_D = \alpha \langle \hat{\mathbf{d}}_1,\mathbf{d} \rangle_D ,$$

$$\langle (\hat{\mathbf{m}}_1+\hat{\mathbf{m}}_2),\mathbf{m} \rangle_M = \langle \hat{\mathbf{m}}_1,\mathbf{m} \rangle_M + \langle \hat{\mathbf{m}}_2,\mathbf{m} \rangle_M ,$$

and

$$\langle \alpha\hat{\mathbf{m}}_1,\mathbf{m} \rangle_M = \alpha \langle \hat{\mathbf{m}}_1,\mathbf{m} \rangle_M ,$$

the dual of a linear vector space is also a linear vector space.

For discrete spaces, it is easy to give a characterization of the elements of the dual: with any linear form $\hat{\mathbf{d}}_1$ over $\boldsymbol{D}$ (resp. $\hat{\mathbf{m}}_1$ over $\boldsymbol{M}$) we can associate constants $\{\hat{d}_1{}^i\}$ (resp. $\{\hat{m}_1{}^\alpha\}$) such that

$$\langle \hat{\mathbf{d}}_1, \mathbf{d}_2 \rangle_D = \sum_{i \in I_D} \hat{d}_1{}^i \; d_2{}^i$$

$$\langle \hat{\mathbf{m}}_1, \mathbf{m}_2 \rangle_M = \sum_{\alpha \in I_M} \hat{m}_1{}^\alpha \; m_2{}^\alpha ,$$

or, using the more compact notations introduced in section 1.1.4,

$$\begin{aligned} \langle \hat{\mathbf{d}}_1, \mathbf{d}_2 \rangle_D &= \hat{\mathbf{d}}_1{}^t \; \mathbf{d}_2 \\ \langle \hat{\mathbf{m}}_1, \mathbf{m}_2 \rangle_M &= \hat{\mathbf{m}}_1{}^t \; \mathbf{m}_2 . \end{aligned} \tag{4.66}$$

The "bracket" notation $\langle\ ,\ \rangle$ corresponds to a general mathematical notation, while the notation $(\ )^t\ (\ )$ has the advantage of allowing easy writing of complicated formulas. Also, when using column matrices to represent the components of a vector, the notation $(\ )^t\ (\ )$ directly corresponds with the usual matricial notation.

It should be noticed that the previous equations imply, in particular, that the components of a dual vector have as physical dimensions the *reciprocals* of the physical dimensions of the components of the primal vector. It is, in general, not possible intuitively to imagine an element of the dual space as belonging to the primal space. Figure 4.2 suggests a geometrical interpretation of a dual element.

Let $\mathbf{d}_0 \in D$ and $\mathbf{m}_0 \in M$. From the definition of covariance operator, it is easy to see that for arbitrary $\mathbf{d} \in D$ and $\mathbf{m} \in M$, the expressions $(\ \mathbf{C}_D{}^{-1}\ \mathbf{d}_0\ )^t\ \mathbf{d}$ and $(\ \mathbf{C}_M{}^{-1}\ \mathbf{m}_0\ )^t\ \mathbf{m}$ make sense, are linear in $\mathbf{d}$ and $\mathbf{m}$, and define real numbers. *We can then identify* $\mathbf{C}_D{}^{-1}\ \mathbf{d}_0$ *and* $\mathbf{C}_M{}^{-1}\ \mathbf{m}_0$ *as members of* $\hat{D}$ *and* $\hat{M}$ respectively: the operators $\mathbf{C}_D{}^{-1}$ and $\mathbf{C}_M{}^{-1}$ map D and M into their respective duals $\hat{D}$ and $\hat{M}$. As $\mathbf{C}_D$ and $\mathbf{C}_M$ are, by definition, positive definite, and thus regular operators, in fact they define bijections between D and M and their respective duals. This can be formally written

$$D = \mathbf{C}_D\ \hat{D} \qquad\qquad \hat{D} = \mathbf{C}_D{}^{-1}\ D \tag{4.67a}$$

$$M = \mathbf{C}_M\ \hat{M} \qquad\qquad \hat{M} = \mathbf{C}_M{}^{-1}\ M\ . \tag{4.67b}$$

These equalities are important because they allow of intuitively interpreting the dual of the data space D (resp. of the model space M) as *the space of data vectors weighted by* C_D^{-1} (resp. *the space of model vectors weighted by* C_M^{-1}). Sometimes, in ℓ_2-norm spaces, the dual space is abusively identified with the original space; we see that, in fact, they are quite different (in particular, they have different physical dimensions), and that they could be identified only in as far as it would make sense to take the identity as covariance operator.

Throughout the rest of this book, the element of $\hat{D}$ (resp. $\hat{M}$) associated to $\mathbf{d}$ (resp. $\mathbf{m}$) through the bijection defined by equation (4.67), will be denoted $\hat{\mathbf{d}}$ (resp. $\hat{\mathbf{m}}$):

$$\mathbf{d} = C_D \, \hat{\mathbf{d}} \qquad\qquad \hat{\mathbf{d}} = C_D^{-1} \, \mathbf{d} \tag{4.68a}$$

$$\mathbf{m} = C_M \, \hat{\mathbf{m}} \qquad\qquad \hat{\mathbf{m}} = C_M^{-1} \, \mathbf{m}. \tag{4.68b}$$

Example 2: Assume a two-dimensional model space with elements $\mathbf{m} = (m^1,m^2)$. If, for instance, m^1 represents a density, and m^2 a velocity, the vector (4 g cm^{-3} , 6 cm s^{-1}) belongs to M . The vector (3 g^{-1} cm^3 , 12 cm^{-1} s) belongs to the dual of M . ■

Example 3: Sometimes it happens that the components of $\mathbf{d}$ (resp. m) represent digitized values of some continuous field (see, for instance, example 42 of Chapter 1). The covariance operator C_D (resp. C_M) is then generally a smoothing operator (see, for instance, Pugachev, 1965), and the elements of D (resp. M) are *smooth* (i.e., are the discretized versions of smooth functions). The inverse C_D^{-1} (resp. C_M^{-1}) is then a "roughing" operator. Equations (4.67)-(4.68) then show that the elements of $\hat{D}$ (resp. $\hat{M}$) functions). ■

The introduction of the dual spaces helps in solving inverse problems essentially because the *gradient* of the misfit function S is *not* an element of the model space, but is an element of the dual of the model space. This cannot be ignored if we wish to use gradient methods of minimization, in which "gradient" cannot be taken as synonymous with "direction of steepest ascent" (see Box 4.7).

From (4.63) and (4.68) the following identities can be deduced:

$$\langle \, \hat{\mathbf{d}}_1 \, , \mathbf{d}_2 \, \rangle_D = (\, \mathbf{d}_1 \, , \mathbf{d}_2 \,)_D \tag{4.69a}$$

$$\langle \, \hat{\mathbf{m}}_1 \, , \mathbf{m}_2 \, \rangle_M = (\, \mathbf{m}_1 \, , \mathbf{m}_2 \,)_M \, . \tag{4.69b}$$

While (,) is termed a scalar product, $\langle \; , \; \rangle$ is called a *duality product.*

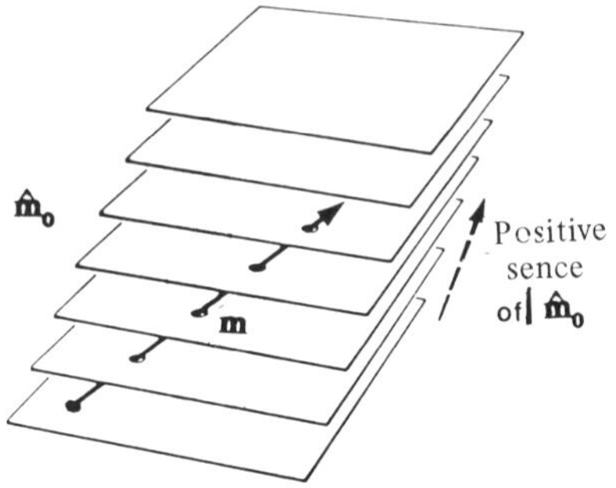
$\hat{m}_0$
Positive
sence
of $\hat{m}_0$
m

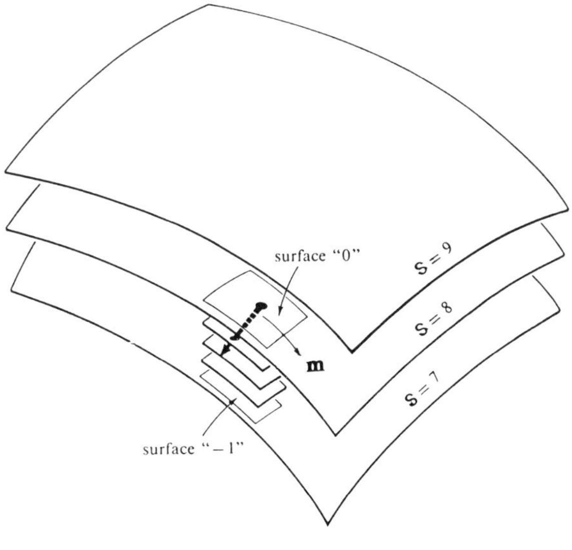
surface "0"
S = 9
S = 8
m
S = 7
surface "−1"

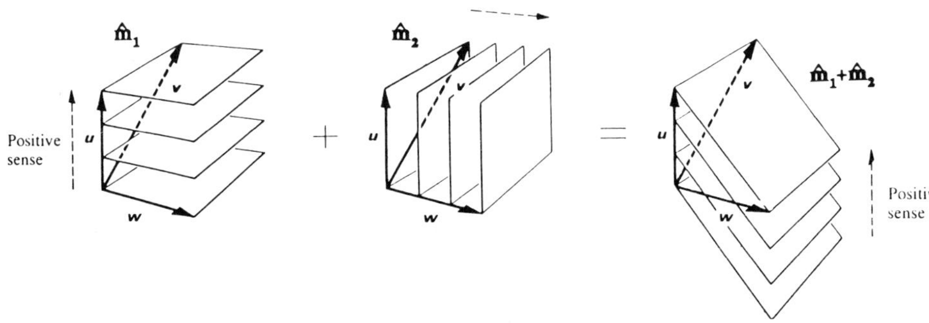
Positive sense
$\hat{m}_1$
$\hat{m}_2$
$\hat{m}_1+\hat{m}_2$
v
u
w
Positive
sense
+
=
Positive
sense

← *Figure 4.2:* Adapted from Misner, Thorne, and Wheeler (1973).

Top: Let M be a linear vector space, and $\hat{m}_0$ a particular element of the dual of M, i.e., a linear *form* over $\boldsymbol{M}$. By definition, the form $\hat{\mathbf{m}}_0$ asociates the real number $\langle \hat{\mathbf{m}}_0, \mathbf{m} \rangle$ to any vector $\mathbf{m}$ of M. If the vectors of M are intuitively represented by "arrows" in a 3D space, the correct representation of a linear form is a "millefeuilles", that associates any vector $\mathbf{m}$ with the number of "feuilles" (flakes) the vector $\mathbf{m}$ goes through. In general, it is not possible to represent an element of the dual of M as an arrow in $\boldsymbol{M}$ because: a) a scalar product is not necessarily defined over M; b) the physical dimensions of the components of $\hat{\mathbf{m}}_0$ are different from those of the components of the elements of $\boldsymbol{M}$ (see text). Of course, an element of the dual of M can be represented as an arrow in the dual space $\hat{M}$, which is in fact very different from M.

Middle: Let $\hat{\gamma}_0$ denote the *gradient* at $\mathrm{m} = \mathbf{m}_0$ of a real nonlinear functional $S(\mathbf{m})$ defined over M. By definition, $\hat{\gamma}_0$ is a form over M. The "millefeuilles" representing the gradient is tangential to the surfaces of constant S value, the flakes are flat, and their density equals the density of surfaces of constant S value at $\mathbf{m}_0$. In the figure, Misner et al. have drawn extra flakes, to show more clearly the structure of the millefeuilles. In this example, $\langle \hat{\gamma}_0, \mathbf{m} \rangle = -0.5$.

Bottom: Geometrical interpretation of the addition of two linear forms.

Defining a scalar product over the dual spaces by

$$(\hat{\mathbf{d}}_1 , \hat{\mathbf{d}}_2)_{\hat{D}} = \hat{\mathbf{d}}_1{}^t \ \mathbf{C}_D \ \hat{\mathbf{d}}_2 \tag{4.70a}$$

$$(\hat{\mathbf{m}}_1 , \hat{\mathbf{m}}_2)_{\hat{M}} = \hat{\mathbf{m}}_1{}^t \ \mathbf{C}_M \ \hat{\mathbf{m}}_2 , \tag{4.70b}$$

gives

$$(\mathbf{d}_1 , \mathbf{d}_2)_D = (\hat{\mathbf{d}}_1 , \hat{\mathbf{d}}_2)_{\hat{D}} \tag{4.71a}$$

$$(\mathbf{m}_1 , \mathbf{m}_2)_M = (\hat{\mathbf{m}}_1 , \hat{\mathbf{m}}_2)_{\hat{M}} . \tag{4.71b}$$

As the scalar products of isomorphic elements are identical, it is said that the covariance operators define an *isometric* isomorphism between the spaces $\boldsymbol{D}$ and $\boldsymbol{M}$ and their respective duals.

Traditionally, mathematical textbooks ignore the fact that the spaces under consideration do have physical dimensions and only consider the linear vector space $\boldsymbol{R}^n$ with the euclidean scalar product. This is misleading, because the dual of $\boldsymbol{R}^n$ is $\boldsymbol{R}^n$ itself, and we have already seen that the duals of the physical spaces $\boldsymbol{D}$ and $\boldsymbol{M}$ are quite different from $\boldsymbol{D}$ and $\boldsymbol{M}$. To make the link with the conventional mathematical point of view, we can introduce the *square root* of $C_M{}^{-1}$ and $C_D{}^{-1}$ (see Box 4.1):

$$C_D = C_D{}^{1/2}\ C_D{}^{t/2} \tag{4.72a}$$

$$C_M = C_M{}^{1/2}\ C_M{}^{t/2} \ , \tag{4.72b}$$

and define new spaces, without physical dimensions, by

$$\tilde{\boldsymbol{D}} = C_D{}^{-1/2}\ \boldsymbol{D} \tag{4.73a}$$

$$\tilde{\boldsymbol{M}} = C_M{}^{-1/2}\ \boldsymbol{M} \ . \tag{4.73b}$$

The scalar products over $\tilde{\boldsymbol{D}}$ and $\tilde{\boldsymbol{M}}$ induced by the least-squares scalar products over $\boldsymbol{D}$ and $\boldsymbol{M}$ are euclidean,

$$(\tilde{\mathbf{d}}_1,\tilde{\mathbf{d}}_2)_{\tilde{\mathrm{D}}} = \tilde{\mathbf{d}}_1{}^t\ \tilde{\mathbf{d}}_2 \tag{4.74a}$$

$$(\tilde{\mathbf{m}}_1,\tilde{\mathbf{m}}_2)_{\tilde{\mathrm{M}}} = \tilde{\mathbf{m}}_1{}^t\ \tilde{\mathbf{m}}_2 \ , \tag{4.74b}$$

the associated norm is the usual ℓ_2 norm,

$$\|\ \tilde{\mathbf{d}}\ \|_{\tilde{\mathrm{d}}} = (\ \tilde{\mathbf{d}}^t\ \tilde{\mathbf{d}}\ )^{1/2} \tag{4.75a}$$

$$\|\ \tilde{\mathbf{m}}\ \|_{\tilde{\mathrm{m}}} = (\ \tilde{\mathbf{m}}^t\ \tilde{\mathbf{m}}\ )^{1/2} \ , \tag{4.75b}$$

and the duals of $\tilde{\boldsymbol{D}}$ and $\tilde{\boldsymbol{M}}$ are $\tilde{\boldsymbol{D}}$ and $\tilde{\boldsymbol{M}}$ themselves. Although this approach is totally equivalent to the one developed here, it has the disadvantage of hiding both operational and intuitive aspects of least-squares inversion.

Box 4.4: The different meanings of the word "kernel".

When one asks clever people what a kernel is, the answer invariably is that it is the stone of a fruit. Mathematicians and physicists invariably give less poetical definitions. The two main ones are:

a) A Kernel may be a subspace. Let $\mathbf{M}$ and D be two metric spaces (i.e., spaces with a definition of distance), $\mathbf{m} \rightarrow \mathbf{g}(\mathbf{m})$ represent an application from M into D, δM_0 the tangent linear space to M at $\mathbf{m} = \mathbf{m}_0$, and δD the tangent linear space to D at $\mathbf{d} = \mathbf{g}(\mathbf{m}_0)$. The *tangent linear application* to $\mathbf{g}$ at $\mathbf{m} = \mathbf{m}_0$, denoted $\delta\mathbf{m} \rightarrow \mathbf{G}_0\, \delta\mathbf{m}$ may be defined by:

$$\mathbf{g}(\mathbf{m}) = \mathbf{g}(\mathbf{m}_0) + \mathbf{G}_0\,(\mathbf{m}-\mathbf{m}_0) + O(\|\mathbf{m}-\mathbf{m}_0\|^2) .$$

The subspace of elements $\delta\mathbf{m} \in \delta M_0$ such that

$$\mathbf{G}_0\, \delta\mathbf{m} = 0 \qquad (1)$$

is termed the *kernel* of $\mathbf{G}_0$ (or also the *null space* of $\mathbf{G}_0$). More generally, the (possibly nonlinear) subspace of elements $\mathbf{m} \in M$ such that

$$\mathbf{g}(\mathbf{m}) = \mathbf{g}(\mathbf{m}_0) \qquad (2)$$

is called the *kernel* of $\mathbf{g}$ at $\mathbf{m}_0$. Clearly, this concept of kernel is associated with the inversibility of an operator.

b) A kernel may be the representation of a linear operator Let M and D be two arbitrary spaces, and $\mathbf{G} : M \rightarrow D$ a *linear* operator from M into D. According to whether M or D are discrete or continuous spaces, the abstract linear equation

$$\mathbf{d} = \mathbf{G}\,\mathbf{m} \qquad (3)$$

may take one of the following explicit representations

$$d^i = \sum_{\alpha \in I_M} G^{i\alpha}\; m^\alpha \qquad (i \in I_D) \qquad (4a)$$

(...)

$$d(\mathbf{y}) = \sum_{\alpha \in I_M} G^\alpha(\mathbf{y}) \; m^\alpha \qquad (\mathbf{y} \in V_\mathbf{y}) \tag{4b}$$

$$d^i = \int_{V_\mathbf{x}} d\mathbf{x} \;\; G^i(\mathbf{x}) \;\; m(\mathbf{x}) \qquad (i \in I_D) \tag{4c}$$

$$d(\mathbf{y}) = \int_{V_\mathbf{x}} d\mathbf{x} \;\; G(\mathbf{y},\mathbf{x}) \; m(\mathbf{x}) \qquad (\mathbf{y} \in V_\mathbf{y}) . \tag{4d}$$

The array of constants $G^{i\alpha}$, the arrays of functions $G^\alpha(\mathbf{y})$ or $G^i(\mathbf{x})$, or the function $G(\mathbf{y},\mathbf{x})$ are respectively termed the *kernel* of the linear operator $\mathbf{G}$.

By extension, if $\mathbf{m}_0$ and $\mathbf{d}_0$ are given elements of linear spaces $\boldsymbol{M}$ and $\boldsymbol{D}$, the arrays of constants $m_0{}^\alpha$ or $d_0{}^i$, or the functions $m_0(\mathbf{x})$ or $d_0(\mathbf{y})$ may also be named the kernels of $\mathbf{m}$ and $\mathbf{d}$.

Box 4.5: Transposed operator and adjoint operator.

Let $\boldsymbol{M}$ and $\boldsymbol{D}$ represent two arbitrary vector spaces, and $\hat{\boldsymbol{M}}$ and $\hat{\boldsymbol{D}}$ the respective dual spaces. The duality product of $\hat{\mathbf{d}}_1 \in \hat{\boldsymbol{D}}$ by $\mathbf{d}_2 \in \boldsymbol{D}$ (resp. of $\hat{\mathbf{m}}_1 \in \hat{\boldsymbol{M}}$ by $\mathbf{m}_2 \in \boldsymbol{M}$) is denoted $\langle \; \hat{\mathbf{d}}_1 , \mathbf{d}_2 \; \rangle_D$ (resp. $\langle \; \hat{\mathbf{m}}_1 , \mathbf{m}_2 \; \rangle_M$) .

Let $\mathbf{G}$ be an arbitrary linear operator mapping $\boldsymbol{M}$ into $\boldsymbol{D}$. The *transpose* of $\mathbf{G}$ is denoted $\mathbf{G}^t$ and is a linear operator, mapping $\hat{\boldsymbol{D}}$ into $\hat{\boldsymbol{M}}$, and defined by

$$\left.\begin{array}{l} \mathbf{G} \;\; \text{linear from } \boldsymbol{M} \text{ into } \boldsymbol{D} \\ \mathbf{G}^t \;\; \text{linear from } \hat{\boldsymbol{D}} \text{ into } \hat{\boldsymbol{M}} \\ \text{for any } \hat{\mathbf{d}} \in \hat{\boldsymbol{D}} \text{ and any } \mathbf{m} \in \boldsymbol{M} : \\ \langle \; \mathbf{G}^t \hat{\mathbf{d}} , \mathbf{m} \; \rangle_M = \langle \; \hat{\mathbf{d}} , \mathbf{G} \, \mathbf{m} \; \rangle_D . \end{array}\right. \tag{1}$$

(...)

Using the pseudo-matricial notation

$$\langle\ \hat{\mathbf{d}}_1\ ,\ \mathbf{d}_2\ \rangle_D\ =\ \hat{\mathbf{d}}_1{}^t\ \mathbf{d}_2$$

$$\langle\ \hat{\mathbf{m}}_1\ ,\ \mathbf{m}_2\ \rangle_M\ =\ \hat{\mathbf{m}}_1{}^t\ \mathbf{m}_2\ ,$$

definition (1) can be written

$$(\ \mathbf{G}^t\ \hat{\mathbf{d}}\ )^t\ \mathbf{m}\ =\ \hat{\mathbf{d}}^t\ (\ \mathbf{G}\ \mathbf{m}\ )\ , \qquad (2)$$

which is the form to be used for numerical computations. For discrete spaces,

$$\langle\ \hat{\mathbf{d}}_1\ ,\ \mathbf{d}_2\ \rangle_D\ =\ \hat{\mathbf{d}}_1{}^t\ \mathbf{d}_2\ =\ \sum_{i\in I_D} \hat{d}_1{}^i\ d_2{}^i$$

$$\langle\ \hat{\mathbf{m}}_1\ ,\ \mathbf{m}_2\ \rangle_M\ =\ \hat{\mathbf{m}}_1{}^t\ \mathbf{m}_2\ =\ \sum_{\alpha\in I_M} \hat{m}_1{}^\alpha\ m_2{}^\alpha\ ,$$

and denoting the elements of $\mathbf{G}$ by $G^{i\alpha}$ $(i\in I_D)$ $(\alpha\in I_M)$:

$$(\ \mathbf{G}\ \mathbf{m}\ )^i\ =\ \sum_{\alpha\in I_M} G^{i\alpha}\ m^\alpha\ ,$$

and denoting the elements of $\mathbf{G}^t$ by $\psi^{\alpha i}$ $(\alpha\in I_M)$ $(i\in I_D)$:

$$(\ \mathbf{G}^t\ \mathbf{m}\ )^\alpha\ =\ \sum_{i\in I_D} \psi^{\alpha i}\ d^i\ ,$$

the definition of transpose operator readily gives

$$\psi^{\alpha i}\ =\ G^{i\alpha}\ . \qquad (3)$$

In the particular case where the kernel of $\mathbf{G}$ is an ordinary (two-dimensional) matrix, this equation shows that the matrix representing the operator $\mathbf{G}^t$ is the transpose of that representing $\mathbf{G}$.

In defining the transpose of a linear operator, it is not assumed that the vector spaces in consideration have a scalar product. If they do have, then it is possible to define the "adjoint" of a linear operator.

(...)

Let G represent an arbitrary linear operator mapping $\mathcal{M}$ into $\mathcal{D}$, and let $(\ ,\ )_D$ and $(\ ,\ )_M$ represent the scalar product in $\mathcal{D}$ and $\mathcal{M}$ respectively. The *adjoint* of G is denoted $\mathbf{G}^*$ and is a linear operator, mapping $\mathcal{D}$ into $\mathcal{M}$, and defined by

$$\begin{array}{l} \text{G linear from } \mathcal{M} \text{ into } \mathcal{D} \\ \mathbf{G}^* \text{ linear from } \mathcal{D} \text{ into } \mathcal{M} \\ \text{for any } \mathbf{d} \in \mathcal{D} \text{ and any } \mathbf{m} \in \mathcal{M} : \\ (\mathbf{G}^*\, \mathbf{d}\, ,\, \mathbf{m})_M \;=\; (\mathbf{d}\, ,\, \mathbf{G}\, \mathbf{m})_D \;. \end{array} \tag{4}$$

Let C_M and C_D be the covariance operators defining the bijection between the primal spaces $\mathcal{M}$ and $\mathcal{D}$ and the dual spaces $\hat{\mathcal{M}}$ and $\hat{\mathcal{D}}$ (section 4.4.2) :

$$\hat{\mathbf{m}} = \mathbf{C_M}^{-1}\ \mathbf{m} \qquad\qquad \langle\ \hat{\mathbf{m}}_1\, ,\, \mathbf{m}_2\ \rangle_M = (\ \mathbf{m}_1\, ,\, \mathbf{m}_2\ )_M$$

$$\hat{\mathbf{d}} = \mathbf{C_D}^{-1}\ \mathbf{d} \qquad\qquad \langle\ \hat{\mathbf{d}}_1\, ,\, \mathbf{d}_2\ \rangle_D = (\ \mathbf{d}_1\, ,\, \mathbf{d}_2\ )_D$$

We have successively

$$\begin{aligned} (\ \mathbf{G}^*\ \mathbf{d}\, ,\, \mathbf{m}\ )_M &= (\ \mathbf{d}\, ,\, \mathbf{G}\ \mathbf{m}\ )_D = \langle\ \hat{\mathbf{d}}\, ,\, \mathbf{G}\ \mathbf{m}\ \rangle_D \\ &= \langle\ \mathbf{G}^t\ \hat{\mathbf{d}}\, ,\, \mathbf{m}\ \rangle_M = (\ \mathbf{C_M}\ \mathbf{G}^t\ \hat{\mathbf{d}}\, ,\, \mathbf{m}\ )_M \\ &= (\ \mathbf{C_M}\ \mathbf{G}^t\ \mathbf{C_D}^{-1}\ \mathbf{d}\, ,\, \mathbf{m}\ )_M . \end{aligned}$$

This shows the relationship between $\mathbf{G}^*$ and $\mathbf{G}^t$:

$$\mathbf{G}^* = \mathbf{C_M}\ \mathbf{G}^t\ \mathbf{C_D}^{-1} \tag{5}$$

Sometimes, the terms *adjoint* and *transpose* are abusively used as synonyms. The last equation shows that they are not.

If a linear operator L maps a space $\mathcal{E}$ into its dual $\hat{\mathcal{E}}$, then, by definition, the transpose $\mathbf{L}^t$ also maps $\mathcal{E}$ into $\hat{\mathcal{E}}$. If, in that case, $\mathbf{L} = \mathbf{L}^t$, the operator is *symmetric* .

(...)

If $\mathbf{L}$ maps $\mathbf{E}$ into $\hat{\mathbf{E}}$ and for any $\mathbf{e} \in \mathbf{E}$, $\mathbf{e} \neq 0$

$$\langle \mathbf{L}\,\mathbf{e} , \mathbf{e} \rangle \quad > \quad 0 ,$$

$\mathbf{L}$ is *positive-definite* .

If a linear operator $\mathbf{L}$ maps a space E into itself, then, by definition, the adjoint $\mathbf{L}$ also maps E into itself. If, in that case, $\mathbf{L} = \mathbf{L}^*$, the operator is *self-adjoint* .

If $\mathbf{L}$ maps $\mathbf{E}$ into itself and for any $\mathbf{e} \in \mathbf{E}$, $\mathbf{e} \neq 0$

$$(\mathbf{L}\,\mathbf{e} , \mathbf{e}) \quad > \quad 0 ,$$

$\mathbf{L}$ is *positive-definite* .

In inverse problem theory, one only needs to define the inverse ($\mathbf{L}\,\mathbf{L}^{-1} = \mathbf{L}^{-1}\,\mathbf{L} = \mathbf{I}$) of operators which are either symmetric positive-definite, or self-adjoint positive-definite. These inverses always exist.

The reader will easily give sense to (and verify) the following equalities

$$(\mathbf{A}\,\mathbf{B})^t \; = \; \mathbf{B}^t\,\mathbf{A}^t$$

$$(\mathbf{A}\,\mathbf{B})^* \; = \; \mathbf{B}^*\,\mathbf{A}^*$$

$$(\mathbf{A}\,\mathbf{B})^{-1} \; = \; \mathbf{B}^{-1}\,\mathbf{A}^{-1} .$$

4.4.3: Gradient, Hessian, Steepest ascent direction, and Curvature of $S(\mathbf{m})$:

Our problem here is to obtain effective methods for the minimization of

$$S(\mathbf{m}) = \frac{1}{2}\Big[(\mathbf{g}(\mathbf{m})-\mathbf{d}_{obs})^t\,\mathbf{C}_D^{-1}\,(\mathbf{g}(\mathbf{m})-\mathbf{d}_{obs}) + (\mathbf{m}-\mathbf{m}_{prior})^t\,\mathbf{C}_M^{-1}\,(\mathbf{m}-\mathbf{m}_{prior})\Big] . \tag{4.8 again}$$

In Box 4.7 the gradient and the Hessian of $S(\mathbf{m})$ at a given point $\mathbf{m}_n$ are respectively denoted $\hat{\boldsymbol{\gamma}}_n$ and $\hat{\mathbf{H}}_n$ and are defined by the second-order development

$$S(\mathbf{m}_n+\delta\mathbf{m}) \; = \; S(\mathbf{m}_n) + \langle \; \hat{\boldsymbol{\gamma}}_n \, , \delta\mathbf{m} \; \rangle \;\; \frac{1}{2}\,\langle \; \hat{\mathbf{H}}_n\,\delta\mathbf{m} \, , \delta\mathbf{m} \; \rangle \;\; O(\,\|\,\delta\mathbf{m}\,\|^3), \tag{4.76}$$

which gives

$$\hat{\gamma}_n{}^\alpha = \left(\frac{\partial S}{\partial m^\alpha} \right)_{\mathbf{m}_n} \tag{4.77a}$$

and

$$\hat{H}_n{}^{\alpha\beta} = \left(\frac{\partial^2 S}{\partial m^\alpha \partial m^\beta} \right)_{\mathbf{m}_n} = \left(\frac{\partial \hat{\gamma}^\alpha}{\partial m^\beta} \right)_{\mathbf{m}_n} , \tag{4.78a}$$

or, for short,

$$\hat{\gamma}_n = \left(\frac{\partial S}{\partial \mathbf{m}} \right)_n \tag{4.77b}$$

and

$$\hat{H}_n = \left(\frac{\partial^2 S}{\partial \mathbf{m}^2} \right)_n = \left(\frac{\partial \hat{\gamma}}{\partial \mathbf{m}} \right)_n . \tag{4.78b}$$

Equation (4.76) allows an abstract understanding of what gradient and Hessian are; in particular, it is seen that $\hat{\gamma}_n$ is an element of $\hat{\boldsymbol{M}}$, dual of $\boldsymbol{M}$, and that $\hat{H}_n$ is an operator mapping $\boldsymbol{M}$ into $\hat{\boldsymbol{M}}$. Equations (4.77) and equations (4.78) allow the effective computation of the components of their kernels.

As outlined in section 4.4.2, the gradient of $S(\mathbf{m})$ at $\mathbf{m}_n$ is not a vector of the model space $\boldsymbol{M}$, but can be intuitively interpreted as a "mille-feuille", located at $\mathbf{m}_n$, which associates the number of flakes the vector $\delta\mathbf{m}$ goes through with any vector $\delta\mathbf{m}$ at $\mathbf{m}_n$.

Using the results of section 4.2.5, we have

$$\hat{\gamma}_n = \mathbf{G}_n{}^t \ \mathbf{C}_D{}^{-1} \ (\mathbf{g}(\mathbf{m}_n)-\mathbf{d}_{obs}) + \mathbf{C}_M{}^{-1} \ (\mathbf{m}_n-\mathbf{m}_{prior}) . \tag{4.79}$$

$$\hat{H}_n \simeq \mathbf{G}_n{}^t \ \mathbf{C}_D{}^{-1} \ \mathbf{G}_n + \mathbf{C}_M{}^{-1} , \tag{4.80}$$

where

$$G_n{}^{i\alpha} = \left(\frac{\partial g^i}{\partial m^\alpha} \right)_{\mathbf{m}_n} , \tag{4.81a}$$

or, for short,

$$\mathbf{G}_n = \left(\frac{\partial \mathbf{g}}{\partial \mathbf{m}} \right)_n , \tag{4.81b}$$

and where the second-order derivatives of $g^i(\mathbf{m})$ have been neglected in (4.80). With a little training, equations (4.79) and (4.80) can directly be obtained from equation (4.8) using equations (4.77b) and (4.78b) and a formal differentiation "with respect to the vector $\mathbf{m}$".

In most problems, the derivatives (4.81) can be obtained analytically. If not, a finite-difference approximation should be used.

It is shown in Box 4.7 that the *direction of steepest ascent* at $\mathbf{m}_n$, with respect to the metric defined through equation (4.63), is given by

$$\gamma_n = \mathbf{C}_M \hat{\gamma}_n , \tag{4.82}$$

while the *curvature operator* is defined by

$$\mathbf{H}_n = \mathbf{C}_M \hat{\mathbf{H}}_n . \tag{4.83}$$

The direction of steepest ascent is an element of M , while the curvature operator maps $\boldsymbol{M}$ into itself.

This gives, for the direction of steepest ascent,

$$\gamma_n = \mathbf{C}_M \mathbf{G}_n^t \mathbf{C}_D^{-1} (\mathbf{g}(\mathbf{m}_n)-\mathbf{d}_{obs}) + (\mathbf{m}_n-\mathbf{m}_{prior}) , \tag{4.84}$$

and, for the curvature operator,

$$\mathbf{H}_n \simeq \mathbf{I} + \mathbf{C}_M \mathbf{G}_n^t \mathbf{C}_D^{-1} \mathbf{G}_n . \tag{4.85}$$

Introducing the *adjoint* of $\mathbf{G}_n$ (see Box 4.5):

$$\mathbf{G}_n^* = \mathbf{C}_M \mathbf{G}_n^t \mathbf{C}_D^{-1} , \tag{4.86}$$

equations (4.84) and (4.85) can be written more simply:

$$\gamma_n = \mathbf{G}_n^* (\mathbf{g}(\mathbf{m}_n)-\mathbf{d}_{obs}) + (\mathbf{m}_n-\mathbf{m}_{prior}) , \tag{4.87}$$

and

$$\mathbf{H}_n \simeq \mathbf{I} + \mathbf{G}_n^* \mathbf{G}_n . \tag{4.88}$$

Box 4.6: The notation $O(\|\delta m \|^r)$.

Let $\mathbf{u}$ be an application from a linear vector space ΔM into a linear vector space ΔD , furnished respectively with norms $\| \cdot \|_M$ and $\| \cdot \|_D$. We say that

$$\mathbf{u}(\delta\mathbf{m}) \textit{ is of order } \| \delta\mathbf{m} \|_M^r , \tag{1}$$

and we write

$$\mathbf{u}(\delta\mathbf{m}) = O(\| \delta\mathbf{m} \|_M^r) , \tag{2}$$

if there exists a real number K such that for $\| \delta\mathbf{m} \|_M$ *small enough*,

$$\| \mathbf{u}(\delta\mathbf{m}) \|_D \leq K \| \delta\mathbf{m} \|_M^r . \tag{3}$$

4.4.4: Introduction to gradient methods

Let $\mathbf{m}_n$ represent the "current point", and $\boldsymbol{\gamma}_n$ the direction of steepest ascent at $\mathbf{m}_n$. A crude steepest descent method defines an "updated point" $\mathbf{m}_{n+1}$ by

$$\mathbf{m}_{n+1} = \mathbf{m}_n - \mu_n \boldsymbol{\gamma}_n , \tag{4.89}$$

where μ_n is an arbitrary positive real number small enough to ensure that $S(\mathbf{m}_{n+1})$ will be smaller than $S(\mathbf{m}_n)$ (its existence is guaranteed because, by definition, $-\boldsymbol{\gamma}_n$ is a direction of descent for S at $\mathbf{m}_n$).

There are many ways of giving a definite value to μ_n . Among them:

i) by trial and error,

ii) by interpolation; for instance, when $S(\mathbf{m}_{n+1})$ is computed for three different values of μ_n , a parabola is fitted to these values, and the value of μ_n giving the minimum of the parabola is used,

iii) by linearization of $\mathbf{g}(\mathbf{m})$ around $\mathbf{m}_n$:

$$\mathbf{g}(\mathbf{m}_n - \mu_n \boldsymbol{\gamma}_n) \simeq \mathbf{g}(\mathbf{m}_n) - \mu_n \mathbf{G}_n \boldsymbol{\gamma}_n .$$

The last choice is the simplest, although not very robust. In all the formulas given below, the value of μ_n corresponding to that linearization is shown. The reader may decide to use other values. In any case, he has to be aware that if the function $\mathbf{g}(\mathbf{m})$ is not linear, the function $S(\mathbf{m})$ is not quadratic, so whatever the method used for evaluating a value for μ_n, the condition $S(\mathbf{m}_{n+1}) < S(\mathbf{m}_n)$ has to be checked, and the computed value of μ_n eventually has to be diminished until the condition is satisfied.

The direction of steepest descent is, by definition, locally optimal, and if infinitely many steps were taken, each one being infinitely small, this would simulate the motion of a drop of rain on the slope of a mountain. In a steepest descent method, however, we wish to take finite steps (in fact, as long as possible), so that the path obtained is generally not optimum (Figure 4.3). The Newton method (section 4.5.3) uses not only the information on the direction of steepest ascent at $\mathbf{m}_n$ for defining a direction of search, but also uses curvature information, as given by the curvature operator $\mathbf{H}_n$. The algorithm is given by

$$\mathbf{m}_{n+1} \simeq \mathbf{m}_n - \mathbf{H}_n^{-1} \gamma_n , \tag{4.90}$$

and it can be seen that the point $\mathbf{m}_{n+1}$ so defined corresponds to the minimum of the "paraboloid" tangent at S at the point $\mathbf{m}_n$ (figures 4.4 and 4.5). Of course, the Newton algorithm converges much more rapidly than the steepest descent algorithm. The trouble is that the operator $\mathbf{H}_n^{-1}$ may be difficult to handle. By comparison of (4.89) and (4.90), we see that if the operator $\hat{\mathbf{S}}_n$ is a reasonable approximation of $\mathbf{H}_n^{-1}$, but more easy to handle, the algorithm

$$\mathbf{m}_{n+1} = \mathbf{m}_n - \mu_n \hat{\mathbf{S}}_n \gamma_n , \tag{4.91}$$

where μ_n is again defined so as to minimize $S(\mathbf{m}_{n+1})$, can greatly improve the convergence with respect to the steepest descent algorithm. The choice $\hat{\mathbf{S}}_n = \mathbf{I}$ gives the steepest descent, while the choice $\hat{\mathbf{S}}_n = \mathbf{H}_n^{-1}$ gives the Newton algorithm. As for everything in life, the best way is obtained by a compromise between different tendencies: here we need an operator which has to approach $\mathbf{H}_n^{-1}$ as closely as possible, but which remains manageable.

The "preconditioned" steepest descent (section 4.5.1) is obtained when the operator $\hat{\mathbf{S}}_n$ takes a constant value $\hat{\mathbf{S}}_0$, independent of the iteration number. The variable metric method (section 4.5.4) gives a simple formula for updating $\hat{\mathbf{S}}_n$ in such a way that $\hat{\mathbf{S}}_n \to \mathbf{H}_n^{-1}$.

The "conjugate gradients" method (section 4.5.2) is based on the following idea: Let $\mathbf{m}_0$ be the starting point, and γ_0 the direction of steepest ascent at $\mathbf{m}_0$. The point $\mathbf{m}_1$ is defined as in the steepest descent method. Let γ_1 be the direction of steepest ascent at $\mathbf{m}_1$. The point $\mathbf{m}_2$ is not

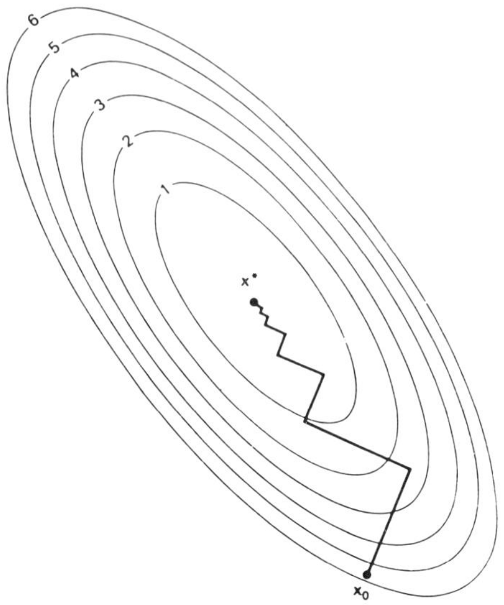

Figure 4.3: The path obtained using the steepest descent method of minimization in a schematic problem with two variables (from Walsh, 1975).

defined as the point minimizing S in the direction given by $\boldsymbol{\gamma}_1$, but as the point minimizing S *in the subspace generated by* $\boldsymbol{\gamma}_0$ *and* $\boldsymbol{\gamma}_1$. The point $\mathbf{m}_3$ is defined as the point minimizing S in the subspace generated by $\boldsymbol{\gamma}_0$, $\boldsymbol{\gamma}_1$ and $\boldsymbol{\gamma}_3$, and so on until convergence. It can be shown (see, for instance, Fletcher, 1980) that this method generally converges at the same rate as the variable metric method (it has quadratic convergence for linear problems). The miracle is that the computations needed to perform this method are not more difficult than those needed in using the steepest descent method. To "precondition" the conjugate gradients means using $\hat{S}_0\, \boldsymbol{\gamma}_n$ instead of $\boldsymbol{\gamma}_n$.

Appendix 4.7 gives more details on gradient methods.

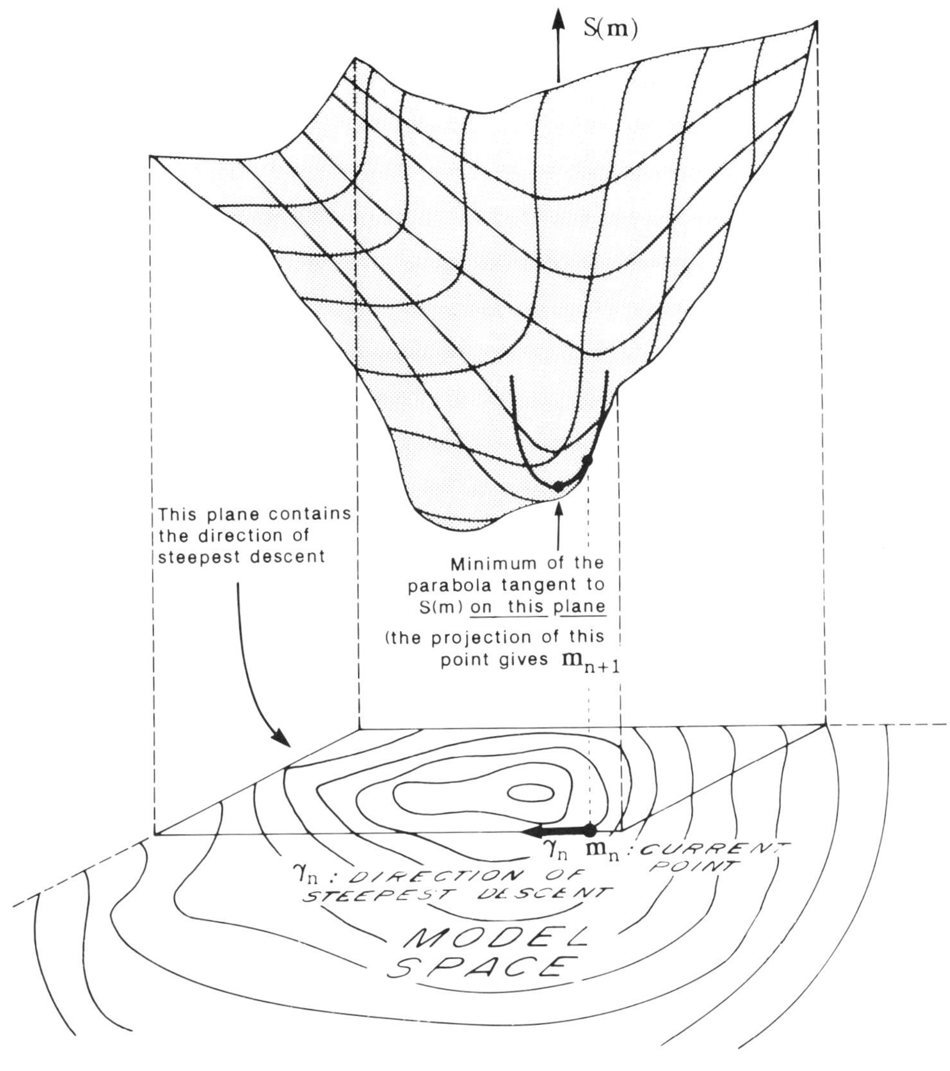

Figure 4.4: In the steepest descent method of minimization, a minimum of $S(\mathbf{m})$ is sought for *in the direction of steepest descent* . This figure illustrates, for a two-dimensional problem, the "level lines" of $S(\mathbf{m})$ (at bottom), and the "surface" representing $S(\mathbf{m})$. The easiest way of defining a step length along the direction of steepest descent is by fitting a parabola at the current point in the intersection of the surface $S(\mathbf{m})$ with the plane containing the direction of steepest descent, and by choosing as updated point the projection of the point at the minimum of the parabola.

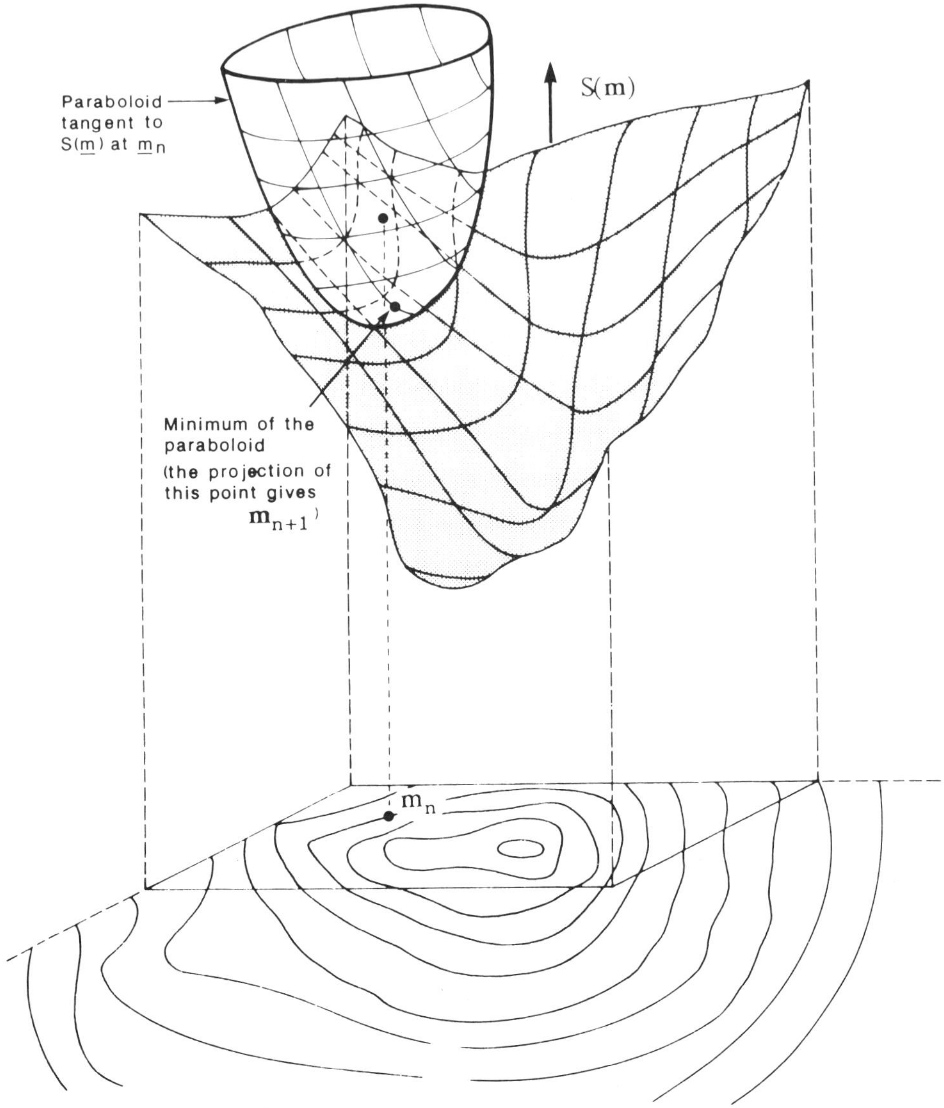

Figure 4.5: In the Newton method, the tangent paraboloid to S(**m**) at the current point is defined (using curvature information), and the updated point is the projection of the minimum of the paraboloid.

Box 4.7: Gradient and direction of steepest ascent.

Let $\mathbf{m} \rightarrow S(\mathbf{m})$ be a nonlinear form over a discrete vector space M (i.e., a nonlinear application from M into R). Let $\mathbf{C}$ be the covariance operator defining the scalar product over M :

$$(\mathbf{m}_1 , \mathbf{m}_2) = \mathbf{m}_1^t \ \mathbf{C}^{-1} \ \mathbf{m}_2 \ . \tag{1}$$

The (ℓ_2) norm over M is defined as usual:

$$\| \mathbf{m} \| = (\mathbf{m} , \mathbf{m})^{1/2} = (\mathbf{m}^t \ \mathbf{C}^{-1} \ \mathbf{m})^{1/2}$$

The *gradient* of S at $\mathbf{m} = \mathbf{m}_n$ is denoted $\hat{\gamma}_n$ and is defined by

$$\hat{\gamma}_n{}^\alpha = \left(\frac{\partial S}{\partial m^\alpha} \right)_{\mathbf{m}_n} \ . \tag{2}$$

The *Hessian* of S at $\mathbf{m} = \mathbf{m}_n$ is denoted $\hat{\mathbf{H}}_n$ and is defined by

$$\hat{H}_n{}^{\alpha\beta} = \left(\frac{\partial^2 S}{\partial m^\alpha \ \partial m^\beta} \right)_{\mathbf{m}_n} \ . \tag{3}$$

A second-order development of $S(\mathbf{m})$ around $\mathbf{m}_n$ gives

$$S(\mathbf{m}_n + \delta\mathbf{m}) = S(\mathbf{m}_n) + \sum_\alpha \left(\frac{\partial S}{\partial m^\alpha} \right)_{\mathbf{m}_n} \delta m^\alpha + \frac{1}{2} \sum_\alpha \left(\frac{\partial^2 S}{\partial m^\alpha \ \partial m^\beta} \right)_{\mathbf{m}_n} \delta m^\alpha \ \delta m^\beta + O(\| \delta\mathbf{m} \|^3) \ ,$$

where $O(\| \delta\mathbf{m} \|^3)$ denotes a term which tends to zero more rapidly than the second order term when $\| \delta\mathbf{m} \| \rightarrow \mathbf{0}$. Using the definitions of gradient and Hessian, the second-order development can be written

$$S(\mathbf{m}_n + \delta\mathbf{m}) = S(\mathbf{m}_n) + \hat{\gamma}_n{}^t \ \delta\mathbf{m} + \frac{1}{2} (\hat{\mathbf{H}}_n \ \delta\mathbf{m})^t \ \delta\mathbf{m} + O(\| \delta\mathbf{m} \|^3) \ . \tag{4a}$$

This equation shows that the gradient $\hat{\gamma}_n$ at a given point $\mathbf{m} = \mathbf{m}_n$ defines a linear application from M into the real line R : it is an element of $\hat{M}$, dual of M . The Hessian $\mathbf{H}_n$ at a given point $\mathbf{m} = \mathbf{m}_n$ can be interpreted as a linear operator mapping M into the dual $\hat{M}$. Let $\hat{\mathbf{f}}$ be an element of $\hat{M}$. Denoting the real number associated by $\hat{\mathbf{f}}$ with $\mathbf{m} \in M$ by $\langle \hat{\mathbf{f}}, \mathbf{m} \rangle$, equation (4a) can be rewritten into the form

(...)

$$S(\mathbf{m}_n+\delta\mathbf{m}) = S(\mathbf{m}_n) + \langle \hat{\gamma}_n , \delta\mathbf{m} \rangle + \frac{1}{2} \langle \hat{H}_n \delta\mathbf{m} , \delta\mathbf{m} \rangle + O(\|\delta\mathbf{m}\|^3) , \tag{4b}$$

which can be taken as an abstract definition of gradient and Hessian.

Consider a "circle" of radius R around a point $\mathbf{m}_n$ (i.e., a set of points $\mathbf{m}$ such that $\|\mathbf{m}-\mathbf{m}_n\| = R$). The point of the circle in which $S(\mathbf{m})$ is maximum when the radius R tends to zero defines the *direction of steepest ascent* at $\mathbf{m}_n$. It should be noticed that this does not define a "vector" of steepest ascent, but only a direction, because the norm remains undefined. Letting $\mathbf{d}_n$ be a vector of fixed norm $\|\mathbf{d}_n\| = N$, the previous geometric definition can be formalized as follows: $\mathbf{d}_n$ is colinear to the direction of steepest ascent at $\mathbf{m}_n$ if

$$S(\mathbf{m}_n+\epsilon\mathbf{d}_n) \qquad \text{is maximum when} \qquad \epsilon \rightarrow 0 .$$

When $\epsilon \rightarrow 0$ a first order development of $S(\mathbf{m})$ gives

$$\begin{aligned} S(\mathbf{m}_n + \epsilon\, \mathbf{d}_n) &= S(\mathbf{m}_n) + \epsilon \sum_{\alpha} \left(\frac{\partial S}{\partial m^\alpha} \right)_{\mathbf{m}_n} d_n^{\alpha} \\ &= S(\mathbf{m}_n) + \epsilon\, \hat{\gamma}_n^{t}\, \mathbf{d}_n , \end{aligned}$$

so that the problem of obtaining $\mathbf{d}_n$ is now equivalent to the problem:

find $\mathbf{d}_n$ such that $\qquad \hat{\gamma}_n^{t}\, \mathbf{d}_n \qquad$ is maximum

with the constraint $\qquad \|\mathbf{d}_n\|^2 = N^2 .$

This can be solved using the method of Lagrange's multipliers (see appendix 4.6). Letting λ be the Lagrange parameter of the problem, we should now maximize the function

$$\begin{aligned} \Psi(\mathbf{d}_n , \lambda) &= \hat{\gamma}_n^{t}\, \mathbf{d}_n - \frac{\lambda}{2} (\|\mathbf{d}_n\|^2 - N^2) , \\ &= \hat{\gamma}_n^{t}\, \mathbf{d}_n - \frac{\lambda}{2} (\mathbf{d}_n^{t}\, \mathbf{C}^{-1}\, \mathbf{d}_n - N^2) . \\ &= \hat{\gamma}_n^{t}\, \mathbf{d}_n - \frac{\lambda}{2} \left[\sum_{\alpha} \sum_{\beta} d_n^{\alpha}\, (C^{-1})^{\alpha\beta}\, d_n^{\beta} - N^2 \right] . \end{aligned}$$

(...)

The condition $\partial S/\partial \lambda = 0$ simply gives the constraint

$$\mathbf{d}_n{}^t\ \mathbf{C}^{-1}\ \mathbf{d}_n = N^2\ , \tag{5}$$

while the condition $\partial S/\partial d_n{}^\alpha = 0$ gives

$$\hat{\gamma}_n{}^\alpha - \lambda\ (\mathbf{C}^{-1}\ \mathbf{d}_n)^\alpha = 0\ ,$$

i.e.,

$$\mathbf{d}_n = \frac{1}{\lambda}\ \mathbf{C}\ \hat{\gamma}_n\ .$$

The value of λ can then be obtained from (5):

$$\lambda = \frac{(\hat{\gamma}_n{}^t\ \mathbf{C}\ \hat{\gamma}_n)^{1/2}}{N}\ .$$

The value of $N = \|\mathbf{d}_n\|$ is, until now, arbitrary. The choice

$$N = (\hat{\gamma}_n{}^t\ \mathbf{C}\ \hat{\gamma}_n)^{1/2}$$

imposes that the norm of $\mathbf{d}_n$ equals the norm of $\hat{\gamma}_n$. It simply gives $\lambda = 1$ and

$$\boxed{\mathbf{d}_n = \mathbf{C}\ \hat{\gamma}_n\ .} \tag{6}$$

This last equation shows then that $\mathbf{d}_n$ and $\hat{\gamma}$ are related by the usual relation between isomorphic elements of $\mathcal{M}$ and $\hat{\mathcal{M}}$, thus justifying the notation

$$\mathbf{d}_n = \gamma_n\ . \tag{7}$$

Let $\phi(\mathbf{m})$ be the tangent linear application to $S(\mathbf{m})$ at $\mathbf{m} = \mathbf{m}_n$:

$$\phi(\mathbf{m}) = S(\mathbf{m}_n) + \hat{\gamma}_n{}^t\ (\mathbf{m} - \mathbf{m}_n)\ .$$

(...)

We have

$$\phi\left(\mathbf{m}_n + \frac{\boldsymbol{\gamma}_n}{\|\boldsymbol{\gamma}_n\|}\right) - \phi(\mathbf{m}_n) = \frac{\hat{\boldsymbol{\gamma}}_n{}^t\, \boldsymbol{\gamma}_n}{\|\boldsymbol{\gamma}_n\|} = \|\boldsymbol{\gamma}_n\| .$$

This shows that the norm of $\boldsymbol{\gamma}_n$ represents the variation of the linear application tangent to $S(\mathbf{m})$ at $\mathbf{m}_n$ per unit norm-length of variation of $\mathbf{m}$ in the direction of the steepest ascent, i.e., the norm of $\boldsymbol{\gamma}_n$ represents the *slope* of S at $\mathbf{m}_n$. See Figure 4.6

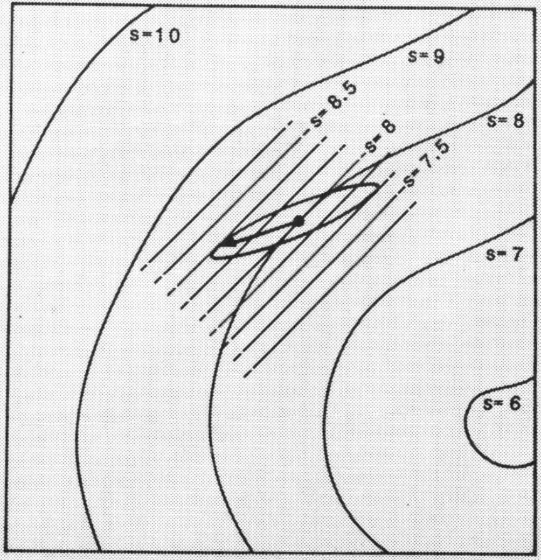

Figure 4.6: Level lines of a nonlinear real function $S(\mathbf{m})$ in a 2D space, the "millefeuilles" representing the gradient of S at a given point $\mathbf{m}_0$, the unit "circle" around $\mathbf{m}_0$ (with respect to some given metric), and the direction of steepest ascent.

(...)

The operator

$$\mathbf{H}_n = \mathbf{C}\,\hat{\mathbf{H}}_n = \left(\frac{\partial \boldsymbol{\gamma}}{\partial \mathbf{m}}\right)_{\mathbf{m}_n} \tag{8}$$

is the *curvature* operator.

So, the terminology is as follows:

$\hat{\boldsymbol{\gamma}}_n = \left(\frac{\partial S}{\partial \mathbf{m}}\right)_{\mathbf{m}_n}$: **gradient of S at $\mathbf{m}_n$**

$\boldsymbol{\gamma}_n = \mathbf{C}\,\hat{\boldsymbol{\gamma}}_n$: **direction of steepest ascent at $\mathbf{m}_n$**

$\hat{\mathbf{H}}_n = \left(\frac{\partial^2 S}{\partial^2 \mathbf{m}}\right)_{\mathbf{m}_n} = \left(\frac{\partial \hat{\boldsymbol{\gamma}}}{\partial \mathbf{m}}\right)_{\mathbf{m}_n}$: **Hessian of S at $\mathbf{m}_n$**

$\mathbf{H}_n = \mathbf{C}\,\hat{\mathbf{H}}_n = \left(\frac{\partial \boldsymbol{\gamma}}{\partial \mathbf{m}}\right)_{\mathbf{m}_n}$: **curvature of S at $\mathbf{m}_n$.**

$\boldsymbol{\gamma}_n$ is sometimes abusively called the gradient, while $\mathbf{H}_n$ is sometimes termed the Hessian. This terminology can be misleading because although these elements are isomorphic, they are by no means identical.

4.4.5: The choice of the optimization method

a) Newton methods versus gradient methods. The use of a Newton or quasi-Newton method implies the resolution of a linear system at each iteration, with the linear operator equal, for instance, to $(\mathbf{G}^t\,\mathbf{C}_D^{-1}\,\mathbf{G} + \mathbf{C}_M^{-1})$, or to $(\mathbf{C}_D + \mathbf{G}\,\mathbf{C}_M\,\mathbf{G}^t)$ (see section 4.5.3). For *small sized* problems, the components of model and data vectors can be arranged into column matrices. The two-dimensional symmetric, positive definite matrices representing $(\mathbf{G}^t\,\mathbf{C}_D^{-1}\,\mathbf{G} + \mathbf{C}_M^{-1})$ or $(\mathbf{C}_D + \mathbf{G}\,\mathbf{C}_M\,\mathbf{G}^t)$ can then be explicitly built, and direct methods of resolution can be used (such as, for instance, the Cholesky's decomposition of appendix 4.5, or the Lanczos decomposition of Box 4.2). The dimension of the matrix corresponding to $(\mathbf{G}^t\,\mathbf{C}_D^{-1}\,\mathbf{G} + \mathbf{C}_M^{-1})$ is (Dimension of model space × Dimension of model space) , while the dimension of the matrix $(\mathbf{C}_D + \mathbf{G}\,\mathbf{C}_M\,\mathbf{G}^t)$ is (Dimension of data space × Dimension of data space) . As these can be very different, the operator of smaller dimension should be chosen.

The use of iterative methods of resolution of the Newton linear system can of course also be used (appendix 4.4), but if one wishes to use iterative methods, it is best to drop the Newton method of optimization completely and use gradient methods which directly avoid the resolution of linear systems.

If the covariance operators $\mathbf{C}_M$ and $\mathbf{C}_D$ are positive definite (non-null or infinite variances, nor perfect correlations), it is easy to see that the operators $(\mathbf{G}^t\ \mathbf{C}_D^{-1}\ \mathbf{G} + \mathbf{C}_M^{-1})$ and $(\mathbf{C}_D + \mathbf{G}\ \mathbf{C}_M\ \mathbf{G}^t)$ are also positive definite, so their inverses always exist. Of course, an operator may be positive definite in theory, but the finite accuracy of computers may transform the operator into one numerically singular. In practical computations when realistic a priori information on the solution is introduced using $\mathbf{C}_M$, no numerical troubles generally arise. If they do, we have to resort to a trick, and diminish variances (or correlations) in $\mathbf{C}_M$ artificially until a good conditioning of the operator is obtained.

The Newton method is not well adapted to large-sized inverse problems, which are more easily solved using gradient methods.

b) The use of gradient methods. The choice of the preconditioning operator $\hat{\mathbf{S}}_0 \simeq (\mathbf{I} + \mathbf{C}_M\ \mathbf{G}_0{}^t\ \mathbf{C}_D^{-1}\ \mathbf{G}_0)^{-1} = (\mathbf{G}_0{}^t\ \mathbf{C}_D^{-1}\ \mathbf{G}_0 + \mathbf{C}_M^{-1})^{-1}\ \mathbf{C}_M^{-1}$ is important if one needs to accelerate the convergence. For instance, taking $\hat{\mathbf{S}}_0 = \mathbf{I}$ in the preconditioned steepest descent method gives the crude steepest descent, which in general converges very slowly.

If the direct estimation of $(\mathbf{I} + \mathbf{C}_M\ \mathbf{G}_0{}^t\ \mathbf{C}_D^{-1}\ \mathbf{G}_0)^{-1}$ is too difficult, a good strategy is the following: the first iterate of a gradient method (either steepest descent or conjugate gradients) is

$$\mathbf{m}_1 = \mathbf{m}_{prior} - \mu_0\ \hat{\mathbf{S}}_0\ \mathbf{C}_M\ \mathbf{G}_0{}^t\ \mathbf{C}_D^{-1}\ (\mathbf{g}(\mathbf{m}_{prior}) - \mathbf{d}_{obs}))$$

Taking a synthetic example, in which $\mathbf{d}_{obs}$ has in fact been computed from a "true" model $\mathbf{m}_{true}$,

$$\mathbf{d}_{obs} = \mathbf{g}(\mathbf{m}_{true}) \ ,$$

gives

$$\mathbf{m}_1 = \mathbf{m}_{prior} + \mu_0\ \hat{\mathbf{S}}_0\ \delta\mathbf{m}_0$$

where

$$\delta\mathbf{m}_0 = \mathbf{C}_M\ \mathbf{G}_0{}^t\ \mathbf{C}_D^{-1}\ (\mathbf{g}(\mathbf{m}_{true}) - \mathbf{g}(\mathbf{m}_{prior})) \ .$$

The idea is to carefully analyze $\delta\mathbf{m}_0$ and to use our physical intuition to infer an ad-hoc "deconvolution" operator $\hat{\mathbf{S}}_0$ such that

$$\hat{\mathbf{S}}_0\ \delta\mathbf{m} \simeq \mathbf{m}_{true} - \mathbf{m}_{prior} \ , \qquad (4.92)$$

which gives $\mu_0 \simeq 1$ and $\mathbf{m}_1 \simeq \mathbf{m}_{true}$.

The algorithm using conjugate directions and the variable metric algorithm usually have a very similar convergence rate (much better than the convergence rate of the steepest descent algorithm). In as far as the conjugate directions are not much more difficult to obtain than the steepest descent directions, there is no reason to prefer the steepest descent method (except, perhaps, its conceptual simplicity).

The choice between conjugate directions and variable metric has to be made by considering that the first needs less computer memory, but that the second gives a direct approximation to the posterior covariance operator (because $\hat{\mathbf{S}}_n \mathbf{C}_M \rightarrow \mathbf{C}_{M'}$).

4.5: Methods of resolution (II)

4.5.1: Preconditioned steepest descent

From appendix 4.7, and the results of the previous section, we obtain the algorithm:

$$\gamma_n = \mathbf{C}_M \mathbf{G}_n^t \mathbf{C}_D^{-1} (\mathbf{g}(\mathbf{m}_n)-\mathbf{d}_{obs}) + (\mathbf{m}_n-\mathbf{m}_{prior}) \qquad (\mathbf{m}_0 \text{ arbitrary})$$

$$\phi_n = \hat{\mathbf{S}}_0 \gamma_n \qquad (\hat{\mathbf{S}}_0 \text{ arbitrary}) \qquad (4.93)$$

$$\mathbf{m}_n+1 = \mathbf{m}_n - \mu_n \phi_n \qquad (\text{obtain } \mu_n \text{ by linear search})$$

The starting point $\mathbf{m}_0$ is arbitrary, the simplest choice being $\mathbf{m}_0 = \mathbf{m}_{prior}$, although the use of different starting points helps to check the existence of secondary solutions.

The simplest choice for $\hat{\mathbf{S}}_0$ is $\hat{\mathbf{S}}_0 = \mathbf{I}$. Usually a good approximation of the initial inverse Hessian

$$\hat{\mathbf{S}}_0 \simeq \left(\mathbf{I} + \mathbf{C}_M \mathbf{G}_0^t \mathbf{C}_D^{-1} \mathbf{G}_0\right)^{-1} \qquad (4.94)$$

gives good results.

An adequate value for μ_n has to be obtained by linear search. Alternatively, a linearization of $\mathbf{g}(\mathbf{m})$ around $\mathbf{m}_n$ gives

$$\mu_n \simeq \frac{\gamma_n{}^t \; C_M{}^{-1} \; \phi_n}{\phi_n{}^t \; C_M{}^{-1} \; \phi_n \; + \; b_n{}^t \; C_D{}^{-1} \; b_n} , \tag{4.95}$$

where

$$b_n = G_n \; \phi_n \; . \tag{4.96}$$

4.5.2: Preconditioned conjugate directions:

From appendix 4.7, and the results of the previous section, we obtain the algorithm:

$$\begin{array}{lll} \gamma_n = C_M \, G_n{}^t \, C_D{}^{-1} \, (g(m_n)-d_{obs}) + (m_n - m_{prior}) & & (m_0 \text{ arbitrary}) \\ \lambda_n = \hat{S}_0 \; \gamma_n & (\hat{S}_0 \text{ arbitrary}) & \\ \phi_n = \lambda_n + \alpha_n \; \phi_{n-1} & (\phi_0 = \lambda_0) & (\alpha_n \text{ defined below}) \\ m_{n+1} = m_n - \mu_n \; \phi_n & (\text{obtain } \mu_n \text{ by linear search}) & \end{array} \tag{4.97}$$

The starting point m_0 is arbitrary, the simplest choice being $m_0 = m_{prior}$, although the use of different starting points helps to check the existence of secondary solutions.

The simplest choice for $\hat{S}_0$ is $\hat{S}_0 = I$. Usually a good approximation of the initial inverse Hessian

$$\hat{S}_0 \simeq \left(I + C_M \; G_0{}^t \; C_D{}^{-1} \; G_0 \right)^{-1} \tag{4.98}$$

gives good results.

There are different common choices for α_n , for instance, those in Fletcher and Reeves (1964), Polak and Ribière (1969), and Hestenes and Stiefel (1952) which are all equivalent for quadratic optimization problems (see appendix 4.7). Powell (1977) suggests that for non quadratic optimization problems, the Polak-Ribière formula sometimes gives superior results. From the corresponding formula of appendix 4.7 we then obtain

$$\alpha_n = \frac{\omega_n - \gamma_{n-1}{}^t \; C_M{}^{-1} \; \lambda_n}{\omega_{n-1}} \tag{4.99}$$

where

$$\omega_n = \gamma_n^t \; C_M^{-1} \; \lambda_n \; . \tag{4.100}$$

The value μ_n has to be obtained by linear search. Alternatively, a linearization of $\mathbf{g(m)}$ around $\mathbf{g(m_n)}$ gives

$$\mu_n \simeq \frac{\gamma_n^t \; C_M^{-1} \; \phi_n}{\phi_n^t \; C_M^{-1} \; \phi_n + \mathbf{b}_n^t \; C_D^{-1} \; \mathbf{b}_n} \tag{4.101}$$

where

$$\mathbf{b}_n = \mathbf{G}_n \; \phi_n \; . \tag{4.102}$$

Notice that the numerator of (4.101) can be written

$$\gamma_n^t \, C_M^{-1} \, \phi_n = \gamma_n^t \, C_M^{-1} \, \lambda_n - \alpha_n \, \gamma_n^t \, C_M^{-1} \, \phi_{n-1} \; . \tag{4.103}$$

If the linear searches are accurate, the steepest ascent direction γ_n is approximately orthogonal to the previous search direction ϕ_{n-1} :

$$\gamma_n^t \; C_M^{-1} \; \phi_{n-1} \simeq 0 \; , \tag{4.104}$$

and the following simplification can be used:

$$\mu_n \simeq \frac{\omega_n}{\phi_n^t \; C_M^{-1} \; \phi_n + \mathbf{b}_n^t \; C_D^{-1} \; \mathbf{b}_n} \; . \tag{4.105}$$

4.5.3: Quasi-Newton method

The Newton method gives

$$\mathbf{m}_{n+1} = \mathbf{m}_n - \hat{\mathbf{H}}_n^{-1} \; \hat{\gamma}_n \; , \tag{4.106a}$$

or, equivalently,

$$\mathbf{m}_{n+1} = \mathbf{m}_n - \mathbf{H}_n^{-1} \; \gamma_n \; . \tag{4.106b}$$

Equation (4.106a) uses the Hessian $\hat{\mathbf{H}}_n$ and the gradient $\hat{\gamma}_n$, while (4.106b) uses the curvature $\mathbf{H}_n$ and the direction of steepest descent γ_n .

As explained in section 4.2.5, it is usual in least squares to make the approximation

$$\hat{\mathrm{H}}_n \simeq \left[\mathbf{G}_n{}^t \, \mathbf{C}_D{}^{-1} \, \mathbf{G}_n + \mathbf{C}_M{}^{-1} \right] , \tag{4.107a}$$

$$\mathrm{H}_n \simeq \left[\mathbf{I} + \mathbf{C}_M \, \mathbf{G}_n{}^t \, \mathbf{C}_D{}^{-1} \, \mathbf{G}_n \right] , \tag{4.107b}$$

where second-order derivatives of $\mathbf{g}(\mathbf{m})$ are neglected. This is the reason why, strictly speaking, the formulas in this section are "quasi-Newton". In the particular case where there is no a priori information in the model space ($\mathbf{C}_M \rightarrow \infty \, \mathbf{I}$), they are known as the "Gauss-Newton" formulas.

We directly obtain

$$\mathbf{m}_{n+1} = \mathbf{m}_n - \mu_n \left(\mathbf{G}_n{}^t \, \mathbf{C}_D{}^{-1} \, \mathbf{G}_n + \mathbf{C}_M{}^{-1} \right)^{-1} \left[\mathbf{G}_n{}^t \, \mathbf{C}_D{}^{-1} \, (\mathbf{g}(\mathbf{m}_n) - \mathbf{d}_{obs}) + \mathbf{C}_M{}^{-1} \, (\mathbf{m}_n - \mathbf{m}_{prior}) \right]$$

$$\mathbf{m}_{n+1} = \mathbf{m}_{prior} - \mu_n \left(\mathbf{G}_n{}^t \, \mathbf{C}_D{}^{-1} \, \mathbf{G}_n + \mathbf{C}_M{}^{-1} \right)^{-1} \mathbf{G}_n{}^t \, \mathbf{C}_D{}^{-1} \left[(\mathbf{g}(\mathbf{m}_n) - \mathbf{d}_{obs}) - \mathbf{G}_n \, (\mathbf{m}_n - \mathbf{m}_{prior}) \right]$$

$$\mathbf{m}_{n+1} = \mathbf{m}_{prior} - \mu_n \, \mathbf{C}_M \, \mathbf{G}_n{}^t \left(\mathbf{C}_D + \mathbf{G}_n \, \mathbf{C}_M \, \mathbf{G}_n{}^t \right)^{-1} \left[(\mathbf{g}(\mathbf{m}_n) - \mathbf{d}_{obs}) - \mathbf{G}_n \, (\mathbf{m}_n - \mathbf{m}_{prior}) \right]$$

$$\mathbf{m}_{n+1} = \mathbf{m}_n - \mu_n \left(\mathbf{I} + \mathbf{C}_M \, \mathbf{G}_n{}^t \, \mathbf{C}_D{}^{-1} \, \mathbf{G}_n \right)^{-1} \left[\mathbf{C}_M \, \mathbf{G}_n{}^t \, \mathbf{C}_D{}^{-1} \, (\mathbf{g}(\mathbf{m}_n) - \mathbf{d}_{obs}) + (\mathbf{m}_n - \mathbf{m}_{prior}) \right]$$

$$\mathbf{m}_{n+1} = \mathbf{m}_{prior} - \mu_n \left(\mathbf{I} + \mathbf{C}_M \, \mathbf{G}_n{}^t \, \mathbf{C}_D{}^{-1} \, \mathbf{G}_n \right)^{-1} \mathbf{C}_M \, \mathbf{G}_n{}^t \, \mathbf{C}_D{}^{-1} \left[(\mathbf{g}(\mathbf{m}_n) - \mathbf{d}_{obs}) - \mathbf{G}_n \, (\mathbf{m}_n - \mathbf{m}_{prior}) \right]$$

$$\mathbf{m}_{n+1} = \mathbf{m}_{prior} - \mu_n \, \mathbf{C}_M \, \mathbf{G}_n{}^t \, \mathbf{C}_D{}^{-1} \left(\mathbf{I} + \mathbf{G}_n \, \mathbf{C}_M \, \mathbf{G}_n{}^t \, \mathbf{C}_D{}^{-1} \right)^{-1} \left[(\mathbf{g}(\mathbf{m}_n) - \mathbf{d}_{obs}) - \mathbf{G}_n \, (\mathbf{m}_n - \mathbf{m}_{prior}) \right] , \tag{4.108}$$

where the equivalences between different formulas can easily be shown from the identities of problem 1.19.

These formulas were first derived by Rodgers (1976), and rediscovered by Tarantola and Valette (1982b).

As usual, the starting point $\mathbf{m}_0$ is arbitrary, the simplest choice is $\mathbf{m}_0 = \mathbf{m}_{prior}$, but the use of different starting points helps to check the existence of secondary solutions.

The optimum value for μ_n has to be obtained by linear search. A linearization of $\mathbf{g}(\mathbf{m})$ around $\mathbf{g}(\mathbf{m}_n)$ gives

$$\mu_n \simeq 1 \ . \tag{4.109}$$

4.5.4: Variable metric method

As explained in appendix 4.7, in a variable metric method, the preconditioning operator $\hat{\mathbf{S}}$ is initialized as an arbitrary approximation of $\mathbf{H}_0^{-1}$,

$$\hat{\mathbf{S}}_0 \simeq \mathbf{H}_0^{-1} \simeq \left(\mathbf{I} + \mathbf{C}_M \ \mathbf{G}_0^t \ \mathbf{C}_D^{-1} \ \mathbf{G}_0 \right)^{-1} , \tag{4.110}$$

and is updated in such a way that, at least for linear functions $\mathbf{g}(\mathbf{m})$ (i.e., for quadratic functions $S(\mathbf{m})$),

$$\hat{\mathbf{S}}_n \rightarrow \mathbf{H}^{-1} \ . \tag{4.111}$$

The advantage of this method is that it starts out by behaving like a preconditioned steepest descent, and ends up by behaving like the Newton method, with its rapid termination. In addition, as the posterior covariance operator in the model space is given by (section 4.3)

$$\mathbf{C}_{M'} \simeq \mathbf{H}_n^{-1} , \tag{4.112}$$

the variable metric method is potentially well adapted to the resolution of least-squares inverse problems because it gives, as a by-product, a direct estimation of posterior errors (without requiring the inversion of an operator).

Using the formulas of appendix 4.7, we obtain the following algorithm:

$$\begin{aligned}
\gamma_n &= C_M\, G_n^t\, C_D^{-1}\, (g(\mathbf{m}_n)-\mathbf{d}_{obs}) + (\mathbf{m}_n-\mathbf{m}_{prior}) && (\mathbf{m}_0 \text{ arbitrary}) \\
\phi_n &= \hat{S}_n\, \gamma_n && (\hat{S}_0 \text{ arbitrary}) \\
\mathbf{m}_{n+1} &= \mathbf{m}_n - \mu_n\, \phi_n && (\text{obtain } \mu_n \text{ by linear search}) \\
\hat{S}_{n+1} &= \hat{S}_n + \delta\hat{S}_n && (\delta\hat{S}_n \text{ defined below})
\end{aligned} \qquad (4.113)$$

Many different formulas exist for the updating of $\hat{S}_n$. The best known are (see appendix 4.7) the "rank one formula"

$$\hat{S}_{n+1} = \hat{S}_n + \frac{\mathbf{u}_n\, \mathbf{u}_n^t\, C_M^{-1}}{\mathbf{u}_n^t\, C_M^{-1}\, \delta\gamma_n^t} \qquad (4.114)$$

due to Davidon (1959), the "DFP formula"

$$\hat{S}_{n+1} = \hat{S}_n + \frac{\delta\mathbf{m}_n\, \delta\mathbf{m}_n^t\, C_M^{-1}}{\delta\mathbf{m}_n^t\, C_M^{-1}\, \delta\gamma_n} - \frac{\mathbf{v}_n\, \mathbf{v}_n^t\, C_M^{-1}}{\mathbf{v}_n^t\, C_M^{-1}\, \delta\gamma_n} \qquad (4.115)$$

due to Davidon (1959) and to Fletcher and Powell (1963), and the "BFGS formula"

$$\hat{S}_{n+1} = \left(I - \frac{\delta\mathbf{m}_n\, \delta\gamma_n^t\, C_M^{-1}}{\delta\gamma_n^t\, C_M^{-1}\, \delta\mathbf{m}_n} \right) \hat{S}_n \left(I - \frac{\delta\mathbf{m}_n\, \delta\gamma_n^t\, C_M^{-1}}{\delta\gamma_n^t\, C_M^{-1}\, \delta\mathbf{m}_n} \right) + \frac{\delta\mathbf{m}_n\, \delta\mathbf{m}_n^t\, C_M^{-1}}{\delta\gamma_n^t\, C_M^{-1}\, \delta\mathbf{m}_n} \qquad (4.116)$$

due to Broyden (1970), Fletcher (1970), Goldfarb (1970), and Shanno (1970). In these equations,

$$\delta\mathbf{m}_n = \mathbf{m}_{n+1} - \mathbf{m}_n \ , \qquad (4.117)$$

$$\delta\gamma_n = \gamma_{n+1} - \gamma_n \ , \qquad (4.118)$$

$$\mathbf{v}_n = \hat{S}_n\, \delta\gamma_n \ , \qquad (4.119)$$

and

$$\mathbf{u}_n \;=\; \delta\mathbf{m}_n - \mathbf{v}_n \; . \tag{4.120}$$

Although it is easy to see that all the operators $\mathbf{S}_n$ thus defined are definite positive, they may become numerically singular. It seems that the BFGS formula has a greater tendency to keep the definiteness of $\mathbf{S}_n$. Maybe it is for this reason that it is today the most widely used updating formula.

The kernels (matrices) representing the operators $\hat{\mathbf{S}}_n$ should only be explicitly computed for small-sized problems. For large-sized problems, it should be noticed that all that we need is to be able to compute the result of the action of $\hat{\mathbf{S}}_n$ on an arbitrary model vector $\mathbf{f}$. Using, for instance, the rank-one formula, we have

$$\hat{\mathbf{S}}_n\,\mathbf{f} \;=\; \hat{\mathbf{S}}_0\,\hat{\mathbf{f}} \;+\; \sum_{p=0}^{n-1} \frac{\mathbf{u}_p{}^t\,\mathbf{C}_M{}^{-1}\,\mathbf{f}}{\nu_p}\,\mathbf{u}_p \tag{4.121}$$

where ν_p are the real numbers

$$\nu_p \;=\; \mathbf{u}_p{}^t\,\mathbf{C}_M{}^{-1}\,\delta\gamma_p \; . \tag{4.122}$$

This shows that, in order to operate with $\hat{\mathbf{S}}_n$, we only have to store in the computer memory the vectors $\mathbf{u}_0 ,\dots , \mathbf{u}_n$ and the scalars $\nu_0 ,\dots , \nu_n$.

This remark also applies for the estimation of the posterior covariance operator in large-sized problems. One interesting property of the variable metric method is that it allows, at least in principle, an inexpensive estimate of the posterior covariance operator: the property

$$\hat{\mathbf{S}}_n \;\rightarrow\; \left(\mathbf{I} + \mathbf{C}_M\,\mathbf{G}_\infty{}^t\,\mathbf{C}_D{}^{-1}\,\mathbf{G}_\infty\right)^{-1} \tag{4.123}$$

gives

$$\mathbf{C}_{M'} \;\simeq\; \hat{\mathbf{S}}_N\,\mathbf{C}_M \; , \tag{4.124}$$

where the index N represents the value of n for which iterations are stopped. Using, for instance, the rank-one formula gives, using (4.114),

$$(\mathbf{C}_{M'})^{\alpha\beta} \;=\; (\;\hat{\mathbf{S}}_N\,\mathbf{C}_M\;)^{\alpha\beta} \;=\; (\;\hat{\mathbf{S}}_0\,\mathbf{C}_M\;)^{\alpha\beta} \;+\; \sum_{n=0}^{N-1} \frac{u_n{}^{\alpha}\,u_n{}^{\beta}}{\phi_n} \; . \tag{4.125}$$

For large-sized problems, the kernel ($\mathbf{C}_{M'}$) is too large to be computed explicitly. A useful strategy is to compute only the standard deviations

$$(\sigma')^{\alpha} = \sqrt{\quad N')^{\alpha\alpha}} \; , \tag{4.126}$$

where

$$(C_{M'})^{\alpha\alpha} = (\hat{S}_0 \, C_M)^{\alpha\alpha} + \sum_{n=0}^{N-1} \frac{(u_n{}^{\alpha})^2}{\phi_n} \; , \tag{4.127}$$

and, for selected values of α_0 , the posterior covariances of the α_0-th parameter with all other parameters:

$$(C_{M'})^{\alpha_0\beta} = (\hat{S}_0 \, C_M)^{\alpha_0\beta} + \sum_{n=0}^{N-1} \frac{u_n{}^{\alpha_0} \, u_n{}^{\beta}}{\phi_n} \tag{4.128}$$

(see Chapter 7 for an example of such an analysis of covariances).

For large-sized problems, good approximations of the solution are usually obtained after a few iterations. Unfortunately, very little is known about the accuracy of the covariance operator obtained after a few iterations of a variable metric method.

The value μ_n has to be obtained by linear search. Alternatively, a linearization of $\mathbf{g}(\mathbf{m})$ around $\mathbf{g}(\mathbf{m}_n)$ gives

$$\mu_n \simeq \frac{\gamma_n{}^t \, C_M{}^{-1} \, \phi_n}{\phi_n{}^t \, C_M{}^{-1} \, \phi_n + \mathbf{b}_n{}^t \, C_D{}^{-1} \, \mathbf{b}_n} \tag{4.129}$$

where

$$\mathbf{b}_n = \mathbf{G}_n \, \phi_n \; . \tag{4.130}$$

Box 4.8: What is ART?

The Algebraic Reconstruction Techniques (ART) originated in the problem of image reconstruction from projections (see Herman, 1980). Mathematically, ART algorithms correspond to the use of "row-generation" methods for optimization problems (Censor and Herman, 1979). Let us see here the associated physical interpretation.

Assume that a data set can be divided into subsets with independent uncertainties,

(...)

$$\mathbf{d}_{obs} = \begin{bmatrix} \mathbf{d}_{1obs} \\ \mathbf{d}_{2obs} \\ \cdots \end{bmatrix} \qquad \mathbf{C}_D = \begin{bmatrix} \mathbf{C}_1 & 0 & \cdots \\ 0 & \mathbf{C}_2 & \cdots \\ \cdots & \cdots & \cdots \end{bmatrix} .$$

As it is shown in problem 4.2, it is possible to perform the inversion of each data subset separately, provided that the prior values of $\mathbf{m}$ and $\mathbf{C}_M$ used at each step equal the posterior values of the previous step. After the inversion of the last data subset, the posterior values of $\mathbf{m}$ and $\mathbf{C}_M$ equal those that could be obtained from the global inversion of the whole data set.

Generalizing somewhat, each iteration of ART proceeds as just described, excepted in that the model covariance operator $\mathbf{C}_M$ is not updated. Consequently, after the inversion of the last data subset the final solution is not yet obtained, and the loop over the data subsets has to be iterated, until convergence. To ensure this convergence, each model update predicted by the described calculation has to be dumped, i.e., multiplied by an ad-hoc factor < 1 .

This concept of data partition may help convergence in linear as well as in nonlinear problems, because at any stage of the computations the data residuals contain information on all the data subsets already used.

In image reconstruction from projections, the data partition corresponds to use each ray independently. In general I would say: *update your model as soon as you can.*

4.6: Particular formulas for linear problems

For a strictly linear problem:

$$\mathbf{g}(\mathbf{m}) = \mathbf{G}\,\mathbf{m} ,$$

where, as usual, $\mathbf{G}$ denotes a linear operator. The misfit function S is given by

$$S = \frac{1}{2} \left[(\mathbf{Gm}-\mathbf{d}_{obs})^t\, \mathbf{C}_D^{-1}\, (\mathbf{Gm}-\mathbf{d}_{obs}) + (\mathbf{m}-\mathbf{m}_{prior})^t\, \mathbf{C}_M^{-1}\, (\mathbf{m}-\mathbf{m}_{prior}) \right], \tag{4.131}$$

a functional which is quadratic in the unknown $\mathbf{m}$.

Linear problems are much easier to solve than nonlinear problems:

i) The function S has no secondary minima.

ii) The Newton method converges in only one iteration.

iii) The method of conjugate directions and the variable metric method converge in a finite number of iterations (if we disregard truncation errors).

Specializing the algorithms of section 4.5 for linear problems, we obtain:

4.6.1: Preconditioned steepest descent

$$\gamma_n = C_M\, G^t\, C_D^{-1}\, (Gm_n - d_{obs}) + (m_n - m_{prior}) \qquad (m_0 \text{ arbitrary})$$

$$\phi_n = \hat{S}_0\, \gamma_n \qquad\qquad \hat{S}_0 \simeq \left(I + C_M\, G^t\, C_D^{-1}\, G\right)^{-1}$$

$$b_n = G\, \phi_n \tag{4.132}$$

$$\mu_n = \frac{\gamma_n{}^t\, C_M^{-1}\, \phi_n}{\phi_n{}^t\, C_M^{-1}\, \phi_n + b_n{}^t\, C_D^{-1}\, b_n}$$

$$m_{n+1} = m_n - \mu_n\, \phi_n\, .$$

4.6.2: Preconditioned conjugate directions:

$$\gamma_n = C_M\ G^t\ C_D^{-1}\ (G\mathbf{m}_n - \mathbf{d}_{obs}) + (\mathbf{m}_n - \mathbf{m}_{prior}) \qquad (m_0 \text{ arbitrary})$$

$$\lambda_n = \hat{\mathbf{S}}_0\ \gamma_n \qquad\qquad \hat{\mathbf{S}}_0 \simeq \left(\mathbf{I} + C_M\ G^t\ C_D^{-1}\ G\right)^{-1}$$

$$\begin{aligned}
\omega_n &= \lambda_n{}^t\ C_M^{-1}\ \gamma_n \\
\phi_n &= \lambda_n + \frac{\omega_n}{\omega_{n-1}}\ \phi_{n-1} \qquad\qquad (\phi_0 = \lambda_0) \\
\mathbf{b}_n &= G\ \phi_n \\
\mu_n &= \frac{\omega_n}{\phi_n{}^t\ C_M^{-1}\ \phi_n + \mathbf{b}_n{}^t\ C_D^{-1}\ \mathbf{b}_n} \\
\mathbf{m}_{n+1} &= \mathbf{m}_n - \mu_n\ \phi_n
\end{aligned} \qquad (4.133)$$

4.6.3: Newton method

$$\mathbf{m} = \left(G^t\ C_D^{-1}\ G + C_M^{-1}\right)^{-1} (G^t\ C_D^{-1}\ \mathbf{d}_{obs} + C_M^{-1}\ \mathbf{m}_{prior}) \qquad (1.89 \text{ again})$$

$$\begin{aligned}
\mathbf{m} &= \mathbf{m}_{prior} - \left(G^t\ C_D^{-1}\ G + C_M^{-1}\right)^{-1} G^t\ C_D^{-1}\ (G\ \mathbf{m}_{prior} - \mathbf{d}_{obs}) \\
\mathbf{m} &= \mathbf{m}_{prior} - C_M\ G^t \left(C_D + G\ C_M\ G^t\right)^{-1} (G\ \mathbf{m}_{prior} - \mathbf{d}_{obs}) \qquad (4.134) \\
\mathbf{m} &= \mathbf{m}_{prior} - \left(I + C_M\ G^t\ C_D^{-1}\ G\right)^{-1} C_M\ G^t\ C_D^{-1}\ (G\ \mathbf{m}_{prior} - \mathbf{d}_{obs}) \\
\mathbf{m} &= \mathbf{m}_{prior} - C_M\ G^t\ C_D^{-1} \left(\mathbf{I} + G\ C_M\ G^t\ C_D^{-1}\right)^{-1} (G\ \mathbf{m}_{prior} - \mathbf{d}_{obs})
\end{aligned}$$

4.6.4: Variable metric method

$$\gamma_n = \mathbf{C_M}\, \mathbf{G}^t\, \mathbf{C_D}^{-1}\, (\mathbf{G}\, \mathbf{m}_n - \mathbf{d}_{obs}) + (\mathbf{m}_n - \mathbf{m}_{prior}) \qquad (\mathbf{m}_0 \text{ arbitrary})$$

$$\phi_n = \hat{\mathbf{S}}_n\, \gamma_n \qquad\qquad \hat{\mathbf{S}}_0 \simeq \left(\mathbf{I} + \mathbf{C_M}\, \mathbf{G}^t\, \mathbf{C_D}^{-1}\, \mathbf{G}\right)^{-1}$$

$$\mathbf{b}_n = \mathbf{G}\, \phi_n$$

$$\mu_n = \frac{\gamma_n{}^t\, \mathbf{C_M}^{-1}\, \phi_n}{\phi_n{}^t\, \mathbf{C_M}^{-1}\, \phi_n + \mathbf{b}_n{}^t\, \mathbf{C_D}^{-1}\, \mathbf{b}_n} \tag{4.135}$$

$$\mathbf{m}_{n+1} = \mathbf{m}_n - \mu_n\, \phi_n$$

$$\mathbf{S}_{n+1} = \mathbf{S}_n + \delta\mathbf{S}_n \qquad \text{(see section 4.5)}$$

If a number of iterations equal to the dimension of the model space are effectively performed, at the last iteration

$$\hat{\mathbf{S}}_n \equiv \left(\mathbf{I} + \mathbf{C_M}\, \mathbf{G}^t\, \mathbf{C_D}^{-1}\, \mathbf{G}\right)^{-1} .$$

4.7: Particular formulas for linearizable problems

It often happens that the computation of $\mathbf{g}(\mathbf{m})$ (i.e., the resolution of the forward problem) is difficult for an arbitrary $\mathbf{m}$, but that there exists a *reference* model $\mathbf{m}_{ref}$ for which the computation of $\mathbf{g}(\mathbf{m}_{ref})$ is easier than that of $\mathbf{g}(\mathbf{m})$, and for which the linearized approximation

$$\mathbf{g}(\mathbf{m}) \simeq \mathbf{g}(\mathbf{m}_{ref}) + \mathbf{G}_{ref}\, (\mathbf{m} - \mathbf{m}_{ref}) \tag{4.136}$$

holds with adequate accuracy. This greatly simplifies the computations, essentially because the misfit function $S(\mathbf{m})$ becomes a quadratic function of $\mathbf{m}$. Then all gradient algorithms necessarily converge, the conjugate gradient algorithm converges in a finite number of iterations, and the Newton algorithm converges in only one iteration.

The main trouble with linearized problems is that modelization errors (see section 1.3) usually become important and difficult to account for, so that it is often difficult to assess the reliability of the solution.

To simplify the notations, I assume here that the reference model $\mathbf{m}_{ref}$, the starting model $\mathbf{m}_0$, and the a priori model $\mathbf{m}_{prior}$ coincide:

$$\mathbf{m}_{ref} = \mathbf{m}_0 = \mathbf{m}_{prior} . \tag{4.137}$$

Defining

$$\delta\mathbf{m} = \mathbf{m} - \mathbf{m}_0 \tag{4.138}$$

$$\delta\mathbf{d} = \mathbf{g}(\mathbf{m}) - \mathbf{g}(\mathbf{m}_0) \tag{4.139}$$

$$\delta\mathbf{d}_{obs} = \mathbf{d}_{obs} - \mathbf{g}(\mathbf{m}_0) , \tag{4.140}$$

we have, for the misfit function S :

$$S = \frac{1}{2} \left[(\mathbf{G}_0\, \delta\mathbf{m} - \delta\mathbf{d}_{obs})^t\, \mathbf{C}_D^{-1}\, (\mathbf{G}_0\, \delta\mathbf{m} - \delta\mathbf{d}_{obs}) + \delta\mathbf{m}^t\, \mathbf{C}_M^{-1}\, \delta\mathrm{m} \right] , \tag{4.141}$$

a functional which is quadratic in the unknown $\delta\mathbf{m}$.

The formulas corresponding to a linearized problem can be obtained from the formulas of nonlinear problems using the approximation (4.136). But as linearized problems are much closer to linear problems than to nonlinear problems, it is simpler to imagine an ad-hoc linear problem with data $\delta\mathbf{d}_{obs}$, data covariance operator C_D , forward equation

$$\delta\mathbf{d} = \mathbf{G}_0\, \delta\mathbf{m} ,$$

a priori model $\delta\mathbf{m} \equiv \mathbf{0}$, and a priori model covariance operator C_M . The corresponding formulas can then be obtained from those corresponding to linear problems (section 4.6).

4.7.1: Preconditioned steepest descent

$$\gamma_n = C_M\, G_0{}^t\, C_D{}^{-1}\, (\, G_0\, \delta m_n - \delta d_{obs}\,) + \delta m_n \qquad (\delta m_0 \text{ arbitrary})$$

$$\phi_n = \hat{S}_0\, \gamma_n \qquad\qquad \hat{S}_0 \simeq \left(I + C_M\, G_0{}^t\, C_D{}^{-1}\, G_0 \right)^{-1}$$

$$b_n = G_0\, \phi_n \tag{4.142}$$

$$\mu_n = \frac{\gamma_n{}^t\, C_M{}^{-1}\, \phi_n}{\phi_n{}^t\, C_M{}^{-1}\, \phi_n + b_n{}^t\, C_D{}^{-1}\, b_n}$$

$$\delta m_{n+1} = \delta m_n - \mu_n\, \phi_n\, .$$

4.5.2: Preconditioned conjugate directions:

$$\gamma_n = C_M\, G_0{}^t\, C_D{}^{-1}\, (\, G_0\, \delta m_n - \delta d_{obs}\,) + \delta m_n \qquad (\delta m_0 \text{ arbitrary})$$

$$\lambda_n = \hat{S}_0\, \gamma_n \qquad\qquad \hat{S}_0 \simeq \left(I + C_M\, G_0{}^t\, C_D{}^{-1}\, G_0 \right)^{-1}$$

$$\omega_n = \lambda_n{}^t\, C_M{}^{-1}\, \gamma_n$$

$$\phi_n = \lambda_n + \frac{\omega_n}{\omega_{n-1}}\, \phi_{n-1} \qquad\qquad (\phi_0 = \lambda_0)$$

$$b_n = G_0\, \phi_n \tag{4.143}$$

$$\mu_n = \frac{\omega_n}{\phi_n{}^t\, C_M{}^{-1}\, \phi_n + b_n{}^t\, C_D{}^{-1}\, b_n}$$

$$\delta m_{n+1} = \delta m_n - \mu_n\, \phi_n$$

4.7.3: Newton method

$$\delta\mathbf{m} = \left(\mathbf{G}_0^t \, \mathbf{C}_D^{-1} \, \mathbf{G}_0 + \mathbf{C}_M^{-1} \right)^{-1} \mathbf{G}_0^t \, \mathbf{C}_D^{-1} \, \delta\mathbf{d}_{obs}$$

$$\delta\mathbf{m} = \mathbf{C}_M \, \mathbf{G}_0^t \left(\mathbf{C}_D + \mathbf{G}_0 \, \mathbf{C}_M \, \mathbf{G}_0^t \right)^{-1} \delta\mathbf{d}_{obs} \qquad (4.144)$$

$$\delta\mathbf{m} = \left(\mathbf{I} + \mathbf{C}_M \, \mathbf{G}_0^t \, \mathbf{C}_D^{-1} \, \mathbf{G}_0 \right)^{-1} \mathbf{C}_M \, \mathbf{G}_0^t \, \mathbf{C}_D^{-1} \, \delta\mathbf{d}_{obs}$$

$$\delta\mathbf{m} = \mathbf{C}_M \, \mathbf{G}_0^t \, \mathbf{C}_D^{-1} \left(\mathbf{I} + \mathbf{G}_0 \, \mathbf{C}_M \, \mathbf{G}_0^t \, \mathbf{C}_D^{-1} \right)^{-1} \delta\mathbf{d}_{obs}$$

4.7.4: Variable metric method

$$\gamma_n = \mathbf{C}_M \, \mathbf{G}_0^t \, \mathbf{C}_D^{-1} \left(\mathbf{G}_0 \, \delta\mathbf{m}_n - \delta\mathbf{d}_{obs} \right) + \delta\mathbf{m}_n \qquad (\delta\mathbf{m}_0 \text{ arbitrary})$$

$$\phi_n = \hat{\mathbf{S}}_n \, \gamma_n \qquad \hat{\mathbf{S}}_0 \simeq \left(\mathbf{I} + \mathbf{C}_M \, \mathbf{G}_0^t \, \mathbf{C}_D^{-1} \, \mathbf{G}_0 \right)^{-1}$$

$$\mathbf{b}_n = \mathbf{G}_0 \, \phi_n$$

$$\mu_n = \frac{\gamma_n^t \, \mathbf{C}_M^{-1} \, \phi_n}{\phi_n^t \, \mathbf{C}_M^{-1} \, \phi_n + \mathbf{b}_n^t \, \mathbf{C}_D^{-1} \, \mathbf{b}_n} \qquad (4.145)$$

$$\delta\mathbf{m}_{n+1} = \delta\mathbf{m}_n - \mu_n \, \phi_n$$

$$\mathbf{S}_{n+1} = \mathbf{S}_n + \delta\mathbf{S}_n \qquad \text{(see section 4.5)}$$

If a number of iterations equal to the dimension of the model space are effectively performed, at the last iteration

$$\hat{\mathbf{S}}_n \equiv \left(\mathbf{I} + \mathbf{C}_M \, \mathbf{G}_0^t \, \mathbf{C}_D^{-1} \, \mathbf{G}_0 \right)^{-1} .$$

Appendix 4.1: Tensor Notations for Discrete Inverse Problems

The interest of tensor notation is twofold: first, the formulas thus obtained have a form well adapted to the translation into usual computer language, and second, the "rules" of tensor notation allow one to obtain easily formulas which would require, otherwise, a more technical demonstration.

Let $\mathbf{d}$ represent an element of the data vector space $\boldsymbol{D}$, and $\mathbf{m}$ an element of the model vector space $\boldsymbol{M}$. Numerical computations are not performed with the *vectors* themselves, but with their "components" : $\{d^i\}$ $(i \in \mathbf{I}_D)$ and $\{m^\alpha\}$ $(\alpha \in \mathbf{I}_M)$, where $\mathbf{I}_D$ and $\mathbf{I}_M$ denote the respective index sets of the data and model spaces.

When we speak of *components* of a vector, we are referring to a given *basis* of the space. Let us assume that bases $\{e_i\}$ $\{\mathbf{e}_\alpha\}$ are given in the data and model space such that we can write

$$\mathbf{d} = d^i \, \mathbf{e}_i$$
$$\mathbf{m} = m^\alpha \, e_\alpha \; , \tag{1}$$

where d^i and m^α are the usual components of data and model vectors, and where an implicit sum is assumed over a term containing repeated indexes: *one upper* index, and *one lower* index. The components defined by (1) are called *contravariant*, by opposition to the components over the dual basis (to be introduced below), which are called *covariant.*

The Kronecker "symbols" can be introduced by

$$\delta^i{}_j = 1 \quad \text{if} \quad i \equiv j \quad \text{or} \quad 0 \quad \text{otherwise}$$

$$\delta^\alpha{}_\beta = 1 \quad \text{if} \quad \alpha \equiv \beta \quad \text{or} \quad 0 \quad \text{otherwise} \; .$$

Let C^{ij} and $C^{\alpha\beta}$ represent the elements of given *covariance operators* (supposed regular) in the data and model space respectively. We define C_{ij} and $C_{\alpha\beta}$ by

$$C^{ij} \, C_{jk} = \delta^i{}_k$$
$$C^{\alpha\beta} \, C_{\beta\gamma} = \delta^\alpha{}_\gamma \; , \tag{2}$$

so that the elements with lower indexes are the components of the *inverse* of the covariance operator. C_{ij} and $C_{\alpha\beta}$ can be used to introduce a *metric* over $\boldsymbol{D}$ and $\boldsymbol{M}$, by defining the scalar products by

$$(d\ ,\ d') = C_{ij}\ d^i\ d^{j'}$$
$$(m\ ,\ m') = C_{\alpha\beta}\ m^\alpha\ m^{\beta'}\ , \tag{3}$$

In particular, covariance operators are such that: i) the above expressions define real (adimensional) numbers, whatever the physical dimensions of the components $\{d^i\}$ and $\{m^\alpha\}$ may be; ii) the above expressions are symmetric; and iii) they are positive definite (if we assume that variances are not null or infinite, or that there are no perfect correlations). As it can be seen from equation (3), C_{ij} and $C_{\alpha\beta}$ are the *metric tensors* of the data and model space.

It should be noted that the same symbol $\mathbf{C}$ is used to denote the metric tensor in data and model space; the indexes specify which tensor we are using. Also, in equation (3), the traditional notation consisting in writing $d^{j'}$ and $m^{\beta'}$ rather than d'^j and m'^β is used. It can be noticed that, as the metrics C_{ij} and $C_{\alpha\beta}$ are independent of the coordinates, the spaces $\boldsymbol{D}$ and $\boldsymbol{M}$ are *flat* (i.e., they have no curvature).

We have

$$(e_k\ ,\ e_l) = C_{ij}\ (e_k)^i\ (e_l)^j = C_{ij}\ \delta^i{}_k\ \delta^j{}_l\ ,$$

$$(e_\gamma\ ,\ e_\delta) = C_{\alpha\beta}\ (e_\gamma)^\alpha\ (e_\delta)^\beta = C_{\alpha\beta}\ \delta^\alpha{}_\gamma\ \delta^\beta{}_\delta\ ,$$

i.e.,

$$(e_k\ ,\ e_l) = C_{kl}\ .$$
$$(e_\gamma\ ,\ e_\delta) = C_{\gamma\delta}\ . \tag{4}$$

We have seen that there exists an isometric bijection between a space and its dual. The simplest interpretation of tensor equations is obtained when a vector of the space is considered as an abstract entity representing both isomorphic elements. The "dual basis" is then a basis on the same vector space, denoted $\{\mathbf{e}^i\}$ (respectively $\{\mathbf{e}^\alpha\}$). The components of a given vector on the primal basis are named *contravariant components* and correspond to the usual vector components. The components of the vector on the dual basis are named *covariant components* and correspond to the usual components of the dual vector in the basis of the dual space. The terms "contravariant" and "covariant" will be justified later.

In as far as the dual basis is considered to belong to the considered space, the scalar product between elements of the primal and dual basis can be considered. This leads to the formal definition of the dual basis:

$$(\mathbf{e}^i , \mathbf{e}_j) = \delta^i{}_j$$
$$(\mathbf{e}^\alpha , \mathbf{e}_\beta) = \delta^\alpha{}_\beta \,. \tag{5}$$

The *covariant components* of **d** and **m** are then defined by:

$$\mathbf{d} = d_i \; \mathbf{e}^i$$
$$\mathbf{m} = m_\alpha \; \mathbf{e}^\alpha \,. \tag{6}$$

The following properties can now easily be demonstrated:

$$(\mathbf{d} , \mathbf{e}_i) = d_i$$
$$(\mathbf{m} , \mathbf{e}_\alpha) = m_\alpha \tag{7}$$

$$(\mathbf{d} , \mathbf{e}^i) = d^i$$
$$(\mathbf{m} , \mathbf{e}^\alpha) = m^\alpha \tag{8}$$

$$C_{ij} \; d^j = d_i$$
$$C_{\alpha\beta} \; m^\beta = m_\alpha \tag{9}$$

$$C^{ij} \; d_j = d^i$$
$$C^{\alpha\beta} \; m_\beta = m^\alpha \,. \tag{10}$$

Equations (9) and (10) show that the metric tensor (i.e., the covariance tensor) can be used for "raising and lowering indexes". In particular, it is interesting to note the property of the components of the metric tensor when we take mixed contravariant and covariant components:

$$C^i{}_j = \delta^i{}_j$$
$$C^\alpha{}_\beta = \delta^\alpha{}_\beta \ . \qquad (11)$$

Then, raising and lowering indexes,

$$C^{ij} = \delta^{ij}$$
$$C^{\alpha\beta} = \delta^{\alpha\beta} \ . \qquad (12a)$$

$$C_{ij} = \delta_{ij}$$
$$C_{\alpha\beta} = \delta_{\alpha\beta} \ , \qquad (12b)$$

but, of course, δ^{ij} , δ_{ij} , $\delta^{\alpha\beta}$, and $\delta_{\alpha\beta}$ do not satisfy the definition of Kronecker symbols.

Let now G represent a linear operator mapping ***M*** into ***D*** . We can write

$$d^i = G^i{}_\alpha \ m^\alpha \ . \qquad (13)$$

The expression

$$d_i \rightarrow \psi_\alpha = G^i{}_\alpha \ d_i \qquad (14)$$

represents a linear application from the dual of ***D*** (because the index of d is a lower one) into the dual of ***M*** (because the index of ψ is Greek and low): it can be identified with the *transposed* operator defined in Box 4.5. Comparing (13) and (14) it is evident that the transposed of an operator has components which are identical to those of the original operator. The expression

$$d^i \rightarrow \phi^\alpha = G_i{}^\alpha \ d^i \qquad (15)$$

represents a linear application from ***D*** into ***M*** : it can be identified with the *adjoint* operator defined in Box 4.5. Raising and lowering indexes, we easily obtain

$$G_i{}^\alpha = C_{ij} \; G^j_\beta \; C^{\beta\alpha} \tag{16},$$

which corresponds to the formula

$$\mathbf{G}^* = \mathbf{C_M} \; \mathbf{G}^t \; \mathbf{C_D}^{-1} ,$$

which has been demonstrated in Box 4.5. We thus see that, as usual, tensor notations allow us to obtain trivially expressions that, otherwise, require more sophisticated demonstrations.

The function $S(\mathbf{m})$ to be minimized in least-squares inverse problems (equation 4.8) can now be written:

$$2\,S(m) = C_{ij}\,(g(\mathbf{m})-\mathbf{d}_{obs})^i\,(g(\mathbf{m})-\mathbf{d}_{obs})^j + C_{\alpha\beta}\,(\mathbf{m}-\mathbf{m}_{prior})^\alpha\,(m-m_{prior})^\beta$$

$$= (g(\mathbf{m})-\mathbf{d}_{obs})^i\,(g(\mathbf{m})-\mathbf{d}_{obs})_i + (\mathbf{m}-\mathbf{m}_{prior})^\alpha\,(\mathbf{m}-\mathbf{m}_{prior})_\alpha \;. \tag{17}$$

Using the fact that the derivative with respect to a contravariant component gives a covariant component (see, for instance, Lichnérowicz, 1960), we obtain for the gradient and the Hessian the expressions:

$$\gamma_\alpha = \left(\frac{\partial S}{\partial m^\alpha}\right) = G^i_\alpha \; C_{ij} \; (g(m)-d_{obs})^j + C_{\alpha\beta} \; (\mathbf{m}-\mathbf{m}_{prior})^\beta \tag{18}$$

$$= G^i_\alpha \; (g(m)-d_{obs})_i + (\mathbf{m}-\mathbf{m}_{prior})_\alpha$$

$$H_{\alpha\beta} = \left(\frac{\partial^2 S}{\partial m^\alpha \partial m^\beta}\right) = G^i_\alpha \; C_{ij} \; G^j_\beta + C_{\alpha\beta} + K^i_{\alpha\beta} \; C_{ij} \; (g(\mathbf{m})-\mathbf{d}_{obs})^j \; , \tag{19}$$

where

$$G^i_\alpha = \frac{\partial g^i}{\partial m^\alpha} ,$$

and

$$K^i_{\alpha\beta} = \frac{\partial G^i_\alpha}{\partial m^\beta} = \frac{\partial^2 g^i}{\partial m^\alpha \partial m^\beta} .$$

Defining $S^{\alpha\beta}$ by

$$S^{\alpha\beta} \; H_{\beta\gamma} = \delta^{\alpha}_{\gamma} ,$$

the first of the Newton formulas in (4.108) becomes

$$(m_{n+1})^{\alpha} = (\mathbf{m}_n)^{\alpha} - (S_n)^{\alpha\beta} \; (\gamma_n)_{\beta} , \tag{20}$$

while the fourth Newton formula in (4.108) becomes simply

$$(m_{n+1})^{\alpha} = (\mathbf{m}_n)^{\alpha} - (S_n)^{\alpha}{}_{\beta} \; (\gamma_n)^{\beta} . \tag{21}$$

The steepest descent algorithm is written

$$(m_{n+1})^{\alpha} = (\mathbf{m}_n)^{\alpha} - \mu_n \; (\gamma_n)^{\alpha} . \tag{22}$$

It should be noticed that the rule that free indexes have to be in the same position at both sides of an equality, protects the user of tensor notations against writing senseless formulas, such as for instance

$$(m_{n+1})^{\alpha} = (\mathbf{m}_n)^{\alpha} - \mu_n \; (\gamma_n)_{\alpha} , \tag{23}$$

which would correspond, in the notations of section 4.4, to the formula

$$\mathbf{m}_{n+1} = \mathbf{m}_n + \mu_n \; \hat{\gamma}_n , \tag{24}$$

the error being there that $\hat{\gamma}_n$ is an element of the dual of M , and the addition of elements of M and of its dual is not defined. Someone used to tensor notations would never make the mistake of using equation (23), but it is possible to forget the "detail" that $\hat{\gamma}_n$ is not an element of the model space (and, as computers usually ignore the physical dimensions of the variables, a program based on such an incorrect formula will work, although its performances will be very dependent on the choice of physical units made by the user).

Let us finally see the reason for the contravariant-covariant terminology. Let $\{e_{i'}\}$ and $\{e_{\alpha'}\}$ respectively represent new choices of basis vectors in our spaces:

$$\begin{aligned} e_{i'} &= \Lambda_{i'}{}^{i} \; e_i \\ e_{\alpha'} &= \Lambda_{\alpha'}{}^{\alpha} \; \mathbf{e}_{\alpha} . \end{aligned} \tag{25}$$

The reader will easily verify the following equalities

$$\mathbf{e}^{i'} = \Lambda^{i'}{}_{i} \ \mathbf{e}^{i}$$

$$\mathbf{e}^{\alpha'} = \Lambda^{\alpha'}{}_{\alpha} \ \mathbf{e}^{\alpha} \tag{26}$$

$$d_{i'} = \Lambda_{i'}{}^{i} \ d_{i}$$

$$m_{\alpha'} = \Lambda_{\alpha'}{}^{\alpha} \ m_{\alpha} \tag{27}$$

$$d^{i'} = \Lambda^{i'}{}_{i} \ d^{i}$$

$$m^{\alpha'} = \Lambda^{\alpha'}{}_{\alpha} \ m^{\alpha} . \tag{28}$$

Equation (28) concerns the components over the primal basis. They do not change through the "matrix" $\Lambda_{x'}{}^{x}$ as the primal basis, but they change through the "matrix" $\Lambda^{x'}{}_{x}$, as the dual basis: they are "contravariant". Equation (27) concerns the components over the dual basis. They change through the same "matrix" as the primal basis: they are "covariant".

A complete development of tensor calculus can be found, for instance, in Lichnérowicz (1960).

Appendix 4.2: More notations

Let $\mathbf{d}$ denote a generic element of the data space $\boldsymbol{D}$, and $\mathbf{m}$ a generic element of the model space $\boldsymbol{M}$. The dual spaces $\hat{\boldsymbol{D}}$ and $\hat{\boldsymbol{M}}$ have elements which are generically denoted $\hat{\mathbf{d}}$ and $\hat{\mathbf{m}}$. The notations so far defined are such that the real number associated by the linear form $\hat{\mathbf{d}}_1$ (resp. $\hat{\mathbf{m}}_1$) with the element $\mathbf{d}_2$ (resp. $\mathbf{m}_2$) is denoted $\hat{\mathbf{d}}_1{}^t \ \mathbf{d}_2$ (resp. $\hat{\mathbf{m}}_1{}^t \ \mathbf{m}_2$) . This notation is justified because it recalls usual matricial notations. In that sense, if the elements of $\boldsymbol{D}$ and $\boldsymbol{M}$ are imagined as column matrices, the elements of $\hat{\boldsymbol{D}}$ and $\hat{\boldsymbol{M}}$ are also column matrices. These notations are coherent and allow an intuitive use of the formulas. But they do not satisfy me completely. I prefer to imagine the elements of the dual spaces as *row matrices* instead of column matrices. Let us see how this works.

Using first a purely notational point of view, let me denote

$$\hat{\mathbf{d}}^t \equiv \mathbf{d}^* ,$$
$$\hat{\mathbf{m}}^t \equiv \mathbf{m}^* , \tag{1}$$

then

$$\hat{\mathbf{d}}_1{}^t\, \mathbf{d}_2 = \mathbf{d}_1{}^*\, \mathbf{d}_2 ,$$
$$\hat{\mathbf{m}}_1{}^t\, \mathbf{m}_2 = \mathbf{m}_1{}^*\, \mathbf{m}_2 . \tag{2}$$

Given a linear operator $\mathbf{G}$ mapping $\mathcal{M}$ into $\mathcal{D}$, I have defined in box 4.5 the operators $\mathbf{G}^t$ (mapping $\hat{\mathcal{D}}$ into $\hat{\mathcal{M}}$) and $\mathbf{G}^*$ (mapping $\mathcal{D}$ into $\mathcal{M}$). I recall that if $\mathbf{C}_D$ and $\mathbf{C}_M$ are the covariance operators defining the scalar products in the data and model space, respectively, then

$$\mathbf{G}^* = \mathbf{C}_M\, \mathbf{G}^t\, \mathbf{C}_D{}^{-1} . \tag{3}$$

I now introduce a new operator, $\hat{\mathbf{G}}$, mapping $\hat{\mathcal{M}}$ into $\hat{\mathcal{D}}$, defined by

$$\hat{G} = (\mathbf{G}^*)^t = (\mathbf{G}^t)^* . \tag{4}$$

This gives

$$\hat{\mathbf{G}} = \mathbf{C}_D{}^{-1}\, \mathbf{G}\, \mathbf{C}_M . \tag{5}$$

The reader will easily verify the following sets of identities

$$\mathbf{d} = \mathbf{G}\,\mathbf{m} \;\Leftrightarrow\; \hat{\mathbf{d}} = \hat{\mathbf{G}}\,\hat{\mathbf{m}} \;\Leftrightarrow\; \mathbf{d}^t = \mathbf{m}^t\,\mathbf{G}^t \;\Leftrightarrow\; \mathbf{d}^* = \mathbf{m}^*\,\mathbf{G}^* . \tag{6}$$

$$\mathbf{m} = \mathbf{G}^*\,\mathbf{d} \;\Leftrightarrow\; \hat{\mathbf{m}} = \mathbf{G}^t\,\hat{\mathbf{d}} \;\Leftrightarrow\; \mathbf{m}^t = \mathbf{d}^t\,\hat{\mathbf{G}} \;\Leftrightarrow\; \mathbf{m}^* = \mathbf{d}^*\,\mathbf{G} . \tag{7}$$

I now formally introduce the spaces $\mathcal{D}^*$ and $\mathcal{M}^*$, respectively containing the elements denoted $\mathbf{d}^*$ and $\mathbf{m}^*$, and the spaces $\mathcal{D}^t$ and $\mathcal{M}^t$ respectively containing the elements denoted $\mathbf{d}^t$ and $\mathbf{m}^t$. Intuitively, the spaces $\mathcal{D}^t$ and $\mathcal{M}^t$ are simply the transposes of the spaces $\mathcal{D}$ and $\mathcal{M}$, and the spaces $\mathcal{D}^*$ and $\mathcal{M}^*$ are the transposes of $\hat{\mathcal{D}}$ and $\hat{\mathcal{M}}$. If the elements of $\mathcal{D}$ and $\mathcal{M}$ are imagined as column matrices, and we wish to consider the elements of the dual spaces as *row* matrices, then we will call $\mathcal{D}^*$ and $\mathcal{M}^*$ the duals of $\mathcal{D}$ and $\mathcal{M}$, instead of $\hat{\mathcal{D}}$ and $\hat{\mathcal{M}}$.

The denomination is then as follows:

$$\begin{aligned} &\boldsymbol{D} \ , \ \boldsymbol{M} &&: \text{ original (primal) spaces.} \\ &\boldsymbol{D}^t \ , \ \boldsymbol{M}^t &&: \text{ transposed spaces.} \\ &\boldsymbol{D}^* \ , \ \boldsymbol{M}^* &&: \text{ dual spaces.} \\ &\hat{\boldsymbol{D}} \ , \ \hat{\boldsymbol{M}} &&: \text{ transposes of the duals.} \end{aligned} \tag{8}$$

Looking to expressions (6) and (7), we may wish to consider the operators as acting indifferently on vectors on their right or left hand. They then map the following spaces:

$$\begin{aligned} &G && \text{acting on the right: maps } \boldsymbol{M} \text{ into } \boldsymbol{D} \\ &\hat{G} && \text{acting on the right: maps } \hat{\boldsymbol{M}} \text{ into } \hat{\boldsymbol{D}} \\ &G^* && \text{acting on the left: maps } \boldsymbol{M}^* \text{ into } \boldsymbol{D}^* \\ &G^t && \text{acting to the left: maps } \boldsymbol{M}^t \text{ into } \boldsymbol{D}^t \ , \end{aligned} \tag{9}$$

or, alternatively,

$$\begin{aligned} &G && \text{acting on the left: maps } \boldsymbol{D}^* \text{ into } \boldsymbol{M}^* \\ &\hat{G} && \text{acting on the left: maps } \boldsymbol{D}^t \text{ into } \boldsymbol{M}^t \\ &G^* && \text{acting on the right: maps } \boldsymbol{D} \text{ into } \boldsymbol{M} \\ &G^t && \text{acting on the right: maps } \hat{\boldsymbol{D}} \text{ into } \hat{\boldsymbol{M}} \ . \end{aligned} \tag{10}$$

Appendix 4.3: Rate of convergence of an iterative sequence.

Let $\mathbf{m}_0, \mathbf{m}_1, \ldots$ be an iterative sequence converging on $\mathbf{m}^*$. Assume for simplicity that $\mathbf{m}_n \neq \mathbf{m}_{n'}$ for $n \neq n'$, and that for no index n , $\mathbf{m}_n = \mathbf{m}^*$. The sequence $\mathbf{m}_0, \mathbf{m}_1, \ldots$ is said to converge *with order* r to $\mathbf{m}^*$ if r is the largest number such that

$$0 \leq \lim_{n\to\infty} \frac{\| m_{n+1} - m^* \|}{\| m_n - m^* \|^r} < \infty ,$$

where $\| \cdot \|$ denotes a norm. For $r = 1$ it is said that the convergence is *linear*, for $r = 2$, it is *quadratic*, etc. If the sequence $m_0, m_1, \ldots$ converges with order r , the limit

$$e = \lim_{n\to\infty} \frac{\| m_{n+1} - m^* \|}{\| m_n - m^* \|^r}$$

is named the *asymptotic error constant.* If $e = 0$, it is said that the convergence is *super* order r (superlinear, superquadratic, ...).

As an example, consider the sequences obtained when using a Newton method for minimizing the functions

$S_1(x) = x^4$, (case 1)

$S_2(x) = x^2 + x^3$, (case 2)

$S_3(x) = x^2 + x^4$, (case 3)

$S_4(x) = x^2$. (case 4)

This gives respectively the sequences

$x_{n+1} = \frac{2}{3} x_n$ linear convergence (case 1)

$x_{n+1} = \frac{3}{2 + 6 x_n} x_n^2$ quadratic convergence (case 2)

$x_{n+1} = \frac{4}{1 + 6 x_n^2} x_n^3$ cubic convergence (case 3)

$x_{n+1} = 0$ infinite order convergence (case 4).

Figure 4.7 illustrates these sequences. In particular, it can be seen that for a linear (respectively quadratic, cubic) convergence, the number of significant digits increases linearly (respectively doubles, trebles) with the iteration number (for a sufficiently high iteration number).

5.000000000000000	5.0D+00	5.000000000000000	5.0D+00	5.000000000000000	5.0D+00	5.000000000000000	5.0D+00
3.333333333333333	3.3D+00	2.343750000000000	2.3D+00	3.311258278145695	3.3D+00	0.000000000000000	0.0D+00
2.222222222222222	2.2D+00	1.025960603112840	1.0D+00	2.174452392758933	2.2D+00	0.000000000000000	0.0D+00
1.481481481481481	1.5D+00	0.387184526803539	3.9D-01	1.400276344265300	1.4D+00	0.000000000000000	0.0D+00
0.987654320987654	9.9D-01	0.104030632750452	1.0D-01	0.860384490513769	8.6D-01	0.000000000000000	0.0D+00
0.658436213991770	6.6D-01	0.012372272740441	1.2D-02	0.468180783168134	4.7D-01	0.000000000000000	0.0D+00
0.438957475994513	4.4D-01	0.000221392320626	2.2D-04	0.177304529707353	1.8D-01	0.000000000000000	0.0D+00
0.292638317329675	2.9D-01	0.000000073473040	7.3D-08	0.018757542851859	1.9D-02	0.000000000000000	0.0D+00
0.195092211553117	2.0D-01	0.000000000000008	8.1D-15	0.000026343408864	2.6D-05	0.000000000000000	0.0D+00
0.130061474368745	1.3D-01	0.000000000000000	9.8D-29	0.000000000000073	7.3D-14	0.000000000000000	0.0D+00
0.086707649579163	8.7D-02	0.000000000000000	1.5D-56	0.000000000000000	1.6D-39	0.000000000000000	0.0D+00
0.057805099719442	5.8D-02	0.000000000000000	3.2D-112	0.000000000000000	1.5D-117	0.000000000000000	0.0D+00
0.038536733146295	3.9D-02						
0.025691155430863	2.6D-02						
0.017127436953909	1.7D-02						
0.011418291302606	1.1D-02						
0.007612194201737	7.6D-03						
0.005074796134491	5.1D-03						
0.003383197422994	3.4D-03						
0.002255464948663	2.3D-03						
0.001503643299109	1.5D-03						
0.001002428866072	1.0D-03						
0.000668285910715	6.7D-04						
0.000445523940477	4.5D-04						
0.000297015960318	3.0D-04						
0.000198010640212	2.0D-04						
0.000132007093475	1.3D-04						
0.000088004728983	8.8D-05						
0.000058669819322	5.9D-05						
0.000039113212881	3.9D-05						
0.000026075475254	2.6D-05						
0.000017383650169	1.7D-05						
0.000011589100113	1.2D-05						
0.000007726066742	7.7D-06						
0.000005150711161	5.2D-06						
0.000003433807441	3.4D-06						
0.000002289204961	2.3D-06						
0.000001526136640	1.5D-06						
0.000001017424427	1.0D-06						
0.000000678282951	6.8D-07						
0.000000452188634	4.5D-07						
0.000000301459089	3.0D-07						
0.000000200972726	2.0D-07						
0.000000133981818	1.3D-07						
0.000000089321212	8.9D-08						
0.000000059547474	6.0D-08						
0.000000039698316	4.0D-08						
0.000000026465544	2.6D-08						

← *Figure 4.7:* Illustration of the rate of convergence of an iterative sequence. For a better understanding, each sequence is written twice, first in decimal notation, and second in powers of 10. The first sequence corresponds to the Newton minimization of $S(x) = x^4$, is given by $x_{n+1} = 2/3\ x_n$, and has linear convergence. The second sequence corresponds to the Newton minimization of $S(x) = x^2 + |x|^3$, is given by $x_{n+1} = 3/(2 + 6\ x_n)\ x_n^2$, and has quadratic convergence. The third corresponds to the Newton minimization of $S(x) = x^2 + x^4$, is given by $x_{n+1} = 4/(1 + 6\ x_n^2)\ x_n^3$, and has cubic convergence. The fourth corresponds to the Newton minimization of $S(x) = x^2$, is given by $x_{n+1} = 0$, and has an infinite order rate of convergence. All sequences converge to $x = 0$, and start at $x = 5$. Notice that for a sufficiently high iteration number, the number of significant digits increases linearly for the linear example, doubles for the quadratic example, and trebles for the cubic example.

Appendix 4.4: Iterative resolution of linear systems

Let us consider here the problem of obtaining the vector $\mathbf{x}$ solution of

$$\mathbf{A}\,\mathbf{x} = \mathbf{y}\ , \tag{1}$$

where $\mathbf{y}$ is a given vector and $\mathbf{A}$ represents a *positive definite* linear operator, either symmetric or self-adjoint (see Box 4.5).

Let $\mathbf{Q}$ be a positive definite operator which is as close as possible to $\mathbf{A}^{-1}$, but which is more easy to obtain (or to handle) than $\mathbf{A}^{-1}$ (we will later see how "close" it has to be):

$$\mathbf{Q}\,\mathbf{A} \simeq \mathbf{I}\ . \tag{2}$$

We have successively

$$\mathbf{Q}\,\mathbf{A}\,\mathbf{x} = \mathbf{Q}\,\mathbf{y}$$

$$(\mathbf{I} - (\mathbf{I} - \mathbf{Q}\,\mathbf{A}))\,\mathbf{x} = \mathbf{Q}\,\mathbf{y}$$

$$\mathbf{x} = \mathbf{Q}\,\mathbf{y} + (\mathbf{I} - \mathbf{Q}\,\mathbf{A})\ \mathbf{x}\ .$$

This last equation can now be solved using a fixed point algorithm:

$$\mathbf{x}_{n+1} = \mathbf{Q}\,\mathbf{y} + (\mathbf{I} - \mathbf{Q}\,\mathbf{A})\,\mathbf{x}_n \qquad (\mathbf{x}_0 = \mathbf{0}), \tag{3}$$

i.e.,

$$\boxed{\mathbf{x}_{n+1} = \mathbf{x}_n + \mathbf{Q}\,(\mathbf{y} - \mathbf{A}\,\mathbf{x}_n) \qquad (\mathbf{x}_0 = \mathbf{0}) .} \tag{4}$$

If the iterative scheme effectively converges into a fixed point ($\mathbf{x}_{(n+1)} \simeq \mathbf{x}_n$), then

$$\mathbf{Q}\,(\mathbf{y} - \mathbf{A}\,\mathbf{x}_n) \simeq \mathbf{0} ,$$

and, as $\mathbf{Q}$ is positive definite,

$$\mathbf{A}\,\mathbf{x}_n \simeq \mathbf{y} ,$$

so that the linear system (1) has effectively been solved.

It often happens that the operator $\mathbf{A}$ is "diagonally dominant". In that case, a good choice for $\mathbf{Q}$ is to take a diagonal operator containing as elements the inverse of the diagonal elements of $\mathbf{A}$ (Jacobi's method). Sometimes, it is possible to take for $\mathbf{Q}$ the inverse of an operator identical to $\mathbf{A}$ over a "diagonal band", and zero elsewhere. The better the choice of $\mathbf{Q}$, the better the convergence.

From (4) we deduce

$$\mathbf{x}_{n+1} = \left[\sum_{p=0}^{n} (\mathbf{I} - \mathbf{Q}\,\mathbf{A})^p \right] \mathbf{Q}\ \mathbf{y} ,$$

and it can be shown that a necessary and sufficient condition for the convergence of this series is that

$$0 < \max \lambda_\alpha < 2 ,$$

where λ_α are the eigenvalues of $\mathbf{Q}\,\mathbf{A}$. Unfortunately, this condition is too difficult to check in general, and often we can only use physical arguments to guess if a given approximation of $\mathbf{A}^{-1}$ will be good enough. The proof is only obtained when the algorithm is tested and the convergence numerically established.

More generally, the reader can apply the algorithms of section 4.6 to the minimization of $\| \mathbf{A}\,\mathbf{x} - \mathbf{y} \|$.

This iterative method can of course also be used for the inversion of a linear operator. Replacing equation (1) by

$$\mathbf{A}\ \mathbf{A}^{-1} = \mathbf{I} \tag{5}$$

leads to

$$\mathbf{A}^{-1}_{n+1} = \mathrm{A}^{-1}_{n} + \mathrm{Q}\left(\mathrm{I} - \mathbf{A}\ \mathbf{A}^{-1}_{n}\right) \qquad (\mathbf{A}^{-1}{}_{0} = \mathbf{0})\ . \tag{6}$$

Further details in these methods can be found in any text book on Numerical Analysis, such as for instance, Ciarlet (1982).

Appendix 4.5: The Cholesky decomposition.

A being a symmetric positive definite matrix, it can be shown that there exists a unique *lower triangular* matrix **B** with positive diagonal elements, and such that

$$\mathbf{A} = \mathbf{B}\ \mathbf{B}^t\ . \tag{1}$$

Writing

$$\mathbf{A} = \begin{bmatrix} a_{11} & a_{12} & a_{13} & \dots \\ a_{21} & a_{22} & a_{23} & \dots \\ a_{31} & a_{32} & a_{33} & \dots \\ \dots & \dots & \dots & \dots \end{bmatrix} \qquad \mathbf{B} = \begin{bmatrix} b_{11} & 0 & 0 & \dots \\ b_{21} & b_{22} & 0 & \dots \\ b_{31} & b_{32} & b_{33} & \dots \\ \dots & \dots & \dots & \dots \end{bmatrix},$$

equation (1) directly gives

$$a_{11} = (b_{11})^2 \qquad \Rightarrow \qquad b_{11} = \sqrt{a_{11}}$$

$$a_{12} = b_{11}\, b_{21} \qquad \Rightarrow \qquad b_{21} = \frac{a_{12}}{b_{11}}$$

(...)

and, more generally,

$$a_{ii} = b_{i1}\, b_{i1} + \ldots + b_{ii}\, b_{ii} \quad \Rightarrow \quad b_{ii} = \sqrt{a_{ii} - \sum_{k=1}^{i-1} (b_{ik})^2} \tag{2a}$$

$$a_{ij} = b_{i1}\, b_{j1} + \ldots + b_{ii}\, b_{ji} \quad \Rightarrow \quad b_{ji} = \frac{a_{ij} - \sum_{k=1}^{i-1} b_{ik}\, b_{jk}}{b_{ii}} \; . \tag{2b}$$

a) Resolution of a linear system.

We now consider the problem of obtaining the column matrix $\mathbf{x}$ solution of

$$\mathbf{A}\,\mathbf{x} = \mathbf{y} \, , \tag{3}$$

where the column matrix $\mathbf{y}$ is given. Using Cholesky's decomposition, we have

$$\mathbf{B}\,\mathbf{B}^t\,\mathbf{x} = \mathbf{y} \, ,$$

equation equivalent to the system

$$\begin{aligned} \mathbf{B}\,\mathbf{w} &= \mathbf{y} \\ \mathbf{B}^t\,\mathbf{x} &= \mathbf{w} \, . \end{aligned} \tag{4}$$

Cholesky's method for solving the linear system consists in computing $\mathbf{B}$, then in solving the last two systems successively (which is very easy because $\mathbf{B}$ is lower triangular).

b) Inversion of a matrix.

Let $\mathbf{C}$ denote the inverse of $\mathbf{A}$. The equation

$$\mathbf{A}\,\mathbf{C} = \mathbf{I}$$

is equivalent to the whole set of linear equations

$$\mathbf{A} \begin{bmatrix} C^{11} \\ C^{21} \\ C^{31} \\ \cdots \end{bmatrix} = \begin{bmatrix} 1 \\ 0 \\ 0 \\ \cdots \end{bmatrix} , \qquad \mathbf{A} \begin{bmatrix} C^{12} \\ C^{22} \\ C^{32} \\ \cdots \end{bmatrix} = \begin{bmatrix} 0 \\ 1 \\ 0 \\ \cdots \end{bmatrix} ,$$

$$\mathbf{A} \begin{bmatrix} C^{13} \\ C^{23} \\ C^{33} \\ \cdots \end{bmatrix} = \begin{bmatrix} 0 \\ 0 \\ 1 \\ \cdots \end{bmatrix} , \qquad (...)$$

The successive resolution of all these linear equations, using the method described in (a), gives the inverse of the matrix $\mathbf{A}$.

In particular, we see that if the dimension of the matrix $\mathbf{A}$ is $(M \times M)$, the computation of $\mathbf{A}^{-1}$ is M times more expensive than the resolution of the linear system $\mathbf{A}\ \mathbf{x} = \mathbf{y}$ (in fact, slightly less, because Cholesky's decomposition $\mathbf{A} = \mathbf{B}\ \mathbf{B}^t$ has to be performed only once).

c) Computer programs.

Figure 4.8 shows a computer code for the inversion of a positive definite symmetric matrix, and Figure 4.9 shows a test-run of this program. Figure 4.10 shows a computer code for the resolution of a linear system, and Figure 4.11 shows a test-run of this program.

Appendix 4.6: The Lagrange multipliers.

Let the problem be that of minimizing the real functional

$$S = S(x^\alpha) \qquad (\alpha \in \mathbf{I}_X) \tag{1a}$$

under the nonlinear constraints

$$\Psi^i(x^\alpha) = 0 \qquad (i \in \mathbf{I}_\Psi), \tag{1b}$$

where $\mathbf{I}_X$ and $\mathbf{I}_\Psi$ represent discrete index sets.

The Lagrange method consists in introducing unknown parameters λ^i and defining a new functional $S'(x^\alpha,\lambda^i)$ by

```
100 REM ***********************************************************************
101 REM *      INVERSION OF A MATRIX USING THE CHOLESKY'S DECOMPOSITION.       *
102 REM *      The input matrix is A(N,N). The output matrix is AI(N,N).       *
103 REM *      W(N) is a working vector. DT is the determinant of A.           *
104 REM *      INPUTS: N, A(1,1), ... , A(N,N)                                 *
105 REM *      OUTPUTS: AI(1,1), ... , AI(N,N), DT.                            *
106 REM ***********************************************************************
107 REM
108 REM ***********************************************************************
109 REM *      Definition of dimensions.                                       *
110 REM ***********************************************************************
111 REM
112      DIM A(10,10): DIM AI(10,10): DIM W(10): DIM I(10,10)
113 REM
114 REM ***********************************************************************
115 REM *      Introduction of the test example.                               *
116 REM ***********************************************************************
117 REM
118 N = 9
119 FOR I = 1 TO N: FOR J = 1 TO N: A(I,J) = EXP(-ABS(I-J)/2): NEXT J: NEXT I
120 FOR I=1 TO N
121 FOR J=1 TO N: PRINT USING"####.###"; A(I,J),: NEXT J
122 PRINT: NEXT I: PRINT
123 REM
124 REM ***********************************************************************
125 REM *      Calculation begins.                                             *
126 REM ***********************************************************************
127 REM
128 DT = A(1,1)
129 AI(1,1) = SQR(A(1,1))
130      FOR I = 2 TO N
131      AI(I,1) = A(I,1)/AI(1,1)
132      NEXT I
133           FOR J = 2 TO N
134           L = J - 1
135           M = J + 1
136           S = A(J,J)
137                FOR K = 1 TO L
138                S = S - AI(J,K)^2
139                NEXT K
140           DT = DT * S
141           AI(J,J) = SQR(S)
142           IF J = N GOTO 151
143                FOR I = M TO N
144                S = A(I,J)
145                     FOR K = 1 TO L
146                     S = S - AI(J,K) * AI(I,K)
147                     NEXT K
148                AI(I,J) = S/AI(J,J)
149                NEXT I
150           NEXT J
151      N1 = N - 1
152      FOR I = 1 TO N1
153      W(I) = 1/AI(I,I)
154      M = I + 1
155           FOR J = M TO N
156           S = 0
157           L = J - 1
158                FOR K = I TO L
159                S = S - AI(J,K) * W(K)
160                NEXT K
161           W(J) = S/AI(J,J)
162           NEXT J
```

```
163          AI(I,N) = W(N)/AI(N,N)
164          M = N - 1
165               FOR J1 = 1 TO M
166               J = N - J1
167               L = J + 1
168               S = W(J)
169                    FOR K = L TO N
170                    S = S - AI(K,J) * AI(I,K)
171                    NEXT K
172               AI(I,J) = S/AI(J,J)
173               NEXT J1
174          NEXT I
175 AI(N,N) = 1/AI(N,N)/AI(N,N)
176          FOR I = 1 TO N
177               FOR J = 1 TO N
178               AI(J,I) = AI(I,J)
179               NEXT J
180          NEXT I
181 REM
182 REM **********************************************************************
183 REM     Calculation ends. Printing of the results begins.                *
184 REM **********************************************************************
185 REM
186 PRINT DT: PRINT
187 FOR I = 1 TO N
188 FOR J = 1 TO N: PRINT USING"####.###"; AI(I,J),: NEXT J
189 PRINT: NEXT I: PRINT
190 REM
191 REM **********************************************************************
192 REM     Test: A*AI is computed and printed.                              *
193 REM **********************************************************************
194 REM
195 FOR I = 1 TO N
196 FOR J = 1 TO N
197 I(I,J) = 0
198 FOR K = 1 TO N
199 I(I,J) = I(I,J) + A(I,K) * AI(K,J)
200 NEXT K: NEXT J: NEXT I
201 FOR I = 1 TO N
202 FOR J = 1 TO N: PRINT USING "####.###"; I(I,J),:NEXT J
203 PRINT: NEXT I: PRINT
204 REM
205 REM **********************************************************************
206 REM     End of the program                                               *
207 REM **********************************************************************
```

Figure 4.8: Computer code for the inversion of a positive definite symmetric matrix using the Cholesky decomposition. A BASIC-like notation has been used.

```
run
 1.000   0.607   0.368   0.223   0.135   0.082   0.050   0.030   0.018
 0.607   1.000   0.607   0.368   0.223   0.135   0.082   0.050   0.030
 0.368   0.607   1.000   0.607   0.368   0.223   0.135   0.082   0.050
 0.223   0.368   0.607   1.000   0.607   0.368   0.223   0.135   0.082
 0.135   0.223   0.368   0.607   1.000   0.607   0.368   0.223   0.135
 0.082   0.135   0.223   0.368   0.607   1.000   0.607   0.368   0.223
 0.050   0.082   0.135   0.223   0.368   0.607   1.000   0.607   0.368
 0.030   0.050   0.082   0.135   0.223   0.368   0.607   1.000   0.607
 0.018   0.030   0.050   0.082   0.135   0.223   0.368   0.607   1.000

 2.549172E-02

 1.582  -0.960  -0.000   0.000   0.000  -0.000  -0.000   0.000  -0.000
-0.960   2.164  -0.960   0.000  -0.000   0.000   0.000  -0.000   0.000
-0.000  -0.960   2.164  -0.960  -0.000   0.000  -0.000   0.000  -0.000
 0.000   0.000  -0.960   2.164  -0.960  -0.000   0.000  -0.000   0.000
 0.000  -0.000  -0.000  -0.960   2.164  -0.960  -0.000   0.000   0.000
-0.000   0.000   0.000  -0.000  -0.960   2.164  -0.960   0.000  -0.000
-0.000   0.000  -0.000   0.000  -0.000  -0.960   2.164  -0.960   0.000
 0.000  -0.000   0.000  -0.000   0.000   0.000  -0.960   2.164  -0.960
-0.000   0.000  -0.000   0.000   0.000  -0.000   0.000  -0.960   1.582

 1.000   0.000   0.000   0.000  -0.000   0.000   0.000  -0.000   0.000
-0.000   1.000   0.000  -0.000  -0.000   0.000   0.000  -0.000   0.000
 0.000  -0.000   1.000  -0.000   0.000  -0.000  -0.000  -0.000   0.000
-0.000   0.000   0.000   1.000   0.000   0.000   0.000  -0.000   0.000
 0.000   0.000   0.000  -0.000   1.000   0.000  -0.000   0.000   0.000
-0.000   0.000   0.000  -0.000  -0.000   1.000   0.000   0.000   0.000
-0.000   0.000  -0.000   0.000  -0.000  -0.000   1.000  -0.000   0.000
 0.000   0.000  -0.000  -0.000  -0.000   0.000  -0.000   1.000  -0.000
-0.000   0.000   0.000  -0.000  -0.000  -0.000   0.000  -0.000   1.000
Ok
```

Figure 4.9: Test-run of the previous program. The inverse of the matrix defined in line 119 is strictly tridiagonal.

$$S'(x^\alpha,\lambda^i) \;=\; S(x^\alpha) - \sum_{i\in I_\Psi} \lambda^i \;\Psi^i(x^\alpha)\;. \tag{2}$$

The conditions $\partial S'/\partial\lambda^i = 0$ give

$$\Psi^i(x^\alpha) \;=\; 0 \qquad\qquad (i \in \mathbf{I}_\Psi)\;, \tag{3}$$

while the conditions $\partial S'/\partial x^\alpha = 0$ give

```
100 REM ***********************************************************************
101 REM *     CHOLESKY'S RESOLUTION OF A LINEAR SYSTEM.                        *
102 REM *     This program solves  Y = A * X , where  A  is an (NxN) symmetric*
103 REM *     positive definite matrix. Factorizes  A = B * BT (Lower triang).*
104 REM *     It also computes the determinant of A (named DT).               *
105 REM *     INPUTS: N, A(1,1), ... , A(N,N), Y(1), ... , Y(N)               *
106 REM *     OUTPUTS: DT, X(1), ... , X(N), B(1,1), ... , B(N,N)             *
107 REM ***********************************************************************
108 REM
109 REM ***********************************************************************
110 REM *     Definition of dimensions.                                       *
111 REM ***********************************************************************
112 REM
113     DIM A(10,10): DIM B(10,10): DIM Y(10): DIM X(10)
114 REM
115 REM ***********************************************************************
116 REM *     Introduction of the test example.                               *
117 REM ***********************************************************************
118 REM
119 N = 9
120 FOR I = 1 TO N: FOR J = 1 TO N: A(I,J) = EXP(-ABS(I-J)/2): NEXT J: NEXT I
121 FOR I = 1 TO N
122 FOR J = 1 TO N: PRINT USING"####.###"; A(I,J),: NEXT J
123 PRINT: NEXT I: PRINT
124 FOR I = 1 TO N: Y(I) = 0: NEXT I
125 Y(3) = 1
126 FOR I = 1 TO N: PRINT USING"####.###"; Y(I),: NEXT I: PRINT
127 PRINT
128 REM
129 REM ***********************************************************************
130 REM *     Calculation begins.                                             *
131 REM ***********************************************************************
132 REM
133 REM
134 DT = A(1,1)
135 B(1,1) = SQR(A(1,1))
136       FOR I = 2 TO N
137       B(I,1) = A(I,1)/B(1,1)
138       NEXT I
139             FOR J = 2 TO N
140             L = J-1
141             M = J+1
142             S = A(J,J)
143                   FOR K = 1 TO L
144                   S = S - B(J,K)^2
145                   NEXT K
146       DT = DT * S
147       B(J,J) = SQR(S)
148       IF J = N THEN GOTO 157
149             FOR I = M TO N
150             S = A(I,J)
151                   FOR  K = 1 TO L
152                   S = S - B(J,K)*B(I,K)
153                   NEXT K
154             B(I,J) = S/B(J,J)
155             NEXT I
156       NEXT J
157 X(1) = Y(1)/B(1,1)
158       FOR J = 2 TO N
159       S = Y(J)
160       L = J - 1
161             FOR K = 1 TO L
162             S = S - B(J,K) * X(K)
```

```
163              NEXT K
164         X(J) = S/B(J,J)
165         NEXT J
166         X(N) = X(N)/B(N,N)
167         M = N - 1
168              FOR J1 = 1 TO M
169              J = N - J1
170              L = J + 1
171              S = X(J)
172                   FOR K = L TO N
173                   S = S - B(K,J) * X(K)
174                   NEXT K
175              X(J) = S/B(J,J)
176              NEXT J1
177 REM
178 REM ***********************************************************************
179 REM *     Calculation ends. Printing of the results begins.               *
180 REM ***********************************************************************
181 REM
182 PRINT DT: PRINT
183 FOR I = 1 TO N: PRINT USING"####.###"; X(I),: NEXT I
184 REM
185 REM ***********************************************************************
186 REM *     End of the program.                                             *
187 REM ***********************************************************************
```

Figure 4.10: Computer code for the resolution of a linear system with a positive definite symmetric matrix using Cholesky's decomposition. The resolution of a linear system is much less expensive than the inversion of a matrix.

$$\frac{\partial S}{\partial x^\alpha} - \sum_{i \in \mathbf{I}_\Psi} \lambda^i \frac{\partial \Psi^i}{\partial x^\alpha} = 0 \qquad (\alpha \in \mathbf{I}_X) . \tag{4}$$

Equations (3) show that at the minimum, the constraints (1b) will be satisfied. It follows that the *constrained* minimization of (1a) is equivalent to the *unconstrained* minimization of (2).

The system (3)-(4) has as many equations as unknowns (the x^α and the λ^i). Its resolution gives the solution of the problem.

```
run
   1.000     0.607     0.368     0.223     0.135     0.082     0.050     0.030     0.018
   0.607     1.000     0.607     0.368     0.223     0.135     0.082     0.050     0.030
   0.368     0.607     1.000     0.607     0.368     0.223     0.135     0.082     0.050
   0.223     0.368     0.607     1.000     0.607     0.368     0.223     0.135     0.082
   0.135     0.223     0.368     0.607     1.000     0.607     0.368     0.223     0.135
   0.082     0.135     0.223     0.368     0.607     1.000     0.607     0.368     0.223
   0.050     0.082     0.135     0.223     0.368     0.607     1.000     0.607     0.368
   0.030     0.050     0.082     0.135     0.223     0.368     0.607     1.000     0.607
   0.018     0.030     0.050     0.082     0.135     0.223     0.368     0.607     1.000

   0.000     0.000     1.000     0.000     0.000     0.000     0.000     0.000     0.000

 2.549172E-02

  -0.000    -0.960     2.164    -0.960    -0.000     0.000    -0.000     0.000    -0.000
Ok
```

Figure 4.11: Test-run of the previous program. The inverse of the matrix defined in line 119 is strictly tridiagonal. Notice that, as the vector used in this example is (0,0,1,0,0,0,0,0,0), the corresponding solution equals the third column of the inverse matrix obtained in Figure 4.9 .

Appendix 4.7: Gradient methods of minimization in normed linear spaces

Most of the material in this appendix has already been presented in sections 4.4 and 4.5, but here I do not assume that a scalar product has been defined.

Let $\mathcal{M}$ represent a discrete finite dimensional (vector) linear space, $\mathbf{m} \rightarrow \| \mathbf{m} \|$ a norm over $\mathcal{M}$, and $\mathbf{m} \rightarrow S(\mathbf{m})$ an arbitrary nonlinear form over $\mathcal{M}$. We are here interested in the gradient methods allowing us to obtain the point $\mathbf{m}$ at which $S(\mathbf{m})$ reaches its minimum.

I assume that the norm $\| \mathbf{m} \|$ is arbitrary (it is *not* necessarily an ℓ_2 norm). This will allow to use the results of this box in subsequent chapters.

Let $\hat{\mathcal{M}}$ be the dual of $\mathcal{M}$. The elements of $\hat{\mathcal{M}}$ (i.e., the linear forms over $\mathcal{M}$) are denoted with a "hat", like $\hat{\mathbf{f}}$. As usual, the notation $\langle \hat{\mathbf{f}} , \mathbf{m} \rangle$ is used to denote the real number associated to $\mathbf{m}$ by the linear form $\hat{\mathbf{f}}$.

Let $\mathbf{m}_0$ be an arbitrary point of $\mathcal{M}$. A second order development of S around $\mathbf{m}_0$ can be written

$$S(\mathbf{m}_0+\delta\mathbf{m}) = S(\mathbf{m}_0) - \langle \hat{\gamma}_0 , \delta\mathbf{m} \rangle + \frac{1}{2} \langle \hat{\mathbf{H}}_0 \, \delta\mathbf{m} , \delta\mathbf{m} \rangle + O(\| \delta\mathbf{m} \|^3), \tag{1}$$

thus defining the *gradient*, $\hat{\gamma}_0$, and the *Hessian*, $\hat{\mathbf{H}}_0$, of S at $\mathbf{m} = \mathbf{m}_0$. As usual, $O(\| \delta\mathbf{m} \|^3)$ represents a term which is at least third order in $\| \delta\mathbf{m} \|$ We are here interested in problems where gradient and Hessian are defined everywhere.

As discussed in Box 4.7, if $\mathbf{m}$ has components $\{ m^\alpha \}$ $(\alpha \in \mathbf{I}_M)$ then gradient and Hessian have components

$$\hat{\gamma}_0{}^\alpha = \left(\frac{\partial S}{\partial m^\alpha} \right)_{\mathbf{m}_0} \qquad (\alpha \in \mathbf{I}_M) \tag{2}$$

$$\hat{H}_0{}^{\alpha\beta} = \left(\frac{\partial \hat{\gamma}^\alpha}{\partial m^\beta} \right)_{\mathbf{m}_0} = \left(\frac{\partial^2 S}{\partial m^\alpha \partial m^\beta} \right)_{\mathbf{m}_0} \qquad (\alpha \in \mathbf{I}_M) \quad (\beta \in \mathbf{I}_M) . \tag{3}$$

The gradient of S at a point $\mathbf{m}_0$, $\hat{\gamma}_0$, is a form over M (it is *not* an element of M). The Hessian of S at $\mathbf{m}_0$, $\hat{\mathbf{H}}_0$, is a linear operator mapping M into $\hat{M}$.

Equation (2) shows that the gradient of S at a point $\mathbf{m}_0$ is, in fact, *independent* of the particular norm chosen (in this sense, all the norms over a discrete space are equivalent). On the contrary, the *direction of steepest ascent* at $\mathbf{m}_0$ (which is an element of M) *depends* on the particular norm considered (see box 4.7). In the following, *gradient* and *direction of steepest ascent* are not to be taken as synonymous.

Let $\hat{\mathbf{f}} \in \hat{M}$ and $\mathbf{m} \in M$. Using the notation

$$\langle \hat{\mathbf{f}} , \mathbf{m} \rangle = \hat{\mathbf{f}}^t \, \mathbf{m} , \tag{4}$$

it is possible to give the formulas of this appendix a form readily interpretable in terms of matrix operations. As I prefer here to use notations allowing an easy understanding of the *type* of objects under manipulation, I will make use of that possibility only occasionally.

There exist many good books on gradient methods. As usual, English books (Walsh, 1975; Fletcher, 1980; Powell, 1981; Scales, 1985) are excellent for their empirical taste, but suffer sometimes from a lack of generality. French books (Céa, 1971; Ciarlet, 1982) have opposite qualities. Many of the results of this box are only suggested, and the reader is referred to classical references for more details.

a) Minimization along a given direction:

Let $\mathbf{m}_n$ be the *current point*. We wish to obtain an *updated point*

$\mathbf{m}_{(n+1)}$. Let ϕ_n be a *direction of search* at $\mathbf{m}_n$:

$$\mathbf{m}_{n+1} = \mathbf{m}_n - \mu_n \phi_n , \tag{5}$$

where μ_n is a real number to be chosen such that $S(\mathbf{m}_{n+1})$ takes a minimum along the direction defined by ϕ_n.

Usual methods for obtaining adequate values for μ_n are, for instance, trial and error, or parabolic interpolation (the value of S is computed for three values of μ_n , a parabola is fitted to these three points, and the value of μ_n giving the minimum of the parabola is chosen). These methods, although robust, need the computation of S at some points, and for large-dimensioned problems, this can be too expensive. If the functional S is sufficiently well behaved, a useful approximation for μ_n can be obtained by considering not S , but its second-order approximation around $\mathbf{m}_n$:

$$\begin{aligned} S(\mathbf{m}_{n+1}) &= S(\mathbf{m}_n - \mu_n \phi_n) \\ &\simeq S(\mathbf{m}_n) - \mu_n \langle \hat{\gamma}_n , \phi_n \rangle + \frac{\mu^2_n}{2} \langle \hat{\mathbf{H}}_n \phi_n , \phi_n \rangle . \end{aligned}$$

The condition $\partial S/\partial \mu_n = 0$ then gives

$$\mu_n \simeq \frac{\langle \hat{\gamma}_n , \phi_n \rangle}{\langle \hat{\mathbf{H}}_n \phi_n , \phi_n \rangle} . \tag{6}$$

The symbol $\simeq$ in this last equation means that as the value of μ_n has been obtained by a second order development of S , it is optimum only for quadratic functions. For arbitrary functions, it can only be considered as a guess to be consequently checked and, eventually, modified. Appendix 5.1 suggests that if the minimization of $S(\mathbf{m})$ corresponds to the minimization of an ℓ_p norm, a better approximation is

$$\mu_n \simeq (p-1) \frac{\langle \hat{\gamma}_n , \phi_n \rangle}{\langle \hat{\mathbf{H}}_n \phi_n , \phi_n \rangle} . \tag{7}$$

For an ℓ_2 norm, $p-1 = 1$, and the usual expression is found.

b) The Newton Method

Let $\mathbf{m}_n$ represent the current point, and $\delta\mathbf{m}_n$ the correction to be computed for obtaining the updated point $\mathbf{m}_{n+1}$:

$$\mathbf{m}_{n+1} = \mathbf{m}_n - \delta\mathbf{m}_n \ .$$

A second order development of S around $\mathbf{m}_n$ gives

$$S(\mathbf{m}_{n+1}) = S(\mathbf{m}_n - \delta\mathbf{m}_n) = S(\mathbf{m}_n) - \langle \hat{\gamma}_n , \delta\mathbf{m}_n \rangle + \frac{1}{2} \langle \hat{\mathbf{H}}_n \delta\mathbf{m}_n , \delta\mathbf{m}_n \rangle + O(\| \delta\mathbf{m}_n \|^3) \ .$$

The Newton solution is obtained by the condition

$$S(\mathbf{m}_n) - \langle \hat{\gamma}_n , \delta\mathbf{m}_n \rangle + \frac{1}{2} \langle \hat{\mathbf{H}}_n \delta\mathbf{m}_n , \delta\mathbf{m}_n \rangle \qquad \text{MINIMUM} \ .$$

If $\hat{\mathbf{H}}_n$ is positive definite, this gives

$$\hat{\mathbf{H}}_n \delta\mathbf{m}_n = \hat{\gamma}_n \ ,$$

and

$$\mathbf{m}_{n+1} = \mathbf{m}_n - \hat{\mathbf{H}}_n^{-1} \hat{\gamma}_n \ . \tag{8}$$

As (8) is obtained by a second-order development, the Newton step is only optimum for quadratic functions (in which case it gives the minimum in only one iteration). If $S(\mathbf{m})$ is not quadratic, the Newton step is replaced by

$$\mathbf{m}_{n+1} = \mathbf{m}_n - \mu_n \hat{\mathbf{H}}_n^{-1} \hat{\gamma}_n \ , \tag{9}$$

where μ_n is a real number ensuring the condition $S(\mathbf{m}_{n+1}) < S(\mathbf{m}_n)$, and which has to be obtained by linear search.

Let us define the *Newton direction* at $\mathbf{m} = \mathbf{m}_n$ by

$$\phi_n = \hat{\mathbf{H}}_n^{-1} \hat{\gamma}_n \ .$$

It is well known that the resolution of the linear system

$$\hat{\mathbf{H}}_n \phi_n = \hat{\gamma}_n$$

does not necessarily require the computation of the kernel of the inverse of $\hat{\mathbf{H}}_n$ (see for instance Appendix 4.5). Newton's algorithm is then written

$$\begin{aligned} \hat{\mathbf{H}}_n \phi_n &= \hat{\gamma}_n && \text{(solve for } \phi_n \text{)} \\ \mathbf{m}_{n+1} &= \mathbf{m}_n - \mu_n \phi_n && \text{(obtain } \mu_n \text{ by linear search)} \ . \end{aligned} \tag{10}$$

Using the approximation (7) gives

$$\mu_n \simeq p-1 . \tag{11}$$

For sufficiently nonlinear problems, this approximation may not be adequate, and a more elaborate linear search along the Newton direction has to be made.

c) The "preconditioning" operator

The Newton direction

$$\phi_n = \hat{\mathbf{H}}_n^{-1} \hat{\gamma}_n$$

although a good direction of ascent in general, is not always easy to obtain, in particular for large-dimension problems. Let $\mathbf{S}_0$ be a symmetric positive definite operator arbitrarily approximating $\hat{\mathbf{H}}_0^{-1}$:

$$\mathbf{S}_0 \simeq \hat{\mathbf{H}}_0^{-1} . \tag{12}$$

The better $\mathbf{S}_0$ aproximates $\hat{\mathbf{H}}_0^{-1}$, the better the direction

$$\phi_0 = \mathbf{S}_0 \gamma_0$$

will approximate the Newton direction at $\mathbf{m}_0$. In practical applications, it is often possible (using as much ingenuity as possible) to define operators $\mathbf{S}_0$ which are simple to handle and which, in some sense, well "simulate" the action of $\hat{\mathbf{H}}_0^{-1}$.

Variable metric methods start with such an operator $\mathbf{S}_0$ and update it at each iteration

$$\mathbf{S}_{n+1} = \mathbf{S}_n + \delta\mathbf{S}_n$$

in order to approach $\hat{\mathbf{H}}_n^{-1}$, as closely as possible, as iterations proceed (see below).

It is also possible to keep the same operator $\mathbf{S}_0$ all through the iterations, using as direction of search at the n-th iteration

$$\phi_n = \mathbf{S}_0 \hat{\gamma}_n , \tag{13}$$

where $\hat{\gamma}_n$ is the gradient at $\mathbf{m}_n$. It generally takes more iterations to converge, but each iteration is simpler.

As discussed in Appendix 5.2, the direction ϕ_n defined by (13) is the direction of steepest descent at $\mathbf{m} = \mathbf{m}_n$ *with respect to* the metric induced by the ℓ_2-norm

$$\| \mathbf{m} \|_2 = \left(\mathbf{m}^t \; \mathbf{S}_0^{-1} \; \mathbf{m} \right)^{1/2} \qquad (14)$$

It should be noticed that, although arbitrary, $\mathbf{S}_0$ is assumed to work over the same spaces as $\hat{\mathbf{H}}_0^{-1}$, i.e., it maps $\hat{M}$, the dual of M , into M ; this means in particular that the components of the kernel of $\mathbf{S}_0$ must have the same physical dimensions as those of $\hat{\mathbf{H}}_0\text{-}1$; for instance, taking $\mathbf{S}_0 = \mathbf{I}$ does not necessarily make sense (and generally it does not). Instead, if we are considering an ℓ_2 norm over M , deriving from a scalar product

$$\| \mathbf{m} \|_2 = \left(\mathbf{m}^t \; \mathbf{C}_M^{-1} \; \mathbf{m} \right)^{1/2} , \qquad (15)$$

then the choice $\mathbf{S}_0 = \mathbf{C}_M$ gives directions ϕ_n which are directions of steepest descent in the usual sense.

The name of "preconditioning operator" given to $\mathbf{S}_0$ takes its source in the theory of resolution of linear systems: letting $\mathbf{A}\ \mathbf{x} = \mathbf{y}$ represent a system to be solved, and $\mathbf{S}_0 \simeq \mathbf{A}^{-1}$, the system $\mathbf{S}_0\ \mathbf{A}\ \mathbf{x} = \mathbf{S}_0\ \mathbf{y}$ is equivalent to the first, but if $\mathbf{S}_0$ is astutely chosen, it has a lower *condition number* (see problem 1.10), so its numerical resolution is more stable: the system has been "preconditioned".

d) The method of steepest descent (with preconditioning)

Letting $\mathbf{S}_0$ be a symmetric positive definite operator, hopefully a good approximation of the inverse of the initial Hessian $\hat{\mathbf{H}}_0$, the method of steepest descent chooses as direction of search at the current point $\mathbf{m}_n$ the direction of steepest descent (with respect to the metric induced by $\mathbf{S}_0$). This gives

$$\begin{aligned} \phi_n &= \mathbf{S}_0 \; \hat{\gamma}_n & & (\mathbf{S}_0 \simeq \hat{\mathbf{H}}_0^{-1}) \\ \mathbf{m}_{n+1} &= \mathbf{m}_n - \mu_n \; \phi_n & & (\text{obtain } \mu_n \text{ by linear search}) . \end{aligned} \qquad (16)$$

Using the approximation (7) gives

$$\mu_n \simeq (p-1) \frac{\langle \hat{\gamma}_n , \phi_n \rangle}{\langle \hat{H}_n \phi_n , \phi_n \rangle} . \tag{17}$$

e) The method of Conjugate Directions (with preconditioning):

This method starts as the Steepest Descent method: letting $\mathbf{m}_0$ be the starting point, and λ_0 the direction of steepest ascent at $\mathbf{m}_0$ (for a given metric), the point $\mathbf{m}_1$ is obtained by minimization of S in the direction defined by λ_0 (and, of course, in the sense defined by $-\boldsymbol{\lambda}_0$). Once the point $\mathbf{m}_1$ and the direction of steepest ascent $\boldsymbol{\lambda}_1$ have been obtained, the point $\mathbf{m}_2$ is *not* defined by minimization of S along $\boldsymbol{\lambda}_1$, but *over the subspace* spanned by λ_0 and λ_1. More generally, letting $\mathbf{m}_n$ be the current point, and $\lambda_0 , \lambda_1 ,..., \lambda_n$ respectively the directions of steepest ascent at $\mathbf{m}_0 , \mathrm{m}_2 ,..., \mathrm{m}_n$, the point $\mathbf{m}_{n+1}$ is defined as the point minimizing S in the subspace spanned by $\lambda_0 , \boldsymbol{\lambda}_1 ,..., \lambda_n$.

Thus defined, it is can be guessed that an algorithm based on such a method will converge much more rapidly than one based on a steepest descent method (which is actually the case), but what is much less evident is that the operations needed per iteration are not more difficult than those needed in the steepest descent method.

The general algorithm for the method of conjugate directions is (see Céa, 1971; Walsh, 1975; Fletcher, 1980; Powell, 1981; Ciarlet, 1982; or Scales, 1985):

$$\lambda_n = \mathbf{S}_0 \hat{\gamma}_n \qquad (\mathbf{S}_0 \simeq \hat{H}_0^{-1})$$

$$\phi_n = \lambda_n + \alpha_n \phi_{n-1} \quad (\phi_0 = \lambda_0) \qquad (\alpha_n \text{ defined below}) \tag{18}$$

$$\mathbf{m}_{n+1} = \mathbf{m}_n - \mu_n \phi_n \qquad (\text{obtain } \mu_n \text{ by linear search})$$

Different expressions can be obtained for α_n, such as for instance

$$\alpha_n = \frac{\langle \hat{\gamma}_n , \lambda_n \rangle}{\langle \hat{\gamma}_{n-1} , \lambda_{n-1} \rangle} , \tag{19}$$

due to Fletcher and Reeves (1964), or

$$\alpha_n = \frac{\langle \hat{\gamma}_n - \hat{\gamma}_{n-1} , \lambda_n \rangle}{\langle \hat{\gamma}_{n-1} , \lambda_{n-1} \rangle} , \tag{20}$$

due to Polak and Ribière (1969), or

$$\alpha_n = \frac{\langle \hat{\gamma}_n - \hat{\gamma}_{n-1} , \lambda_n \rangle}{\langle \hat{\gamma}_n - \hat{\gamma}_{n-1} , \phi_{n-1} \rangle} , \tag{21}$$

due to Hestenes and Stiefel (1952). Although these expressions are perfectly equivalent for quadratic functions, they are not equivalent for general problems. The Fletcher-Reeves formula is still probably the most widely used. Nevertheless, Powell (1977) has suggested that in some situtations, the Polak-Ribière formula may give superior results. For instance, in non-quadratic minimizations it may happen that $\boldsymbol{\phi}_n$ becomes almost orthogonal to the direction of steepest descent $\boldsymbol{\lambda}_n$. In that case, $\mathbf{m}_{n+1} \simeq \mathbf{m}_n$, $\hat{\gamma}_{n+1} \simeq \hat{\gamma}_n$, and $\lambda_{n+1} \simeq \lambda_n$. The Fletcher-Reeves formula then gives $\alpha_{n+1} \simeq 1$ and

$$\phi_{n+1} \simeq \lambda_{n+1} + \phi_n ,$$

while the Polak-Ribière formula gives $\alpha_n \simeq 0$ and

$$\phi_{n+1} \simeq \lambda_{n+1} .$$

This shows that in critical situations, when small advance can be made, the Polak-Ribière method is more robust, because it has a tendency to take the steepest descent direction as a direction of search.

Let NM represent the dimension of the space $\boldsymbol{M}$. For quadratic problems, it can be shown that if NM iterations are effectively performed, the actual minimum is attained.

Using the approximation (7) for μ_n gives

$$\mu_n \simeq (p-1) \frac{\langle \hat{\gamma}_n , \phi_n \rangle}{\langle \hat{H}_n \phi_n , \phi_n \rangle} . \tag{22}$$

f) Variable metric methods

The central idea of variable metric methods is to allow the preconditioning operator S_0 to vary from iteration to iteration, and to find an updating formula

$$S_{n+1} = S_n + \delta S_n$$

such that, in some sense,

$$S_n \to \hat{H}_n^{-1} , \tag{23}$$

so that a variable metric method will start behaving like a steepest descent method, but will finish behaving like a Newton method (with its rapid termination).

Let $\mathbf{S}_0$ be an arbitrary symmetric positive definite operator, hopefully a good approximation of the inverse on the initial Hessian

$$S_0 \simeq \hat{H}_0^{-1} .$$

The general structure of a variable metric method is (see Céa, 1971; Walsh, 1975; Fletcher, 1980; Powell, 1981; Ciarlet, 1982; Scales, 1985):

$$\phi_n = \mathbf{S}_n \hat{\gamma}_n \qquad (\mathbf{S}_0 \simeq \hat{\mathbf{H}}_0^{-1})$$

$$\mathbf{m}_{n+1} = \mathbf{m}_n - \mu_n \phi_n \qquad \text{(obtain } \mu_n \text{ by linear search)} \tag{24}$$

$$S_{n+1} = S_n + \delta S_n \qquad (\delta S_n \text{ defined below})$$

In numerical applications, the kernels of the operators S_1 , S_2 ,... do not need to be explicitly computed. All that is needed is the possibility of computing the result of the action of the operators on arbitrary vectors. In the following, $S_n(\cdot)$ represents the result of the application of S_n over a generic vector represented by " . ".

Let me define

$$\begin{aligned} \delta\mathbf{m}_n &= \mathbf{m}_{n+1} - \mathbf{m}_n \\ \delta\hat{\gamma}_n &= \hat{\gamma}_{n+1} - \hat{\gamma}_n \\ \mathbf{v}_n &= S_n\, \delta\hat{\gamma}_n \\ \mathbf{u}_n &= \delta\mathbf{m}_n - \mathbf{v}_n . \end{aligned} \tag{25}$$

Many different formulas exist for the updating of $\mathbf{S}_n$. For instance, the "symmetric rank-one formula"

$$S_{n+1}(\cdot) = S_n(\cdot) + \frac{\langle \cdot , \mathbf{u}_n \rangle}{\langle \delta\hat{\gamma}_n , \mathbf{u}_n \rangle} \mathbf{u}_n \tag{26}$$

due to Davidon (1959), the "DFP formula"

$$S_{n+1}(\cdot) = S_n(\cdot) + \frac{\langle \cdot , \delta\mathbf{m}_n \rangle}{\langle \delta\hat{\gamma}_n , \delta\mathbf{m}_n \rangle} \delta\mathbf{m}_n - \frac{\langle \cdot , \mathbf{v}_n \rangle}{\langle \delta\hat{\gamma}_n , \mathbf{v}_n \rangle} \mathbf{v}_n , \tag{27}$$

due to Davidon (1959) and to Fletcher and Powell (1963), and the "BFGS formula"

$$S_{n+1}(\cdot) = S_n(\cdot) + \frac{1}{\langle \delta\hat{\gamma}_n , \delta\mathbf{m}_n \rangle} \times$$

$$\times \left(\beta_n \langle \cdot , \delta \mathbf{m}_n \rangle \ \delta \mathbf{m}_n - \langle \ \delta \hat{\gamma}_n , \mathbf{S}_n(\cdot) \rangle \ \delta \mathbf{m}_n - \langle \ \cdot , \delta \mathbf{m}_n \rangle \ \ \mathbf{v}_n \right), \tag{28}$$

where

$$\beta_n = 1 + \frac{\langle \ \delta\hat{\gamma}_n , \mathbf{v}_n \rangle}{(\langle \ \delta\hat{\gamma}_n , \delta\mathbf{m}_n \rangle)},$$

due to Broyden (1970), Fletcher (1970), Goldfarb (1970), and Shanno (1970).

Although it is easy to see that all the operators $\mathbf{S}_n$ thus defined are definite positive, they may become numerically singular. It seems that the BFGS formula has a greater tendency to keep the definiteness of $\mathbf{S}_n$. Maybe this is the reason why it is today the most widely used updating formula.

Let NM denote the dimension of the space $\boldsymbol{M}$. When using these formulas for quadratic problems, it can be shown that if NM iterations are effectively performed, the actual minimum is attained, and

$$S_{NM} = H^{-1} , \tag{29}$$

where $\mathbf{H}$ denotes the (constant) Hessian of S. For large-dimensioned problems, good approximations of the minimum are sometimes obtained after a few iterations. Of course, the better $\mathbf{S}_0$ approximates $\mathbf{H}_0{}^{-1}$, the better the convergence, is in general.

As previously indicated, to operate with S_n we do not need to "build" the kernel of the operator: we only need to store in the computer's memory some vectors and scalars. For instance, letting $\hat{\mathbf{f}}$ be an arbitrary vector, we have, for the rank one formula

$$S_n \hat{\mathbf{f}} = \mathbf{S}_0 \hat{\mathbf{f}} + \sum_{p=0}^{n-1} \frac{\langle \ \hat{\mathbf{f}} , \mathbf{u}_p \rangle}{\nu_p} \ \mathbf{u}_p \tag{30}$$

where ν_p are the real numbers

$$\nu_p = \langle \ \delta\hat{\gamma}_p , \mathbf{u}_p \rangle ,$$

which shows that, in order to operate with $\mathbf{S}_n$ we only have to store in the computer's memory the vectors $\mathbf{u}_0 ,\dots , \mathbf{u}_n$ and the scalars $\nu_0 ,\dots , \nu_n$.

For small-sized problems it is nevertheless possible explicitly to compute the kernels of S_n. Using the notation (4) formulas (26)-(28) can be written

$$S_{n+1} = \mathbf{S}_n + \frac{\mathbf{u}_n \ \mathbf{u}_n{}^t}{\delta\hat{\gamma}_n{}^t \ \mathbf{u}_n} \tag{26 bis}$$

$$S_{n+1} = \mathbf{S}_n + \frac{\delta\mathbf{m}_n \ \delta\mathbf{m}_n{}^t}{\delta\hat{\gamma}^t \ \delta\mathbf{m}_n} - \frac{\mathbf{v}_n \ \mathbf{v}_n{}^t}{\delta\hat{\gamma}_n{}^t \ \mathbf{v}_n} \tag{27 bis}$$

$$S_{n+1} = \left[I - \frac{\delta \mathbf{m}_n \ \delta\hat{\gamma}_n{}^t}{\delta\hat{\gamma}_n{}^t \ \delta \mathbf{m}_n} \right] \mathbf{S}_n \left[I - \frac{\delta \mathbf{m}_n \ \delta\hat{\gamma}_n{}^t}{\delta\hat{\gamma}_n{}^t \ \delta \mathbf{m}_n} \right]^t + \frac{\delta \mathbf{m}_n \ \delta \mathbf{m}_n{}^t}{\delta\hat{\gamma}^t \ \delta \mathbf{m}_n} . \tag{28 bis}$$

Using the approximation (7) for μ_n gives

$$\mu_n \simeq (p-1) \frac{\langle \hat{\gamma}_n , \phi_n \rangle}{\langle \hat{H}_n \phi_n , \phi_n \rangle} . \tag{32}$$

Figure 4.12, due to Gill, Murray, and Wright (1981) compares different methods for a nonquadratic function of two variables.

Appendix 4.8: The spherical harmonics: basis of functional spaces.

a) Functions defined over the surface of a sphere

Let us consider the space of complex functions $f(\theta,\phi)$ defined over the surface of a sphere ($\theta \in [0,\pi]$, $\phi \in [0,2\pi]$). The scalar product of two functions is defined by

$$(f_1 , f_2) = \int_0^{2\pi} d\phi \int_0^{\pi} d\theta \ \sin\theta \ f_1(\theta,\phi) \ \tilde{f}_2(\theta,\phi) , \tag{1}$$

where (~) represents the conjugate complex. For $-\infty \le n \le +\infty$ $-n \le m \le +n$, the sequence of functions

$$Y_{nm}(\theta,\phi) = \tag{2}$$

$$\frac{(-1)^{n+m}}{2^n \ n!} \sqrt{\frac{2n+1}{4\pi} \frac{(n-m)!}{(n+m)!}} \ e^{im\phi} \ (\sin\theta)^m \ \frac{d^{n+m}}{d(\cos\theta)^{n+m}} \ (\sin\theta)^{2n}$$

is orthonormal:

$$(Y_{nm} , Y_{n'm'}) = \int_0^{2\pi} d\phi \int_0^{\pi} d\theta \ \sin\theta \ Y_{nm}(\theta,\phi) \ \tilde{Y}_{n'm'}(\theta,\phi)$$

$$= \delta_{nn'} \ \delta_{mm'} . \tag{3}$$

As the whole sequence is complete,

$$\sum_{n=-\infty}^{n=+\infty} \sum_{m=-n}^{m=+n} Y_{nm}(\theta,\phi) \ \tilde{Y}_{nm}(\theta',\phi') = \delta(\cos\theta - \cos\theta') \ \delta(\phi-\phi') , \tag{4}$$

it constitutes a basis of the functional space. Any function $f(\theta,\phi)$ can then

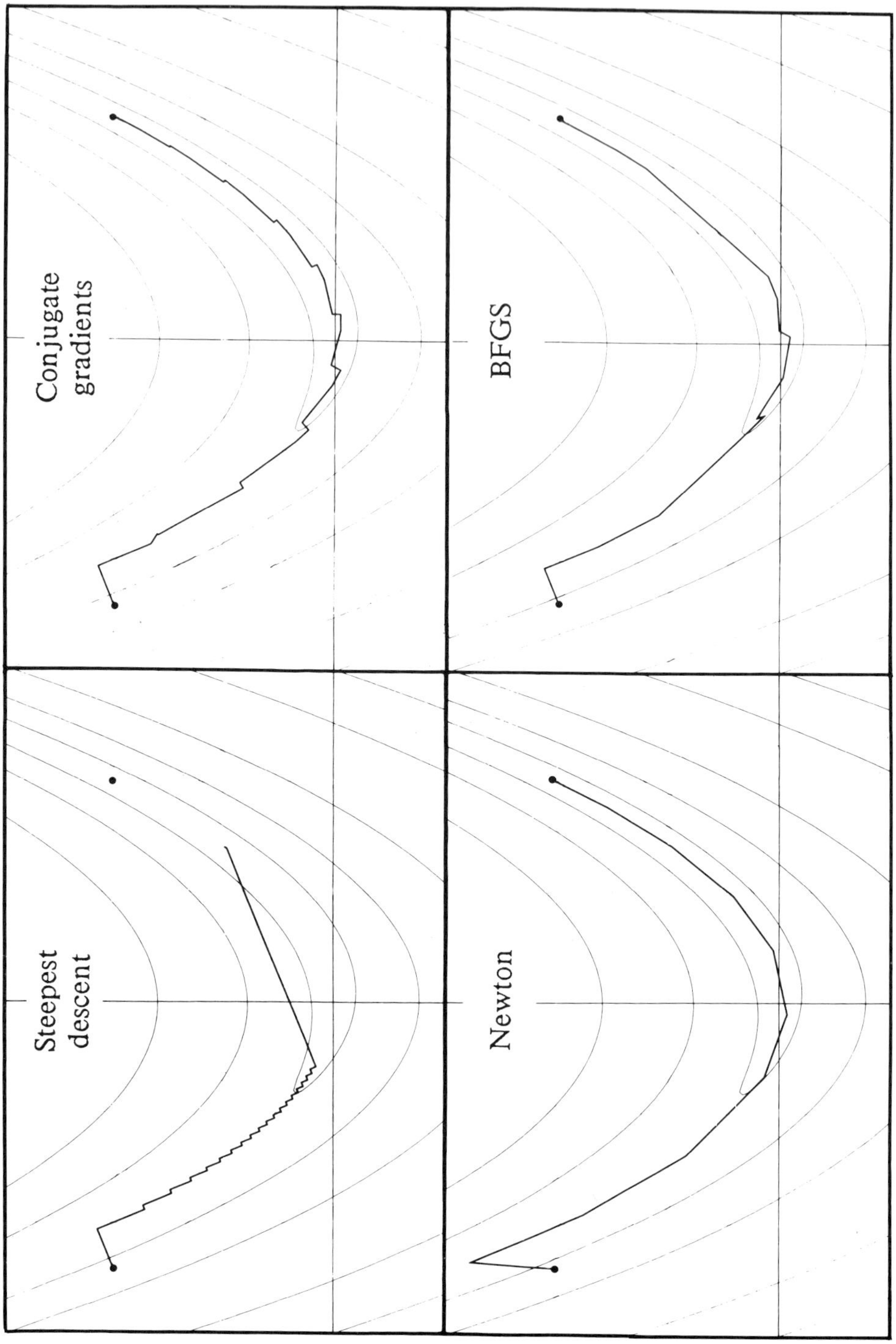
Conjugate gradients
BFGS
Steepest descent
Newton

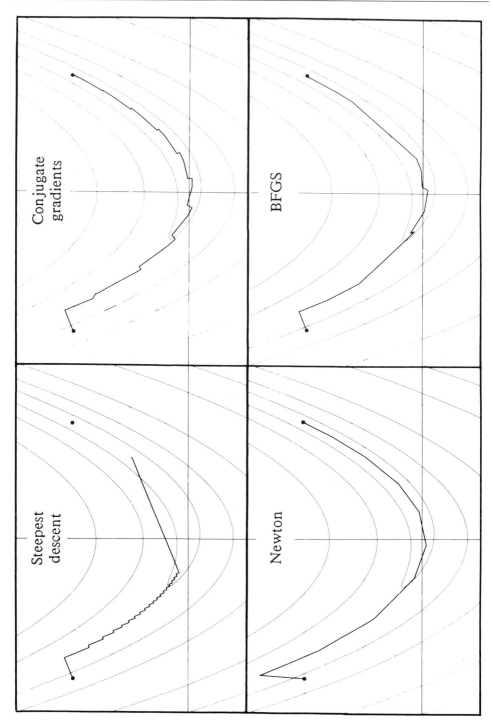

← *Figure 4.12:* The *Rosenbrock function* $S(x,y) = 100\ (y-x^2)^2 + (1-x)^2$ is often used to test optimization algorithms. It has a unique minimum at point $(x,y) = (1,1)$. Gill, Murray, and Wright (1981) show the solution paths obtained with different methods. All algorithms are initialized at $(x,y) = (-1.2,1.0)$.

Top left: Steepest-descent algorithm. Note that the algorithm would have failed in the vicinity of the point $(x,y) = (-0.3,0.1)$ but for the fact that the linear search found, by chance, the second minimum along the search direction. Several hundred iterations were performed close to the new point without any perceptible change. The algorithm was terminated after 1000 iterations. From Gill et al. (1981).

Top right: Conjugate gradient algorithm. Although the method is not intended for problems with such a small number of parameters, the figure is useful in illustrating the cyclic nature of the traditional conjugate gradient method. From Gill et al. (1981).

Bottom left: Modified Newton's algorithm. Except for the first iteration, the method follows the base of the valley in an almost "optimal" number of steps, given that piecewise linear segments are used. From Gill et al. (1981).

Bottom right: BFGS quasi-Newton (variable metric) algorithm. Like Newton's method, the algorithm makes good progress at points remote from the solution. The worst behaviour occurs near the origin, where the curvature is changing most rapidly. From Gill et al. (1981).

be expressed by its components over the basis

$$f(\theta,\phi) = \sum_{n=-\infty}^{n=+\infty} \sum_{m=-n}^{m=+n} c_{nm}\ Y_{nm}(\theta,\phi) , \tag{5}$$

and the components of **f** over the basis can be obtained by the scalar products of **f** with the elements of the basis:

$$c_{nm} = (\mathbf{f} , \mathbf{Y}_{nm}) = \int_0^{2\pi} d\phi \int_0^{\pi} d\theta\ \sin\theta\ f(\theta,\phi)\ Y_{nm}(\theta,\phi) . \tag{6}$$

The explicit expressions of the low order spherical harmonics are:

$$Y_{00}(\theta,\phi) = \sqrt{\frac{1}{4\pi}}$$

$$\begin{bmatrix} Y_{1-1}(\theta,\phi) \\ Y_{1+1}(\theta,\phi) \end{bmatrix} = \pm \sqrt{\frac{3}{8\pi}} \ \sin\theta \ \ e\mp i\phi$$

$$Y_{10}(\theta,\phi) = \sqrt{\frac{3}{4\pi}} \ \cos\phi$$

$$\begin{bmatrix} Y_{2-2}(\theta,\phi) \\ Y_{2+2}(\theta,\phi) \end{bmatrix} = \sqrt{\frac{15}{32\pi}} \ \sin^2\theta \ \ e\mp 2i\phi$$

$$\begin{bmatrix} Y_{2-1}(\theta,\phi) \\ Y_{2+1}(\theta,\phi) \end{bmatrix} = \pm \sqrt{\frac{15}{8\pi}} \ \sin\theta \ \cos\theta \ \ e\mp 2i\phi$$

$$Y_{20}(\theta,\phi) = \sqrt{\frac{5}{16\pi}} \ (3\cos^2\theta - 1)$$

b) Functions defined over the space

Let $f(r,\theta.\phi)$ be an arbitrary complex function defined over the three-dimensional space ($r \in [0,\infty]$, $\theta \in [0,\pi]$, $\phi \in [0,2\pi]$). For each value of r the development (3) holds:

$$f(r,\theta,\phi) = \sum_{n=-\infty}^{n=+\infty} \sum_{m=-n}^{m=+n} c_{nm}(r) \ Y_{nm}(\theta,\phi) \qquad (7)$$

The functions $c_{nm}(r)$ completely characterize $f(r,\theta,\phi)$. They can in turn be developed on a given basis of one-dimensional functions (polynomials, box-car functions,...)

c) Functions defined over a circle

Let us consider the space of complex functions $f(\phi)$ defined over a circle ($\phi \in [0,2\pi]$). The scalar product of two functions is defined by

$$(\mathbf{f}_1 , \mathbf{f}_2) = \int_0^{2\pi} d\phi \ f_1(\phi) \ \tilde{f}_2(\phi) , \tag{8}$$

where (~) represents the conjugate complex. For $-\infty \leq n \leq +\infty$ the sequence of functions

$$Y_n(\phi) = \frac{1}{\sqrt{2\pi}} e^{in\phi} \tag{9}$$

is orthonormal:

$$(\mathbf{Y}_n , \mathbf{Y}_{n'}) = \int_0^{2\pi} d\phi \ Y_n(\phi) \ \tilde{Y}_{n'}(\phi) = \delta_{nn'} \tag{10}$$

As the whole sequence is complete,

$$\sum_{n=-\infty}^{n=+\infty} Y_n(\phi) \ \tilde{Y}_n(\phi') = \delta(\phi-\phi') , \tag{11}$$

it constitutes a basis of the functional space. Any function $f(\phi)$ can then be expressed by its components over the basis

$$f(\phi) = \sum_{n=-\infty}^{n=+\infty} c_n \ Y_n(\phi) , \tag{12}$$

and the components of $\mathbf{f}$ over the basis can be obtained by the scalar products of $\mathbf{f}$ with the elements of the basis:

$$c_n = (\mathbf{f} , \mathbf{Y}_n) = \int_0^{2\pi} d\phi \ f(\phi) \ \tilde{Y}_n(\phi)$$

$$= \frac{1}{\sqrt{2\pi}} \int_0^{2\pi} d\phi \ f(\phi) \ e^{-in\phi} . \tag{13}$$

d) Functions defined over the plane

Let $f(r,\phi)$ be an arbitrary complex function defined over the plane ($r \in [0,\infty]$, $\phi \in [0,2\pi]$). For each value of r the development (8) holds:

$$f(r,\phi) = \sum_{n=-\infty}^{n=+\infty} c_n(r)\ Y_n(\phi) \tag{14}$$

The functions $c_n(r)$ completely characterize $f(r,\phi)$. They can in turn be developed on a given basis of one-dimensional functions (polynomials, box-car functions,...)

Problems for chapter 4:

Problem 4.1: Instead of a theoretical equation of the form $d = g(m)$, consider the more general form $\mathbf{f(x) = 0}$, where $\mathbf{x}$ contains all the parameters of the problem (measurable parameters and model parameters). Give the Newton and steepest descent formulas for the resolution of the general least-squares problem.

Solution: Let $\mathbf{x}_{prior}$ denote the a priori values (prior to the inversion) of all the parameters of the problem. For a measured parameter, its a priori value is of course the measured value. Let $\mathbf{C}_X$ be the covariance operator describing uncertainties in $\mathbf{x}_{prior}$. Let $\mathbf{f(x) = 0}$ be the theoretical relationship between the parameters. And finally let $\mathbf{C}_T$ be the covariance operator describing uncertainties in this theoretical relationship. The maximum likelihood estimate of the true value of $\mathbf{x}$ minimizes (see section 1.7.2)

$$S(\mathbf{x}) = \frac{1}{2} \left[\mathbf{f(x)}^t\ \mathbf{C}_T^{-1}\ \mathbf{f(x)} + (\mathbf{x}-\mathbf{x}_{prior})^t\ \mathbf{C}_X^{-1}\ (\mathbf{x}-\mathbf{x}_{prior}) \right] . \tag{1}$$

Using the Newton method gives (Tarantola and Valette, 1982b):

$$\mathbf{x}_{n+1} = \mathbf{x}_n - \mu_n \left(\mathbf{F}_n^t\ \mathbf{C}_T^{-1}\ \mathbf{F}_n + \mathbf{C}_X^{-1} \right)^{-1}$$

$$\left(\mathbf{F}_n^t\ \mathbf{C}_T^{-1}\ \mathbf{f}(\mathbf{x}_n) + \mathbf{C}_X^{-1}\ (\mathbf{x}_n - \mathbf{x}_{prior}) \right)$$

m= number of data

n= number of layers

$F_{m \times n+m}$

$$\mathbf{x}_{n+1} = \mathbf{x}_{prior} - \mu_n \left(\mathbf{F}_n^t \mathbf{C}_T^{-1} \mathbf{F}_n + \mathbf{C}_X^{-1} \right)^{-1} \mathbf{F}_n^t \mathbf{C}_T^{-1} \left(\mathbf{f}(\mathbf{x}_n) - \mathbf{F}_n (\mathbf{x}_n - \mathbf{x}_{prior}) \right)$$

$$\mathbf{x}_{n+1} = \mathbf{x}_{prior} - \mu_n \mathbf{C}_X \mathbf{F}_n^t \left(\mathbf{C}_T + \mathbf{F}_n \mathbf{C}_X \mathbf{F}_n^t \right)^{-1} \left(\mathbf{f}(\mathbf{x}_n) - \mathbf{F}_n (\mathbf{x}_n - \mathbf{x}_{prior}) \right)$$

$$\mathbf{x}_{n+1} = \mathbf{x}_n - \mu_n \left(\mathbf{I} + \mathbf{C}_X \mathbf{F}_n^t \mathbf{C}_T^{-1} \mathbf{F}_n \right)^{-1} \left(\mathbf{C}_X \mathbf{F}_n^t \mathbf{C}_T^{-1} \mathbf{f}(\mathbf{x}_n) + (\mathbf{x}_n - \mathbf{x}_{prior}) \right)$$

$$\mathbf{x}_{n+1} = \mathbf{x}_{prior} - \mu_n \left(\mathbf{I} + \mathbf{C}_X \mathbf{F}_n^t \mathbf{C}_T^{-1} \mathbf{F}_n \right)^{-1} \mathbf{C}_X \mathbf{F}_n^t \mathbf{C}_T^{-1} \left(\mathbf{f}(\mathbf{x}_n) - \mathbf{F}_n (\mathbf{x}_n - \mathbf{x}_{prior}) \right)$$

$$\mathbf{x}_{n+1} = \mathbf{x}_{prior} - \mu_n \mathbf{C}_X \mathbf{F}_n^t \mathbf{C}_T^{-1} \left(\mathbf{I} + \mathbf{F}_n \mathbf{C}_X \mathbf{F}_n^t \mathbf{C}_T^{-1} \right)^{-1} \left(\mathbf{f}(\mathbf{x}_n) - \mathbf{F}_n (\mathbf{x}_n - \mathbf{x}_{prior}) \right) \tag{2}$$

The optimum value for μ_n has to be obtained by linear search. A linearization of $\mathbf{f}(\mathbf{x})$ around $\mathbf{x}_n$ gives

$$\mu_n \simeq 1 \ .$$

The (preconditioned) steepest descent method leads to the algorithm

$$\gamma_n = \mathbf{C}_X \mathbf{F}_n^t \mathbf{C}_T^{-1} \mathbf{f}(\mathbf{x}_n) + (\mathbf{x}_n - \mathbf{x}_{prior}) \qquad (\mathbf{x}_0 \text{ arbitrary})$$

$$\phi_n = \hat{\mathbf{S}}_0 \gamma_n \qquad (\hat{\mathbf{S}}_0 \text{ arbitrary}) \tag{3}$$

$$\mathbf{x}_{n+1} = \mathbf{x}_n - \mu_n \phi_n \qquad (\text{obtain } \mu_n \text{ by linear search})$$

The simplest choice for $\hat{\mathbf{S}}_0$ is $\hat{\mathbf{S}}_0 = \mathbf{I}$, thus leading to the (unpreconditioned) steepest descent. An arbitrary approximation of the initial Hessian

$$\hat{\mathbf{S}}_0 \simeq \left(\mathbf{I} + \mathbf{C}_X \mathbf{F}_0^t \mathbf{C}_T^{-1} \mathbf{F}_0 \right)^{-1}$$

may be good. An adequate value for μ_n has to be obtained by linear search. Alternatively, a linearization of $\mathbf{g}(\mathbf{m})$ around $\mathbf{m}_n$ gives

$$\mu_n \simeq \frac{\gamma_n{}^t \; C_X{}^{-1} \; \phi_n}{\phi_n{}^t \; C_X{}^{-1} \; \phi_n \; + \; b_n{}^t \; C_T{}^{-1} \; b_n} ,$$

where

$$b_n = F_n \; \phi_n \; .$$

In both algorithms, a good choice for the starting point is $x_0 = x_{prior}$, although other choices may help in checking the existence of secondary minima of (1).

Problem 4.2: Demonstrate that if a data set can be divided into two subsets with independent uncertainties,

$$d_{obs} = \begin{bmatrix} d_{1obs} \\ d_{2obs} \end{bmatrix} \qquad C_D = \begin{bmatrix} C_1 & 0 \\ 0 & C_2 \end{bmatrix} , \tag{1}$$

then the results obtained by the inversion of d_{1obs} , then of d_{2obs} , are identical to those obtained by the simultaneous inversion of the whole data set.

Solution: The least squares solution is

$$m = m_{prior} + \left(G^t \; C_D{}^{-1} \; G + C_M{}^{-1} \right)^{-1} \; G^t \; C_D{}^{-1} \; (d_{obs} - G \; m_{prior}) \; , \tag{2}$$

$$C_{M'} = \left(G^t \; C_D{}^{-1} \; G + C_M{}^{-1} \right)^{-1} \; . \tag{3}$$

Introducing

$$\begin{bmatrix} d_1 \\ d_2 \end{bmatrix} = \begin{bmatrix} G_1 \\ G_2 \end{bmatrix} m \; , \tag{4}$$

we have

$$m = m_{prior} + \left[[\; G_1{}^t \quad G_2{}^t \;] \begin{bmatrix} C_1{}^{-1} & 0 \\ 0 & C_2{}^{-1} \end{bmatrix} \begin{bmatrix} G_1 \\ G_2 \end{bmatrix} + C_M{}^{-1} \right]^{-1}$$

$$[\ G_1^t \quad G_2^t\ ] \begin{bmatrix} C_1^{-1} & 0 \\ 0 & C_2^{-1} \end{bmatrix} \left[\begin{bmatrix} d_{1obs} \\ d_{2obs} \end{bmatrix} - \begin{bmatrix} G_1 \\ G_2 \end{bmatrix} m_{prior} \right]$$

$$= m_{prior} + (\ S_0 + S_1 + S_2\ )^{-1}\ (\ G_1^t\ C_1^{-1}\ (d_{1obs} - G_1\ m_{prior})$$

$$+\ G_2^t\ C_2^{-1}\ (d_{2obs} - G_2\ m_{prior})\ )\ , \tag{5}$$

where

$$S_0 = C_M^{-1}\ ,$$

$$S_1 = G_1^t\ C_1^{-1}\ G_1\ , \tag{6}$$

$$S_2 = G_2^t\ C_2^{-1}\ G_2\ .$$

The following identity holds:

$$(S_0 + S_1 + S_2)^{-1} = (S_0 + S_1 + S_2)^{-1}\ (\ (S_0 + S_1 + S_2) - S_2\ )\ (S_0 + S_1)^{-1}$$

$$= (S_0 + S_1 + S_2)^{-1}\ (S_0 + S_1 + S_2) \left(I - (S_0 + S_1 + S_2)^{-1}\ S_2 \right) (S_0 + S_1)^{-1}$$

$$= \left(I - (S_0 + S_1 + S_2)^{-1}\ S_2 \right) (S_0 + S_1)^{-1}$$

$$= (S_0 + S_1)^{-1} - (S_0 + S_1 + S_2)^{-1}\ S_2\ (S_0 + S_1)^{-1}\ . \tag{7}$$

This gives

$$m = m_A + \left(G_2^t\ C_2^{-1}\ G_2 + C_A^{-1} \right)^{-1}\ G_2^t\ C_2^{-1}\ (\ d_{2obs} - G_2\ m_A)\ , \tag{8a}$$

and

$$C_{M'} = \left(G_2^t\ C_2^{-1}\ G_2 + C_A^{-1} \right)^{-1}\ , \tag{8b}$$

where

$$m_A = m_{prior} + \left(G_1^t\ C_1^{-1}\ G_1 + C_M^{-1} \right)^{-1}\ G_1^t\ C_1^{-1}\ (\ d_{1obs} - G_1\ m_{prior}) \tag{9a}$$

and

$$\mathbf{C}_A = \left(\mathbf{G}_1^t \, \mathbf{C}_1^{-1} \, \mathbf{G}_1 + \mathbf{C}_M^{-1} \right)^{-1} . \tag{9b}$$

The interpretation of this result is as follows. We start with the a priori values $\mathbf{m}_{prior}$ and $\mathbf{C}_M$. After the inversion of the first data subset, $\mathbf{d}_{1obs}$, we obtain the a posteriori values $\mathbf{m}_A$ and $\mathbf{C}_A$ (equations (8)). Using these a posteriori values in (9) as a priori values for the inversion of the second data subset, $\mathbf{d}_{2obs}$, leads to the same solution (2) that should be obtained by simultaneous inversion of the whole data set.

As each data subset may eventually be divided into smaller subsets (if uncertainties remain independent), the results hold with all generality. In particular, if all errors are independent,

$$\mathbf{d}_{obs} = \begin{bmatrix} d^1{}_{obs} \\ d^2{}_{obs} \\ d^3{}_{obs} \\ \cdots \end{bmatrix} \qquad \mathbf{C}_D = \begin{bmatrix} (\sigma^1)^2 & 0 & 0 & \cdots \\ 0 & (\sigma^2)^2 & 0 & \cdots \\ 0 & 0 & (\sigma^3)^2 & \cdots \\ \cdots & \cdots & \cdots & \cdots \end{bmatrix} , \tag{10}$$

a procedure can be devised which inverts *only one datum at a time.* Using equation (4.5c), the only matrix inversion is for a matrix (1x1), and the resulting algorithm is astonishingly simple:

```
SUBROUTINE ONEONE(ND,NP,D0,CD,P0,CP,G,Q)
DIMENSION D0(ND), CD(ND), P0(NP), CP(NP,NP), G(ND,NP), Q(NP)
      DO 1 K=1,ND
      V=D0(K)
          DO 2 I=1,NP
          V=V-G(K,I)*P0(I)
          Q(I)=0.
          DO 2 J=1,NP
2         Q(I)=Q(I)+CP(I,J)*G(K,J)
      A=CD(K)
          DO 3 I=1,NP
3         A=A+G(K,I)*Q(I)
          DO 4 I=1,NP
          P0(I)=P0(I)+Q(I)*V/A
          DO 4 J=1,NP
4         CP(I,J)=CP(I,J)-Q(I)*Q(J)/A
1     CONTINUE
      RETURN
      END
```

This subroutine is self contained, and solves the general linear least-squares inverse problem, for arbitrary a priori information on model parameters, and for data with independent errors.

Problem 4.3: Let $\mathbf{d} = \mathbf{G}\ \mathbf{m}$ be the equation solving the forward problem, and let $\mathbf{d}_{obs}$, $\mathbf{C}_D$, $\mathbf{m}_{prior}$, and $\mathbf{C}_M$ as usual in least-squares problems. The best solution in the least squares sense is given, for instance, by

$$\mathbf{m} \; = \; \mathbf{m}_{prior} + \left(\mathbf{G}^t \; \mathbf{C}_D^{\;-1} \; \mathbf{G} + \mathbf{C}_M^{\;-1} \right)^{-1} \; \mathbf{G}^t \; \mathbf{C}_D^{\;-1} \; (\mathbf{d}_{obs} - \mathbf{G} \; \mathbf{m}_{prior}) \; ,$$

and the posterior covariance operator is given, for instance, by

$$\mathbf{C}_{M'} \; = \; \left(\mathbf{G}^t \; \mathbf{C}_D^{\;-1} \; \mathbf{G} + \mathbf{C}_M^{\;-1} \right)^{-1} \; .$$

Assume that we can partition the model vector into

$$\mathbf{m} \; = \; \begin{bmatrix} \mathbf{m}_1 \\ \mathbf{m}_2 \end{bmatrix} ,$$

with

$$\mathbf{C}_M \; = \; \begin{bmatrix} \mathbf{C}_1 & 0 \\ 0 & \mathbf{C}_2 \end{bmatrix} ,$$

and that we are only interested in $\mathbf{m}_1$. Derive the corresponding formulas.

Solution: Let

$$\Delta\mathbf{d} \; = \; \mathbf{d}_{obs} - \mathbf{G} \; \mathbf{m}_{prior} \; ,$$

$$\Delta\mathbf{m} \; = \; \mathbf{m} - \mathbf{m}_{prior} \; ,$$

$$\mathbf{G} \; = \; [\, \mathbf{G}_1 \quad \mathbf{G}_2 \,] \; .$$

We obtain

$$\begin{bmatrix} \Delta\mathbf{m}_1 \\ \Delta\mathbf{m}_2 \end{bmatrix} \; = \; \begin{bmatrix} \mathbf{S}_{11} & \mathbf{S}_{12} \\ \mathbf{S}_{21} & \mathbf{S}_{22} \end{bmatrix}^{-1} \begin{bmatrix} \mathbf{G}_1^{\;t} \\ \mathbf{G}_2^{\;t} \end{bmatrix} \mathbf{C}_D^{\;-1} \; \Delta\mathbf{d} \; ,$$

where

$$S_{11} = G_1^t \, C_D^{-1} \, G_1 + C_1^{-1} ,$$

$$S_{22} = G_2^t \, C_D^{-1} \, G_2 + C_2^{-1} ,$$

$$S_{12} = G_1^t \, C_D^{-1} \, G_2 ,$$

$$S_{21} = S_{12}^t .$$

Using an inversion per block (see the following problem) gives

$$\begin{bmatrix} \Delta \mathbf{m}_1 \\ \Delta \mathbf{m}_2 \end{bmatrix} = \begin{bmatrix} \mathbf{A} & \mathbf{B} \\ \mathbf{C} & \mathbf{D} \end{bmatrix} \begin{bmatrix} G_1^t \\ G_2^t \end{bmatrix} C_D^{-1} \, \Delta \mathbf{d} ,$$

where

$$\mathbf{A} = \left(S_{11} - S_{12} \, S_{22}^{-1} \, S_{12}^t \right)^{-1} ,$$

$$\mathbf{D} = \left(S_{22} - S_{12}^t \, S_{11}^{-1} \, S_{12} \right)^{-1} ,$$

$$\mathbf{B} = - \mathbf{A} \, S_{12} \, S_{22}^{-1} ,$$

$$\mathbf{C} = \mathbf{B}^t .$$

For $\Delta \mathbf{m}_1$ we obtain

$$\Delta \mathbf{m}_1 = (\mathbf{A} \, G_1^t + \mathbf{B} \, G_2^t) \, C_D^{-1} \, \Delta \mathbf{d} .$$

This gives

$$\Delta \mathbf{m}_1 = (S_{11} - G_1^t \, \mathbf{T}_{22} \, G_1)^{-1} \, G_1^t \, (C_D^{-1} - \mathbf{T}_{22}) \, \Delta \mathbf{d} ,$$

where

$$\mathbf{T}_{22} = C_D^{-1} \, G_2 \, S_{22}^{-1} \, G_2^t \, C_D^{-1} .$$

The a posteriori covariance operator for $\mathbf{m}_1$ clearly equals $\mathbf{A}$.

The previous formulas allow us to take into consideration model parameters which, although not interesting for themselves, are poorly known, and they introduce uncertainties than cannot be ignored.

Problem 4.4: Obtain the inverse of the partitioned matrix

$$\begin{bmatrix} \mathbf{B} & \mathbf{C}^t \\ \mathbf{C} & \mathbf{D} \end{bmatrix} .$$

Solution:

$$\begin{bmatrix} \mathbf{B} & \mathbf{C}^t \\ \mathbf{C} & \mathbf{D} \end{bmatrix}^{-1} = \begin{bmatrix} \mathbf{E} & \mathbf{F}^t \\ \mathbf{F} & \mathbf{G} \end{bmatrix} ,$$

with

$$\mathbf{E} = (\mathbf{B} - \mathbf{C}^t \, \mathbf{D}^{-1} \, \mathbf{C})^{-1} ,$$

$$\mathbf{G} = (\mathbf{D} - \mathbf{C} \, \mathbf{B}^{-1} \, \mathbf{C}^t)^{-1} ,$$

and

$$\mathbf{F} = - \mathbf{G} \, \mathbf{C} \, \mathbf{B}^{-1} = - \mathbf{D}^{-1} \, \mathbf{C} \, \mathbf{E} .$$

Problem 4.5: Let E be a linear space, and $\mathbf{C}$ the covariance operator defining a scalar product over E :

$$(\mathbf{e}_1,\mathbf{e}_2) = \mathbf{e}_1{}^t \, \mathbf{C}^{-1} \, \mathbf{e}_2 . \tag{1}$$

Demonstrate that the adjoint of $\mathbf{C}$ equals its inverse:

$$\mathbf{C}^* = \mathbf{C}^{-1} . \tag{2}$$

Solution: Let us first consider a linear operator $\mathbf{C}$ mapping a linear space F into the linear space E . If the scalar product over F is written

$$(\mathbf{f}_1,\mathbf{f}_2) = \mathbf{f}_1{}^t \, \mathbf{C}_f \, \mathbf{f}_2 , \tag{3}$$

(notice that I use $\mathbf{C}_f$ in (3) instead of $\mathbf{C}_f^{-1}$, for reasons that will become apparent below), then, from the definitions of Box 4.5,

$$\mathbf{C}^* = \mathbf{C}_f^{-1}\, \mathbf{C}^t\, \mathbf{C}^{-1} . \tag{4}$$

A covariance operator over E maps $\hat{E}$, dual of E , into E (see section 4.4.2), so that, using the previous notations, $F \equiv \hat{E}$. If the scalar product over E is defined by (1), then the scalar product over $\hat{E}$ (equation (3)) is (see section 4.4.2):

$$(\hat{\mathbf{e}}_1,\hat{\mathbf{e}}_2) = \hat{\mathbf{e}}^t\, \mathbf{C}\, \hat{\mathbf{e}}_2 . \tag{5}$$

This gives

$$\mathbf{C}^* = \mathbf{C}^{-1}\, \mathbf{C}^t\, \mathbf{C}^{-1} , \tag{6}$$

from which result (2) follows, using the symmetry of $\mathbf{C}$.

Problem 4.6: In a problem involving a density and a velocity as parameters, we obtain the following covariance matrix

$$\mathbf{C} = \begin{bmatrix} 25.0 \text{ g}^2 \text{ cm}^{-6} & 4.0 \text{ g cm}^{-2}\text{s}^{-1} \\ 4.0 \text{ g cm}^{-2}\text{s}^{-1} & 1.0 \text{ cm}^2\text{s}^{-2} \end{bmatrix} .$$

Compute its square root.

Solution: Using the definitions of Box 4.1 gives

$$\boldsymbol{\Sigma} = \begin{bmatrix} 5.0 \text{ g cm}^{-3} & 0 \\ 0 & 1.0 \text{ cm s}^{-1} \end{bmatrix}$$

$$\mathbf{R} = \begin{bmatrix} 1.0 & 0.8 \\ 0.8 & 1.0 \end{bmatrix}$$

$$\mathbf{U} = \begin{bmatrix} 0.707 & 0.707 \\ -0.707 & 0.707 \end{bmatrix}$$

$$\mathbf{\Lambda} = \begin{bmatrix} 0.2 & 0.0 \\ 0.0 & 1.8 \end{bmatrix}$$

$$\mathbf{R}^{1/2} = \mathbf{U}\,\mathbf{\Lambda}^{1/2}\,\mathbf{U}^t = \begin{bmatrix} 0.894 & 0.447 \\ 0.447 & 0.894 \end{bmatrix}$$

(changing the order of the eigenvalues and eigenvalues, changes $\mathbf{U}$ and $\mathbf{\Lambda}$, but does not change $\mathbf{R}^{1/2}$).

$$\mathbf{C}^{1/2} = \mathbf{\Sigma}\,\mathbf{R}^{1/2} = \begin{bmatrix} 4.472 \text{ g cm}^{-3} & 2.236 \text{ g cm}^{-3} \\ 0.447 \text{ cm s}^{-1} & 0.894 \text{ cm s}^{-1} \end{bmatrix}$$

$$\mathbf{C}^{-1/2} = \begin{bmatrix} 0.298 \text{ g}^{-1}\text{cm}^3 & -0.745 \text{ cm}^{-1}\text{s} \\ -0.149 \text{ g}^{-1}\text{cm}^3 & 1.490 \text{ cm}^{-1}\text{s} \end{bmatrix} .$$

Problem 4.7: Solve the problem 1.1 (estimation of the epicentral coordinates of a seismic event) using the Newton method, the steepest descent method, the conjugate directions method, and the variable metric method. Consider different kinds of a priori information. Examine the evolution of posterior errors when eliminating some data. Use only one datum (arrival time) and the a priori information that the x-coordinate of the epicenter is 15km±5km.

Note: As the forward problem is solved by the equation

$$d^i = g^i(X,Y) = \frac{1}{v}\sqrt{(X-x^i)^2 + (Y-y^i)^2} \,, \tag{1}$$

if we order arrival times in a column matrix, the matrix of partial derivatives is

$$\mathbf{G}_n = \begin{bmatrix} \left(\dfrac{\partial g^1}{\partial X}\right)_n & \left(\dfrac{\partial g^1}{\partial Y}\right)_n \\ \left(\dfrac{\partial g^2}{\partial X}\right)_n & \left(\dfrac{\partial g^2}{\partial Y}\right)_n \\ \cdots & \cdots \end{bmatrix} , \tag{2}$$

where

$$\left(\frac{\partial g^i}{\partial X}\right)_n = \frac{X_n - x^i}{v\sqrt{(X_n - x^i)^2 + (Y_n - y^i)^2}} , \tag{3a}$$

and

$$\left(\frac{\partial g^i}{\partial Y}\right)_n = \frac{Y_n - x^i}{v\sqrt{(X_n - x^i)^2 + (Y_n - y^i)^2}} . \tag{3b}$$

Problem 4.8: Solve the problem 1.2 (elementary approach to tomography) using the Newton method, the steepest descent method, the conjugate directions method, and the variable metric method. Consider different kinds of a priori information. Examine the evolution of posterior errors when adding much more data. Solve a problem with $NX \times NY$ blocks, where NX and NY are the number of pixels of the colour output of your computer. Then read Chapter 7.

CHAPTER 5

THE LEAST-ABSOLUTE-VALUES (ℓ_1-norm) CRITERION AND THE MINIMAX (ℓ_∞-norm) CRITERION

When a traveler reaches a fork in the road, the ℓ_1-norm tells him to take either one way or the other, but the ℓ_2-norm instructs him to head off into the bushes.

John F. Claerbout and Francis Muir, 1973.

Because of its simplicity, the least-squares criterion (ℓ_2-norm criterion) is widely used for the resolution of inverse problems. We have seen that least squares is intimately related with the hypothesis of Gaussian uncertainties. For other types of uncertainties, better criteria exist. Among them, those based on an ℓ_p norm ($1 \leq p \leq \infty$) have the advantage of allowing an easy mathematical formulation.

As suggested in Chapter 1, when outliers are suspected in a data set, long-tailed probability density functions should be used to model uncertainties. "Long tailed" means, in fact, functions tending to zero less rapidly than the Gaussian function, when the distance between the variable and any of its central estimators tends to infinity. Two typical long-tailed probability densities are: the Cauchy function $1/(1+x^2)$ and the symmetric exponential function $\exp(-| x |)$. The former is a very tempting function to use because, although being nicely bell-shaped, it has infinite variance, which seems to be adequate for modeling suspected outliers by an unknown amount. The symmetric exponential function, on the other hand, has the advantage of leading to results intimately related with the concept of the ℓ_1 norm, so that a lot of mathematics are already available for solving the problem. The results obtained using the minimum ℓ_1-norm (least-absolute-values) criterion are known to be sufficiently insensitive to outliers (i.e., to be *robust*).

The ℓ_1-norm criterion was already used by Laplace and Gauss. In the words of Gauss (1809), quoted by Plackett (1972), "Laplace made use of another principle for the solution of linear equations, the number of which is greater than the number of unknown quantities, which had been previously proposed by Boscovich, namely that the differences themselves, but all of them taken positively, should make up as small sum as possible". In modern times, Clearbout and Muir (1973) have given a detailed discussion of the robustness of the ℓ_1-norm criterion for the resolution of inverse problems, and have suggested a method of resolution related with the "linear programming" techniques.

The ℓ_∞-norm criterion arises when we use box-car functions to model the probability density for uncertainties. This assumes that we have a strict control on errors, as for instance when errors are mainly due to *rounding* the last digit used (see Problem 1.5).

This chapter starts by recalling the definition of a ℓ_p norm, and by introducing a natural bijection between a ℓ_p-normed space and its dual. For $1 < p < \infty$, the methods for solving inverse problems are similar to the methods used for $p = 2$ (Chapter 4). For $p = 1$ and $p = \infty$, linear programming methods are best adapted. Although the minimum ℓ_1-norm and minimum ℓ_∞-norm criteria are used in almost opposite circumstances, the underlying mathematics are very similar. This justifies including them in the same chapter.

A comprehensive explanation of linear programming methods would require an entire book. Unfortunately, current presentations of linear programming do not make the effort of distinguishing what is intrinsic (linear applications, dual spaces,...) or extrinsic (the matricial representation of a linear application or a vector). But as I have not yet been able to entirely rewrite the theory, I choose to refer the reader to the usual text books.

5.1: The ℓ_p norm

5.1.1: Definition of the weighted ℓ_p norm

Let $\mathbf{x}$ be an element of a discrete vector space X , x^i $(i \in I_X)$ the components of $\mathbf{x}$, and let σ^i $(i \in I_X)$ be given positive constants such that, for any i, σ^i has the same physical dimensions as x^i (so that x^i/σ^i is an adimensional real number). For $1 \le p < \infty$, the (*weighted*) ℓ_p *norm* of $\mathbf{x}$ is denoted $\| \mathbf{x} \|_p$, and is defined by

$$\| \mathbf{x} \|_p = \left(\sum_{i \in I_X} \frac{| x^i |^p}{(\sigma^i)^p} \right)^{1/p} . \tag{5.1}$$

The ℓ_∞ norm is defined as the limit of (5.1) when $p \to \infty$, which gives (e.g., Watson, 1980):

$$\| \mathbf{x} \|_\infty = \underset{i \in I_X}{\text{MAX}} \left(\frac{| x^i |}{\sigma^i} \right) . \tag{5.2}$$

It is well known that this definition verifies the usual properties of a norm:

$$\begin{aligned}
&\| \mathbf{x} \|_p \in R \\
&\mathbf{x} \neq 0 \quad \Rightarrow \quad \| \mathbf{x} \|_p > 0 \\
&\| \alpha\, \mathbf{x} \|_p = | \alpha | \; \| \mathbf{x} \|_p \\
&\| \mathbf{x}_1 + \mathbf{x}_2 \|_p \leq \| \mathbf{x}_1 \|_p + \| \mathbf{x}_2 \|_p .
\end{aligned} \tag{5.3}$$

A *circle* of radius R centered at $\mathbf{x}_0$ is defined as the set of points $\mathbf{x}$ such that $\| \mathbf{x} - \mathbf{x}_0 \|_p = R$. Figure 5.1 shows some 2D "circles" corresponding to different ℓ_p norms.

It should be noticed that, given the constants σ^i $(i \in I_X)$ which define a ℓ_p norm over X , it is possible to define, over the same space, an infinity of different $\ell_{p'}$ norms

$$\| \mathbf{x} \|_{p'} = \left[\sum_{i \in I_X} \frac{| x^i |^{p'}}{(\sigma^i)^{p'}} \right]^{1/p'}$$

with $1 \leq p' < \infty$. Sometimes it is necessary to consider different ℓ_p norms over the same space simultaneously.

5.1.2: The dual of an ℓ_p space.

Let $\hat{\chi}$ denote a linear form over X (i.e., a linear application from X into the real line R). The space of all linear forms over X is named the *dual* of X , and is denoted $\hat{X}$. The result of the action of $\hat{\chi} \in \hat{X}$ on $\mathbf{x} \in X$ is denoted by $< \hat{\chi} , \mathbf{x} >$ or also $\hat{\chi}^t \, \mathbf{x}$. For any form $\hat{\chi}$ over a discrete space, it is possible to find constants $\hat{\chi}^i$ $(i \in I_X)$ such that

$$< \hat{\chi} , \mathbf{x} > = \hat{\chi}^t \, \mathbf{x} = \sum_{i \in I_X} \hat{\chi}^i \, x^i . \tag{5.4}$$

If the elements of X and of $\hat{X}$ are represented by column matrices, the

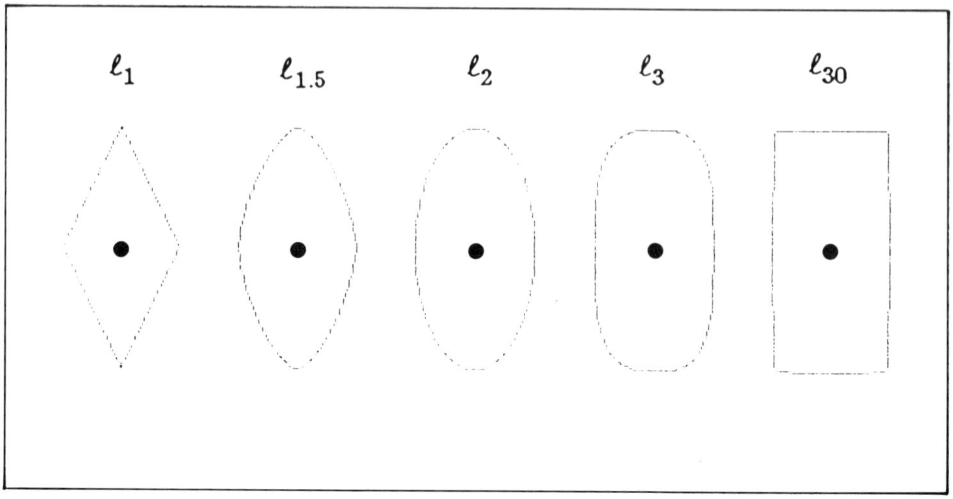

Figure 5.1: Some unit "circles" in the ℓ_p norm sense. The points on the circles are such that $\frac{|x^1-x_0^1|^p}{(\sigma^1)^p} + \frac{|x^2-x_0^2|^p}{(\sigma^2)^p} = 1$, and are drawn for p respectively equal to 1 , 1.5 , 2 , 3 , and 30 .

notation $\hat{\chi}^t x$ corresponds to the usual matrix notation; in the case where the elements of X and of $\hat{X}$ have a more complex structure (in practice: they are represented by multidimensional arrays), the notation $\hat{\chi}^t \mathbf{x}$ is still practical for analytical developments. See also the tensor notations introduced in the Appendix 4.1

The space $\hat{X}$ can be identified with a space of vectors $\hat{\chi}$ whose components are arbitrary excepting that the physical dimension of the i-th component of χ , $\hat{\chi}^i$, is the inverse of the physical dimension of the i-th component of x , x^i (so that $\hat{\chi}^i x^i$ is adimensional).

Given a particular ℓ_p norm over a space X , it is useful to define a bijection between X and its dual $\hat{X}$: for $1 < p < \infty$, with a given $\mathbf{x}_0 \in X$, I associate an element of $\hat{X}$, denoted $\hat{\mathbf{x}}_0$ and defined by

$$\hat{x}_0{}^i = \frac{1}{p} \left[\frac{\partial}{\partial x^i} \| \mathbf{x} \|_p^p \right]_{\mathbf{x}=\mathbf{x}_0} . \tag{5.5}$$

This bijection will be written, for short,

$$\hat{\mathbf{x}} = \frac{1}{p} \frac{\partial}{\partial \mathbf{x}} \| \mathbf{x} \|_p^p . \tag{5.6}$$

Introducing

$$\hat{\sigma}^i = \frac{1}{\sigma^i} , \tag{5.7}$$

we can, for any q , define a ℓ_q norm over the dual space $\hat{X}$:

$$\| \hat{\mathbf{x}} \|_q = \left[\sum_i \frac{| \hat{x}^i |^p}{(\hat{\sigma}^i)^q} \right]^{1/q} \qquad (1 \le q < \infty) , \tag{5.8a}$$

$$\| \hat{\mathbf{x}} \|_\infty = \underset{i \in I_X}{\text{MAX}} \left(\frac{| \hat{x}^i |}{\hat{\sigma}^i} \right) . \tag{5.8b}$$

Let p satisfy $1 < p < \infty$, and let q be defined by

$$\frac{1}{p} + \frac{1}{q} = 1 , \tag{5.9}$$

then the symmetric of equation (5.6) holds (see Problem 5.2):

$$\mathbf{x} = \frac{1}{q} \left(\frac{\partial}{\partial \hat{\mathbf{x}}} \| \hat{\mathbf{x}} \|_q^q \right) , \tag{5.10}$$

and we have the following equalities (see Problem 5.3):

$$\boxed{\| \mathbf{x} \|_p^p = \| \hat{\mathbf{x}} \|_q^q = \| \mathbf{x} \|_p \cdot \| \hat{\mathbf{x}} \|_q = \hat{\mathbf{x}}^t \, \mathbf{x} .} \tag{5.11}$$

The previous equations show, in particular, that if we consider an ℓ_p norm over an space X , it is natural (although not necessary) to furnish its dual $\hat{X}$ with an associated ℓ_q norm , where p and q are related through equation (5.9). By extension, if we consider a ℓ_1 norm, we furnish the dual with an ℓ_∞ norm, and viceversa, but this case is pathological and we do not have a natural bijection between the primal and the dual spaces.

From (5.6) and (5.10) we obtain the explicit representations

$$\hat{x}^i = \frac{sg(x^i) \left| x^i \right|^{p-1}}{(\sigma^i)^p} , \tag{5.12a}$$

and

$$x^i = \frac{sg(\hat{x}^i) \left| \hat{x}^i \right|^{q-1}}{(\hat{\sigma}^i)^q} . \tag{5.12b}$$

To simplify analytical computations, I suggest introducing the following notation: letting u be an arbitrary scalar, and r a real positive number,

$$u^{\{r\}} = sg(u) \; |u|^r . \tag{5.13}$$

The usefulness of this notation comes from the following properties

$$u^{\{r\}} = v \quad \Leftrightarrow \quad u = v^{\{1/r\}} \tag{5.14}$$

$$\frac{\partial}{\partial x} u^{\{r\}} = r \; |u|^{r-1} \; \frac{\partial u}{\partial x} \tag{5.15a}$$

$$\frac{\partial}{\partial x} |u|^r = r \; u^{\{r-1\}} \; \frac{\partial u}{\partial x} . \tag{5.15b}$$

In particular, using (5.14), the bijection between X and $\hat{X}$ can be rewritten in a more compact form:

$$\frac{\hat{x}^i}{\hat{\sigma}^i} = \left(\frac{x^i}{\sigma^i} \right)^{\{p-1\}} \tag{5.16}$$

$$\frac{x^i}{\sigma^i} = \left(\frac{\hat{x}^i}{\hat{\sigma}^i} \right)^{\{q-1\}} . \tag{5.17}$$

Figure 5.2 shows the function $u^{\{r\}}$ for different values of $r > 0$. For positive values of u , it simply corresponds to the function u^r . For negative values of u , it is possible to interpret $u^{\{r\}}$ as an interpolation of the functions u^r obtained when r is an odd integer.

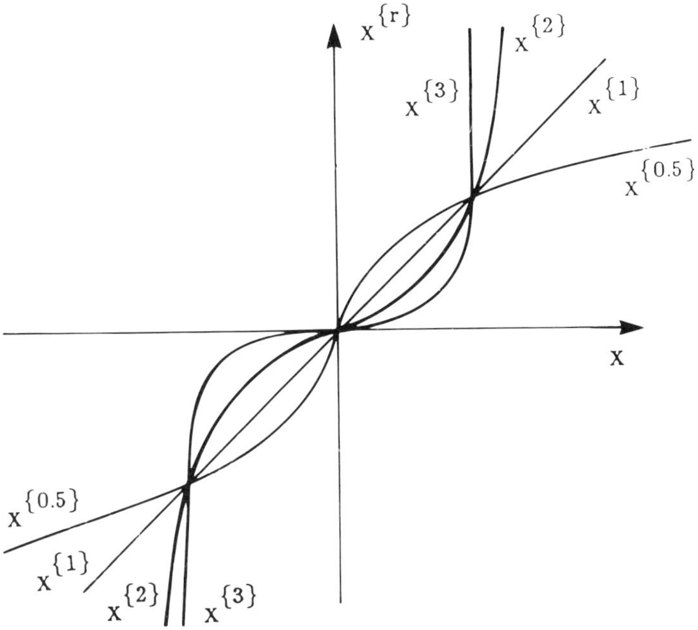

Figure 5.2: Some examples of the function $x^{\{r\}}$. For any r it is a real function, symmetric (with respect to the origin), defined for any value of r (even for negative x), and is, for given x , a continuous function of r .

The bijection defined between X and $\hat{X}$ is nonlinear excepting for $p = 2$. In that case, $q = p = 2$, and

$$\frac{\hat{x}^i}{\hat{\sigma}^i} = \frac{x^i}{\sigma^i} .$$

It should be mentioned that the relationship (5.6) between a space and its dual does not correspond to the usual definition (e.g., Watson, 1980; exercise 1.27):

$$\hat{\mathbf{x}} = \frac{\partial}{\partial \mathbf{x}} \| \mathbf{x} \|_p ,$$

in which case (5.11) is replaced by

$$\| \hat{\mathbf{x}} \|_q = 1 ,$$

and the relationship between X and $\hat{X}$ is no longer a bijection. This justifies the definition given in this book.

Box 5.1: Gradient and direction of steepest ascent in ℓ_p norm spaces.

Let M be a discrete linear vector space with elements $\mathbf{m} = \{ m^\alpha \}$, $(\alpha \in I_M)$, where I_M represents the (discrete) index set. If there are constants σ^α $(\alpha \in I_M)$ such that

$$\| \mathbf{m} \|_p = \left(\sum_{\alpha \in I_M} \frac{| m^\alpha |^p}{(\sigma^\alpha)^p} \right)^{1/p} \tag{1}$$

is defined, then M is a ℓ_p norm linear vector space.

The *dual* of M is denoted $\hat{M}$, and is defined as the space of linear forms over M, (i.e., of linear applications from M into R). Let $\hat{\mu} \in \hat{M}$, and $\mathbf{m} \in M$, the corresponding element of R is then denoted either $< \hat{\mu} , \mathbf{m} >$ or $\hat{\mu}^t \mathbf{m}$. For any $\hat{\mu} \in \hat{M}$ there exist constants $\hat{\mu}^\alpha$ $(\alpha \in I_\mu)$ such that, for any $\mathbf{m} \in M$,

$$< \hat{\mu} , \mathbf{m} > = \hat{\mu}^t \mathbf{m} = \sum_{\alpha \in I_M} \hat{\mu}^\alpha m^\alpha . \tag{2}$$

Let $S(\mathbf{m})$ represent an arbitrary nonlinear form over M. A second-order development of $S(\mathbf{m})$ around a given point $\mathbf{m} = \mathbf{m}_n$ can be written

(...)

$$S(\mathbf{m}_n+\delta\mathbf{m}) = S(\mathbf{m}_n) + < \hat{\gamma}_n , \delta\mathbf{m} > + \frac{1}{2} < \hat{H}_n \, \delta\mathbf{m} , \delta\mathbf{m} > + O(\| \delta\mathbf{m} \|^3) , \tag{3}$$

thus defining $\hat{\gamma}_n$, the *gradient* of $S(\mathbf{m})$ at $\mathbf{m}_n$, and $\hat{H}_n$, the *Hessian* of $S(\mathbf{m})$ at $\mathbf{m}_n$. The norm $\| \, . \, \|$ in (3) is arbitrary. By definition, the gradient $\hat{\gamma}_n$ is a linear application from $\boldsymbol{M}$ into $\boldsymbol{R}$ (i.e., an element of $\hat{\boldsymbol{M}}$), while the Hessian $\hat{H}_n$ is a linear application from $\boldsymbol{M}$ into its dual $\hat{\boldsymbol{M}}$.

The previous definition is general and can be used to define gradient and Hessian for general Banach spaces. For discrete spaces it gives

$$\hat{\gamma}_n{}^\alpha = \left(\frac{\partial S}{\partial m^\alpha} \right)_{\mathbf{m}_n} \tag{4}$$

and

$$\hat{H}_n{}^{\alpha\beta} = \left(\frac{\partial^2 S}{\partial m^\alpha \, \partial m^\beta} \right)_{\mathbf{m}_n} = \left(\frac{\partial \hat{\gamma}^\alpha}{\partial m^\beta} \right)_{\mathbf{m}_n} . \tag{5}$$

It should be noticed that the values thus obtained for the gradient and Hessian are *independent* of any particular choice of norm over $\boldsymbol{M}$. In Chapter 4 we have seen that the right intuitive image of the gradient at a given point is a "millefeuilles" (see Figures 4.2 and 4.6).

The direction of steepest ascent at $\mathbf{m} = \mathbf{m}_n$ is defined as the direction from $\mathbf{m}_n$ to the point of an infinitesimal "circle", centered at $\mathbf{m}_n$, for which the function $S(\mathbf{m})$ takes its maximum value. As for any value of p it is possible to consider the corresponding ℓ_p-norm circle, the direction of steepest ascent depends on the particular norm chosen (see Figure 5.3 for a schematic illustration).

Considering a fixed value for p , the problem of obtaining the direction of steepest ascent at $\mathbf{m}_n$ is equivalent to the problem of obtaining the vector $\mathbf{d}_n \in \boldsymbol{M}$ maximizing $S(\mathbf{m}_n + \epsilon \, \mathbf{d}_n)$ in the limit $\epsilon \to 0$. As when ϵ tends to zero, $S(\mathbf{m}_n + \epsilon \, \mathbf{d}_n)$ can be approximated by $S(\mathbf{m}_n) + \epsilon < \hat{\gamma}_n , \mathbf{d}_n >$, the problem is equivalent to:

$$\begin{aligned} &\text{Maximize} && < \hat{\gamma}_n , \mathbf{d}_n > \\ &\text{under the constraint} && \| \mathbf{d}_n \|_p = N \text{ (arbitrary fixed constant).} \end{aligned} \tag{6}$$

Introducing Lagrange's multiplier λ (see appendix 4.6), this constrained maximization problem is equivalent to the unconstrained problem

(...)

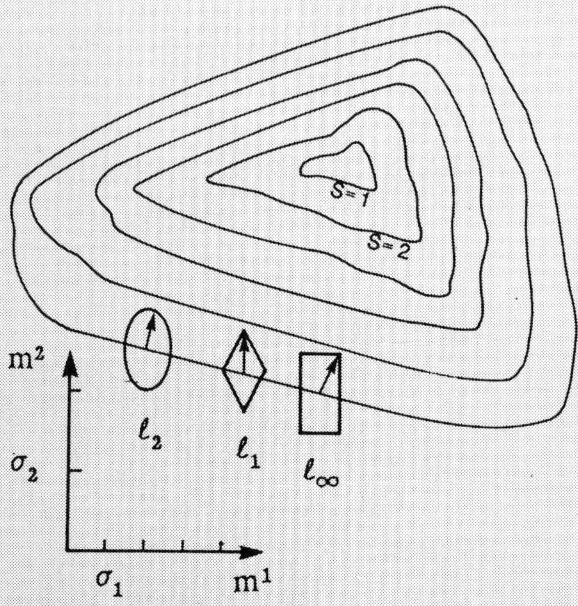

Figure 5.3: The concept of direction of steepest ascent is not intrinsically defined, and only makes sense if referred to a particular definition of norm. In this illustration, the ℓ_p norms

$$\| \mathbf{m} \|_p = \left(\frac{|m^1|^p}{(\sigma^1)^p} + \frac{|m^2|^p}{(\sigma^2)^p} \right)^{1/p}$$

are considered. The geometrical definition of direction of steepest ascent at a given point $\mathbf{m}_0$ is the direction pointing from $\mathbf{m}_0$ to the point of a infinitesimally small unit circle around $\mathbf{m}_0$ which gives the highest value of the function considered. This figure illustrates in which manner the direction of steepest ascent depends on the chosen norm. It can be shown that if an arbitrary direction is a direction of ascent for one particular norm, then it is also a direction of ascent for any other norm.

(...)

Maximize $\quad \Psi(\mathbf{d}_n , \lambda) = < \hat{\gamma}_n , \mathbf{d}_n > - \frac{\lambda}{p} \left(\| \mathbf{d}_n \|^p - N^p \right)$

$$= \sum_{\alpha \in I_M} \hat{\gamma}_n{}^\alpha \; d_n{}^\alpha - \frac{\lambda}{p} \left(\sum_{\alpha \in I_M} \frac{\left| d_n{}^\alpha \right|^p}{(\sigma^\alpha)^p} - N^p \right) .$$

The condition $\partial S / \partial \lambda = 0$ simply gives the constraint $\| \mathbf{d}_n \|_p = N$, while the condition $\partial S / \partial d_n{}^\alpha = 0$ gives

$$\frac{\hat{\gamma}_n{}^\alpha}{\hat{\sigma}^\alpha} = \lambda \left(\frac{d_n{}^\alpha}{\sigma^\alpha} \right)^{\{p-1\}} ,$$

where $\hat{\sigma}^\alpha = 1/\sigma^\alpha$, and where the symbol $u^{\{r\}}$ has been defined in equation (5.13). Using (5.16)-(5.17) gives

$$\frac{d_n{}^\alpha}{\sigma^\alpha} = \frac{1}{\lambda} \left(\frac{\hat{\gamma}_n{}^\alpha}{\hat{\sigma}^\alpha} \right)^{\{q-1\}}$$

where

$$\frac{1}{p} + \frac{1}{q} = 1 .$$

Up to nowm the value of λ is arbitrary, and serves to ensure the constraint $\| \mathbf{d}_n \|_p = N$. Choosing $\lambda = 1$ gives

$$\| \mathbf{d}_n \|_p^p = \| \hat{\gamma}_n \|_q^q \tag{7}$$

and

$$\frac{d_n{}^\alpha}{\sigma^\alpha} = \left(\frac{\hat{\gamma}_n{}^\alpha}{\hat{\sigma}^\alpha} \right)^{\{q-1\}} , \tag{8}$$

thus showing that the direction of steepest ascent $\mathbf{d}_n$ is simply related with the gradient $\hat{\gamma}_n$ through the bijection which has been defined in section 5.1.2 between a normed vector space and its dual. This justifies the notation

(...)

$$\mathbf{d}_n = \boldsymbol{\gamma}_n \, , \qquad (9)$$

where, by dropping the hat, it is meant that $\boldsymbol{\gamma}_n$ is the element of M related with the gradient $\hat{\boldsymbol{\gamma}}_n$ through the usual bijection defined by equations (5.6), (5.12a), or (5.16).

5.1.3: The ℓ_p-norm criterion for the resolution of inverse problems

In Chapter 1 it has been shown that if estimated errors on the observed data values $\mathbf{d}_{obs}$ and estimated errors on the a priori model $\mathbf{m}_{prior}$ are assumed to be conveniently modeled using a generalized Gaussian of order p , then the a posteriori probability density in the model space is given by

$$\sigma_M(\mathbf{m}) = \text{const.} \exp - \frac{1}{p} \left(\sum_{i \in \mathbf{I}_D} \frac{\left| g^i(\mathbf{m}) - d^i_{obs} \right|^p}{\left(\sigma^i_D \right)^p} + \sum_{\alpha \in \mathbf{I}_M} \frac{\left| m^\alpha - m^\alpha_{prior} \right|^p}{\left(\sigma^\alpha_M \right)^p} \right) , \qquad (5.18)$$

where σ^i_D and σ^α_M respectively represent the ℓ_p norm estimators of dispersion for errors in observed data and a priori model. The *maximum likelihood* model maximizes $\sigma_M(\mathbf{m})$, i.e. minimizes the *misfit function* $S(\mathbf{m})$ defined by

$$S(\mathbf{m}) = \frac{1}{p} \left(\sum_{i \in \mathbf{I}_D} \frac{\left| g^i(\mathbf{m}) - d^i_{obs} \right|^p}{\left(\sigma^i_D \right)^p} + \sum_{\alpha \in \mathbf{I}_M} \frac{\left| m^\alpha - m^\alpha_{prior} \right|^p}{\left(\sigma^\alpha_M \right)^p} \right) , \qquad (5.19)$$

where the factor $1/p$ is left for subsequent simplifications. The minimization of $S(\mathbf{m})$ given by (5.19) is clearly equivalent to the minimization of the ℓ_p norm in the space $\boldsymbol{D} \times \boldsymbol{M}$

$$\left\| \begin{matrix} \mathbf{d} - \mathbf{d}_{obs} \\ \mathbf{m} - \mathbf{m}_{prior} \end{matrix} \right\|_p = \left(\sum_{i \in \mathbf{I}_D} \frac{\left| d^i - d^i_{obs} \right|^p}{\left(\sigma^i_D \right)^p} + \sum_{\alpha \in \mathbf{I}_M} \frac{\left| m^\alpha - m^\alpha_{prior} \right|^p}{\left(\sigma^\alpha_M \right)^p} \right)^{1/p} , \qquad (5.20a)$$

under the constraint

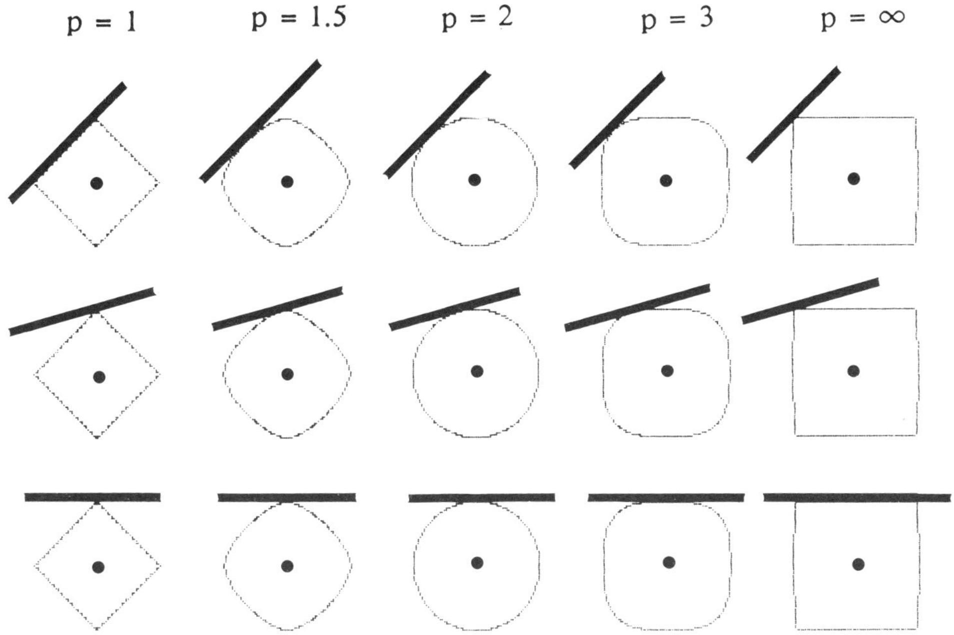

Figure 5.4: Minimization of a ℓ_p norm under a linear constraint. The two-dimensional problem illustrated in this figure is to obtain the point, constrained to lie on a given straight line, which is the closest to a given point. The problem can be solved geometrically by "expanding" the unit circle until it becomes tangent to the line. In the first example (top) the solution is unique excepting for $p = 1$; in the second example (middle) it is always unique; in the third example (bottom) it is unique excepting for $p = \infty$. The generalization of this geometrical illustration is also true (see, for instance, Watson, 1980): the minimization of a ℓ_p norm $\| \mathbf{x}-\mathbf{x}_0 \|_p$ under a linear constraint $\mathbf{F}\,\mathbf{x} = \mathbf{0}$ gives a unique solution excepting for $p = 1$ and $p = \infty$, in which cases the solution may not be unique. When the constraint is nonlinear, the minimization of a ℓ_p norm does not necessarily have a unique solution.

$$\mathbf{d} = \mathbf{g}(\mathbf{m}) \ . \tag{5.20b}$$

To recognize that the minimization of $S(\mathbf{m})$ in fact corresponds to the constrained minimization of a ℓ_p norm is important, because it gives a clear understanding of the uniqueness of the solution. Figure 5.4 suggests (as is indeed the case (Watson, 1980)) that *for a linear constraint, the minimum is unique if* $1 < p < \infty$ (because the unit circle is strictly convex), but it is not necessarily unique for $p = 1$ and $p = \infty$ (the unit circle is still convex, but not strictly convex). For a nonlinear constraint, multiple minima, secondary minima, and saddle points may exist.

The model $\mathbf{m}$ minimizing $S(\mathbf{m})$ can then be regarded either as a maximum likelihood model (from the point of view developed in Chapter 1), or as the "best model" under the minimum ℓ_p-norm "criterion". It has then to be emphasized that the ℓ_p-norm criterion is only justified if the assumption of errors distributed following a generalized Gaussian of order p is acceptable.

This introduction to ℓ_p-norm methods is useful for a better understanding of ℓ_1 and ℓ_∞ methods, but it is not of sufficient practical use to justify giving here the methods of resolution, which are given in Appendices 5.1 and 5.2.

5.2: The ℓ_1-norm criterion for the resolution of inverse problems

In Chapter 1 it has been shown that if estimated errors in the observed data values $\mathbf{d}_{obs}$ and estimated errors in the a priori model $\mathbf{m}_{prior}$ are assumed to be conveniently modeled using a Double Exponential, then the a posteriori probability density in the model space is given by

$$\sigma_M(m) = \exp - \left(\sum_{i \in I_D} \frac{\left| g^i(m) - d^i_{obs} \right|}{\sigma_D{}^i} + \sum_{\alpha \in I_M} \frac{\left| m^\alpha - m^\alpha_{prior} \right|}{\sigma_M{}^\alpha} \right) , \qquad (5.21)$$

where $\sigma_D{}^i$ and $\sigma_M{}^\alpha$ respectively represent the ℓ_1-norm estimators of dispersion (i.e., the mean deviations) for errors in observed data and a priori model. The *maximum likelihood* model maximizes $\sigma_M(\mathbf{m})$, i.e. minimizes the *misfit function* $S(m)$ defined by

$$S(m) = \sum_{i \in I_D} \frac{\left| g^i(m) - d^i_{obs} \right|}{\sigma^i_D} + \sum_{\alpha \in \mathbf{I}_M} \frac{\left| m^\alpha - m^\alpha_{prior} \right|}{\sigma^\alpha_M} . \qquad (5.22)$$

For obvious reasons, the model minimizing $S(\mathbf{m})$ given by (5.22) is called the *best* model with respect to the least-absolute-values (ℓ_1-norm) criterion.

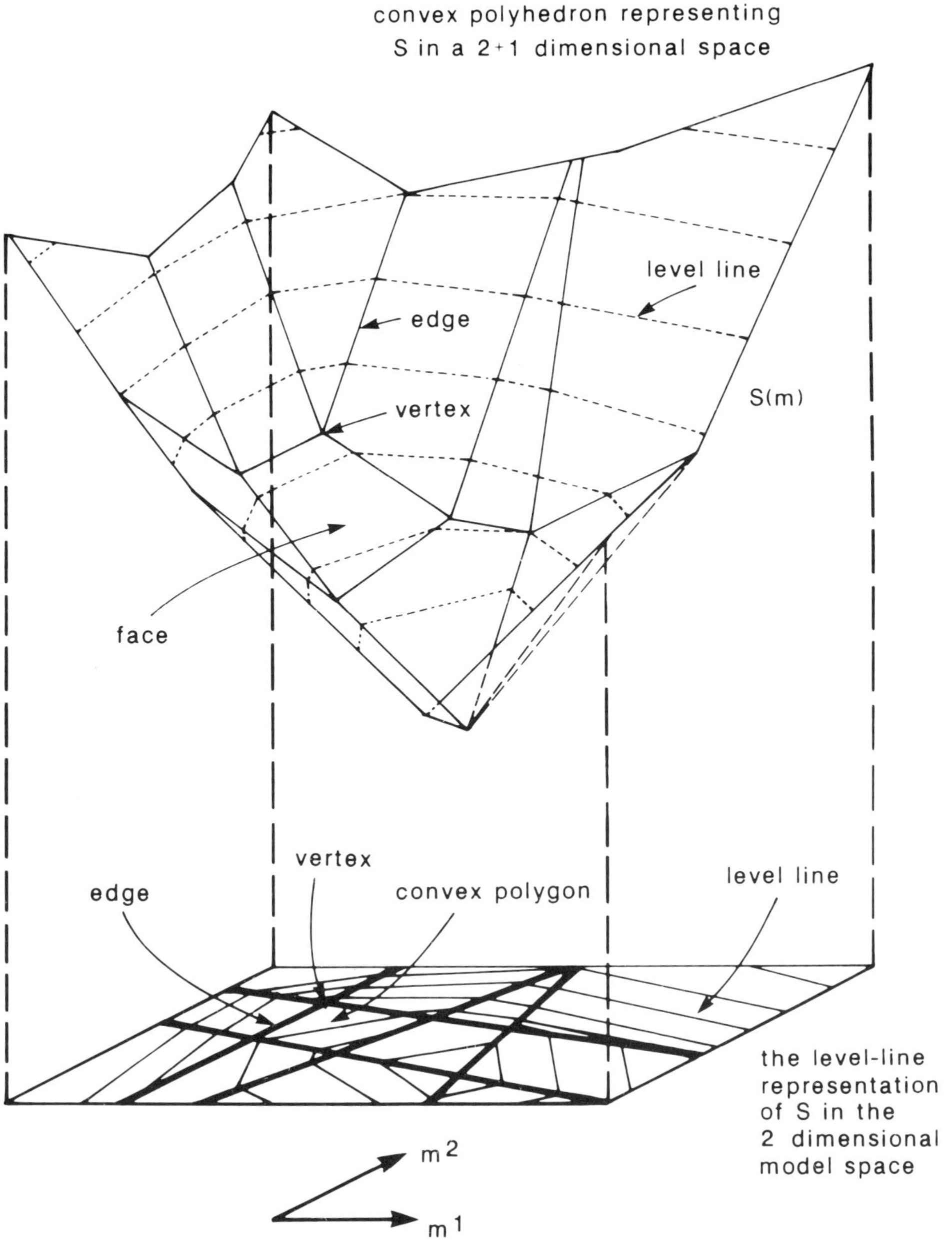

Figure 5.5: When the model space $\boldsymbol{M}$ is two dimensional, the misfit function $S(\mathbf{m})$ can be represented by a convex polyhedron in a 3D space (top). The function $S(\mathbf{m})$ defines a partition of the model space into convex polygons (bottom). The level lines of S ($S(\mathbf{m}) = \text{const}$) are also convex polygons.

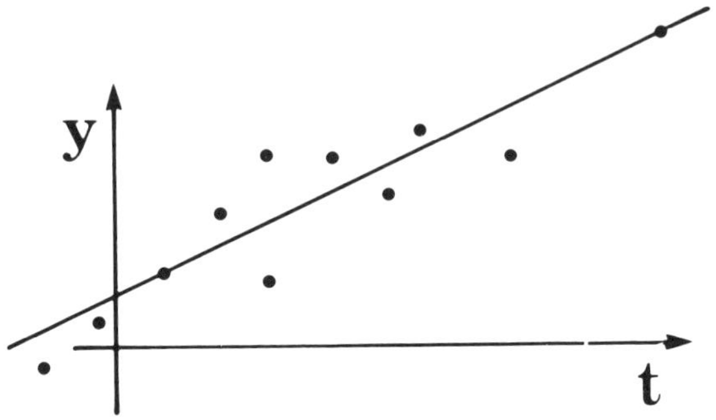

Figure 5.6: The best straight line in the ℓ_1 norm sense minimizes

$$S(a,b) = \sum_i | (a\, t^i + b) - y^i | .$$

It fits at least 2 points exactly.

5.2.1: Linear problems

For a linear problem,

$$\mathbf{g}(\mathbf{m}) = \mathbf{G}\, \mathbf{m} , \tag{5.23}$$

$\mathbf{G}$ representing a linear operator.

To fix our ideas, let us first consider the case where the model space has a dimension of 2 , i.e., we have only two parameters. Figure 5.5 illustrates the shape of the function $S(\mathbf{m})$. It is easy to see that the surface representing $S(\mathbf{m})$ is then a convex polyhedron or, more precisely, the "lower part" of an infinite convex polyhedron. Each edge of the polyhedron clearly corresponds to a null residual in (5.22), i.e.,

$$\mathbf{m} \in \text{edge} \Leftrightarrow \begin{cases} (\mathbf{G}\,\mathbf{m})^i = d^i{}_{obs} & \text{for a given } i \in \mathbf{I}_D \\ \text{or} & \\ m^\alpha = m^\alpha{}_{prior} & \text{for a given } \alpha \in \mathbf{I}_M . \end{cases} \tag{5.24}$$

It is also clear that each vertex of the polyhedron corresponds to the points

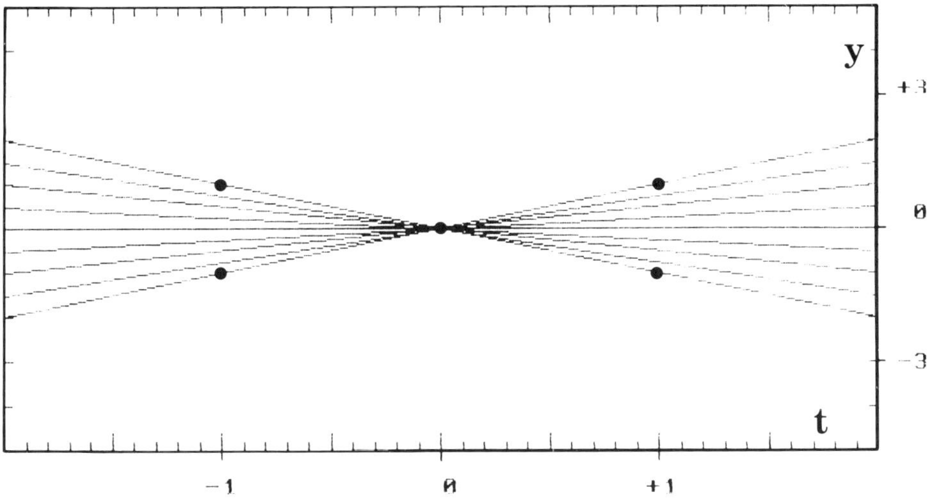

Figure 5.7: Even for linear problems, the best solution in the ℓ_1-norm sense may not be unique. This figure illustrates a highly degenerated problem. We seek for the best line $y = a\, t + b$ fitting the five points (t is the abscissa). In this example, the best line in the ℓ_1-norm sense minimizes the function $S(a,b) = |-a+b+1| + |-a+b-1| + |b| + |a+b-1| + |a+b+1|$. S(a,b) is represented in Figure 5.8. The minimum of S(a,b) is attained by a solution passing through 2 points (as it must always be), but also for all the stright lines represented in the figure.

where *at least two* residuals are null. As the minimum of the convex polyhedron is necessarily attained at a vertex (or edge, face, ...), we arrive at the conclusion that at the minimum of the function $S(\mathbf{m})$ at least two of the equations

$$(\mathbf{G}\,\mathbf{m})^i = d^i_{obs} \qquad (i \in \mathbf{I}_D)$$

$$m^\alpha = m^\alpha_{prior} \qquad (\alpha \in \mathbf{I}_M) \tag{5.25}$$

are exactly satisfied (and only two in the most common situations). Figure 5.6 illustrates the well-known problem of fitting a series of points by a straight line. Following the previous discussion, the best line in the ℓ_1-norm

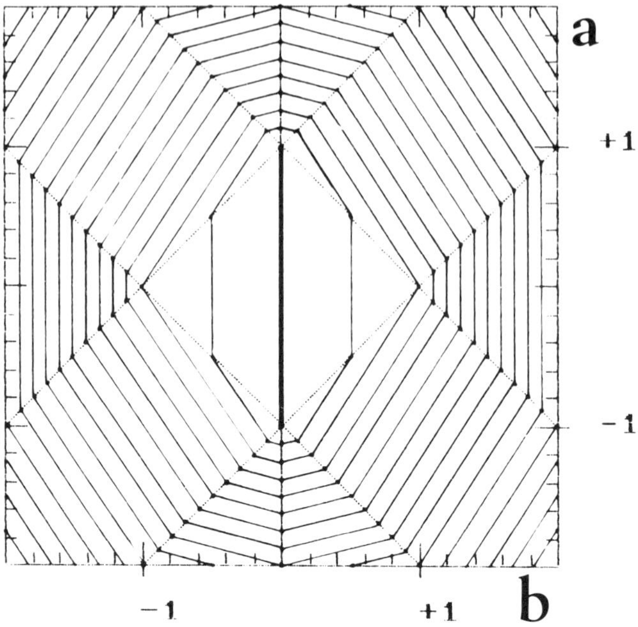

Figure 5.8: The function S(a,b) corresponding to the problem in Figure 5.7. The vertical axis is for the slope a , and the horizontal axis is for b . The minimum of S is attained at an edge (of the polyhedron representing S in a 3D space). The minimum value of S is S = 4 , and the level lines represented are spaced as $\Delta S = 0.5$.

sense will necessarily go through at least 2 points. Of course it may happen that the minimum of S is attained at an "horizontal" edge or face (Figures 5.7 and 5.8). As in that case the minimum of S is not attained at a single point, the solution of the inverse problem based on the least ℓ_1-norm criterion is *not necessarily unique*.

It is helpful to keep in mind a geometric representation of the problem even when the dimension of the model space is greater than 2. To attempt this generalization, I suggest that the reader drops the image of the convex polyhedron in the top of Figure 5.5 and focuses on the representation in the bottom of the figure. Figure 5.9 is the equivalent of the bottom of Figure 5.5, but for a 3D model space. So, if we denote the dimension of the model space by NM , each of the equations (5.25) defines an hyper-plane of dimension N-1 over *M* (a *line* for NM = 2 , an *ordinary plane* for NM = 3 , etc.). The whole set of equations (5.25) defines a partition of the model space into convex NM dimensional hyper-polyhedrons where the function S has constant gradient (convex polygons for NM = 2 , convex polyhedrons for NM = 3 , etc.). The minimum of S is then attained at one vertex of

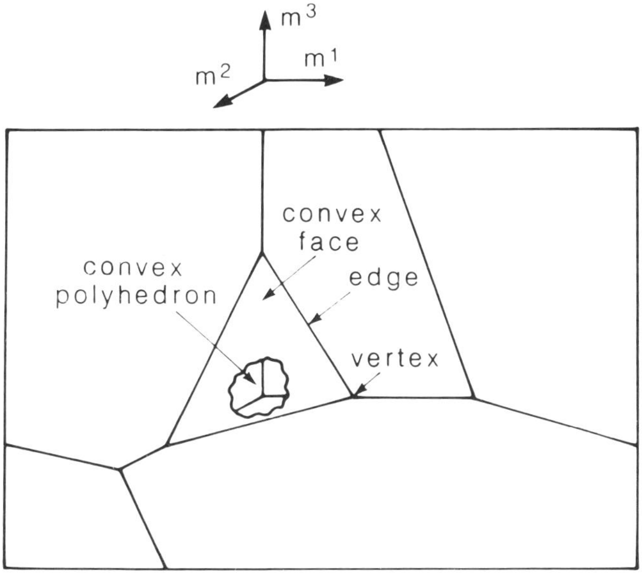

Figure 5.9: If the model space is three dimensional, the function $S(\mathbf{m})$ defines a partition into convex polyhedrons, connected by convex polygons, connected by edges, connected by vertices. In a NM dimensional model space, we have vertices, edges, convex polygons, convex polyhedrons, convex 4D hyper-polyhedrons, ... , and convex NM hyper-polyhedrons. Linear programming defines paths along edges and vertices.

one of these hyper-polyhedrons (or, in case of degeneracy, at one edge, or face, etc.). As the vertices are defined by the intersection of at least NM hyper-planes, we arrive at the conclusion that, at the minimum of S, *at least NM of the equations (5.25) are exactly satisfied* (and often only NM).

This means that the least ℓ_1-norm criterion "selects" among the components of d^i_{obs} $(i \in I_D)$ and m^α_{prior} $(\alpha \in \mathbf{I}_M)$ the NM components whose posterior values will be identical to the prior ones. The posterior values of all other components will be computed from these NM values using the linear system (5.25). The prior values of these other component values are not used (robustness results from this), except in that they have influenced in the choice of the selected NM components.

5.2.2: Nonlinear problems

For nonlinear problems, none of the properties discussed above necessarily remain true. In Figure 1.13 of Chapter 1, the concept of *quasi-linearity* of a problem has been discussed (which is more general than the concept of *linearizability*). A problem is quasi-linear if the functions $g(\mathbf{m})$ can be linearized inside the region of significant posterior probability density (but cannot necessarily be linearized with respect to the a priori model $\mathbf{m}_{\text{prior}}$). All the methods to be described below will only make sense for quasi-linear problems. For strong nonlinear problems, the more general (and expensive) methods of Chapter 1 should be used.

Box 5.2: The mutidimensional exponential probability

a) The unidimensional probability.

Let x be a non negative random variable with probability density $f(x)$, and let $P(\, x \geq x_0\,)$ be the probability

$$P(\, x \geq x_0\,) \;=\; \int_{x_0}^{\infty} dx'\; f(x') \;. \tag{1}$$

If the conditional probability of the event $x \geq x_1 + x_2$, given the event $x \geq x_1$, equals the (unconditional) probability of the event $x \geq x_2$,

$$P(\, x \geq x_1 + x_2 \mid x \geq x_1\,) \;=\; P(\, x \geq x_2\,) \;, \tag{2}$$

it is said that the random process has the *lack of memory* property. For instance, the probability for a radioactive atomic nucleus now alive to remain alive after the next hour equals the probability of remaining alive one hour after its birth: radioactivity lacks its memory.

A random process with lack of memory property has necessarily a probability of the form (e.g., Azlarov and Volodin, 1986)

$$P(\, x \geq x_0) \;=\; \exp\left(-\frac{x_0}{\sigma}\right) \;, \tag{3}$$

where σ is some positive constant. The associated probability density is

$$f(x) \;=\; -\frac{dP}{dx_0}(x) \;=\; \frac{1}{\sigma}\,\exp\left(-\frac{x}{\sigma}\right) \;. \tag{4}$$

(...)

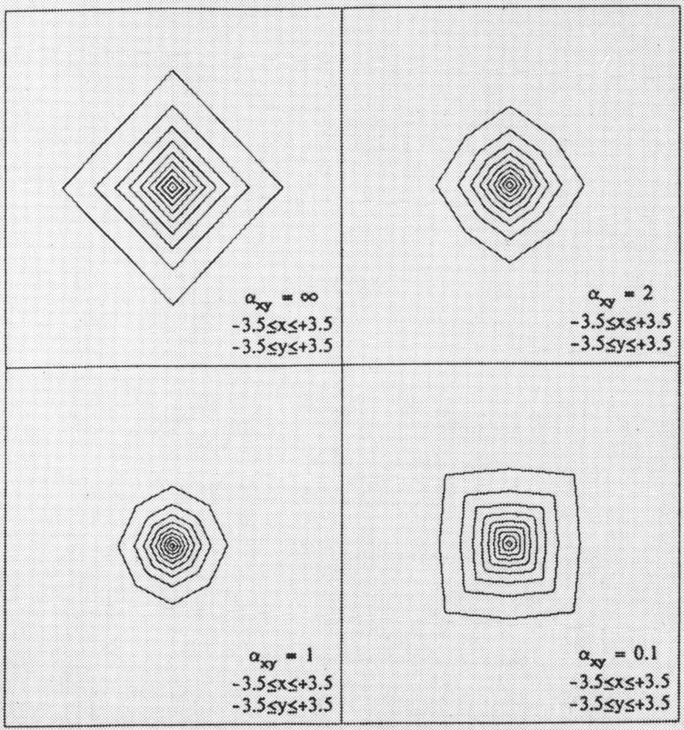

Figure 5.10: The bidimensional probability density

$$f(x,y) = \begin{cases} \dfrac{1}{4\alpha_y}\left(\dfrac{1}{\alpha_x}+\dfrac{1}{\alpha_{xy}}\right)\exp\left[-\left(\dfrac{|x|}{\alpha_x}+\dfrac{|y|}{\alpha_y}+\dfrac{\max(|x|,|y|)}{\alpha_{xy}}\right)\right] & \text{if } |x| > |y| \\ \\ \dfrac{1}{4\alpha_x}\left(\dfrac{1}{\alpha_y}+\dfrac{1}{\alpha_{xy}}\right)\exp\left[-\left(\dfrac{|x|}{\alpha_x}+\dfrac{|y|}{\alpha_y}+\dfrac{\max(|x|,|y|)}{\alpha_{xy}}\right)\right] & \text{if } |x| < |y| \end{cases}$$

is represented for $\alpha_x = \alpha_y = 1$, and for α_{xy} respectively equal to ∞ , 2 , 1 , and 0.1 . Although the variables x and y are not independent (for finite α_{xy}), they are uncorrelated (null correlation coefficient).

b) The bidimensional probability.

For two variables, the lack of memory property is defined by

$$P(\ x \geq x_1 + x_2\ ,\ y \geq y_1 + y_2\ |\ x \geq x_1\ ,\ y \geq y_1\ )\ =\ P(\ x \geq x_2\ ,\ y \geq y_2\ )\ , \tag{2}$$

It can be shown (Azlarov and Volodin, 1986) that if the marginal probabilities of x and y are exponential, then the lack of memory property (5) is satisfied for all positive x_1 , y_1 , $x_2 = y_2$ if and only if

$$P(\ x \geq x_0\ ,\ y \geq y_0\ )\ =\ \exp\left[- \left(\frac{x_0}{\alpha_x} + \frac{y_0}{\alpha_y} + \frac{\max(x_0,y_0)}{\alpha_{xy}} \right) \right]\ , \tag{6}$$

for some non negative α_x , α_y , and α_{xy} . The associated probability density is $f(x,y) = \frac{\partial^2 P}{\partial x \partial y}(x,y)$, i.e.,

$$f(x,y) = \begin{cases} \frac{1}{\alpha_y}\left(\frac{1}{\alpha_x}+\frac{1}{\alpha_{xy}}\right) \exp\left[-\left(\frac{x}{\alpha_x}+\frac{y}{\alpha_y}+\frac{\max(x,y)}{\alpha_{xy}}\right)\right] & \text{if } x > y \\ \text{not necessarily defined} & \text{if } x = y \\ \frac{1}{\alpha_x}\left(\frac{1}{\alpha_y}+\frac{1}{\alpha_{xy}}\right) \exp\left[-\left(\frac{x}{\alpha_x}+\frac{y}{\alpha_y}+\frac{\max(x,y)}{\alpha_{xy}}\right)\right] & \text{if } x < y\ . \end{cases} \tag{7}$$

Replacing x and y by $|x|$ and $|y|$ defines a probability density in the real plane. Figure 5.10 represents $f(x,y)$ for different values of α_{xy} .

c) The multidimensional probability.

A multivariate probability on non negative variables is called of the Marshall-Olkin type if it is given by (Marshall and Olkin, 1967a,b)

$$P(\ \xi_1 \geq x_1\ ,\ldots,\ \xi_n \geq x_n\ ) =$$

$$= \exp\left[- \left(\sum_i \frac{x_i}{\alpha_i} + \sum_{i<j} \frac{\max(x_i,x_j)}{\alpha_{ij}} + \ldots + \frac{\max(x_1,x_2,\ldots,x_n)}{\alpha_{12\ldots n}} \right) \right], \tag{8}$$

with $\alpha_{i_1,\ldots,i_k} \geq 0$, $k = 1,\ldots n$, $1 \leq i_1 < i_2 < \ldots i_k \leq n$.

5.3: The ℓ_1 norm criterion and the method of steepest descent

When direct methods of minimization (systematic exploration of the space, etc.) are too expensive to be used, and if the derivatives

$$G_n{}^{i\alpha} = \left(\frac{\partial g^i}{\partial m^\alpha} \right)_{\mathbf{m}_n} , \tag{5.26}$$

can be obtained, either analytically or by finite differencing, gradient methods of minimizations can be used. I limit the scope of this section to the steepest-descent method of minimization.

The function $S(\mathbf{m})$ does not, in general, have a continuous gradient (see, for instance, figures 5.15 and 5.17). Typically, steepest-descent methods of minimization work fairly well in the first iterations of a ℓ_1-norm problem, but, when approaching the minimum, we often arrive at a hyper-face where the directions of steepest descent at each side are almost opposite (see Figure 5.15). Very small advances are then usually made in subsequent iterations. This is why the linear programming methods in section 5.4 are best suited for ℓ_1-norm minimization problems than gradient methods.

5.3.1: Gradient and direction of steepest ascent

From now on, the terms vertex, edge, ... correspond to the representation of $S(\mathbf{m})$ in the model space (bottom of Figure 5.5 and Figure 5.9) and not to the representation of $S(m)$ as in the top of Figure 5.5.

The gradient of the cost function $S(\mathbf{m})$ can be obtained directly from (5.22):

$$\hat{\gamma}_n{}^\alpha = \left(\frac{\partial S}{\partial m^\alpha} \right)_{\mathbf{m}_n} = \sum_{i \in \mathbf{I}_D} G_n{}^{i\alpha} \frac{1}{\sigma_D^i} \operatorname{sg}\left(g^i(m_n) - d^i_{obs} \right)$$
$$+ \frac{1}{\sigma_M^\alpha} \operatorname{sg}\left(m^\alpha - m^\alpha_{prior} \right) , \tag{5.27}$$

where sg(x) is the sign of x:

$$\operatorname{sg}(x) = \begin{cases} 1 & \text{if} \quad x > 0 \\ 0 & \text{if} \quad x = 0 \\ -1 & \text{if} \quad x < 0 \,. \end{cases} \tag{5.28}$$

In all rigour, the gradient is not defined if one of the residuals $g^i(\mathbf{m})-d^i{}_{obs}$ or $m^\alpha - m^\alpha{}_{prior}$ is null (this corresponds to a hyper-face of dimension NM-1 of one of the convex hyper-polyhedrons of dimension NM representing S). If instead of considering an ℓ_1 norm we consider an ℓ_p norm, it is possible to define the gradient everywhere. For the present purposes, it is helpful to extend the definition of the gradient by defining its value at the hyper-planes of dimension NM-1 as the limit of the value for $p > 1$ when $p \to 1$. This simply corresponds to using equation (5.27) everywhere with the convention $sg(0) = 0$. It is then easy to see that, at any hyper-plane of dimension q , the value thus obtained for the gradient is the mean of the gradient values at each of the adjacent hyper-polyhedrons of dimension q+1 . For instance, the value of the gradient thus obtained at a vertex is the mean of the gradient values at each on the concurrent edges.

To define a direction of steepest ascent, we have to make clear which norm over M we are considering. As discussed in Appendix 5.2, the simplest results are obtained when the ad-hoc ℓ_2 norm

$$\| \mathbf{m} \| = \left(\sum_\alpha \frac{(m^\alpha)^2}{\left(\sigma_M^\alpha\right)^2} \right)^{1/2} , \tag{5.29}$$

is introduced.

Using (5.27), this gives, denoting the direction of steepest ascent at $\mathbf{m}_n$ by γ_n ,

$$\gamma_n{}^\alpha = \left(\sigma_M^\alpha\right)^2 \hat{\gamma}_M^\alpha \tag{5.30}$$

$$= \sigma_M^\alpha \left(\sigma_M^\alpha \sum_{i \in I_D} G_n{}^{i\alpha} \frac{1}{\sigma_D^i} sg\left(g^i(\mathbf{m}_n) - d^i_{obs} \right) + sg\left(m^\alpha - m^\alpha_{prior} \right) \right) ,$$

5.3.2: *The algorithm*

Let $\mathbf{m}_n$ represent the current point we wish to update. In the previous section we have obtained the direction of steepest ascent γ_n . As usual, the steepest descent method is written

$$\mathbf{m}_{n+1} = \mathbf{m}_n - \mu_n \gamma_n , \tag{5.31}$$

where μ_n is an arbitrary positive real number small enough to ensure convergence. One possibility for choosing μ_n is as follows: a linearization of the cost function $S(\mathbf{m})$ in the vicinity of $\mathbf{m}_n$ defines a partition of the model space into convex hyper-polyhedrons (bottom of Figure 5.5 and Figure 5.9). The direction $-\gamma_n$ is a direction of descent at $\mathbf{m}_n$. It is possible (and easy) to obtain the point of the intersection of $-\gamma_n$ with the first hyper-face, and to take this point as the updated point $\mathbf{m}_{n+1}$ (or, in practice, a point slightly "after" the hyper-face), thus fixing the value of μ_n . Let us consider in some detail how this works .

Hyper-faces of the cost function $S(\mathbf{m})$ correspond to changes of sign of the expressions $g^i(\mathbf{m})-\mathbf{d}_{obs}$ and of $\mathbf{m}-\mathbf{m}_{prior}$, which means that the point $\mathbf{m}_{n+1}$ will be at a hyper-face if

$$g^i(\mathbf{m}_{n+1}) - d^i_{obs} = 0 \qquad \text{for a given } i \in \mathbf{I}_D \tag{5.32a}$$

or if

$$m^\alpha_{n+1} - m^\alpha_{prior} = 0 \qquad \text{for a given } \alpha \in \mathbf{I}_M \ . \tag{5.32b}$$

Linearizing $g^i(\mathbf{m})$ around $\mathbf{m}_n$ gives

$$g^i(\mathbf{m}_n - \mu_n \gamma_n) \simeq g^i(\mathbf{m}_n) - \mu_n \sum_{\alpha\in\mathbf{I}_M} G_n{}^{i\alpha} \gamma_n{}^\alpha ,$$

and, using (5.31), the conditions of being at an hyper face (5.32) are written:

$$g^i(\mathbf{m}_n) - d^i_{obs} = \mu_n \sum_{\alpha\in\mathbf{I}_M} G_n{}^{i\alpha} \gamma_n{}^\alpha \quad \text{for a given } i \in \mathbf{I}_D \tag{5.33a}$$

$$m^\alpha_n - m^\alpha_{prior} = \mu_n \gamma_n{}^\alpha \qquad \text{for a given } \alpha \in \mathbf{I}_M \ . \tag{5.33b}$$

From these equations it is easy to compute all the values of μ_n for which we will have hyper-faces of the linearized function:

$$\mu^i_n = \frac{g^i(\mathbf{m}_n) - d^i_{obs}}{\displaystyle\sum_{\alpha\in\mathbf{I}_M} G_n{}^{i\alpha} \gamma^\alpha_n} \qquad (i \in \mathbf{I}_D) \tag{5.34a}$$

$$\mu_n^\alpha = \frac{m_n^\alpha - m_{prior}^\alpha}{\gamma_n^\alpha} \qquad (\alpha \in \mathbf{I}_M) . \tag{5.34b}$$

As $-\gamma_n$ is a direction of descent, negative values of μ_n have to be disregarded, because they would give an increase of S . Among the positive values, it is the smaller which gives the most neighbouring hyper-face:

$$\mu_n = \tag{5.35}$$
$$\mathrm{MIN}\left[\mu_n^{\ i} \ (i \in \mathbf{I}_D) \ , \ \mu_n^{\ \alpha} \ (\alpha \in \mathbf{I}_M) \ , \quad \text{for } \mu_n^{\ i} > 0 \ , \ \mu_n^{\ \alpha} > 0 \right] .$$

Taking a slightly larger value will, in general, avoid dropping *before* the true hyper-face of the nonlinearized function (the chances of dropping at the hyperface are small, due to computer arithmetics).

Very little is known of the acceptability of such a simplistic method for large-sized problems.

5.4: The ℓ_1 norm criterion and the linear programming techniques

"Programming" techniques have been developed to obtain the solution of the problem of optimizing a real function $S(\mathbf{x})$ when the vector variable $\mathbf{x}$ is subject to a given vector constraint $\mathbf{\Psi}(\mathbf{x}) \leq \mathbf{0}$. Very numerous books exist on the subject. The history of programming methods can be read in Dantzig (1963), who first proposed the *Simplex method* in 1947. A good text book is, for instance, Murty (1974). The reader may also refer to Luenberger (1973), Gass (1975), or Bradley, Hax, and Magnanti (1977). A new method developed by Karmarkar (1984) has been reported recently.

When both the cost function $S(\mathbf{x})$ and the constraints $\mathbf{\Psi}(\mathbf{x}) \leq \mathbf{0}$ are linear functions of $\mathbf{x}$, the problem is one of "linear programming". Linear programming techniques are often used by economists (maximizing a benefit for given limitations in available supplies), or by the military (minimizing the time needed for invading a neighbouring country for given limitations on troop transportation).

The central problem in linear programming is always to obtain the minimum of a convex polyhedron in a multidimensional space, and all the methods suggested for solving the problem are very similar: first, one has to manage to obtain a vertex of the polyhedron, then one has to leave the current vertex following a descending edge until the next vertex. After a finite number of moves, the minimum is necessarily attained. The different methods only differ in the way of obtaining the first vertex, or, for a given vertex, in the choice of the edge by which the current vertex has to be left.

The most widely used method for solving linear programming problems is the *Simplex method*, introduced in Box 5.3. For solving inverse problems, the FIFO method of Claerbout and Muir (1973) is simpler.

5.4.1: The FIFO method

The terms vertex, edge, ... correspond to the representation of $S(\mathbf{m})$ in the model space (bottom of Figure 5.5 and Figure 5.9).

The original ℓ_1 problem for solving *linear* inverse problems is the unconstrained minimization of

$$S(\mathbf{m}) = \sum_{i \in I_D} \frac{\left| (\mathbf{G}\,\mathbf{m})^i - d^i_{obs} \right|}{\sigma^i_D} + \sum_{\alpha \in I_M} \frac{\left| m^\alpha - m^\alpha_{prior} \right|}{\sigma^\alpha_M} \, . \qquad (5.22 \text{ again})$$

Let NM denote the dimension of the model space (i.e., the number of model parameters). We have already seen that at the minimum of S, at least NM of the residuals in (5.2) will vanish. This means that we can limit the search of the minimum of S among the models giving at least NM null residuals (i.e., among the vertices of the convex polyhedrons in the model space with constant gradient of S).

Let us first consider the key problem: assume we are at a given vertex, and let us see how we can reach one of the neighbouring vertices with lower value of S. This problem is divided in two subproblems: a) characterization of a vertex, and b) obtaining the new vertex.

a) Characterization of a vertex. Let $\mathbf{m}$ denote an arbitrary vertex of S, which will be called the *current* vertex. If it is a vertex, it has to give NM null residuals, i.e., it has to verify the NM equations

$$m^\alpha = m^\alpha_{prior} \qquad \text{(for some } \alpha \in I_M \text{)} \qquad (5.36a)$$

$$\sum_{\alpha \in I_M} G^{i\alpha} \, m^\alpha = d^i_{obs} \qquad \text{(for some } i \in I_D \text{)} \, , \qquad (5.36b)$$

which are termed the *basic equations.* To simplify the notations, assume that we reclass the components of the model and data vectors so that the equations (5.36a) concern the first components of $\mathbf{m}$, and the equations (5.36b) concern the first components of $\mathbf{d}$. We can then consider the partitions

$$\mathbf{m} = \begin{bmatrix} \mathbf{m}^B \\ \mathbf{m}^N \end{bmatrix} \tag{5.37a}$$

$$\mathbf{d} = \begin{bmatrix} \mathbf{d}^B \\ \mathbf{d}^N \end{bmatrix} \tag{5.37b}$$

where $\mathbf{m}^B$ represent the components of $\mathbf{m}$ appearing in (5.36a) and where $\mathbf{d}^B$ represent the components of $\mathbf{d}$ appearing in (5.37b). This implies the following partition for $\mathbf{G}$:

$$\mathbf{G} = \begin{bmatrix} \mathbf{G}^{BB} & \mathbf{G}^{BN} \\ \mathbf{G}^{NB} & \mathbf{G}^{NN} \end{bmatrix} . \tag{5.38c}$$

The NM basic equations (5.36) can now be written in the more compact form

$$\begin{bmatrix} \mathbf{I} & \mathbf{0} \\ \mathbf{G}^{BB} & \mathbf{G}^{BN} \end{bmatrix} \begin{bmatrix} \mathbf{m}^B \\ \mathbf{m}^N \end{bmatrix} = \begin{bmatrix} \mathbf{m}^B_{prior} \\ \mathbf{d}^B_{obs} \end{bmatrix} \quad \text{(basic equations)} . \tag{5.39}$$

The rest of the equations are termed *nonbasic*, and are not necessarily satisfied

$$\begin{bmatrix} \mathbf{0} & \mathbf{I} \\ \mathbf{G}^{NB} & \mathbf{G}^{NN} \end{bmatrix} \begin{bmatrix} \mathbf{m}^B \\ \mathbf{m}^N \end{bmatrix} = \begin{bmatrix} \mathbf{m}^N_{prior} \\ \mathbf{d}^N_{obs} \end{bmatrix} \quad \text{(nonbasic equations)} . \tag{5.40}$$

From (5.39) it is possible to obtain all the components of $\mathbf{m}$:

$$\begin{bmatrix} \mathbf{m}^B \\ \mathbf{m}^N \end{bmatrix} = \begin{bmatrix} \mathbf{I} & \mathbf{0} \\ \mathbf{G}^{BB} & \mathbf{G}^{BN} \end{bmatrix}^{-1} \begin{bmatrix} \mathbf{m}^B_{prior} \\ \mathbf{d}^B_{obs} \end{bmatrix} . \tag{5.41}$$

Using an inversion by blocks we have

$$\begin{bmatrix} I & 0 \\ G^{BB} & G^{BN} \end{bmatrix}^{-1} = \begin{bmatrix} I & 0 \\ A & B \end{bmatrix} \tag{5.42}$$

where

$$A = -(G^{BN})^{-1}\, G^{BB} \qquad\qquad B = (G^{BN})^{-1} ,$$

which gives

$$m^B = m^B_{prior} \tag{5.43a}$$

$$m^N = (G^{BN})^{-1} \left(d^B_{obs} - G^{BB}\, m^B_{prior} \right) . \tag{5.43b}$$

It is easy to see that the matrix G^{BN} is squared. It can be shown (see below) that the choice of basic and nonbasic equations is always made in such a way that G^{BN} is regular.

b) The equation of a vertex leaving the current point. Consider the following direction in the model space:

$$m'(t) = m + t\, g , \tag{5.44}$$

where, if σ_p denotes the mean deviation appearing in (5.2) for the p-th basis equation in (5.39), g is σ_p times the p-th column of the matrix (5.42), and where t is a real parameter. By definition of g , we have

$$\begin{bmatrix} I & 0 \\ G^{BB} & G^{BN} \end{bmatrix} \begin{bmatrix} g^B \\ g^N \end{bmatrix} = \begin{bmatrix} \mu^B \\ \delta^B \end{bmatrix} , \tag{5.45}$$

where the right-hand vector contains zeroes everywhere except at the p-th component:

$$\begin{bmatrix} \mu^B \\ \delta^B \end{bmatrix} = \begin{bmatrix} 0 \\ 0 \\ \cdots \\ \sigma_p \\ \cdots \\ 0 \\ 0 \end{bmatrix} . \tag{5.46}$$

Using (5.39) and the definition of $\mathbf{g}$, we have

$$\begin{bmatrix} \mathbf{I} & 0 \\ \mathbf{G}^{BB} & \mathbf{G}^{BN} \end{bmatrix} \begin{bmatrix} \mathbf{m}'(t)^B \\ \mathbf{m}'(t)^N \end{bmatrix} = \begin{bmatrix} \mathrm{m}^B_{prior} \\ \mathbf{d}^B_{obs} \end{bmatrix} + t \begin{bmatrix} \mu^B \\ \delta^B \end{bmatrix} . \tag{5.47}$$

As all the components of the last vector are null except the p-th , this equation shows that the point $\mathbf{m}'(t)$ verifies all the equations (5.39) but the p-th : this means that *the direction given by (5.44) defines one of the edges leaving the current vertex, and corresponds to dropping the p-th equation in (5.39).*

c) Obtaining a new vertex. Once we have the equation of the segment leaving the current vertex, one possibility for defining the new vertex is to take the closest vertex in that direction. A better choice is to take as the new vertex the one minimizing S along that direction. We then have

$$\begin{aligned} S(t) &= \sum_{i \in I_D} \frac{\left| (\mathbf{G}\,\mathbf{m}'(t))^i - d^i_{obs} \right|}{\sigma^i_D} + \sum_{\alpha \in I_M} \frac{\left| m'(t)^\alpha - m^\alpha_{prior} \right|}{\sigma^\alpha_M} \\ &= \sum_{i \in I_D} \frac{\left| (\mathbf{G}\,\mathbf{m})^i + t\,(\mathbf{G}\,\mathbf{g})^i - d^i_{obs} \right|}{\sigma^i_D} + \sum_{\alpha \in I_M} \frac{\left| m^\alpha + t\,g^\alpha - m^\alpha_{prior} \right|}{\sigma^\alpha_M} \\ &= \sum_{i \in I_D} \left| w^i\, t - e^i \right| + \sum_{\alpha \in I_M} \left| w^\alpha\, t - e^\alpha \right| , \end{aligned} \tag{5.48}$$

where

$$w^i = \frac{(G\,g)^i}{\sigma_D^i} \qquad e^i = \frac{d^i_{obs} - (G\,m)^i}{\sigma_D^i} \quad (i \in I_D)$$

and (5.49)

$$w^\alpha = \frac{g^\alpha}{\sigma_M^\alpha} \qquad e^\alpha = \frac{m^\alpha_{prior} - m^\alpha_n}{\sigma_M^\alpha} \quad (\alpha \in I_M) .$$

To obtain the value of t giving the minimum of S from (5.48) then simply corresponds to obtaining a weighted median (see, for instance, the Hoare's method quoted by Claerbout and Muir).

For computing the constants appearing in (5.49), the following equalities should be used

$$G\,g = \begin{bmatrix} G^{BB} & G^{BN} \\ G^{NB} & G^{NN} \end{bmatrix} \begin{bmatrix} g^B \\ g^N \end{bmatrix} = \begin{bmatrix} G^{BB}\,g^B + G^{BN}\,g^N \\ G^{NB}\,g^B + G^{NN}\,g^N \end{bmatrix}$$

$$= \begin{bmatrix} \delta^B \\ G^{NB}\,g^B + G^{NN}\,g^N \end{bmatrix} \tag{5.50}$$

and

$$G\,m = \begin{bmatrix} G^{BB} & G^{BN} \\ G^{NB} & G^{NN} \end{bmatrix} \begin{bmatrix} m^B \\ m^N \end{bmatrix} = \begin{bmatrix} d^B_{obs} \\ G^{NB}\,m^B + G^{NN}\,m^N \end{bmatrix} . \tag{5.51}$$

d) Which edge has to be chosen to leave the current vertex? Many choices are possible. For instance, we could choose the edge with maximum slope, but this needs the computation of the slopes of all edges leaving the current vertex. Claerbout and Muir (1973) suggest the following strategy. Leaving the current vertex taking an edge corresponds to droping one of the basic equations (5.39) and replacing it by one of the nonbasic equations (5.40). To choose an edge thus means choosing the basic equation which has to be dropped. The simplest choice consists in dropping the equation which was the first in becoming basic (FIFO: First In, First Out). Of course, in the beginning all the equations have the same "age", so arbitrary ages are given to the basic equations for the first iteration. After NM iterations, the ages are true. If, in trying to leave a vertex following a given edge (i.e., in trying to drop one basic equation), S cannot be diminished, the equation reenters the basis, and its age is set to zero. If this happens consecutively for the NM basic equations, then the minimum is attained.

5.4.2: The FIFO algorithm

Step 01: One makes an arbitrary choice of the starting edge, i.e. one arbitrarily chooses the NM residuals to be taken as null, or, equivalently, the N basic equations to be satisfied. One gives arbitrary ages to the NM first basic equations.

The simplest (and recommended) choice is simply to take in the basis all the equations corresponding to model parameters, i.e., to take

$$\mathbf{m}_0 = \mathbf{m}_{prior} \cdot \tag{5.52}$$

In that case, the components $\mathbf{m}^N$ $\mathbf{d}^B$, and the matrices $\mathbf{G}^{BB}$ and $\mathbf{G}^{BN}$ are nonexistent. Any other choice is possible, but then the user has to verify that the corresponding matrix $\mathbf{G}^{BN}$ thus defined is regular. Let us assume, in what follows that we are placed in that more general case.

Step 02: One drops the oldest equation of the basis (First In, First Out). Assume it is the p-th equation. One obtains the vector $\mathbf{g}$, defined previously by solving the linear equation

$$\begin{bmatrix} \mathbf{I} & \mathbf{0} \\ \mathbf{G}^{BB} & \mathbf{G}^{BN} \end{bmatrix} \begin{bmatrix} \mathbf{g}^B \\ \mathbf{g}^N \end{bmatrix} = \begin{bmatrix} \boldsymbol{\mu}^B \\ \boldsymbol{\delta}^B \end{bmatrix} , \tag{5.53}$$

where the right hand vector is defined in (5.46).

Step 03: Compute

$$\mathbf{d}_1^N = \mathbf{G}^{NB} \mathbf{g}^B + \mathbf{G}^{NN} \mathbf{g}^N \tag{5.54a}$$

$$\mathbf{d}_2^N = \mathbf{G}^{NB} \mathbf{m}^B + \mathbf{G}^{NN} \mathbf{m}^N , \tag{5.54b}$$

set

$$\mathbf{d}_1 = \begin{bmatrix} \boldsymbol{\delta}^B \\ \mathbf{d}_1^N \end{bmatrix} \tag{5.55a}$$

$$\mathbf{d}_2 = \begin{bmatrix} \mathbf{d}^B_{obs} \\ \mathbf{d}_2{}^N \end{bmatrix} ,$$

compute

$$w^i = \frac{d_1{}^i}{\sigma^i_D} \qquad e^i = \frac{d^i_{obs} - d_2{}^i}{\sigma^i_D} \qquad (i \in \mathbf{I}_D) \tag{5.56a}$$

$$w^\alpha = \frac{g^\alpha}{\sigma^\alpha_M} \qquad e^\alpha = \frac{m^\alpha_{prior} - m^\alpha}{\sigma^\alpha_M} \qquad (\alpha \in \mathbf{I}_M) , \tag{5.56b}$$

and obtain the value of t minimizing the sum

$$S(t) = \sum_{i \in \mathbf{I}_D} | \, w^i \, t - e^i \, | + \sum_{\alpha \in \mathbf{I}_M} | \, w^\alpha \, t - e^\alpha \, | \, . \tag{5.57}$$

If $t \neq 0$, a given residual in S becomes null. Sthe p-th basic equation is then drpped, taking as new basic equation the equation giving a null residual. Set its age to zero. GO TO step 02.

If $t = 0$, the p-th equation cannot be dropped (the corresponding vertex is an ascending edge). The age of the p-th equation is reset to zero.

If the NM basic equations have consecutively reentered the basis in such a way, the minimum of S is attained, and the problem is solved.

Problem 5.5 illustrates this method in a very simple example.

5.4.3: Application to nonlinear problems

If the linearization

$$\mathbf{g}(\mathbf{m}) \simeq \mathbf{g}(\mathbf{m}_0) + \mathbf{G}_0 \, (\mathbf{m} - \mathbf{m}_0) \tag{5.58}$$

holds with sufficient accuracy for any model $\mathbf{m}$ to be considered, then the problem is in fact linear, and standard linear programming techniques can be applied. If not, the only easy strategy is to use the linearization (5.58) to obtain a model $\mathbf{m}_1$ using linear programming. Then, use the linearization

$$\mathbf{g}(\mathbf{m}) \simeq \mathbf{g}(\mathbf{m}_1) + \mathbf{G}_1 \, (\mathbf{m} - \mathbf{m}_1) \tag{5.59}$$

and linear programming to obtain a model m_2 , and so on. I do not, however, strongly recommend the use of linear programming for fundamentally nonlinear problems.

Box 5.3: The simplex method of linear programming

Let the following be given:

M a (n x m) rectangular matrix
$\hat{\chi}$ a (m x 1) column matrix
y a (n x 1) column matrix

We wish to obtain a (m x 1) column matrix **x** solving the following problem

$$\text{Minimize} \quad \hat{\chi}^t \, \mathbf{x} \qquad (1a)$$

$$\text{Subject to the constraints} \quad \begin{cases} \mathbf{M}\,\mathbf{x} = \mathbf{y} & (1b) \\ \mathbf{x} \geq 0 & (1c) \end{cases}$$

where by $\mathbf{x} \geq 0$ we mean that all the components ox **x** are non-negative. In the applications we will consider, the matrix **M** will have the following properties:

$m > n$ (more columns than rows) (2a)
M is full rank (rows are linearly independent) (2b)
there is no row with all elements but one null (2c)

Equation (1b) defines a hyperplane with dimension $m - n$. Conditions (2a) and (2b) assure that this hyperplane is not empty and is not reduced to a single point. Condition (2c) ensures that the hyperplane is not parallel to one of the coordinate axes. The set of constraints (1b)-(1c) then always have an infinity of solutions. The set of solutions of (1b)-(1c) can easily be seen to constitute a non-empty convex set (convex means that the straight segment joining two arbitrary points of the set belongs to the set). The problem then reduces to one of obtaining the point **x** of the convex set which minimizes the scalar function $\hat{\chi}^t \, \mathbf{x}$ defined by (1a). As $\hat{\chi}^t \, \mathbf{x}$ is a linear function of **x** , the minimum of $\hat{\chi}^t \, \mathbf{x}$ is attained at a "vertex" of the convex set (or at an edge, if the solution is not unique).

(...)

Let us turn to the "simplex method" for obtaining this solution. The idea of the method is to start at an arbitrary vertex of the convex "polyhedron" and to follow the steepest descending edge to the next vertex. It is clear that the minimum will be attained in a finite number of steps. It can be shown that the vector $\mathbf{x}$ solution of (1) has at most n components different from zero. We can then limit ourselves to looking for the solution in a subspace with n eventually non-null components.

We start by arbitrarily choosing n components (called the *basic* components) among the m components of $\mathbf{x}$ (remember that $m > n$). Setting all the other components to zero, and using equation (1b) will allow us to compute the values of the basic components. This gives a vertex of the polyhedron. If, by chance, we have the minimum of $\hat{\chi}^t \mathbf{x}$, we stop the computations, if not, we drop one of the basic components and replace it by another component. This gives a new vertex connected to the old one by an edge. The simplex method chooses as a new vertex the one for which the corresponding edge has maximum (descending) slope. Let us see how this can be done.

Assume that we have chosen the starting n components of $\mathbf{x}$ arbitrarily. For simplicity, assume that we class the components of $\mathbf{x}$ in such a way that the basic components appear at the first n places of the column matrix representing $\mathbf{x}$:

$$\mathbf{x} = \begin{bmatrix} x^B \\ x^N \end{bmatrix} = \begin{bmatrix} x^1 \\ x^2 \\ \cdots \\ x^n \\ x^{n+1} \\ \cdots \\ x^m \end{bmatrix} = \begin{bmatrix} \text{basic components} \\ \\ \text{other components} \end{bmatrix}$$

The matrix $\mathbf{M}$ can then be written in partitioned form:

$$\mathbf{M} = [\mathbf{M}^B \quad \mathbf{M}^N] .$$

As we have assumed that $\mathbf{M}$ has full rank, we can always choose our basis components in such a way that $\mathbf{M}^B$ is regular. Equation (1b) can be rewritten

(...)

$$[\mathbf{M}^B \quad \mathbf{M}^N] \begin{bmatrix} \mathbf{x}^B \\ \mathbf{x}^N \end{bmatrix} = \mathbf{y}$$

i.e.,

$$\mathbf{M}^B \ \mathbf{x}^B = \mathbf{y} - \mathbf{M}^N \ \mathbf{x}^N$$

which gives

$$\mathbf{x}^B = (\mathbf{M}^B)^{-1} \ (\mathbf{y} - \mathbf{M}^N \mathbf{x}^N) . \tag{3}$$

As we have not yet used the constraint (1c), we have no reason to have $\mathbf{x}_B \geq 0$. We then have to change the choice of basic components until this condition is fulfilled. When it is, we can pass to the study of $\hat{\chi}^t \mathbf{x}$. We have

$$\hat{\chi}^t \ \mathbf{x} = \begin{bmatrix} \mathbf{c}^B & \mathbf{c}^N \end{bmatrix}^t \begin{bmatrix} \mathbf{x}^B \\ \mathbf{x}^N \end{bmatrix} = \mathbf{c}^{B^t} \ \mathbf{x}^B + \mathbf{c}^{N^t} \ \mathbf{x}^N$$

and, using (3),

$$\hat{\chi}^t \ \mathbf{x} = \mathbf{c}^{B^t} \ (\mathbf{M}^B)^{-1} \ \mathbf{y} + \gamma^{N^t} \ \mathbf{x}^N \tag{4}$$

where

$$\gamma^{N^t} = \mathbf{c}^{N^t} - \mathbf{c}^{B^t} \ (\mathbf{M}^B)^{-1} \ \mathbf{M}^N . \tag{5}$$

If all the components of γ^N are ≥ 0 , $\hat{\chi}^t \ \mathbf{x}$ is clearly minimized for $\mathbf{x}^N = \mathbf{0}$, and then the solution obtained using (3):

$$\mathbf{x} = \begin{bmatrix} \mathbf{x}^B \\ \mathbf{x}^N \end{bmatrix} = \begin{bmatrix} (\mathbf{M}^B)^{-1} \ \mathbf{y} \\ \mathbf{0} \end{bmatrix} , \tag{6}$$

(...)

is the solution of the problem. If some of the components of γ^N are negative, we choose the most negative. Let us denote as x^k the corresponding component of $\mathbf{x}$ (which is not in the basic components). If we give increasing positive values to x^k and we compute the corresponding values for $\mathbf{x}^B$ using (3), one of the basis components, say x^j, will first vanish. It can be seen that replacing x^j by x^k in the basic components corresponds to a move from one vertex to the neighbooring vertex whose direction from the previous vertex is the steepest one.

Iterating this procedure, we will end at a vertex for which all the components of γ_N are positive, and the solution will be attained.

Bibliography: For important questions concerning implementation, the reader may refer to Cuer (1984) or to the references there quoted (Bartels, 1971; Bartels and Golub, 1969; Gill and Murray, 1973; Hanson and Wisnieski, 1979; Dickson and Frederick, 1960; Kuhn and Quandt, 1962; Goldfarb, 1976; Reid, 1975; Harris, 1975; Ho, 1978; Bartels, Stoer, and Zenger, 1971; Cuer and Bayer, 1980).

For algorithmic questions, see Klee and Minty, 1972; Jeroslov, 1973; Hacijan, 1979; Gacs and Lovasc, 1981; Kônig and Pallaschke, 1981; Cottle and Dantzig, 1968; Bartels, 1971; Mangasarian, 1981, 1983; Ciarlet and Thomas, 1982; Censor and Elfving, 1982; Cimino, 1938; Magnanti, 1976; Nazareth, 1984; McCall, 1982; Ecker and Kupferschmid, 1985; Karmarkar, 1984; De Ghellinck and Vial, 1985, 1986)

Box 5.4: ℓ_1-norm minimization using Linear Programming.

Assume given the following constants:

$\{d^i_{obs}\}$ $\quad (i \in I_D)$

$\{\sigma^i_D\}$ $\quad (i \in I_D)$

$\{m^\alpha_{prior}\}$ $\quad (\alpha \in I_M)$

$\{\sigma^\alpha_M\}$ $\quad (\alpha \in I_M)$

$\{G^{i\alpha}\}$ $\quad (i \in I_D) \quad (\alpha \in I_M)$,

and consider the problem of obtaining the unknowns:

(...)

$\{u^i\}$ $(i \in I_D)$
$\{v^i\}$ $(i \in I_D)$
$\{a^\alpha\}$ $(\alpha \in I_M)$
$\{b^\alpha\}$ $(\alpha \in I_M)$,

solution of the following problem of constrained minimization:

$$\text{Minimize} \quad \sum_{i \in I_D} \frac{u^i + v^i}{\sigma_D^i} + \sum_{\alpha \in I_M} \frac{a^\alpha + b^\alpha}{\sigma_M^\alpha} \tag{1a}$$

$$\text{Subject to} \quad \begin{cases} \mathbf{G}\,(\mathbf{a} - \mathbf{b}) - (\mathbf{u} - \mathbf{v}) = \mathbf{d}_{obs} - \mathbf{G}\,\mathbf{m}_{prior} & (1b) \\ \mathbf{u}, \mathbf{v}, \mathbf{a}, \mathbf{b} \geq 0 & (1c) \end{cases}$$

Defining

$$w_D^i = 1 / \sigma_D^i \qquad (i \in I_D)$$

$$w_M^\alpha = 1 / \sigma_M^\alpha \qquad (\alpha \in I_M)$$

and setting

$$\mathbf{x} = \begin{bmatrix} \mathbf{a} \\ \mathbf{u} \\ \mathbf{b} \\ \mathbf{v} \end{bmatrix} \qquad \hat{\chi} = \begin{bmatrix} \mathbf{w}_M \\ \mathbf{w}_D \\ \mathbf{w}_M \\ \mathbf{w}_D \end{bmatrix}$$

$$\mathbf{M} = [\, \mathbf{G} \quad -\mathbf{I} \quad -\mathbf{G} \quad \mathbf{I} \,]$$

$$\mathbf{y} = \mathbf{d}_{obs} - \mathbf{G}\,\mathbf{m}_{prior} ,$$

equations (1a)-(1c) write

(...)

$$\text{Minimize} \quad \hat{\chi}^t \ \mathbf{x}$$

$$\text{Subject to} \quad \begin{cases} \mathbf{M}\,\mathbf{x} = \mathbf{y} \\ \mathbf{x} \geq \mathbf{0} \ , \end{cases} \qquad (2)$$

which corresponds to the standard form of the linear programming problem.

To see the equivalence between this problem and the unconstrained ℓ_1 norm minimization problem, let us define $\mathbf{m}$ by

$$\mathbf{a} - \mathbf{b} = \mathbf{m} - \mathbf{m}_{prior} \ . \qquad (3)$$

The constraint (1b) then writes

$$\mathbf{u} - \mathbf{v} = \mathbf{G}\,\mathbf{m} - \mathbf{d}_{obs} \ . \qquad (4)$$

It can be shown that in each vertex of the convex polyhedron defined by (1b)-(1c), and due to the particular structure of system (1b), for any α, both a^α and b^α cannot be $\neq 0$ simultaneously, and for any i, both u^i and v^i can not be $\neq 0$ simultaneously. Then at each vertex,

$$u^i + v^i = | u^i - v^i |$$

$$a^\alpha + b^\alpha = | a^\alpha - b^\alpha | \ ,$$

so that expression (1a) can be written

$$S = \sum_{i \in I_D} \frac{\left| (\mathbf{G}\,\mathbf{m})^i - d^i_{obs} \right|}{\sigma^i_D} + \sum_{\alpha \in I_M} \frac{\left| m^\alpha - m^\alpha_{prior} \right|}{\sigma^\alpha_M} \ , \qquad (5)$$

which corresponds to the standard cost function for the ℓ_1 norm criterion for the resolution of linear problems. We have thus seen that the minimization of (1a) under the constraints (1b)-(1c) is equivalent to the usual unconstrained ℓ_1 norm minimization.

(...)

In numerical analysis, ℓ_1 norm minimization problems have been studied in approximation theory (Barrodale and Young, 1966; Barrodale, 1970; Barrodale and Roberts, 1973, 1974; Bartels, Conn, and Sinclair, 1978; Armstrong and Golfrey, 1979; Watson, 1980).

A useful package of routines has been developed by Cuer and Bayer (1980). They apply to the general problem

$$\text{Minimize} \quad \sum_{A \in I_x} \frac{\left| x^A - x^A_{prior} \right|}{\sigma^A}$$

$$\text{Under the constraints} \quad \begin{cases} \mathbf{A}\,\mathbf{x} = \mathbf{b} \\ \mathbf{0} \leq \mathbf{x} \leq \mathbf{x}_{Max} \,, \end{cases}$$

where $\mathbf{x}_{prior}$, $\mathbf{x}_{Max}$, $\mathbf{A}$, and $\mathbf{b}$ are given, and where some of the constants σ^A may be infinite. These algorithms allow the resolution of the standard linear programming problem.

5.5: Robust inversion using the Cauchy probability density

The Cauchy probability density is defined by

$$f(x) = \frac{1}{\pi\sigma} \frac{1}{1 + \frac{(x-x_0)^2}{\sigma^2}} \,. \tag{5.60}$$

It is centered at $x = x_0$, its variance is infinite, and, for large k ,

$$f(x_0 \pm k\sigma) \simeq \frac{f(x_0)}{k^2} \,, \tag{5.61}$$

i.e., tends to zero as $1/k^2$. As can be seen from Figure 5.11, it represents a bell-shaped curve, quite similar to the Gaussian curve, although having large "tails". This property suggests it as a candidate for robust inversion. Let us see the expressions obtained when assuming Cauchy probability densities for representing a priori data and model parameters uncertainties. Simple results are only obtained when assuming independent uncertainties. This gives

$$\rho_D(\mathbf{d}) = \prod_{i \in I_D} \frac{1}{\pi \sigma_D{}^i} \frac{1}{1 + \dfrac{(d^i - d^i_{obs})^2}{(\sigma_D{}^i)^2}} , \tag{5.62a}$$

$$\rho_M(\mathbf{m}) = \prod_{\alpha \in I_M} \frac{1}{\pi \sigma_M{}^\alpha} \frac{1}{1 + \dfrac{(m^\alpha - m^\alpha_{prior})^2}{(\sigma_M{}^\alpha)^2}} . \tag{5.62b}$$

Assuming an exact resolution of the forward problem, $\mathbf{d} = \mathbf{g}(\mathbf{m})$, and a uniform noninformative probability density $\mu_D(\mathbf{d})$, we obtain from (1.69) the posterior probability density

$$\sigma_M(\mathrm{m}) = \mathrm{const} \left[\prod_{i \in I_D} \frac{1}{1 + \dfrac{(g^i(\mathbf{m}) - d^i_{obs})^2}{(\sigma_D{}^i)^2}} \right] \left[\prod_{\alpha \in I_M} \frac{1}{1 + \dfrac{(m^\alpha - m^\alpha_{prior})^2}{(\sigma_M{}^\alpha)^2}} \right] . \tag{5.63}$$

In some problems with only a few model parameters, $\sigma_M(\mathrm{m})$ can explicitly be computed in a grid of the model space. More often, only the maximum likelihood point $\mathbf{m}_{ML}$ (maximizing $\sigma_M(\mathbf{m})$) is needed. The maximum likelihood point minimizes

$$S(\mathbf{m}) = - \frac{1}{2} \,\mathrm{Log} \frac{\sigma_M(\mathbf{m})}{\mathrm{const}} \tag{5.64}$$

$$= \frac{1}{2} \left(\sum_{i \in I_D} \mathrm{Log} \left(1 + \frac{(g^i(\mathbf{m}) - d^i_{obs})^2}{(\sigma_D{}^i)^2} \right) + \sum_{\alpha \in I_M} \mathrm{Log} \left(1 + \frac{(m^\alpha - m^\alpha_{prior})^2}{(\sigma_M{}^\alpha)^2} \right) \right) .$$

The gradient of $S(\mathbf{m})$ at a point $\mathbf{m} = \mathbf{m}_n$ is easily obtained:

$$\hat{\gamma}_n{}^\alpha = \left(\frac{\partial S}{\partial m^\alpha} \right)_{\mathbf{m}_n}$$

$$= \sum_{i \in \mathbf{I}_D} G_n{}^{i\alpha} \frac{1}{(\sigma_D{}^i)^2} \frac{g^i(\mathbf{m}_n) - d^i_{obs}}{1 + \dfrac{(g^i(\mathbf{m}_n) - d^i_{obs})^2}{(\sigma_D{}^i)^2}}$$

$$+ \frac{1}{(\sigma_M{}^\alpha)^2} \frac{m_n{}^\alpha - m^\alpha_{prior}}{1 + \dfrac{(m_n{}^\alpha - m^\alpha_{prior})^2}{(\sigma_M{}^\alpha)^2}} , \tag{5.65}$$

and the direction of steepest ascent (with respect to an ad-hoc ℓ_2 norm, as discussed in Appendix 5.2) is

$$\gamma_n{}^\alpha = (S_M{}^\alpha)^2 \, \hat{\gamma}_n{}^\alpha$$

$$= (\sigma_M{}^\alpha)^2 \sum_{i \in I_D} G_n{}^{i\alpha} \frac{1}{(\sigma_D{}^i)^2} \frac{g^i(m_n) - d^i_{obs}}{1 + \dfrac{(g^i(\mathbf{m}_n) - d^i_{obs})^2}{(\sigma_D{}^i)^2}}$$

$$+ \frac{m_n{}^\alpha - m^\alpha_{prior}}{1 + \dfrac{(m_n{}^\alpha - m^\alpha_{prior})^2}{(\sigma_M{}^\alpha)^2}} . \tag{5.66}$$

A steepest descent algorithm

$$\mathbf{m}_{n+1} = \mathbf{m}_n - \mu_n \, \gamma_n \tag{5.66}$$

can then easily be used to obtain $\mathbf{m}_{ML}$. In fact, some numerical tests have suggested to me that accurate linear searches (for obtaining adequate values of μ_n) are quite difficult, due to the rapid variation of $S(\mathbf{m})$ in the "valleys". The gradient method can then be used for approaching the region of the minimum of $S(\mathbf{m})$, the minimum itself being difficult to obtain.

Box 5.5: The Cauchy probability density

It is defined by

$$f(x) = \frac{1}{\pi\sigma} \frac{1}{1 + \frac{(x-x_0)^2}{\sigma^2}} , \tag{1}$$

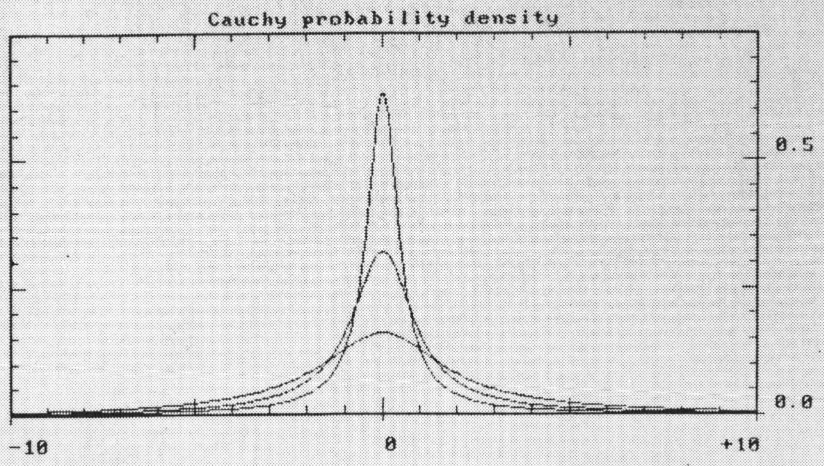

Figure 5.11: The Cauchy probability density for σ respectively equal to 0.5 , 1 , and 2 .

and is represented in Figure 5.11 for σ equal to 0.5 , 1 , and 2 . It has infinite variance. Using $(x-x_0)/\sigma = \text{sh}\,\Theta$ one easily obtains

$$\int_{x_1}^{x_2} dx\ f(x) = \frac{1}{\pi} \left(\text{Arc tg} \frac{x_2-x_0}{\sigma} - \text{Arc tg} \frac{x_1-x_0}{\sigma} \right) . \tag{2}$$

The cumulated probability is then

(...)

$$\Psi(x) = \int_{-\infty}^{x} du\, f(u) = \frac{1}{2} + \frac{1}{\pi}\, \text{Arc tg} \frac{x-x_0}{\sigma} . \tag{3}$$

In particular, $\Psi(\infty) = 1$ (it is normalized).

The variable change $\text{Erf}(x^*) = \Psi(x)$, i.e.,

$$x^* = \text{Erf}^{-1}\left(\frac{1}{2} + \frac{1}{\pi}\, \text{Arc tg}\, \frac{x-x_0}{\sigma}\right) \tag{4a}$$

or

$$\frac{x-x_0}{\sigma} = \text{tg}\left(\pi\left(\text{Erf}(x^*) - \frac{1}{2}\right)\right) , \tag{4b}$$

where $\text{Erf}(u)$ is the error function

$$\text{Erf}(u) = \int_{-\infty}^{u} dt\, \frac{1}{(2\pi)^{1/2}}\, \exp\left(-\frac{t^2}{2}\right) , \tag{5}$$

leads to a Gaussian probability density:

$$f^*(x^*) = \frac{1}{(2\pi)^{1/2}}\, \exp\left(-\frac{(x^*)^2}{2}\right) . \tag{6}$$

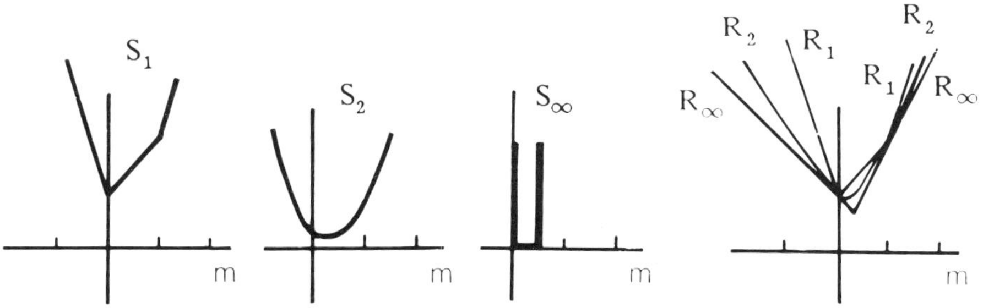

Figure 5.12: The function $S_p(m) = 1/p\ (|2m|^p + |m-1|^p)$ is shown for p respectively equal to 1, 2, and ∞ . Also shown is the function $R_p(m) = (\ p\ S(m)\ )^{1/p}$, which gives $R_1(m) = S_1(m) = |2m| + |m-1|$, $R_2(m) = ((5m^2-2m+1))^{1/2}$, and $R_\infty(m) = \text{MAX}(|2m| , |m-1|)$

5.6: The ℓ_∞ norm criterion for the resolution of inverse problems

In Chapter 1 it has been shown that if uncertainties in the observed data values, $\mathbf{d}_{obs}$, and uncertainties in the prior model, $\mathbf{m}_{prior}$, are assumed to be conveniently modeled using a generalized Gaussian of order p , then the posterior probability density in the model space is given by

$$\sigma_M(m) = \text{const. } \exp - \frac{1}{p} \left(\sum_{i \in I_D} \frac{\left| g^i(\mathbf{m}) - d^i_{obs} \right|^p}{(\sigma^i_D)^p} + \sum_{\alpha \in I_M} \frac{\left| m^\alpha - m^\alpha_{prior} \right|^p}{(\sigma^\alpha_M)^p} \right) . \tag{5.67}$$

In particular, if $p \rightarrow \infty$, the Generalized Gaussian becomes a box-car function (see Box 1.2), which means that this limit corresponds to the assumption of strict error bounds

$$d^i_{obs} - \sigma^i_D \quad \le \quad g^i(\mathbf{m}) \quad \le \quad d^i_{obs} + \sigma^i_D \qquad (i \in I_D) \tag{5.68a}$$

$$m^\alpha_{prior} - \sigma^\alpha_M \quad \le \quad m^\alpha \quad \le \quad m^\alpha_{prior} - \sigma^\alpha_M \qquad (\alpha \in I_M) . \tag{5.68b}$$

When $p \rightarrow \infty$, the posterior probability density gives

$$\sigma_M(m) \quad = \quad \begin{cases} \text{const.} & \text{if the bounds (5.68) are satisfied} \\ 0 & \text{otherwise .} \end{cases} \tag{5.69}$$

It may well happen that no point of the model space satisfies the bounds. This means, in general, that the theoretical equation $\mathbf{d} = \mathbf{g}(\mathbf{m})$ does not model the real world accurately enough, or, more often, that we have been too optimistic in setting the errors bars. In what follows, let us assume that $\sigma_M(m)$ is defined, i.e., that there exists a region M' of the model space satisfying the error bounds and the theoretical equation.

As $\sigma_M(m)$ is uniform over M' , the maximum likelihood point is not defined. For sufficiently regular forward equations, the maximum likelihood point, say $\mathbf{m}_p$, can be defined for any finite value of p . Using the results of Descloux (1963), it can be shown that the sequence m_p is convergent when $p \rightarrow \infty$. The convergence point may be termed *the strict ℓ_∞ solution.* For any finite p , the point maximizing $\sigma_M(\mathbf{m})$ minimizes the function

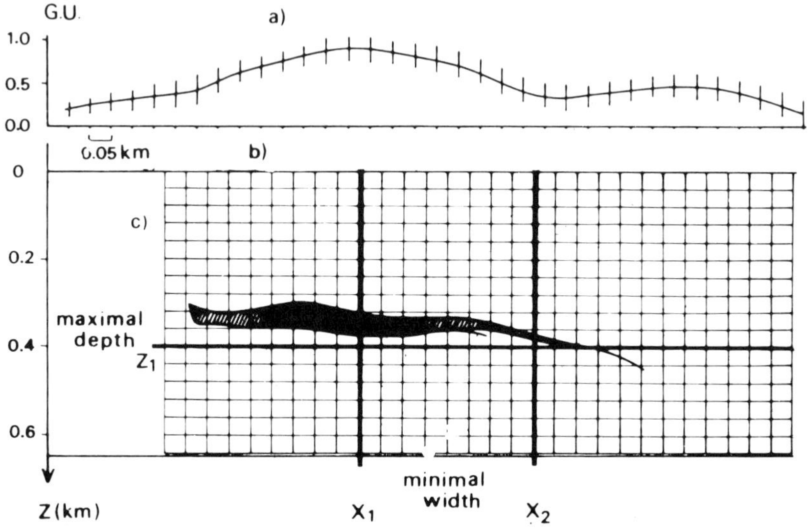

Figure 5.13: The top of the figure shows the gravimetric anomaly measured along a profile on the Earth's surface. It suggests the existence of an orebody. The unknowns of the problem are the densities at each of the cells shown in the figure. Its is assumed a priori that the density lies between 2.5 and 4.2 g cm^{-3}. Using linear programming techniques, Richard, Bayer, and Cuer (1984) show that the data set cannot be fitted unless densities in excess of 2.5 g cm^{-3} are allowed *above* a plane of a depth of 0.4 km . This can be interpreted as the maximal depth of the orebody. They also compute the minimal width. The extent of the actual orebody (estimated by drilling) is shown.

$$S(\mathbf{m}) = \frac{1}{p} \left(\sum_{i \in I_D} \frac{\left| g^i(\mathbf{m}) - d^i_{obs} \right|^p}{(\sigma^i_D)^p} + \sum_{\alpha \in I_M} \frac{\left| m^\alpha - m^\alpha_{prior} \right|^p}{(\sigma^\alpha_M)^p} \right) , \tag{5.70}$$

which has been already used for the values $1 \leq p < \infty$. For any finite p , minimizing $S(\mathbf{m})$ is equivalent to minimizing the function

$$R(\mathbf{m}) = (p\ S(\mathbf{m}))^{1/p} \tag{5.71}$$

i.e.,

$$R(\mathbf{m}) = \left[\sum_{i \in \mathbf{I}_D} \frac{\left| g^i(\mathbf{m}) - d^i_{obs} \right|^p}{(\sigma^i_D)^p} + \sum_{\alpha \in \mathbf{I}_M} \frac{\left| m^\alpha - m^\alpha_{prior} \right|^p}{(\sigma^\alpha_M)^p} \right]^{1/p} . \tag{5.72}$$

When $p \rightarrow \infty$, this gives:

$$R(\mathbf{m}) = \mathrm{MAX} \left(\left(\frac{\left| g^i(\mathbf{m}) - d^i_{obs} \right|}{\sigma^i_D} , (i \in \mathbf{I}_D) \right) , \left(\frac{\left| m^\alpha - m^\alpha_{prior} \right|}{\sigma^\alpha_M} , (\alpha \in \mathbf{I}_M) \right) \right) . \tag{5.73}$$

The advantage of $R(\mathbf{m})$ over $S(\mathbf{m})$ is that $R(\mathbf{m})$ remains finite for any $\mathbf{m}$, while $S(\mathbf{m})$ is either zero (if the bounds (5.68) are satisfied) or infinite (if the bounds are not satisfied). On the contrary, for finite p , $S(\mathbf{m})$ gives simpler expressions for the gradient (for $p = 1$, $R(\mathbf{m}) = S(\mathbf{m})$). Figure 5.12 shows the functions $R(\mathbf{m})$ and $S(\mathbf{m})$ for a one-dimensional example.

Equation (5.73) shows that the minimization of $R(\mathbf{m})$ thus corresponds to minimizing the *maximum weighted residual*. This explains the term **minimax** criterion, often used for the ℓ_∞-norm criterion. Used as synonimous of "ℓ_∞ norm" are: minimax norm, uniform norm, and Chebyshev norm (in honour of the mathematician P.L. Chebyshev, who used the ℓ_∞-norm in the 1850's).

Figure 5.13 shows an interesting example of application of the ℓ_∞ norm to the resolution of a geophysical inverse problem.

5.7: The ℓ_∞ norm and the method of steepest descent

When direct methods of minimization (systematic exploration of the space, etc.) are too expensive to be used, and if the derivatives

$$\mathbf{G}_n{}^{i\alpha} = \left(\frac{\partial g^i}{\partial m^\alpha} \right)_{\mathbf{m}_n} , \tag{5.74}$$

can be obtained either analytically or by finite differencing, gradient methods of minimization can be tried. As for the ℓ_1 norm, gradient methods are less natural than programming methods which are discussed in section 5.8.

5.7.1: Gradient and direction of steepest ascent

At a given point $\mathbf{m}_n$, the maximum of $R(\mathbf{m})$ is attained for one of the (weighted) residuals in (5.73) (if it is attained for more than one residual simultaneously, we are either at a point where the gradient is not defined or the two residuals are identical). Let us distinghuish two cases:

5.7.1.a): The maximum of $R(\mathbf{m})$ is attained for a data residual, say the term $i = K$. Then

$$\hat{\gamma}_n{}^\alpha = G_n{}^{K\alpha} \frac{1}{\sigma_D^K} \, sg(g^K(m_n)-d^K_{obs}) \ . \tag{5.75a}$$

5.7.1b): The maximum of $R(\mathbf{m})$ is attained for a model residual, say the term $\alpha = \Delta$. Then

$$\hat{\gamma}_n{}^\alpha = \delta^{\Delta\alpha} \frac{1}{\sigma_M^\alpha} \, sg(m^\alpha-m^\alpha_{prior}) \ . \tag{5.75b}$$

Using an ℓ_2-norm in the model space for defining the direction of steepest ascent (see Appendix 5.2), we have:

$$\gamma^\alpha = (\sigma_M^\alpha)^2 \ \hat{\gamma}^\alpha \ . \tag{5.76}$$

This gives, respectively, for the case a),

$$\gamma_n{}^\alpha = (\sigma_M^\alpha)^2 \ G_n{}^{K\alpha} \frac{1}{\sigma_D^K} \, sg(g^K(\mathbf{m}_n)-d^K_{obs}) \ , \tag{5.77a}$$

and for the case b),

$$\gamma_n{}^\alpha = \delta^{\Delta\alpha} \ \sigma_M^\alpha \ sg(m^\alpha-m^\alpha_{prior}) \ . \tag{5.77b}$$

5.7.2: The algorithm

The steepest descent algorithm writes

$$\mathbf{m}_{n+1} = \mathbf{m}_n - \mu_n \, \gamma_n \, , \tag{5.78}$$

where the positive real number μ_n can be chosen using any of the methods suggested in Chapter 4 (trial and error, parabola fitting, etc.).

In general, such a crude steepest-descent method is too slowly convergent to be useful. At most, it may help, in the first iterations, in approaching the region to which the minimum belongs.

5.8: The ℓ_∞ norm criterion and the linear programming techniques

The following is adapted from Watson (1980). The problem of minimizing R as defined by (5.77) is equivalent to the problem of minimizing R subject to the constraints

$$-R \leq \frac{(G\,\mathbf{m} - \mathbf{d}_{obs})^i}{\sigma_D^i} \leq +R \tag{5.79a}$$

$$-R \leq \frac{(\mathrm{m} - \mathbf{m}_{prior})^\alpha}{\sigma_M^\alpha} \leq +R \, , \tag{5.79b}$$

which can be rewritten as

$$\begin{aligned}
(G\,\mathbf{m})^i + \sigma_D^i \, R &\geq \mathbf{d}^i_{obs} \\
(G\,\mathbf{m})^i - \sigma_D^i \, R &\leq \mathbf{d}^i_{obs} \\
m^\alpha + \sigma_M^\alpha \, R &\geq m^\alpha_{prior} \\
m^\alpha - \sigma_M^\alpha \, R &\leq m^\alpha_{prior} \, .
\end{aligned} \tag{5.80}$$

The problem can then be written, in matricial form:

$$\text{Minimize} \qquad \begin{bmatrix} 0 \\ 1 \end{bmatrix}^t \begin{bmatrix} \mathbf{m} \\ R \end{bmatrix} \tag{5.81a}$$

$$\text{Subject to} \quad \begin{bmatrix} \mathbf{G} & \sigma_D \\ \mathbf{I} & \sigma_M \\ -\mathbf{G} & \sigma_D \\ -\mathbf{I} & \sigma_M \end{bmatrix} \begin{bmatrix} \mathbf{m} \\ R \end{bmatrix} \geq \begin{bmatrix} \mathbf{d}_{obs} \\ \mathbf{m}_{prior} \\ -\mathbf{d}_{obs} \\ -\mathbf{m}_{prior} \end{bmatrix} , \tag{5.81b}$$

where 0 denotes a vector of zeroes, and where σ_D and σ_M are vectors containing estimated data errors and estimated a priori model errors:

$$\sigma_D = \{ \sigma_D^i \quad (i \in I_D) \} \tag{5.82a}$$

$$\sigma_M = \{ \sigma_M^\alpha \quad (\alpha \in I_M) \} . \tag{5.82b}$$

The problem (5.81) is a Linear Programming problem, but it cannot be solved by direct application of the Simplex method. The dual problem writes (see Box 5.6):

$$\text{Maximize} \quad \begin{bmatrix} \mathbf{d}_{obs} \\ \mathbf{m}_{prior} \\ -\mathbf{d}_{obs} \\ -\mathbf{m}_{prior} \end{bmatrix}^t \begin{bmatrix} \hat{\mathbf{d}}_1 \\ \hat{\mathbf{m}}_1 \\ \hat{\mathbf{d}}_2 \\ \hat{\mathbf{m}}_2 \end{bmatrix} \tag{5.83a}$$

$$\text{Subject to} \quad \begin{cases} \begin{bmatrix} \mathbf{G} & \sigma_D \\ \mathbf{I} & \sigma_M \\ -\mathbf{G} & \sigma_D \\ -\mathbf{I} & \sigma_M \end{bmatrix}^t \begin{bmatrix} \hat{\mathbf{d}}_1 \\ \hat{\mathbf{m}}_1 \\ \hat{\mathbf{d}}_2 \\ \hat{\mathbf{m}}_2 \end{bmatrix} \geq \begin{bmatrix} 0 \\ 1 \end{bmatrix} \\ \hat{\mathbf{d}}_1 , \hat{\mathbf{m}}_1 , \hat{\mathbf{d}}_2 , \hat{\mathbf{m}}_2 \quad \geq \quad \mathbf{0} . \end{cases} \tag{5.83b}$$

This last formulation corresponds to the canonical form of the Linear Programming problem, and can be solved using the standard Simplex method described in Box 5.3. Once the solution of the problem in the dual variables $\hat{\mathbf{d}}_1$, $\hat{\mathbf{m}}_1$, $\hat{\mathbf{d}}_2$, and $\hat{\mathbf{m}}_2$ has been obtained, the values of the variables $\mathbf{m}$ and R can be obtained using the equations of Box 5.6.

Using more compact notations, the problem (5.83) can be written

$$\text{Maximize} \quad (\mathbf{d}_{obs})^t \, (\hat{\mathbf{d}}_1 - \hat{\mathbf{d}}_2) \; + \; (\mathbf{m}_{prior})^t \, (\hat{\mathbf{m}}_1 - \hat{\mathbf{m}}_2) \tag{5.84a}$$

$$
\text{Subject to} \quad \begin{cases} G^t\,(\hat{d}_1-\hat{d}_2) + (\hat{m}_1-\hat{m}_2) = 0 \\ \sigma_D{}^t\,(\hat{d}_1+\hat{d}_2) + \sigma_M{}^t\,(\hat{m}_1+\hat{m}_2) = 1 \\ \hat{d}_1\,,\ \hat{m}_1\,,\ \hat{d}_2\,,\ \hat{m}_2 \ \geq\ \mathbf{0}\,. \end{cases} \tag{5.84b}
$$

Barrodale and Phillips (1975a, 1975b) give a modification of the standard Simplex method which is well adapted to the special structure of this problem. For more details, see Watson (1980).

Box 5.6: Dual problems in Linear Programming

Let $\mathcal{X}$ and $\mathcal{Y}$ be two abstract vector spaces, and $\mathbf{M}$ be a linear operator mapping $\mathcal{X}$ into $\mathcal{Y}$. Assume that the dimension of $\mathcal{X}$ is greater than the dimension of $\mathcal{Y}$, and that $\mathbf{M}$ is full rank. Let $\mathbf{y}$ denote a given vector of $\mathcal{Y}$, and $\hat{\chi}$ a given form over $\mathcal{X}$, i.e., an element of $\hat{\mathcal{X}}$, dual of $\mathcal{X}$. The problem of obtaining the vector $\mathbf{x}$ of $\mathcal{X}$ satisfying the conditions

Minimize	$\hat{\chi}^t\,\mathbf{x}$	(1a)
Subject to the constraints	$\mathbf{M}\,\mathbf{x} = \mathbf{y}$	(1b)
	$\mathbf{x} \geq 0$	(1c)

is called the *standard* problem of Linear Programming. As usual, $\mathbf{x} \geq 0$ means that all the components of $\mathbf{x}$ are ≥ 0.

Let $\mathbf{M}$, $\mathbf{y}$, and $\hat{\chi}$ be the same elements as above. As $\mathbf{M}$ maps $\mathcal{X}$ into $\mathcal{Y}$, by definition of transpose operator (see Box 4.2), $\mathbf{M}^t$ maps $\hat{\mathcal{Y}}$, dual of $\mathcal{Y}$, into $\hat{\mathcal{X}}$, dual of $\mathcal{X}$. The element $\mathbf{y}$ of $\mathcal{Y}$ defines a linear form over $\hat{\mathcal{Y}}$. The problem of obtaining the vector $\hat{\upsilon}$ of $\hat{\mathcal{Y}}$ satisfying the conditions

Maximize	$\mathbf{y}^t\,\hat{\upsilon}$	(2a)
Subject to the constraints	$\mathbf{M}^t\,\hat{\upsilon} \leq \hat{\chi}$	(2b)

is called the *canonical* problem of Linear Programming.

(...)

The inputs of the two problems are the same (namely $\mathbf{M}$, $\mathbf{y}$ and $\hat{\chi}$). The unknown of (1) is an element of X , while the unknown of (2) is an element of the dual of Y . Any of the problems (1)-(2) is termed the *dual* of the other problem, called *primal*. This definition is useful because the solution of one problem also gives the solution to the dual problem.

The basic duality theorem (Dantzig, 1963; Gass, 1975) is: if feasible solutions to both the primal and dual problems exist, there exists an optimum solution to both problems, and

$$\text{Minimum of } \hat{\chi}\, \mathbf{x} = \text{Maximum of } \mathbf{y}^t\, \hat{\mathbf{v}} . \tag{3}$$

The solution of the standard problem (1) can be written (see Box 5.3)

$$\mathbf{x}_{sol} = \begin{bmatrix} \mathbf{x}_B \\ \mathbf{x}_N \end{bmatrix} = \begin{bmatrix} (\mathbf{M}_B)^{-1}\, \mathbf{y} \\ \mathbf{0} \end{bmatrix} , \tag{4}$$

where $\mathbf{x}_B$ is the column matrix of *basic components*, and where $\mathbf{M}_B$ is the *basic submatrix* of $\mathbf{M}$. Let

$$\hat{\chi} = \begin{bmatrix} \hat{\chi}_B \\ \hat{\chi}_N \end{bmatrix} \tag{5}$$

represent the partition of $\hat{\chi}$ into basic and nonbasic components. It can be shown (see, for instance, Gass, 1975) that the solution of the dual problem (2) is then given by

$$\hat{\mathbf{v}}_{sol} = (\mathbf{M}_B)^{-t}\, \hat{\chi}_B , \tag{6}$$

where $(\mathbf{M}_B)^{-t}$ denotes the transpose of the inverse of $\mathbf{M}_B$. The property (3) is then easily verified:

$$\hat{\chi}^t\, \mathbf{x}_{sol} = \hat{\chi}_B{}^t\, \mathbf{x}_B = \hat{\chi}_B{}^t\, (\mathbf{M}_B)^{-1}\, \mathbf{y} = \mathbf{y}^t\, (\mathbf{M}_B)^{-t}\, \hat{\chi}_B = \mathbf{y}^t\, \hat{\mathbf{v}}_{sol} .$$

The standard problem (1) has been arbitrarily assumed to be a minimization problem. The dual of a maximization problem can be obtained simply by changing the signs of $\mathbf{M}$, $\mathbf{y}$, and χ . This gives:

(...)

$$\text{Maximize} \quad \hat{\chi}^t \, \mathbf{x} \tag{7a}$$

$$\text{Subject to the constraints} \quad \begin{cases} \mathbf{M} \, \mathbf{x} = \mathbf{y} & (7b) \\ \mathbf{x} \geq 0 & (7c) \end{cases}$$

$$\text{Minimize} \quad \mathbf{y}^t \, \hat{\upsilon} \tag{8a}$$

$$\text{Subject to the constraints} \quad \mathbf{M}^t \, \hat{\upsilon} \geq \hat{\chi} \tag{8b}$$

where the inequality constraints of the dual problem are now $\geq$, instead of the inequalities $\leq$ of (2).

Box 5.7: Slack variables in Linear Programming

The most general form of a linear programming problem is

$$\text{Minimize (resp. maximize)} \quad \mathbf{a}^t \, \mathbf{b} \tag{1a}$$

$$\text{Under the constraints} \quad \begin{cases} \mathbf{L}_1 \, \mathbf{b} = \mathbf{c}_1 & (1b) \\ \mathbf{L}_2 \, \mathbf{b} \geq \mathbf{c}_2 & (1c) \\ \mathbf{L}_3 \, \mathbf{b} \leq \mathbf{c}_3 \, , & (1d) \end{cases}$$

where $\mathbf{b}$ is the unknown vector, $\mathbf{a}$, $\mathbf{c}_1$, $\mathbf{c}_2$ and $\mathbf{c}_3$ are given vectors, and $\mathbf{L}_1$, $\mathbf{L}_2$, and $\mathbf{L}_3$ are given linear operators.

The particular choice $\mathbf{a} = \hat{\chi}$, $\mathbf{b} = \mathbf{x}$, $\mathbf{L}_1 = \mathbf{M}$, $\mathbf{c}_1 = \mathbf{y}$, $\mathbf{L}_2 = \mathbf{I}$, $\mathbf{c}_2 = \mathbf{0}$, $\mathbf{L}_3 = \mathbf{0}$, and $\mathbf{c}_3 = \mathbf{0}$, leads to the *standard* problem

$$\text{Minimize (resp. maximize)} \quad \hat{\chi}^t \, \mathbf{x} \tag{2a}$$

$$\text{Subject to the constraints} \quad \begin{cases} \mathbf{M} \, \mathbf{x} = \mathbf{y} & (2b) \\ \mathbf{x} \geq 0 \, , & (2c) \end{cases}$$

which is the standard problem of Box 5.6.

The particular choice $\mathbf{a} = -\mathbf{y}$, $\mathbf{b} = \hat{\upsilon}$, $\mathbf{L}_1 = \mathbf{0}$, $\mathbf{c}_1 = \mathbf{0}$, $\mathbf{L}_2 = \mathbf{0}$, $\mathbf{c}_2 = \mathbf{0}$, $\mathbf{L}_3 = \mathbf{M}^t$, and $\mathbf{c}_3 = \hat{\chi}$, leads to the *canonical* problem

(...)

Maximize (resp. minimize)	$\mathbf{y}^t\ \hat{\mathbf{\upsilon}}$	(3a)
Subject to the constraints	$\mathbf{M}^t\ \hat{\mathbf{\upsilon}} \le \hat{\chi}$.	(3b)

We see that problems (2) and (3) are special cases of (1). We will now see that the reciprocal is also true.

Problem (1) can be written

Minimize (resp. maximize) $\quad \mathbf{a}^t\ \mathbf{b}$ (1a)

Under the constraints
$$\begin{cases} \mathbf{L}_1\ \mathbf{b} \le \mathbf{c}_1 \\ -\mathbf{L}_1\ \mathbf{b} \le -\mathbf{c}_1 \\ -\mathbf{L}_2\ \mathbf{b} \le -\mathbf{c}_2 \\ \mathbf{L}_3\ \mathbf{b} \le \mathbf{c}_3\ , \end{cases}$$

and using $\mathbf{a} = -\mathbf{y}$, $\mathbf{b} = \hat{\mathbf{\upsilon}}$, $\mathbf{M}^t = \begin{bmatrix} \mathbf{L}_1 \\ -\mathbf{L}_1 \\ -\mathbf{L}_2 \\ \mathbf{L}_3 \end{bmatrix}$, $\hat{\chi} = \begin{bmatrix} \mathbf{c}_1 \\ -\mathbf{c}_1 \\ -\mathbf{c}_2 \\ \mathbf{c}_3 \end{bmatrix}$, leads to the canonical form (3).

Introducing the *slack variables* $\mathbf{b}' \ge \mathbf{0}$, $\mathbf{b}'' \ge \mathbf{0}$, $\mathbf{c}_2' \ge \mathbf{0}$, and $\mathbf{c}_3' \ge \mathbf{0}$, and writing $\mathbf{b} = \mathbf{b}'-\mathbf{b}''$, the problem (1) becomes

Minimize (resp. maximize) $\quad \mathbf{a}^t\ (\mathbf{b}'-\mathbf{b}'')$ (4a)

Under the constraints
$$\begin{cases} \mathbf{L}_1\ (\mathbf{b}'-\mathbf{b}'') = \mathbf{c}_1 & (4b) \\ \mathbf{L}_2\ (\mathbf{b}'-\mathbf{b}'') - \mathbf{c}_2' = \mathbf{c}_2 & (4c) \\ \mathbf{L}_3\ (\mathbf{b}'-\mathbf{b}'') + \mathbf{c}_3' = \mathbf{c}_3 & (4d) \\ \mathbf{b}' \ge \mathbf{0}\ ,\ \mathbf{b}'' \ge \mathbf{0}\ ,\ \mathbf{c}_2' \ge \mathbf{0}\ ,\ \mathbf{c}_3' \ge \mathbf{0}\ , & (4e) \end{cases}$$

and using $\mathbf{x} = \begin{bmatrix} \mathbf{b}' \\ \mathbf{b}'' \\ \mathbf{c}_2' \\ \mathbf{c}_3' \end{bmatrix}$, $\hat{\chi} = \begin{bmatrix} \mathbf{a} \\ -\mathbf{a} \\ \mathbf{0} \\ \mathbf{0} \end{bmatrix}$, $\mathbf{y} = \begin{bmatrix} \mathbf{c}_1 \\ \mathbf{c}_2 \\ \mathbf{c}_3 \end{bmatrix}$, and

(...)

$$\mathbf{M} = \begin{bmatrix} \mathbf{L}_1 & -\mathbf{L}_1 & \mathbf{0} & \mathbf{0} \\ \mathbf{L}_2 & -\mathbf{L}_2 & -\mathbf{I} & \mathbf{0} \\ \mathbf{L}_3 & -\mathbf{L}_3 & \mathbf{0} & \mathbf{I} \end{bmatrix}$$ **, leads to the standard form (2).**

Appendix 5.1: The Newton step for the minimization of a ℓ_p norm.

Let $S(\mathbf{m})$ denote a real nonlinear convex functional, and let $\hat{\boldsymbol{\gamma}}_n$ and $\hat{\mathbf{H}}_n$ respectively denote the gradient and Hessian of $S(\mathbf{m})$ at $\mathbf{m} = \mathbf{m}_n$ (see section 4.4.3). If $S(\mathbf{m})$ can be approximated by a quadratic function near its minimum, then a traditional Newton method can be used (see Figure 5.14):

$$\mathbf{m}_{n+1} = \mathbf{m}_n - \mu_n \hat{\mathbf{H}}_n \hat{\boldsymbol{\gamma}}_n , \tag{1}$$

where

$$\mu_n \simeq 1 . \tag{2}$$

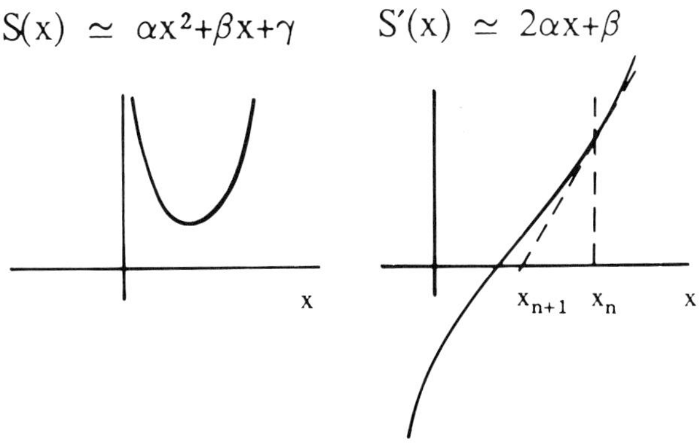

Figure 5.14: The minimum of a sufficiently regular function $S(x)$ is attained at the zero of its derivative, $S'(x)$. In Newton's method, the zero of the derivative is approximated by the zero of the tangent at the current point x_n :

$$S'(x_n) + S''(x_n)\,(x_{n+1}-x_n) = 0 \Rightarrow \quad x_{n+1} = x_n - \frac{1}{S''(x_n)}\, S'(x_n) .$$

Newton's method performs best when the function to be minimized is well approximated, near its minimum, by a quadratic function.

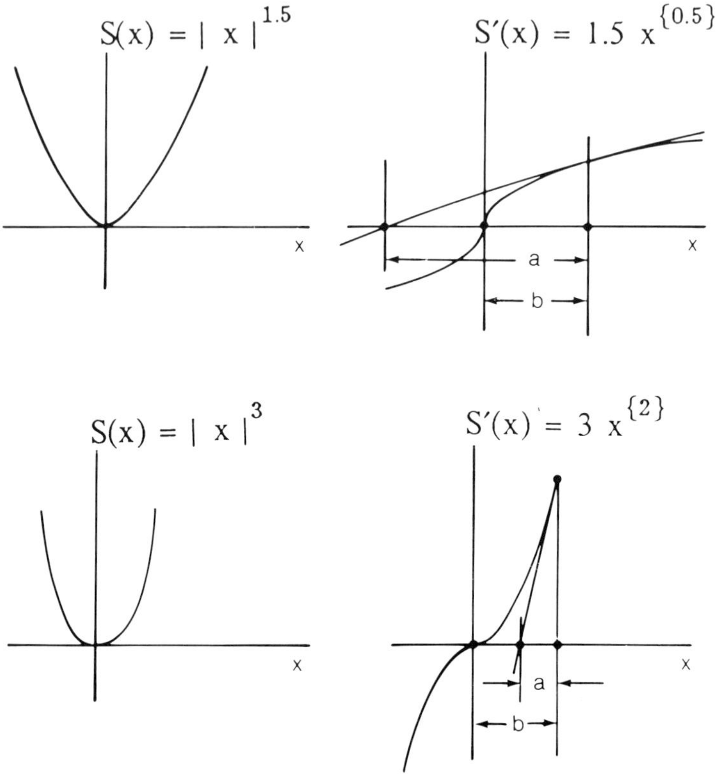

Figure 5.15: The functions S(x) to be minimized when solving inverse problems based on a least ℓ_p-norm criterion are generally best approximated, near their minimum, by a function proportional to $|x|^p$ rather than by a quadratic function (in an illustrative one-dimensional problem). The trouble is that for $p \neq 2$, the derivative of S(x) becomes tangent to one of the axes, and the Newton step is not necessarily very good. Figure a) illustrates the behaviour of a Newton iteration in the case $p = 1.5$, where the Newton step $S'(x_n) / S''(x_n)$ is too big. Figure b) illustrates the case $p = 3$, where the Newton step is too small. Letting a denote the Newton step, and b the optimum step (leading to the minimum of S(x)), we clearly have, for any value of p, $(S'(b))/a = S''(b)$, and using $S(x) = \text{const}\, |x|^p$, this gives $b/a = (p-1)$. This suggests that for minimizing a function behaving like $|x|^p$ near its minimum, Newton's formula $x_{n+1} = x_n - \frac{1}{S''(x_n)} S'(x_n)$ should be replaced by $x_{n+1} = x_n - (p-1) \frac{1}{S''(x_n)} S'(x_n)$.

The functionals S(**m**) encountered in ℓ_p-norm optimization problems are not approximately parabolic near their minimum, and the Newton step (2) is in general not good.

For consider the one-dimensional example

$$S(x) = \alpha \left| x \right|^p \tag{3}$$

in the case $p \neq 2$. The Newton method gives

$$x_{n+1} = x_n - \mu_n \frac{1}{S''(x_n)} S'(x_n) , \tag{4}$$

and we obtain (assuming $x > 0$)

$$x_{n+1} = x_n - \frac{\mu_n}{p-1} x_n . \tag{5}$$

As illustrated in Figure 5.15, for $\mu_n = 1$, the Newton step is too large if $p < 2$ and too small if $p > 2$. Equation (5) suggests the choice

$$\mu_n = \mu = p-1 , \tag{6}$$

which gives convergence in only one step.

More generally, if ϕ_n denotes a direction of search:

$$\mathbf{m}_{n+1} = \mathbf{m}_n - \mu_n \phi_n , \tag{7}$$

this suggest replacing the linearized approximation

$$\mu_n \simeq \frac{< \hat{\gamma}_n , \phi_n >}{< \hat{H}_n \phi_n , \phi_n >} \tag{8}$$

suggested in Chapter 4 by

$$\mu_n \simeq (p-1) \frac{< \hat{\gamma}_n , \phi_n >}{< \hat{H}_n \phi_n , \phi_n >} . \tag{9}$$

Appendix 5.2: Gradient methods for ℓ_p norms

It is helpful for computational developments to introduce some compact notations. Let C_D and C_M be two diagonal operators defined by

$$C_D{}^{ij} = (\sigma_D^i)^p \ \delta^{ij} \tag{1a}$$

$$C_M{}^{\alpha\beta} = (\sigma_M^\alpha)^p \ \delta^{\alpha\beta} \ , \tag{1b}$$

let Q_D and Q_M be two diagonal operators given, when they are defined, by

$$Q_D{}^{ij} = \left| g^i(\mathbf{m}) - d^i_{obs} \right|^{1-p/2} \delta^{ij} \tag{2a}$$

$$Q_M{}^{\alpha\beta} = \left| m^\alpha - m^\alpha_{prior} \right|^{1-p/2} \delta^{\alpha\beta} \ , \tag{2b}$$

and finally let $\tilde{C}_D$ and $\tilde{C}_M$ be two diagonal operators given by

$$\tilde{C}_D = Q_D{}^t \ C_D \ Q_D \tag{3a}$$

$$\tilde{C}_M = Q_M{}^t \ C_M \ Q_M \ . \tag{3b}$$

With these definitions, the misfit function $S(\mathbf{m})$ takes the form

$$S(\mathbf{m}) = \tag{4}$$

$$= \frac{1}{p} \left[(\mathbf{g}(\mathbf{m}) - \mathbf{d}_{obs})^t \ \tilde{C}_D{}^{-1} \ (\mathbf{g}(\mathbf{m}) - \mathbf{d}_{obs}) + (\mathbf{m} - \mathbf{m}_{prior})^t \ \tilde{C}_M{}^{-1} \ (\mathbf{m} - \mathbf{m}_{prior}) \right] ,$$

which is identical to the least-squares misfit function except in the fact that the "covariance operators" $\tilde{C}_D$ and $\tilde{C}_M$ are here diagonal operators whose diagonal elements are given by

$$(\tilde{C}_D{}^{-1})^{ii} = \frac{\left| g^i(\mathbf{m}) - d^i_{obs} \right|^{p-2}}{(\sigma_D^i)^p} \tag{5a}$$

$$(\tilde{C}_M^{-1})^{\alpha\alpha} = \frac{\left| m^\alpha - m^\alpha_{prior} \right|^{p-2}}{(\sigma^\alpha_M)^p} , \qquad (5b)$$

and which depend on the current value of $\mathbf{m}$. This suggests an interesting interpretation of the ℓ_p-norm minimization criterion:

The elements of the diagonal operators $\tilde{\mathbf{C}}_D^{-1}$ and $\tilde{\mathbf{C}}_M^{-1}$ appearing in (4) clearly act as "weights" for the corresponding residuals $(g^i(m)-d^i_{obs})$ or $(m^\alpha - m^\alpha_{prior})$. In particular, for $p = 2$,

$$(\tilde{C}_D^{-1})^{ii} = \frac{1}{(\sigma^i_D)^2}$$

$$(\tilde{C}_M^{-1})^{\alpha\alpha} = \frac{1}{(\sigma^\alpha_M)^2}$$

and the weights are independent of the values of the residuals. For $p > 2$, (5) shows that the importance of large residuals is increased with respect to that of small residuals, while for $p < 2$, it is the importance of the small residuals which is increased with respect to that of large residuals (in the limit $p \to 1$, null residuals have infinite weight). Roughly speaking, a ℓ_p-norm criterion will not tolerate large residuals for $p \to \infty$, will give equal consideration to large and small residuals for $p = 2$, and will be insensitive to large residuals for $p \to 1$. This latter property is named the "robustness" of the minimum ℓ_1-norm criterion with respect to blunders.

The posterior probability density in the model space is given by (5.18). In principle, analysis and error and resolution may be made from this probability density, but for problems other than academical, it turns out to be too difficult (or too expensive) to be practical. For $p \simeq 2$, the inverse Hessian (see below)

$$\hat{\mathbf{H}}_\infty^{-1} \simeq \left[\mathbf{G}_\infty{}^t \; \tilde{\mathbf{C}}_D^{-1} \; \mathbf{G}_\infty + \tilde{\mathbf{C}}_M^{-1} \right]^{-1}$$

approximates the posterior ℓ_2 norm covariance operator, but it is clear that for values of p significantly different from 2 , the property no longuer holds. No simple result is known in the general case that could simplify the analysis.

The gradient methods developed below are likely to work only for sufficiently smooth functions $S(\mathbf{m})$. When p tends to 1 or to ∞ , the function $S(\mathbf{m})$ tends to be a "polyhedron", with sharp edges (see Problem 5.1), thus the numerical methods to be used will greatly depend on the chosen value of p . As suggested in the introduction to this chapter, it seems reasonable to limit the spectrum of the values effectively used for p to the three values $p = 1$, $p = 2$, and $p = \infty$. For $p = 2$, the methods in Chapter 4 should be used, while for $p = 1$ and $p = \infty$, Linear (and Nonlinear) Programming methods should be used. Nevertheless, let me try to comment on some of the numerical difficulties awaiting intrepid users of general ℓ_p norms.

First, as can be seen from equation (10b) below, and contrarily to the ℓ_2-norm case, the usual approximation to the Hessian is not defined, for $1 \leq p < 2$, each time a data residual $g^i(\mathbf{m})-d^i{}_{obs}$ or a model residual $m^\alpha - m^\alpha{}_{prior}$ becomes null. Watson (1980) suggests avoiding the difficulty practically by substituting the null residuals by non-zero values which are small (in the sense of a computer's precision arithmetic). Ekblom (1973) and Wolfe (1979) suggest more rigorous (and difficult) ways for going around the problem. For $2 > p \geq \infty$, the Hessian is defined but it may be singular.

In fact, the Hessian is only needed when using Newton's method, and the preconditioned steepest descent or the preconditioned conjugate gradient method only use a given approximation to the Hessian. The approximation suggested in section 5.3, namely

$$S_0{}^{\alpha\beta} = (p-1)\ (\sigma_M^\alpha)^2\ \delta^{\alpha\beta}\ ,$$

is positive definite, and is not too bad in general.

To my knowledge, the behaviour of the variable metric method for values of p very close to 1 or to ∞ has not yet been studied, but it will presumably lead to operators which are numerically singular, unless a very precautious use of the method is made.

More generally, all Newton and gradient methods of resolution can be expected to degradate in their performances rapidly when p takes values increasingly different from $p = 2$.

The definition of the ℓ_p-norm used in section 5.1 is

$$\|\ \mathbf{x}\ \|_p = \left(\sum_{i \in \mathbf{I}_x} \frac{|\ x^i\ |^p}{(\sigma^i)^p} \right)^{1/p} , \qquad (5.1\ \text{again})$$

where the constants σ^i represent some "estimated errors". For $p = 2$ this gives

$$\| \mathbf{x} \|_2 = \left[\sum_{i \in \mathbf{I}_x} \frac{(x^i)^2}{(\sigma^i)^2} \right]^{1/2} .$$

Letting **C** be a covariance operator (in the usual sense), not necessarily diagonal, the general definition of a ℓ_2-norm is

$$\| \mathbf{x} \|_2 = \left[\sum_{i \in \mathbf{I}_x} \sum_{j \in \mathbf{I}_x} x^i \; (\mathbf{C}^{-1})^{ij} \; x^j \right]^{1/2} = (\mathbf{x}^t \; \mathbf{C}^{-1} \; \mathbf{x})^{1/2} .$$

As discussed in Chapter 4, the use of covariance operators which are *not* diagonal is essential for solving inverse problems with a convenient degree of generality: experimental errors and/or errors in the a priori model may not be uncorrelated. Unfortunately, to my knowledge, no coherent theory exists allowing the generalization of the definition (5.1) to include the possibility of correlated errors. Box 5.2 suggests a way of research.

Methods of resolution

For small-sized problems, methods not using the derivatives of $S(\mathbf{m})$, like

i) systematic exploration of the model space
ii) trial and error
iii) grid search
iv) Monte-Carlo

can of course be used. As they have been already described elsewhere, let us now focus on *gradient methods* of resolution.

The Gradient and Hessian of $S(m)$

If the functions $g^i(\mathbf{m})$ solving the forward problem are differentiable, i.e., if the derivatives

$$G_n{}^{i\alpha} = \left(\frac{\partial g^i}{\partial m^\alpha} \right)_{\mathbf{m}_n} \qquad (6)$$

exist at any point $\mathbf{m}_n$ (or at "almost" any point), and if they can easily be computed (either analytically or by finite-differencing), then, gradient methods can be used for the minimization of $S(\mathbf{m})$.

The *gradient* of S at $\mathbf{m}_n$ is denoted $\hat{\gamma}_n$ and is defined by (see Box 5.1):

$$\hat{\gamma}_n{}^\alpha = \left(\frac{\partial S}{\partial m^\alpha} \right)_{\mathbf{m}_n} \quad . \tag{7}$$

We easily obtain

$$\hat{\gamma}_n{}^\alpha = \sum_{i \in I_D} G_n{}^{i\alpha} \frac{1}{\sigma_D^i} \left(\frac{g^i(m_n) - d^i_{obs}}{\sigma_D^i} \right)^{\{p-1\}} + \frac{1}{\sigma_M^\alpha} \left(\frac{m_n^\alpha - m_{priori}^\alpha}{\sigma_M^\alpha} \right)^{\{p-1\}} , \tag{8}$$

where the symbol $u^{\{r\}}$ has been defined by

$$u^{\{r\}} = sg(u) \; |u|^r \; . \tag{5.13 again}$$

Using the definition of the operators $\tilde{\mathbf{C}}_D$ and $\tilde{\mathbf{C}}_M$ introduced in (3), the expression giving the gradient can be written under the form

$$\hat{\gamma}_n = \mathbf{G}_n{}^t \; \tilde{\mathbf{C}}_D{}^{-1} \; (\mathbf{g}(\mathbf{m}_n) - \mathbf{d}_{obs}) + \tilde{\mathbf{C}}_M{}^{-1} \; (\mathbf{m}_n - \mathbf{m}_{prior}) \; , \tag{9}$$

which is formally identical to the expression obtained for ℓ_2 norm inverse problems, but where the operators $\tilde{\mathbf{C}}_D$ and $\tilde{\mathbf{C}}_M$ depend on $\mathbf{m}_n$ if $p \neq 2$.

The Hessian of S(m) may be obtained from (see Box 5.1)

$$\hat{H}_n{}^{\alpha\beta} = \left(\frac{\partial \hat{\gamma}^\alpha}{\partial m^\beta} \right)_{\mathbf{m}_n} ,$$

which gives

$$\hat{H}_n{}^{\alpha\beta} = (p-1) \left(\sum_{i \in \mathbf{I}_D} G_n{}^{i\alpha} \, G_n{}^{i\beta} \frac{\left| g^i(\mathbf{m}_n) - d^i_{obs} \right|^{p-2}}{(\sigma_D^i)^p} + \delta^{\alpha\beta} \frac{\left| m_n^\alpha - m_{prior}^\alpha \right|^{p-2}}{(\sigma_M^\alpha)^p} \right)$$

$$+ \sum_{i \in \mathbf{i}_D} \left(\frac{\partial^2 g^i}{\partial m^\alpha \partial m^\beta} \right)_n \frac{\left(g^i(\mathbf{m}_n) - d^i_{obs} \right)^{\{p-1\}}}{(\sigma_D^i)^p} \; . \tag{10a}$$

As the Hessian $\hat{\mathbf{H}}_n$ is never to be known with great accuracy, and as the last term in (10a) is generally: i) not predominant (at least near the minimum), and ii) difficult to compute, it is generally neglected, thus giving the approximation

$$\hat{H}_n{}^{\alpha\beta} \simeq (p-1) \left(\sum_{i \in I_D} G_n{}^{i\alpha}\, G_n{}^{i\beta} \frac{\left| g^i(\mathbf{m}_n) - d^i_{obs} \right|^{p-2}}{\left(\sigma^i_D\right)^p} + \delta^{\alpha\beta} \frac{\left| m^\alpha_n - m^\alpha_{prior} \right|^{p-2}}{\left(\sigma^\alpha_M\right)^p} \right) \tag{10b}$$

Using the operators $\tilde{\mathbf{C}}_D$ and $\tilde{\mathbf{C}}_M$ we have

$$\hat{\mathbf{H}}_n \simeq (p-1)\,(\mathbf{G}_n{}^t\, \tilde{\mathbf{C}}_D{}^{-1}\, \mathbf{G}_n + \tilde{\mathbf{C}}_M{}^{-1})\ , \tag{11}$$

which looks similar to the expression of the Hessian in ℓ_2 norm inverse problems, but where, although it is not explicitly indicated, it should not be forgotten that the operators $\tilde{\mathbf{C}}_D$ and $\tilde{\mathbf{C}}_M$ depend on $\mathbf{m}_n$.

The direction of steepest ascent

The problem here is the minimization of the misfit function $S(\mathbf{m})$ defined in equation (4). The gradient of $S(m)$ at a point $\mathbf{m}_n$ is given by equation (8) and it can be deduced from (4), using (7), independently of any particular definition of norm over M . This is not the case for the direction of steepest ascent at $\mathbf{m}_n$, which is different for different norms over M (see Figure 5.3). As the function $S(m)$ is obtained from an ℓ_p norm defined over the space $D \times M$ (see equation (5.20)), it may seem natural to consider the ℓ_p norm over M defined by the restriction over M of the ℓ_p norm over $D \times M$:

$$\| \mathbf{m} \|_p = \left(\sum_{\alpha \in I_M} \frac{\left| m^\alpha \right|^p}{(\sigma^\alpha_M)^p} \right)^{1/p} .$$

This choice of norm over M although natural, is not necessary, and arbitrary norms can be considered for minimizing $S(m)$, such as for instance, the $\ell_{p'}$ norm

$$\| \mathbf{m} \|_{p'} = \left(\sum_{\alpha \in \mathbf{I}_M} \frac{| m^\alpha |^{p'}}{(\sigma_M^\alpha)^{p'}} \right)^{1/p'} ,$$

with $p' \neq p$. In Box 5.1 it is shown that the direction of steepest ascent at a point m_n , with respect to an $\ell_{p'}$ norm defined over M , is given by the vector in M which is related with the gradient at $\mathbf{m}_n$, $\hat{\gamma}_n$, trough the bijection induced between M and $\hat{M}$ by the $\ell_{p'}$ norm over M (equation 5.17):

$$\frac{\gamma_n{}^\alpha}{\sigma_M^\alpha} = \left[\sigma_M^\alpha \ \hat{\gamma}_n{}^\alpha \right]^{\{q'-1\}} \qquad \left(\frac{1}{p'} + \frac{1}{q'} = 1 \right) , \tag{12}$$

which justifies using for the direction of steepest ascent the symbol γ_n , i.e., the same symbol as for the gradient, without the "hat". Using (8) this gives

$$\gamma_n{}^\alpha =$$

$$= \sigma_M^\alpha \left[\sigma_M^\alpha \sum_{i \in I_D} G_n{}^{i\alpha} \frac{1}{\sigma_D^i} \left[\frac{g^i(\mathbf{m}_n) - d^i_{obs}}{\sigma_D^i} \right]^{\{p-1\}} + \left[\frac{m^\alpha - m^\alpha_{prior}}{\sigma_M^\alpha} \right]^{\{p-1\}} \right]^{\{q'-1\}} . \tag{13}$$

For instance, the direction of steepest descent of S(m) at $\mathbf{m}_n$ with respect to the ℓ_2 norm

$$\| \mathbf{m} \|_2 = \left(\sum_{\alpha \in \mathbf{I}_M} \frac{(m^\alpha)^2}{(\sigma_M^\alpha)^2} \right)^{1/2}$$

is

$$\gamma_n{}^\alpha = \sigma_M^\alpha \left[\sigma_M^\alpha \sum_{i \in I_D} G_n{}^{i\alpha} \frac{1}{\sigma_D^i} \left[\frac{g^i(\mathbf{m}_n) - d^i_{obs}}{\sigma_D^i} \right]^{\{p-1\}} + \left[\frac{m^\alpha - m^\alpha_{prior}}{\sigma_M^\alpha} \right]^{\{p-1\}} \right] . \tag{14}$$

Figure 5.3 illustrates, for a functional S(**m**) defined over a two-dimensional space M , the multiplicity of directions of steepest descent for different norms considered over M .

In "preconditioned" gradient methods to be studied below, a linear, symmetric, positive definite operator $\mathbf{S}_0$ mapping $\hat{M}$ into M is introduced (which corresponds to an arbitrary approximation to H_0^{-1} , the inverse of the Hessian at a given point $\mathbf{m}_0$), and a vector $\phi_n \in M$ is defined by

$$\phi_n = \mathbf{S}_0 \hat{\gamma}_n \tag{15}$$

(which corresponds to a "direction of search"). From the hypothesis made on S_0 , the expression

$$\| \mathbf{m} \|_2 = \langle \mathbf{S}_0^{-1} \mathbf{m} , \mathbf{m} \rangle^{1/2} = (\mathbf{m}^t \mathbf{S}_0^{-1} \mathbf{m})^{1/2} \tag{16}$$

makes sense (for any $\mathbf{m}$) and defines a particular ℓ_2 norm over M . It is easy to see (for instance by using the results in Chapter 4) that ϕ_n defined by (15) is the *direction of steepest ascent* with respect to the ad-hoc ℓ_2 norm defined by (16). So, it is interesting to note that, in using preconditioned gradient methods for minimizing a functional $S(\mathbf{m})$ which has been defined from a ℓ_p-norm over the space $D \times M$, we are, in fact, using directions of steepest descent for a given ℓ_2 norm over M .

Gradient methods

This section applies the results of appendix 4.7 to the particular values of gradient and Hessian obtained above.

a) The choice of the preconditioning operator

Gradient methods use as direction of search the direction

$$\phi_n = \mathbf{S}_0 \hat{\gamma}_n , \tag{15 again}$$

where the "preconditioning" operator $\mathbf{S}_0$ is a linear, symmetric, positive definite operator arbitrarily approximating the inverse of the Hessian at the point $\mathbf{m}_0$ where iterations will start:

$$\mathbf{S}_0 \simeq \hat{\mathbf{H}}_0 .$$

We have (equation (10b)

$$\hat{H}_n{}^{\alpha\beta} \simeq (p-1) \left\{ \sum_{i \in \mathbf{i}_D} G_n{}^{i\alpha} \, G_n{}^{i\beta} \, \frac{\left| g^i(\mathbf{m}_n) - d^i_{obs} \right|^{p-2}}{(\sigma^i_D)^p} + \delta^{\alpha\beta} \, \frac{\left| m^\alpha_n - m^\alpha_{prior} \right|^{p-2}}{(\sigma^\alpha_M)^p} \right\} .$$

The importance of astute choices of the preconditioning operator which may tremendously improve the speed of convergence of gradient algorithms has been emphasized in Chapter 4 . On the other hand, it is always necessary to bear in mind which is the *simplest* approximation. In the present case it is

$$S_0{}^{\alpha\beta} = (\sigma^\alpha_M)^2 \; \delta^{\alpha\beta} , \tag{17}$$

which, in fact, simply corresponds to a proper scaling of the variables.

Let me make here a brief digression. In naive approaches to optimization, the implicit assumption that direction of steepest descent is synonymous to gradient is used. Let us see how this can work. First, to identify the gradient with a direction in the model space it is necessary to assume that the model space can be identified with its dual. This is possible if all the variables of the problem have been made adimensional using

$$\tilde{m}^\alpha = \frac{m^\alpha}{\sigma^\alpha_M} , \tag{18}$$

Second, it is necessary to give sense to the assertion that the gradient is "perpendicular" to the "level lines" of $S(\mathbf{m})$. This contains the implicit assumption that the euclidean (ℓ_2) norm

$$\| \tilde{\mathbf{m}} \|_2 = \left[\sum_{\alpha \in I_M} (\tilde{m}^\alpha)^2 \right]^{1/2} \tag{19}$$

is considered over the (adimensional) model space. We have already seen that the direction $\phi_n = S_0 \, \hat{\gamma}_n$ is the direction of steepest ascent for the ad-hoc ℓ_2 norm

$$\| \mathbf{m} \|_2 = \left[\sum_{\alpha \in I_M} \sum_{\beta \in I_M} m^\alpha \; (\mathbf{S}_0{}^{-1})^{\alpha\beta} \; m^\beta \right]^{1/2} .$$

Using the approximation (19) to the inverse Hessian gives

$$\| \mathbf{m} \|_2 = \left[\sum_{\alpha \in I_M} \frac{(m^\alpha)^2}{(\sigma_M^\alpha)^2} \right]^{1/2} ,$$

which is identical to (19). This shows that the approximation (17) to the inverse Hessian gives the same results as a traditional naive approach.

b) Preconditioned Steepest Descent. We obtain

$$\begin{aligned} \hat{\gamma}_n &= \mathbf{G}_n{}^t\, \tilde{C}_D{}^{-1}\, (\mathbf{g}(\mathbf{m}_n) - \mathbf{d}_{obs}) + \tilde{C}_M{}^{-1}\, (\mathbf{m}_n - \mathbf{m}_{prior}) && (\ \mathbf{m}_0 \text{ arbitrary}) \\ \phi_n &= S_0\, \hat{\gamma}_n && (\ S_0 \text{ arbitrary}) \\ \mathbf{m}_{n+1} &= \mathbf{m}_n - \mu_n\, \phi_n && (\text{obtain } \mu_n \text{ by linear search}) \end{aligned} \qquad (20)$$

The starting point $\mathbf{m}_0$ is arbitrary. The simplest choice for S_0 is (17), although a better estimation of the inverse of the intial Hessian

$$S_0 \simeq \frac{1}{p-1} \left(G_0{}^t\, \tilde{C}_D{}^{-1}\, G_0 + \tilde{C}_M{}^{-1} \right)^{-1} \qquad (21)$$

may give good results. An adequate value for μ_n has to be obtained by linear search. Alternatively, a linearization of $\mathbf{g}(\mathbf{m})$ around $\mathbf{m}_n$ gives

$$\mu_n \simeq (p-1) \frac{\hat{\gamma}_n{}^t\, \phi_n}{\phi_n{}^t\, \tilde{C}_M{}^{-1}\, \phi_n + b_n{}^t\, \tilde{C}_D{}^{-1}\, b_n} , \qquad (22)$$

where

$$b_n = G_n\, \phi_n . \qquad (23)$$

c) Preconditioned conjugate directions. We obtain

$$\begin{aligned} \hat{\gamma}_n &= \mathbf{G}_n{}^t\, \tilde{C}_D{}^{-1}\, (\mathbf{g}(\mathbf{m}_n) - \mathbf{d}_{obs}) + C_M{}^{-1}\, (\mathbf{m}_n - \mathbf{m}_{prior}) && (\ \mathbf{m}_0 \text{ arbitrary}) \\ \lambda_n &= S_0\, \hat{\gamma}_n && (\ S_0 \text{ arbitrary}) \\ \phi_n &= \lambda_n + \alpha_n\, \phi_{n-1} && (\phi_0 = \lambda_0) \quad (\alpha_n \text{ defined below}) \\ \mathbf{m}_{n+1} &= \mathbf{m}_n - \mu_n\, \phi_n && (\text{obtain } \mu_n \text{ by linear search}) \end{aligned} \qquad (24)$$

$\mathbf{m}_0$ and $\mathbf{S}_0$ are as in b). The value of α_n proposed by Polak and Ribière (1969) (see appendix 4.7) is

$$\alpha_n = \frac{\omega_n - \hat{\gamma}_{n-1}{}^t \ \lambda_n}{\omega_{n-1}} \tag{25}$$

where

$$\omega_n = \hat{\gamma}_n{}^t \ \lambda_n \ . \tag{26}$$

The value μ_n has to be obtained by linear search. Alternatively, a linearization of $g(\mathbf{m})$ around $g(\mathbf{m}_n)$ gives

$$\mu_n \simeq (p-1) \frac{\hat{\gamma}_n{}^t \ \phi_n}{\phi_n{}^t \ \tilde{C}_M{}^{-1} \ \phi_n + \mathbf{b}_n{}^t \ \tilde{C}_D{}^{-1} \ \mathbf{b}_n} , \tag{27}$$

where

$$\mathbf{b}_n = \mathbf{G}_n \ \boldsymbol{\phi}_n \ . \tag{28}$$

The numerator of the expression giving μ_n can be written

$$\hat{\gamma}_n{}^t \ \phi_n = \hat{\gamma}_n{}^t \ \lambda_n - \alpha_n \ \hat{\gamma}_n{}^t \ \phi_{n-1} \ . \tag{29}$$

If the linear searches are accurate, we have:

$$\hat{\gamma}_n{}^t \ \phi_{n-1} \simeq 0 \ ,$$

and the following simplification can be used:

$$\mu_n \simeq \frac{\omega_n}{\phi_n{}^t \ \tilde{C}_M{}^{-1} \ \phi_n + \mathbf{b}_n{}^t \ \tilde{C}_D{}^{-1} \ \mathbf{b}_n} \ . \tag{30}$$

d) Variable metric method We obtain

$$\begin{aligned}
&\hat{\gamma}_n = \mathbf{G}_n{}^t \ \tilde{C}_D{}^{-1} \ (g(\mathbf{m}_n)-\mathbf{d}_{obs}) + \tilde{C}_M{}^{-1} \ (\mathbf{m}_n - \mathbf{m}_{prior}) && (\ \mathbf{m}_0 \text{ arbitrary}) \\
&\phi_n = \mathbf{S}_n \ \hat{\gamma}_n && (\ \mathbf{S}_0 \text{ arbitrary}) \\
&\mathbf{m}_{n+1} = \mathbf{m}_n - \mu_n \ \phi_n && (\text{obtain } \mu_n \text{ by linear search}) \\
&\mathbf{S}_{n+1} = \mathbf{S}_n + \delta\mathbf{S}_n && (\delta\mathbf{S}_n \text{ defined below})
\end{aligned} \tag{31}$$

$\mathbf{m}_0$ and $\mathbf{S}_0$ are as in b). Using the BFGS updating formula for $\mathbf{S}_n$ (see Appendix 4.7) gives

$$\mathbf{S}_{n+1} = \left[\mathbf{I} - \frac{\delta\mathbf{m}_n \ \delta\hat{\gamma}_n{}^t}{\delta\hat{\gamma}_n{}^t \ \delta\mathbf{m}_n} \right] \mathbf{S}_n \left[\mathbf{I} - \frac{\delta\mathbf{m}_n \ \delta\hat{\gamma}_n{}^t}{\delta\hat{\gamma}_n{}^t \ \delta\mathbf{m}_n} \right]^t + \frac{\delta\mathbf{m}_n \ \delta\mathbf{m}_n{}^t}{\delta\hat{\gamma}_n{}^t \ \delta\mathbf{m}_n} \tag{32}$$

where

$$\delta\mathbf{m}_n = \mathbf{m}_{n+1} - \mathbf{m}_n \tag{33}$$

and

$$\delta\hat{\gamma}_n = \hat{\gamma}_{n+1} - \hat{\gamma}_n \ . \tag{34}$$

e) Newton's method We obtain

$$\hat{\gamma}_n = \mathbf{G}_n{}^t \ \tilde{\mathbf{C}}_D{}^{-1} \ (\mathbf{g}(\mathbf{m}_n)\text{-}\mathbf{d}_{obs}) + \tilde{\mathbf{C}}_M{}^{-1} \ (\mathbf{m}_n\text{-}\mathbf{m}_{prior}) \qquad (\ \mathbf{m}_0 \text{ arbitrary})$$

$$\hat{\mathbf{H}}_n \ \phi_n = \hat{\gamma}_n \qquad (\text{solve for } \phi_n\ ) \tag{35}$$

$$\mathbf{m}_{n+1} = \mathbf{m}_n - \mu_n \ \phi_n \qquad (\text{obtain } \mu_n \text{ by linear search})\ .$$

The Hessian is given by (11):

$$\hat{\mathbf{H}}_n = (p\text{-}1) \left(\mathbf{G}_n{}^t \ \tilde{\mathbf{C}}_D{}^{-1} \ \mathbf{G}_n + \tilde{\mathbf{C}}_M{}^{-1} \right) . \tag{36}$$

A linear approximation of $\mathbf{g}(\mathbf{m})$ gives

$$\mu_n \simeq p\text{-}1 \ . \tag{37}$$

It should be noticed that the factor $1/(p\text{-}1)$ in (36) compensates the value $p\text{-}1$ in (37), so, in fact, these values can simply be dropped.

Problems for chapter 5:

Problem 5.1: Consider a schematic problem with two model parameters m^1 and m^2, and a single datum d^1. The datum is theoretically related to model parameters through the linear equation

$$d^1 = g^1(m^1,m^2) = m^1 - m^2 . \tag{1}$$

Its observed value is

$$d^1{}_{obs} = 0 \pm 1 . \tag{2}$$

The a priori values of the model parameters are

$$m^1{}_{prior} = 3 \pm 1 , \tag{3}$$

and

$$m^2{}_{prior} = 1 \pm 2 . \tag{4}$$

Represent the misfit function $S(m^1,m^2)$ in the three following cases:

i) The symbols $\pm$ in (2)-(4) represent ℓ_1 norm error bars.

ii) The symbols $\pm$ in (2)-(4) represent ℓ_2 norm error bars.

iii) The symbols $\pm$ in (2)-(4) represent ℓ_∞ norm error bars.

Solution for the ℓ_1 norm: Using a ℓ_p norm, the misfit function is (equation (5.19)):

$$S(\mathbf{m}) = \frac{1}{p} \left(\sum_{i \in I_D} \frac{\left| g^i(\mathbf{m}) - d^i{}_{obs} \right|^p}{(\sigma^i_D)^p} + \sum_{\alpha \in I_M} \frac{\left| m^\alpha - m^\alpha{}_{prior} \right|^p}{(\sigma^\alpha_M)^p} \right) . \tag{5}$$

For $p = 1$ we have

$$S(\mathbf{m}) = \sum_{i \in \mathbf{I}_D} \frac{\left| g^i(\mathbf{m}) - d^i_{obs} \right|}{\sigma^i_D} + \sum_{\alpha \in \mathbf{I}_M} \frac{\left| m^\alpha - m^\alpha_{prior} \right|}{\sigma^\alpha_M} . \qquad (6)$$

For a reason that will become apparent in the case $p \to \infty$, instead of $S(\mathbf{m})$ I introduce the function

$$R(\mathbf{m}) = (p\, S(\mathbf{m}))^{1/p}$$

$$= \left[\sum_{i \in \mathbf{I}_D} \frac{\left| g^i(\mathbf{m}) - d^i_{obs} \right|^p}{(\sigma^i_D)^p} + \sum_{\alpha \in \mathbf{I}_M} \frac{\left| m^\alpha - m^\alpha_{prior} \right|^p}{(\sigma^\alpha_M)^p} \right]^{1/p} . \qquad (7)$$

Of course, for $p = 1$,

$$R(\mathbf{m}) = S(\mathbf{m}) . \qquad (8)$$

In our present example we have

$$R(m^1,m^2) = |m^1-m^2| + |m^1-3| + \frac{|m^2-1|}{2} . \qquad (9)$$

This function is represented in Figure 5.16. The level lines are polygons. In a 3D space with axes (m^1,m^2,R), the function $R(m^1,m^2)$ is clearly a convex polyhedron.

The minimum of R is attained at $(m^1,m^2) = (3,3)$. Some ℓ_1-norm "circles" of radius 1/8 have been drawn. The associated steepest descent directions have a pathology along the segment joining the point (1,1) to the point (3,3).

Solution for the ℓ_2 norm: We have

$$R(m^1,m^2) = \left[(m^1-m^2)^2 + (m^1-3)^2 + \frac{(m^2-1)^2}{4} \right]^{1/2} . \qquad (10)$$

This function is represented in Figure 5.17. Its minimum is attained at $(m^1,m^2) = (8/3\ ,\ 7/3)$.

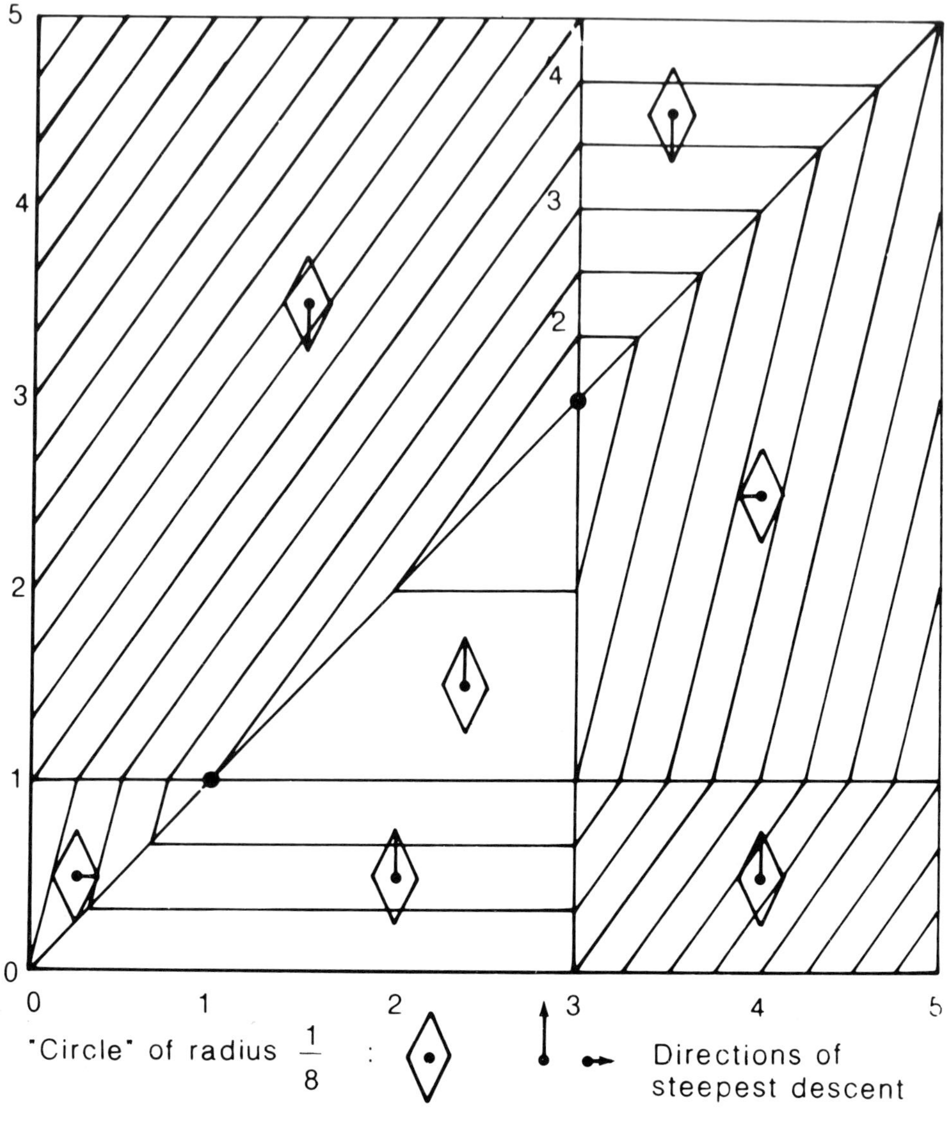
5
4
3
2
1
0
0
1
2
3
4
5
4
3
2
"Circle" of radius $\frac{1}{8}$:
Directions of steepest descent

← *Figure 5.16:* The function $R(\mathbf{m})$ is defined by

$$(R(\mathbf{m}))^p = p\, S(\mathbf{m}) = \sum_{i \in I_D} \frac{\left| g^i(\mathbf{m}) - d^i_{obs} \right|^p}{(\sigma^i_D)^p} + \sum_{\alpha \in I_M} \frac{\left| m^\alpha - m^\alpha_{prior} \right|^p}{(\sigma^\alpha_M)^p} . \tag{1}$$

The constants $\sigma_M{}^\alpha$ allow the introduction of a natural definition of distance over the model space:

$$\| \mathbf{m} \|^p = \sum_{\alpha \in I_M} \frac{\left| m^\alpha - m^\alpha_{prior} \right|^p}{(\sigma^\alpha_M)^p} . \tag{2}$$

For $p = 1$, this gives

$$R(\mathbf{m}) = S(\mathbf{m}) = \sum_{i \in I_D} \frac{\left| g^i(\mathbf{m}) - d^i_{obs} \right|}{\sigma^i_D} + \sum_{\alpha \in I_M} \frac{\left| m^\alpha - m^\alpha_{prior} \right|}{\sigma^\alpha_M} , \tag{3}$$

and

$$\| \mathbf{m} \| = \sum_{\alpha \in I_M} \frac{\left| m^\alpha - m^\alpha_{prior} \right|}{\sigma^\alpha_M} . \tag{4}$$

The figure shows the function $R(m^1,m^2)$ defined by the data given in the text. The level lines are polygons. In a 3D space with axes (m^1,m^2,R) , the function $R(m^1,m^2)$ is a convex polyhedron.

Some "circles" (for the distance (4)) or radius 1/8 are shown. The points at each side of the segment joining the point (1,1) to the point (3,3) have directions of steepest descent which are opposite. An algorithm of steepest descent will oscillate indefinitely between the two sides of the segment. Defining circles with respect to the ad-hoc ℓ_2 distance

$$\| \mathbf{m} \|^2 = \sum_{\alpha \in I_M} \frac{(m^\alpha - m^\alpha_{prior})^2}{(\sigma^\alpha_M)^2} , \tag{5}$$

will slightly ameliorate the convergence.

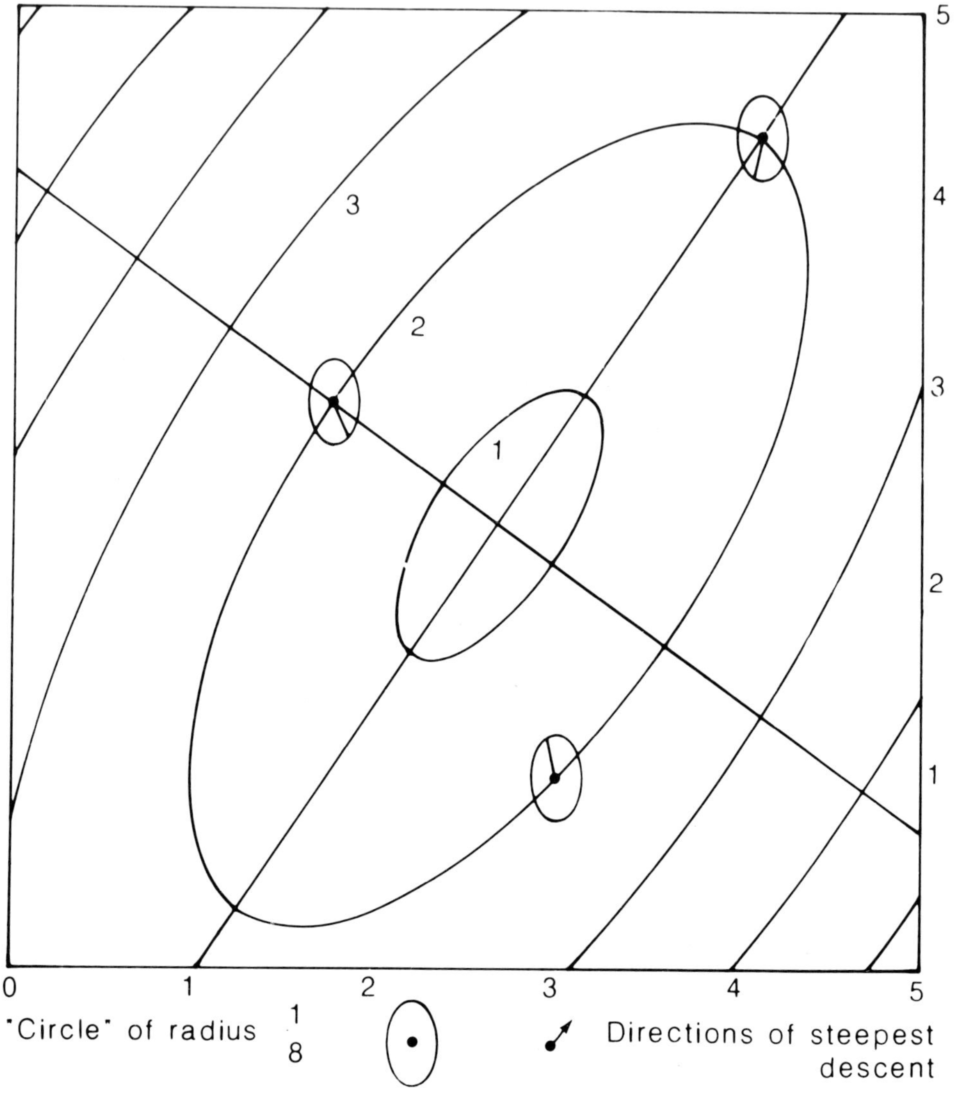

Figure 5.17: Same as Figure 5.16 for $p = 2$.

Solution for the ℓ_∞ norm: For $p \to \infty$ we have (see text)

$$R(\mathbf{m}) = \\ \mathrm{MAX}\left(\left(\frac{\left| g^i(\mathbf{m}) - d^i_{obs} \right|}{\sigma^i_D} \ , \ (i \in \mathbf{I}_D) \right) \ , \ \left(\frac{\left| m^\alpha - m^\alpha_{prior} \right|}{\sigma^\alpha_M} \ , \ (\alpha \in \mathbf{I}_M) \right) \right) \ . \tag{11}$$

In our present example, this gives

$$R(m^1,m^2) = \mathrm{MAX}\left(|m^1-m^2| \ , \ |m^1-3| \ , \ \frac{|m^2-1|}{2} \right) \ . \tag{12}$$

This function is represented in Figure 5.18. As for the ℓ_1 case, the level lines of $R(m^1,m^2)$ are polygons, and in a 3D space with axes (m^1,m^2,R) , the function $R(m^1,m^2)$ is a convex polyhedron.

The minimum of R is attained at $(m^1,m^2) = (5/2 \ , \ 2)$. Some ℓ_∞ norm "circles" of radius 1/8 have been drawn. The associated directions of steepest descent are not uniquely defined in all the polyhedron faces.

Problem 5.2: Demonstrate the equivalence

$$\hat{\mathbf{x}} = \frac{1}{p} \frac{\partial}{\partial \mathbf{x}} \| \mathbf{x} \|_p^p \qquad \Leftrightarrow \qquad \mathbf{x} = \frac{1}{q} \frac{\partial}{\partial \hat{\mathbf{x}}} \| \hat{\mathbf{x}} \|_q^q \ , \tag{1}$$

where

$$\frac{1}{p} + \frac{1}{q} = 1 \ . \tag{2}$$

Solution: The norm $\| \mathbf{x} \|_p$ is defined by

$$\| \mathbf{x} \|_p = \left(\sum_i \frac{\left| x^i \right|^p}{(\sigma^i)^p} \right)^{1/p} \ . \tag{3}$$

From

$$\hat{x}^i = \frac{1}{p} \left(\frac{\partial}{\partial x^i} \| \mathbf{x} \|_p^p \right) \tag{4}$$

it follows that

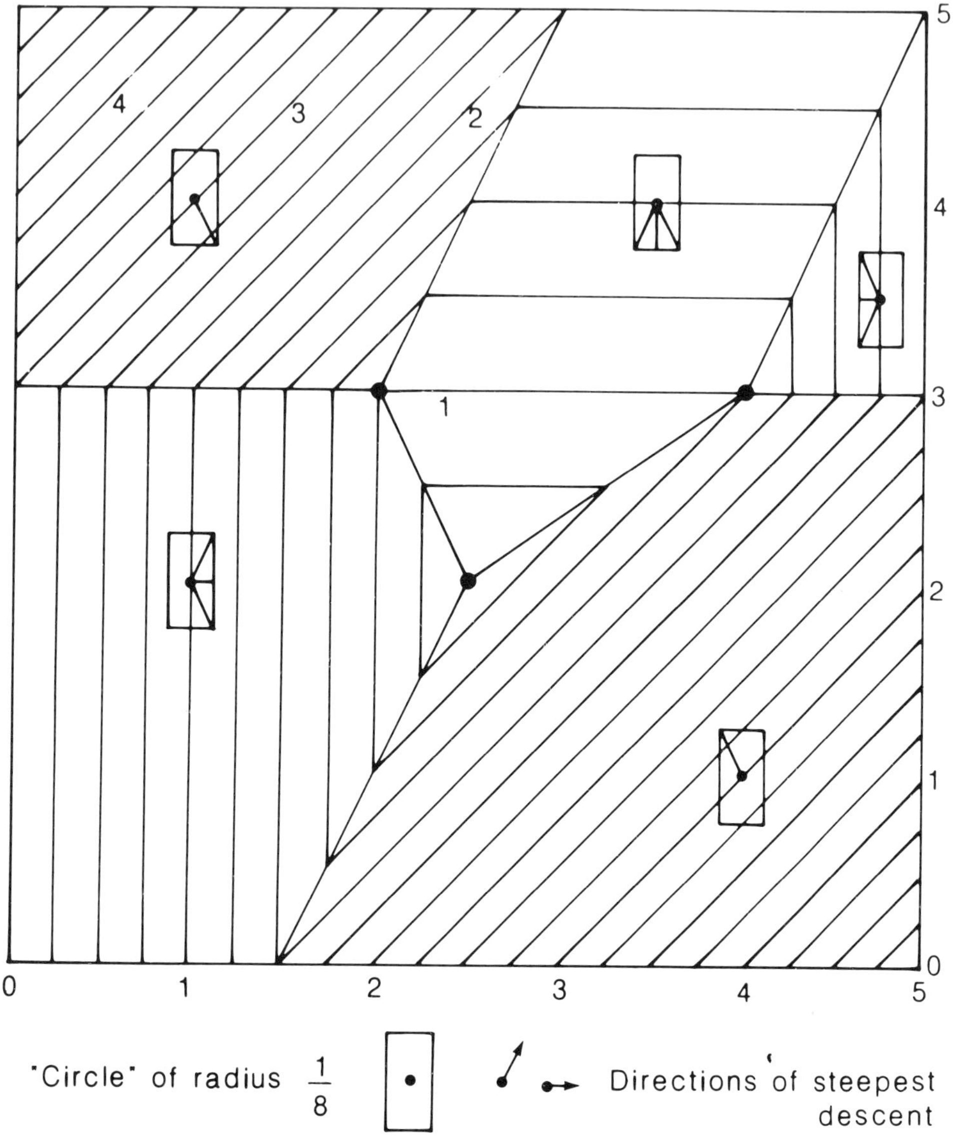
4
3
2
1
0
1
2
3
4
5
0
1
2
3
4
5
"Circle" of radius $\frac{1}{8}$
Directions of steepest descent

← *Figure 5.18:* Same as Figure 5.16 for $p = \infty$. This gives

$$R(\mathbf{m}) = \mathrm{MAX}\left(\left(\frac{\left| g^i(\mathbf{m})-d^i_{obs} \right|}{\sigma^i_D} \ , (i\in\mathbf{I}_D) \right) , \left(\frac{\left| m^\alpha - m^\alpha_{prior} \right|}{\sigma^\alpha_M} \ , (\alpha\in\mathbf{I}_M) \right) \right) .$$

Some "circles" for the distance

$$\| \mathbf{m} \| = \mathrm{MAX}\left(\frac{\left| m^\alpha - m^\alpha_{prior} \right|}{\sigma^\alpha_M} \ , (\alpha\in\mathbf{I}_M) \right) ,$$

are shown. The direction of steepest descent is not uniquely defined everywhere. Again, the use of an ad-hoc ℓ_2 norm distance simplifies the problem.

$$\hat{x}^i = \frac{sg(x^i) \left| x^i \right|^{p-1}}{(\sigma^i)^p} , \qquad (5)$$

equivalent to the two equations

$$sg(\hat{x}^i) = sg(x^i) \qquad (6a)$$

$$\left| \hat{x}^i \right| = \frac{\left| x^i \right|^{p-1}}{(\sigma^i)^p} . \qquad (6b)$$

From (6b) it follows that

$$\left| x^i \right| = (\sigma^i)^{p/(p-1)} \left| \hat{x}^i \right|^{1/(p-1)} = (\sigma^i)^q \left| \hat{x}^i \right|^{q-1} , \qquad (7)$$

and, using (6a),

$$x^i = \frac{sg(\hat{x}^i) \left| \hat{x}^i \right|^{q-1}}{(\hat{\sigma}^i)^q} , \tag{8}$$

where

$$\hat{\sigma}^i = \frac{1}{\sigma^i} . \tag{9}$$

Defining

$$\| \hat{\mathbf{x}} \|_q = \left(\sum_i \frac{\left| \hat{x}^i \right|^p}{(\hat{\sigma}^i)^q} \right)^{1/q} , \tag{10}$$

gives

$$x^i = \frac{1}{q} \left(\frac{\partial_i}{\partial \hat{x}} \| \hat{\mathbf{x}} \|_q^q \right) . \tag{11}$$

Problem 5.3: Demonstrate the identities

$$\hat{\mathbf{x}}^t \mathbf{x} = \| \mathbf{x} \|_p \cdot \| \hat{\mathbf{x}} \|_q = \| \mathbf{x} \|_p^p = \| \hat{\mathbf{x}} \|_q^q . \tag{1}$$

Solution: From equation (5) of the previous problem it directly follows that

$$\hat{\mathbf{x}}^t \mathbf{x} = \sum_i \hat{x}^i \; x^i = \| \mathbf{x} \|_p^p , \tag{2}$$

and from equation (8),

$$\hat{\mathbf{x}}^t \mathbf{x} = \| \hat{\mathbf{x}} \|_q^q . \tag{3}$$

From the identity

$$\| \mathbf{x} \|_p^p = \| \hat{\mathbf{x}} \|_q^q \tag{4}$$

already demonstrated, it follows that

$$\| \hat{\mathbf{x}} \|_q = \| \mathbf{x} \|_p^{p/q} = \| \mathbf{x} \|_p^{p-1} , \tag{5}$$

i.e.,

$$\| \hat{\mathbf{x}} \|_q \cdot \| \mathbf{x} \|_p = \| \mathbf{x} \|_p^p . \tag{6}$$

Problem 5.4: Use the simplex method of Box 5.3 to obtain the minimum of the function

$$\phi(x^1,x^2,x^3) = 2x^1 + x^2 + 2x^3 , \tag{1a}$$

under the constraints

$$\begin{aligned} x^1 + x^2 + x^3 &= 5 , \\ x^1 + 2x^2 &= 3 , \end{aligned} \tag{1b}$$

and

$$x^1 > 0 \qquad x^2 > 0 \qquad x^3 > 0 . \tag{1c}$$

Solution: In the following, the notations of Box 5.3 are used. In compact notation, the problem is to minimize

$$\phi(\mathbf{x}) = \hat{\chi}^t \mathbf{x} \tag{2a}$$

under the constraints

$$\mathbf{M} \mathbf{x} = \mathbf{y} \tag{2b}$$

$$\mathbf{x} \geq \mathbf{0} , \tag{2c}$$

where

$$\mathbf{x} = \begin{bmatrix} x^1 \\ x^2 \\ x^3 \end{bmatrix} \qquad \hat{\chi} = \begin{bmatrix} 2 \\ 1 \\ 2 \end{bmatrix} , \tag{3a}$$

and

$$\mathbf{M} = \begin{bmatrix} 1 & 1 & 1 \\ 1 & 2 & 0 \end{bmatrix} \qquad \mathbf{y} = \begin{bmatrix} 5 \\ 3 \end{bmatrix} . \tag{3b}$$

In this problem, a basis contains two parameters. We have three different possibilities for choosing the basic parameters:

First possible choice: $$\mathbf{x}_B = \begin{bmatrix} x^1 \\ x^2 \end{bmatrix} \qquad \mathbf{x}_N = [x^3] \tag{4}$$

Second possible choice: $$\mathbf{x}_B = \begin{bmatrix} x^2 \\ x^3 \end{bmatrix} \qquad \mathbf{x}_N = [x^1] \tag{5}$$

Third possible choice: $$\mathbf{x}_B = \begin{bmatrix} x^1 \\ x^3 \end{bmatrix} \qquad \mathbf{x}_N = [x^2] \; . \tag{6}$$

As discussed in Box 5.3, we will try to take the non basic parameters as null.

I arbitrarily choose first the choice (4). This gives

$$\mathbf{M}_B = \begin{bmatrix} 1 & 1 \\ 1 & 2 \end{bmatrix} \qquad \mathbf{M}_N = \begin{bmatrix} 1 \\ 0 \end{bmatrix} . \tag{7}$$

I must first check that in taking the non basic parameters as null, I satisfy the positivity constraints for the basic parameters. From equation (3) of Box 5.3,

$$\mathbf{x}_B = \mathbf{M}_B^{-1} \, \mathbf{y} = \begin{bmatrix} 7 \\ -2 \end{bmatrix} . \tag{8}$$

As the positivity constraints are not satisfied, the choice (4) is not acceptable.

Turning to the choice (5), we have

$$\mathbf{M}_B = \begin{bmatrix} 1 & 1 \\ 2 & 0 \end{bmatrix} \qquad \mathbf{M}_N = \begin{bmatrix} 1 \\ 1 \end{bmatrix} , \tag{9}$$

which gives

$$\mathbf{x}_B = \mathbf{M}_B^{-1} \, \mathbf{y} = \begin{bmatrix} 3/2 \\ 7/2 \end{bmatrix} . \tag{10}$$

As the positivity constraints are satisfied, we can now check if this possible solution is the optimum solution. We have

$$\mathbf{c}_B = \begin{bmatrix} 1 \\ 2 \end{bmatrix} \qquad \mathbf{c}_N = [2] , \tag{11}$$

and, using equation (5) of Box 5.3,

$$\gamma_N = \mathbf{c}_N - \mathbf{M}_N{}^t \, \mathbf{M}_B{}^{-t} \, \mathbf{c}_B = [1/2] . \tag{12}$$

As all the components of γ_N (we only have one) are positive, the function ϕ is effectively minimized taking $\mathbf{x}_N = \mathbf{0}$, and (10) is the solution of the problem.

The problem is now completely solved, but let us see what happens when using the choice (6). This gives

$$\mathbf{M_B} = \begin{bmatrix} 1 & 1 \\ 1 & 0 \end{bmatrix} \qquad \mathbf{M_N} = \begin{bmatrix} 1 \\ 2 \end{bmatrix} , \tag{13}$$

and

$$\mathbf{x_B} = \mathbf{M_B}^{-1} \mathbf{y} = \begin{bmatrix} 3 \\ 2 \end{bmatrix} , \tag{14}$$

which is acceptable. We have

$$\mathbf{c_B} = \begin{bmatrix} 2 \\ 2 \end{bmatrix} \qquad \mathbf{c_N} = [1] , \tag{15}$$

and, using equation (5) of box 6.1,

$$\gamma_N = \mathbf{c_N} - \mathbf{M_N}^t \, \mathbf{M_B}^{-t} \, \mathbf{c_B} = [-1] . \tag{16}$$

As all the components of γ_N are not positive, (14) is not the solution of the problem. The parameter associated with the most negative component of γ_N must leave the basis (as we only have the parameter, x^2 , in $\mathbf{x_N}$, this is the parameter). Equation (3a) of Box 5.3 gives

$$\mathbf{x_B} = \mathbf{M_B}^{-1} (\mathbf{y} - \mathbf{M_N} \, \mathbf{x_N}) = \begin{bmatrix} 3-2x^2 \\ 2+x^2 \end{bmatrix} , \tag{17}$$

As the first component of $\mathbf{x_B}$, x^1 , first becomes negative when increasing x^2 , it is the parameter x^1 that must replace x^2 in the basis. This leads directly to the choice (5), which has already been explored (and which gives the solution).

Problem 5.5 (problem 1.9 revisited): Two variables y and t are related through a linear relationship

$$y = a\,t + b . \tag{1}$$

In order to estimate the parameters a and b , the 11 experimental points (y^i, t^i) : (10,1), (11,3), (11,5), (12,7), (13,10), (14,12), (14,13), (15,15), (15,16), (16,18), (2,31), shown in Figure 5.19, have been obtained. Find the straight line which best fits the points in the ℓ_1-norm sense.

Solution: We wish to minimize

$$S(a,b) = \sum_{i=1}^{i=11} \frac{\left| y^i_{cal} - y^i_{obs} \right|}{\sigma^i} , \tag{2}$$

where

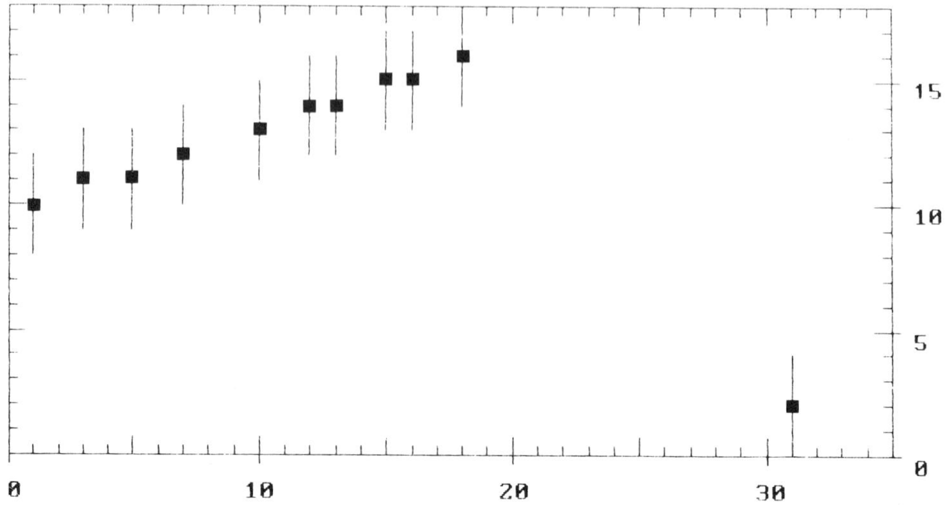

Figure 5.19: We wish to find the straight line fitting these 11 points (same as Figure 1.26). Notice the outlier.

$$y^i_{cal} = a\,t^i + b \;, \tag{3}$$

and $\sigma^i = \sigma = 2$.

Figure 5.20 shows the projection on the plane (a,b) of the convex polyhedron representing S(a,b) . Figures 5.21 and 5.22 zoom the region of interest. The level lines S(a,b) = 12.5 and S(a,b) = 25 are shown in Figure 5.20 and the level line S(a,b) = 12.5 is shown in Figure 5.21.

Each line of the net corresponds to an experimental point. For instance, the most horizontal line corresponds to the (probable outlier) point (2,31). The minimum of S is necessarily attained at a vertex of the polyhedron, i.e., at a knot in the mesh in Figures 5.20-5.22. At least two experimental points are exactly fitted at each knot.

To solve the problem using linear programing techniques, we first choose one knot, i.e., two of the equations

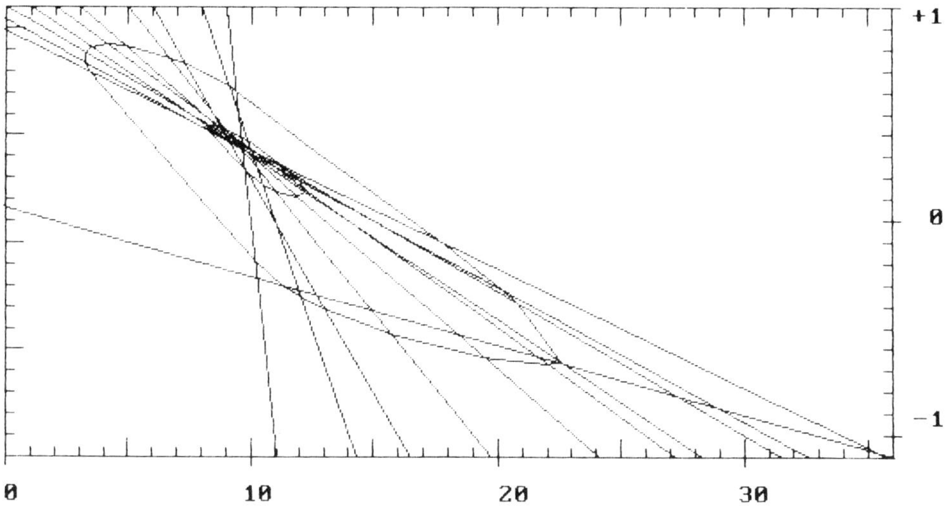

Figure 5.20: Projection on the parameter space (a,b) of the convex polyhedron representing the misfit function S(a,b). The level lines S = 12.5 and S = 25. are shown.

$$y^i_{obs} = a\, t^i + b \ , \qquad (4)$$

for instance, the first two. This gives the point (0.5,9.5) . To leave the knot, we have to drop one of the basis equations. The FIFO method drops the "oldest" equation. At the first round equations have arbitrary ages. Take for instance ages decreasing from the first to the last equation. Dropping the first equation gives the line shown in Figure 5.21. The minimum along the line is attained at the point (0.308,10.08) , where the 9-th equation enters the basis. Dropping the oldest equation (the second) we get a minimum at the point (0.333,9.67) , where equations 1, 4, and 7 are exactly satisfied and they can all enter the basis. Whatever the choice we make, it leaves the basis again, because we cannot make S diminish; we are at the minimum.

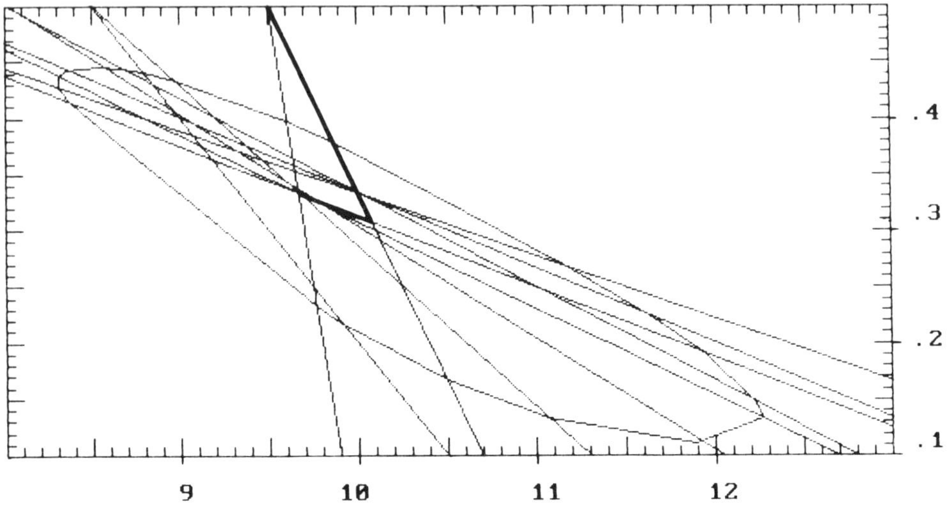

Figure 5.21: Zoom of figure 5.20. The solution path is shown (see text).

Problem 5.6: Figure 5.23 shows 5 points on a euclidian plane. Ten different distances between these points have been measured experimentally. The the results are as follows.

Observed distance	Estimated error	True (unknown) error
D^0 = 9 486.843 0 m	±2 cm	+1 cm
D^1 = 15 000.010 0 m	±2 cm	+1 cm
D^2 = 12 727.902 1 m	±2 cm	-2 cm
D^3 = 6 708.203 9 m	±2 cm	0 cm
D^4 = 6 708.193 9 m	±2 cm	-1 cm
D^5 = 11 998.000 0 m	±2 cm	-200 cm
D^6 = 6 708.193 9 m	±2 cm	-1 cm
D^7 = 10 816.653 8 m	±2 cm	0 cm
D^8 = 9 486.853 0 m	±2 cm	+2 cm
D^9 = 6 708.223 9 m	±2 cm	+2 cm

Observe that estimated errors are uniform, $\sigma^i = 2$ cm , and that distance D^5

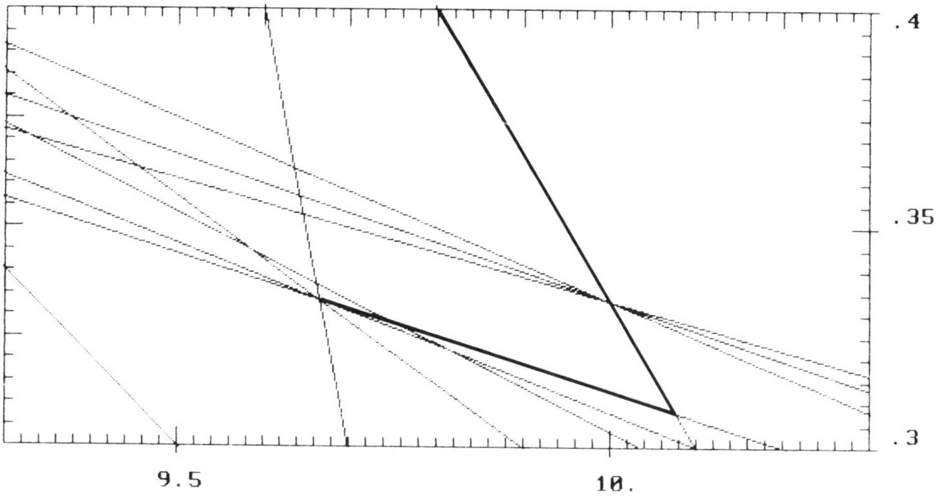

Figure 5.22: Zoom of Figure 5.21.

is an outlier.

To define the geometric figure perfectly, only 7 distances are needed, but, as is usual in geodetic measurements, some redundant measurements have been made, in order to minimize posterior true errors. Due to the observational errors, the 10 distances obtained are not compatible with the geometric constraints. Obtain the new set of distances D^i , compatible with these geometric constraints, minimizing the ℓ_1 norm

$$S_1 = \sum_{i=0}^{i=9} \frac{\left| D^i - D^i_{obs} \right|}{\sigma^i} .$$

Compare with the solution of the minimization of the ℓ_2 norm

$$S_2 = \sum_{i=0}^{i=9} \frac{\left(D^i - D^i_{obs} \right)^2}{(\sigma^i)^2} .$$

Solution: As the expected corrections are small, the problem can be linearized. The ℓ_1-norm minimization problem can be solved using the linear

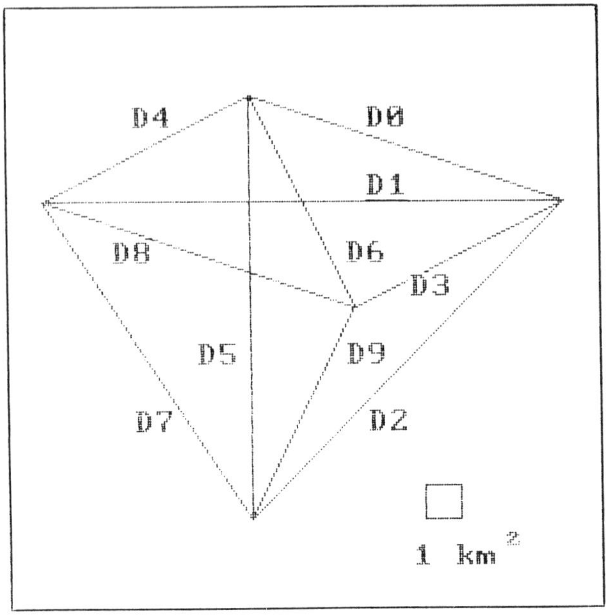

Figure 5.23: Between 5 points of a euclidean plane, the 10 distances shown in the figure have been measured. One is an outlier (see text). Estimate the true distances between the points.

programming methods, and the ℓ_2-norm minimization problem using a Newton's method. The following table compares the true errors with the corrections predicted by the two methods.

True errors	ℓ_1-norm corrections	ℓ_2-norm corrections
+1.00 cm	0.00 cm	+14.22 cm
+1.00 cm	-0.89 cm	-19.21 cm
-2.00 cm	0.00 cm	+4.64 cm
0.00 cm	0.00 cm	+2.72 cm
-1.00 cm	0.00 cm	+39.79 cm
-200.00 cm	-195.70 cm	-66.49 cm
-1.00 cm	0.00 cm	+49.41 cm
0.00 cm	0.00 cm	+35.94 cm
+2.00 cm	0.00 cm	-38.28 cm
+2.00 cm	+4.31 cm	+37.24 cm

Notice that the ℓ_1-norm solution exactly satisfies 7 distances (the number of independent data). As expected, the ℓ_1-norm criterion allows an easy identification of the outlier, while the ℓ_2-norm criterion smears the error across the whole geodetic network.

CHAPTER 6

THE GENERAL PROBLEM

When in doubt, smooth.

Sir Harold Jeffreys (Quoted by Moritz, 1980)

Many inverse problems involve functions: the data set sometimes consists in recordings as a function of time or space, and the main unknown in the parameter set sometimes consist in a function of the spatial coordinates and/or of time. Nevertheless, there are two kinds of arguments suggesting that the Inverse Theory could be limited to finite-dimensional problems:

i) The "technological" argument: data "functions" are recorded digitally (or, if they are recorded analogically, they have a finite bandwidth and, thus, a finite amount of information (Shannon, 1948)). The "functions" used to describe the model are handled by digital computers which can only consider a finite amount of information.

ii) The "mathematical" argument: central in the theory of inverse problems is the concept of probability. We will see in this chapter that the concept of probability density function, so essential for finite-dimensional spaces, cannot be generalized to infinite-dimensional spaces. Also, in a function space, a probability can only be defined over certain subsets, named cylinder sets, whose elements have, in fact, a finite number of degrees of freedom.

For the general inverse problem, these arguments prevail. From both the technological and mathematical points of view, after a proper understanding of all the subtleties of infinite-dimensional spaces, practical applications can only be developed over discretized problems (with finite numbers of degrees of freedom). It is only for particular cases (e.g., Gaussian assumption) that the functional approach reveals itself to be tremendously more powerful (both practically and intellectually) than the discretized approach, as demonstrated in Chapter 7.

After introducing the concept of a (infinite-dimensional) random process, I examine in this chapter the extent to which the general theory of Chapter 1 can be generalized for function spaces.

6.1: Random processes in function spaces

6.1.1: Description of a random function.

Looking at Figure 6.1, we can understand what is generally meant by a *random function* . The figure represents some of its *realizations*. Each realization is a (ordinary) function $t \rightarrow x(t)$ of the real variable t , for $t_{min} \leq t \leq t_{max}$. Clearly, to characterize an arbitrary realization of such a random function perfectly, an infinity of values $x(t)$ have to be defined. Each realization of a random function may be viewed abstractly as a point in a properly defined infinite-dimensional space (named a *function space*, or *space of functions*).

In the first chapter we saw the usefulness of the introduction of probability density functions over finite dimensional parameter spaces. Unfortunately, there does not exist any simple generalization of the concept of probability density functions over infinite dimensional spaces.

Let us denote the random function under consideration by $t \rightarrow X(t)$, and any particular realization by $t \rightarrow x(t)$. For given t , $X(t)$ is an ordinary random variable. Its probability density function is denoted $f(x;t)$.

If the probability density functions $f(x;t)$ are known for any t , central estimators or estimators of dispersion of the random function can be computed. For instance, the *mean value* of the random function is the (nonrandom) function $m_2(t)$ defined by

$$m_2(t) = \int dx \ x \ f(x;t) \ , \qquad (6.1)$$

while the *median value* $m_1(t)$ verifies

$$\int_{-\infty}^{m_1(t)} dx \ f(x;t) = \int_{m_1(t)}^{\infty} dx \ f(x;t) = \frac{1}{2} \ . \qquad (6.2)$$

The standard deviation $\sigma_2(t)$ and the mean deviation $\sigma_1(t)$ are then respectively given by (see Chapter 1)

$$\sigma_2(t)^2 = \int dx \ (x-m_2(t))^2 \ f(x;t) \qquad (6.3)$$

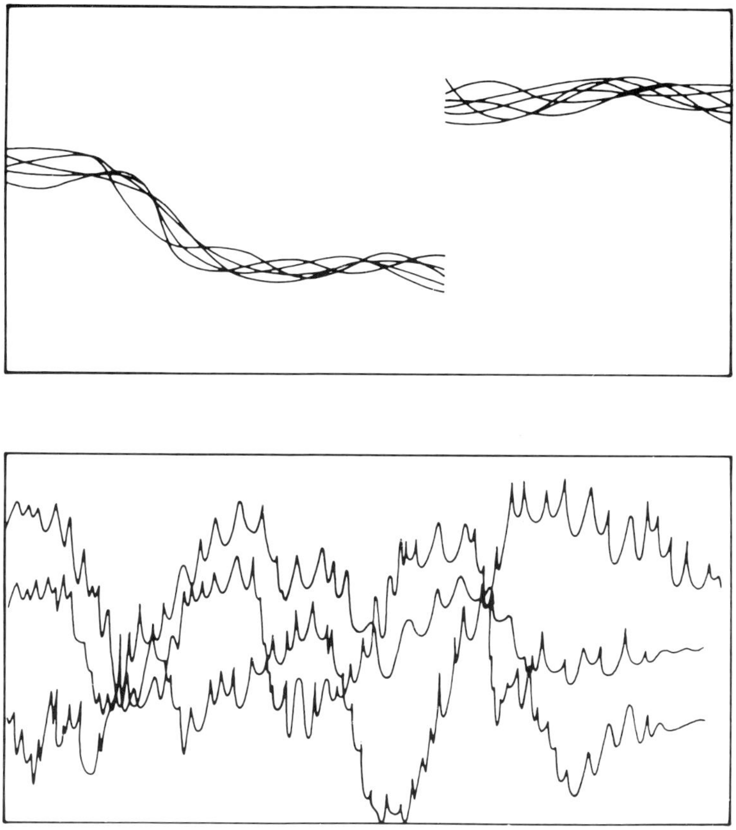

Figure 6.1: Schematic illustration of some realizations of two different random functions.

$$\sigma_1(t) \;=\; \int dx \;\; | \; x-m_1(t) \; | \;\; f(x;t) \; . \qquad (6.4)$$

Let us now consider the behaviour of the random function at two different points t_1 and t_2 . Clearly, the knowledge of the probability density function $f(x;t)$ for any t does *not* give any information on the *correlations* between the random variables $X(t_1)$ and $X(t_2)$. This information is given by the *two-dimensional* joint probability density function $f(x_1,x_2;t_1,t_2)$. If the two-dimensional probability density function $f(x_1,x_2;t_1,t_2)$ is known for any t_1 and t_2 , the correlations are known perfectly. For instance, the *covariance function* $C_2(t,t')$ of the random function is defined as the covariance (in the usual sense) between the random variables $X(t_1)$ and $X(t_2)$:

$$C(t,t') = \int dx \int dx' \ (x-m_2(t)) \ (x'-m_2(t')) \ f(x,x';t,t') . \tag{6.5}$$

It can be shown that a covariance function is symmetric:

$$C(t,t') = C(t',t) , \tag{6.6}$$

and definite nonnegative:

$$\int_{t_{min}}^{t_{max}} dt \int_{t_{min}}^{t_{max}} dt' \ \phi(t) \ C(t,t') \ \phi(t') \geq 0 \qquad \text{for any function } \phi(t) . \tag{6.7}$$

The knowledge of the two-dimensional probability density function $f(x,x';t,t')$ does not give any information on moments of order higher than two. For a complete characterization of a random function, we need to know joint probability densities of any order. Thus, from a mathematical point of view,

> a random function $t \rightarrow X(t)$ is perfectly characterized
> if the n-dimensional probability density function
> $f(x_1,...,x_n;t_1,...,t_n)$
> is known, *for any points* $t_1,...,t_n$ *and for any value of n.*

From a practical point of view, a random function $t \rightarrow X(t)$ is conveniently characterized if the n-dimensional probability density function $f(x_1,...,x_n;t_1,...,t_n)$ is known for, say, equally spaced points $t_1,...,t_n$ and for a sufficiently high value of *n*. Practically, the value of n corresponds to the number of points which are needed for a reasonably accurate representation of any realization of the random function. This implies, of course, that the realizations are sufficiently regular. For instance, a realization of pure white noise (see Section 7.1.3) cannot be described by a finite number of points. Physical white noise is always somewhat colored (finite bandwidth) and can always be described by a finite (sufficiently high) number of points.

Example 6.1: Gaussian random function with given mean value and covariance function. A random function $t \rightarrow X(t)$ is termed Gaussian (or normal) with mean value $m(t)$ and covariance function $C(t,t')$ if, for any points $t_1,...,t_n$, and for any n , the joint probability density function of the random variables $X(t_1),...,X(t_n)$ is

$$f(x_1,...,x_n;t_1,...,t_n) = \frac{1}{(2\pi)^{n/2} \det^{1/2} C} (\mathbf{x}-\mathbf{m})^t \mathbf{C}^{-1} (\mathbf{x}-\mathbf{m})$$

where

$$\mathbf{x} = \begin{bmatrix} x_1 \\ \cdots \\ x_n \end{bmatrix} \quad \mathbf{m} = \begin{bmatrix} m(t_1) \\ \cdots \\ m(t_n) \end{bmatrix} \quad \mathbf{C} = \begin{bmatrix} C(t_1,t_1) & \cdots & \cdots \\ \cdots & \cdots & \cdots \\ \cdots & \cdots & C(t_n,t_n) \end{bmatrix} .$$

■

6.1.2: Computing probabilities.

Let $t \to X(t)$ represent a random function, and assume that we have a perfect description of it, i.e., that for any points $t_1,...,t_n$ and for any n, the joint probability density function $f(x_1,...,x_n;t_1,...,t_n)$ of the random variables $X(t_1),...,X(t_n)$ is known. We wish here to compute the probability of a realization $x(t)$ of the random function $X(t)$ being within certain limits. More precisely, for each value of t, let us introduce a numerical set $E(t)$. We wish to compute the probability for $x(t)$ to verify (see Figure 6.2)

$$x(t) \in E(t) . \tag{6.8}$$

The set of all the possible realizations of the random function verifying equation (6.8) will be denoted by $\mathbf{A}$. Following Pugachev (1965), let us first consider the subset $\mathbf{A}_n$ of the set of all the possible realizations of the random function for which the values associated with the points $t_1,...,t_n$ belong to the numerical sets $\mathbf{E}(t_1),...,\mathbf{E}(t_n)$. As explained in Box 6.2, $\mathbf{A}_n$ is a "cylinder set". By definition of a probability density, the probability of the set A_n is clearly

$$P(A_n) = \int_{E(t_1)} dx_1 \cdots \int_{E(t_n)} dx_n \; f(x_1,...,x_n;t_1,...,t_n) . \tag{6.9}$$

For any positive integer value of n we now consider a partition of the range of variation of the argument t into n sub-intervals such that the length of each of the n subintervals tends to zero as $n \to \infty$. Now, from each subinterval we choose a value of t such that the points $t_1,...,t_n$ contain, for every n, all the points of the preceding partitions. Thus, we get a sequence $\mathbf{A}_1$, $\mathbf{A}_2$, ... of sets of realizations of X each including all the subsequent sets. If the realizations of the random function are sufficiently regular, we can assume that the product of all the sets A_n coincides with the initial set A . The numbers $P(A_1)$, $P(A_2)$,... form a monotonic non-increasing progression of non-negative numbers. Then the limit

$$P(A) = \lim_{n\to\infty} P(A_n) \tag{6.10}$$

exists and corresponds to the probability of a realization of the random function verifying (6.8).

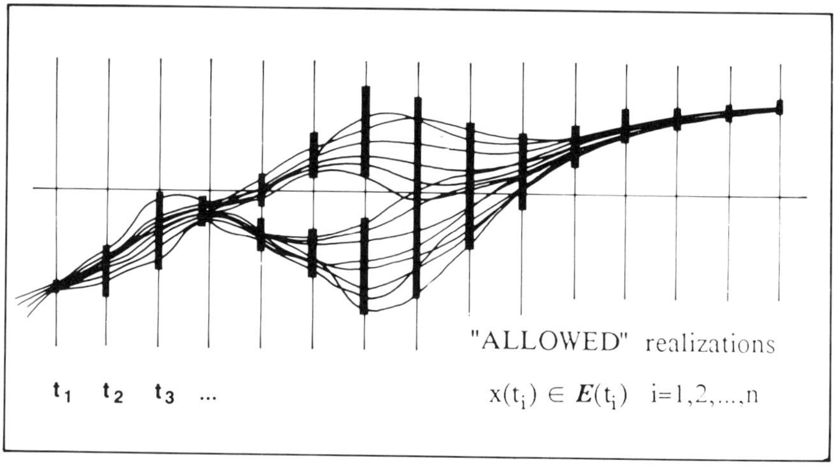

Figure 6.2: Given a random function, it is possible to compute the probability of a realization being within given bounds (see text).

Box 6.1: The covariance operator as a scalar product over the dual space.

A random function $t \to X(t)$ defines, for each value of t, the random variable $X(t)$. Another useful notation for a random function is $x(t;\omega)$, where ω is a random variable each realization of which, say $\omega = \omega_0$, gives the realization $x(t,\omega_0)$.

Let $\mathcal{X}$ denote the functional space of the realizations of the random function and $\hat{\mathcal{X}}$ its dual. For any $\hat{\chi}\in\hat{\mathcal{X}}$, $\langle\ \hat{\chi}\ ,\ \mathbf{x}(\omega)\ \rangle$ defines a random variable.

(...)

The *mathematical expectation* $E_x(t)$ of a random function $x(t;\omega)$ is the (non-random) function defined by

$$\forall\ \hat{\chi} \in \hat{X} \qquad \langle\ \hat{\chi}\ ,\ E_x\ \rangle\ = E(\langle\ \hat{\chi}\ ,\ \mathbf{x}(\omega)\ \rangle\ \ , \tag{1}$$

where $E(\ \bullet\ )$ denotes the mathematical expectation of an ordinary random variable. Formally,

$$E(\ \langle\ \hat{\chi}\ ,\ \mathbf{x}(\omega)\ \rangle\ \ = \int_{\Omega} dP(\omega) \int_{t_{min}}^{t_{max}} dt\ \hat{\chi}(t)\ x(t;\omega)$$

$$= \int_{t_{min}}^{t_{max}} dt\ \hat{\chi}(t) \int_{\Omega} dP(\omega)\ x(t;\omega)$$

i.e.,

$$E_x(t) = \int_{\Omega} dP(\omega)\ x(t;\omega)\ . \tag{2}$$

A *scalar product* over $\hat{X}$, denoted $(\hat{x}_1,\hat{x}_2)$ or $C(\hat{x}_1,\hat{x}_2)$ may be defined by

$$(\hat{x}_1,\hat{x}_2) = C(\hat{x}_1,\hat{x}_2) = E\{\langle\ \hat{x}_1\ ,\ (\mathbf{x}(\omega)-E_x)\ \rangle\ \langle\ \hat{x}_2\ ,\ (\mathbf{x}(\omega)-E_x)\ \rangle\ \}\ . \tag{3}$$

This gives

$C(\hat{x}_1,\hat{x}_2) =$

$$= \int_{\Omega} dP(\omega) \int_{t_{min}}^{t_{max}} dt\ \hat{x}_1(t)\ (x(t;\omega)-E_x(t)) \int_{t_{min}}^{t_{max}} dt'\ \hat{x}_2(t')\ (x(t';\omega)-E_x(t'))$$

$$= \int_{t_{min}}^{t_{max}} dt\ \hat{x}_1(t) \int_{t_{min}}^{t_{max}} dt'\ \hat{x}_2(t') \int_{\Omega} dP(\omega)\ (x(t;\omega)-E_x(t))\ (x(t';\omega)-E_x(t'))\ .$$

Defining the *covariance function* $C(t,t')$ by

(...)

$$C(t,t') = \int_{\Omega} dP(\omega)\ (x(t;\omega)-E_x(t))\ (x(t';\omega)-E_x(t')) \tag{4}$$

gives

$$C(\hat{x}_1,\hat{x}_2) = \int_{t_{min}}^{t_{max}} dt \int_{t_{min}}^{t_{max}} dt'\hat{x}_1(t)\ \ C(t,t')\ \hat{x}_2(t')\ . \tag{5}$$

Box 6.2: Cylindrical measures.

In this box, I closely follow Balakrishnan (1976), but I consider general linear spaces instead of Hilbert spaces, and I use a simpler (and less rigorous) language.

Let $\boldsymbol{X}$ be a (separable) real vector space, $\hat{\boldsymbol{X}}$ its dual, $\hat{x}_1,...,\hat{x}_n$ n elements (distinct or not) in $\hat{\boldsymbol{X}}$, and $\boldsymbol{B}$ a (Borel) set in $\boldsymbol{R}^n$. The duality product of $\hat{x}_1 \in \hat{\boldsymbol{X}}$ and $x_2 \in \boldsymbol{X}$ is denoted $\langle\ \hat{x}_1\ ,\ x_2\ \rangle$.

A *cylinder set of base* $\boldsymbol{B}$ is the set of all $x \in \boldsymbol{X}$ such that $\langle\ \hat{x}_1\ ,\ x\ \rangle\ ,...,\ \langle\ \hat{x}_n\ ,\ x\ \rangle$ belongs to $\boldsymbol{B}$. Figure 6.3 explains the use of the term "cylinder" for such a set.

Let $\hat{\boldsymbol{X}}^n$ denote the finite-dimensional subspace generated by the elements $\hat{x}_1,...,\hat{x}_n$ and let $\nu_n(\ \bullet\ )$ be a countably additive probability measure on the (σ-algebra of Borel) subsets of $\hat{\boldsymbol{X}}^n$ (i.e., an ordinary probability over the finite-dimensional space $\hat{\boldsymbol{X}}^n$). Let Z be an arbitrary set in $\boldsymbol{X}$ with base $\boldsymbol{B}$ in $\hat{\boldsymbol{X}}^n$. Given $\hat{\boldsymbol{X}}^n$ and $\nu_n(\ \bullet\ )$, a *cylinder set measure* $\mu(\ \bullet\ )$ is defined over the infinite-dimensional space $\boldsymbol{X}$ by

$$\mu(Z) = \nu_n(\boldsymbol{B})\ . \tag{1}$$

A cylinder set measure is sometimes called a *weak distribution.*

One of the properties of a measure (over a finite-dimensional space) is that if $\boldsymbol{A}_1$, $\boldsymbol{A}_2$, ... represents a disjoint sequence of sets, then

$$P(\cup\ \boldsymbol{A}_i) = \Sigma\ P(\boldsymbol{A}_i)\ .$$

It can be shown that a cylindrical measure only has this property for a *finite* number of sets.

(...)

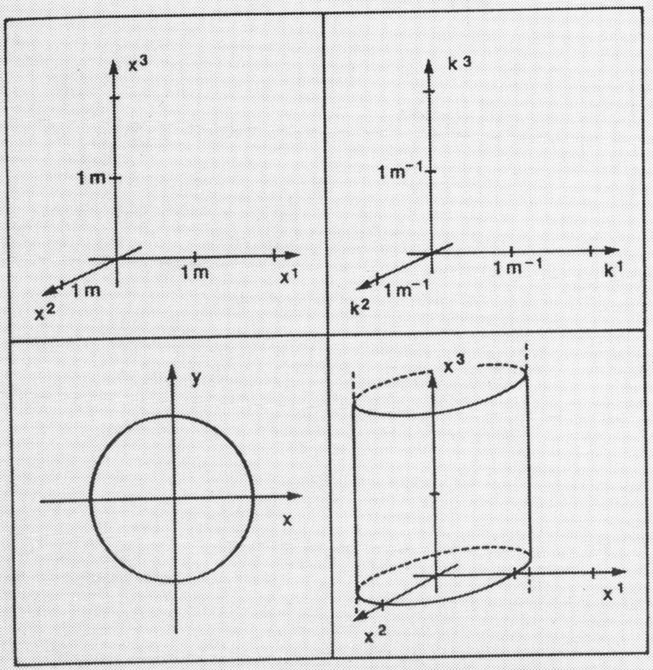

Figure 6.3: This figure illustrates the abstract definition of a "cylinder set". The space in consideration is the euclidean three-dimensional linear vector space E^3. Each vector $\mathbf{x} \in E^3$ has components $\mathbf{x} = (x^1,x^2,x^3)$ with the physical dimension of a length. The dual of E^3 is K^3, a three-dimensional linear vector space with elements $\mathbf{k} = (k^1,k^2,k^3)$ such that $\langle \mathbf{k} , \mathbf{x} \rangle = k^1x^1+k^2x^2+k^3x^3$ is a real number. The physical dimensions of the components of $\mathbf{k}$ are the inverse of a length ($\mathbf{k}$ can be seen as the Fourier dual variable associated with $\mathbf{x}$). I choose in K^3 the two particular vectors $\mathbf{k}_1 = (1\ m^{-1},0,0)$ and $\mathbf{k}_2 = (0,1\ m^{-1},0)$ and, letting (x,y) denote a pair of real numbers, I choose in R^2 the Borel set B defined by $x^2+y^2 = 1$. The corresponding cylinder set in E^3 is the set of the vectors $\mathbf{x}$ such that $(\langle \mathbf{k}_1 , \mathbf{x} \rangle , \langle \mathbf{k}_2 , \mathbf{x} \rangle) \in B$, i.e., $(x^1)^2+(x^2)^2 = 1\ m^2$. The top of the figure represents the physical spaces E^3 and K^3. The bottom left represents the Borel set B in R^2, and the bottom right represents the corresponding cylinder set in E^3.

(...)

Characteristic function: The concept of probability density function in finite-dimensional spaces does not generalize into infinite-dimensional spaces, but the concept of characteristic function does. Let $\mu(\,\cdot\,)$ denote a cylinder set measure over a linear vector space $\boldsymbol{X}$. For any (Borel) set $\boldsymbol{B}$ in $\boldsymbol{R}$ and for any $\hat{\mathbf{x}} \in \hat{\boldsymbol{X}}$ (dual of $\boldsymbol{X}$), we can define the countably additive probability measure on $\boldsymbol{R}$

$$\nu(\boldsymbol{B}) = \mu([\,\mathbf{x}' \mid \langle \hat{\mathbf{x}},\mathbf{x}' \rangle \in \boldsymbol{B}\,])\,. \tag{2}$$

For any $\hat{\mathbf{x}} \in \hat{\boldsymbol{X}}$ we can then define the integral

$$\kappa(\hat{\mathbf{x}}) = \int_{\boldsymbol{B}} \mathrm{d}\mu \;\; e^{i\langle \hat{\mathbf{x}},\mathbf{x}' \rangle} \tag{3}$$

by

$$\kappa(\hat{\mathbf{x}}) = \int_{\boldsymbol{B}} \mathrm{d}\mu \;\; e^{i\langle \hat{\mathbf{x}},\mathbf{x}' \rangle} = \int_{-\infty}^{\infty} \mathrm{d}\nu \;\; e^{i\langle \hat{\mathbf{x}},\mathbf{x}' \rangle}\,. \tag{4}$$

$\kappa(\,\cdot\,)$ is termed the *characteristic function* corresponding to the measure μ. It can be shown that a characteristic function completely characterizes the measure μ. For instance, the characteristic function of a Gaussian cylinder set measure with mean $\mathbf{m}$ and covariance operator $\mathbf{C}$ is (Balakrishnan, 1976)

$$\kappa(\hat{\mathbf{x}}) = e^{\,i\langle \hat{\mathbf{x}},\mathbf{m} \rangle - \frac{1}{2}\langle \hat{\mathbf{x}},\mathbf{C}\hat{\mathbf{x}} \rangle}\,, \tag{5}$$

which is identical to the expression corresponding to the usual finite-dimensional case.

Examples of cylinder sets: Choosing as n elements in the dual space the "functions"

$$\hat{\mathbf{x}}_i = \delta(t-t_i)\,,\; (i=1,\ldots,n)\,,$$

gives

$$\langle \hat{\mathbf{x}}_i,\mathbf{x} \rangle = \int_{t_{min}}^{t_{max}} dt \;\; \delta(t-t_i)\;\; x(t) = x(t_i)\,,$$

(...)

and choosing for each value of i a numerical set $E(t_i)$, defines the cylinder set

$$h(t_i) \in E(t_i) . \tag{6}$$

Choosing as n elements in the dual space the box car functions

$$\hat{x}_i = f_i(t) = \begin{cases} 1 & \text{if } \alpha_i \le t < \beta_i \\ 0 & \text{otherwise} \end{cases} \qquad (i=1,...,n)$$

gives

$$\langle \hat{x}_i , x \rangle = \int_{t_{min}}^{t_{max}} dt \ \hat{x}_i(t) \ x(t) = \int_{\alpha_i}^{\beta_i} dt \ x(t)$$

and defines the cylinder set

$$\int_{\alpha_i}^{\beta_i} dt \ x(t) \in E(t_i) . \tag{7}$$

We see thus that a choice of a sequence of elements of the dual space defines, in fact, a particular discretization of the (primal) space under consideration.

6.1.3: Does a covariance function give a convenient description of a random function?

The response is:

in general, **NO** .

Pugachev (1965), for instance, shows two examples of random functions with completely different realizations but having exactly the same mean function and covariance function (examples 2 and 3, pages 202 and 203).

This should not be surprising, because this is only the equivalent, in infinite-dimensional spaces, of the well-known fact that one-dimensional random variables with completely different probability densities may well have the same mean and the same variance. The problem is that when the number of dimensions under consideration is high, it is not easy to obtain an intuitive idea of what the probability densities look like.

As we will see in the next chapter, least-squares methods may be used when all the probabilities under consideration are Gaussian. A Gaussian probability is defined when the mean function and the covariance function of the random function are known. When *only* the mean function and the covariance function of a random function are known, it is tempting to use least squares, but this may be very misleading. Unfortunately, for highly-dimensioned problems, only least-squares lead to calculable solutions (with present-day computers).

6.1.4: General random processes.

We have seen that at each value of t, a random function $t \to X(t)$, defines a random variable. Sometimes we have to consider two random functions $t \to X^1(t)$ and $t \to X^2(t)$ simultaneously. Each random function is individually characterized by a joint probability density $f^1(x^1{}_1,...,x^1{}_n;t_1,...,t_n)$ and $f^2(x^2{}_1,...,x^2{}_m;t'_1,...,t'_m)$ as indicated in Section 6.1.1. In addition, we need now to characterize the correlations between the random variables $X^1(t)$ and $X^2(t)$. This is made by the joint probability density

$$f^{12}(x^1{}_1,...,x^1{}_n,x^2{}_1,...,x^2{}_m;t_1,...,t_n,t'_1,...,t'_m)$$

from which the marginal probability densities

$$f^1(x^1{}_1,...,x^1{}_n;t_1,...,t_n) =$$

$$= \int dx^2{}_1 \dots \int dx^2{}_m \; f^{12}(x^1{}_1,...,x^1{}_n,x^2{}_1,...,x^2{}_m;t_1,...,t_n,t'_1,...,t'_m) \qquad (6.11)$$

$$f^2(x^2{}_1,...,x^2{}_m;t'_1,...,t'_m) =$$

$$= \int dx^1{}_1 \dots \int dx^1{}_n \; f^{12}(x^1{}_1,...,x^1{}_n,x^2{}_1,...,x^2{}_m;t_1,...,t_n,t'_1,...,t'_m) \qquad (6.12)$$

can be computed. The reader will easily generalize to that case the definitions of mean value, median value, standard deviation, or mean deviation as given in Section 6.1.1. The only new definition which is needed is the cross-covariance between the two random functions:

$$C^{12}(t,t') = \int dx^1 \int dx^2 \; (\, x^1 - m_2{}^1(t)\,)\; (x^2 - m_2{}^2(t')\,)\; f(x^1,x^2;t,t') \,. \qquad (6.13)$$

We then have a matrix of covariance functions:

$$C(t,t') = \begin{bmatrix} C^{11}(t,t') & C^{12}(t,t') \\ C^{21}(t,t') & C^{22}(t,t') \end{bmatrix} \tag{6.14}$$

which is symmetric

$$C^{ij}(t,t') = C^{ji}(t',t) , \tag{6.15}$$

and definite non-negative

$$\sum_i \sum_j \int_{t_{min}}^{t_{max}} dt \int_{t_{min}}^{t_{max}} dt' \; \phi^i(t) \; C^{ij}(t,t') \; \phi^j(t') \geq 0 \qquad \text{for any} \quad \phi^i(t) . \tag{6.16}$$

In more general problems, instead of two random functions, we have to consider an arbitrary number. Their variables, in turn, may be multidimensional, and may be different. We may have to consider not only random functions, but simultaneously also some discrete random variables as well. The notations may become complicated in detail, but there is no conceptual difficulty.

From an abstract point of view, a general *random process* is defined over a general linear space $\mathcal{X}$. Let us assume in the following that it is infinite-dimensional. For instance, its elements may be of the form

$$\mathbf{x} = \begin{bmatrix} \mathbf{a}(s) \\ \mathbf{b}(t) \\ \mathbf{c} \end{bmatrix} ,$$

where $\mathbf{a}(s)$ and $\mathbf{b}(t)$ are vector functions of different vector variables, and where $\mathbf{c}$ is a discrete vector. A random process is defined over $\mathcal{X}$ if, for any finite-dimensional subspace $\mathcal{X}^n$ a usual probability is defined. Obviously, a compatibility condition is needed, namely, that if a subspace $\mathcal{X}^m$ is bigger than a subspace $\mathcal{X}^n$, the probability over $\mathcal{X}^n$ has to be obtained as a marginal probability from the probability over $\mathcal{X}^m$. Such a system of probabilities over the finite-dimensional subspaces of a linear space $\mathcal{X}$ is said to define a *cylinder set probability* over $\mathcal{X}$ (see Box 6.2 for details).

Let $\mathcal{X}^n$ be an n-dimensional subspace of $\mathcal{X}$, and $\mathbf{x}^n = (x^1,...,x^n)$ a generic point of $\mathcal{X}^n$. If a cylinder set probability has been defined over $\mathcal{X}$, we can associate the probability $P(\mathcal{A}^n)$ with any subset $\mathcal{A}^n$ of $\mathcal{X}^n$. As usual, the associated probability density is defined by

$$P(\mathcal{A}^n) = \int_{\mathcal{A}^n} d\mathbf{x}^n \; f(\mathbf{x}^n) \qquad \text{for any} \quad \mathcal{A}^n \subset \mathcal{X}^n . \tag{6.17}$$

One of the properties of a probability is that if $\mathcal{A}_1, \mathcal{A}_2,...$ represents a disjoint sequence of sets with probabilities $P(\mathcal{A}_1), P(\mathcal{A}_2),...$ then

$$P(\cup A_i) = \Sigma P(A_i) , \tag{6.18}$$

(see Chapter 1). In general, a cylinder set probability only verifies this property for a *finite* number of sets. In mathematical language, a cylinder set probability is (generally) not a *measure*, and this implies that integration cannot be defined such that, for any $A \subset X$,

$$P(A) = \int_A P(dx) . \tag{6.19}$$

Only particular cylinder set probabilities over infinite-dimensional spaces are true probabilities (such as, for instance, Gaussian probabilities with nuclear covariance operators (e.g., Balakrishnan, 1976)).

In any case, cylinder set probabilities or true probabilities, when defined over an infinite-dimensional space, do not have a probability density. The reason is that the equivalent of the Lebesgue measure does not exist in infinite-dimensional spaces.

6.2: *Solution of general inverse problems*

In Chapter 1, I introduced the data space D , the parameter space M , and the space $X = D \times M$. I have postulated that (in the finite-dimensional case) the most general way of describing any state of information is by defining a probability. The reference state of information is defined by the probability $M(\cdot)$, with density $\mu(\cdot)$, the state of information describing measurements and a priori information is defined by the probability $R(\cdot)$, with density $\rho(\cdot)$, and the theoretical state of information is defined by the probability $T(\cdot)$, with density $\theta(\cdot)$. The final state of information, $S(\cdot)$, obtained by combining these states of information has been defined by conditions (1.35), and the corresponding probability density is

$$\sigma(\mathbf{x}) = \frac{\rho(\mathbf{x})\,\theta(\mathbf{x})}{\mu(\mathbf{x})} . \tag{6.20}$$

It is not clear at present which is the right way in which these notions should be properly generalized for infinite-dimensional problems. Should we postulate that a general state of information is represented by a general cylinder set probability or by a true probability? If $R(\cdot)$ and $T(\cdot)$ are cylinder-set probabilities (resp. probabilities), do the conditions (1.35) define a cylinder-set probability (resp. a probability)?

These questions are important from both philosophical and mathematical points of view. But as they remain unsolved, I am forced here to take a more particular and operational approach.

Some very general results exists concerning the solution of very special infinite-dimensional inverse problems. For instance, in a stochastic (Gaussian) context, Franklin (1970) gave the least-squares solution to inverse problems where both the model and data space can be infinite-dimensional, but where data parameters are *linearly* related to model parameters, and Backus (1970,a,b,c) (see Box 6.2) gave the general Bayesian solution to problems where, if the model space can be infinite-dimensional, it is assumed that the data space is finite-dimensional, that the relationship between data and model parameters is linear, and where one is only interested in the prediction of a *finite* number of properties of the model (which are also *linear* functions of model parameters). These papers are historically important, but they are far from giving to the infinite-dimensional problem a solution with a degree of generality comparable to that given to finite-dimensional problems in the first chapter of this book.

For the time being, my suggestion is to take a practical (i.e., numerical) point of view, and to limit ourselves to the consideration of sufficiently regular functions (i.e., functions which can be well approximated by a finite number of points). We will then always be able to obtain solutions of our inverse problems numerically (using, for instance, equation (6.20)). The solutions thus obtained will probably be very close to the solutions that could be obtained from a rigorous, infinite-dimensional approach.

Box 6.3: The Bayesian viewpoint of Backus (1970).

In his paper "Inference from inadequate and inaccurate data", Backus (1970) made the first effort in formalizing the probabilistic approach to inverse problems. Although indigestible, the paper is historically important. The essentials of the theory are as follows.

The infinite-dimensional linear model space M is assumed to be a Hilbert space with scalar product denoted by $(\bullet \ , \bullet \ )$. The true (unknown) model is denoted by $\mathbf{m}_{true}$. n measurements of physical properties of $\mathbf{m}_{true}$ give the n real quantities $d^1, d^2, ..., d^n$ which are assumed linearly related with $\mathbf{m}_{true}$. Then there exist n vectors $\mathbf{g}^1, \mathbf{g}^2, ..., \mathbf{g}^n$ of M , called "data vectors", such that if $\mathbf{m}$ was $\mathbf{m}_{true}$, we could predict the n real quantities $d^1, d^2, ..., d^n$ by the scalar products

$$d^1 = (\mathbf{g}^1, \mathbf{m}) \qquad d^2 = (\mathbf{g}^2, \mathbf{m}) \qquad (...) \ d^n = (\mathbf{g}^n, \mathbf{m}) \ . \tag{1}$$

(In fact, introducing $\mathbf{g}^1, \mathbf{g}^2, ..., \mathbf{g}^n$ as elements of the dual of M , and considering duality products instead of scalar products allows dropping the assumption that M is a Hilbert space, which seems unnecessary).

(...)

The n actual measurements give the values $d^1_{obs}, d^2_{obs}, ..., d^n_{obs}$, as well as an estimation of the probabilistic distribution of experimental errors.

Backus is not interested in the description of $\mathbf{m}_{true}$ itself, but only in the prediction of m numerical properties of $\mathbf{m}_{true}$, which are also assumed to be linear functions of the model :

$$\delta^1 = (\gamma^1,\mathbf{m}) \qquad \delta^2 = (\gamma^2,\mathbf{m}) \qquad (...) \; \delta^m = (\gamma^m,\mathbf{m}) \; , \tag{2}$$

where $\gamma^1,\gamma^2,...,\gamma^m$ are the "prediction vectors".

The following finite-dimensional spaces are introduced:

G : n-dimensional linear vector space generated by the data vectors $\mathbf{g}^1,\mathbf{g}^2,...,\mathbf{g}^n$. It is a subspace of M .

D : n-dimensional linear vector space where the measurements $d^1,...,d^n$ may take their values (in fact, R^n)

Γ : m-dimensional linear vector space generated by the prediction vectors $\gamma^1,\gamma^2,...,\gamma^m$. It is a subspace of M .

Δ : m-dimensional linear vector space where the predictions $\delta^1,...,\delta^m$ may take their values (in fact, R^m)

S : arbitrary *finite-dimensional* subspace of M *containing* G and Γ as subspaces.

The n measurements only give information on the projection of $\mathbf{m}_{true}$ over G . As we are only interested in the m predictions, we only need information on the projection of $\mathbf{m}_{true}$ over Γ . We can then drop the infinite-dimensional model space M from our attention and only consider the finite-dimensional space S , which contains both the projections of $\mathbf{m}_{true}$ over G and over Γ . As S is finite-dimensional, the standard Bayesian inference can be used.

For instance, let us denote by $\mathbf{s}$ a generic element of S . We can introduce the probability density $\rho_{meas}(\mathbf{d}|\mathbf{s})$ over D representing the density of probability of obtaining $\mathbf{d}$ as the result of our measurements if (the projection of) the true model was $\mathbf{s}$. This probability density is practically obtained from the knowledge of the data vectors $\mathbf{g}^1,\mathbf{g}^2,...,\mathbf{g}^m$ and knowledge of the statistics of errors of our measuring instruments

(...)

(see for instance Chapter 1). The a priori information over $\boldsymbol{S}$ is described using a probability density $\rho_{prior}(\mathbf{s})$. The Bayes' rule then gives the posterior probability density over $\boldsymbol{S}$:

$$\rho_{post}(\mathbf{s}|\mathbf{d}_{obs}) = k \ \rho_{meas}(\mathbf{d}_{obs}|\mathbf{s}) \ \rho_{prior}(\mathbf{s}) \ , \tag{3}$$

where k is a norming constant.

Once this posterior probability has been defined over $\boldsymbol{S}$, the corresponding probability over $\boldsymbol{\Gamma}$ is obtained as a marginal probability. As we know how to associate the m-dimensional vector $\delta^1,...,\delta^m$ with any element of $\boldsymbol{\Gamma}$ (equation 2), it is then easy to deduce the corresponding posterior probability over $\boldsymbol{\Delta}$, the space of predictions.

The conceptually important result proved by Backus (theorem 29, page 54) is that if M_{prior} is a *cylinder measure* over $\boldsymbol{M}$, and for each choice of the finite-dimensional space $\boldsymbol{S}$ the a priori probability over $\boldsymbol{S}$ is the corresponding marginal probability of the cylinder measure M_{prior} , then all the posterior probability distributions over $\boldsymbol{S}$ obtained from the Bayesian solution are marginal distributions of a single cylinder measure M_{post} over $\boldsymbol{M}$.

The results are then independent of the particular choice of $\boldsymbol{S}$. The simplest results are obtained for the smallest $\boldsymbol{S}$, i.e., $\boldsymbol{S} = \boldsymbol{G} + \boldsymbol{\Gamma}$.

This theory can be applied to linearized inverse problems, but does not generalize to true nonlinear problems.

CHAPTER 7

THE LEAST-SQUARES CRITERION IN FUNCTIONAL SPACES

Tu quoque, fili!.

Caius Julius Caesar
(when he saw that even Brutus discretized least-squares problems).

Many objects in the physical world are "fields", i.e., functions defined over spatial coordinates and/or time and taking values in an arbitrary space. For instance, seismologists describe an Earth model using the fields $\rho(\mathbf{x})$, $\kappa(\mathbf{x})$, $\mu(\mathbf{x})$,... (density, bulk modulus, shear modulus,...) as functions of the spatial coordinates $\mathbf{x} = (x^1,x^2,x^3)$, and describe a seismic wave as the field $\mathbf{u}(\mathbf{x},t)$ (displacement at the point $\mathbf{x}$ at time t). Given the boundary and initial conditions, the (macroscopic) relationships between a wavefield and an Earth model are described by a differential equation (the wave equation). As analytic solutions of the wave equation are usually not known (except for some trivial nonrealistic examples), the problem of obtaining the wavefield corresponding to a given Earth model is solved numerically, for instance, by *discretizing* the fields representing the Earth and the wavefield, and by replacing the differential equation by the corresponding finite-difference equation. But even if all the computations of wavefields that we might perform in our whole life were discretized, the functional language would remain useful, because of its compactness.

The same thing happens in least-squares inversion for problems involving fields. A theory can be built up which is essentially functional, and which considers the fields as abstract elements of a conveniently defined infinite-dimensional space. The resolution of the problem implies the use of differential and integral operators. Nevertheless, for all non-trivial applications, numerical results are always obtained after discretization. As the theory of least squares is considerably more abstract in infinite-dimensional spaces, the question arises whether we could limit ourselves to the consideration of discrete least squares. Unambiguously, the response is **no**, as, again, the functional language allows an indispensable compactness of the discussion. For instance, to obtain the differential equations giving the inverse solution of a forward problem involving a differential equation is much easier than for the discretized version of the forward problem. Once the final (functional) equation giving the solution of the inverse problem has been obtained, and in order to perform numerical computations, then (and only then) do we have to discretize everything, the original differential equation, its adjoint, and all the integral/differential equations obtained.

Of course, there exist inverse problems which are essentially discrete. Of course, an astute parameterization of our model and/or our data can allow of using discrete inverse theory instead of the functional theory. But I suggest that the reader should first consider the forward problem alone: when you imagine a model or a data set, do you imagine a discrete sequence of parameters or a field? If you imagine a discrete sequence, go ahead with discrete inverse theory. If you imagine fields, do not discretize your model or your data set for the purposes of inversion: use the functional approach.

In writing a chapter like this one, it is a very difficult problem to choose an adequate level of mathematical rigour. Any user of the functional least-squares theory acknowledges that least-squares formulas have a much larger domain of validity than one for which rigorous mathematical proofs already exist. So, in that field, to be rigorous means enormously restricting the class of spaces we can deal with. My personal philosophy in that matter is not to be overpreoccupied with pure mathematical aspects: common sense will guide the physicist well, with low probability of error. What is important, is to learn the intuitive meaning of a wide enough set of concepts and of the accompanying usual language (linear and nonlinear functionals, dual spaces, transposes of linear operators,...).

As we saw in the first part of the book, least-squares theory is justified only when data uncertainties and a priori model uncertainties can be described using Gaussian statistics. As this is not always the case in actual problems involving functions, I emphasize again that the solutions obtained using least squares are usually only approximations (in a badly defined sense) to more optimal solutions described in Chapters 1 and 6.

7.1: Covariance operators and working spaces

7.1.1: Representation of the dual of a linear space

Let δE be a linear space, eventually infinite dimensional, and let $\delta\hat{E}$ be its dual, i.e., the space of the linear real functionals over δE (a real *functional* over δE is an application from δE into the real line R). If $\delta \mathbf{e}_1$ is an element of δE , and $\delta\hat{\mathbf{e}}_2$ an element of $\delta\hat{E}$, the real number associated with $\delta \mathbf{e}_1$ by $\delta\hat{\mathbf{e}}_2$ is denoted $\langle\ \delta\hat{\mathbf{e}}_2 \, , \delta \mathbf{e}_1 \rangle$. It follows from

Riesz's representation theorem (see, for instance, Taylor and Lay, 1980) that any linear functional $\delta\hat{e}_2$ can be associated with an "object" with the same indexes or variables as the elements of δE , also denoted $\delta\hat{e}_2$, and such that if the matricial-like notation $\delta\hat{\mathbf{e}}_2{}^t\,\delta\mathbf{e}_1$ represents the real (adimensional) number obtained by summing or integrating over all the (common) variables of $\delta\hat{e}_2$ and δe_1 , then

$$\langle\ \delta\hat{e}_2\ ,\ \delta e_1\ \rangle\ =\ \delta\hat{e}_2{}^t\ \delta e_1 \tag{7.1}$$

Example 7.1: Let δE be the space of all couples of functions

$$\delta e\ =\ \begin{bmatrix}\delta\rho(\mathbf{x})\\ \delta\kappa(\mathbf{x})\end{bmatrix}\ ,$$

where $\mathbf{x}$ represents an arbitrary point inside the Earth, and where $\delta\rho(\mathbf{x})$ and $\delta\kappa(\mathbf{x})$ are respectively matter density and bulk modulus perturbations at point $\mathbf{x}$. Any element of $\delta\hat{E}$ can then be (uniquely) represented by another couple of functions (or, more generally, distributions):

$$\delta\hat{e}\ =\ \begin{bmatrix}\delta\hat{\rho}(\mathbf{x})\\ \delta\hat{\kappa}(\mathbf{x})\end{bmatrix}\ ,$$

such that

$$\langle\ \delta\hat{e}_2\ ,\ \delta e_1\ \rangle\ =\ \delta\hat{e}_2{}^t\ \delta e_1$$

$$=\ \int_{\mathbf{V}} dV\mathbf{x}\ \ \delta\hat{\rho}_2(\mathbf{x})\ \delta\rho_1(\mathbf{x})\ +\ \int_{\mathbf{V}} dV\mathbf{x}\ \ \delta\hat{\kappa}_2(\mathbf{x})\ \delta\kappa_1(\mathbf{x})\ .$$

It should be noticed that if the physical dimensions of $\delta\rho$ and $\delta\kappa$ are respectively a density and a pressure, as $\langle\ \delta\hat{e}_2\ ,\ \delta e_1\ \rangle$ has to be a real number, the physical dimensions of $\delta\hat{\rho}$ and $\delta\hat{\kappa}$ are respectively (density × volume)$^{-1}$, and (pressure × volume)$^{-1}$. We will say that the "components" of δe_1 and $\delta\hat{e}_2$ have *dual* physical dimensions. ■

7.1.2: Definition of a covariance operator

A *covariance operator* over δE is, by definition, a linear symmetric definite non-negative operator (see Box 7.1) mapping $\delta\hat{E}$ into δE .

Example 7.2: Let $\mathbf{C}$ be a covariance operator over the space δE of example 7.1, and let

$$\delta e\ =\ \mathbf{C}\ \delta\hat{e}\ . \tag{7.2}$$

An explicit representation of the operator **C** will be given by

$$\delta\rho(\mathbf{x}) = \int_{\mathbf{V}} dV(\mathbf{x})\ C_{\rho\rho}(\mathbf{x},\mathbf{x}')\ \delta\hat{\rho}(\mathbf{x}) + \int_{\mathbf{V}} dV(\mathbf{x})\ C_{\rho\kappa}(\mathbf{x},\mathbf{x}')\ \delta\hat{\kappa}(\mathbf{x})$$

$$\delta\kappa(\mathbf{x}) = \int_{\mathbf{V}} dV(\mathbf{x})\ C_{\kappa\rho}(\mathbf{x},\mathbf{x}')\ \delta\hat{\rho}(\mathbf{x}) + \int_{\mathbf{V}} dV(\mathbf{x})\ C_{\kappa\kappa}(\mathbf{x},\mathbf{x}')\ \delta\hat{\kappa}(\mathbf{x}) ,$$

or, for short,

$$\begin{bmatrix} \delta\rho \\ \delta\kappa \end{bmatrix} = \begin{bmatrix} C_{\rho\rho} & C_{\rho\kappa} \\ C_{\kappa\rho} & C_{\kappa\kappa} \end{bmatrix} \begin{bmatrix} \delta\hat{\rho} \\ \delta\hat{\kappa} \end{bmatrix} .$$

We see that the kernel of the operator is here a 2×2 matrix of (covariance) functions. From the physical dimensions of $\delta\rho$, $\delta\kappa$, $\delta\hat{\rho}$, and $\delta\hat{\kappa}$, we can easily see that the physical dimensions of the covariance functions are:

$C_{\rho\rho}(\mathbf{x},\mathbf{x}')$: $(\text{density})^2$
$C_{\rho\kappa}(\mathbf{x},\mathbf{x}')$: (density $\times$ pressure)
$C_{\kappa\rho}(\mathbf{x},\mathbf{x}')$: (density $\times$ pressure)
$C_{\kappa\kappa}(\mathbf{x},\mathbf{x}')$: $(\text{pressure})^2$,

i.e., the same dimensions as usual covariance functions (see example 7.3). ■

Example 7.3: Assume that we have N models of density and bulk modulus (N >> 1)

$$\begin{bmatrix} \rho_1 \\ \kappa_1 \end{bmatrix} , \begin{bmatrix} \rho_2 \\ \kappa_2 \end{bmatrix} , \ldots$$

Using the classical definitions of statistics, the mean model $\begin{bmatrix} \langle \rho \rangle \\ \langle \kappa \rangle \end{bmatrix}$ is defined by

$$\langle \rho \rangle (\mathbf{x}) = \frac{1}{N} \sum_{i=1}^{N} \rho_i(\mathbf{x}) \tag{7.3a}$$

$$\langle \kappa \rangle (\mathbf{x}) = \frac{1}{N} \sum_{i=1}^{N} \kappa_i(\mathbf{x}) , \tag{7.3b}$$

and the following covariance functions are defined:

$$C_{\rho\rho}(\mathbf{x},\mathbf{x}') = \frac{1}{N} \sum_{i=1}^{N} (\ \rho_i(\mathbf{x}) - \langle\ \rho\ \rangle\ (\mathbf{x})\ )\ (\ \rho_i(\mathbf{x}') - \langle\ \rho\ \rangle\ (\mathbf{x}')\ ) \tag{7.4a}$$

$$C_{\rho\kappa}(\mathbf{x},\mathbf{x}') = \frac{1}{N} \sum_{i=1}^{N} (\ \rho_i(\mathbf{x}) - \langle\ \rho\ \rangle\ (\mathbf{x})\ )\ (\ \kappa_i(\mathbf{x}') - \langle\ \kappa\ \rangle\ (\mathbf{x}')\ ) \tag{7.4b}$$

$$C_{\kappa\rho}(\mathbf{x},\mathbf{x}') = \frac{1}{N} \sum_{i=1}^{N} (\ \kappa_i(\mathbf{x}) - \langle\ \kappa\ \rangle\ (\mathbf{x})\ )\ (\ \rho_i(\mathbf{x}') - \langle\ \rho\ \rangle\ (\mathbf{x}')\ ) \tag{7.4c}$$

$$C_{\kappa\kappa}(\mathbf{x},\mathbf{x}') = \frac{1}{N} \sum_{i=1}^{N} (\ \kappa_i(\mathbf{x}) - \langle\ \kappa\ \rangle\ (\mathbf{x})\ )\ (\ \kappa_i(\mathbf{x}') - \langle\ \kappa\ \rangle\ (\mathbf{x}')\ )\ . \tag{7.4d}$$

It is easy to see that the physical dimensions of the covariance function are as in example 7.2.

A well-known result (e.g., Pugachev, 1965) is that the covariance functions are symmetric:

$$C_{\rho\rho}(\mathbf{x},\mathbf{x}') = C_{\rho\rho}(\mathbf{x}',\mathbf{x}) \quad C_{\rho\kappa}(\mathbf{x},\mathbf{x}') = C_{\kappa\rho}(\mathbf{x}',\mathbf{x}) \quad C_{\kappa\kappa}(\mathbf{x},\mathbf{x}') = C_{\kappa\kappa}(\mathbf{x}',\mathbf{x})\ , \tag{7.5}$$

and definite non-negative:

$$\int_{\mathbf{V}} dV(\mathbf{x})\ \phi_\rho(\mathbf{x})\ C_{\rho\rho}(\mathbf{x},\mathbf{x}')\ \phi_\rho(\mathbf{x}') + \int_{\mathbf{V}} dV(\mathbf{x})\ \phi_\rho(\mathbf{x})\ C_{\rho\kappa}(\mathbf{x},\mathbf{x}')\ \phi_\kappa(\mathbf{x}') +$$

$$\int_{\mathbf{V}} dV(\mathbf{x})\ \phi_\kappa(\mathbf{x})\ C_{\kappa\rho}(\mathbf{x},\mathbf{x}')\ \phi_\rho(\mathbf{x}') + \int_{\mathbf{V}} dV(\mathbf{x})\ \phi_\kappa(\mathbf{x})\ C_{\kappa\kappa}(\mathbf{x},\mathbf{x}')\ \phi_\kappa(\mathbf{x}')\ \geq\ 0$$

$$\text{for any } \phi_\rho\ ,\ \phi_\kappa\ . \tag{7.6}$$

All these properties justify the abstract definition used for introducing covariance operators. ■

7.1.3: Some examples of covariance functions and associated random realizations

It is very useful to have in mind what is exactly meant by a Gaussian random process with given mathematical expectation and given covariance operator. The following examples give some illustrations.

Example 7.4: Exponential covariance function. Let M be the space of all functions

$$m = [m(z)] ,$$

where z represents a spatial or temporal variable. A covariance operator over M will have as kernel a covariance function $C(z,z')$.

One covariance function often encountered in practical applications is the exponential covariance function

$$C(z,z') = \sigma^2 \exp\left(- \frac{|z-z'|}{L} \right) . \tag{7.7}$$

Figure 7.1 shows a (pseudo) random realization of a Gaussian random function with zero expectation and exponential covariance. In the example, $\sigma = 1$ and $L = 1$.

It can be shown that the realizations of a (sufficiently regular) stationnary random function are n times differentiable if the correlation function $C(z,0)$ is differentiable 2n times at $z = 0$ (see section 7.1.4). Thus, the realizations of Figure 7.1 have *discontinuous derivative at every point.* They are *fractals* (Mandelbrot, 1977), in the sense that if a small portion of the "curve" is zoomed with adequate horizontal and vertical scales, we obtain another "curve" which looks like the original one. ■

Example 7.5: Gaussian covariance function. Another covariance function often encountered in practical applications is the Gaussian covariance function:

$$C(z,z') = \sigma^2 \exp\left(- \frac{1}{2} \frac{(z-z')^2}{L^2} \right) . \tag{7.8}$$

The bottom of Figure 7.2, due to Frankel and Clayton (1986), shows a (pseudo) random realization of a two-dimensional Gaussian random function with Gaussian covariance. For the construction of the pseudo-random realizations, Frankel and Clayton first used a pseudo-random number generator with Gaussian probability density to assign sequentially (and independently) a number at each point of the 2D grid. The random field was then Fourier transformed in its two dimensions, filtered to achieve the desired spectrum (see section 7.1.4) and transformed back to the spatial domain. ■

Example 7.6: "White-noise" covariance function. The covariance "function"

$$C(z,z') = \beta \, \delta(z-z') , \tag{7.9}$$

where β is a finite constant, and where $\delta(z)$ represents the Dirac distribu-

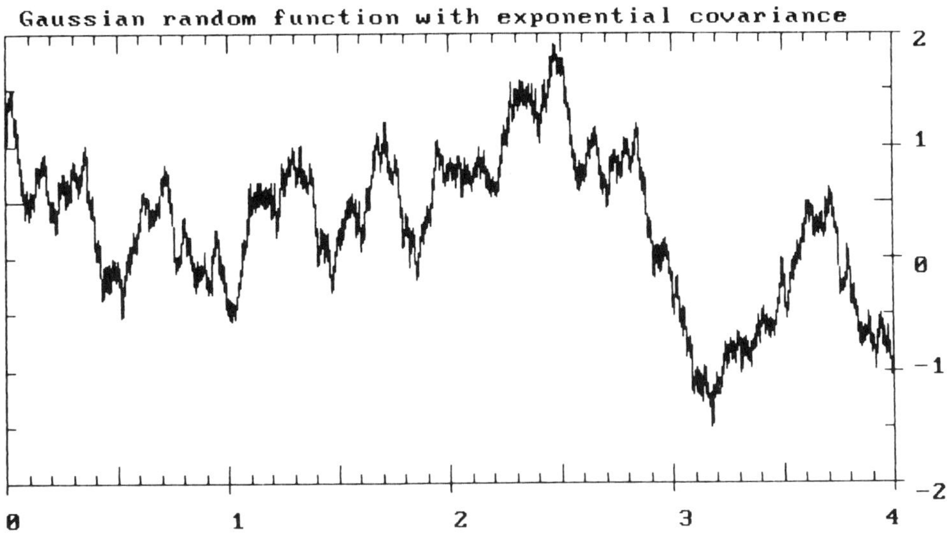

Figure 7.1: Example of pseudo-random realization of a Gaussian random function with zero mean and exponential covariance function

$$C(z,z') = \sigma^2 \exp\left(- \frac{|z-z'|}{L} \right) .$$

In this example, $L = 1$ and $\sigma = 1$. Notice that the function has discontinuous derivative at every point. It has fractal characteristics (Mandelbrot, 1977). The value of the function has been computed in 40 000 points (the graphic printer device has much less resolution). Some artifacts are due to periodicities in the pseudorandom generator used.

tion, corresponds to white noise (see example 7.8). Figure 7.3 suggests what a realization of a white noise random function may be. ■

Example 7.7: Random-walk covariance function. Also interesting for practical applications is the covariance function

$$C(z,z') = \beta \operatorname{Min}(z,z') \qquad (\text{for } z \geq 0 \text{ and } z' \geq 0) . \tag{7.10}$$

The variance at point z is $\sigma^2 = \beta z$. It corresponds to an (unidimensional) random walk, which, in turn, corresponds to the primitive of a white noise. Figure 7.4 shows a (pseudo) random realization of a Gaussian random func-

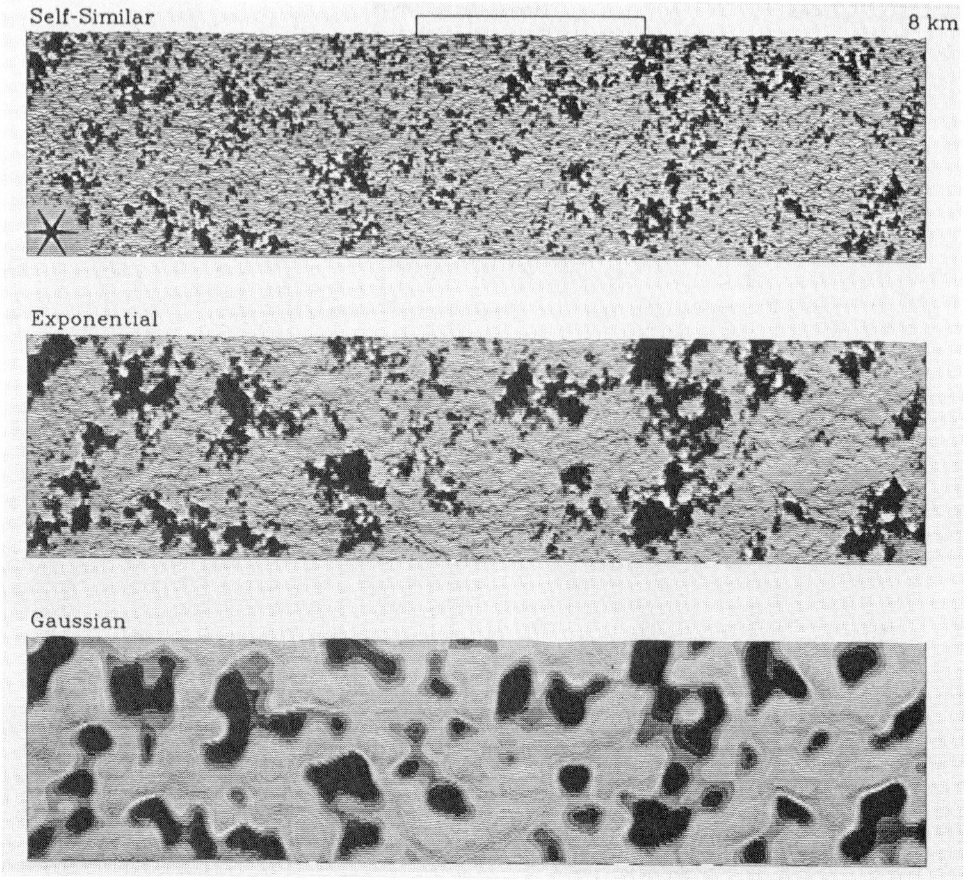
Self-Similar
8 km
Exponential
Gaussian

← *Figure 7.2:* From Frankel and Clayton (1986). Three types of two-dimensional pseudo-random realizations. Areas higher than average are shaded. Top: Von Karman covariance function (Tatarski, 1961):

$$C(x,y;x',y') = K_0\left(\frac{\sqrt{(x-x')^2+(y-y')^2}}{L} \right) ,$$

where $K_0(\bullet)$ is the modified Bessel function of order zero. Middle: Exponential covariance function:

$$C(x,y;x',y') = \sigma^2 \exp\left(- \frac{\sqrt{(x-x')^2+(y-y')^2}}{L} \right) .$$

Bottom: Gaussian covariance function:

$$C(x,y;x',y') = \sigma^2 \exp\left(- \frac{(x-x')^2+(y-y')^2}{L^2} \right) .$$

tion with zero expectation and such a covariance function, with $\beta = 1$ (which gives $\sigma(z=40) \simeq 6.32$). ■

Example 7.8: The use of White Noise for modeling errors. A Gaussian random function $t \rightarrow X(t)$ is named *white noise* if its covariance function is

$$C(t,t') = w(t)\ \delta(t-t') , \tag{7.11}$$

where $w(t)$ is termed the *intensity* of the white noise. In particular, it should be noticed that, for any t , the variance of the random variable $X(t)$ is infinite, and that, for any t and t' , the random variables $X(t)$ and $X(t')$ are uncorrelated. The easiest way to understand white noise is to consider it as a limit of a Gaussian random process where the correlation length tends to vanish. Take for instance a Gaussian random function with exponential covariance

$$C(t,t') = \sigma^2 \exp\left(- \frac{|t-t'|}{T} \right) .$$

In the limit when $T \rightarrow 0$, if the product $\sigma^2\ T$ remains finite (and so, if the variance tends to infinity), the function $C(t,t')$ tends to the particular white-noise distribution

$$C(t,t') = 2\ \sigma^2\ T\ \delta(t-t') . \tag{7.12}$$

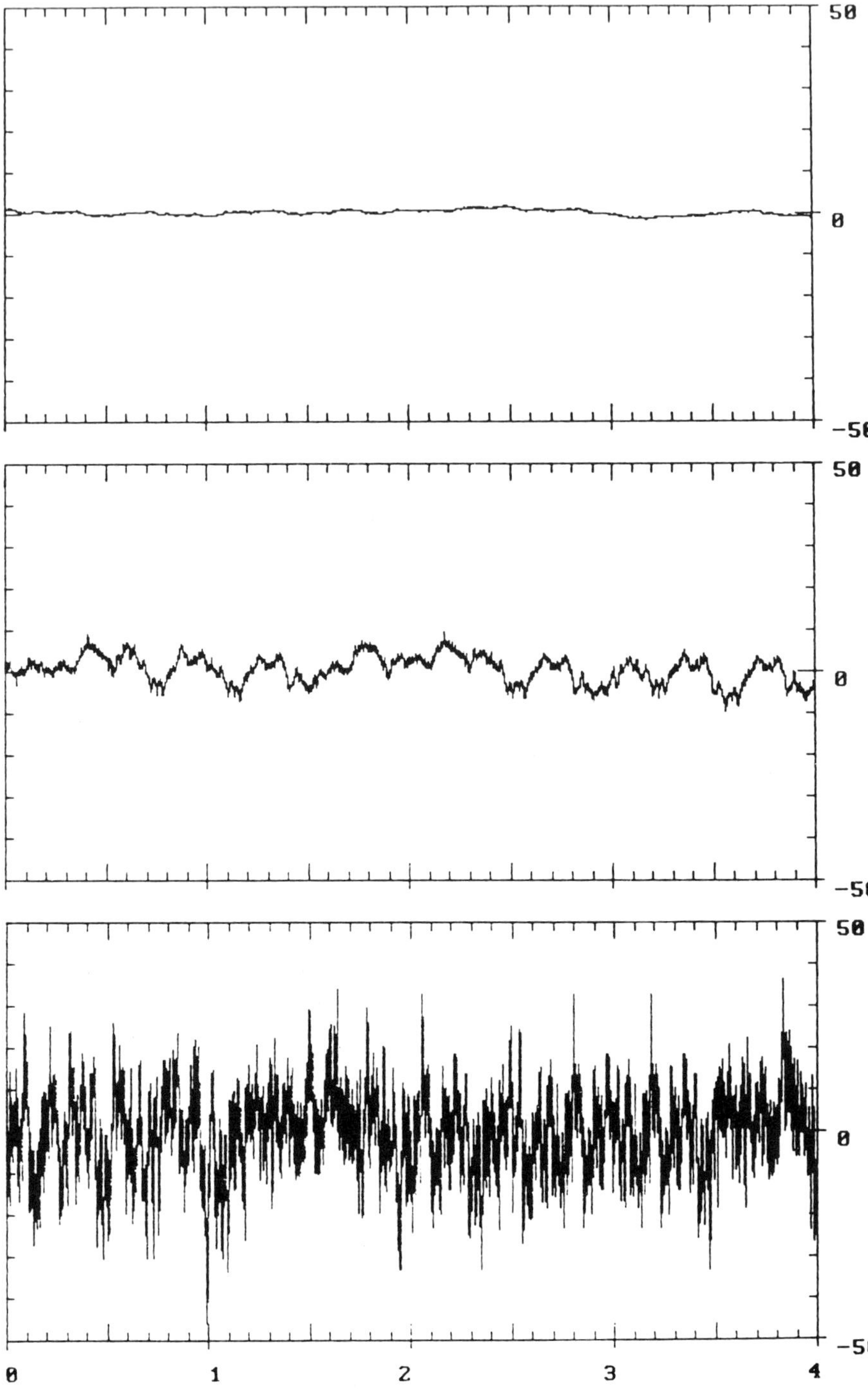
50
0
-50
50
0
-50
50
0
-50
0
1
2
3
4

← *Figure 7.3:* A pseudo-random realization of a pure White Noise random function cannot be represented, because the variance of the random function at any point is infinite. But white noise can be seen, for instance, as the limit of an ordinary Gaussian random function when the correlation length L tends to zero and the variance σ^2 tends to infinity, the product $L\sigma^2$ remaining constant (see example 7.8). This figure illustrates such a limit. The three functions are pseudorandom realizations of a Gaussian random function with null mathematical expectation and exponential covariance function $C(z,z') = \sigma^2 \exp(-|z-z'|/L)$. The product $L\sigma^2$ is constant for the three realizations. At top, $L = 1$ and $\sigma^2 = 1$, at middle, $L = 0.1$ and $\sigma^2 = 10$, and at bottom, $L = 0.01$ and $\sigma^2 = 100$ (the top realization is the same as in Figure 7.1, but represented here with a much larger vertical scale). The reader can intuitively take the visual limit of the process, and apprehend what a realization of a white-noise random process may be. The value of the function has been computed in 40 000 points (the graphic printer device has much less resolution). Some artifacts are due to periodicities in the pseudo-random generator used.

The adjective "white" comes from the fact that a realization of such a random function has a flat frequency spectrum (i.e., flat Fourier transform), like white light; the noun "noise" comes from the fact that a sound wave with such a spectrum really sounds as noise). The term $2\,\sigma^2\,T$ in the last equation can be interpreted as the area under the curve $C(t,t')$:

$$\int_{-\infty}^{\infty} dt'\; C(t,t') \;=\; 2\,\sigma^2\,T \qquad\qquad \text{(for any } t \text{) .}$$

It is not possible to represent a realization of a true white-noise random function, because the values at any point are $+\infty$ or $-\infty$ (with probability 1), but it is possible to obtain a good intuitive feeling by a direct consideration of the concept of limit of random functions whose correlation lengths tend to zero. Figure 7.3 shows an illustration.

If instead of using an exponential covariance function we use a Gaussian covariance function:

$$C(t,t') \;=\; \sigma^2 \exp\left(-\,\frac{1}{2}\,\frac{(t-t')^2}{T^2} \right) ,$$

then, in the limit $T \to 0$, $\sigma^2 \to \infty$, $T\,\sigma^2$ constant ,

$$C(t,t') \;=\; (2\pi)^{1/2}\;\sigma^2\;T\;\;\delta(t-t') \; , \qquad (7.13)$$

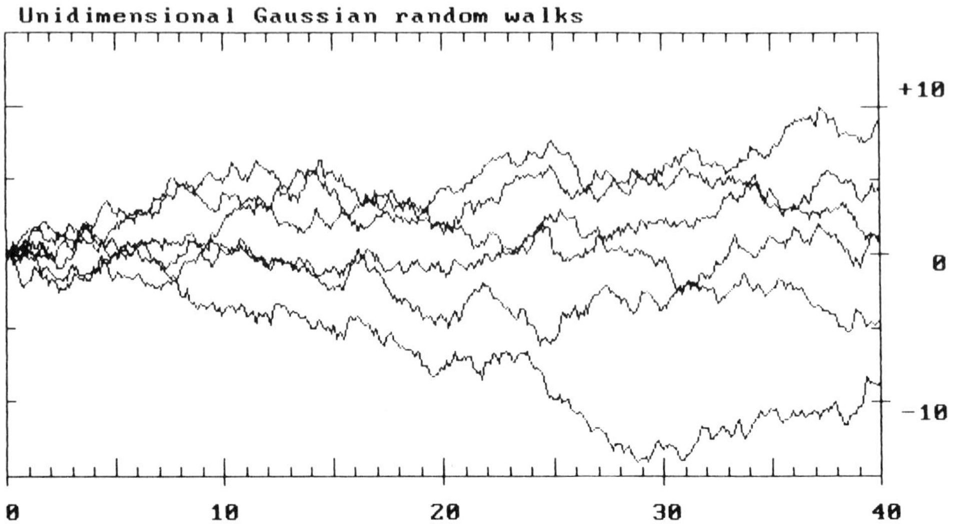

Figure 7.4: Some example of pseudo-random realizations of a Gaussian random function with zero mean and covariance function $C(z,z') = \beta \operatorname{Min}(z,z')$. In this example, $\beta = 1$. Notice that the variance depends on z : $\sigma^2(z) = C(z,z) = \beta z$. For $z = 40$ this gives $\sigma \simeq 6.32$. Only 400 points have been used to generate this pseudorandom realizations. The derivative of a random walk is white noise.

where, again, the factor of the delta function represents the area under the covariance function. Comparing (7.12) with (7.13), we see that the factor of $\delta(t-t')$ differs when obtaining white noise as the limit of different random functions.

Assume that some data $d(t)$ consist in signal $s(t)$ plus noise $n(t)$ (Figure 7.5):

$$d(t) = s(t) + n(t) .$$

Usually, the $d(t)$ data are sampled. If the correlation length of the signal is much larger than the correlation length of the noise, the last may be under-sampled, and then, for all numerical purposes, its correlation length may be taken as null. The covariance function describing uncertainties in the observed data may then be approximated by (7.11), where

$w(t) \simeq$ (True variance of the noise) $\times$ (True correlation length of the noise) .

This justifies the conventional use of white noise to represent some experimental uncertainties. ■

Example 7.9: Experimental estimation of a covariance function for describing experimental uncertainties. Figure 7.6 shows an experimentally estimated covariance function for representing data uncertainties. ■

7.1.4: The power spectrum of a random realization equals the spectral amplitude of the correlation function.

Let $M(z)$ be a random function with mathematical expectation $\langle m \rangle (z)$ and covariance $C(z,z')$. The random function $M(z)$ is *stationnary* if the mathematical expectation is constant,

$$\langle m \rangle (z) = \text{const.} ,$$

and the covariance function $C(z,z')$ depends only on the distance $z-z'$,

$$C(z,z') = \Gamma(z-z') ,$$

where $\Gamma(z)$ is the *correlation function.* Let $m(z)$ be a realization of $M(z)$. Without loss of generality, assume $\langle m \rangle (z) = 0$. We accept without demonstration the intuitive result that the autocorrelation of $m(z)$ gives an estimation of the correlation function $\Gamma(z)$:

$$\Gamma(z) = \lim_{Z \to \infty} \frac{1}{2Z} \int_{-Z}^{+Z} dz' \; m(z') \; m(z'+z) ,$$

or, for short,

$$\Gamma(z) = m(z) * m(-z) .$$

In the Fourier domain,

$$\tilde{\Gamma}(k) = \tilde{m}(k) \; \tilde{m}(k)^* ,$$

where $(\;)^*$ means complex conjugate. We can introduce the *spectral ampli-*

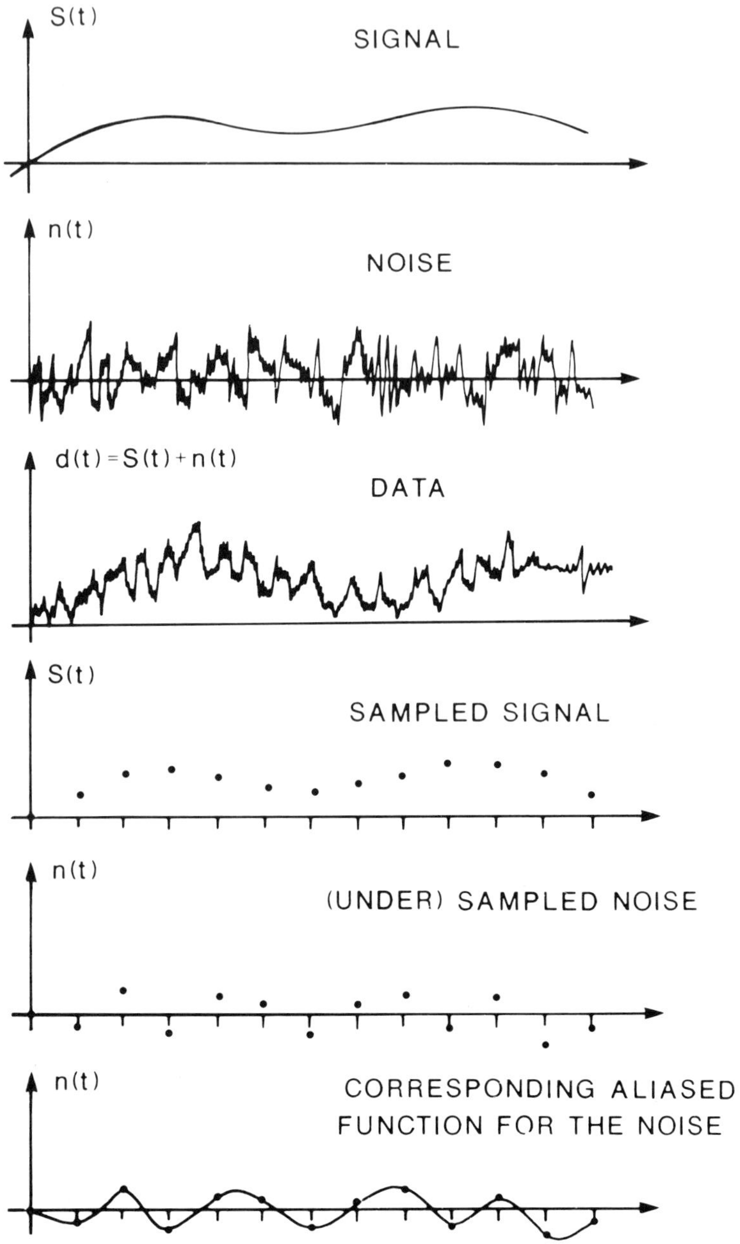
S(t)
SIGNAL
n(t)
NOISE
d(t) = S(t) + n(t)
DATA
S(t)
SAMPLED SIGNAL
n(t)
(UNDER) SAMPLED NOISE
n(t)
CORRESPONDING ALIASED
FUNCTION FOR THE NOISE

← *Figure 7.5:* Observed data can sometimes be interpreted as signal plus (data-independent) noise. Assume that data are sampled. If the signal has a correlation length much larger than the noise, it may happen that the noise is "undersampled". Let T be the true correlation length of the noise, and σ^2 its variance. If the sampling length is much larger than T , the noise "function" will be strongly aliased. A good enough representation of the noise covariance function is then $C(t,t') = G\ \delta(t-t')$, where $G \simeq T\sigma^2$. True white noise does not exist, because the correlation length is never strictly null, but this example shows that white noise may sometimes be used as a correct simplification.

tude and *phase* of $\tilde{m}(k)$ and $\tilde{\Gamma}(k)$,

$$\tilde{m}(k) = A_m(k)\ e^{i\,\phi_m(k)} ,$$

$$\tilde{\Gamma}(k) = A_\Gamma(k)\ e^{i\,\phi_\Gamma(k)} .$$

As the correlation function $\Gamma(z)$ is symmetric,

$$\phi_\Gamma(k) = 0 .$$

We then have

$$A_\Gamma(k) = \tilde{\Gamma}(k) = \tilde{m}(k)\ \tilde{m}(k)^* = (A_m(k))^2 , \tag{7.14}$$

which shows that the spectral amplitude of the correlation function equals the power spectrum (square of the spectral amplitude) of the random realization.

An important classical result concerning a stationnary (sufficiently regular) random function is that its realizations are differentiable n times if, and only if, the correlation function $\Gamma(z) = C(z,0)$ is differentiable at $z = 0$ up to the order $2n$. For instance, for the realizations to be differentiable, the second derivative of $\Gamma(z)$ must exist at $z = 0$. This is not the case in the example of Figure 7.1, so that the realizations shown are not differentiable. On the contrary, the realizations of a random function with Gaussian covariance are infinitely differentiable (they are *very* smooth).

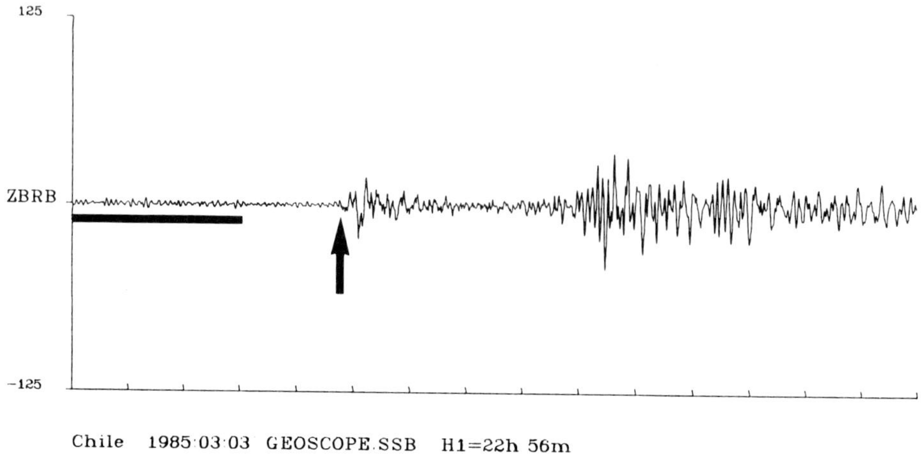

125
ZBRB
-125
Chile 1985 03 03 GEOSCOPE.SSB H1=22h 56m

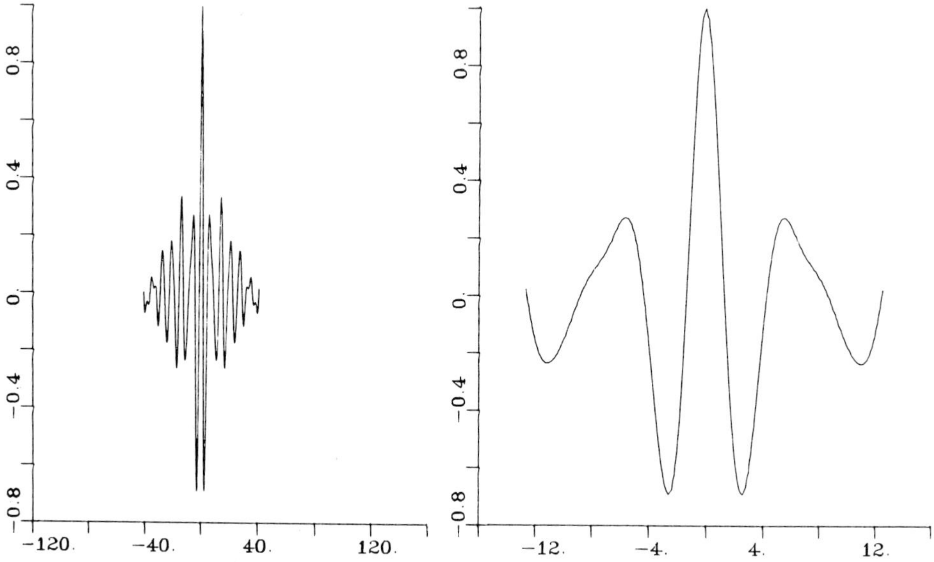

0.8
0.4
0.
-0.4
-0.8
-120.
-40.
40.
120.
0.8
0.4
0.
-0.4
-0.8
-12.
-4.
4.
12.

Inversion Ideas:

1. Use constrained optimization with Lagrangian multipliers to find a "smooth" model

2. Use $\log v_s$ instead of v_s and $\log c$ instead of c to provide more realistic probability distributions. This requires a major effort to recalculate the derivatives. Both v_s and c should be converted to log values so that the relative values remain the same and the inversion will not favor one over the other. (p 42 - 53 of Tarantola)

3. Add thicknesses ($\log h$?) to the inversion

4. Weight the data

5. Use discrete layer matrices. May need a complex discretization

$S_{n+1} < S_n$

✓ 8. Select y_n to minimize S_{n+1} (and $S_{n+1} < S_n$).

9. Use second derivatives to help convergence. (p.194 of Tarantola)

10. Use χ^2 distribution to check solution

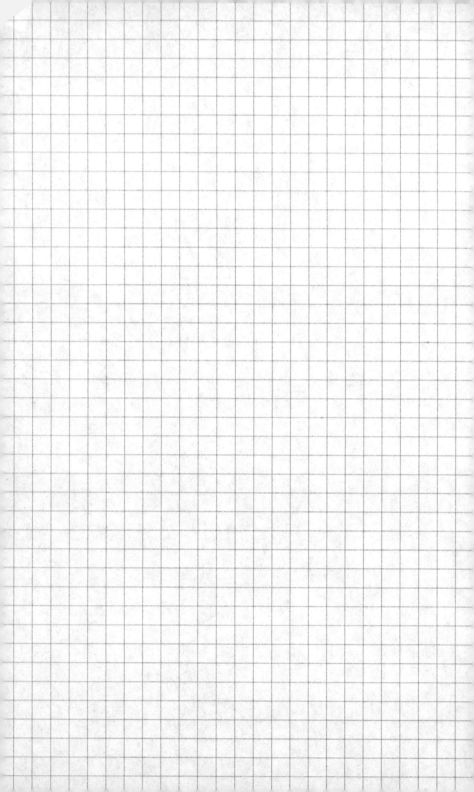

← Figure 7.6: 900 seconds of recording of the displacement of a point on the Earth's surface south of France (top). One can see the arrival of the seismic waves produced by the big Chilean earthquake of March 3, 1985 (arrow). To estimate the noise contaminating the signal after the first wave arrival, 180 seconds of recording before the arrival have been selected (thick line). The normalized autocorrelation of this part of the recording is shown at the bottom in two different scales (in seconds). This can be taken directly as correlation function for representing data uncertainties. Data from the GEOSCOPE seismic network (G. Levy and B. Romanowicz, personal communication).

7.1.5: Weighting operators, scalar products, and "least-squares" norms

By definition, a covariance operator $\mathbf{C}$ is definite non-negative and maps a linear vector space $\hat{\boldsymbol{\delta E}}$ into its (anti) dual $\boldsymbol{\delta E}$. If $\boldsymbol{C}$ is positive definite, then its inverse C^{-1} can be defined and maps δE into $\hat{\delta E}$. C^{-1} is named the *weighting operator* and is often denoted by $\mathbf{W}$.

The expression

$$(\delta e_1 , \delta e_2) = \delta e_1{}^t \ C^{-1} \ \delta e_2 \tag{7.15}$$

can be shown to define a *scalar product* over $\boldsymbol{\delta E}$ (see Appendix 7.1). The expression

$$\| \delta e \| = (\delta e , \delta e)^{1/2} = (\delta e^t \ C^{-1} \ \delta e)^{1/2} \tag{7.16}$$

then defines a *norm* over $\boldsymbol{\delta E}$. It is sometimes called an L_2 norm, although this terminology is improper (see section 7.1.7). The name "least-squares norm" is more familiar, although a proper terminology would be "covariance-related norm".

A space $\boldsymbol{\delta E}$ furnished with a scalar product is named a *Hilbert space* (in fact, it is only a *pre-Hilbert space*, because it is not necessarily complete (see Appendix 7.1), but for all interesting choices of covariance functions it will be (see, for instance, example 7.11).

The reader may wonder why I use $\boldsymbol{\delta E}$ instead of $\boldsymbol{E}$ as a notation for the space over which I define covariance operators and norms. An example will show the reason:

Let, for instance, $\boldsymbol{M}_A$ be the space of all conceivable Earth models of density and bulk modulus, as defined in example 7.3. Again, assume that, by some random process, we have generated a large number of realizations $\mathbf{m}_1$, $\mathbf{m}_2$, ... Once the corresponding mean model $\langle \mathbf{m} \rangle$ and the covariance operator $\mathbf{C}_M$ have been computed, if the random realizations are adequately described using Gaussian statistics, we can restrict the space in consideration to the subspace $\boldsymbol{M}_B$ of all possible realizations of a Gaussian random process with mathematical expectation $\langle \mathbf{m} \rangle$ and covariance operator $\mathbf{C}_M$. With the obvious definitions of sum of two models and of multiplication of a model by a real number, $\boldsymbol{M}_B$ is a linear vector space. This space is still very large, and it is sometimes useful to introduce a further restriction, and to consider only the subspace $\boldsymbol{M}_C$ compounded by the models $\mathbf{m}$ for which the *distance* to the mathematical expectation is finite:

$$\| \mathbf{m} - \langle \mathbf{m} \rangle \| = \left[(\mathbf{m} - \langle \mathbf{m} \rangle)^t \, \mathbf{C}_m^{-1} \, (\mathbf{m} - \langle \mathbf{m} \rangle) \right]^{1/2} \quad \text{FINITE} . \tag{7.17}$$

The space $\boldsymbol{M}_C$ thus defined is *not* a linear vector space. For the null vector, $\mathbf{m} = \mathbf{0}$ does not necessarily belong to $\boldsymbol{M}_C$. The space $\boldsymbol{M}_C$ is an *affine linear space* (see Appendix 7.1), i.e., a space obtained by translation of a linear vector subspace of $\boldsymbol{M}_B$. We can define a new space $\delta\boldsymbol{M}$ whose elements are the *deviations* with respect to the mathematical expectation $\langle \mathbf{m} \rangle$. Formally, $\delta\boldsymbol{M} = \boldsymbol{M}_C - \langle \mathbf{m} \rangle$. $\delta\boldsymbol{M}$ is a linear vector space, and it has a norm:

$$\| \delta\mathbf{m} \| = \left[\delta\mathbf{m}^t \, \mathbf{C}_M^{-1} \, \delta\mathbf{m} \right]^{1/2} . \tag{7.18}$$

The space $\delta\boldsymbol{M}$ is central in least-squares theory. Following the situations, it is called the space of *perturbations*, or *corrections*, or *deviations* (with respect to $\langle \mathbf{m} \rangle$). This space and its dual $\delta\hat{\boldsymbol{M}}$ are the domains of definition and the images of the operators $\mathbf{C}_M$ and $\mathbf{C}_M^{-1}$.

Example 7.10: Let $\delta\boldsymbol{E}$ be the space of all functions

$$\delta\mathbf{e} = [\, \delta e(z) \,] ,$$

and let $\delta\hat{\mathbf{e}}$ be an element of $\delta\hat{\boldsymbol{E}}$, dual of $\delta\boldsymbol{E}$, related with an element $\delta\mathbf{e}$ of $\delta\boldsymbol{E}$ through

$$\delta\mathbf{e} = \mathbf{C} \, \delta\hat{\mathbf{e}} , \tag{7.19}$$

where **C** is a covariance operator over δE with integral kernel (i.e., covariance function) $C(z,z')$:

$$\delta e(z) = \int dz \; C(z,z') \; \delta\hat{e}(z') .$$

By definition of the weighting operator C^{-1} ,

$$\delta\hat{e} = C^{-1} \, \delta e . \tag{7.20}$$

Formally, the integral kernel of the weighting operator C^{-1} can be introduced by

$$\delta\hat{e}(z) = \int dz \; C^{-1}(z,z') \; \delta e(z') . \tag{7.21}$$

This gives the formal equation

$$\int dz \int dz' \; C(z,z') \; C^{-1}(z',z'') = \delta(z-z'') . \tag{7.22}$$

In fact, usual covariance functions are smooth functions, and the linear operator defined by (7.19)-(7.20) is a true integral operator. Its inverse is then a differential operator, and its integral representation (7.21) makes sense only if we interpret $C^{-1}(z,z')$ as a distribution (see example 7.11). Nevertheless, by languistic abuse, $C^{-1}(z,z')$ is named the *weighting function.* ■

Example 7.11: Let $C(z,z')$ be the covariance function considered in example 7.4:

$$C(z,z') = \sigma^2 \exp\left(- \frac{|z-z'|}{L} \right) . \tag{7.23}$$

σ^2 is the *variance* of the random function, L is the *correlation length*, and if the random process is Gaussian, it corresponds to the length along which successive values of any realization are correlated.

The covariance operator **C** corresponding with the integral kernel (7.23) to any function $\delta\hat{e}(z)$ associates the function

$$\delta e(z) = \int_{z_1}^{z_2} dz' \; C(z,z') \; \delta\hat{e}(z') \tag{7.24}$$

As shown in problem 7.7, we have

$$C^{-1}(z,z') = \frac{1}{2\sigma^2} \left(\frac{1}{L}\, \delta(z-z') - L\, \delta''(z-z') \right) , \tag{7.25}$$

where I have used the definition of the derivative of a distribution (see Appendix 7.2). In particular, for Dirac's Delta "function",

$$\int dz'\, \delta^{(n)}(z-z')\, \mu(z') = (-1)^n \frac{d^n}{dz^n} \mu(z) .$$

For the norm of an element $\delta\mathbf{e}$ this gives (see problem 7.7)

$$\| \delta e \|^2 = \frac{1}{2\sigma^2} \left(\frac{1}{L} \int_{z_1}^{z_2} dz \left[\delta e(z)\right]^2 + L \int_{z_1}^{z_2} dz \left[\frac{\partial \delta e}{\partial z}(z) \right]^2 \right) . \tag{7.26}$$

This corresponds to the usual norm in the Sobolev space H^1 (see Appendix 7.2), which equals the sum of the usual L_2 norm of the function and of the L_2 norm of its derivative. The Sobolev spaces are *complete*. In fact, the norm associated with the exponential covariance function may also contain boundary terms. See problem 7.7 for details. ■

Example 7.12: For $0 \le z \le L$, let $C(z,z')$ be the covariance function

$$C(z,z') = \beta\, \mathrm{Min}(z,z') . \tag{7.27}$$

The covariance operator $\mathbf{C}$ whose kernel is the covariance function (7.27) associates to any function $\delta\hat{e}(z)$ the function

$$\delta e(z) = \int_0^L dz'\, C(z,z')\, \delta\hat{e}(z') . \tag{7.28}$$

As shown in problem 7.8, we have

$$C^{-1}(z,z') = -\frac{1}{\beta}\, \delta''(z-z') . \tag{7.29}$$

For the norm of an element $\delta\mathbf{e}$, this gives (see problem 7.8)

$$\| \delta e \|^2 = \frac{1}{\beta} \int_0^L dz \left(\frac{\partial \delta e}{\partial z}(z) \right)^2 . \tag{7.30}$$

This result is interesting, because we see that a criterion of least norm associated with the covariance function (7.27) imposes that the *derivative* of the function (and not the function itself) is small. ■

Example 7.13: The covariance function

$$C(z,z') = w(z)\ \delta(z-z') \tag{7.31}$$

represents white noise (see Example 7.8). w(z) is termed the *intensity* of the white noise. It is easy to see that if

$$\delta e(z) = \int dz'\ C(z,z')\ \delta\hat{e}(z') ,$$

then,

$$\delta\hat{e}(z) = \frac{1}{w(z)}\ \delta e(z) . \tag{7.32}$$

This corresponds to the kernel

$$C^{-1}(z,z') = \frac{1}{w(z)}\ \delta(z-z') . \tag{7.33}$$

The associated norm is the usual weighted L_2 norm:

$$\|\ \delta e\ \|^2 = \int dz\ \frac{(\delta e(z))^2}{w(z)} . \tag{7.34}$$

■

7.1.6: Relationship between the "least-squares" norm and the L_2 norm

Let **C** be an arbitrary positive definite covariance operator over an space *δE* . The norm over *δE* has been defined by

$$\|\ \delta e\ \|^2 = \delta e^t\ \mathbf{C}^{-1}\ \delta e ,$$

and we have already seen that it may correspond to very different usual norms, such as for instance the usual L_2 norm (example 7.8) or the usual H^1 Sobolev norm (example 7.11). Why then use the the generic name of L_2 for such "covariance-related" norms?

The reason is the following. Using the method of Box 4.1 (conveniently generalized to the infinite-dimensional case) it is possible to introduce the square root of the covariance operator $\mathbf{C}$ by

$$\mathbf{C} = \mathbf{C}^{1/2} \ \mathbf{C}^{t/2} . \tag{7.35}$$

The space $\delta \tilde{E}$ with elements

$$\delta\tilde{\mathbf{e}} = \mathbf{C}^{-1/2} \ \delta\mathbf{e} \tag{7.36}$$

is isomorphic to δE . The norm over δE is then given by

$$\| \delta\tilde{\mathbf{e}} \|^2 = \| \delta\mathbf{e} \|^2 = \delta\mathbf{e}^t \ \mathbf{C}^{-1} \ \delta\mathbf{e} = (\mathbf{C}^{1/2} \ \delta\tilde{\mathbf{e}})^t \ \mathbf{C}^{-1} \ (\mathbf{C}^{1/2} \ \delta\tilde{\mathbf{e}})$$

$$= \delta\tilde{\mathbf{e}}^t \ \mathbf{C}^{t/2} \ \mathbf{C}^{-1} \ \mathbf{C}^{1/2} \ \delta\tilde{\mathbf{e}} = \delta\tilde{\mathbf{e}}^t \ \delta\tilde{\mathbf{e}} , \tag{7.37}$$

which is clearly the usual L_2 norm.

The situation is as follows: a covariance-related norm in an arbitrary working space may be very general; nevertheless, it is always possible (in principle) to introduce a (complicated) space, isomorphic to the working space, where the norm becomes L_2 . This allows mathematicians to make demonstrations by reduction to the well-known L_2 norm case, but it is of no interest for practical applications. This is why I prefer to talk about covariance-related norms (in the plural) rather than about the L_2 norm (in the singular).

7.2: Derivative operators and transposed operators in functional spaces

7.2.1: The derivative operator

Let us consider the (possibly nonlinear) operator $\mathbf{m} \rightarrow g(\mathbf{m})$ associating with any model $\mathbf{m}$ of the model space $\boldsymbol{M}$ a data vector $\mathbf{d} = g(\mathbf{m})$ of the data space $\boldsymbol{D}$. The *tangent linear application* to g at a point $\mathbf{m} = \mathbf{m}_0$ is the linear operator $\mathbf{G}_0$ defined by the first order development

$$g(\mathbf{m}_0+\delta\mathbf{m}) = g(\mathbf{m}_0) + \mathbf{G}_0 \ \delta\mathbf{m} + O(\| \delta\mathbf{m} \|^2) . \tag{7.38}$$

When applied to a scalar function $x \rightarrow g(x)$ of a one dimensional variable, this expression defines the *tangent* (in the usual geometrical sense) to the graph $(x , g(x))$ at x_0 (see Figure 7.7), thus justifying the terminology.

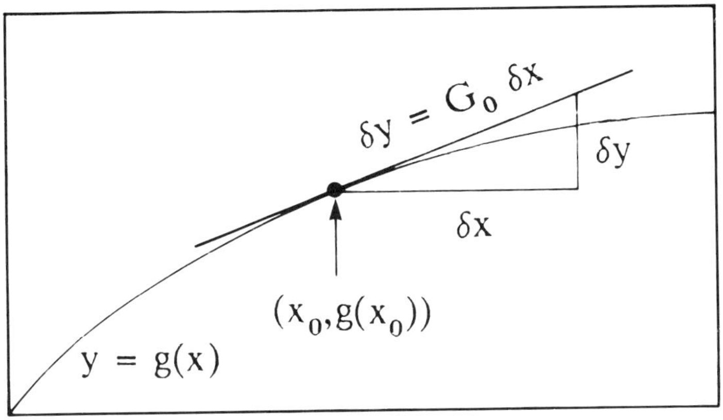

Figure 7.7: The *tangent linear application* to the nonlinear application $y = g(x)$ at the point $(x_0 , g(x_0))$ is the linear application whereby any δx is associated with $\delta y = G_0\, \delta x$, where G_0 is the *slope* of the curve $y = g(x)$ at x_0 . This definition generalizes to a nonlinear application $\mathbf{d} = g(\mathbf{m})$ between functional spaces (see text). The tangent linear application is then usually called the *Fréchet derivative* or simply the *derivative* at $\mathbf{m}_0$.

When $\boldsymbol{M}$ and/or $\boldsymbol{D}$ are functional spaces, the tangent linear application $\mathbf{G}_0$ is usually named the *Fréchet derivative* (or simply the *derivative*) of $\mathbf{g}$ at $\mathbf{m}_0$. Note that this terminology may be misleading, because for a one dimensional scalar function $x \rightarrow g(x)$, what is generally termed the derivative at a given point is the slope of the tangent (i.e., a number), and not the tangent itself (i.e., an application).

By definition, the operator $\mathbf{g}$ in (7.38) maps the space $\boldsymbol{M}$ into the space $\boldsymbol{D}$. As explained in section 7.1.5, these are usually affine vector spaces, and not necessarily linear vector spaces. The spaces $\Delta\boldsymbol{M} = \boldsymbol{M} - \mathbf{m}_0$ and $\Delta\boldsymbol{D} = \boldsymbol{D} - \mathbf{g}(\mathbf{m}_0)$ are then linear vector spaces. In all rigor, if $\mathbf{g}$ maps the affine vector space $\boldsymbol{M}$ into the affine vector space $\boldsymbol{D}$, the tangent linear application $\mathbf{G}_0$ maps the linear vector space $\Delta\boldsymbol{M}$ into the linear vector space $\Delta\boldsymbol{D}$.

Example 7.14: If $\boldsymbol{M}$ and $\boldsymbol{D}$ are discrete,

$$d^i = g^i(m^1, m^2, ...) \qquad (i=1,2,...) \ . \tag{7.39}$$

From the definition (7.38) it follows that the kernel of the operator $\mathbf{G}_0$ is a matrix. Its elements are given by

$$G_0^{\,i\alpha} = \left(\frac{\partial g^i}{\partial m^\alpha} \right)_{\mathbf{m}=\mathbf{m}_0} . \blacksquare \tag{7.40}$$

Example 7.15: Let $\boldsymbol{M}$ be a space of functions $z \rightarrow m(z)$, for $0 \leq z \leq Z$, $\boldsymbol{D}$ be a discrete N-dimensional space, and $\mathbf{m} \rightarrow g(\mathbf{m})$ be the nonlinear operator from $\boldsymbol{M}$ to $\boldsymbol{D}$ that associates the function $z \rightarrow m(z)$ with the following element of $\boldsymbol{D}$:

$$d^i = g^i(\mathbf{m}) = \int_0^Z dz\ \beta^i(z)\ (\ m(z)\ )^2 \qquad (i=1,2,...,N)\ , \tag{7.41}$$

where $\beta^i(z)$ are given functions. Let $z \rightarrow m_0(z)$ be a particular element of $\boldsymbol{M}$, denoted $\mathbf{m}_0$. We wish to compute the derivative of g at the point $\mathbf{m}_0$.

We have

$$g^i(\mathbf{m}_0+\delta\mathbf{m}) = \int_0^Z dz\ \beta^i(z)\ (\ m_0(z) + \delta m(z)\ )^2$$

$$= \int_0^Z dz\ \beta^i(z)\ (\ (m_0(z))^2 + 2\ m_0(z)\ \delta m(z) + (\delta m(z))^2\ )$$

$$= g^i(\mathbf{m}_0) + 2 \int_0^Z dz\ \ \beta^i(z)\ \ m_0(z)\ \delta m(z)\ \ + O^i(\,\|\,\delta\mathbf{m}\,\|^2)\ ,$$

so, using the definition (7.38),

$$(G_0\ \delta\mathbf{m})^i = 2 \int_0^Z dz\ \ \beta^i(z)\ \ m_0(z)\ \ \delta m(z)\ . \tag{7.42}$$

In words, the derivative at the point $\mathbf{m} = \mathbf{m}_0$ of the nonlinear operator $\mathbf{m} \rightarrow g(\mathbf{m})$ is the linear operator that associates any function $\delta\mathbf{m}$ (i.e., any function $z \rightarrow \delta m(z)$) with the vector given by (7.42).

Introducing an integral representation of G_0 :

$$(G_0\ \delta\mathbf{m})^i = \int_0^Z dz\ \ G_0^{\,i}(z)\ \ \delta m(z)\ , \tag{7.43}$$

gives the integral kernels of G_0 :

$$G_0^i(z) = 2\,\beta^i(z)\, m_0(z)\ . \ \blacksquare \tag{7.44}$$

Example 7.16: The Fréchet derivatives in the problem of X-ray tomography. A useful technique for obtaining images of the interior of a human body is (computerized) X-ray tomography. As the etymology of the word indicates, tomography consists in obtaining graphics of a *section* of a body. Typically, X-rays are sent between a point source and a point receiver which counts the number of photons not absorbed by the medium, thus giving an indication of the integrated attenuation coefficient along that particular ray path (see Figure 7.8). Repeating the measurement for many different ray paths, conveniently sampling the medium, the bidimensional structure of the attenuation coefficient can be inferred, and so, an image of the medium be obtained. Ray paths of X-rays through an animal body can be assimilated to straight lines with an excellent approximation.

More precisely, let ρ^i be the *transmittance* along the i-th ray, i.e., the *probability* of a photon being transmitted. Clearly, this can be obtained directly from the measurement of the flux of photons sent by the source and the flux arriving at the receiver. It can be shown (see, for instance, Herman, 1980) that ρ^i is given by

$$\rho^i = \exp\left(- \int_{R^i} ds^i \ m(\mathbf{x}^i) \right) , \tag{7.45}$$

where $m(\mathbf{x})$ represents the *linear attenuation coefficient* at point $\mathbf{x}$, and corresponds to the probability per unit length of path of a photon arriving at $\mathbf{x}$ being absorbed, R^i denotes the ray path, ds^i is the element of length along the ray path, and $\mathbf{x}^i$ denotes the current point considered in the line integral along R^i . In a three-dimensional experience, $\mathbf{x}$ is represented by the cartesian coordinates (x,y,z) or the spherical coordinates (r,θ,ϕ) .

The relationship between the data ρ^i and the unknown function $m(\mathbf{x})$ in (7.45) is not linear, but defining the new data

$$d^i = - \operatorname{Log} \rho^i \tag{7.46}$$

gives the linear relationship

$$d^i = \int_{R^i} ds^i \ m(\mathbf{x}^i) \ . \tag{7.47}$$

For more compactness, let me denote by $\mathbf{d} = \mathbf{g}(\mathbf{m})$ the resolution of this forward problem. As the operator $\mathbf{g}$ is linear, the definition (7.38) of the derivative of $\mathbf{g}$ at a point $\mathbf{m}_0$ gives

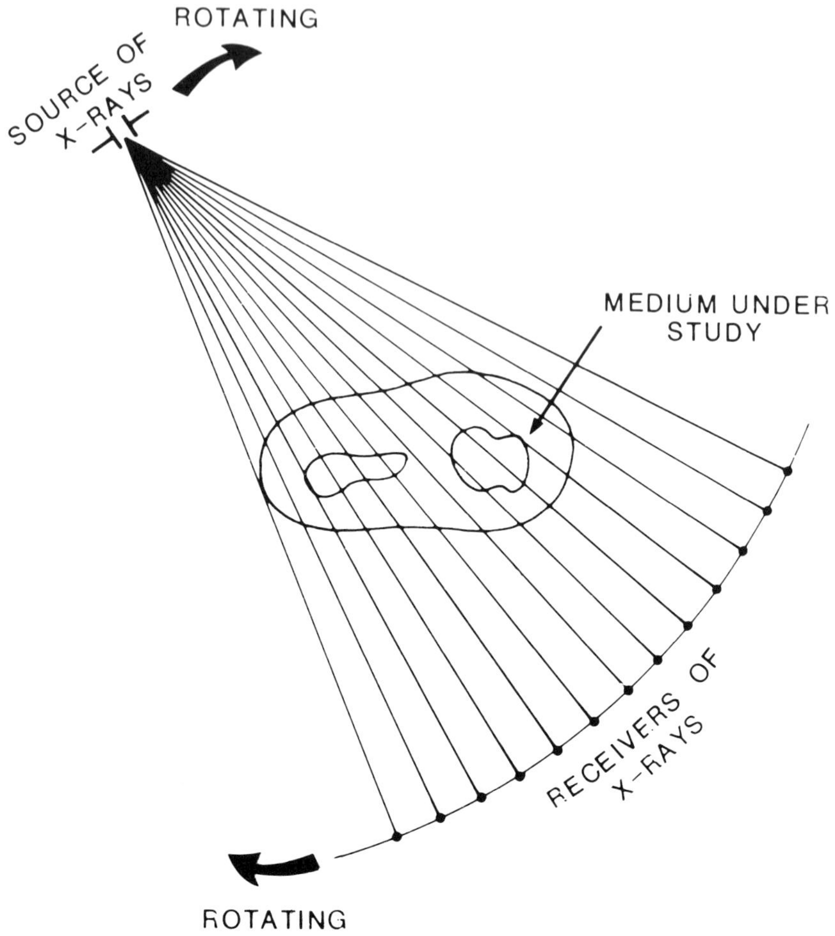

Figure 7.8: Schematical representation of a X-ray tomography experiment. Typically, a source sends a beam of X-rays, and the number of photons arriving at each of the receivers is counted. With a sufficient accuracy, photons can be assumed to propagate along stright lines. In the example in the figure, source and receivers are slowly rotating around the medium under study. Each particular measurement of number of photons arriving at a source brings information about the integrated attenuation along a particular line across the medium. When sufficiently many integrated attenuations have been measured, with sufficiently different azimuths, the bidimensional structure of the attenuation coefficient can be inferred, thus obtaining a "tomography" (i.e., an image of a section of the 3D medium).

$$\mathbf{G}_0 \, \delta\mathbf{m} \; = \; \mathbf{g}(\delta\mathbf{m}) \; , \tag{7.48}$$

which simply means that the tangent linear application to a linear application is the linear application itself. As $\mathbf{G}_0$ does not depend on $\mathbf{m}_0$, the index $(_0)$ can be dropped. For linear problems like this one, the notation

$$\mathbf{d} \; = \; \mathbf{g}(\mathbf{m})$$

is always replaced by the notation

$$\mathbf{d} \; = \; \mathbf{G} \, \mathbf{m} \; . \tag{7.49}$$

Let us see how can we introduce the kernels, $G^i(\mathbf{x})$, of $\mathbf{G}$. By definition,

$$d^i \; = \; \int_{\mathbf{V}} d\mathbf{x} \;\; G^i(\mathbf{x}) \;\; m(\mathbf{x}) \; , \tag{7.50}$$

where $\mathbf{V}$ denotes the volume under study. I introduce a "delta-like" function $\Delta^i(\mathbf{x})$ which is null everywhere in the space except along the i-th ray path. It is formally defined by

$$\int_{R^i} ds^i \;\; m(\mathbf{x}^i) \; = \; \int_{\mathbf{V}} d\mathbf{x} \;\; \Delta^i(\mathbf{x}) \;\; m(\mathbf{x}) \; . \tag{7.51}$$

Then,

$$d^i \; = \; (\, \mathbf{G} \, \mathbf{m} \,)^i \; = \; \int_{\mathbf{V}} d\mathbf{x} \;\; \Delta^i(\mathbf{x}) \;\; m(\mathbf{x}) \; ,$$

and

$$G^i(\mathbf{x}) \; = \; \Delta^i(\mathbf{x}) \; . \; \blacksquare \tag{7.52}$$

Example 7.17: The Fréchet derivatives in Geophysical acoustic tomography. To infer the velocity structure of a medium, acoustic waves are generated by some sources, and the travel times to some receivers are measured. The main difference (and difficulty) with respect to the problem of X-ray tomography is that geophysical media are often highly heterogeneous, and the ray-paths depend on the velocity structure (see Figure 7.9), so that there is no way to make the problem strictly linear.

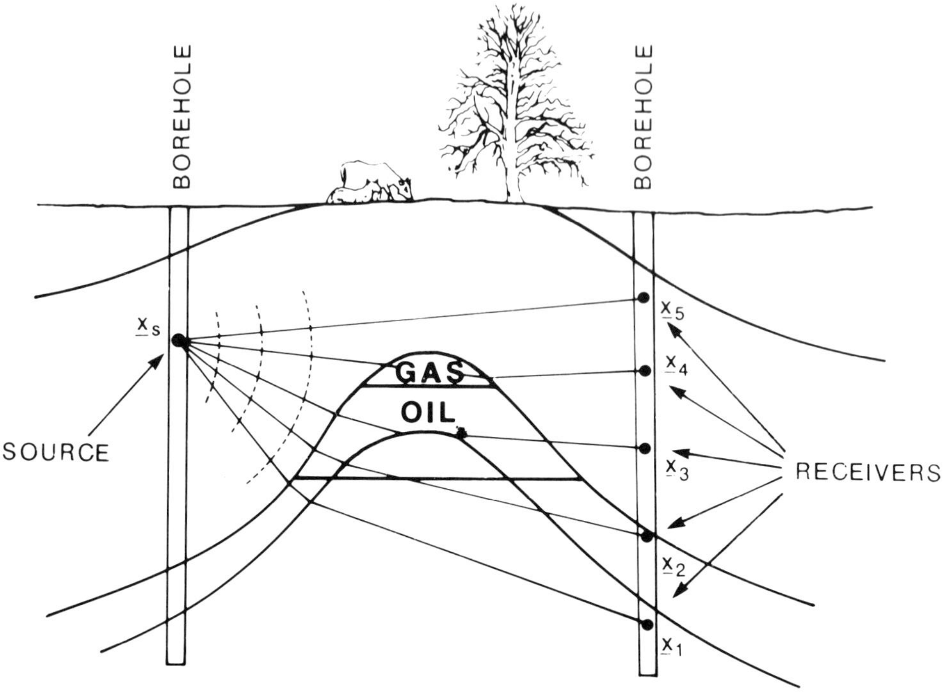

Figure 7.9: Typical experiment of geophysical tomography. A source of seismic waves is "shot" at different positions inside a borehole. For each shot position, some receivers are placed in boreholes around the region under study which record the time of arrival of the first wave front. The purpose of the experiment is to infer the acoustic structure of the medium. This problem is very similar to that of X-ray tomography, except that the ray-paths are a priori unknown, because they depend on the actual structure of the medium. From the point of view of inverse theory, the equation solving the forward problem is, in this case, essentially nonlinear.

Here we assume that the high-frequency limit is acceptable, i.e., ray theory can be used instead of wave theory. If $c(\mathbf{x})$ denotes the celerity of the waves at point $\mathbf{x}$, let $m(\mathbf{x})$ be the slowness

$$m(\mathbf{x}) = \frac{1}{c(\mathbf{x})} . \tag{7.53}$$

The i-th datum is the travel time for the i-th ray:

$$d^i = g^i(\mathbf{m}) = \int_{R^i(\mathbf{m})} ds^i \ m(\mathbf{x}^i) \ , \tag{7.54}$$

where $R^i(\mathbf{m})$ denotes the i-th ray path. As this ray path depends on $\mathbf{m}$, d^i is a nonlinear function of $\mathbf{m}$. Given a medium $\mathbf{m}$, the actual ray path is obtained using Fermat's theorem (or, equivalently, the eikonal equation) and some numerical method. Here we wish to obtain the derivative of the nonlinear operator g^i at a point $\mathbf{m}_0$. We have

$$g^i(\mathbf{m}_0+\delta\mathbf{m}) = \int_{R^i(\mathbf{m}_0+\delta\mathbf{m})} ds^i \ (\ m_0(\mathbf{x}^i) + \delta m(\mathbf{x}^i) \ ) \ .$$

The travel time being stationary along the actual ray path (Fermat's theorem),

$$\int_{R^i(\mathbf{m}_0+\delta\mathbf{m})} ds^i \ (\ m_0(\mathbf{x}^i) + \delta m(\mathbf{x}^i) \ ) =$$

$$= \int_{R^i(\mathbf{m}_0)} ds^i \ (\ m_0(\mathbf{x}^i) + \delta m(\mathbf{x}^i) \ ) + O^i(\| \delta\mathbf{m} \|^2) \ . \tag{7.55}$$

This gives

$$g^i(\mathbf{m}_0+\delta\mathbf{m}) = g^i(\mathbf{m}_0) + \int_{R^i(\mathbf{m}_0)} ds^i \ \delta m(\mathbf{x}^i) + O^i(\| \delta\mathbf{m} \|^2) \ . \tag{7.56}$$

Comparison with definition (7.38) directly gives the derivative operator:

$$(\mathbf{G}_0 \ \delta\mathbf{m})^i = \int_{R^i(\mathbf{m}_0)} ds^i \ \delta m(\mathbf{x}^i) \ . \tag{7.57}$$

The derivative of $\mathbf{g}$ at the point $\mathbf{m}_0$ associates the travel-time perturbations (7.57) to the model perturbation $\delta\mathbf{m}$.

Equation (7.57) is well adapted to all numerical computations involving the derivative operator $\mathbf{G}_0$, but, for analytic developments it is sometimes useful to introduce the kernel of $\mathbf{G}_0$. Introducing a "delta-like" function as in the previous example by

$$\int_{R^i(\mathbf{m}_0)} ds^i \ \delta m(\mathbf{x}^i) = \int_{\mathbf{V}} d\mathbf{x} \ \Delta_0{}^i(\mathbf{x}) \ \delta m(\mathbf{x}) \ , \tag{7.58}$$

and from the integral representation

$$(G_0 \ \delta\mathbf{m})^i = \int_{\mathbf{V}} d\mathbf{x} \ G_0{}^i(\mathbf{x}) \ \delta m(\mathbf{x}) \ , \tag{7.59}$$

we directly obtain

$$(G_0 \ \delta\mathbf{m})^i = \int_{\mathbf{V}} d\mathbf{x} \ \Delta_0{}^i(\mathbf{x}) \ \delta m(\mathbf{x}) \ ,$$

and

$$G_0{}^i(\mathbf{x}) = \Delta_0{}^i(\mathbf{x}) \ . \ \blacksquare \tag{7.60}$$

Example 7.18: The Fréchet derivatives in the problem of inversion of acoustic waveforms. Consider a three-dimensional unbounded medium supporting the propagation of acoustic waves. The waves are assumed to propagate according to the wave equation

$$\frac{1}{c^2(\mathbf{x})} \frac{\partial^2 p}{\partial t^2}(\mathbf{x},t) - \nabla^2 p(\mathbf{x},t) = S(\mathbf{x},t) \tag{7.61a}$$

where $\mathbf{x}$ denotes a point of the medium, t is time, $p(\mathbf{x},t)$ is the pressure (variation) at point $\mathbf{x}$ and time t , $S(\mathbf{x},t)$ is the source function, and $c(\mathbf{x})$ represents the velocity (or, more properly, the *celerity*) of the pressure waves at point $\mathbf{x}$. The pressure field $p(\mathbf{x},t)$ is supposed to be at rest at the initial time:

$$p(\mathbf{x},0) = 0 \tag{7.61b}$$

$$\dot{p}(\mathbf{x},0) = 0 \ , \tag{7.61c}$$

Assume that we have a complete control of the source $S(\mathbf{x},t)$, so that we can send into the medium any type of wave we wish (usually the source has limited spatial extension, and we send quasi-spherical waves), and that we have some sensors which are able to measure the actual value of the pressure at arbitrary points in the medium.

Assume we start an experiment at $t = 0$, sending waves into the medium by an appropiate choice of the source function $S(\mathbf{x},t)$. From $t = 0$ to $t = T$ we record the pressure at some (finite number of) points $\mathbf{x}_r$ $(r=1,2,...)$. We wish to use the actual results of our measurements to infer the actual value of the function $c(\mathbf{x})$ characterizing the medium (see Figure 7.10). This is, of course, an inverse problem and will be solved in the next sections. Let us first consider the resolution of the forward problem.

Our data space $\boldsymbol{D}$ consists in all conceivable realizations for the results of our measurements:

$$\mathbf{d} = [\ p(\mathbf{x}_r,t) \qquad \text{for} \quad 0 \le t \le T \qquad \text{and} \quad r=1,2,...\ ] .$$

For more simplicity in the analytical derivations, let us consider the *squared slowness* instead of the celerity:

$$m(\mathbf{x}) = \frac{1}{c^2(\mathbf{x})} \tag{7.62}$$

Our model space $\boldsymbol{M}$ consists then in all possible models $m(\mathbf{x})$.

The original differential system becomes

$$m(\mathbf{x}) \frac{\partial^2 p}{\partial t^2}(\mathbf{x},t) - \nabla^2 p(\mathbf{x},t) = S(\mathbf{x},t) \qquad t \in [0,T] \tag{7.63a}$$

$$p(\mathbf{x},0) = 0 \tag{7.63b}$$

$$\dot{p}(\mathbf{x},0) = 0 . \tag{7.63c}$$

Given a model $\mathbf{m}$, to solve the forward problem means to compute the data vector $\mathbf{d}$ predicted from the knowledge of $\mathbf{m}$. This is denoted by

$$\mathbf{d} = \mathbf{g}(\mathbf{m}) . \tag{7.64}$$

The most general way of solving the forward problem is by replacing the differential equation (7.63a) by its finite-difference approximation (see, for instance, Altermann and Karal, 1968) and to compute numerically the approximated values of the pressure at time $t+\Delta t$ from the knowledge of the approximated values of the pressure at times t and $t-\Delta t$. This gives the pressure field everywhere. The forward problem is solved by just picking the values obtained at the receiver locations $\mathbf{x}_r$.

Let $\mathbf{m}_0$ denote a particular model. We wish to compute the derivative of $\mathbf{g}$ at $\mathbf{m} = \mathbf{m}_0$. By definition,

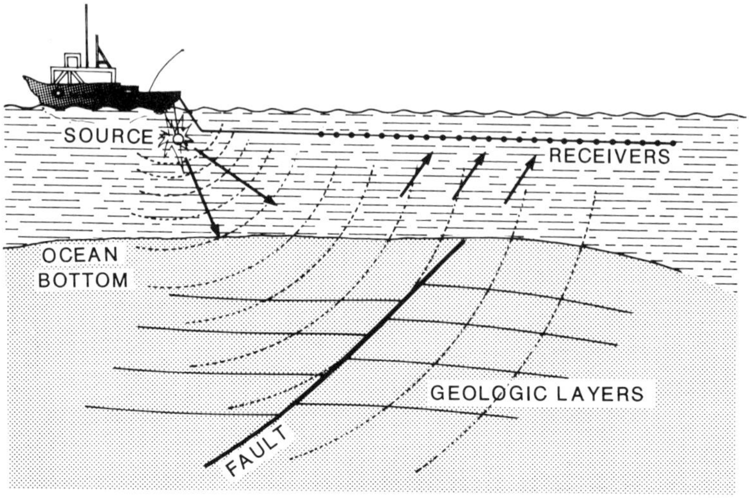
SOURCE
RECEIVERS
OCEAN
BOTTOM
GEOLOGIC LAYERS
FAULT

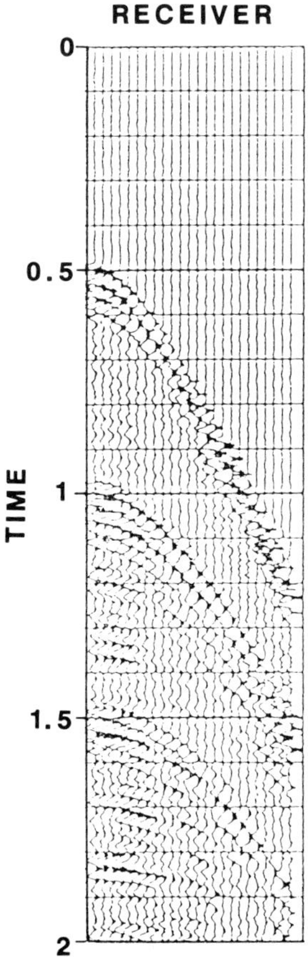
RECEIVER
TIME
0
0.5
1
1.5
2

← *Figure 7.10:* Typical marine seismic reflection experiment. *Top:* Pressure waves are generated by some source, they interact with the medium, and the total pressure field is recorded by some receivers. The ship advances regularly, and shots a source from time to time (typically one shot every 20 meters). The problem is to infer the seismic structure of the medium. *Bottom:* A real example of recorded seismograms for a given shot point. A typical marine seismic reflection experiment produces as many as 10^8 bits of measurements if only a 2D vertical section of the medium is sought, and as many as 10^{11} bits for a 3D experiment. In example 7.18, an (incorrect) acoustic assumption has been made. The real Earth is elastic, and the corresponding theory is developed in problem 7.3.

$$\mathbf{g}(\mathbf{m}_0+\delta\mathbf{m}) = \mathbf{g}(\mathbf{m}_0) + \mathbf{G}_0\,\delta\mathbf{m} + O(\delta\mathbf{m}^2)\ . \tag{7.65}$$

Let us denote

$$\delta\mathbf{p} = \mathbf{G}_0\,\delta\mathbf{m}\ . \tag{7.66}$$

As demontrated in problem 7.2, $\delta p(\mathbf{x}_r,t)$ is the value at $\mathbf{x}_r$ and time t of the field $\delta p(\mathbf{x},t)$ solution of

$$m_0(\mathbf{x})\,\frac{\partial^2\delta p}{\partial t^2}(\mathbf{x},t) - \nabla^2\delta p(\mathbf{x},t) = -\,\delta m(\mathbf{x})\,\frac{\partial^2 p_0}{\partial t^2}(\mathbf{x},t) \qquad t\in[0,T] \tag{7.67a}$$

$$\delta p(\mathbf{x},0) = 0 \tag{7.67b}$$

$$\delta\dot{p}(\mathbf{x},0) = 0\ . \tag{7.67c}$$

where $p_0(\mathbf{x},t)$ is the field propagating in the model $\mathbf{m}_0$:

$$m_0(\mathbf{x})\,\frac{\partial^2 p_0}{\partial t^2}(\mathbf{x},t) - \nabla^2 p_0(\mathbf{x},t) = S(\mathbf{x},t) \qquad t\in[0,T] \tag{7.68a}$$

$$p_0(\mathbf{x},0) = 0 \tag{7.68b}$$

$$\dot{p}_0(\mathbf{x},0) = 0\ . \tag{7.68c}$$

To compute $\delta p(\mathbf{x},t)$ we then essentially need to solve a forward problem (and we have already seen how to do this) using $-\delta m(\mathbf{x})\ \partial^2 p_0/\partial t^2(\mathbf{x},t)$ as source term and propagating the waves in the "unperturbed" medium

$m_0(\mathbf{x})$. If we are able to compute $\delta p(\mathbf{x},t)$, then, from (7.66) we know the derivative operator $\mathbf{G}_0$. In words, $\mathbf{G}_0$ is the linear operator that associates the values $\delta p(\mathbf{x}_r,t)$, obtained by the resolution of (7.67), to the model perturbation $\delta m(\mathbf{x})$.

Let me point out here that to estimate data $\mathbf{g}(\mathbf{m}_0+\delta\mathbf{m})$ using the first order approximation

$$\mathbf{g}(\mathbf{m}_0+\delta\mathbf{m}) \simeq \mathbf{g}(\mathbf{m}_0) + \mathbf{G}_0\,\delta\mathbf{m} \tag{7.69}$$

corresponds to *Born's approximation* (see Box 7.3). We are not using here Born's approximation. We are simply computing Fréchet derivatives.

As shown in problem 7.2, the kernel $G_0(\mathbf{x}_r,t;\mathbf{x})$ introduced by the integral representation

$$\delta p(\mathbf{x}_r,t) = \int_V d\mathbf{x}\ G_0(\mathbf{x}_r,t;\mathbf{x})\ \delta m(\mathbf{x}) \tag{7.70}$$

is given by

$$G_0(\mathbf{x}_r,t;\mathbf{x}) = -\,\Gamma_0(\mathbf{x}_r,t;\mathbf{x},0)\ \frac{\partial^2 p_0}{\partial t^2}(\mathbf{x},t)\ , \tag{7.71}$$

where $\Gamma_0(\mathbf{x},t;\mathbf{x}',0)$ is Green's function, defined as the solution of

$$m_0(\mathbf{x})\ \frac{\partial^2\Gamma_0}{\partial t^2}(\mathbf{x},t;\mathbf{x}',t') - \nabla^2\Gamma_0(\mathbf{x},t;\mathbf{x}',t') = \delta(\mathbf{x}-\mathbf{x}')\ \delta(t-t') \tag{7.72a}$$

$$\Gamma_0(\mathbf{x},t;\mathbf{x}',t') = 0 \qquad \text{for}\ \ t < t' \tag{7.72b}$$

$$\frac{\partial\Gamma_0}{\partial t}(\mathbf{x},t;\mathbf{x}',t') = 0 \qquad \text{for}\ \ t < t'\ . \tag{7.72c}$$

■

7.2.2: The transposed operator

Let $\mathbf{G}$ represent a linear operator from a linear space δM into a linear space $\boldsymbol{\delta D}$. Its *transposed*, $\mathbf{G}^t$, is a linear operator mapping $\delta\hat{D}$ into $\delta\hat{M}$ defined by

$$\langle\ \mathbf{G}^t\,\delta\hat{\mathbf{d}}\ ,\ \delta\mathbf{m}\ \rangle_M = \langle\ \delta\hat{\mathbf{d}}\ ,\ \mathbf{G}\,\delta\mathbf{m}\ \rangle_D \qquad (\text{for any}\ \ \delta\hat{\mathbf{d}}\ \ \text{and}\ \ \delta\mathbf{m}\ )\ . \tag{7.73}$$

The reader should refer to Box 7.1 for (important) details.

Notice that the transposed of a linear operator is defined independently of any particular choice of scalar products over $\delta\boldsymbol{M}$ and $\delta\boldsymbol{D}$.

Again, let G represent a linear operator from a linear space $\delta\boldsymbol{M}$ into a linear space $\delta\boldsymbol{D}$, and assume now that $\delta\boldsymbol{M}$ and $\delta\boldsymbol{D}$ are scalar product vector spaces. The *adjoint* of G , G^* , is a linear operator mapping $\delta\boldsymbol{D}$ into $\delta\boldsymbol{M}$ defined by the equality of scalar products

$$(\mathbf{G}^* \ \delta\mathbf{d} \ , \delta\mathbf{m} \ )_M = (\ \delta\mathbf{d} \ , \mathbf{G} \ \delta\mathbf{m} \ )_D \qquad \text{(for any } \delta\mathbf{d} \text{ and } \delta\mathbf{m} \text{) .} \tag{7.74}$$

We successively have

$$(\mathbf{G}^* \ \delta\mathbf{d} \ , \delta\mathbf{m} \ )_M = (\ \delta\mathbf{d} \ , \mathbf{G} \ \delta\mathbf{m} \ )_D = \langle \ \delta\hat{\mathbf{d}} \ , \mathbf{G} \ \delta\mathbf{m} \ \rangle_D$$

$$= \langle \ \mathbf{G}^t \ \delta\hat{\mathbf{d}} \ , \delta\mathbf{m} \ \rangle_M = (\ \mathbf{C}_M \ \mathbf{G}^t \ \delta\hat{\mathbf{d}} \ , \delta\mathbf{m} \ )_M$$

$$= (\ \mathbf{C}_M \ \mathbf{G}^t \ \mathbf{C}_D^{-1} \ \delta\mathbf{d} \ , \delta\mathbf{m} \ )_M \ ,$$

i.e.,

$$\mathbf{G}^* = \mathbf{C}_M \ \mathbf{G}^t \ \mathbf{C}_D^{-1} \ . \tag{7.75}$$

By linguistic abuse, the terms *adjoint* and *transpose* are sometimes used as synonymous. The last equation shows that they are not.

Let us see some examples practically illustrating the operational aspects of the definition of the transposed of a linear operator. In discrete spaces, the linear equation

$$\delta\mathrm{d}_1 = \mathrm{G} \ \delta\mathrm{m}_1$$

is explicitly written

$$\delta\mathrm{d}_1{}^i = \sum_\alpha \mathrm{G}^{i\alpha} \ \delta\mathrm{m}_1{}^\alpha \ , \tag{7.76a}$$

and the array $(G^{i\alpha})$ is termed the *kernel* of G . If the indexes i and α are simple integers, then the array $(G^{i\alpha})$ is a usual (two-dimensional) matrix. The linear equation

$$\delta\hat{\mathbf{m}}_2 = \mathbf{G}^t \ \delta\hat{\mathbf{d}}_2$$

is explicitly written

$$\delta m_2{}^\alpha = \sum_i (G^t)^{\alpha i} \, \delta d_2{}^i \ , \tag{7.76b}$$

and the definition (7.73) gives directly

$$(G^t)^{\alpha i} = (G)^{i\alpha} \ , \tag{7.76c}$$

which simply shows that the kernel of an operator and the kernel of the corresponding transposed operator are essentially the same, modulo a "transposition" of the variables. In particular, if the kernel of the linear operator $\mathbf{G}$ is a matrix, the kernel of the transposed operator is simply the transposed of the matrix. If we consider more general spaces, the linear equation $\boldsymbol{\delta}\mathbf{d}_1 = \mathbf{G}\ \boldsymbol{\delta}\mathbf{m}_1$ may take, for instance, the explicit form

$$\delta d_1{}^{ij\cdots}(u,v,...) = \tag{7.77a}$$

$$= \sum_\alpha \sum_\beta \cdots \int dx \int dy \ ... \quad G^{ij\ldots\alpha\beta\cdots}(u,v,...,x,y,...) \ \ \delta m_1{}^{\alpha\beta\cdots}(x,y,...)$$

The linear equation $\boldsymbol{\delta}\hat{\mathbf{m}}_2 = \mathbf{G}^t\ \boldsymbol{\delta}\hat{\mathbf{d}}_2$ is then explicitly written

$$\delta\hat{m}_2{}^{\alpha\beta\cdots}(x,y,...) = \tag{7.77b}$$

$$= \sum_i \sum_j \cdots \int du \int dv \ ... \quad (G^t)(\alpha\beta...ij...)(x,y,...,u,v,...) \ \ \delta\hat{d}_2{}^{ij\cdots}(u,v,...)$$

and the definition of $\mathbf{G}^t$ gives then

$$(G^t)^{\alpha\beta\ldots ij\cdots}(x,y,...,u,v,...) = G^{ij\ldots\alpha\beta\cdots}(u,v,...,x,y,...) \ , \tag{7.77c}$$

thus generalizing the notion of "variable transposition". In words, "if the kernel of a linear operator is $G^{ij\ldots\alpha\beta\cdots}(u,v,...,x,y,...)$, and the application of $\mathbf{G}$ implies sums (or integrals) over the variables $\alpha,\beta,...,x,y,...$, then the kernel of the transposed operator is essentially the same, and the application of $\mathbf{G}^t$ implies sums over the other variables $i,j,...,u,v,...$ ".

As a further example, consider the operator $\hat{\mathbf{G}}$ mapping the space $\delta \mathcal{M}$ into the space $\boldsymbol{\delta \mathcal{D}}$. $\mathbf{G}^t$ then maps $\boldsymbol{\delta}\hat{\boldsymbol{\mathcal{D}}}$ into $\boldsymbol{\delta}\hat{\boldsymbol{\mathcal{M}}}$. If $\mathbf{C}$ is a covariance operator over $\boldsymbol{\delta \mathcal{M}}$, it maps $\boldsymbol{\delta}\hat{\boldsymbol{\mathcal{M}}}$ into $\boldsymbol{\delta \mathcal{M}}$, and it makes sense to consider the operator

$$\mathbf{U} = \mathbf{C}\ \mathbf{G}^t \ , \tag{7.78}$$

mapping $\delta\hat{D}$ into δM . Let $G^{ij...\alpha\beta...}(u,v,...,x,y,...)$ and $C^{\alpha\beta...\alpha'\beta'...}(x,y,...,x',y',...)$ be respectively the kernels of **G** and **C** . As can easily be verified, the kernel of **U** is then

$$U^{\alpha\beta...ij...}(x,y,...,u,v,...) = \tag{7.79}$$

$$= \sum_{\alpha'}\sum_{\beta'} ... \int dx' \int dy' ... G^{\alpha\beta...\alpha'\beta'...}(x,y,...,x',y',...)\ G^{ij...\alpha'\beta'...}(u,v,...,x',y',...)\ .$$

In inverse problems, we always have to consider the forward equation $\mathbf{d} = \mathbf{g}(\mathbf{m})$, and the operator $\mathbf{G}_n$, the derivative of $\mathbf{g}$ at some point $\mathbf{m}_n$. In all the formulas for least squares inversion the transposed operator $\mathbf{G}_n{}^t$ appears (see next section), and, when using gradient methods of resolution, is in fact an iterative application of $\mathbf{G}_n{}^t$ to the residuals which performs the data inversion (see Chapter 4). The understanding of the meaning of the transposed of a linear operator is very important for functional inverse problems. In particular it is important to understand that to compute a vector $\mathbf{G}^t\ \delta\mathbf{d}$ it is *not* necessary to explicitly use the kernel of **G** (see example 7.19 below).

Example 7.19: We wish here to obtain the transposed of the operator **G** obtained in the problem of X-ray tomography (example 7.16). We have

$$(\ \mathbf{G}\ \mathbf{m}\ )^i = \int_{R^i} ds^i\ \ m(\mathbf{x}^i)\ . \tag{7.47 again}$$

The definition

$$\langle\ \mathbf{G}^t\ \hat{\mathbf{d}}\ ,\ \mathbf{m}\ \rangle_M = \langle\ \hat{\mathbf{d}}\ ,\ \mathbf{G}\ \mathbf{m}\ \rangle_D\ , \tag{7.73 again}$$

is written, explicitly,

$$\int_{\mathbf{V}} d\mathbf{x}\ \ (\ \mathbf{G}^t\ \hat{\mathbf{d}}\ )(\mathbf{x})\ \ m(\mathbf{x}) = \sum_i \hat{d}^i \int_{R^i} ds^i\ \ m(\mathbf{x}^i)\ . \tag{7.80}$$

We will see later (problem 7.2) that this expression characterizes the transposed operator sufficiently well for practical computations in inversion. But let us try here to obtain a more compact representation. Using the definition of the delta-like function $\Delta^i(\mathbf{x})$

$$\int_{R^i} ds^i\ \ m(\mathbf{x}) = \int_{\mathbf{V}} d\mathbf{x}\ \ \Delta^i(\mathbf{x})\ \ m(\mathbf{x})\ , \tag{7.51 again}$$

equation (7.80) can be rewritten

$$\int_{\mathbf{V}} d\mathbf{x} \;\; m(\mathbf{x}) \left[(\mathbf{G}^t \; \hat{\mathbf{d}})(\mathbf{x}) \; - \; \sum_i \hat{d}^i \; \Delta^i(\mathbf{x}) \right] \;=\; 0 \; ,$$

and, this being satisfied for any $\mathbf{m}$, it follows that

$$(\mathbf{G}^t \; \hat{\mathbf{d}})(\mathbf{x}) \;=\; \sum_i \hat{d}^i \; \Delta^i(\mathbf{x}) \; . \tag{7.81}$$

Notice that this last result can also be obtained directly from the knowledge of the kernel of $\mathbf{G}$:

$$G^i(\mathbf{x}) \;=\; \Delta^i(\mathbf{x}) \; , \qquad \text{(7.52 again)}$$

and the rule of variable transposition. ■

Example 7.20: In this example we wish to obtain the transposed of the operator $\mathbf{G}_0$ appearing in the problem of inversion of pressure waveforms (example 7.18). The kernel of $\mathbf{G}_0$ was

$$G_0(\mathbf{x}_r,t;\mathbf{x}) \;=\; - \; \Gamma_0(\mathbf{x}_r,t;\mathbf{x},0) \; \frac{\partial^2 p_0}{\partial t^2}(\mathbf{x},t) \; . \qquad \text{(7.71 again)}$$

The linear equation

$$\delta \mathbf{d}_1 \;=\; \mathbf{G}_0 \;\; \delta \mathbf{m}_1$$

is written explicitly

$$\delta d_1(\mathbf{x}_r,t) \;=\; \int d\mathbf{x} \;\; G_0(\mathbf{x}_r,t,\mathbf{x}) \;\; \delta m_1(\mathbf{x}) \; ,$$

then the equation

$$\delta \hat{\mathbf{m}} \;=\; \mathbf{G}_0{}^t \;\; \delta \hat{\mathbf{d}}$$

involving $\mathbf{G}_0{}^t$ will be written explicitly

$$\delta \hat{m}(\mathbf{x}) \;=\; \sum_r \int_0^T dt \;\; (G_0{}^t)(\mathbf{x},\mathbf{x}_r,t) \;\; \delta \hat{p}(\mathbf{x}_r,t)$$

(receiver positions are assumed discrete), and, using the rule that an operator and its transposed have same kernels,

$$\delta\hat{m}(\mathbf{x}) = \sum_r \int_0^T dt \ G_0(\mathbf{x}_r,t,\mathbf{x}) \ \delta\hat{p}(\mathbf{x}_r,t) \ .$$

As shown in problem 7.2 this finally gives

$$\delta\hat{m}(\mathbf{x}) = \int_0^T dt \ \dot{\Psi}_0(\mathbf{x},t) \ \dot{p}_0(\mathbf{x},t) \ , \tag{7.82}$$

where $\Psi_0(\mathbf{x},t)$ is the field defined by

$$m_0(\mathbf{x}) \ \frac{\partial^2 \Psi_0}{\partial t^2}(\mathbf{x},t) - \nabla^2\Psi_0(\mathbf{x},t) = \sum_r \delta(\mathbf{x}-\mathbf{x}_r) \ \delta\hat{p}(\mathbf{x}_r,t) \tag{7.83a}$$

$$\Psi_0(\mathbf{x},T) = 0 \tag{7.83b}$$

$$\dot{\Psi}_0(\mathbf{x},T) = 0 \ , \tag{7.83c}$$

where it should be noticed that there are *final* (instead of initial) conditions.

We see thus that to evaluate the action of the operator $G_0{}^t$ over $\delta\hat{p}(x_r,t)$, we first have to solve a propagation problem (7.83) *reversed in time*, then to compute the time correlation (7.82). ■

Box 7.1: Transposed and adjoint of a differential operator

Let $\mathcal{M}$ and $\mathcal{D}$ represent two vector spaces, and $\hat{\mathcal{M}}$ and $\hat{\mathcal{D}}$ their respective duals. The duality product of $\hat{\mathbf{d}}_1 \in \hat{\mathcal{D}}$ by $\mathbf{d}_2 \in \mathcal{D}$ (resp. of $\hat{\mathbf{m}}_1 \in \hat{\mathcal{M}}$ by $\mathbf{m}_2 \in \mathcal{M}$) is denoted $\langle \hat{\mathbf{d}}_1 , \mathbf{d}_2 \rangle_D$ (resp. $\langle \hat{\mathbf{m}}_1 , \mathbf{m}_2 \rangle_M$).

Let $\mathbf{G}$ be a linear operator mapping $\mathcal{M}$ into $\mathcal{D}$. If the spaces $\mathcal{M}$ and $\mathcal{D}$ are finite dimensional discrete spaces, or if $\mathbf{G}$ is an ordinary *integral operator* between functional spaces, the definitions of transposed and adjoint are as in Box 4.5. If $\mathbf{G}$ is a *differential operator* between functional spaces, these definitions need some generalization.

Let $\mathbf{G}$ be a differential operator mapping the functional space $\mathcal{M}$ into the functional space $\mathcal{D}$, $\mathbf{x} = (x^1,x^2,x^3,...)$ the (common) variables of these functional spaces, $\mathbf{V}$ the (generalized) volume into consideration, $\mathbf{S}$ its boundary, and $n^i(\mathbf{x})$ the (contravariant) components of the outward normal unit vector on $\mathbf{S}$. The *formal transposed* of $\mathbf{G}$, $\mathbf{G}^t$, is

(...)

the unique operator mapping $\hat{D}$ into $\hat{M}$ such that the difference $\langle\, \hat{\mathbf{d}} , \mathbf{G}\,\mathbf{m} \,\rangle_D - \langle\, \mathbf{G}^t\, \hat{\mathbf{d}} , \mathbf{m} \,\rangle_M$ equals the volume integral of the divergence of a certain bilinear form $\mathbf{P}[\hat{\mathbf{d}},\mathbf{m}]$:

$$\langle\, \hat{\mathbf{d}} , \mathbf{G}\,\mathbf{m} \,\rangle_D - \langle\, \mathbf{G}^t\, \hat{\mathbf{d}} , \mathbf{m} \,\rangle_M = \int_V dV(\mathbf{x})\ \frac{DP^i[\hat{\mathbf{d}},\mathbf{m}]}{Dx^i}(\mathbf{x}) \ , \qquad (1a)$$

where an implicit sum over i is assumed. Using Green's theorem, this difference can be written as the boundary integral of the flux of the bilinear form:

$$\langle\, \hat{\mathbf{d}} , \mathbf{G}\,\mathbf{m} \,\rangle_D - \langle\, \mathbf{G}^t\, \hat{\mathbf{d}} , \mathbf{m} \,\rangle_M = \int_S dS(\mathbf{x})\ n_i(\mathbf{x})\ P^i[\hat{\mathbf{d}},\mathbf{m}](\mathbf{x}) \ . \qquad (1b)$$

Although $\mathbf{G}^t$ is uniquely defined by (1a) or (1b), the vector $\mathbf{P}[\,\cdot\,,\,\cdot\,]$ is defined to the addition of a divergenceless vector.

Example: The transposed of the gradient operator is minus the divergence operator. Let V denote a volume in the physical (euclidean) 3D space, bounded by a surface S , and let (x,y,z) denote cartesian coordinates. We consider a space of functions $m(x,y,z)$ defined inside (and at the surface of) V . The gradient operator

$$\boldsymbol{\nabla} = \begin{bmatrix} \partial/\partial x \\ \partial/\partial y \\ \partial/\partial z \end{bmatrix} \qquad (2)$$

associates to $m(x,y,z)$ its gradient

$$\mathbf{d} = \boldsymbol{\nabla}\,\mathbf{m} = \begin{bmatrix} \partial m/\partial x \\ \partial m/\partial y \\ \partial m/\partial z \end{bmatrix} . \qquad (3)$$

Let us verify that the transposed of the gradient operator equals the divergence operator with reversed sign:

$$\boldsymbol{\nabla}^t = -\,[\partial/\partial x \quad \partial/\partial y \quad \partial/\partial z] = -\,\boldsymbol{\nabla}\cdot \ . \qquad (4)$$

For any $\hat{\mathbf{d}} \in \hat{D}$ and any $\mathbf{m} \in M$ we have

(...)

$$\langle\, \hat{\mathbf{d}} \,,\, \nabla\, \mathbf{m} \,\rangle_D - \langle\, \nabla^t\, \hat{\mathbf{d}} \,,\, \mathbf{m} \,\rangle_M =$$

$$= \int_V dV(\mathbf{x})\ \hat{d}^i(\mathbf{x})\ \frac{\partial m}{\partial x^i}(\mathbf{x}) + \int_V dV(\mathbf{x})\ \frac{\partial \hat{d}^i}{\partial x^i}(\mathbf{x})\ m(\mathbf{x})$$

$$= \int_V dV(\mathbf{x})\ \frac{\partial}{\partial x^i}\left(\hat{d}^i(\mathbf{x})\ m(\mathbf{x})\right) = \int_S dS(\mathbf{x})\ n_i(\mathbf{x})\ \hat{d}^i(\mathbf{x})\ m(\mathbf{x}) \ ,$$

and the components of the bilinear form are

$$P^i[\hat{d},m](\mathbf{x}) = \hat{d}^i(\mathbf{x})\ m(\mathbf{x}) \ . \ \blacksquare \qquad (5)$$

Example: Demonstration of $(\mathbf{AB})^t = \mathbf{B}^t\ \mathbf{A}^t$ *:* Let $\mathbf{B} : \boldsymbol{E} \rightarrow \boldsymbol{F}$ and $\mathbf{A} : \boldsymbol{F} \rightarrow \boldsymbol{G}$. Then, $\mathbf{AB} : \mathbf{E} \rightarrow \mathbf{G}$. From

$$\langle\, \hat{\mathbf{g}} \,,\, \mathbf{A}\, \mathbf{f} \,\rangle_G - \langle\, \mathbf{A}^t\, \hat{\mathbf{g}} \,,\, \mathbf{f} \,\rangle_F = \int dV\ D_i\ a^i$$

$$\langle\, \hat{\mathbf{f}} \,,\, \mathbf{B}\, \mathbf{e} \,\rangle_F - \langle\, \mathbf{B}^t\, \hat{\mathbf{f}} \,,\, \mathbf{e} \,\rangle_E = \int dV\ D_i\ b^i$$

it follows, setting $\mathbf{f} = \mathbf{B}\ \mathbf{e}$ and $\hat{\mathbf{f}} = \mathbf{A}^t\ \hat{\mathbf{g}}$,

$$\langle\, \hat{\mathbf{g}} \,,\, \mathbf{A}\, \mathbf{B}\, \mathbf{e} \,\rangle_G - \langle\, \mathbf{A}^t\, \hat{\mathbf{g}} \,,\, \mathbf{B}\, \mathbf{e} \,\rangle_F = \int dV\ D_i\ a^i$$

$$\langle\, \mathbf{A}^t\, \hat{\mathbf{g}} \,,\, \mathbf{B}\, \mathbf{e} \,\rangle_F - \langle\, \mathbf{B}^t\, \mathbf{A}^t\, \hat{\mathbf{e}} \,,\, \mathbf{e} \,\rangle_E = \int dV\ D_i\ b^i \ ,$$

and

$$\langle\, \hat{\mathbf{g}} \,,\, \mathbf{A}\, \mathbf{B}\, \mathbf{e} \,\rangle_G - \langle\, \mathbf{B}^t\, \mathbf{A}^t\, \hat{\mathbf{e}} \,,\, \mathbf{e} \,\rangle_E = \int dV\ D_i\ (a^i + b^i) \ ,$$

i.e.,

$$(\mathbf{AB})^t = \mathbf{B}^t\ \mathbf{A}^t \ . \ \blacksquare \qquad (6)$$

(...)

If the right-hand side of (1a)-(1b) vanishes for any $\mathbf{m}$ and $\hat{\mathbf{d}}$,

$$\int_V dV(\mathbf{x}) \; \frac{DP^i[\hat{d},m]}{Dx^i}(\mathbf{x}) \; = \; \int_S dS(\mathbf{x}) \; n_i(\mathbf{x}) \; P^i[\hat{d},m](\mathbf{x}) \; = \; 0 \; , \tag{7}$$

then the *formal transposed* is simply termed the *transposed*, and we have

$$\langle \; \hat{\mathbf{d}} \, , \mathbf{G} \, \mathbf{m} \; \rangle_D \; = \; \langle \; \mathbf{G}^t \, \hat{\mathbf{d}} \, , \mathbf{m} \; \rangle_M \; . \tag{8}$$

In that case, we can harmlessly use all the equations involving transpositions as if we were dealing with discrete spaces.

Usually, the domains of definition of $\mathbf{G}$ and $\mathbf{G}^t$ are restricted so as to satisfy (7). It is then said that these domains of definition satisfy *dual boundary conditions*. See below for an example.

In the special case where a linear operator $\mathbf{W}$ maps a space E into its dual $\hat{E}$, then $\mathbf{W}^t$ also maps E into $\hat{E}$. In that case, it may happen that $\mathbf{W}^t = \mathbf{W}$, and the operator $\mathbf{W}$ is *symmetric*. Let us come to this definition with some care.

Assume that a linear operator $\mathbf{W}$ maps $E_0 \subset E$ into Y, and define its transpose $\mathbf{W}^t$ as mapping $\hat{Y}_1 \subset \hat{Y}$ into $\hat{E}$. If the subspaces E_0 and $\hat{Y}_1$ satisfy *dual boundary conditions* (see above), then

$$\forall \mathbf{e}_0 \in E_0 \;\; \forall \hat{\mathbf{y}}_1 \in \hat{Y}_1 \qquad \langle \; \hat{\mathbf{y}}_1 \, , \mathbf{W} \, \mathbf{e}_0 \; \rangle_Y \; = \; \langle \; \mathbf{W}^t \, \hat{\mathbf{y}}_1 \, , \mathbf{e}_0 \; \rangle_E \; .$$

If $Y = \hat{E}$ and we identify the bidual of E to E, then $\mathbf{W} : E_0 \to \hat{E}$ and $\mathbf{W}^t : E_1 \to \hat{E}$, and, by definition of transpose,

$$\forall \mathbf{e}_0 \in E_0 \;\; \forall \mathbf{e}_1 \in E_1 \qquad \langle \; \mathbf{e}_1 \, , \mathbf{W} \, \mathbf{e}_0 \; \rangle_{\hat{E}} \; = \; \langle \; \mathbf{W}^t \, \mathbf{e}_1 \, , \mathbf{e}_0 \; \rangle_E \; .$$

Now, if

$$\forall \mathbf{e} \in E_0 \cap E_1 \;\; \forall \mathbf{e}' \in E_0 \cap E_1 \qquad \langle \; \mathbf{W} \, \mathbf{e} \, , \mathbf{e}' \; \rangle_E \; = \; \langle \; \mathbf{W}^t \, \mathbf{e} \, , \mathbf{e}' \; \rangle_E \, , \tag{9}$$

the operator $\mathbf{W}$ is termed *symmetric*, the notation

$$\mathbf{W} \; = \; \mathbf{W}^t \tag{10}$$

is used, and the identities

(...)

$$\langle \mathbf{e} , \mathbf{W}^t \mathbf{e}' \rangle_{\hat{E}} = \langle \mathbf{W} \mathbf{e} , \mathbf{e}' \rangle_E = \langle \mathbf{W}^t \mathbf{e} , \mathbf{e}' \rangle_E = \langle \mathbf{e} , \mathbf{W} \mathbf{e}' \rangle_{\hat{E}} \quad (11)$$

hold $\forall\ \mathbf{e},\mathbf{e}' \in E_0 \cap E_1$.

In defining the transposed of a linear operator, it is not assumed that the vector spaces into consideration have a scalar product. If they do, then it is possible to define the adjoint of a linear operator.

Let M and D represent two scalar product vector spaces. The scalar product of $\mathbf{d}_1$ by $\mathbf{d}_2$ (resp. of $\mathbf{m}_1$ by $\mathbf{m}_2$ is denoted $(\mathbf{d}_1 , \mathbf{d}_2)_D$ (resp. $(\mathbf{m}_1 , \mathbf{m}_2)_M$) .

Using the same notations as above, the *formal adjoint* of $\mathbf{G}$, $\mathbf{G}^*$, is the unique operator mapping D into M such that

$$(\mathbf{d} , \mathbf{G}\, \mathbf{m})_D - (\mathbf{G}^* \mathbf{d} , \mathbf{m})_M = \int_V dV(\mathbf{x}) \ \frac{DP^i[\mathbf{d},\mathbf{m}]}{Dx^i}(\mathbf{x}) \ , \quad (12a)$$

or, using Green's theorem,

$$(\mathbf{d} , \mathbf{G}\, \mathbf{m})_D - (\mathbf{G}^* \mathbf{d} , \mathbf{m})_M = \int_S dS(\mathbf{x}) \ n_i(\mathbf{x}) \ P^i[\mathbf{d},\mathbf{m}](\mathbf{x}) \ . \quad (12b)$$

Again, although $\mathbf{G}^*$ is uniquely defined by (12a) or (12b), the vector $\mathbf{P}[\ \cdot\ ,\ \cdot\ ]$ is defined to the addition of a divergenceless vector.

Let $\mathbf{C}_M$ and $\mathbf{C}_D$ be the covariance operators defining the scalar products over M and D respectively (and, thus, the natural isomorphisms between M and D and their respective duals):

$$(\mathbf{d}_1 , \mathbf{d}_2)_D = \langle \mathbf{C}_D^{-1} \mathbf{d}_1 , \mathbf{d}_2 \rangle_D \ , \quad (13a)$$

$$(\mathbf{m}_1 , \mathbf{m}_2)_M = \langle \mathbf{C}_M^{-1} \mathbf{m}_1 , \mathbf{m}_2 \rangle_M \ . \quad (13b)$$

Using for instance (1a), we obtain

$$(\mathbf{d} , \mathbf{G}\, \mathbf{m})_D - (\mathbf{C}_M \mathbf{G}^t \mathbf{C}_D^{-1} \mathbf{d} , \mathbf{m})_M = \int_V dV(\mathbf{x}) \ \frac{DP^i[\mathbf{d},\mathbf{m}]}{Dx^i}(\mathbf{x}) \ ,$$

or, for short,

$$\mathbf{G}^* = \mathbf{C}_M \mathbf{G}^t \mathbf{C}_D^{-1} \ . \quad (14)$$

(...)

The reader will easily give sense to and demonstrate the property

$$(\mathbf{A}\ \mathbf{B})^* = \mathbf{B}^*\ \mathbf{A}^* . \tag{15}$$

If the right-hand side of (12a)-(12b) vanishes for any $\mathbf{m}$ and $\mathbf{d}$,

$$\int_V dV(\mathbf{x})\ \frac{DP^i[\mathbf{d},\mathbf{m}]}{Dx^i}(\mathbf{x}) = \int_S dS(\mathbf{x})\ n_i(\mathbf{x})\ P^i[\mathbf{d},\mathbf{m}](\mathbf{x}) = 0 , \tag{16}$$

then the *formal adjoint* is simply termed the *adjoint*, and we have

$$(\mathbf{d} , \mathbf{G}\ \mathbf{m})_D = (\mathbf{G}^*\ \mathbf{d} , \mathbf{m})_M . \tag{17}$$

In the special case where a linear operator $\mathbf{L}$ maps a space E into itself, then the adjoint operator $\mathbf{L}^*$ also maps E into itself. In that case, it may happen that $\mathbf{L}^* = \mathbf{L}$, and the operator $\mathbf{L}$ is *self-adjoint*. Let us replace duality products by scalar products in the equations above.

Assume that a linear operator $\mathbf{W}$ maps $E_0 \subset E$ into Y, and define its adjoint $\mathbf{W}^*$ as mapping $Y_1 \subset Y$ into E. If the subspaces E_0 and Y_1 satisfy *dual boundary conditions* (see above), then

$$\forall \mathbf{e}_0 \in E_0 \quad \forall \mathbf{y}_1 \in Y_1 \qquad (\mathbf{y}_1 , \mathbf{W}\ \mathbf{e}_0)_Y = (\mathbf{W}^*\ \mathbf{y}_1 , \mathbf{e}_0)_E .$$

If $Y = E$, then $\mathbf{W} : E_0 \to E$ and $\mathbf{W}^* : E_1 \to E$, and, by definition of adjoint,

$$\forall \mathbf{e}_0 \in E_0 \quad \forall \mathbf{e}_1 \in E_1 \qquad (\mathbf{e}_1 , \mathbf{W}\ \mathbf{e}_0)_E = (\mathbf{W}^*\ \mathbf{e}_1 , \mathbf{e}_0)_E .$$

Now, if

$$\forall \mathbf{e} \in E_0 \cap E_1 \quad \forall \mathbf{e}' \in E_0 \cap E_1 \qquad (\mathbf{W}\ \mathbf{e} , \mathbf{e}')_E = (\mathbf{W}^*\ \mathbf{e} , \mathbf{e}')_E , \tag{18}$$

the operator $\mathbf{W}$ is termed *self-adjoint*, the notation

$$\mathbf{W} = \mathbf{W}^* \tag{19}$$

is used, and the identities

$$(\mathbf{e} , \mathbf{W}^*\ \mathbf{e}')_E = (\mathbf{W}\ \mathbf{e} , \mathbf{e}')_E = (\mathbf{W}^*\ \mathbf{e} , \mathbf{e}')_E = (\mathbf{e} , \mathbf{W}\ \mathbf{e}')_E \tag{20}$$

hold $\forall \mathbf{e},\mathbf{e}' \in E_0 \cap E_1$.

(...)

It is easy to see that Identities (20) define a scalar product, denoted $\mathbf{W(e,e')}$:

$$\mathbf{W(e,e')} = (\mathbf{e} , \mathbf{W^* e'})_E = (\mathbf{W\, e} , \mathbf{e'})_E = (\mathbf{W^* e} , \mathbf{e'})_E = (\mathbf{e} , \mathbf{W\, e'})_E \quad (21)$$

If $\mathbf{W}$ is the operator defining the original scalar product over E ,

$$\mathbf{W(e,e')} = (\mathbf{e} , \mathbf{e'})_E , \quad (22)$$

it is named the *weighting* operator over E , or the *inverse of the covariance* operator over E , and the following notation is used:

$$\mathbf{W(e,e')} = (\mathbf{e} , \mathbf{e'})_E = \mathbf{e}^t \, \mathbf{W} \, \mathbf{e'} . \quad (23)$$

Example: The transposed of the elastodynamics operator. Let $\mathbf{L}$ denote the elastic wave equation operator that to a displacement field $\mathbf{u}(\mathbf{x},t)$ associates its sources $\boldsymbol{\phi}(\mathbf{x},t)$ (volume density of external forces):

$$\mathbf{L\, u} = \boldsymbol{\phi} . \quad (24a)$$

Explicitly, using tensor notations,

$$\rho(\mathbf{x}) \frac{\partial^2 u^i}{\partial t^2}(\mathbf{x},t) - \frac{\partial}{\partial x^j}\left[c^{ijkl}(\mathbf{x}) \frac{\partial u_k}{\partial x^l}(\mathbf{x},t) \right] = \phi^i(\mathbf{x},t) . \quad (24b)$$

Let $\mathbf{u} \in U$ and $\boldsymbol{\phi} \in \boldsymbol{\Phi}$. Each source vector $\boldsymbol{\phi}$ can be considered as a linear form over the space of displacements:

$$\langle \boldsymbol{\phi} , \mathbf{u} \rangle_U = \int_V dV(\mathbf{x})\, \phi^i(\mathbf{x},t)\, u_i(\mathbf{x},t) = \int_V dV(\mathbf{x})\, \phi_i(\mathbf{x},t)\, u^i(\mathbf{x},t) , \quad (25)$$

The physical dimension of $\langle \boldsymbol{\phi} , \mathbf{u} \rangle_U$ is an *action*. Equation (11) allows to identify $\boldsymbol{\Phi}$ as $\hat{U}$, dual of U . Furthermore, identifying the bidual of U with U , we see that $\mathbf{L}$ and $\mathbf{L}^t$ both map $U = \hat{\boldsymbol{\Phi}}$ into $\boldsymbol{\Phi} = \hat{U}$. We will now check under which conditions the wave equation operator is symmetric:

$$\mathbf{L}^t = \mathbf{L} . \quad (26)$$

For any $\hat{\boldsymbol{\phi}} \in \hat{\boldsymbol{\Phi}} = U$ and any $\mathbf{u} \in U = \hat{\boldsymbol{\Phi}}$,

(...)

$$\langle \hat{\phi} , L\, u \rangle_\Phi - \langle L^t \hat{\phi} , u \rangle_U$$

$$= \int_V dV(\mathbf{x}) \int_{t_0}^{t_1} dt\, \hat{\phi}^i \left\{ \rho \frac{\partial^2 u^i}{\partial t^2} - \frac{\partial}{\partial x^j} \left[c^{ijkl} \frac{\partial u_k}{\partial x^l} \right] \right\}$$

$$- \int_V dV(\mathbf{x}) \int_{t_0}^{t_1} dt \left\{ \rho \frac{\partial^2 \phi_i}{\partial t^2} - \frac{\partial}{\partial x_j} \left[c_{ijkl} \frac{\partial \hat{\phi}^k}{\partial x_l} \right] \right\} u^i$$

$$= \int_V dV(\mathbf{x}) \int_{t_0}^{t_1} dt\, \frac{\partial}{\partial t} \left\{ \rho \left[\hat{\phi}_i \frac{\partial u^i}{\partial t} - \frac{\partial \hat{\phi}_i}{\partial t} u^i \right] \right\}$$

$$- \int_V dV(\mathbf{x}) \int_{t_0}^{t_1} dt\, \frac{\partial}{\partial x^j} \left\{ c^{ijkl} \left[\hat{\phi}^i \frac{\partial u_k}{\partial x^l} - \frac{\partial \hat{\phi}_k}{\partial x^l} u^i \right] \right\} , \tag{27}$$

and the components of the bilinear form are

$$P^t[\hat{\phi},u](\mathbf{x},t) = \rho(\mathbf{x}) \left[\hat{\phi}_i(\mathbf{x},t) \frac{\partial u^i}{\partial t}(\mathbf{x},t) - \frac{\partial \hat{\phi}_i}{\partial t}(\mathbf{x},t)\, u^i(\mathbf{x},t) \right] \tag{28a}$$

$$P^j[\hat{\phi},u](\mathbf{x},t) = c^{ijkl}(\mathbf{x}) \left[\frac{\partial \hat{\phi}_k}{\partial x^l}(\mathbf{x},t)\, u^i(\mathbf{x},t) - \hat{\phi}^i(\mathbf{x},t) \frac{\partial u_k}{\partial x^l}(\mathbf{x},t) \right] . \tag{28b}$$

Using Green's theorem we obtain

$$\langle \hat{\phi} , L\, u \rangle_\Phi - \langle L^t \hat{\phi} , u \rangle_U$$

$$= \int_V dV(\mathbf{x})\, \rho \left[\hat{\phi}_i \frac{\partial u^i}{\partial t} - \frac{\partial \hat{\phi}_i}{\partial t} u^i \right] \Bigg|_{t_0}^{t_1}$$

$$- \int_{t_0}^{t_1} dt \int_S dS(\mathbf{x})\, n_j\, c^{ijkl} \left[\hat{\phi}^i \frac{\partial u_k}{\partial x^l} - \frac{\partial \hat{\phi}_k}{\partial x^l} u^i \right] . \tag{29}$$

Assume for instance that the fields $\mathbf{u}(\mathbf{x},t)$ and $\hat{\phi}(\mathbf{x},t)$ satisfy the boundary conditions

(...)

$$u^i(\mathbf{x},t_0) = 0$$

$$\frac{\partial u^i}{\partial t}(\mathbf{x},t_0) = 0 \, , \qquad (30a)$$

$$\hat{\phi}^i(\mathbf{x},t_1) = 0$$

$$\frac{\partial \hat{\phi}^i}{\partial t}(\mathbf{x},t_1) = 0 \, , \qquad (30b)$$

$$n_j(\mathbf{x}) \; c^{ijkl}(\mathbf{x}) \; \frac{\partial u_k}{\partial x^l}(\mathbf{x},t) = 0 \, , \qquad (31a)$$

$$n_j(\mathbf{x}) \; c^{ijkl}(\mathbf{x}) \; \frac{\partial \hat{\phi}_k}{\partial x^l}(\mathbf{x},t) = 0 \, . \qquad (31b)$$

As expression (29) then vanishes, these are dual boundary conditions. If we only consider fields $\mathbf{u}(\mathbf{x},t)$ and $\hat{\phi}(\mathbf{x},t)$ satisfying these conditions, then

$$\langle \; \hat{\phi} \, , \mathbf{L} \, \mathbf{u} \; \rangle_{\Phi} = \langle \; \mathbf{L}^t \, \hat{\phi} \, , \mathbf{u} \; \rangle_{U} \, , \qquad (31)$$

and the symmetry property (26) holds. We will see in Section 7.8 that when solving usual inverse problems we are actually faced with wave-fields satisfying dual boundary conditions.

Notice that usual text books (Morse and Feshbach, 1953; Courant and Hilbert, Dautray and Lions, 1984) talk about the "self-adjointness" of the wave equation operator. This assumes that a scalar product can be defined over U , which is not necessarily the case (for instance, U may be a general Banach space). Only the symmetry of the wave equation operator is needed.

7.3: General least-squares inversion

7.3.1: Linear problems

I start by recalling some results from Chapter 1 for finite-dimensional problems. Assume that the forward problem is (exactly) solved by the linear equation

$$\mathbf{d} = \mathbf{G}\,\mathbf{m}\ ,$$

that the results of the observations are described by a Gaussian probability with mathematical expectation $\mathbf{d}_{obs}$ and covariance operator C_D , and that the a priori information is described by a Gaussian probability with mathematical expectation $\mathbf{m}_{prior}$ and covariance operator C_M . Then the posterior probability in the model space is also Gaussian, with mathematical expectation

$$\langle\ \mathbf{m}\ \rangle\ = \mathbf{m}_{prior} + (\mathbf{G}^t\ C_D^{-1}\ \mathbf{G} + C_M^{-1})^{-1}\ \mathbf{G}^t\ C_D^{-1}\ (\mathbf{d}_{obs} - \mathbf{G}\ \mathbf{m}_{prior}) \quad (7.84a)$$

$$= \mathbf{m}_{prior} + C_M\ \mathbf{G}^t\ (\mathbf{G}\ C_M\ \mathbf{G}^t + C_D)^{-1}\ (\mathbf{d}_{obs} - \mathbf{G}\ \mathbf{m}_{prior})\ , \quad (7.84b)$$

and covariance operator

$$C_{M'}\ =\ (\mathbf{G}^t\ C_D^{-1}\ \mathbf{G} + C_M^{-1})^{-1} \quad (7.85a)$$

$$=\ C_M - C_M\ \mathbf{G}^t\ (\mathbf{G}\ C_M\ \mathbf{G}^t + C_D)^{-1}\ \mathbf{G}\ C_M\ . \quad (7.85b)$$

Using either the arguments of Franklin (1970) or Backus (1970), it can be shown that this result remains true for infinite dimensional problems: if priors are Gaussian, and the forward equation is linear, the posterior is Gaussian, with mathematical expectation and covariance operator as given above. This result shows that, when using adequate notations, the results obtained in Chapter 4 are valid in a much more general context (i.e., the context of functional analysis).

The major difference between finite and infinite dimensional problems is that in the finite-dimensional case, probability densities can be introduced, while they cannot be introduced in the infinite-dimensional case.

A minor difference is that in finite-dimensional problems there are never fundamental problems for solving linear systems with positive definite operators, while in the infinite-dimensional case it may well happen that a linear system with positive definite operator has no solution belonging to the spaces under consideration. If I call this difficulty "minor" it is because it can be usually solved by introducing larger functional spaces.

Let us consider the misfit function

$$S(\mathbf{m}) = \frac{1}{2}\left[(\mathbf{G}\ \mathbf{m} - \mathbf{d}_{obs})^t\ \mathbf{C}_D^{-1}\ (\mathbf{G}\ \mathbf{m} - \mathbf{d}_{obs}) + (\mathbf{m}-\mathbf{m}_{prior})^t\ \mathbf{C}_M^{-1}\ (\mathbf{m}-\mathbf{m}_{prior}) \right] . \tag{7.86}$$

For arbitrary $\mathbf{m}$ there is no reason to have a finite value of S . We will have finite values only if the data residuals $\mathbf{G}\ \mathbf{m} - \mathbf{d}_{obs}$ and the model residuals $\mathbf{m} - \mathbf{m}_{prior}$ have finite norm. It is clear that there may exist forward operators $\mathbf{G}$ and data covariance operators $\mathbf{C}_D$ such that for no $\mathbf{m}$ this happens. We are then in trouble. This means that the forward modelization is not coherent with our estimation of errors by $\mathbf{C}_D$, and we have to change one or the other. If the forward operator $\mathbf{G}$ and the data covariance operator $\mathbf{C}_D$ are coherent, then for some models $\mathbf{m}$ the misfit function $S(\mathbf{m})$ will take finite values. It is then easy to show that the minimum value is obtained when $\mathbf{m}$ is given by (7.84).

7.3.2: Nonlinear problems

Let us now turn to the case where the equation solving the forward problem is nonlinear. As explained in Chapter 1, by nonlinear I mean that, if $\mathbf{m}$ is any model in the region of significant posterior probability, the linearization

$$\mathbf{g}(\mathbf{m}) \simeq \mathbf{g}(\mathbf{m}_{prior}) + \mathbf{G}_0\ (\mathbf{m} - \mathbf{m}_{prior}) ,$$

where $\mathbf{G}_0$ denotes the derivative of $\mathbf{g}$ at $\mathbf{m} = \mathbf{m}_{prior}$, is *not* assumed to be good enough for approximating the solution of the forward problem (see Section 1.7.1 for more details).

For discrete problems, we have seen in Chapter 1 that if the results of the observations can be described by a Gaussian probability with mathematical expectation $\mathbf{d}_{obs}$ and covariance operator $\mathbf{C}_D$, the a priori information can be described by a Gaussian probability with mathematical expectation $\mathbf{m}_{prior}$ and covariance operator $\mathbf{C}_M$, and if the equation solving the forward problem is quasi-linear in the region of the model space with significant a posteriori probability, then the a posteriori probability in the model space is approximately Gaussian. The mathematical expectation, $\langle\ \mathbf{m}\ \rangle$, of the approximated Gaussian probability minimizes the misfit function

$$S(\mathbf{m}) = \tag{7.87a}$$
$$\frac{1}{2}\left[(\mathbf{g}(\mathbf{m})-\mathbf{d}_{obs})^t\ \mathbf{C}_D^{-1}\ (\mathbf{g}(\mathbf{m})-\mathbf{d}_{obs}) + (\mathbf{m}-\mathbf{m}_{prior})^t\ \mathbf{C}_M^{-1}\ (\mathbf{m}-\mathbf{m}_{prior})\right] ,$$

can be obtained using some iterative process: $\mathbf{m}_{prior}=\mathbf{m}_0$, $\mathbf{m}_1$, $\mathbf{m}_2$, ... , $\mathbf{m}_\infty=\langle\ \mathbf{m}\ \rangle$, and the covariance operator of the approximated Gaussian probability is given by

$$\mathbf{C}_{M'} = (\mathbf{G}_\infty{}^t\ \mathbf{C}_D^{-1}\ \mathbf{G}_\infty + \mathbf{C}_M^{-1})^{-1}$$
$$= \mathbf{C}_M - \mathbf{C}_M\ \mathbf{G}_\infty{}^t\ (\mathbf{G}_\infty\ \mathbf{C}_M\ \mathbf{G}_\infty{}^t + \mathbf{C}_D)^{-1}\ \mathbf{G}_\infty\ \mathbf{C}_M\ , \tag{7.87b}$$

where $\mathbf{G}_\infty$ denotes the derivative of the operator $\mathbf{g}$ at the point $\langle\ \mathbf{m}\ \rangle = \mathbf{m}_\infty$.

We accept without demonstration that these results remain true for infinite dimensional problems.

7.3.3: Linearizable problems

As defined in Chapter 1, a problem is linearizable if, for any model in the region of significant posterior probability, the approximation

$$\mathbf{g}(\mathbf{m}) \simeq \mathbf{g}(\mathbf{m}_0) + \mathbf{G}_0\ (\mathbf{m}-\mathbf{m}_0) \tag{7.88}$$

holds with sufficient accuracy, where $\mathbf{m}_0$ is some "initial" model (usually, $\mathbf{m}_0 = \mathbf{m}_{prior}$). Defining then the *data residuals*

$$\delta\mathbf{d} = \mathbf{d} - \mathbf{G}_0\ \mathbf{m}_0 \tag{7.89}$$

and the *model residuals*

$$\delta\mathbf{m} = \mathbf{m} - \mathbf{m}_0\ , \tag{7.90}$$

the equation solving the forward problem can be replaced by the linearized equation

$$\delta\mathbf{d} = \mathbf{G}_0\ \delta\mathbf{m}\ . \tag{7.91}$$

The observations are then described by the "observed residuals"

$$\delta\mathbf{d}_{obs} = \mathbf{d}_{obs} - \mathbf{G}_0\ \mathbf{m}_0\ . \tag{7.92}$$

If $\mathbf{C}_D$ is the covariance operator describing experimental uncertainties in $\mathbf{d}_{obs}$, it also describes uncertainties in $\delta\mathbf{d}_{obs}$, because $\mathbf{m}_0$ is an arbitrary, but perfectly defined model. The a priori information in the model space is described by the a priori model residuals

$$\delta\mathbf{m}_{prior} = \mathbf{m}_{prior} - \mathbf{m}_0 . \tag{7.93}$$

If $\mathbf{C}_M$ is the covariance operator describing our confidence in $\mathbf{m}_{prior}$, it also describes our confidence in $\delta\mathbf{m}_{prior}$ (for the same reason as before).

The inputs of a linear problem were $\mathbf{G}$, $\mathbf{d}_{obs}$, $\mathbf{C}_D$, $\mathbf{m}_{prior}$, and $\mathbf{C}_M$. Replacing these respectively by $\mathbf{G}_0$, $\delta\mathbf{d}_{obs}$, $\mathbf{C}_D$, $\delta\mathbf{m}_{prior}$, and $\mathbf{C}_M$, allows us to reformulate the inverse problem corresponding to a linearized forward problem as has been done for linear problems.

7.4: Methods of resolution

In the previous sections of this chapter I have reviewed the main concepts of functional analysis. In particular, it has been emphasized in which sense the usual formulas of matricial calculus have to be reinterpreted in terms of abstract linear operators and their kernels. Let me recall here some of the analogies and differences between matricial and functional calculus.

First of all, our present-day computers are digital, and this implies that their inputs and outputs are not functions but discrete and finite sets of numbers. Our brains know that, but — as our senses have finite resolution — it is often impossible to *feel* this essential discretization (think, for instance, of the output of a tomographic section on a cathodic screen with very high resolution). I usually name an approach "functional" if, say, multiplying the number of points used both to exhibit a function (on each of the axes of the graphic display) and to represent the function in the computer's memory by 10, increases the computing time but does not change — perceptibly — the solution.

An important question arising in the practical resolution of functional least-squares problems concerns the operational significance of the resolution of a linear equation. Let, for instance, $\delta\mathbf{d}$ be a vector of residuals (note that I use the word "vector" for "element of a linear vector space", not for "column matrix"). In least-squares computations, we often need to evaluate the weighted residuals $\delta\hat{\mathbf{d}} = \mathbf{C}^{-1}\,\delta\mathbf{d}$, i.e., to solve for $\delta\hat{\mathbf{d}}$ the linear equation

$$\mathbf{C}\,\delta\hat{\mathbf{d}} = \delta\mathbf{d} , \tag{7.94}$$

where $\mathbf{C}$ is a given covariance operator. To compute $\delta\hat{\mathbf{d}}$ we can, for instance, use an iterative algorithm (see Appendix 4.4)

$$\delta\hat{\mathbf{d}}_{n+1} = \delta\hat{\mathbf{d}}_n + \mathbf{Q}\,(\,\delta\mathbf{d} - \mathbf{C}\,\delta\hat{\mathbf{d}}_n\,)\,, \tag{7.95}$$

where $\mathbf{Q}$ is an arbitrary operator suitably chosen for accelerating convergence. Usually, a good choice for $\mathbf{Q}$ is a diagonal operator proportional to the inverse of the variances in $\mathbf{C}$ (the reader will easily generalize the concept of diagonal matrix into the concept of diagonal operator). Equation (7.95) shows that, for numerical computations, we do not need explicitly to introduce the inverse of the operator $\mathbf{C}$. Usually, a covariance operator is an integral operator, and computing $\mathbf{C}\,\delta\hat{\mathbf{d}}_n$ implies some numerical method, such as, for instance, a Runge-Kutta method. Of course any numerical method will imply a discretization of the working space (here the data space), but it is important to acknowledge that to discretize the working space does *not* imply considering the operator $\mathbf{C}$ as a matrix and effectively building the matrix in the computer's memory (for problems other than academic this would be practically impossible).

Another important reason not to look on (7.94) as a matricial equation is that, as we have seen in section 7.1.6, for some particularly simple choices of covariance function, the use of functional concepts allows an *analytic resolution* of the equation.

A second important problem arising in least-squares computations is the computation of

$$\delta\hat{\mathbf{m}} = \mathbf{G}^t\ \delta\hat{\mathbf{d}}\,, \tag{7.96}$$

where $\delta\hat{\mathbf{d}}$ is a weighted data vector, $\mathbf{G}$ a linear operator from the model into the data space, and $\mathbf{G}^t$ its transposed. We have seen in section 7.2.2 that the abstract definition of the transposed operator leads to a very deep physical understanding of the computations to be performed: backprojection of the weighted residuals in an X-ray tomography problem, backpropagation of the weighted residuals plus a time correlation with the current predicted field in a problem of inversion of acoustic waveforms. Again, to perform these operations numerically, some discretization of the working spaces has to be used, but the naïve approach consisting in introducing a matrix representing the discretized version of $\mathbf{G}$, and interpreting (7.96) as a matrix multiplication equation, not only destroys the physical interpretation, but forces us to effectively compute the elements of the matrix $\mathbf{G}$, which is usually prohibitive because it takes too much computing time and core memory space.

We see thus in which sense the discretization of functional problems — necessary for numerical computations — has to be understood.

This being so, in Chapter 4 I have developed Newton and gradient methods for the minimization of the least-squares misfit function $S(\mathbf{m})$. As the notation used is general, these methods can be directly used in the more general context of this chapter, without change. The corresponding equations will therefore not be repeated here.

The examples in the two following sections illustrate extensively how the least-squares formulas have to be used in problems involving functions.

Box 7.2: The method of Backus and Gilbert.

In a famous paper, Backus and Gilbert (1970) gave a conceptually simple philosophy for dealing with linear, essentially underdeterminated, problems.

Assume that a model is described by a *function*, $m(r)$, and that we consider a finite amount of *discrete data*, $d^1, d^2, ..., d^n$, which are linear functionals of $m(r)$ through kernels $G^i(r)$:

$$d^i = \int dr \ G^i(r) \ m(r) \ . \tag{1}$$

The kernels $G^i(r)$ are assumed regular enough for equation (1) to be an ordinary integral equation.

The true (unknown) model is denoted $m_{true}(r)$. The observed data are denoted d^i_{obs} and are assumed error free. Then,

$$d^i_{obs} = \int dr \ G^i(r) \ m_{true}(r) \ . \tag{2}$$

The problem is to obtain a good estimator of $m_{true}(r)$ at a given point $r = r_0$. Let us denote this estimator by $m_{est}(r_0)$. As the forward problem is linear we impose that the value $m_{est}(r_0)$ is a *linear function of the observed data*, i.e, we assume the form

$$m_{est}(r_0) = \sum_i Q^i(r_0) \ d^i_{obs} \ , \tag{3}$$

where, at a given point r_0 , $Q^i(r_0)$ are some constants. The problem now is to obtain the best constants $Q^i(r_0)$. Replacing (2) into (3) gives

$$m_{est}(r_0) = \sum_i Q^i(r_0) \int dr \ G^i(r) \ m_{true}(r) \ ,$$

and defining

(...)

$$R(r_0,r) = \sum_i Q^i(r_0)\ G^i(r) \tag{4}$$

gives

$$m_{est}(r_0) = \int dr\ R(r_0,r)\ m_{true}(r)\ . \tag{5}$$

This last equation shows that our estimate at $r = r_0$ will be a filtered version of the true value, with filter $R(r_0,r)$. This filter is called the "resolving kernel". We are only able to see the true world through this filter. The "sharper" the filter is around r_0 , the better our estimate (see Figure 7.18, 7.19, and 7.23). We can arbitrarily choose the coefficients $Q^i(r_0)$. The "deltaness criterion" consists in choosing these coefficients in such a way that the resulting resolving kernel is the closest to a delta function,

$$R(r_0,r) \simeq \delta(r_0-r)\ , \tag{6}$$

in which case

$$m_{est}(r_0) \simeq m_{true}(r_0)\ . \tag{7}$$

Using, for instance, a least-squares deltaness criterion

$$\int dr\ (\ R(r_0,r) - \delta(r_0-r)\ )^2 \qquad \textbf{MINIMUM} \tag{8}$$

gives (see problem 7.6)

$$Q^i(r_0) = \sum_j (S^{-1})^{ij}\ G^j(r_0)\ , \tag{9}$$

where

$$\boxed{S^{ij} = \int dr\ G^i(r)\ G^j(r)\ .} \tag{10}$$

This gives

(...)

$$m_{est}(r_0) = \sum_i \sum_j G^i(r_0) \; (S^{-1})^{ij} \; d^j_{obs} \; , \tag{11}$$

which corresponds to the Backus and Gilbert solution for the estimation problem (the reader will easily verify that the matrix S^{ij} is regular if the $G^i(r)$ are linearly independent, i.e, each datum depends differently on model parameters). Using (9), the resolving kernel is given by

$$R(r_0,r) = \sum_i \sum_j G^i(r_0) \; (S^{-1})^{ij} \; G^j(r) \; . \tag{12}$$

It is easy to see that the data predicted from $m_{est}(r)$ exactly verify the observations:

$$d^i_{cal} = \int dr \; G^i(r) \; m_{est}(r) = \sum_j \sum_k S^{ij} \; (S^{-1})^{jk} \; d^k_{obs} = d^i_{obs} \; .$$

Using more compact and general notations, all previous equations can be rewritten as follows:

$$\mathbf{d} = \mathbf{G} \, \mathbf{m} \; , \tag{1'}$$

$$\mathbf{d}_{obs} = \mathbf{G} \, \mathbf{m}_{true} \; , \tag{2'}$$

$$\mathbf{m}_{est} = \mathbf{Q}^t \, \mathbf{d}_{obs} \; , \tag{3'}$$

$$\mathbf{R} = \mathbf{Q}^t \, \mathbf{G} \; , \tag{4'}$$

$$\mathbf{m}_{est} = \mathbf{R} \, \mathbf{m}_{true} \; , \tag{5'}$$

$$\mathbf{R} \simeq \mathbf{I} \; , \tag{6'}$$

$$\mathbf{m}_{est} \simeq \mathbf{m}_{true} \; , \tag{7'}$$

$$\| \mathbf{R} - \mathbf{I} \|^2 \qquad \textbf{MINIMUM} \; , \tag{8'}$$

$$\mathbf{Q} = (\mathbf{G}\,\mathbf{G}^t)^{-1} \, \mathbf{G} \; , \tag{9'}$$

(...)

$$\mathbf{m}_{est} = \mathbf{G}^t \, (\mathbf{G}\,\mathbf{G}^t)^{-1} \, \mathbf{d}_{obs} \, , \tag{11'}$$

and

$$\mathbf{R} = \mathbf{G}^t \, (\mathbf{G}\,\mathbf{G}^t)^{-1} \, \mathbf{G} \, . \tag{12'}$$

Of course, there is no reason for the estimate $\mathbf{m}_{est}$ to equal the true model $\mathbf{m}_{true}$, which is generally not attainable with a finite amount of data. But as Backus and Gilbert assume exact data, it is easy to show that the true model is necessarily of the form

$$\mathbf{m} = \mathbf{m}_{est} + (\mathbf{I} - \mathbf{R}) \, \mathbf{m}_0 \, , \tag{13}$$

where $\mathbf{m}_0$ is an arbitrary model. For it is sufficient to verify, using (12'), that $(\mathbf{I} - \mathbf{R}) \, \mathbf{m}_0$ belongs to the null space of $\mathbf{G}$, i.e., it is such that

$$\mathbf{G} \, (\, (\mathbf{I} - \mathbf{R}) \, \mathbf{m}_0 \,) = \mathbf{0} \, .$$

Then

$$\mathbf{G}\,\mathbf{m} = \mathbf{G}\,\mathbf{m}_{est} = \mathbf{G}\,\mathbf{m}_{true} = \mathbf{d}_{obs} \, . \tag{14}$$

For Backus and Gilbert, equation (11') gives a *particular solution* of the inverse problem, while equation (13) gives the *general solution*.

Let us make the comparison between Backus and Gilbert's philosophy and the probabilistic approach. For Backus and Gilbert, $\mathbf{m}_{est}$ is the best estimate of $\mathbf{m}_{true}$, and turns out to be a filtered version of it. From a probabilistic point of view, we have some a priori information on $\mathbf{m}_{true}$ described through the a priori model $\mathbf{m}_{prior}$ and the a priori covariance operator $\mathbf{C}_M$. Observations $\mathbf{d}_{obs}$ have estimated errors described by $\mathbf{C}_D$. If a priori information and observational errors are adequately described using the Gaussian hypothesis, then the posterior probability in the model space is also Gaussian, with mathematical expectation (equation (7.84b))

(...)

$$\mathbf{m} = \mathbf{m}_{prior} + \mathbf{C_M}\, \mathbf{G}^t \,(\, \mathbf{C_D} + \mathbf{G}\, \mathbf{C_M}\, \mathbf{G}^t \,)^{-1} \,(\, \mathbf{d}_{obs} - \mathbf{G}\, \mathbf{m}_{prior}) \tag{15}$$

and posterior covariance operator (equation (7.85b))

$$\mathbf{C_{M'}} = \mathbf{C_M} - \mathbf{C_M}\, \mathbf{G}^t \,(\, \mathbf{C_D} + \mathbf{G}\, \mathbf{C_M}\, \mathbf{G}^t \,)^{-1}\, \mathbf{G}\, \mathbf{C_M} \,. \tag{16}$$

Backus and Gilbert do not use a priori information in the model space. This corresponds in a probabilistic context to the particular assumption of white noise (infinite variances and null correlations):

$$\mathbf{C_M} \simeq k\, \mathbf{I} \qquad (k \rightarrow \infty) \,. \tag{17}$$

Equations (15)-(16) give then respectively

$$\mathbf{m} = \mathbf{G}^t \,(\mathbf{G}\, \mathbf{G}^t)^{-1}\, \mathbf{d}_{obs} + (\mathbf{I} - \mathbf{R})\, \mathbf{m}_{prior} \,, \tag{18}$$

$$\mathbf{C_{M'}} = (\mathbf{I} - \mathbf{R})\, \mathbf{C_M} \,. \tag{19}$$

Equation (18) is identical to equation (13), where $\mathbf{m}_{prior}$ replaces the arbitrary $\mathbf{m}_0$. Note that if $\mathbf{R} \simeq \mathbf{I}$, in the Backus and Gilbert context $\mathbf{m}_{est} \simeq \mathbf{m}_{true}$, while in the probabilistic context, $\mathbf{C_{M'}} \simeq \mathbf{0}$ (no error in the a posteriori solution), which means the same. I feel the probabilistic approach to be much richer than the mathematical approach of Backus and Gilbert, but they probably feel the contrary.

7.5: Example: The inverse problem of X-ray tomography

This problem has been introduced in examples 7.16 and 7.20. The datum d^i represents the (measurable) transmittance for an X-ray beam along the ray R^i , and the model parameter $m(\mathbf{x})$ represents the attenuation coefficient at point $\mathbf{x} = (x^1,x^2,x^3)$. Given the attenuation $m(\mathbf{x})$, the forward problem is solved by the linear equation

$$d^i = \int_{R^i} ds^i \;\; m(\, \mathbf{x}(s^i)\,) \,. \tag{7.47 again}$$

As usual, for more compactness, this equation is written

$$\mathbf{d} = \mathbf{G}\, \mathbf{m} \,.$$

Introducing the integral kernel of **G** by

$$d^i = \int_{\mathbf{V}} d\mathbf{x} \;\; G^i(\mathbf{x}) \;\; m(\mathbf{x}) \; , \qquad (7.50 \text{ again})$$

where V denotes the volume under study (I assume here 3D tomography), and defining a delta-like function $\Delta^i(\mathbf{x})$ by

$$\int_{\mathbf{V}} d\mathbf{x} \;\; G^i(\mathbf{x}) \; m(\mathbf{x}) = \int_{R^i} ds^i \;\; m(\; \mathbf{x}(s^i) \;) \; , \qquad (7.51 \text{ again})$$

gives

$$G^i(\mathbf{x}) = \Delta^i(\mathbf{x}) \; . \qquad (7.52 \text{ again})$$

Let $d^i{}_{obs}$ (i=1,2.,,,) be the observed data, $C_D{}^{ij}$ the elements of the covariance matrix describing our estimate of experimental errors, $m_{prior}(\mathbf{x})$ our a priori model of attenuation, and $C_M(\mathbf{x},\mathbf{x}')$ the covariance operator describing our confidence in our a priori model. For instance, assuming uncorrelated errors

$$C_D{}^{ij} = (\sigma_D{}^i)^2 \;\; \delta^{ij} \; . \qquad (7.97)$$

The a priori model of attenuation can, for instance, be taken constant

$$m_{prior}(x) = \text{constant} \qquad \text{(e.g., the attenuation of water)} \; , \qquad (7.98)$$

and a choice of covariance function leading to easy computations is

$$C_M(\mathbf{x},\mathbf{x}') = \sigma_M{}^2 \;\; \exp\left(- \frac{\| \; \mathbf{x}-\mathbf{x}' \; \|}{L} \right) \; , \qquad (7.99)$$

where $\| \; \mathbf{x}-\mathbf{x}' \; \|$ denotes the euclidean distance. The parameter L plays the role of a smoothing distance.

Whatever the choice of $\mathbf{C}_D$, $\mathbf{m}_{prior}$, and $\mathbf{C}_M$ may be, the least-squares solution is defined by the minimization of

$$S(\mathbf{m}) = \qquad (7.86 \text{ again})$$

$$\frac{1}{2}\left[(\mathbf{G} \; \mathbf{m} \; - \; \mathbf{d}_{obs})^t \; \mathbf{C}_D{}^{-1} \; (\mathbf{G} \; \mathbf{m} \; - \; \mathbf{d}_{obs}) + (\mathbf{m}-\mathbf{m}_{prior})^t \; \mathbf{C}_M{}^{-1} \; (\mathbf{m}-\mathbf{m}_{prior}) \right] .$$

For the particular choice of $\mathbf{C}_D$ and $\mathbf{C}_M$ just mentioned, it can easily be shown, using the results of problem 7.9, that this gives

$$2\,S(m) \;=\; \sum_i \left(\frac{(\mathbf{G}\ \mathbf{m}\ -\mathbf{d}_{obs})^i}{\sigma_D{}^i} \right)^2$$

$$+\ \frac{1}{8\pi L^3 \sigma_M{}^2} \int_{\mathbf{V}} dV(\mathbf{x})\ \left(\ m(\mathbf{x})\ -\ m(\mathbf{x})_{prior}\ \right)^2$$

$$+\ \frac{1}{4\pi L \sigma_M{}^2} \int_{\mathbf{V}} dV(\mathbf{x})\ \left(\ \mathbf{grad}\ m(\mathbf{x})\ -\ \mathbf{grad}\ m(\mathbf{x})_{prior}\ \right)^2$$

$$+\ \frac{L}{8\pi \sigma_M{}^2} \int_{V} dV(\mathbf{x})\ \left(\ \nabla^2 m(\mathbf{x})\ -\ \nabla^2 m(\mathbf{x})_{prior}\right)^2\ ,$$

thus showing that the model obtained will predict data close to the observed data (first term), will be close everywhere to the a priori model (second term), and will be as smooth as it (last terms). The values of σ_M and L can be heuristically interpreted as adjustable parameters weighting the relative importance of these three terms.

We will now see that the algorithms leading to the solution of this inverse problem are very simple. I will examine two different algorithms, iterative steepest descent, and explicit Newton.

7.5.1: Iterative solution.

Letting $\mathbf{m}_0 = \mathbf{m}_{prior}$, and $\hat{\mathbf{S}}_0$ to be an arbitrary approximation to the inverse of the positive definite operator $\mathbf{I} + \mathbf{C}_M\, \mathbf{G}^t\, \mathbf{C}_D{}^{-1}\, \mathbf{G}$:

$$\hat{\mathbf{S}}_0 \ \simeq\ (\ \mathbf{I} + \mathbf{C}_M\ \mathbf{G}^t\ \mathbf{C}_D{}^{-1}\ \mathbf{G}\ )^{-1} \tag{7.100}$$

(approximation to be discussed below), the preconditioned steepest descent algorithm (see section 4.4) gives

$$\mathbf{m}_{n+1} \ =\ \mathbf{m}_n\ -\ \mu_n\ \boldsymbol{\phi}_n\ , \tag{7.101}$$

where the "search direction" $\boldsymbol{\phi}_n$ is given by

$$\boldsymbol{\phi}_n \ =\ \hat{\mathbf{S}}_0\ \boldsymbol{\gamma}_n\ , \tag{7.102}$$

the "direction of steepest ascent" $\boldsymbol{\gamma}_n$ is given by

$$\gamma_n = \mathbf{C_M} \mathbf{G}^t \mathbf{C_D}^{-1} (\mathbf{G}\, \mathbf{m}_n - \mathbf{d}_{obs}) + (\mathbf{m}_n - \mathbf{m}_{prior}) \ , \tag{7.103}$$

and μ_n is the real constant

$$\mu_n = \frac{\gamma_n^t \mathbf{C_M}^{-1} \phi_n}{\phi_n^t \mathbf{C_M}^{-1} \phi_n + (\mathbf{G}\, \phi_n)^t \mathbf{C_D}^{-1} (\mathbf{G}\, \phi_n)}$$

$$= \frac{(\gamma_n \, , \phi_n)}{\| \phi_n \|^2 + \| \mathbf{G}\, \phi_n \|^2} \ . \tag{7.104}$$

To interpret these equations in some detail, it is useful to introduce the following partial steps:

$$\delta\mathbf{d}_n = \mathbf{G}\, \mathbf{m}_n - \mathbf{d}_{obs} \ , \tag{7.105}$$

$$\delta\hat{\mathbf{d}}_n = \mathbf{C_D}^{-1}\, \delta\mathbf{d}_n \ , \tag{7.106}$$

$$\delta\hat{\mathbf{m}}_n = \mathbf{G}^t\, \delta\hat{\mathbf{d}}_n \ , \tag{7.107}$$

$$\delta\mathbf{m}_n = \mathbf{C_M}\, \delta\hat{\mathbf{m}}_n \ , \tag{7.108}$$

$$\gamma_n = \delta\mathbf{m}_n + (\mathbf{m}_n - \mathbf{m}_{prior}) \ , \tag{7.109}$$

$$\phi_n = \hat{\mathbf{S}}_0\, \gamma_n \ , \tag{7.110}$$

and

$$\mathbf{m}_{m+1} = \mathbf{m}_n - \mu_n\, \phi_n \tag{7.111}$$

(for the present, let us not worry about the real number μ_n).

If $G^i(x)$, $C_M(\mathbf{x},\mathbf{x}')$, $\hat{S}_0(\mathbf{x},\mathbf{x}')$ and $C_D{}^{ij}$ are the kernels of $\mathbf{G}$, $\mathbf{C_M}$, $\hat{\mathbf{S}}_0$ and $\mathbf{C_D}$, respectively, equations (7.105)-(7.111) are written, in explicit notation,

$$\delta d_n{}^i = \int_V dV(\mathbf{x})\ G^i(\mathbf{x})\ m_n(\mathbf{x}) - d^i{}_{obs} \ , \tag{7.112}$$

$$\delta\hat{d}_n{}^i = \sum_j (C_D{}^{-1})^{ij}\ \delta d_n{}^j \ , \tag{7.113}$$

$$\delta\hat{m}_n(\mathbf{x}) = \sum_i G^i(\mathbf{x}) \; \delta\hat{d}_n^{\,i} , \tag{7.114}$$

$$\delta m_n(\mathbf{x}) = \int_{\mathbf{V}} dV(\mathbf{x}') \; C_M(\mathbf{x},\mathbf{x}') \; \delta\hat{m}_n(\mathbf{x}') , \tag{7.115}$$

$$\gamma_n(\mathbf{x}) = \delta m_n(\mathbf{x}) + (m_n(\mathbf{x}) - m_{prior}(\mathbf{x})) , \tag{7.116}$$

$$\phi_n(\mathbf{x}) = \int_{\mathbf{V}} dV(\mathbf{x}') \; \hat{S}_0(\mathbf{x},\mathbf{x}') \; \gamma_n(\mathbf{x}') , \tag{7.117}$$

and

$$m_{n+1}(\mathbf{x}) = m_n(\mathbf{x}) - \mu_n \; \phi_n(\mathbf{x}) . \tag{7.118}$$

Using now the particular form of the kernel $G^i(\mathbf{x})$ of this problem (equations (7.51)-(7.52)), gives

$$\delta d_n^{\,i} = \int_{R^i} ds^i \; m_n(\, x(s^i) \,) - d^i_{obs} , \tag{7.119}$$

$$\delta\hat{d}_n^{\,i} = \sum_j (\mathbf{C}_D^{\;-1})^{ij} \; \delta d_n^{\,j} , \tag{7.120}$$

$$\delta m_n(\mathbf{x}) = \sum_i \delta\hat{d}_n^{\,i} \; \Psi^i(\mathbf{x}) , \tag{7.121}$$

$$\gamma_n(\mathbf{x}) = \delta m_n(\mathbf{x}) + (m_n(\mathbf{x}) - m_{prior}(\mathbf{x})) , \tag{7.122}$$

$$\phi_n(\mathbf{x}) = \int_{\mathbf{V}} dV(\mathbf{x}') \; \hat{S}_0(\mathbf{x},\mathbf{x}') \; \gamma_n(\mathbf{x}') , \tag{7.123}$$

and

$$m_{n+1}(\mathbf{x}) = m_n(\mathbf{x}) - \mu_n \; \phi_n(\mathbf{x}) , \tag{7.124}$$

where $\Psi^i(\mathbf{x})$ is defined by

$$\Psi^i(\mathbf{x}) \;=\; \int_{R^i} ds^i \;\; C_M(\mathbf{x},\, \mathbf{x}(s^i)\,) \;. \tag{7.125}$$

The interpretation of these formulas is as follows.

Equation (7.119): the values $\delta d_n{}^i$ are simply (minus) the data residuals corresponding to the current model $m_n(\mathbf{x})$.

Equation (7.120): the values $\delta\hat{d}_n{}^i$ are the data residuals weighted with the usual least squares weights. The data covariance matrix is usually diagonal, so that small effort is needed to evaluate the $\delta\hat{d}_n{}^i$.

Equation (7.121): this is the most important of the equations. As explained in the interpretation for equation (7.125), $\Psi^i(\mathbf{x})$ represents the "coefficient of influence" on the point $\mathbf{x}$ of the i-th ray. We see that the contribution of the i-th ray to the value at point $\mathbf{x}$ of the "model perturbation" $\delta\mathbf{m}_n$ is proportional to the weighted residual for the ray and to the coefficient of influence of the ray at point $\mathbf{x}$. This corresponds to a "back-projection" of the weighted residual along (the immediate vicinity of) the ray (see Figure 7.11). This concept of back-projection is usual in algebraic reconstruction techniques (A.R.T.) (see for instance, Herman, 1980). Here they are generalized to the case where the physical space is not discretized. The reader should notice that it is fundamentally the iterative use of simple back-projections which solves the inverse problem.

Equation (7.122): this equation simply corrects the back-projection of the weighted residuals for the information contained in the a priori model. The reader will easily verify, using (7.125) and (7.121), that if we do not have much a priori information (i.e., if the variances in $C_M(\mathbf{x},\mathbf{x}')$ are high), the second term in (7.122) can be neglected.

Equation (7.123): the direction of steepest ascent $\hat{\gamma}_n(\mathbf{x})$ is here "preconditioned". The preconditioning operator $\hat{S}_0$ is arbitrary, although, ideally, it should equal the expression in (7.100), in which case the algorithm would converge in only one iteration (because the steepest descent algorithm then becomes a Newton algorithm). Practically, $\hat{S}_0$ is chosen as a simple operator of approximate deconvolution. One easy way to obtain an adequate deconvolution is as follows. A synthetic model is used, consisting in a few attenuating points in a homogeneous medium. The a priori model and the starting model are equal. The value of the backprojected perturbation $\hat{\gamma}_1(\mathbf{x})$ is computed. The kernel $\hat{S}_0(\mathbf{x},\mathbf{x}')$ is taken space-invariant:

$$\hat{S}_0(\mathbf{x},\mathbf{x}') = D(\mathbf{x}-\mathbf{x}') \ , \tag{7.126}$$

so that equation (7.123) becomes a (de)convolution equation:

$$\phi_n(\mathbf{x}) = \int_{\mathbf{V}} dV(\mathbf{x}') \ D(\mathbf{x}-\mathbf{x}') \ \gamma_n(\mathbf{x}') \ , \tag{7.127}$$

and an adequate function $D(\mathbf{x})$, simply enough for the computation (7.127) to be inexpensive, is sought for, such that the model $\phi_1(\mathbf{x})$ then obtained is as close as possible to the true model we started with.

Equation (7.124): it simply corresponds to the model updating.

Equation (7.125): usually, a covariance function decreases rapidly away from its diagonal ($\| \mathbf{x}-\mathbf{x}' \| >> 0 \ \Rightarrow \ C(\mathbf{x},\mathbf{x}') \simeq 0$). The function $\Psi^i(\mathbf{x})$ then takes significant values only near the i-th ray: it defines a "tube" along the ray. The closer the point $\mathbf{x}$ is to the ray, the greater is the value $\Psi^i(\mathbf{x})$, which represents a sort of *coefficient of influence* on the point $\mathbf{x}$ of the i-th ray.

Practically, the model $m(\mathbf{x})$ is numerically defined on a grid of points, one point per pixel of the graphic device used to plot the model. A covariance function like the one defined in (7.99) is used, with a correlation length L equal to the characteristic size of the holes between rays (so that the model will be smooth). The integrals along rays in (7.119) is evaluated using any (accurate) numerical method. The function $\Psi^i(\mathbf{x})$ may simply be grossly approximated by a reasonably chosen function of the distance between the point $\mathbf{x}$ and the (nearest point of the) i-th ray.

The problem of the computation of the real constant

$$\mu_n = \frac{\gamma_n{}^t \ C_M{}^{-1} \ \phi_n}{\phi_n{}^t \ C_M{}^{-1} \ \phi_n \ + \ (G \ \phi_n)^t \ C_D{}^{-1} \ (G \ \phi_n)}$$

$$= \frac{(\gamma_n \ , \phi_n)}{\| \phi_n \|^2 + \| G \ \phi_n \|^2} \ . \tag{7.104 again}$$

has not yet been addressed. The results given so far are valid for any choice of covariance operators C_D and C_M . Let us now assume the particular choices

$$C_D{}^{ij} = (\sigma_D{}^i)^2 \ \delta^{ij} \tag{7.97 again}$$

$$C_M(\mathbf{x},\mathbf{x}') = \sigma_M{}^2 \ \exp\left(- \frac{\| \mathbf{x}-\mathbf{x}' \|}{L} \right) . \tag{7.99 again}$$

From (7.97) we obtain

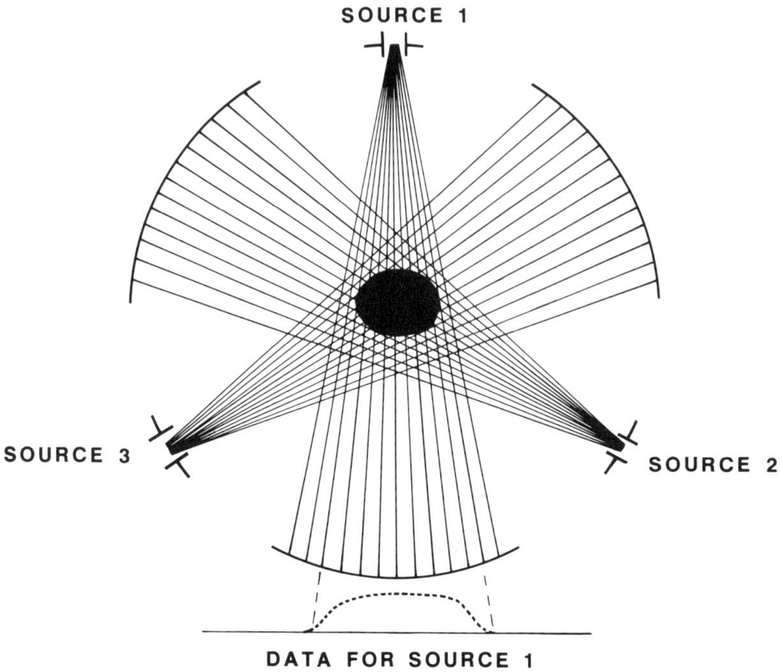
SOURCE 1
SOURCE 3
SOURCE 2
DATA FOR SOURCE 1

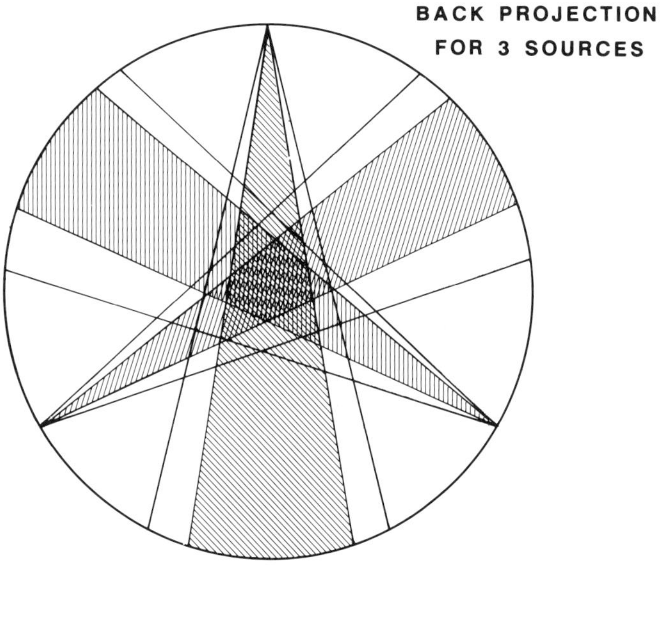
BACK PROJECTION
FOR 3 SOURCES

← *Figure 7.11:* The "backprojection" of data. *Top:* The "true" model and the data corresponding to three different incidences. *Bottom:* Backprojection of the three different incidences. Iterative backprojection (of the residuals) gives the solution of the inverse problem (see text).

$$\| \mathbf{G}\, \phi_n \|^2 = \sum_i \left(\frac{(\mathbf{G}\, \phi_n)^i}{\sigma_D{}^i} \right)^2$$

while from (7.99) we obtain, using the results of problem 7.9,

$$\| \phi_n \|^2 = \frac{1}{8\pi\sigma_M{}^2} \times$$

$$\left[\frac{1}{L^3} \int_{\mathbf{V}} dV(\mathbf{x})\, (\phi(\mathbf{x})_n)^2 + \frac{2}{L} \int_{\mathbf{V}} dV(\mathbf{x})\, (\mathbf{grad}\ \phi(\mathbf{x})_n)^2 + L \int_{\mathbf{V}} dV(\mathbf{x})\, (\Delta\phi(\mathbf{x})_n)^2 \right] ,$$

and

$$(\gamma_n, \phi_n) = \frac{1}{8\pi\sigma_M{}^2} \times$$

$$\left(\frac{1}{L^3} \int_{\mathbf{V}} dV(\mathbf{x})\, \gamma(\mathbf{x})_n\, \phi(\mathbf{x})_n + \frac{2}{L} \int_{\mathbf{V}} dV(\mathbf{x})\, \mathbf{grad}\ \gamma(\mathbf{x})_n \cdot \mathbf{grad}\ \phi(\mathbf{x})_n \right.$$

$$\left. + L \int_{\mathbf{V}} dV(\mathbf{x})\, \Delta\gamma(\mathbf{x})_n\, \Delta\phi(\mathbf{x})_n \right) ,$$

expressions easy to compute numerically.

7.5.2: Newton's noniterative solution.

As seen in Section 4.4, the Newton method gives an explicit (i.e., non-iterative) solution:

$$\mathbf{m} = \mathbf{m}_{prior} + \mathbf{C}_M\, \mathbf{G}^t\, (\mathbf{C}_D + \mathbf{G}\, \mathbf{C}_M\, \mathbf{G}^t)^{-1}\, (\mathbf{d}_{obs} - \mathbf{G}\, \mathbf{m}_{prior}) . \qquad (7.127)$$

In order to interpret this equation, I introduce the following partial steps:

$$\delta \mathbf{d} = \mathbf{d}_{obs} - \mathbf{G}\ \mathbf{m}_{prior} , \tag{7.128}$$

$$\mathbf{S} = \mathbf{C}_D + \mathbf{G}\ \mathbf{C}_M\ \mathbf{G}^t , \tag{7.129}$$

$$\delta \hat{\mathbf{w}} = \mathbf{S}^{-1}\ \delta \mathbf{d} , \tag{7.130}$$

$$\delta \hat{\mathbf{m}} = \mathbf{G}^t\ \delta \hat{\mathbf{w}} , \tag{7.131}$$

and

$$\mathbf{m} = \mathbf{m}_{prior} + \mathbf{C}_M\ \delta \hat{\mathbf{m}} . \tag{7.132}$$

If $G^i(\mathbf{x})$, $C_M(\mathbf{x},\mathbf{x}')$, $\hat{S}_0(\mathbf{x},\mathbf{x}')$ and $C_D{}^{ij}$ are the kernels of $\mathbf{G}$, $\mathbf{C}_M$, $\hat{\mathbf{S}}_0$, and $\mathbf{C}_D$, respectively, equations (7.128)-(7.132) are written, in explicit notation,

$$\delta d^i = d^i{}_{obs} - \int_{\mathbf{V}} d\mathbf{x}\ \ G^i(\mathbf{x})\ \ m_{prior}(\mathbf{x}) , \tag{7.133}$$

$$S^{ij} = C_D{}^{ij} + \int_{\mathbf{V}} d\mathbf{x} \int_{\mathbf{V}} d\mathbf{x}'\ \ G^i(\mathbf{x})\ \ C_M(\mathbf{x},\mathbf{x}')\ \ G^j(\mathbf{x}') , \tag{7.134}$$

$$\delta \hat{w}^i = \sum_j (S^{-1})^{ij}\ \ \delta d^j , \tag{7.135}$$

and

$$m(\mathbf{x}) = m_{prior}(\mathbf{x}) + \int_{\mathbf{V}} d\mathbf{x}' \sum_i C_M(\mathbf{x},\mathbf{x}')\ \ G^i(\mathbf{x}')\ \ \delta \hat{w}^i . \tag{7.136}$$

Using now the particular form of the kernel $G^i(\mathbf{x})$ of this problem (equations (7.51)-(7.52)), gives

$$\delta d^i = d^i{}_{obs} - \int_{R^i} ds^i\ \ m_{prior}(\ \mathbf{x}(s^i)\ ) , \tag{7.137}$$

$$S^{ij} = C_D{}^{ij} + \int_{R^i} ds^i \int_{R^j} ds^j\ \ C_M(\ \mathbf{x}(s^i)\ ,\ \mathbf{x}(s^j)\ ) , \tag{7.138}$$

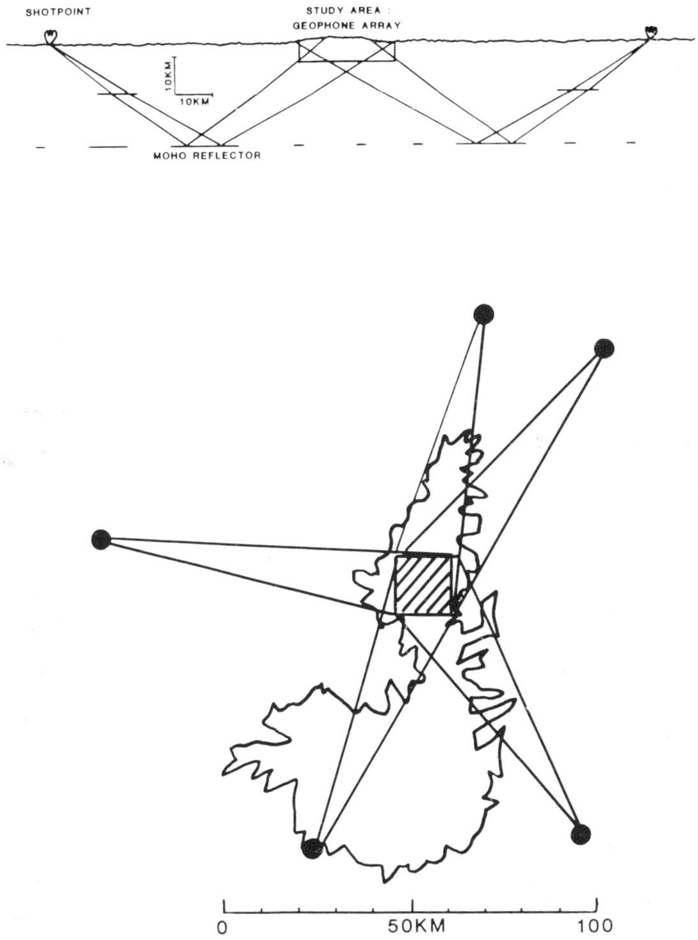

Figure 7.12: A seismic transmission experiment on a volcano (Mont Dore, France). Seismic waves were generated at the surface of the Earth (using explosions). After reflection at a deep discontinuity (the "Moho"), the travel-times of the waves were observed at an array of seismic stations ($\simeq$100 per each shot). The observed travel times were used for inferring the velocity structure for seismic waves in a 3D region under the volcano.

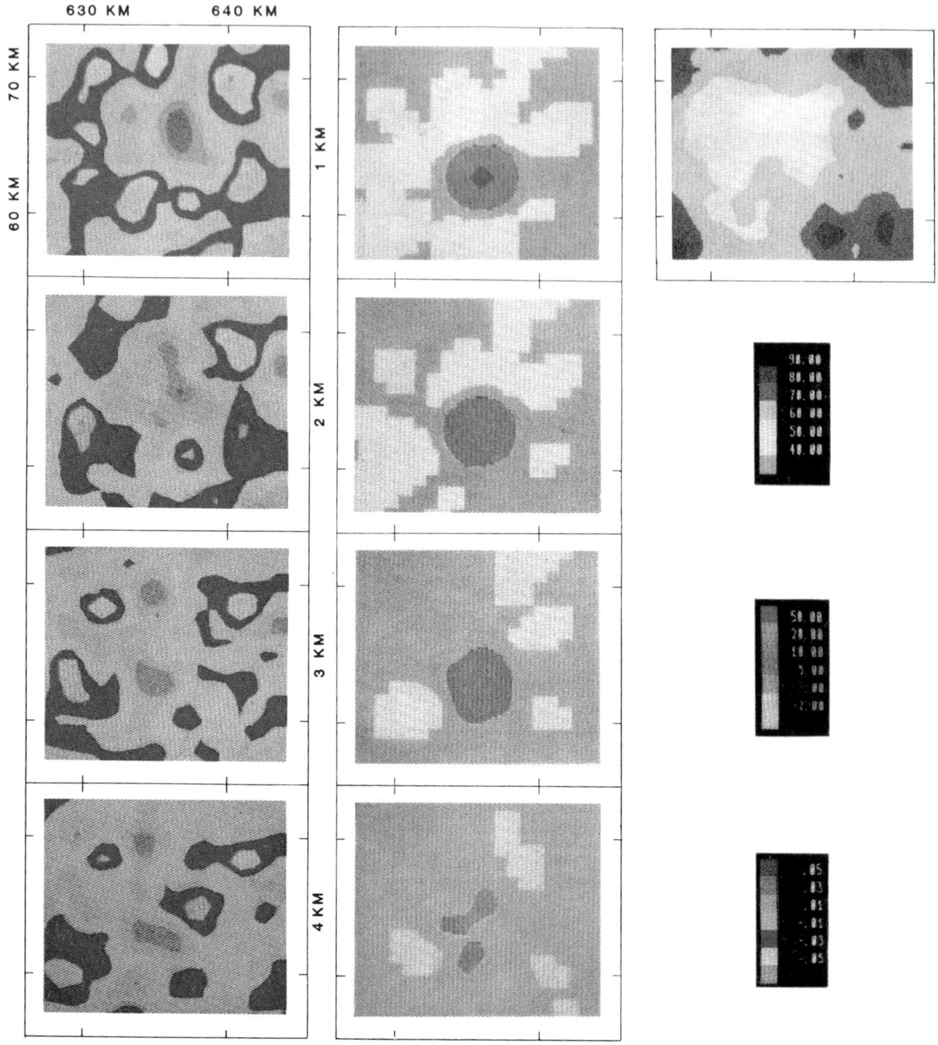
630 KM
640 KM
70 KM
60 KM
1 KM
2 KM
3 KM
4 KM
90.00
80.00
70.00
60.00
50.00
40.00

← *Figure 7.13:* Tomographic results of the Mont Dore seismic transmission experiment, from Nercessian et al., 1984. *Left:* Horizontal sections of the velocity structure under the volcano, at 1, 2, 3, and 4 km depth. Warm colours represent low velocity and cold colours represent high velocity. *Middle:* The posterior covariance function $C(\mathbf{x}_0,\mathbf{x})$, for a particular given point $\mathbf{x}_0$ situated at $x = 636$ km, $y = 63$ km, $z = 1$km . Scale is in percentage of *correlation.* This figure shows in particular that the spatial *resolution* around $\mathbf{x}_0$ attained with this data set is of the order of 1 km . *Right:* A posteriori standard deviations $\sigma(\mathbf{x}) = C(\mathbf{x},\mathbf{x})^{1/2}$ for the horizontal section 1km deep. The scale is in percentage of a priori stardard deviation. Central regions are best resolved.

$$\delta\hat{w}^i = \sum_j (S^{-1})^{ij} \ \delta d^j , \tag{7.139}$$

and

$$m(\mathbf{x}) = m_{prior}(\mathbf{x}) + \sum_i \delta\hat{w}^i \ \Psi^i(\mathbf{x}) , \tag{7.140}$$

where

$$\Psi^i(\mathbf{x}) = \int_{R^i} ds^i \ C_M(\mathbf{x}, \mathbf{x}(s^i)) . \qquad \text{(7.125 again)}$$

The interpretation of these formulas is as follows.

Equation (7.137): the values δd^i are the data residuals corresponding to the prior model $m_{prior}(\mathbf{x})$.

Equation (7.138): the values S^{ij} define the interdependence of each ray with respect all other rays. In particular, they contain information on the geometrical structure of the problem.

Equation (7.139): the values $\delta\hat{w}^i$ are geometrically weighted residuals, where the weighting is such that the back projection of these weighted residuals will give the solution (see below).

Equation (7.140): back-projection (as in equation (7.121)) of the geometrically weighted residuals.

Now, the question arises of which of the two algorithms shown in sections 7.5.1 and 7.5.2 is to be preferred. There is no universal answer. It depends on the available computer hardware. In the steepest descent algorithm the main task is the back projection (7.121), which has to be performed iteratively. In the Newton algorithm, we need to perform a double integral along ray-paths (7.138), to solve a large linear system corresponding to (7.139), and to perform one back propagation (7.140). Current computer technology performs best using the steepest descent.

Figures 7.12 - 7.13 give an illustration of the results of a tomographic (geophysical) method.

7.6: Example: The inversion of acoustic waveforms

This problem has been introduced in examples 7.18 and 7.20. A data vector $\mathbf{d}$ is a set of seismograms

$$\mathbf{d} = [\ p(\mathbf{x}_r,t) \quad ; \quad 0 \le t \le T \quad ; \quad r=1,2,...\ ] \ , \tag{7.141}$$

while a model vector $\mathbf{m}$ is a three-dimensional function

$$\mathbf{m} = [\ m(\mathbf{x}) \quad ; \quad \mathbf{x} \in \mathbf{V}\ ] \ , \tag{7.142}$$

where $m(\mathbf{x})$ represents the squared slowness at point $\mathbf{x}$. The resolution of the forward problem is written

$$\mathbf{d} = \mathbf{g}(\mathbf{m}) \ ,$$

where the operator $\mathbf{g}$ denotes the resolution of the acoustic wave equation (see example 7.18). Let $\mathbf{d}_{obs}$ be the observed data values, $\mathbf{C}_D$ the covariance operator describing data uncertainties, $\mathbf{m}_{prior}$ the a priori model, and $\mathbf{C}_M$ the covariance operator describing our confidence in $\mathbf{m}_{prior}$. The best model is defined by the minimization of

$$S(\mathbf{m}) = \tag{7.143}$$
$$\frac{1}{2} \left[(\mathbf{g}(\mathbf{m})-\mathbf{d}_{obs})^t \ \mathbf{C}_D^{-1} \ (\mathbf{g}(\mathbf{m})-\mathbf{d}_{obs}) + (\mathbf{m}-\mathbf{m}_{prior})^t \ \mathbf{C}_M^{-1} \ (\mathbf{m}-\mathbf{m}_{prior}) \right] ,$$

and can, for instance, be obtained through the preconditioned steepest descent algorithm (see Section 4.4)

$$\mathbf{m}_{n+1} = \mathbf{m}_n - \mu_n \ \hat{\mathbf{S}}_0 \left[\mathbf{C}_M \ \mathbf{G}_n{}^t \ \mathbf{C}_D^{-1} \ (\mathbf{g}(\mathbf{m})-\mathbf{d}_{obs}) + (\mathbf{m}-\mathbf{m}_{prior}) \right] , \tag{7.144}$$

where $\mathbf{G}_n$ is the Fréchet derivative of the nonlinear operator $\mathbf{g}$ at the point $\mathbf{m}_n$, and where $\hat{\mathbf{S}}_0$ is an arbitrary positive definite operator (hopefully a not too bad approximation to $(\mathbf{I}+\mathbf{C}_M\mathbf{G}_0{}^t\mathbf{C}_D{}^{-1}\mathbf{G}_0)^{-1}$).

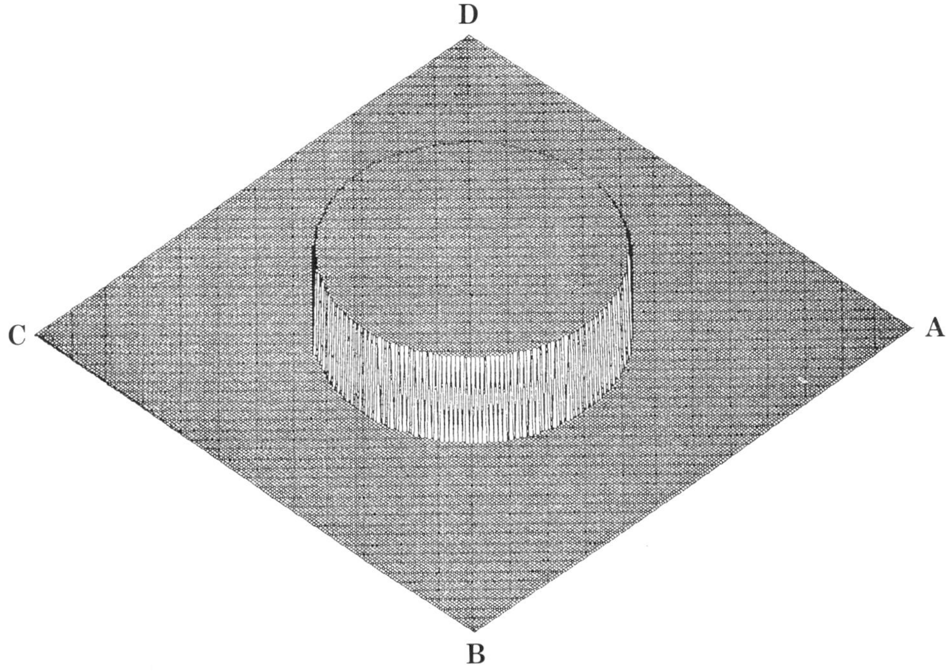

Figure 7.14: The "true model" for the inversion of acustic waveforms. Eight sources have been simulated (stars in Figure 7.16), each being recorded all around the medium (100 receivers per side). The model is defined in a 200 x 200 grid, thus giving 40000 unknowns for the inversion. From Gauthier et al., 1986.

In order to give a physical interpretation of the main operations involved in the algorithm (7.144), I introduce the following partial steps:

$$\delta\mathbf{d}_n = \mathbf{g}(\mathbf{m}_n) - \mathbf{d}_{obs} , \tag{7.145}$$

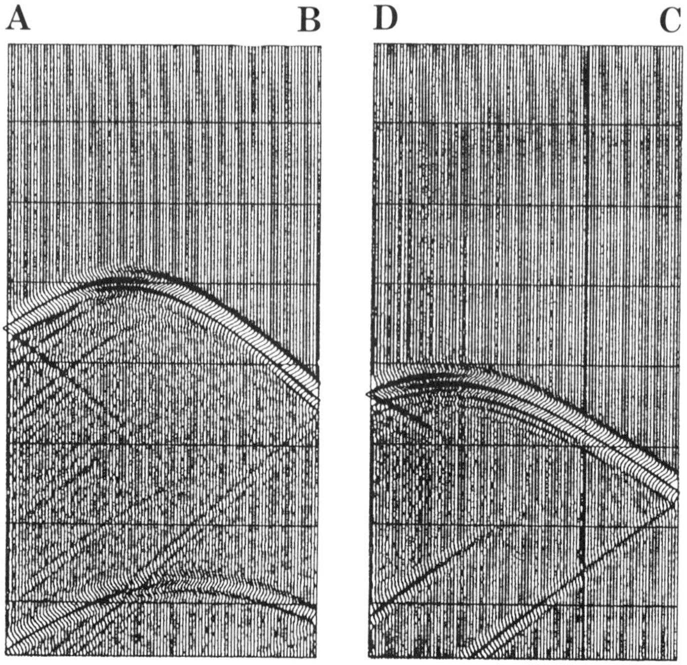

Figure 7.15: Some of the "true seismograms" corresponding to the source at corner A. From Gauthier et al., 1986.

$$\delta\hat{\mathbf{d}}_n = \mathbf{C}_D{}^{-1}\,\delta \mathrm{d}_n \,, \tag{7.146}$$

$$\delta\hat{\mathbf{m}}_n = \mathbf{G}_n{}^t\,\delta\hat{\mathbf{d}}_n \,, \tag{7.147}$$

$$\delta\mathbf{m}_n = \mathbf{C}_M\;\delta\mathbf{m}_n + \mathbf{m}_n - \mathrm{m}_{prior} \,, \tag{7.148}$$

$$\Delta \mathrm{m}_n = \hat{\mathbf{S}}_0\;\delta\mathbf{m}_n \,, \tag{7.149}$$

and

$$\mathbf{m}_{n+1} = \mathbf{m}_n - \mu_n\;\Delta\mathbf{m}_n \,. \tag{7.150}$$

Using the results of examples 7.13 and 7.15, and of problem 7.2, this gives, explicitly,

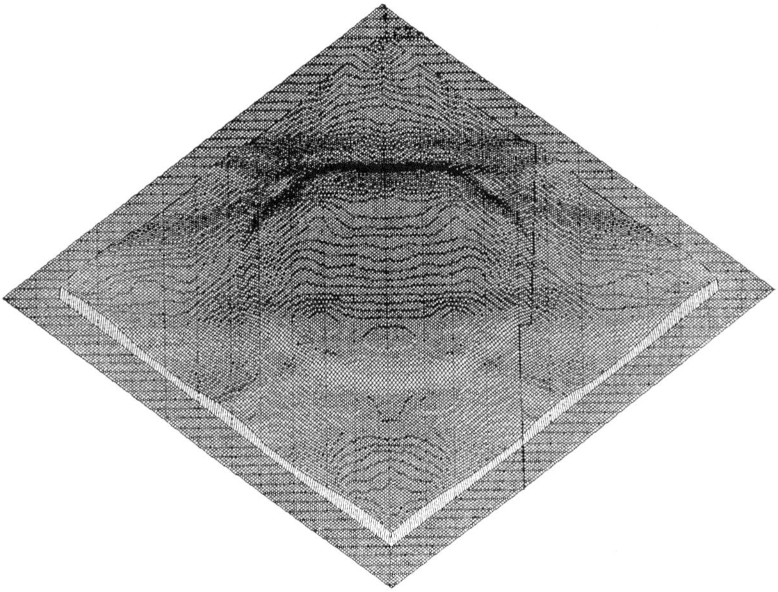

Figure 7.16: The model obtained after one iteration of the steepest descent algorithm. From Gauthier et al., 1986.

$$\delta p(\mathbf{x}_r,t)_n = p(\mathbf{x}_r,t)_n - p(\mathbf{x}_r,t)_{obs} , \tag{7.151}$$

$$\delta\hat{p}(\mathbf{x}_r,t)_n = \sum_r \int_0^T dt' \; C_D^{-1}(\mathbf{x}_r,t;\mathbf{x}_r',t') \; \delta p(\mathbf{x}_r,t)_n , \tag{7.152}$$

$$\delta\hat{m}(\mathbf{x})_n = \int_0^T dt \;\; \dot{\Psi}(\mathbf{x},t)_n \;\; \dot{p}(\mathbf{x},t)_n , \tag{7.153}$$

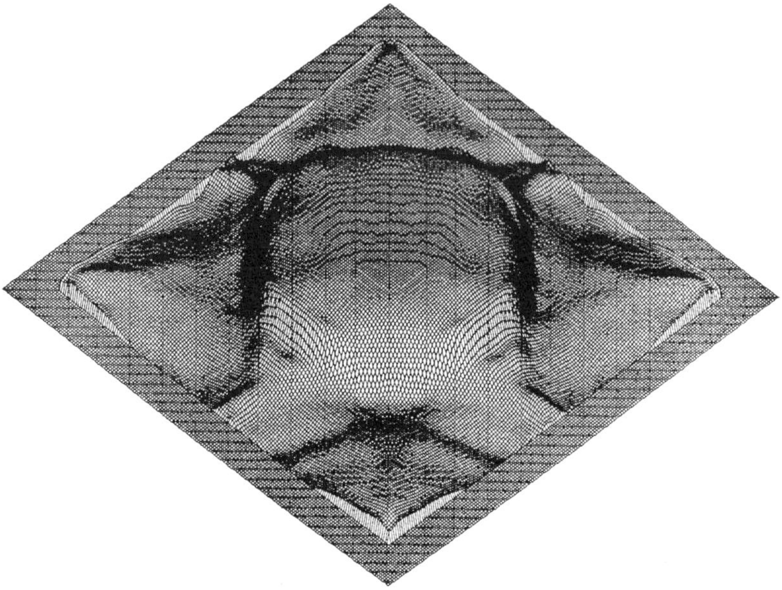

Figure 7.17: The model obtained after five iterations of the steepest descent algorithm. From Gauthier et al., 1986.

$$\delta m(\mathbf{x})_n = \int_V d\mathbf{x}' \; C_M(\mathbf{x},\mathbf{x}') \; \delta\hat{m}(x')_n + m(\mathbf{x})_n - m(\mathbf{x})_{prior} , \tag{7.154}$$

$$\Delta m(x)_n = \int_V d\mathbf{x}' \; \hat{S}_0(\mathbf{x},\mathbf{x}') \; \delta m(\mathbf{x}')_n , \tag{7.155}$$

and

$$m(\mathbf{x})_{n+1} = m(\mathbf{x})_n - \mu_n \; \Delta m(\mathbf{x})_n , \tag{7.156}$$

where $p(\mathbf{x},t)_n$ is the field propagating in the current model $m(\mathbf{x})_n$, and due to the actual sources, i.e., the field solution of

$$m(\mathbf{x})_n \frac{\partial^2}{\partial t^2} p(\mathbf{x},t)_n - \nabla^2 p(\mathbf{x},t)_n = S(\mathbf{x},t) \qquad t\in[0,T] \tag{7.157a}$$

$$p(\mathbf{x},0)_n = 0 \tag{7.157b}$$

$$\dot{p}(\mathbf{x},0)_n = 0 \ , \tag{7.157c}$$

and where $\Psi(\mathbf{x},t)_n$ is the field obtained by propagation *backward in time* of the weighted residuals $\delta\hat{p}(\mathbf{x}_r,t)_n$ (acting as if they were point sources), i.e, the field solution of

$$m(\mathbf{x})_n \frac{\partial^2}{\partial t^2} \Psi(\mathbf{x},t)_n - \nabla^2 \Psi(\mathbf{x},t)_n = \sum_r \delta(\mathbf{x}-\mathbf{x}_r)\ \delta\hat{p}(\mathbf{x}_r,t)_n \qquad t\in[0,T] \tag{7.158a}$$

$$\Psi(x,T)_n = 0 \tag{7.158b}$$

$$\dot{\Psi}(\mathbf{x},T)_n = 0 \tag{7.158c}$$

(note that $\Psi(\mathbf{x},t)_n$ satisfies *final* (instead of initial) conditions.

The field $p(\mathbf{x},t)_n$ is the *current predicted field.* As the field $\Psi(\mathbf{x},t)_n$ is obtained by propagating the current residuals backwards in time, it corresponds to a sort of *current missing field* (a field, that if it were predicted by the current medium, would give null residuals). The important equation (7.153) can then be interpreted: at each point $\mathbf{x}$ where the velocity of the missing field $\Psi(\mathbf{x},t)_n$ is time correlated with the velocity of the current field $p(\mathbf{x},t)_n$, we add a diffractor in our current model, which will try to create this missing field.

Central in a problem involving ray concepts (as in the problem of X-ray tomography) was the concept of *back-projection.* We see that in a problem involving a wave equation, we encounter the concept of *back-propagation.*

For a practical solution of the problem we proceed as follows. Assume we have some numerical code (as for instance a finite-differencing code, e.g, Alterman and Karal, 1968) allowing of solving the wave equation for arbitrary configuration of sources and receivers. Let $m(\mathbf{x})_n$ be the current model. We start solving the wave equation using the actual source $S(x,t)$, (equation (7.157)) thus obtaining the field $p(\mathbf{x},t)_n$, whose time derivative $\dot{p}(\mathbf{x},t)_n$ we store in the computer's memory. In particular, this gives the predicted data $p(\mathbf{x}_r,t)_n$ at the receiver locations. The difference with the observed data $p(\mathbf{x}_r,t)_{obs}$ gives the residuals (7.151), which are weighted with the corresponding weighting operator (this operation is trivial if we use a diagonal operator). These weighted residuals are now used as sources for a new resolution of the wave equation, but now *reversed in time* (equations (7.158). It is very easy to run a finite-difference code backwards in time, just change $t_i = i\ \Delta t$ into $t_i = T - i\ \Delta t$. Once the field $\Psi(\mathbf{x},t)_n$ has been obtained, and as the field $\dot{p}(\mathbf{x},t)_n$ was stored in the computer's memory, we can compute the time correlations (7.153), giving $\delta\hat{m}(x)_n$. This field is smoothed by the covariance operator $C_M(x,x')$ (if the covariance function is stationary, $C_M(\mathbf{x},\mathbf{x}') = K(\mathbf{x}-\mathbf{x}')$, this corresponds to a spatial convolution), and the term $m(\mathbf{x})_n - m(\mathbf{x})_{prior}$ is added, thus giving $\delta m(\mathbf{x})_n$ (equation (7.154)). If we wish to use a preconditioning operator (in order to accelerate convergence) we do that in (7.155). The real number μ_n in (7.156) can be estimated by a linearized approximation (see problem 7.2) but let us simply admit here that we fix its value by trial and error. This finally gives the updated model $m_{n+1}(\mathbf{x})$. Figures 7.14 - 7.17 illustrate the method with a numerical example.

Box 7.3: The first Born and Rytov approximations.

Let $p(\mathbf{x},t)$ represent an arbitrary wave field, solution of a wave equation, for given sources, and with given initial and boundary conditions. The solution $p(\mathbf{x},t)$ will depend on some medium properties, represented by the abstract vector $\mathbf{m}$. For short, we write

$$\mathbf{p} = \mathbf{g}(\mathbf{m}) . \qquad (1)$$

Let $\mathbf{p}_0$ denote the solution of the wave equation for a particular medium $\mathbf{m}_0$ (usually a simple medium: homogeneous, or spherically symmetric, ...),

$$\mathbf{p}_0 = \mathbf{g}(\mathbf{m}_0) . \qquad (2)$$

A Taylor development around $\mathbf{m}_0$ gives

$$\mathbf{g}(\mathbf{m}) = \mathbf{g}(\mathbf{m}_0) + \mathbf{G}_0\,(\mathbf{m}-\mathbf{m}_0) + O(\|\mathbf{m}-\mathbf{m}_0\|^2) , \qquad (3)$$

where $\mathbf{G}_0$ is the Fréchet derivative of $\mathbf{g}$ at $\mathbf{m} = \mathbf{m}_0$. The first Born approximation to the wavefield $\mathbf{p}$ consists in the first-order approximation to (3):

(...)

$$\boxed{\mathbf{p} \simeq \mathbf{g}(\mathbf{m}_0) + \mathbf{G}_0\,(\mathbf{m}-\mathbf{m}_0)} \qquad \textbf{(Born approximation)} . \tag{4}$$

Definition (4) also applies to the Fourier-transformed field

$$\mathbf{p}(\mathbf{x},\omega) = \int_{-\infty}^{\infty} dt\; e^{i\omega t}\; p(\mathbf{x},t) . \tag{5}$$

In an inverse problem where the values of the wavefield $\mathbf{p}$ are data, and the model parameters $\mathbf{m}$ are unknowns, the Born approximation (4) clearly corresponds to a linearization of the forward problem. The Born approximation may be a bad approximation in many circumstances. In particular, if the travel times of the waves are not adequately modeled by the reference model $\mathbf{m}_0$. In that case, a different approximation, named Rytov, may be used. It can be introduced as follows.

Let $p_0(\mathbf{x},\omega)$ be the reference field (corresponding to the reference medium $\mathbf{m}_0$). With any field $p(\mathbf{x},\omega)$ we associate the "Rytov field", or "logarithmic field"

$$\Psi(\mathbf{x},\omega) = \operatorname{Log}\left(\frac{p(\mathbf{x},\omega)}{p_0(\mathbf{x},\omega)} \right) . \tag{5}$$

In terms of this new field, the forward equation (1) becomes

$$\mathbf{\Psi} = \mathrm{h}(\mathrm{m}) , \tag{6}$$

where, clearly,

$$\mathbf{h}(\mathbf{m}) = \operatorname{Log}\left(\frac{\mathbf{g}(\mathbf{m})}{p_0} \right) . \tag{7}$$

A Taylor development gives

$$\mathbf{h}(\mathbf{m}) = \mathbf{h}(\mathbf{m}_0) + \mathbf{H}_0\,(\mathbf{m}-\mathbf{m}_0) + O(\|\mathbf{m}-\mathbf{m}_0\|^2) , \tag{8}$$

and the first-order approximation to the logarithmic field is

$$\mathbf{\Psi} \simeq \mathrm{h}(\mathrm{m}_0) + \mathrm{H}_0\,(\mathrm{m}-\mathrm{m}_0) . \tag{9}$$

As

$$\mathbf{h}(\mathbf{m}_0) = \operatorname{Log}\left(\frac{g(\mathbf{m}_0)}{p_0} \right) = 0 , \tag{10}$$

(...)

and

$$\mathbf{H_0} = \frac{1}{\mathbf{p_0}} \mathbf{G_0} , \qquad (11)$$

we obtain

$$\boldsymbol{\Psi} \simeq \frac{1}{\mathbf{p_0}} \mathbf{G_0} (\mathbf{m}-\mathbf{m_0}) . \qquad (12)$$

In terms of the wavefield **p** this gives, using (5)

$$\boxed{\mathbf{p} \simeq \mathbf{p_0} \exp\left[\frac{1}{\mathbf{p_0}} \mathbf{G_0}(\mathbf{m}-\mathbf{m_0}) \right]} \quad \textbf{(Rytov approximation)} . \qquad (13)$$

Equation (12) is the (first) Rytov approximation, to be compared with that of Born (4). The two approximations can be written in terms of the logarithmic field. This gives respectively:

$$\boldsymbol{\Psi} \simeq \mathrm{Log}(1 + \mathbf{H_0} (\mathbf{m}-\mathbf{m_0})) \qquad \textbf{(Born approximation)} , \qquad (14)$$

and

$$\boldsymbol{\Psi} \simeq \mathbf{H_0} (\mathbf{m}-\mathbf{m_0}) \qquad \textbf{(Rytov approximation)} . \qquad (15)$$

Notice that a first-order development of the exponential in Rytov (13) leads to Born (4), and that a first-order development of the logarithm in Born (14) leads to Rytov (15).

The interest of the logarithmic field can be seen as follows. If $p(\mathbf{x},\omega)$ is the complex spectrum of the wave field $p(\mathbf{x},t)$, we can introduce as usual the real fields $A(\mathbf{x},\omega)$ and $\phi(\mathbf{x},\omega)$, corresponding respectively to the *spectral amplitude* and *phase* of $p(\mathbf{x},\omega)$:

$$p(\mathbf{x},\omega) = A(\mathbf{x},\omega)\ e^{i\phi(\mathbf{x},\omega)} . \qquad (16)$$

Then

$$\Psi(\mathbf{x},\omega) = \mathrm{Log}\left[\frac{A(\mathbf{x},\omega)}{A_0(\mathbf{x},\omega)} \right] + i \, (\, \phi(\mathbf{x},\omega) - \phi_0(\mathbf{x},\omega) \,) . \qquad (17)$$

This justifies the definition of the logarithmic field: it is a linear function of the phase. This is why, Rytov approximates the phase better than Born.

(...)

Assume now that we wish to solve the inverse problem of estimating the medium parameters $\mathbf{m}$ from the observation of the field $p(\mathbf{x},\omega)$ at some points $\mathbf{x}_r$ $(r=1,2,...)$ of the space. Usually, we seek the model $\mathbf{m}$ minimizing

$$S(\mathbf{m}) = \sum_r \int_{-\infty}^{\infty} d\omega \; \left| \, p(\mathbf{x}_r,\omega)_{cal} - p(\mathbf{x}_r,\omega)_{obs} \, \right|^2 . \tag{18}$$

Instead, we may seek the model minimizing

$$\tilde{S}(\mathbf{m}) = \sum_r \int_{-\infty}^{\infty} d\omega \; \left| \, \Psi(\mathbf{x}_r,\omega)_{cal} - \Psi(\mathbf{x}_r,\omega)_{obs} \, \right|^2 , \tag{19}$$

An easy computation gives

$$\tilde{S}(\mathbf{m}) = \sum_r \int_{-\infty}^{\infty} d\omega \left\{ \mathrm{Log}\left(\frac{A(\mathbf{x}_r,\omega)_{cal}}{A(\mathbf{x}_r,\omega)_{obs}} \right) \right\}^2$$

$$+ \sum_r \int_{-\infty}^{\infty} d\omega \; (\, \phi(\mathbf{x}_r,\omega)_{cal} - \phi(\mathbf{x}_r,\omega)_{obs} \,)^2 . \tag{20}$$

Criterion (20) is *quadratic* in phase, and this may be useful in some problems where differences of phase between the the starting wave fields and the observed wave fields are significant.

Example: Consider the acoustic wave equation

$$\omega^2 \, n^2(\mathbf{x}) \, p(\mathbf{x},\omega) + \nabla^2 p(\mathbf{x},\omega) = - \, S(\mathbf{x},\omega) \, , \tag{21}$$

where $p(\mathbf{x},\omega)$ is a pressure wave field, $n(\mathbf{x})$ is the slowness (inverse of celerity) of waves at point $\mathbf{x}$, and $S(\mathbf{x},\omega)$ represents the source of waves. Let $p_0(\mathbf{x},\omega)$ be the pressure field corresponding to the reference medium $n_0(\mathbf{x})$. The (first) Born approximation to the field corresponding to the medium $n_0(\mathbf{x}) + \delta n(\mathbf{x})$ is

$$p(\mathbf{x},\omega) \simeq p_0(\mathbf{x},\omega) + 2\omega^2 \int dV(\mathbf{x}') \, \Gamma_0(\mathbf{x},\omega;\mathbf{x}') \, p_0(\mathbf{x}',\omega) \, n_0(\mathbf{x}') \, \delta n(\mathbf{x}') \, , \tag{22}$$

where $\Gamma_0(\mathbf{x},\omega,\mathbf{x}')$ is Green's function solution of

(...)

$$\omega^2\, n_0^2(\mathbf{x})\, \Gamma_0(\mathbf{x},\omega;\mathbf{x}') + \nabla^2\Gamma_0(\mathbf{x},\omega;\mathbf{x}') = -\,\delta(\mathbf{x}-\mathbf{x}')\ . \qquad (23)$$

The (first) Rytov approximation is

$$p(\mathbf{x},\omega) \simeq \qquad (24)$$

$$p_0(\mathbf{x},\omega)\, \exp\left[\frac{2\omega^2}{p_0(\mathbf{x},\omega)} \int dV(\mathbf{x}')\, \Gamma_0(\mathbf{x},\omega;\mathbf{x}')\, p_0(\mathbf{x}',\omega)\, n_0(\mathbf{x}')\, \delta n(\mathbf{x}') \right] .$$

Some details on Born approximation can be found in Morse and Feshbach (1953). Chernov (1960) and Sobzyk (1985) compare the Born and Rytov approximations. Devaney (1984) uses the Rytov approximation in diffraction acoustic tomography.

7.7: Analysis of error and resolution

As explained in Section 7.3, if measurement uncertainties and a priori model uncertainties may be described using Gaussian probabilities, and if the equation solving the forward problem is strictly linear, the posterior probability in the model space is also Gaussian, with mathematical expectation

$$\langle\, m \,\rangle \ =$$

$$= \mathbf{m}_{prior} + (\mathbf{G}^t\, \mathbf{C}_D^{-1}\, \mathbf{G} + \mathbf{C}_M^{-1})^{-1}\, \mathbf{G}^t\, \mathbf{C}_D^{-1}\, (\mathbf{d}_{obs} - \mathbf{G}\, \mathbf{m}_{prior}) \qquad \text{(7.84a again)}$$

$$= \mathbf{m}_{prior} + \mathbf{C}_M\, \mathbf{G}^t\, (\mathbf{G}\, \mathbf{C}_M\, \mathbf{G}^t + \mathbf{C}_D)^{-1}\, (\mathbf{d}_{obs} - \mathbf{G}\, \mathbf{m}_{prior})\ , \qquad \text{(7.84b again)}$$

and covariance operator

$$\mathbf{C}_{M'} = (\mathbf{G}^t\, \mathbf{C}_D^{-1}\, \mathbf{G} + \mathbf{C}_M^{-1})^{-1} \qquad \text{(7.85a again)}$$

$$= \mathbf{C}_M - \mathbf{C}_M\, \mathbf{G}^t\, (\mathbf{G}\, \mathbf{C}_M\, \mathbf{G}^t + \mathbf{C}_D)^{-1}\, \mathbf{G}\, \mathbf{C}_M\ . \qquad \text{(7.85b again)}$$

If the equation solving the forward problem is not linear, but is not strongly nonlinear, then the a posteriori probability is approximately Gaussian, and its mathematical expectation minimizes the misfit function

$S(\mathbf{m}) =$ (7.87a again)

$$\frac{1}{2}\left[(\mathbf{g}(\mathbf{m})-\mathbf{d}_{obs})^t\ \mathbf{C}_D^{-1}\ (\mathbf{g}(\mathbf{m})-\mathbf{d}_{obs}) + (\mathbf{m}-\mathbf{m}_{prior})^t\ \mathbf{C}_M^{-1}\ (\mathbf{m}-\mathbf{m}_{prior})\right] ,$$

and can be obtained using some iterative process: $\mathbf{m}_0$, $\mathbf{m}_1$, $\mathbf{m}_2$, ... , $\mathbf{m}_\infty = \langle\ \mathbf{m}\ \rangle$. The corresponding covariance operator is then

$$\mathbf{C}_{M'} = (\mathbf{G}_\infty{}^t\ \mathbf{C}_D^{-1}\ \mathbf{G}_\infty + \mathbf{C}_M^{-1})^{-1}$$
$$= \mathbf{C}_M - \mathbf{C}_M\ \mathbf{G}_\infty{}^t\ (\mathbf{G}_\infty\ \mathbf{C}_M\ \mathbf{G}_\infty{}^t + \mathbf{C}_D)^{-1}\ \mathbf{G}_\infty\ \mathbf{C}_M\ , \qquad \text{(7.87b again)}$$

where $\mathbf{G}_\infty$ denotes the derivative of $\mathbf{g}$ at $\langle\ \mathbf{m}\ \rangle = \mathbf{m}_\infty$. All the information we may need about a posteriori uncertainties can be extracted from the a posteriori covariance operator $\mathbf{C}_{M'}$. The simplest analysis is that of "error and resolution". Let us start by the first.

7.7.1: Analysis of error

To fix ideas, I assume here that we examine a problem where the model paremeter is a scalar function of the spatial coordinates, $\mathbf{x} \rightarrow m(\mathbf{x})$, where $\mathbf{x}$ is a point on the plane. The reader will easily generalize to the case where we need more than one function to describe the model (as is, for instance, the case in the elastodynamic waveform inversion of problem 7.3). The kernel of the posterior covariance operator is then a single covariance function $C_{M'}(\mathbf{x},\mathbf{x}')$.

Let us consider a particular point $\mathbf{x}_0$ in the physical medium. What information do we have on the true value of the parameter m at $\mathbf{x}_0$? As the a posteriori probability is Gaussian, the marginal probability for the parameter $m(\mathbf{x}_0)$ is also Gaussian. The corresponding Gaussian probability density is clearly centered at $\langle\ m\ \rangle\ (\mathbf{x}_0)$, and the standard deviation is

$$\sigma(\mathbf{x}_0) = \sqrt{C_{M'}(\mathbf{x}_0,\mathbf{x}_0)}\ . \qquad (7.159)$$

If we are able to compute $\sigma(\mathbf{x})$ for all points $\mathbf{x}$ numerically (see section 7.7.3), we may plot the estimated uncertainty $\sigma(\mathbf{x})$ together with the solution $\langle\ m\ \rangle\ (\mathbf{x})$. This has been done in the example of figures 7.12 - 7.13.

Sometimes we may not be interested in the uncertainty on the a posteriori value of our parameter at $\mathbf{x}_0$, but rather in the uncertainty on the mean value of the parameter over a given ball around $\mathbf{x}_0$:

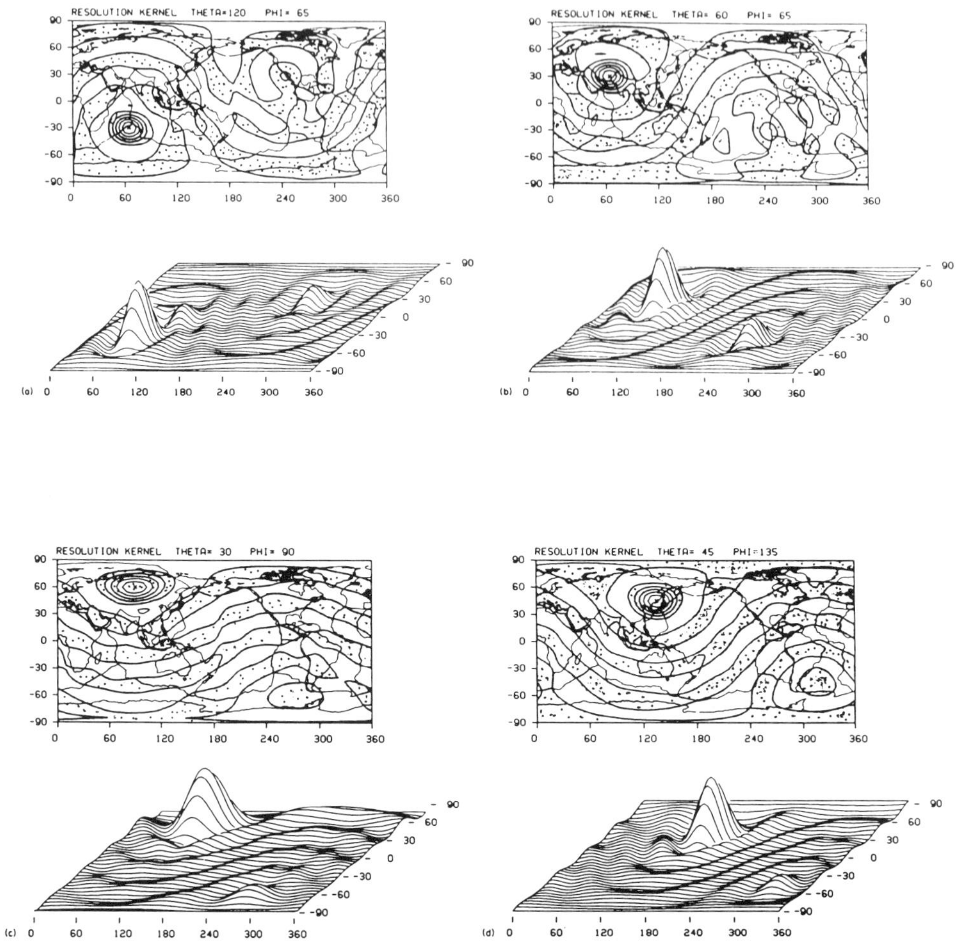

Figure 7.18: The resolving kernels obtained by Tanimoto (1985) in a problem of inversion of seismic waveforms. Rayleigh waves of 200 s period were used. The resolving kernel $R(\theta_0,\phi_0;\theta,\phi)$ is drawn for 4 different positions of (θ_0,ϕ_0) . The figures show the shapes of the resolving kernels in a contoured map at the top and in a relief map at the bottom. In the contoured map, the maximum value at the target point is normalized as 1 and the lines corresponding to 0.8, 0.6, 0.4, 0.2, and 0 are drawn. Positive regions are shaded. Quite localized peaks with radii of about 2000 km are produced for all maps. Note the existence of antipodal peaks. This means that with the kind of data used, the value of the solution obtained at any point on the Earth is (slightly) contaminated with the true values at the antipodes.

$$\tilde{m}(\mathbf{x}_0) = \int_{\mathbf{V}} dV\mathbf{x} \;\; D(\mathbf{x}_0,\mathbf{x}) \;\; \langle \; m \; \rangle \; (\mathbf{x}) \; , \tag{7.160a}$$

or, for short,

$$\tilde{\mathbf{m}} = \mathbf{D} \; \langle \; \mathbf{m} \; \rangle \; . \tag{7.160b}$$

Using the definition of covariance it follows

$$\tilde{\mathbf{C}}_M = \mathbf{D} \; \mathbf{C}_{M'} \; \mathbf{D}^t \; , \tag{7.161a}$$

i.e.,

$$\tilde{C}_M(\mathbf{x},\mathbf{x}') = \int_{\mathbf{V}} dV(\mathbf{x}'') \int_{\mathbf{V}} dV(\mathbf{x}''') \; D(\mathbf{x},\mathbf{x}'') \; C_{M'}(\mathbf{x}'',\mathbf{x}''') \; D(\mathbf{x}',\mathbf{x}''') \; . \tag{7.161b}$$

The uncertainty on the mean value $\tilde{m}(\mathbf{x}_0)$ is then

$$\tilde{\sigma}(\mathbf{x}_0) = \sqrt{\tilde{C}_M(\mathbf{x}_0,\mathbf{x}_0)} \; . \tag{7.162}$$

7.7.2: Analysis of resolution

The concept of resolution corresponds to the possibility of identifying individual points in the posterior model which, in the true model, are close together.

It is important at this point to have good intuition as to what a covariance function represents. In all rigour, the "solution" $\langle \; \mathbf{m} \; \rangle$ and the posterior covariance operator $\mathbf{C}_{M'}$ simply describe a Gaussian random process, allowing of computing the probabilities of different eventualities. For instance, from equation (7.162), and using elementary calculus, we can compute the probability of the mean value of the parameter on a disk defined by (7.160a) belonging to a given interval.

More fundamentally, from $\langle \; \mathbf{m} \; \rangle$ and $\mathbf{C}_{M'}$, we can in principle generate pseudo-random realizations of the Gaussian posterior probability. My personal philosophy is that we could intuitively learn much more from the consideration of a few pseudo-random realizations than from a heavy analysis of error and of resolution. Unfortunately, the pseudorandom realization

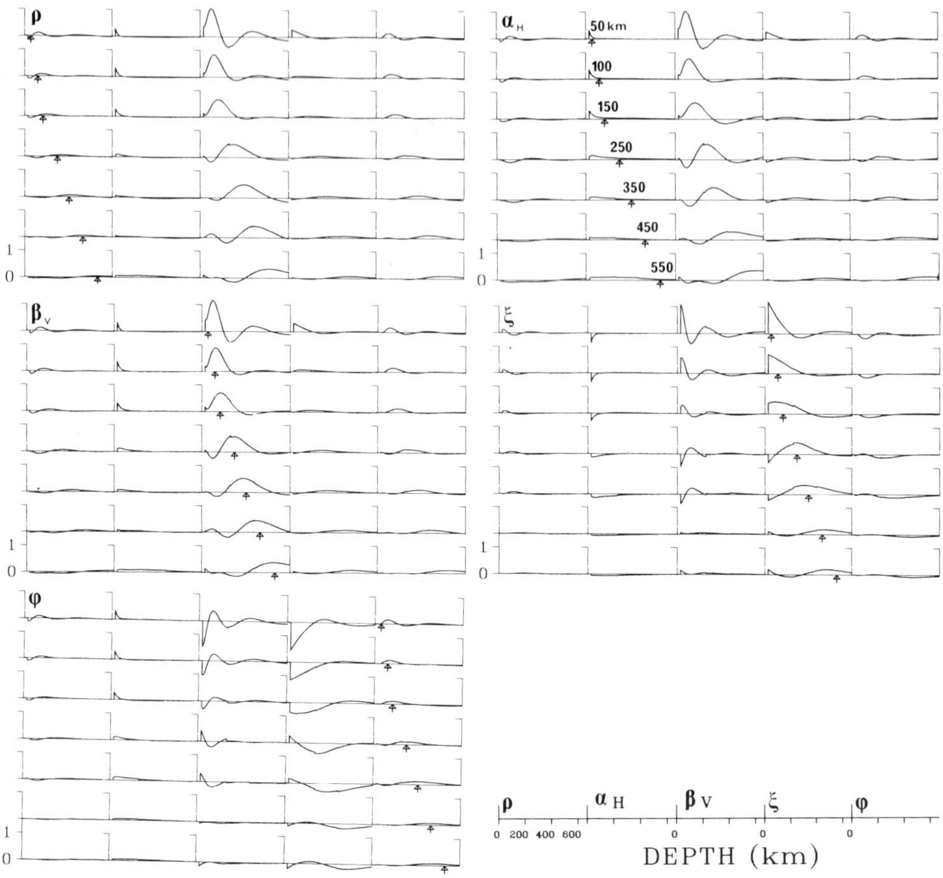
ρ
α_H
50 km
100
150
250
350
450
550
β_V
ξ
φ
1
0
0 200 400 600
DEPTH (km)

← *Figure 7.19:* In a multiparameter problem, the resolution operator has a complex structure. Nataf, Nakanishi and Anderson (1986) use seismic data to infer the values of 5 mecanical parameters in the Earth as a function of depth. The parameters are the density $\rho(z)$, the horizontal celerity of longitudinal waves $\alpha_H(z)$, the vertical celerity of transversal waves $\beta_V(z)$, and the anisotropy of longitudinal $\phi(z)$ and transversal waves $\xi(z)$. The kernel of the resolution operator (7.164) then has the form

$$R(z,z')=\begin{bmatrix} R\rho\rho(z,z') & R\rho\alpha(z,z') & R\rho\beta(z,z') & R\rho\xi(z,z') & R\rho\phi(z,z') \\ R\alpha\rho(z,z') & R\alpha\alpha(z,z') & R\alpha\beta(z,z') & R\alpha\xi(z,z') & R\alpha\phi(z,z') \\ R\beta\rho(z,z') & R\beta\alpha(z,z') & R\beta\beta(z,z') & R\beta\xi(z,z') & R\beta\phi(z,z') \\ R\xi\rho(z,z') & R\xi\alpha(z,z') & R\xi\beta(z,z') & R\xi\xi(z,z') & R\xi\phi(z,z') \\ R\phi\rho(z,z') & R\phi\alpha(z,z') & R\phi\beta(z,z') & R\phi\xi(z,z') & R\phi\phi(z,z') \end{bmatrix} .$$

The full kernel is represented in the figure. Each line is the line of the resolution function that corresponds to the tested parameter at the tested depth (marked by an arrow). Units are 10^{-3} km^{-1} . Each of the five segments spans a depth of 670 km. Each segment corresponds to a different parameter. Note that β_V and ξ are the best resolved parameters. See Nataf et al. (1986) for more details.

of random functions is too expensive in computing time to be practical. But it should be remembered that individual probable realizations are in general quite different from the mathematical expectation $\langle \mathbf{m} \rangle$ (they are much rougher).

Let us try to obtain some measures of what can be called "resolution length". The simplest one is given by the correlation length of the covariance function. In Figures 7.12 - 7.13 we have seen one example of such correlation length: the data set used in that example allows a spatial resolution of the order of 1 km. Another measure of resolution length uses the important concept of "resolving operator", which has been discussed in Section 4.3.3.

Let us here only recall the result corresponding to a linear linear forward equation

$$\mathbf{d} = \mathbf{G}\, \mathbf{m}$$

If $\mathbf{m}_{true}$ is the (unknown) true model, and, per chance, the observed data vector $\mathbf{d}_{obs}$ was error free,

$$\mathbf{d}_{obs} = \mathbf{G}\, \mathbf{m}_{true} ,$$

our inverse solution $\langle \mathbf{m} \rangle$ would verify (using (7.84))

$$\langle \mathbf{m} \rangle - \mathbf{m}_{prior} = \mathbf{R} (\mathbf{m}_{true} - \mathbf{m}_{prior}) , \tag{7.163}$$

where the *resolving operator* $\mathbf{R}$ is given by

$$\begin{aligned} \mathbf{R} &= (\mathbf{G}^t \mathbf{C}_D^{-1} \mathbf{G} + \mathbf{C}_M^{-1})^{-1} \mathbf{G}^t \mathbf{C}_D^{-1} \mathbf{G} \\ &= \mathbf{C}_M \mathbf{G}^t (\mathbf{G} \mathbf{C}_M \mathbf{G}^t + \mathbf{C}_D)^{-1} \mathbf{G} . \end{aligned} \tag{7.164}$$

Equation (7.163) is very important. It shows that we, mortal beings, are only able to see the truth *through a filter* $\mathbf{R}$. In a favourable case, $\mathbf{R} \simeq \mathbf{I}$, and $\langle \mathbf{m} \rangle \simeq \mathbf{m}_{true}$. The more $\mathbf{R}$ differs from the identity, the more deformed is our vision of the true world. The kernel of $\mathbf{R}$ was termed the *resolving kernel* by Backus and Gilbert (1970) in a slightly different context (see Box 7.2).

Let $R(\mathbf{x},\mathbf{x}')$ represent the resolving kernel. If $\mathbf{R} = \mathbf{I}$, then $R(\mathbf{x},\mathbf{x}') = \delta(\mathbf{x},\mathbf{x}')$. The more $R(\mathbf{x},\mathbf{x}')$ differs from $\delta(\mathbf{x}-\mathbf{x}')$, the worse the attained resolution will be. Usually, $R(\mathbf{x},\mathbf{x}')$ is peaked (not necessarily around $\mathbf{x} \simeq \mathbf{x}'$). The width of the peak also gives a measure of the resolution length. Figures 7.18 and 7.19 illustrate the concept.

From (7.85) and (7.164) we deduce the following relationship between the a priori and a posteriori covariance operators:

$$\mathbf{C}_{M'} = (\mathbf{I} - \mathbf{R}) \mathbf{C}_M . \tag{7.165}$$

If $\mathbf{R} \simeq \mathbf{I}$, then $\mathbf{C}_{M'} \simeq \mathbf{0}$, and we closely approach the truth.

7.7.3: Practical computation of the resolving kernel and the posterior covariance function

As the resolving kernel $R(\mathbf{x},\mathbf{x}')$ and the covariance function $C_{M'}(\mathbf{x},\mathbf{x}')$ depend on the two variables $\mathbf{x}$ and $\mathbf{x}'$, they are cumbersome to handle. The interesting information can be obtained by plotting $R(\mathbf{x}_0,\mathbf{x})$ or $C_{M'}(\mathbf{x}_0,\mathbf{x})$, as a function of $\mathbf{x}$, for some chosen values of $\mathbf{x}_0$ (as is done in Figures 7.13, 7.18, and 7.19). Usually, one or two values of $\mathbf{x}_0$ adequately chosen will suffice. As discussed in Section 4.3.3, this computation can be performed using the numerical code used for computing $\langle \mathbf{m} \rangle$, and at the same cost.

7.8: Do it formally.

In the example of section 7.6 and problem 7.2, and in the example of problem 7.3, no effort has been made for indentifying all spaces and operators implicitly used. Furthermore, the reciprocity of the Green function has not been demonstrated. In this section I show that the demonstrations can be carried out in a more formal — and general — way.

Many of the fundamental equations of physics take the form

$$\mathbf{L}\,\mathbf{u} = \boldsymbol{\phi} \,, \tag{7.166}$$

where $u(x,t)$ is a tensor field (defined in some domain of the space-time), $\phi(\mathbf{x},t)$ represents the generalized source of the field $\mathbf{u}$, and $\mathbf{L}$ is a local (differential) operator, possiblly containing boundary terms. Let $\mathbf{u} \in U$ and $\phi \in \boldsymbol{\Phi}$. The transposed operator $\mathbf{L}^t$ maps then $\hat{\boldsymbol{\Phi}}$ into $\hat{U}$.

Example 7.21: Let be an elastic medium inside a volume V , bounded by a surface S , and let $\mathbf{x}$ denote a point in V or on S . $\rho(\mathbf{x})$ and $c^{ijkl}(\mathbf{x})$ are respectively the density and stiffnesses at point $\mathbf{x}$. An elastic wave is described by $u^i(\mathbf{x},t)$, the i-th component of the displacement at point $\mathbf{x}$ and time t . Assume $t \in [t_0,t_1]$. If the functional spaces ito consideration are properly defined, the displacement field $\mathbf{u}$ is uniquely defined by the differential system

$$\rho \frac{\partial^2 u^i}{\partial t^2} - \frac{\partial}{\partial x^j}\left(c^{ijkl} \frac{\partial u_l}{\partial x^k} \right) = f^i \qquad \mathbf{x} \in V \;\; t \in [t_0,t_1] \tag{7.167a}$$

$$n_j\, c^{ijkl} \frac{\partial u_j}{\partial x^k} = \tau^i \qquad \mathbf{x} \in S \;\; t \in [t_0,t_1] \tag{7.167b}$$

$$u^i = \alpha^i \qquad \mathbf{x} \in V \;\; t = t_0 \tag{7.167c}$$

$$\frac{\partial u^i}{\partial t} = \beta^i \qquad \mathbf{x} \in V \;\; t = t_0 \,, \tag{7.167c}$$

where $\mathbf{f}$ is the volume density of external forces, $\mathbf{n}$ the unit normal at the surface, τ the surface traction, and $\boldsymbol{\alpha}$ and $\boldsymbol{\beta}$ the initial displacement and velocity. For short, equations (7.167) can be written

$$\mathbf{L}_f\, \mathbf{u} = \mathbf{f} \tag{7.168a}$$

$$\mathbf{L}_\tau \ \mathbf{u} \ = \ \boldsymbol{\tau} \tag{7.168b}$$

$$\mathbf{L}_\alpha \ \mathbf{u} \ = \ \alpha \tag{7.168c}$$

$$\mathbf{L}_\beta \ \mathbf{u} \ = \ \beta \ , \tag{7.168d}$$

where the operators $\mathbf{L}_f$, $\mathbf{L}_\tau$, $\mathbf{L}_\alpha$, and $\mathbf{L}_\beta$, map U into the spaces of volume sources, surface tractions, and initial conditions respectively. Defining

$$\mathbf{L} \ = \ \begin{bmatrix} \mathbf{L}_f \\ \mathbf{L}_\tau \\ \mathbf{L}_\alpha \\ \mathbf{L}_\beta \end{bmatrix} \qquad \boldsymbol{\phi} \ = \ \begin{bmatrix} \mathbf{f} \\ \boldsymbol{\tau} \\ \alpha \\ \beta \end{bmatrix} , \tag{7.169}$$

leads to (7.166).

Let us verify that the operator $\mathbf{L}^t$ defined by

$$\mathbf{L}^t \ \hat{\boldsymbol{\phi}} \ = \ [\ \mathbf{L}_f{}^t \quad \mathbf{L}_\tau{}^t \quad \mathbf{L}_\alpha{}^t \quad \mathbf{L}_\beta{}^t\ ] \begin{bmatrix} \hat{\mathbf{f}} \\ \hat{\boldsymbol{\tau}} \\ \hat{\alpha} \\ \hat{\beta} \end{bmatrix}$$

$$= \ \mathbf{L}_f{}^t \ \hat{\mathbf{f}} + \mathbf{L}_\tau{}^t \ \hat{\boldsymbol{\tau}} + \mathbf{L}_\alpha{}^t \ \hat{\alpha} + \mathbf{L}_\beta{}^t \ \hat{\beta} \ , \tag{7.170}$$

where

$$(\mathbf{L}_f{}^t \ \hat{\mathbf{f}})^i(\mathbf{x},t) \ = \ \rho(\mathbf{x}) \ \frac{\partial^2 \hat{f}^i}{\partial t^2}(\mathbf{x},t) \ - \ \frac{\partial}{\partial x^j} \left(c^{ijkl}(\mathbf{x}) \ \frac{\partial \hat{f}_l}{\partial x^k}(\mathbf{x},t) \right) \tag{7.171a}$$

$$(\mathbf{L}_\tau{}^t \ \hat{\boldsymbol{\tau}})^l(\mathbf{x},t) \ = \ - \int_S dS(\mathbf{x}') \ n_j(\mathbf{x}') \ c^{ijkl}(\mathbf{x}') \ \hat{\tau}_i(\mathbf{x}',t) \ \frac{\partial}{\partial x^k} \delta(\mathbf{x}-\mathbf{x}') \tag{7.171b}$$

$$(\mathbf{L}_\alpha{}^t \ \hat{\alpha})^i(\mathbf{x},t) \ = \ \hat{\alpha}^i(\mathbf{x}) \ \delta(t-t_0) \tag{7.171c}$$

$$(\mathbf{L}_\beta{}^t \ \hat{\beta})^i(\mathbf{x},t) \ = \ - \ \hat{\beta}^i(\mathbf{x}) \ \delta'(t-t_0) \ . \tag{7.171d}$$

is the transposed of $\mathbf{L}$. Indeed, a computation similar to that made in Box 7.1 gives

$$\langle \ \hat{\boldsymbol{\phi}} \ , \mathbf{L} \ \mathbf{u} \ \rangle_{\Phi} \ - \ \langle \ \mathbf{L}^t \ \hat{\boldsymbol{\phi}} \ , \ \mathbf{u} \ \rangle_U$$

$$= \int_V dV(\mathbf{x}) \int_{t_0}^{t_1} dt \frac{\partial}{\partial t}\left\{\rho \left[\hat{\phi}_i \frac{\partial u^i}{\partial t} - \frac{\partial \hat{\phi}_i}{\partial t} u^i\right]\right\}$$

$$- \int_V dV(\mathbf{x}) \int_{t_0}^{t_1} dt \frac{\partial}{\partial x^j}\left\{c^{ijkl} \left[\hat{\phi}^i \frac{\partial u_k}{\partial x^l} - \frac{\partial \hat{\phi}_k}{\partial x^l} u^i\right]\right\} ,$$

$$= \int_V dV(\mathbf{x}) \, \rho \left[\hat{\phi}_i \frac{\partial u^i}{\partial t} - \frac{\partial \hat{\phi}_i}{\partial t} u^i\right]\Bigg|_{t_0}^{t_1}$$

$$- \int_{t_0}^{t_1} dt \int_S dS(\mathbf{x}) \, n_j \, c^{ijkl} \left[\hat{\phi}^i \frac{\partial u_k}{\partial x^l} - \frac{\partial \hat{\phi}_k}{\partial x^l} u^i\right] . \qquad (7.172)$$

Usually, the field $\mathbf{u}(\mathbf{x},t)$ is at rest at the *initial* time $t = t_0$, and propagates with a free surface (null tractions $n_j \, c^{ijkl} \, \partial u_k/\partial x^l$). Then, if we restrict our consideration to fields $\hat{\boldsymbol{\phi}}(\mathbf{x},t)$ which are at rest at the *final* time $t = t_1$, and also propagate with a free surface, the sums in (7.172) vanish, and

$$\langle \, \hat{\boldsymbol{\phi}} \, , \mathbf{L} \, \mathbf{u} \, \rangle_{\Phi} = \langle \, \mathbf{L}^t \, \hat{\boldsymbol{\phi}} \, , \mathbf{u} \, \rangle_{U} \, . \qquad (7.173)$$

■

The boundary conditions and the functional spaces into consideration are usually defined in such a way that, given $\boldsymbol{\phi} \in \boldsymbol{\Phi}$, (7.166) uniquely defines $\mathbf{u} \in \boldsymbol{U}$. This means that $\mathbf{L}^{-1}$, inverse of $\mathbf{L}$, exists. It is named the *Green operator*, and is denoted by $\mathbf{G}$:

$$\mathbf{G} = \mathbf{L}^{-1} \, . \qquad (7.174)$$

Then, (7.166) can be written

$$\mathbf{u} = \mathbf{G} \, \boldsymbol{\phi} \, . \qquad (7.175)$$

The kernel of the Green operator is termed *Green's function* (see, for instance, Morse and Feshbach, 1953; or Roach, 1982).

Example 7.22: In the elastodynamic problem of the previous example, equation (7.175) gives

$$\mathbf{u} \;=\; \mathbf{G}\,\boldsymbol{\phi} \;=\; \mathbf{G}_f\,\mathbf{f} + \mathbf{G}_\tau\,\boldsymbol{\tau} + \mathbf{G}_\alpha\,\boldsymbol{\alpha} + \mathbf{G}_\beta\,\boldsymbol{\beta}\;, \tag{7.176}$$

or, introducing the kernels,

$$\begin{aligned} u^i(\mathbf{x},t) \;=\; & \int_V dV(\mathbf{x}') \int_{t_0}^{t_1} dt'\; G_f^{\,ij}(\mathbf{x},t;\mathbf{x}',t')\; f_j(\mathbf{x}',t') \\ & + \int_S dS(\mathbf{x}') \int_{t_0}^{t_1} dt'\; G_\tau^{\,ij}(\mathbf{x},t;\mathbf{x}',t')\; \tau_j(\mathbf{x}',t') \\ & + \int_V dV(\mathbf{x}')\; G_\alpha^{\,ij}(\mathbf{x},t;\mathbf{x}',t_0)\; \alpha_j(\mathbf{x}') \\ & + \int_V dV(\mathbf{x}')\; G_\beta^{\,ij}(\mathbf{x},t;\mathbf{x}',t_0)\; \beta_j(\mathbf{x}')\;. \end{aligned} \tag{7.177}$$

These Green's functions are not independent. Using for instance the results of Aki and Richards (1980), or Morse and Feshbach (1953) we obtain

$$G_\tau^{\,ij}(\mathbf{x},t;\mathbf{x}',t') \;=\; G_f^{\,ij}(\mathbf{x},t;\mathbf{x}',t') \tag{7.178a}$$

$$G_\alpha^{\,ij}(\mathbf{x},t;\mathbf{x}',t_0) \;=\; -\,\frac{\partial G_f^{\,ij}}{\partial t}(\mathbf{x},t;\mathbf{x}',t_0)\;\rho(\mathbf{x}') \tag{7.178b}$$

$$G_\beta^{\,ij}(\mathbf{x},t;\mathbf{x}',t_0) \;=\; G_f^{\,ij}(\mathbf{x},t;\mathbf{x}',t_0)\;\rho(\mathbf{x}')\;. \tag{7.178c}$$

Taking $\boldsymbol{\tau} = \mathbf{0}$, $\boldsymbol{\alpha} = \mathbf{0}$, $\boldsymbol{\beta} = \mathbf{0}$, and $f^i(\mathbf{x},t) = \delta^{ip}\;\delta(\mathbf{x}-\mathbf{x}')\;\delta(t-t')$ in (7.177) gives the intuitive interpretation of the Green function as the response to a unit impulse. ■

The use of the Green operator is useful for some analytic developments (sse problems 7.2 and 7.3), but if for given $\boldsymbol{\phi}$, the field $\mathbf{u}$ solution of $\mathbf{L}\,\mathbf{u} = \boldsymbol{\phi}$ is sought for, a numerical method (e.g., integration by finite-differencing) is normally used, rather than an explicit use of the Green operator: $\mathbf{u} = \mathbf{G}\,\boldsymbol{\phi}$. The Green operator also allows to introduce Born's series: if $\mathbf{G}_0$ is known for some $\mathbf{L}_0$, then, setting

$$\mathbf{L} = \mathbf{L}_0 + \delta\mathbf{L} \tag{7.179}$$

we have

$$(\mathbf{L}_0 + \delta\mathbf{L})\ \mathbf{u} \ = \ \boldsymbol{\phi} \ ,$$

i.e.,

$$\mathbf{L}_0\ \mathbf{u} \ = \ \boldsymbol{\phi} - \delta\mathbf{L}\ \mathbf{u} \ ,$$

and, left multiplying by $\mathbf{G}_0$, we obtain the implicit equation

$$\mathbf{u} \ = \ \mathbf{G}_0\ \boldsymbol{\phi} - \mathbf{G}_0\ \delta\mathbf{L}\ \mathbf{u} \ . \tag{7.180}$$

Using a fixed-point method of resolution leads to the Born series

$$\mathbf{u}_0 \ = \ \mathbf{G}_0\ \boldsymbol{\phi} \ , \tag{7.181a}$$

$$\mathbf{u}_{n+1} \ = \ \mathbf{u}_0 - \mathbf{G}_0\ \delta\mathbf{L}\ \mathbf{u}_n \ . \tag{7.181b}$$

The *first Born approximation* (7.181a) is identical to a linearized approximation of $\mathbf{u}$ (see Box 7.3), but the Born series is not identical to the Taylor series development of $\mathbf{u}$. The convergence rate of the Born series is usually very slow.

Illustration of the operator $\delta\mathbf{L}$ is given in example 7.26.

Associated to the *primal problem*

$$\text{Given } \ \boldsymbol{\phi}\in\boldsymbol{\Phi} \ , \ \mathbf{L}: U\to\boldsymbol{\Phi} \ , \text{ solve for } \ \mathbf{u}\in U : \qquad \mathbf{L}\ \mathbf{u} = \boldsymbol{\phi} \ , \tag{7.182}$$

is the *dual problem*

$$\text{Given } \ \hat{\mathbf{u}}\in\hat{U} \ , \ \mathbf{L}: U\to\boldsymbol{\Phi} \ , \text{ solve for } \ \hat{\boldsymbol{\phi}}\in\hat{\boldsymbol{\Phi}} : \qquad \mathbf{L}^t\ \hat{\boldsymbol{\phi}} = \hat{\mathbf{u}} \ . \tag{7.183}$$

As far as $\boldsymbol{\Phi}$ and $\hat{U}$ are restricted to satisfy dual boundary conditions (see Box 7.1), the resolution of the dual problem involves the same numerical methods as the resolution of the primal problem.

Example 7.23: We have seen in example 7.21 that if $\boldsymbol{\phi}$ in the primal problem (7.182) is such that it imposes to $\mathbf{u}$ to be at rest at $t = t_0$, then the dual boundary conditions are satisfied by $\hat{\mathbf{u}}$ in the dual problem (7.183) if it imposes to $\hat{\boldsymbol{\phi}}$ to be at rest at $t = t_1$, the surface tractions associated to both $\mathbf{u}$ and $\hat{\boldsymbol{\phi}}$ being identically null. Then, $\hat{\boldsymbol{\phi}}$ has the form

$$\hat{\boldsymbol{\phi}} = \begin{bmatrix} \hat{\mathbf{f}} \\ \hat{\boldsymbol{\tau}} \\ \hat{\boldsymbol{\alpha}} \\ \hat{\boldsymbol{\beta}} \end{bmatrix} = \begin{bmatrix} \hat{\mathbf{f}} \\ \mathbf{0} \\ \mathbf{0} \\ \mathbf{0} \end{bmatrix} , \tag{7.184}$$

and from (7.170) it follows

$$\mathbf{L}^t \, \hat{\boldsymbol{\phi}} = \mathbf{L}_f{}^t \, \hat{\mathbf{f}} , \tag{7.185}$$

where $\mathbf{L}_f$ is the wave equation operator defined in (7.168), not containing boundary terms. The sums in (7.172) were produced by this operator. As we assume dual boundary conditions, from (7.173) follows the symmetry of $\mathbf{L}_f$:

$$\mathbf{L}_f = \mathbf{L}_f{}^t . \tag{7.186}$$

Practically, if solving the primal problem (7.182) implies taking null initial conditions and finite-differencing forwards in time: $t = t_0$, $t_0+\Delta t$, $t_0+2\Delta t$, ..., t_1 , solving the dual problem (7.183) implies taking null *final* conditions and finite-differencing backwards in time: $t = t_1$, $t_1-\Delta t$, $t_1-2\Delta t$, ..., t_0 . ■

Example 7.24: Tarantola et al. (1987) consider the problem of depth extrapolation of elastic wavefields. The primal boundary conditions are displacements and tractions given at the *top* surface of an elastic slab. The dual boundary conditions are displacements and tractions given at the *bottom* surface of the slab. To solve the dual problem, we can use the same computer code as for solving the primal problem, changing $z_0+\Delta z$, $z_0+2\Delta z$, ..., z_1 into z_1, $z_1-\Delta z$, $z_1-2\Delta z$, ..., z_0 . ■

For given model parameters $\mathbf{m}$ belonging to a model space $\mathcal{M}$ (density and stiffnesses in the example 7.21), the differential operator $\mathbf{L}$ maps $\mathcal{U}$ into $\boldsymbol{\Phi}$. When we wish to make explicit the (nonlinear) dependence of $\mathbf{L}$ and $\mathbf{G}$ on the model parameters we replace (7.166) and (7.171) by

$$\mathbf{L}[\mathbf{m}] \cdot \mathbf{u} = \boldsymbol{\phi} , \tag{7.187}$$

and

$$\mathbf{u} = \mathbf{G}[\mathbf{m}] \cdot \boldsymbol{\phi} . \tag{7.188}$$

The inverse problem usually consists in using the values of the tensor field $\mathbf{u}$ observed at some points of the space-time, to infer the values of the model parameters $\mathbf{m}$. The data space D is then usually the restriction of U to some subspace. Let $\mathbf{P}$ be operator (projector) performing this restriction. Then, the equation solving the forward problem can be written

$$\mathbf{d}_{cal} = \mathbf{g}(\mathbf{m}) = \mathbf{P}\ \mathbf{G}[\mathbf{m}] \cdot \boldsymbol{\phi} . \tag{7.189}$$

$\mathbf{P}^t$ denotes the transposed of $\mathbf{P}$.

Example 7.25: Let us consider the elastic Earth. In a typical seismic inverse problem, the data consists in the displacements $u^i(\mathbf{x}_r,t)$ recorded at some points $\mathbf{x}_r$ (r=1,2,...) of the Earth surface during some time interval ΔT . Let U denote the space of wave fields, and D the data space. The projector $\mathbf{P}$ maps U into D and is defined by

$$(\mathbf{P}\ \mathbf{u})^i(\mathbf{x}_r,t) = u^i(\mathbf{x}_r,t) . \tag{7.190}$$

The transposed operator maps $\hat{D}$ into $\hat{U}$, and is defined by

$$(\mathbf{P}^t\ \hat{\mathrm{d}})^i(\mathbf{x},t) = \sum_r \hat{\mathrm{d}}^i(\mathbf{x}_r,t)\ \delta(\mathbf{x}-\mathbf{x}_r) . \tag{7.191}$$

This gives

$$\langle\ \hat{\mathbf{d}}\ ,\ \mathbf{P}\ \mathbf{u}\ \rangle_D = \langle\ \mathbf{P}^t\ \hat{\mathbf{d}}\ ,\ \mathbf{u}\ \rangle_U . \tag{7.192}$$

■

The least-squares misfit function is then

$$S(\mathbf{m}) = \frac{1}{2}\left(\|\ \mathbf{d}_{cal}-\mathbf{d}_{obs}\ \|^2 + \|\ \mathbf{m}-\mathbf{m}_{prior}\ \|^2\right)$$

$$= \frac{1}{2}\ \langle\ \mathbf{C}_D^{-1}\ (\mathbf{d}_{cal}-\mathbf{d}_{obs})\ ,\ (\mathbf{d}_{cal}-\mathbf{d}_{obs})\ \rangle$$
$$+\ \frac{1}{2}\ \langle\ \mathbf{C}_M^{-1}\ (\mathbf{m}-\mathbf{m}_{prior})\ ,\ (\mathbf{m}-\mathbf{m}_{prior})\ \rangle$$

$$= \frac{1}{2}\ \langle\ \mathbf{C}_D^{-1}\ (\mathbf{P}\ \mathbf{G}[\mathbf{m}] \cdot \boldsymbol{\phi} - \mathbf{d}_{obs})\ ,\ (\mathbf{P}\ \mathbf{G}[\mathbf{m}] \cdot \boldsymbol{\phi} - \mathbf{d}_{obs})\ \rangle$$
$$+\ \frac{1}{2}\ \langle\ \mathbf{C}_M^{-1}\ (\mathbf{m}-\mathbf{m}_{prior})\ ,\ (\mathbf{m}-\mathbf{m}_{prior})\ \rangle\ , \tag{7.193}$$

where $\mathbf{C}_D$ and $\mathbf{C}_M$ are symmetric positive definite (covariance) operators mapping $\hat{\boldsymbol{D}}$ and $\hat{\boldsymbol{M}}$ into $\boldsymbol{D}$ and $\boldsymbol{M}$ respectively. A perturbation $\mathbf{m} \to \mathbf{m} + \delta\mathbf{m}$ gives (using the symmetry of $\mathbf{C}_D$ and $\mathbf{C}_M$)

$$S + \delta S = S + \langle \mathbf{C}_D^{-1} (\mathbf{P}\ \mathbf{G}\ \boldsymbol{\phi} - \mathbf{d}_{obs}) \ , (\mathbf{P}\ \delta\mathbf{G}\ \boldsymbol{\phi} \rangle$$

$$+ \langle \mathbf{C}_M^{-1} (\mathbf{m} - \mathbf{m}_{prior}) \ , \delta\mathbf{m} \rangle \quad + \quad O(\|\delta\mathbf{m}\|^2) \ .$$

From

$$\mathbf{G} = \mathbf{L}^{-1}$$

it follows, successively,

$$\mathbf{G}+\delta\mathbf{G} = (\mathbf{L}+\delta\mathbf{L})^{-1}$$

$$(\mathbf{G}+\delta\mathbf{G})\ (\mathbf{L}+\delta\mathbf{L}) = \mathbf{I}$$

$$\mathbf{I} + \mathbf{G}\ \delta\mathbf{L} + \delta\mathbf{G}\ \mathbf{L} + O(\|\delta\mathbf{m}\|^2) = \mathbf{I}$$

$$\delta\mathbf{G}\ \mathbf{L} = -\ \mathbf{G}\ \delta\mathbf{L} + O(\|\delta\mathbf{m}\|^2)$$

$$\delta\mathbf{G} = -\ \mathbf{G}\ \delta\mathbf{L}\ \mathbf{G} + O(\|\delta\mathbf{m}\|^2) \ . \tag{7.194}$$

Then,

$$\delta S = -\ \langle\ \mathbf{C}_D^{-1} (\mathbf{P}\ \mathbf{G}\ \boldsymbol{\phi} - \mathbf{d}_{obs}) \ , \mathbf{P}\ \mathbf{G}\ \delta\mathbf{L}\ \mathbf{G}\ \boldsymbol{\phi} \rangle$$

$$+ \langle\ \mathbf{C}_M^{-1} (\mathbf{m} - \mathbf{m}_{prior}) \ , \delta\mathbf{m} \rangle \quad + \quad O(\|\delta\mathbf{m}\|^2) \ . \tag{7.195}$$

Example 7.26: In the problem defined in example 7.21, the operator $\delta\mathbf{L}$ is defined by

$$\delta\mathbf{L}\ \mathbf{u} = \begin{bmatrix} \delta\mathbf{f} \\ \delta\boldsymbol{\tau} \\ \delta\boldsymbol{\alpha} \\ \delta\boldsymbol{\beta} \end{bmatrix} , \tag{7.196}$$

where

$$\delta f^i = \delta\rho\ \frac{\partial^2 u^i}{\partial t^2} - \frac{\partial}{\partial x^j}\left(\delta c^{ijkl}\ \frac{\partial u_l}{\partial x^k} \right) \qquad x \in V \quad t \in [t_0, t_1] \tag{7.197a}$$

$$\delta\tau^i = n_j \, \delta c^{ijkl} \, \frac{\partial u_j}{\partial x^k} \qquad \mathbf{x}\in\mathcal{S} \quad t\in[t_0,t_1] \qquad (7.197b)$$

$$\delta\alpha^i = 0 \qquad \mathbf{x}\in\mathcal{V} \qquad (7.197c)$$

$$\delta\beta^i = 0 \qquad \mathbf{x}\in\mathcal{V} \; . \qquad (7.197c)$$

■

Using equation (7.192), equation (7.195) becomes

$$\delta S = - \langle \; \mathbf{P}^t \, \mathbf{C_D}^{-1} \, (\mathbf{P} \, \mathbf{G} \, \boldsymbol{\phi} - \mathbf{d}_{obs}) \; , \; \mathbf{G} \, \delta\mathbf{L} \, \mathbf{G} \, \boldsymbol{\phi} \; \rangle$$

$$+ \langle \; \mathbf{C_M}^{-1} \, (\mathbf{m} - \mathbf{m}_{prior}) \; , \; \delta\mathbf{m} \; \rangle \; + \; O(\| \delta\mathbf{m} \|^2) \; . \qquad (7.198)$$

If $\delta\mathbf{L} \, \mathbf{G} \, \boldsymbol{\phi}$ and $\mathbf{P}^t \, \mathbf{C_D}^{-1} \, (\mathbf{P} \, \mathbf{G} \, \boldsymbol{\phi} - \mathbf{d}_{obs})$ satisfy the dual boundary conditions associated to the current problem, then

$$\delta S = - \langle \; \mathbf{G}^t \, \mathbf{P}^t \, \mathbf{C_D}^{-1} \, (\mathbf{P} \, \mathbf{G} \, \boldsymbol{\phi} - \mathbf{d}_{obs}) \; , \; \delta\mathbf{L} \, \mathbf{G} \, \boldsymbol{\phi} \; \rangle$$

$$+ \langle \; \mathbf{C_M}^{-1} \, (\mathbf{m} - \mathbf{m}_{prior}) \; , \; \delta\mathbf{m} \; \rangle \; + \; O(\| \delta\mathbf{m} \|^2) \; . \qquad (7.199)$$

I now define

$$\delta\mathbf{L} \, \mathbf{G} \, \boldsymbol{\phi} = - \mathbf{A} \, \delta\mathbf{m} \; . \qquad (7.200)$$

Then

$$\delta S = \langle \; \mathbf{G}^t \, \mathbf{P}^t \, \mathbf{C_D}^{-1} \, (\mathbf{P} \, \mathbf{G} \, \boldsymbol{\phi} - \mathbf{d}_{obs}) \; , \; \mathbf{A} \, \delta\mathbf{m} \; \rangle$$

$$+ \langle \; \mathbf{C_M}^{-1} \, (\mathbf{m} - \mathbf{m}_{prior}) \; , \; \delta\mathbf{m} \; \rangle \; + \; O(\| \delta\mathbf{m} \|^2) \; . \qquad (7.201)$$

If $\mathbf{G}^t \, \mathbf{P}^t \, \mathbf{C_D}^{-1} \, (\mathbf{P} \, \mathbf{G} \, \boldsymbol{\phi} - \mathbf{d}_{obs})$ and $\delta\mathbf{m}$ satisfy the dual boundary conditions associated to $\mathbf{A}$, then

$$\delta S = \langle \; \mathbf{A}^t \, \mathbf{G}^t \, \mathbf{P}^t \, \mathbf{C_D}^{-1} \, (\mathbf{P} \, \mathbf{G} \, \boldsymbol{\phi} - \mathbf{d}_{obs}) \; , \; \delta\mathbf{m} \; \rangle$$

$$+ \langle \; \mathbf{C_M}^{-1} \, (\mathbf{m} - \mathbf{m}_{prior}) \; , \; \delta\mathbf{m} \; \rangle \; + \; O(\| \delta\mathbf{m} \|^2) \; . \qquad (7.202)$$

Example 7.27: In the current example of this section, the operator $\mathbf{A}$ has the partitioned structure

$$\mathbf{A}\ \delta\mathbf{m} \ = \ \begin{bmatrix} \mathbf{A}_{f\rho} & \mathbf{A}_{fc} \\ \mathbf{A}_{\tau\rho} & \mathbf{A}_{\tau c} \\ \mathbf{A}_{\alpha\rho} & \mathbf{A}_{\alpha c} \\ \mathbf{A}_{\beta\rho} & \mathbf{A}_{\beta c} \end{bmatrix} \begin{bmatrix} \delta\rho \\ \delta c \end{bmatrix} \ = \ \begin{bmatrix} \mathbf{A}_{f\rho} & \mathbf{A}_{fc} \\ 0 & \mathbf{A}_{\tau c} \\ 0 & 0 \\ 0 & 0 \end{bmatrix} \begin{bmatrix} \delta\rho \\ \delta c \end{bmatrix} . \tag{7.203}$$

Using equations (7.196)-(7.197) gives

$$(\mathbf{A}_{f\rho}\ \delta\rho)^i(\mathbf{x},t) \ = \ -\ \delta\rho(\mathbf{x})\ \ddot{u}^i(\mathbf{x},t) \ , \tag{7.204a}$$

$$(\mathbf{A}_{fc}\ \delta c)^i(\mathbf{x},t) \ = \ \frac{\partial}{\partial x^j}\left[\delta c^{ijkl}(\mathbf{x})\ \frac{\partial u_l}{\partial x^k}(\mathbf{x},t) \right] , \tag{7.204b}$$

and

$$(\mathbf{A}_{\tau\rho}\ \delta c)^i(\mathbf{x},t) \ = \ -\ n_j(\mathbf{x})\ \delta c^{ijkl}(\mathbf{x})\ \frac{\partial u_l}{\partial x^k}(\mathbf{x},t) \ . \tag{7.204c}$$

The corresponding transposed operators are defined by

$$(\mathbf{A}_{f\rho}{}^t\ \delta\hat{\mathbf{f}})(\mathbf{x}) \ = \ -\int_{t_0}^{t_1} dt\ \ddot{u}^i(\mathbf{x},t)\ \delta\hat{f}_i(\mathbf{x},t) \ , \tag{7.205a}$$

$$(\mathbf{A}_{fc}{}^t\ \delta\hat{\mathbf{f}})^{ijkl}(\mathbf{x}) \ = \ -\int_{t_0}^{t_1} dt\ \frac{\partial u^k}{\partial x_l}(\mathbf{x},t)\ \frac{\partial \delta\hat{f}^i}{\partial x_j}(\mathbf{x},t) \ , \tag{7.205b}$$

$$(\mathbf{A}_{\tau c}{}^t\ \delta\hat{\tau})^{ijkl}(\mathbf{x}) \ = \ -\int_{t_0}^{t_1} dt\ \frac{\partial u^k}{\partial x_l}(\mathbf{x},t)\ n^j(\mathbf{x})\ \delta\hat{\tau}^i(\mathbf{x},t)\ \delta_S(\mathbf{x}) \ , \tag{7.205c}$$

where $\delta_S(\mathbf{x})$ is defined by

$$\int_V dV(\mathbf{x})\ \mathbf{Q}(\mathbf{x})\ \delta_S(\mathbf{x}) \ = \ \int_S dS(\mathbf{x})\ \mathbf{Q}(\mathbf{x}) \ , \tag{7.206}$$

for any $\mathbf{Q}(\mathbf{x})$.

Then,

$$\langle\ \delta\hat{\phi}\ ,\ \mathbf{A}\ \delta\mathbf{m}\ \rangle_{\Phi} - \langle\ \mathbf{A}^t\ \delta\hat{\phi}\ ,\ \delta\mathbf{m}\ \rangle_{M}$$

$$= \int_{t_0}^{t_1} dt \int_V dV(\mathbf{x}) \frac{\partial}{\partial x^j} \left\{ \delta\hat{f}_i(\mathbf{x},t)\ \delta c^{ijkl}(\mathbf{x})\ \frac{\partial u_k}{\partial x^l}(\mathbf{x},t) \right\} . \tag{7.207a}$$

$$= \int_{t_0}^{t_1} dt \int_S dS(\mathbf{x})\ n_j(\mathbf{x})\ \delta\hat{f}_i(\mathbf{x},t)\ \delta c^{ijkl}(\mathbf{x})\ \frac{\partial u_k}{\partial x^l}(\mathbf{x},t) , \tag{7.207b}$$

and, if $n_j\ \delta\hat{f}_i\ \delta c^{ijkl}\ \partial u_k/\partial x^l$ vanishes at the surface, then

$$\langle\ \delta\hat{\phi}\ ,\ \mathbf{A}\ \delta\mathbf{m}\ \rangle_{\Phi} = \langle\ \mathbf{A}^t\ \delta\hat{\phi}\ ,\ \delta\mathbf{m}\ \rangle_{M} . \tag{7.208}$$

■

As the gradient of S with respect to $\mathbf{m}$, $\hat{\gamma}$, is defined by

$$\delta S\ =\ \langle\ \hat{\gamma}\ ,\ \delta\mathbf{m}\ \rangle\ +\ O(\|\ \delta\mathbf{m}\ \|^2)\ , \tag{7.209}$$

equation (7.202) gives

$$\hat{\gamma}\ =\ \mathbf{A}^t\ \mathbf{G}^t\ \mathbf{P}^t\ \mathbf{C}_D^{-1}\ (\mathbf{P}\ \mathbf{G}\ \phi\ -\ \mathbf{d}_{obs})\ +\ \mathbf{C}_M^{-1}\ (\mathbf{m}\ -\ \mathbf{m}_{prior})\ . \tag{7.210}$$

Defining the direction of steepest ascent, γ , with respect to the metric defined by $\mathbf{C}_M$ gives

$$\gamma\ =\ \mathbf{C}_M\ \hat{\gamma}\ , \tag{7.211}$$

i.e.,

$$\gamma\ =\ \mathbf{C}_M\ \mathbf{A}^t\ \mathbf{G}^t\ \mathbf{P}^t\ \mathbf{C}_D^{-1}\ (\mathbf{P}\ \mathbf{G}\ \phi\ -\ \mathbf{d}_{obs})\ +\ (\mathbf{m}\ -\ \mathbf{m}_{prior})\ . \tag{7.212}$$

As we have the direction of steepest ascent, we can use any "gradient" method for the minimization of the misfit function. Using for instance a crude steepest descent method gives

$$\mathbf{m}_{n+1}\ =\ \mathbf{m}_n\ -\ \mu_n\ \gamma_n\ . \tag{7.213}$$

Dividing (7.212)-(7.213) into elementary steps gives

$$
\begin{aligned}
&(1) \quad \mathbf{u}_n = \mathbf{G}[\mathbf{m}_n] \cdot \boldsymbol{\phi} \\
&(2) \quad \mathbf{d}_n = \mathbf{P}\, \mathbf{u}_n \\
&(3) \quad \delta\mathbf{d}_n = \mathbf{d}_n - \mathbf{d}_{obs} \\
&(4) \quad \delta\hat{\mathbf{d}}_n = \mathbf{C}_D^{-1}\, \delta\mathbf{d}_n \\
&(5) \quad \delta\hat{\mathbf{u}}_n = \mathbf{P}^t\, \delta\hat{\mathbf{d}}_n \\
&(6) \quad \delta\hat{\boldsymbol{\phi}}_n = \mathbf{G}[\mathbf{m}_n]^t \cdot \delta\hat{\mathbf{u}}_n \\
&(7) \quad \delta\hat{\mathbf{m}}_n = \mathbf{A}^t\, \delta\hat{\boldsymbol{\phi}}_n \\
&(8) \quad \boldsymbol{\gamma}_n = \mathbf{C}_M\, \delta\hat{\mathbf{m}}_n + (\mathbf{m}_n - \mathbf{m}_{prior}) \\
&(9) \quad \mathbf{m}_{n+1} = \mathbf{m}_n - \mu_n\, \boldsymbol{\gamma}_n \ .
\end{aligned}
\tag{7.214}
$$

This implies the following computations:

$$
\begin{aligned}
&(1) \quad \text{Solve} && : \quad \mathbf{L}[\mathbf{m}_n] \cdot \mathbf{u}_n = \boldsymbol{\phi} \\
&(2) \quad \text{Select the recording locations} && : \quad \mathbf{d}_n = \mathbf{P}\, \mathbf{u}_n \\
&(3) \quad \text{Compute the data residuals} && : \quad \delta\mathbf{d}_n = \mathbf{d}_n - \mathbf{d}_{obs} \\
&(4) \quad \text{Weight the data residuals} && : \quad \delta\hat{\mathbf{d}}_n = \mathbf{C}_D^{-1}\, \delta\mathbf{d}_n \\
&(5) \quad \text{Consider these as sources} && : \quad \delta\hat{\mathbf{u}}_n = \mathbf{P}^t\, \delta\hat{\mathbf{d}}_n \\
&(6) \quad \text{Solve the dual problem} && : \quad \mathbf{L}[\mathbf{m}_n]^t \cdot \delta\hat{\boldsymbol{\phi}}_n = \delta\hat{\mathbf{u}}_n \\
&(7) \quad \text{Compute} && : \quad \delta\hat{\mathbf{m}}_n = \mathbf{A}^t\, \delta\hat{\boldsymbol{\phi}}_n \\
&(8) \quad \text{Get} && : \quad \boldsymbol{\gamma}_n = \mathbf{C}_M\, \delta\hat{\mathbf{m}}_n + (\mathbf{m}_n - \mathbf{m}_{prior}) \\
&(9) \quad \text{Update} && : \quad \mathbf{m}_{n+1} = \mathbf{m}_n - \mu_n\, \boldsymbol{\gamma}_n \ .
\end{aligned}
\tag{7.216}
$$

The algorithm (7.216) solves a quite general class of problems involving fundamental equations of physics. The reader will make the necessary corrections for problems not contained in this formulation.

Example 7.28: Compare the result (7.126) with the results obtained in problem 7.3 using a more empirical approach. ■

Important questions have been avoided in this section (characterization of the dual spaces, compatibility of the domains of definition of operators, ...). These questions remain open for further playing.

Appendix 7.1: Spaces and Operators: basic definitions and properties.

This appendix brings together the very basic definitions and properties the reader should know if intending to explore the literature. Useful textbooks are Taylor and Lay (1980) or Dautray and Lions (1984).

Basic terminology:

Let $\mathcal{S}_0$ and $\mathcal{T}$ be arbitrary sets, and $\mathcal{S}$ a subset of $\mathcal{S}_0$. A "rule" that associates each s in $\mathcal{S}$ with a unique element f(s) in $\mathcal{T}$ is termed a *function from* $\mathcal{S}$ *into* $\mathcal{T}$. Such a function is properly denoted by f , or by the expression $s \rightarrow f(s)$, although we often say "the function f(s) ".

To allow suppleness in the discussions, the terms *mapping*, *application*, *transformation* or *operator* are used as synonymous of *function*.

The subset $\mathcal{S}$ of $\mathcal{S}_0$ is the *domain of definition* of f . If $\mathcal{A}$ is a subset of $\mathcal{S}$, the set $f(\mathcal{A})$ (i.e., the subset of $\mathcal{T}$ which can be attained by f from elements of $\mathcal{A}$) is termed the *image* of $\mathcal{A}$ (through f). The image of $\mathcal{S}$, $f(\mathcal{S})$, is called the *range* of f .

If for each t in $f(\mathcal{S})$ there exists only one $s \in \mathcal{S}$ such that $f(s) = t$, the function f is *one-to-one*, or *injective*. We then write $s = f^{-1}(t)$, thus defining the *inverse* of f on $f(\mathcal{S})$. If $f(\mathcal{S}) = \mathcal{T}$, f is termed *surjective*; it is also said that f maps $\mathcal{S}$ *onto* $\mathcal{T}$. When f is both injective and surjective it is named *bijective*; then $g^{-1}(\mathcal{T}) = \mathcal{S}$.

If, $\mathcal{S}$ and $\mathcal{T}$ are vector linear spaces (see below) and if for any s_1 and s_2 , $f(\lambda s_1+\mu s_2) = \lambda f(s_1) + \mu f(s_2)$, where λ and μ are arbitrary real numbers, f is *linear* . A linear bijection between linear vector spaces is an *isomorphism* . If there exists an isomorphism between two linear vector spaces, they are termed *isomorphic* .

Topological space:

A *topological space* is a space S in which a collection of subsets of S has been defined, called the *open subsets* of S , verifying that

i) $\emptyset$ (the empty subset) and S are open subsets ,
ii) any reunion of open subsets is an open subset , (1)
iii) Any finite intersection of open subsets is an open subset .

Example 1: Let R be the real line and $a < b$. An *open interval* (a,b) is defined as the subset of real numbers r verifying $a < r < b$. Any reunion of open intervals is named an open subset. This defines a topology over R . ■

Let S be a topological space, and A an open subset of S . A subset of the form $S - A$ is called a *closed subset* .

Example 2: Let R be the real line. For $a < b$, a *closed interval* [a,b] is defined as the subset of real numbers r verifying $a \leq r \leq b$. A closed interval is a closed subset. ■

Let S be a topological space. The following properties can be demonstrated:

i) $\emptyset$ and S are closed subsets ,
ii) any intersection of closed subsets is a closed subset , (2)
iii) Any finite reunion of closed subsets is a closed subset .

In particular, we see that the sets $\emptyset$ and S are at the same time open and closed. This is exceptional: in general, a subset is neither open or closed, and if it is open, it is not closed, and viceversa.

Let S be a topological space, and s an element of S . A *neighbourhood* of s is an open subset containing s .

Let S be a topological space, and $(s_1, s_2, ...)$ a sequence of elements of S . This sequence *tends to* s if for any neighbourhood A of s in S , there exists an integer N such that

$$n \geq N \quad \Rightarrow \quad s_n \in A \ . \tag{3}$$

In that case, the following notations are used

$$s_n \rightarrow s \tag{4}$$

$$\lim_{n \to \infty} s_n = s . \tag{5}$$

Let S and T be two topological spaces, and f an application from S into T . f is called *continuous* at s_0 if

$$\lim_{s \to s_0} f(s) = f(s_0) , \tag{6}$$

i.e., if for any neighbourhood B of $f(s_0)$ in T , there exists a neighbourhood A in T such that

$$f(A) \subset B . \tag{7}$$

Manifold:

Let M be a topological space, and let M_1, M_2,... be a collection of open subsets of M such that they cover all M. Any bijection ϕ_i from one of the M_i into a space K^n , isomorphic to R^n , is termed a *chart* of M_i . A collection of charts defined for each of the M_i is called an *atlas* of M . If for any (i,j) , the image of the open subset $M_i \cap M_j$ is an open subset of K^n , and if the image of $M_i \cap M_j$ obtained respectively by ϕ_i and ϕ_j are related by an isomorphism, then it is said that the set M is a (n-dimensional) *manifold*. If the isomorphism is p times differentiable, it is named a C^p-manifold.

Metric space:

Let M be an arbitrary set. A *distance* over M associates any couple (m_1,m_2) of elements of M with a *positive real number* denoted $D(m_1,m_2)$ verifying the following conditions

$$D(m_1,m_2) = 0 \quad \Leftrightarrow \quad m_1 = m_2 \tag{8a}$$

$$D(m_1,m_2) = D(m_2,m_1) \qquad \text{for any } m_1 \text{ and } m_2 \tag{8b}$$

$$D(m_1,m_3) \leq D(m_1,m_2) + D(m_2,m_3) \qquad \text{for any } m_1 , m_2 \text{ and } m_3 . \tag{8c}$$

A set furnished with a distance is termed a *metric space*. Each element of a metric space M is called a *point* of M .

Example 3: Let M be the surface of a sphere in a euclidean space. A distance between two points of the sphere can for instance be defined as the length of the (smaller) arc of great circle passing through the points. Alternatively, the distance can be defined as the length of the straight segment joining the two points. With any of these definitions, the surface of a sphere is a metric space. ■

Example 4: Let M be a space of n-dimensional column matrices:

$$\mathbf{m} \in M \quad \Rightarrow \quad \mathbf{m} = \begin{bmatrix} m^1 \\ m^2 \\ \cdots \\ m^n \end{bmatrix} ,$$

each component representing a physical parameter with its own physical dimensions, and $\sigma^1, \sigma^2, ..., \sigma^n$ some positive "error bars". For any $\mathbf{m}_1$ and $\mathbf{m}_2$, the expression

$$D(\mathbf{m}_1,\mathbf{m}_2) = \left(\sum_{i=1}^{n} \frac{\left| m_1^{\ i} - m_2^{\ i} \right|^p}{(\sigma^i)^p} \right)^{1/p} \tag{9}$$

defines a real number. It is a distance over M . In fact, M is a linear vector space, which can be normed by (see below)

$$\| \mathbf{m} \| = D(\mathbf{m},0) \ . \ \blacksquare \tag{10}$$

Example 5: Let E^3 be the usual three-dimensional euclidean space. Let $\mathbf{x}$ represent a generic point of E^3 , and let $\mathbf{x} \rightarrow m(\mathbf{x})$ be a function from E^3 into an space of scalars K . For instance, $\mathbf{x}$ may represent the cartesian coordinates of a point inside a star, and $m(\mathbf{x})$ may represent the temperature at the point $\mathbf{x}$. Let M_0 be the space of all such functions. Letting $m_0(\mathbf{x})$ be a particular function of M_0 , a new, smaller space M can be defined by the condition

$$\int_{E^3} dV(\mathbf{x}) \ \frac{\left| m(\mathbf{x}) - m_0(\mathbf{x}) \right|^p}{s(\mathbf{x})^p} \qquad \textit{is finite} \ , \tag{11}$$

where $s(\mathbf{x})$ represents a positive function with physical dimensions ensuring the adimensionality of the previous expression. Let $\mathbf{m}_1$ and $\mathbf{m}_2$ be two elements of M . The expression

$$D(\mathbf{m}_1,\mathbf{m}_2) = \left[\int_{E^3} d\mathbf{x} \frac{\left| m_1(x) - m_2(\mathbf{x}) \right|^p}{s(\mathbf{x})^p} \right]^{1/p} \tag{12}$$

defines a distance over M . M is then a metric space. We will see later that the space thus defined is, in fact, a linear affine space. It is *not* a linear vector space, because if m_1 and $\mathbf{m}_2$ belong to the space, their sum $\mathbf{m}_1+\mathbf{m}_2$ does not necessarily belong to the space (for the null element $m(\mathbf{x}) \equiv 0$ does not necessarily belong to the space, i.e, it does not necessarily render expression (11) finite). ■

Let M be a metric space, and $(\mathbf{m}_1, \mathbf{m}_2, \dots)$ a sequence of points of M . This sequence *tends to* the point $\mathbf{m}$ of M if

$$D(\mathbf{m}_n,\mathbf{m}) \to 0 \quad \text{when} \quad n \to \infty , \tag{13}$$

i.e., if for any $\epsilon > 0$ there exists an integer N such that

$$n \geq N \quad \Rightarrow \quad D(\mathbf{m}_n,\mathbf{m}) \leq \epsilon . \tag{14}$$

Let M be a metric space, and $(\mathbf{m}_1, \mathbf{m}_2, \dots)$ a sequence of points of M . This sequence is called a *Cauchy sequence* if for any $\epsilon > 0$ there exists an integer N such that

$$n \geq N \quad \text{and} \quad m \geq N \quad \Rightarrow \quad D(\mathbf{m}_n,\mathbf{m}_m) \leq \epsilon . \tag{15}$$

Example 6: Let Q be the set of rational numbers. The sequence (1. , 1.4 , 1.41 , 1.414 , 1.4142 , ...) defined by the decimal development of the number π is a Cauchy sequence. ■

Let M be a metric space. If the limit of every Cauchy sequence of points of M belongs to M , the metric space M is said to be *complete.*

Example 7: The sequence in example 6 does not converge to an element of Q : the set of rational numbers is not complete. The real line R (defined in fact by the "completion" of Q) is complete. ■

Let M and D be metric spaces, and $\mathbf{g}$ an operator from M into D . We say that $\mathbf{g}(m)$ *tends to* $\mathbf{d}_0$ when m tends to $\mathbf{m}_0$ if for any $\epsilon > 0$ there exists a real number r such that

$$D(\mathbf{m},\mathbf{m}_0) \leq r \quad \Rightarrow \quad D(\mathbf{g}(\mathbf{m}),\mathbf{d}_0) \leq \epsilon . \tag{16}$$

Let M and D be metric spaces, and $\mathbf{g}$ an operator from M into D. We say that $\mathbf{g}$ is *continuous* at $\mathbf{m}_0$ if the limit of $\mathbf{g}(\mathbf{m})$ when $\mathbf{m} \to \mathbf{m}_0$ equals $\mathbf{g}(\mathbf{m}_0)$.

Let M be a metric space, a subset A of S is an *metric open subset* if for any point $\mathbf{m}_0 \in A$ there exists $\epsilon > 0$ such that every point $\mathbf{m}$ of M verifying $D(\mathbf{m},\mathbf{m}_0) < \epsilon$ belongs to A. These metric open subsets verify the axioms of the (topological) open subsets as defined in (b): a metric space is always a topological space. It is said that a metric *induces* a topology. The topology induced by the metric is termed the *natural topology*.

It can be shown that the concepts of limit and of continuity defined by the metric or by the natural topology are equivalent.

Let M be a (topological) metric space. It can be shown that if $A \subset M$ is a *closed subset*, every point of M which is the limit of a sequence of points of A belongs to A.

Linear vector space:

Let M be a set, and let $\mathbf{m}$ denote a generic element of M. If we can define the sum $\mathbf{m}_1 + \mathbf{m}_2$ of two elements of M, and the multiplication $\lambda\,\mathbf{m}$ of an element of M by a real number verifying the following conditions

$$\mathbf{m}_1 + \mathbf{m}_2 = \mathbf{m}_2 + \mathbf{m}_1 \tag{17a}$$

$$(\mathbf{m}_1 + \mathbf{m}_2) + \mathbf{m}_3 = \mathbf{m}_1 + (\mathbf{m}_2 + \mathbf{m}_3) \tag{17b}$$

$$\text{There exists } \mathbf{0} \in M \text{ such that } \mathbf{m} + \mathbf{0} = \mathbf{m} \quad \text{for any } \mathbf{m} \tag{17c}$$

$$\text{To each } \mathbf{m} \text{ there corresponds } (-\mathbf{m}) \text{ such that }, \ \mathbf{m} + (-\mathbf{m}) = \mathbf{0} \tag{17d}$$

$$\lambda(\mathbf{m}_1 + \mathbf{m}_2) = \lambda\mathbf{m}_1 + \lambda\mathbf{m}_2 \tag{17e}$$

$$(\lambda + \mu)\mathbf{m} = \lambda\mathbf{m} + \mu\mathbf{m} \tag{17f}$$

$$(\lambda\mu)\mathbf{m} = \lambda(\mu\mathbf{m}) \tag{17g}$$

$$1\mathbf{m} = \mathbf{m}, \tag{17h}$$

then M is called a (real) *linear space*, or *vector space*, or *linear vector space*. The elements of M are called *vectors*.

Example 8: Let E^3 be the usual euclidean space, $\mathbf{x}$ a generic point of E^3, and $\mathbf{x} \to m(\mathbf{x})$ an arbitrary function from E^3 into an space of scalars

K . With the definitions

$$(\mathbf{m}_1 + \mathbf{m}_2)(\mathbf{x}) = \mathbf{m}_1(\mathbf{x}) + \mathbf{m}_2(\mathbf{x}) \tag{18a}$$

$$(\lambda \mathbf{m})(\mathbf{x}) = \lambda\ \mathrm{m}(\mathbf{x}) , \tag{18b}$$

the set of all such functions is a linear vector space, denoted $\boldsymbol{F}$. The subspace of F formed by the *continuous* functions is also a linear vector space, denoted C . The subspace of F formed by n-times differentiable functions is also a linear vector space, denoted C^n . ∎

Let M be a linear vector space. It is a *topological* linear vector space if it is furnished with a topological structure compatible with the structure of linear vector space, i.e., such that the applications $(\mathbf{m}_1,\mathbf{m}_2) \to \mathbf{m}_1+\mathbf{m}_2$ and $(\lambda,\mathbf{m}) \to \lambda\mathbf{m}$ are continuous (with respect to the topology).

Let M be a linear vector space. A *norm* over M associates any element f of M with a *positive real number* denoted $\| \mathbf{f} \|$ verifying the following conditions

$$\| \mathbf{f} \| = 0 \quad \Leftrightarrow \quad \mathbf{f} = \mathbf{0} \tag{19a}$$

$$\| \lambda\mathbf{f} \| = | \lambda | \ \| \mathbf{f} \| \qquad \text{for any } \mathbf{f} \tag{19b}$$

$$\| \mathbf{f} + \mathbf{g} \| \le \| \mathbf{f} \| + \| \mathbf{g} \| \qquad \text{for any } \mathbf{f} \text{ and } \mathbf{g} . \tag{19c}$$

A linear vector space furnished with a norm is named a *normed linear vector space.*

Example 9: Let us consider the space F of arbitrary functions $\mathbf{x} \to \mathrm{m}(\mathbf{x})$ from E^3 into K , and, for $1 \le p < \infty$, the subspace of functions of F , denoted $\mathbf{x} \to \delta\mathrm{m}(\mathbf{x})$, for which the expression

$$\| \delta\mathbf{m} \| = \left(\int_{E^3} dV(\mathbf{x}) \frac{| \delta\mathrm{m}(\mathbf{x}) |^p}{s(\mathbf{x})^p} \right)^{1/p} \tag{19}$$

makes sense and is finite, where $1/s(\mathbf{x})^p$ is a given "weight function" (ensuring in particular the physical adimensionality of $\| \delta\mathbf{m} \|$). It is well known that the space of functions thus defined is a linear vector space and that $\| \cdot \|$ defines a norm. This space is denoted L_p , and plays an important role in mathematical physics (more precisely, an element of the space L_p is not a function, but the *class of functions* which are identical "almost everywhere", i.e., such that the norm of their difference is null). ∎

Example 10: For $p = 2$, the previous definition can be generalized. Let $C(\mathbf{x},\mathbf{x}')$ be a positive definite function, i.e., a function such that for any $\phi(\mathbf{x})$, and any $V \subset E^3$ the sum

$$\int_{\mathbf{V}} dV(\mathbf{x}) \int_{V} dV(\mathbf{x}') \;\; \phi(\mathbf{x}) \;\; C(\mathbf{x},\mathbf{x}') \;\; \phi(\mathbf{x}') \tag{20}$$

is defined and is finite. A distribution (see Appendix 7.2) $C^{-1}(\mathbf{x},\mathbf{x}')$ can then be defined by

$$\int_{E^3} dV(\mathbf{x}') \;\; C(\mathbf{x},\mathbf{x}') \;\; C^{-1}(\mathbf{x}',\mathbf{x}'') \;=\; \delta(\mathbf{x}-\mathbf{x}'') \ , \tag{21}$$

and a linear vector space $\boldsymbol{\Delta M}$ can be defined as the set of functions $\mathbf{x} \rightarrow \delta m(\mathbf{x})$ for which the sum

$$\| \; \delta \mathbf{m} \; \| \;=\; \left[\int_{E^3} dV(\mathbf{x}) \int_{E^3} dV(\mathbf{x}') \;\; \delta m(\mathbf{x}) \;\; C^{-1}(\mathbf{x},\mathbf{x}') \;\; \delta m(\mathbf{x}') \right]^{1/2} \tag{22}$$

is defined and is finite. It is easy to show that $\| \bullet \|$ defines a norm, which is called a *least squares norm*. The space $\boldsymbol{\Delta M}$ is not L_2 , but can be shown to be isomorphic with L_2 (introduce the positive definite operator $\mathbf{C}$ whose kernel is $C(\mathbf{x},\mathbf{x}')$, define the square root of $\mathbf{C}$, verifying $\mathbf{C} = \boldsymbol{\Psi} \, \boldsymbol{\Psi}^t$, and introduce a new space by

$$\Delta \tilde{M} \;=\; \Psi^{-1} \; \Delta M \; ;$$

this defines an isomorphism, and induces the L_2 norm over $\Delta \tilde{M}$). ■

Example 11: It has been mentioned in example 7.11 in the text that the covariance function

$$C(t,t') \;=\; \sigma^2 \;\; \exp\left(- \frac{| \; t-t' \; |}{L} \right) \tag{23}$$

induces the Sobolev H_1 norm. ■

It is easy to see that the expression

$$D(\mathbf{m}_1,\mathbf{m}_2) \;=\; \| \; \mathbf{m}_1 \; - \; \mathbf{m}_2 \; \| \tag{24}$$

defines a distance over $\boldsymbol{M}$: a normed linear vector space is always a metric space. In particular, this allows us to define a Cauchy sequence of vectors. If every Cauchy sequence of vectors of $\boldsymbol{M}$ converges into an element of $\boldsymbol{M}$,

M is *complete.* A complete linear vector space is termed a *Banach space.*

Example 12: The spaces L_p are Banach spaces (i.e., they are complete). The Sobolev spaces H_p (see Appendix 7.2) are complete. ■

Let M be a real linear vector space. A subset S of M is called a *linear vector subspace* if the following conditions are satisfied:

$$\begin{aligned} &\mathbf{m}_1 , \mathbf{m}_2 \in S \quad \Rightarrow \quad \mathbf{m}_1 + \mathbf{m}_2 \in S \\ &\mathbf{m} \in S \quad \Rightarrow \quad \lambda \mathbf{m} \in S \end{aligned} \tag{25}$$

It is easy to see that a linear vector subspace is itself a linear vector space.

Example 13: The space of continuous functions is a linear vector subspace of the space of arbitrary functions. ■

Affine subspace:

Let M_0 be a linear vector space. A subset M of M_0 obtained by translation of a linear vector subspace of M_0 is called an *affine subspace* of M_0. In other words, an affine subspace of M_0 is a set of the form $M = \Delta M + \mathbf{m}_0$, where ΔM is a linear vector subspace of M and $\mathbf{m}_0$ a given element of M_0 (see Figure 7.20).

Example 14: Take as M_0 the set F of arbitrary functions from E^3 into K (example 8), and define a subspace M by the condition

$$\| \mathbf{m} - \mathbf{m}_0 \| = \left(\int_V dx \, \frac{| m(\mathbf{x}) - m_0(\mathbf{x}) |^p}{s(\mathbf{x})^p} \right)^{1/p} \quad \textit{finite} , \tag{26}$$

where $m_0(\mathbf{x})$ is a given element of M_0. The subspace M thus defined is *not* a linear vector space, because the sum of two elements does not always belong to the space (the null element $m(\mathbf{x}) \equiv 0$ may not belong to the space). With any function $\mathbf{m}$ we can associate the function

$$\Delta\mathbf{m} = \mathbf{m} - \mathbf{m}_0 , \tag{27}$$

and it is easy to see that the space ΔM of functions thus defined is the linear vector space L_p. As any element of M can be obtained as

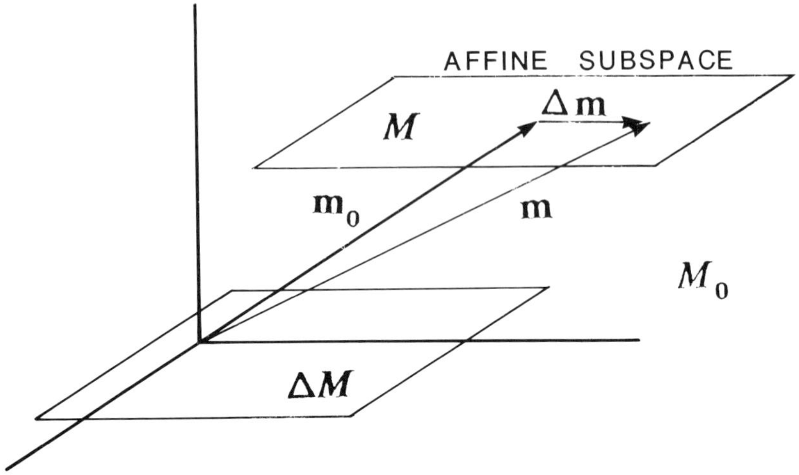

Figure 7.20: An *affine space* M is obtained by translation of a linear vector subspace ΔM of a linear vector space M_0 .

$$\mathbf{m} = \Delta\mathbf{m} + \mathbf{m}_0 ,$$

where $\mathbf{m}_0$ is an element of M_0 and $\Delta\mathbf{m}$ an element of a linear vector subspace of M_0 , M is a linear affine subspace of M_0 . ■

Dimension of a linear vector space. Basis of a linear vector space.:

Let M be a linear vector space, and $\mathbf{m}_i$ ($i \in I$) elements of M . Here, I represents a discrete index set, which can either be finite or infinite. If none of the $\mathbf{m}_i$ can be obtained as a linear combination of the others, the $\mathbf{m}_i$ are *linearly independent*.

Let M be a linear vector space. If there is some positive integer N such that M contains a set of N vectors that are linearly independent, while every set of N+1 vectors are linearly dependent, then M is called *finite dimensional*, and N is called the *dimension* of M . If M is not finite dimensional, then it is called *infinite dimensional*.

A set of vectors of a linear vector space M is called a *basis* if it is linearly independent and if it can generate the whole M by linear combination.

Example 15: Let M be the space of periodical $[\ m(t+2\pi) = m(t)\ ]$, symmetrical $[\ m(-t) = m(t)\ ]$ functions. As this space is infinite dimensional, any basis has an infinite number of elements. A first example is the countable basis

$$b_n(t) = \cos nt \qquad\qquad n = 0,1,... \tag{28}$$

Using the well-known Fourier decomposition, any function of M can be written

$$m(t) = \sum_{n=0}^{\infty} c_n \; b_n(t) \; , \tag{29}$$

with

$$c_0 = \frac{1}{2\pi} \int_0^{2\pi} dt \;\; m(t) \tag{30a}$$

and

$$c_n = \frac{1}{\pi} \int_0^{2\pi} dt \;\; m(t) \;\; \cos nt \qquad\qquad (n=1,2,...) \; . \tag{30b}$$

By languistic abuse, a (noncountable) basis (of distributions) can be considered:

$$b(\nu,t) = \delta(\nu-t) \qquad\qquad -\infty < \nu < +\infty \; . \tag{31}$$

where δ represents Dirac's delta "function". Any function $m(t)$ can be developed into that "basis":

$$m(t) = \int_{-\infty}^{\infty} d\nu \;\; c(\nu) \;\; b(\nu,t) \; , \tag{32}$$

with

$$c(\nu) = m(\nu) \; . \; \blacksquare \tag{33}$$

Linear operator:

Let $\boldsymbol{M}$ and $\boldsymbol{D}$ be two topological linear vector spaces, and $\mathbf{m} \rightarrow \mathbf{G}(\mathbf{m})$ an operator from $\boldsymbol{M}$ into $\boldsymbol{D}$. $\mathbf{G}$ is *linear* if

$$\begin{aligned} \mathbf{G}(\mathbf{m}_1 + \mathbf{m}_2) &= \mathbf{G}(\mathbf{m}_1) + \mathbf{G}(\mathbf{m}_2) \\ \mathbf{G}(\lambda\mathbf{m}) &= \lambda\, \mathbf{G}(\mathbf{m}) \ , \end{aligned} \tag{34}$$

whenever λ is a scalar and $\mathbf{m}$, $\mathbf{m}_1$, $\mathbf{m}_2$ are vectors of $\boldsymbol{M}$. Usually, linear operators are represented by capital letters. If $\mathbf{G}$ is linear, the notation $\mathbf{G}\,\mathbf{m}$ is preferred to $\mathbf{G}(\mathbf{m})$:

$$\mathbf{G}(\mathbf{m}) = \mathbf{G}\,\mathbf{m} \ . \tag{35}$$

The linear operator $\mathbf{G}$ is called *continuous* if $\mathrm{m} \rightarrow \mathbf{m}_0$ (in the topology of $\boldsymbol{M}$) implies $\mathbf{G}\,\mathbf{m} \rightarrow \mathbf{G}\,\mathbf{m}_0$ (in the topology of $\boldsymbol{D}$).

A linear operator defined over a finite dimensional space is always continuous. A linear operator over an infinite dimensional space may be discontinuous.

Example 16: The derivative operator is linear. The derivative of the null function is the null function. But it is easy to define a sequence of functions tending to the null function but such that the limit of their derivatives does not tend to the null function. This implies that the derivative operator, although linear, is not continuous. ∎

The space of all continuous linear operators from $\boldsymbol{M}$ into $\boldsymbol{D}$ is denoted by $\mathrm{L}(\boldsymbol{M},\boldsymbol{D})$. Defining the sum of two operators and the multiplication of an operator by a scalar by

$$\begin{aligned} (\mathbf{G}_1 + \mathbf{G}_2)(\mathbf{m}) &= \mathbf{G}_1\,\mathbf{m} + \mathbf{G}_2\,\mathbf{m} \\ (\lambda\,\mathbf{G})(\mathbf{m}) &= \lambda\,\mathbf{G}\,\mathbf{m} \ , \end{aligned} \tag{36}$$

the space $\mathrm{L}(\boldsymbol{M},\boldsymbol{D})$ is a linear vector space.

Let $\boldsymbol{M}$ and $\boldsymbol{D}$ be two normed vector spaces, and $\mathbf{G}$ a linear operator from $\boldsymbol{M}$ into $\boldsymbol{D}$. It can be shown that $\mathbf{G}$ is continuous if and only if there exists a constant $c > 0$ such that for any $\mathbf{m} \in \boldsymbol{M}$,

$$\|\,\mathbf{G}\,\mathbf{m}\,\| \ \le\ c\ \|\,\mathbf{m}\,\| \ . \tag{37}$$

Then, the *norm* of a continuous operator can be defined by

$$\| \mathbf{G} \| = \sup_{\mathbf{m} \in M,\ \mathbf{m} \neq 0} \frac{\| \mathbf{G\,m} \|}{\| \mathbf{m} \|} . \tag{38}$$

Let M_1 and M_2 be two Banach spaces, and $\mathbf{L}$ a continuous linear operator from M_1 into M_2. If L is a bijection, then $\mathbf{L}^{-1}$ is also a continuous linear operator.

A bijection between the vector linear spaces M_1 and M_2 is said to be an *isomorphism* of M_1 *onto* M_2. Two linear vector spaces M_1 and M_2 are said to be *isomorphic* is there exists an isomorphism of M_1 onto M_2.

Let $\mathbf{G}$ be a linear operator from M into D. The linear vector subspace K of M for which

$$\mathbf{m} \in K \quad \Rightarrow \quad \mathbf{G\,m} = 0 \tag{39}$$

is called the *kernel* , or the *null space*, of $\mathbf{G}$.

Let $\mathbf{G}$ be a linear operator from M into D. The dimension of $\mathbf{G}\,M$ (image of M through $\mathbf{G}$) is called the *rank* of $\mathbf{G}$. It can be shown that it equals the difference between the dimension of M and the dimension of the kernel of $\mathbf{G}$.

A sequence of linear operators $\mathbf{H}_1, \mathbf{H}_2, \ldots$ is called *uniformly convergent* if it converges in norm, i.e., if

$$\lim_{n \to \infty} \| \mathbf{H}_n - \mathbf{H} \| = 0 . \tag{40}$$

Let $\mathbf{H}$ be a continuous linear operator from the Banach space M into itself. Then the limit

$$r(\mathbf{H}) = \lim_{n \to \infty} \| \mathbf{H}^n \|^{1/n} \tag{41}$$

exists, and is called the *spectral radius* of $\mathbf{H}$. If the spectral radius of H is less than one, then $(\mathbf{I} - \mathbf{H})^{-1}$ exists, is a continuous linear operator, and

$$(\mathbf{I} - \mathbf{H})^{-1} = \sum_{n=0}^{\infty} \mathbf{H}^n \qquad \text{(Neumann series)} . \tag{42}$$

Dual of a linear vector space:

Let M be a real topological linear vector space. A *form* over M is an operator from M into the real line R . A *linear form* is a linear operator from M into R .

The space of all continuous linear forms over M is termed the (topological) *dual* space of M, and is denoted by $\hat{M}$. A generic element of $\hat{M}$ is denoted by $\hat{\mathbf{m}}$. The result of the action of $\hat{\mathbf{m}} \in \hat{M}$ over an $\mathbf{m} \in M$ is denoted by either of the two notations

$$\langle \hat{\mathbf{m}},\mathbf{m} \rangle = \hat{\mathbf{m}}^t \mathbf{m} . \tag{43}$$

The second notation is useful for numerical computations, because it recalls matricial notations.

Let $\hat{\mathbf{m}} \in \hat{M}$. The expression

$$\| \hat{\mathbf{m}} \| = \sup_{\mathbf{m}\in M \, , \, \mathbf{m}\neq 0} \frac{| \langle \hat{\mathbf{m}},\mathbf{m} \rangle |}{\| \mathbf{m} \|} \tag{44}$$

defines a norm over $\hat{M}$. It can be shown that with such a norm, $\hat{M}$ is a Banach space (i.e., it is complete).

Example 17: For $1 < p < \infty$, the dual of L_p is L_q, with $\frac{1}{p} + \frac{1}{q} = 1$. ■

Let M be a normed vector space, and $\hat{M}$ its dual. It can be shown that for any non-null $\mathbf{m} \in M$, there exists $\hat{\mathbf{m}} \in \hat{M}$ verifying

$$\begin{aligned} \| \hat{\mathbf{m}} \| &= \| \mathbf{m} \| \\ \langle \hat{\mathbf{m}},\mathbf{m} \rangle &= \| \mathbf{m} \|^2 = \| \hat{\mathbf{m}} \|^2 \end{aligned} \tag{45}$$

Let M be a normed vector space, and $\hat{\hat{M}}$ its bidual (dual of the dual). Then there exists a continuous linear application $\mathbf{J}$ from M into its bidual such that

$$\begin{aligned} &\text{i)} \quad \mathbf{J} \text{ is injective} \\ &\text{ii)} \quad \| \mathbf{J}\,\mathbf{m} \| = \| \mathbf{m} \| \qquad \text{for any } \mathbf{m} \in M . \end{aligned} \tag{46}$$

It is then possible to identify M and its image $\mathbf{J}\, M \subset \hat{\hat{M}}$. We will see later (Riesz theorem) that in Hilbert spaces, there exists a bijection between a space M and its dual $\hat{M}$. But they remain fundamentally different spaces, and the identification of a space and its dual, although perhaps useful for pure mathematical developments, is of no interest for practical applications). On the contrary, the identification between a space and (a subset of) its the bidual is always useful (for the purposes of inverse theory). If M equals its bidual, we say that M is *reflexive*.

Example 18: For $1 < p < \infty$, L_p is a reflexive Banach space. The dual of L_1 is L_∞ , but L_1 is not reflexive. The Sobolev spaces W_p^m (see Appendix 7.2) are reflexive Banach spaces (for $1 < p < \infty$). ■

Let M be a topological linear vector space, and $\hat{M}$ its dual. The sequence $\mathbf{m}_1, \mathbf{m}_2, \ldots$ of elements of M *converges weakly* to $\mathbf{m} \in M$ if

$$\lim_{n \to \infty} \langle \hat{m}, m_n \rangle = \langle \hat{m}, m \rangle \qquad \text{for any } \hat{m} \in \hat{M} . \tag{47}$$

It *converges strongly* if

$$\lim_{n \to \infty} \| \mathbf{m}_n - \mathbf{m} \| = 0 \tag{48}$$

It can be shown that if M has finite dimension, then $\hat{M}$ has the same dimension.

Transposed operator:

Let M and D be two linear vector spaces, and $\mathbf{G}$ a linear operator from M into D . The *transpose* of $\mathbf{G}$ is denoted $\mathbf{G}^t$ and is the linear operator from $\hat{D}$ into $\hat{M}$ defined by

$$\langle \mathbf{G}^t \hat{\mathbf{d}} , \mathbf{m} \rangle = \langle \hat{\mathbf{d}} , \mathbf{G} \mathbf{m} \rangle , \tag{49a}$$

or, using the notations of (43),

$$(\mathbf{G}^t \hat{\mathbf{d}})^t \mathbf{m} = \hat{\mathbf{d}}^t \mathbf{G} \mathbf{m} . \tag{49b}$$

The reader will easily give sense to and demonstrate the equivalences

$$(\mathbf{G}_1 + \mathbf{G}_2)^t = \mathbf{G}_1^t + \mathbf{G}_2^t \tag{50}$$

$$(\mathbf{G}_1 \mathbf{G}_2)^t = \mathbf{G}_2^t \mathbf{G}_1^t \tag{51}$$

$$(\mathbf{G}^t)^{-1} = (\mathbf{G}^{-1})^t \tag{52}$$

$$(\mathbf{G}^t)^t = \mathbf{G} . \tag{53}$$

For more details, see Box 7.1.

Hilbert spaces:

Let $\mathcal{M}$ be a real linear vector space. A *bilinear form* over $\mathcal{M}$ is an application $(m_1,m_2) \to W(m_1,m_2)$ from $\mathcal{M} \times \mathcal{M}$ into R such that

$$W(m_1+m_2,m) = W(m_1,m) + W(m_2,m)$$

$$W(m,m_1+m_2) = W(m,m_1) + W(m,m_2) \tag{54}$$

$$W(\lambda m_1,m_2) = W(m_1,\lambda m_2) = \lambda\, W(m_1,m_2) \; .$$

A bilinear form is *symmetrical* if

$$W(m_1,m_2) = W(m_2,m_1) \; . \tag{55}$$

For a bilinear symmetric form, the following notation is used

$$W(m_1,m_2) = m_1{}^t\, W\, m_2 = m_2{}^t\, W\, m_1 \; . \tag{56}$$

A bilinear form is *positive definite* if

$$m \neq 0 \quad \Rightarrow \quad m^t\, W\, m > 0 \; . \tag{57}$$

It can be shown that if a bilinear form is positive definite, then it is also symmetrical.

It can be shown that if $\mathbf{W}$ is positive definite, then the expression

$$\| \, m \, \| = (\, m^t\, W\, m \,)^{1/2} \tag{58}$$

is a norm over $\mathcal{M}$.

Let $\mathcal{M}$ be a real linear vector space, and $\mathbf{W}$ a positive definite bilinear form over $\mathcal{M}$. The pair $(\mathcal{M},\mathbf{W})$ is called a real *pre-Hilbert* space. A pre-Hilbert space which is complete is termed a *Hilbert* space. If $(\mathcal{M},W)$ is a Hilbert space, the application $m_1{}^t\, W\, m_2$ is called a *scalar product* over $\mathcal{M}$ (and is sometimes denoted (m_1,m_2)).

Let $\mathcal{M}$ be a Banach space. It can be shown that if the norm over $\mathcal{M}$ verifies

$$\| m_1 \|^2 + \| m_2 \|^2 = \frac{1}{2} (\| m_1 + m_2 \|^2 + \| m_1 - m_2 \|^2) , \tag{59}$$

then $\boldsymbol{M}$ is a Hilbert space (i.e., the norm can be defined from a scalar product through (58)). Then, conversely,

$$m_1^t \, W \, m_2 = \frac{1}{4} (\| m_1 + m_2 \|^2 - \| m_1 - m_2 \|^2) . \tag{60}$$

Example 19: L_2 is a Hilbert space. The Sobolev space H^{2m} is a Hilbert space. (See Appendix 7.2). ■

Riesz theorem: Let M be a Hilbert space, and $\hat{M}$ its dual. Let $\hat{m} \in \hat{M}$. Then there exists a unique $m \in \boldsymbol{M}$ such that

$$\begin{aligned} \hat{m}^t \, m' &= m^t \, W \, m' \qquad \text{for any } m' \in \boldsymbol{M} \\ \| \hat{m} \| &= \| m \| . \end{aligned} \tag{61}$$

The reciprocal also holds: letting $m \in \boldsymbol{M}$, there exists a unique $\hat{m} \in \hat{\boldsymbol{M}}$ such that

$$\begin{aligned} \hat{m} \, m' &= m^t \, W \, m' \qquad \text{for any } m' \in \boldsymbol{M} \\ \| \hat{m} \| &= \| m \| . \end{aligned} \tag{62}$$

The Riesz theorem shows that there exists an *isometric isomorphism* between M and its dual $\hat{M}$. Nevertheless, as in inverse problem theory a space and its dual play very different roles, they should never be identified.

It can be shown that a Hilbert space is a reflexive Banach space.

Example 20: Let $\boldsymbol{M}$ be the L_2 space of functions defined over R^n, and D the L_2 space of functions defined over R^m . Let $G(\mathbf{y},\mathbf{x})$ be a function from $R^m \times R^n$ into R such that

$$\int_{R^m} dy \int_{R^n} d\mathbf{x} \; | \, G(y,x) \, |^2 < \infty . \tag{63}$$

Then, the operator $\mathbf{G}$ from $\boldsymbol{M}$ into $\boldsymbol{D}$ defined by

$$d(y) = \int_{R^n} d\mathbf{x} \;\; G(\mathbf{y},\mathbf{x}) \;\; m(\mathbf{x}) , \tag{64}$$

is a continuous linear operator. ■

Adjoint operator:

Let *M* and *D* be two Hilbert spaces, with scalar products denoted respectively $W_m(\bullet , \bullet)$ and $W_d(\bullet , \bullet)$, and let **G** be a linear operator from *M* into *D* . The *adjoint* of **G** is denoted $\mathbf{G}^*$ and is the linear operator from *D* into *M* defined by

$$W_m(\mathbf{G}^* \mathbf{d} , \mathbf{m}) = W_d(\mathbf{d} , \mathbf{G}\, \mathbf{m}) . \tag{65}$$

Using the notations of (56) and the definition (49) of transpose operator, it follows that

$$\mathbf{G}^* = \mathbf{W_m}^{-1} \mathbf{G}^t \mathbf{W_m} . \tag{66}$$

It should be emphasized that the terms "adjoint" and "transpose" are not synonymous. The transpose operator is defined for arbitrary linear vector spaces, irrespectively of the existence of any scalar product (think for instance of the transpose of a matrix), while the adjoint operator is defined only for scalar product (Hilbert) spaces.

The reader will easily give sense to and demonstrate the equivalences

$$(\mathbf{G}_1 + \mathbf{G}_2)^* = \mathbf{G}_1^* + \mathbf{G}_2^* \tag{67}$$

$$(\mathbf{G}_1\, \mathbf{G}_2)^* = \mathbf{G}_2^*\ \mathbf{G}_1^* \tag{68}$$

$$(\mathbf{G}^*)^{-1} = (\mathbf{G}^{-1})^* \tag{69}$$

$$(\mathbf{G}^*)^* = \mathbf{G} . \tag{70}$$

Appendix 7.2: Usual functional spaces.

In the following, $\mathbf{x}$ denotes a point of a euclidean space E, with cartesian coordinates $\mathbf{x} = (x^1,x^2,...) = (x,y,...)$, $\mathbf{x} \to f(\mathbf{x})$ denotes a function from an open subset $\mathbf{V}$ of E into R. The symbol $\int d\mathbf{x}$ denotes the volume integral over $\mathbf{V}$, where $d\mathbf{x}$ denotes the volume element.

The spaces L_p $(1 \le p < \infty)$: The space of functions such that the expression

$$\| \mathbf{f} \| = \left(\int d\mathbf{x} \frac{| f(\mathbf{x}) |^p}{s(\mathbf{x})^p} \right)^{1/p} \tag{1}$$

is defined and is finite, where $1/s(\mathbf{x})$ is a given positive "weighting function", is called the space L_p. Two functions $\mathbf{f}_1$ and $\mathbf{f}_2$, such that $\| \mathbf{f}_1 - \mathbf{f}_2 \| = 0$, are said "equal almost everywhere", and are identified.

The space L_∞ is the space of functions for which the expression

$$\| \mathbf{f} \| = \sup \frac{| f(\mathbf{x}) |}{s(\mathbf{x})} \tag{2}$$

is finite.

If two functions which are equal "almost everywhere" are identified, the previous expressions define a norm over L_p. With such a norm the L_p spaces are Banach spaces (i.e., they are complete).

The space L_2 : In the particular case $p = 2$, a scalar product can be introduced by

$$(\mathbf{f},\mathbf{g}) = \int d\mathbf{x} \frac{f(\mathbf{x})\, g(\mathbf{x})}{s(\mathbf{x})^2} . \tag{3}$$

Then

$$\| \mathbf{f} \| = (\mathbf{f},\mathbf{f})^{1/2} . \tag{4}$$

L_2 is a Hilbert space (i.e., it is complete).

If $f(\mathbf{x})$ is "vector valued" (i.e., if it takes values in $\mathbf{R}^n$), it has components $f^1(\mathbf{x})$, $f^2(\mathbf{x})$,... Then the scalar product is defined by

$$(f,g) = \sum_i \int d\mathbf{x} \; \frac{f^i(\mathbf{x})\, g^i(\mathbf{x})}{s^i(\mathbf{x})^2} \tag{5}$$

The corresponding space is a Hilbert space and is also denoted L_2 .

The Sobolev spaces H^m : By definition, H^m is the space of L_2 functions whose partial derivatives up to order m are L_2 functions. For instance, H^0 is L_2 , H^1 is the space of functions $\mathbf{f} \in L_2$ such that $\partial f/\partial x^i \in L_2$ (for any i), etc. Formally, H^m is the space of functions such that

$$\frac{\partial^{\alpha_1+...+\alpha_n} f}{\partial(x^1)^{\alpha_1} \ldots \partial(x^n)^{\alpha_n}} \in L_2 \quad \text{for } \alpha_i \in N \quad \text{and} \quad 0 \leq (\alpha_1+...+\alpha_n) \leq m \tag{6}$$

Let $\mathbf{f}$ and $\mathbf{g}$ be two functions of H^m . A *scalar product* is defined by

$$(f,g) = \sum_{0\leq(\alpha_1+...+\alpha_n)\leq m} \int d\mathbf{x} \; \frac{\partial^{\alpha_1+...+\alpha_n} f}{\partial(x^1)^{\alpha_1} \ldots \partial(x^n)^{\alpha_n}} \; \frac{\partial^{\alpha_1+...+\alpha_n} g}{\partial(x^1)^{\alpha_1} \ldots \partial(x^n)^{\alpha_n}} \tag{7}$$

(to shorten notations, weighting factors have been omitted). With such a scalar product, the Sobolev spaces H^m are complete.

The corresponding (squared) norm is given by

$$\| f \|^2 = \sum_{0\leq(\alpha_1+...+\alpha_n)\leq m} \int dx \left(\frac{\partial^{\alpha_1+...+\alpha_n} f}{\partial(x^1)^{\alpha_1} \ldots \partial(x^n)^{\alpha_n}} \right)^2 . \tag{8}$$

Example 1: Let $x \rightarrow f(x)$ be a one-dimensional function of a one-dimensional variable. The corresponding H^1 space is the space of L_2 functions such that $\partial f/\partial x$ is L_2 . The scalar product is

$$(f,g) = \int dx \; f(x)\; g(x) + \int dx \; \frac{\partial f}{\partial x} \frac{\partial g}{\partial x} , \tag{9}$$

and the (squared) norm is given by

$$\| f \|^2 = \int dx \; f(x)^2 + \int dx \left(\frac{\partial f}{\partial x} \right)^2 . \tag{10}$$

The space H^2 is the space of L_2 functions such that $\partial f/\partial x$ and $\partial^2 f/\partial x^2$ are L_2 . The scalar product is

$$(f,g) = \int dx \; f(x)\, g(x) + \int dx \; \frac{\partial f}{\partial x}(x) \; \frac{\partial g}{\partial x}(x)$$

$$+ \int dx \; \frac{\partial^2 f}{\partial x^2}(x) \; \frac{\partial^2 g}{\partial x^2}(x) \; , \qquad (11)$$

and the (squared) norm is given by

$$\| \mathbf{f} \|^2 = \int dx \; f(x)^2 + \int dx \left(\frac{\partial f}{\partial x}(x) \right)^2 + \int dx \left(\frac{\partial^2 f}{\partial x^2}(x) \right)^2 . \qquad (12)$$

■

Example 2: Let $(x,y) \to f(x,y)$ be a one-dimensional function of a two-dimensional variable. H^0 is the space L_2 . The space H^1 is the space of L_2 functions such that $\partial f/\partial x$ and $\partial f/\partial y$ are L_2 . The scalar product is

$$(f,g) = \int dx \int dy \; f(x,y)\; g(x,y) + \int dx \int dy \; \frac{\partial f}{\partial x}(x,y) \; \frac{\partial g}{\partial x}(x,y)$$

$$+ \int dx \int dy \; \frac{\partial f}{\partial y}(x,y) \; \frac{\partial g}{\partial y}(x,y) \qquad (13)$$

$$= \int dx \int dy \; f(x,y)\; g(x,y) + \int dx \int dy \; \mathbf{grad}\, f(x,y) \cdot \mathbf{grad}\, g(x,y) \; . \qquad (14)$$

The (squared) norm is given by

$$\| \mathbf{f} \|^2 = \int dx \int dy \; (\, f(x,y)\,)^2 + \int dx \int dy \left(\frac{\partial f}{\partial x}(x,y) \right)^2$$

$$+ \int dx \int dy \left(\frac{\partial f}{\partial y}(x,y) \right)^2 \qquad (15)$$

$$= \int dx \int dy \; (\, f(x,y)\,)^2 + \int dx \int dy \; (\, \mathbf{grad}\, f(x,y)\,)^2 \; . \qquad (16)$$

The space H^2 is the space of L_2 functions such that $\partial f/\partial x$, $\partial f/\partial y$, $\partial^2 f/\partial x \partial y$, $\partial^2 f/\partial x^2$, and $\partial^2 f/\partial y^2$ are L_2 . The scalar product is

$$(f,g) = \int dx \int dy \; f(x,y)\; g(x,y) + \int dx \int dy \; \frac{\partial f}{\partial x}(x,y)\; \frac{\partial g}{\partial x}(x,y)$$

$$+ \int dx \int dy \; \frac{\partial f}{\partial y}(x,y)\; \frac{\partial g}{\partial y}(x,y) + \int dx \int dy \; \frac{\partial^2 f}{\partial x \partial y}(x,y)\; \frac{\partial^2 g}{\partial x \partial y}(x,y)$$

$$+ \int dx \int dy \; \frac{\partial^2 f}{\partial x^2}(x,y)\; \frac{\partial^2 g}{\partial x^2}(x,y) + \int dx \int dy \; \frac{\partial^2 f}{\partial y^2}(x,y)\; \frac{\partial^2 g}{\partial y^2}(x,y) \tag{17}$$

■

The Sobolev spaces $W_p{}^m$: $W_p{}^m$ is the space of L_p functions whose partial derivatives up to order m are L_p functions. For instance, $W_2{}^m$ is H^m , $W_p{}^0$ is L_p , $W_p{}^1$ is the space of functions $\mathbf{f} \in L_p$ such that $\partial f/\partial x^i \in L_p$ (for i=1,...,n). Formally, $W_p{}^m$ is the space of functions such that

$$\frac{\partial^{\alpha_1+\ldots+\alpha_n} f}{\partial (x^1)^{\alpha_1} \ldots \partial (x^n)^{\alpha_n}} \in L_p \quad \text{for} \quad \alpha_i \in N \quad \text{and} \quad 0 \leq (\alpha_1+\ldots+\alpha_n) \leq m \, . \tag{18}$$

The spaces $W_p{}^m$ are Banach spaces with the norm

$$\| \mathbf{f} \| = \left(\sum_{0 \leq (\alpha_1+\ldots+\alpha_n) \leq m} \int \left| \frac{\partial^{\alpha_1+\ldots+\alpha_n} f}{\partial (x^1)^{\alpha_1} \ldots \partial (x^n)^{\alpha_n}} \right|^p \right)^{1/p} . \tag{19}$$

Problems for Chapter 7

Problem 7.1: Travel-time tomography. To infer the velocity structure of a medium, waves are generated by some sources, and the travel times to some receivers are measured (see example 7.17 in text). Solve the inverse problem of estimating the velocity structure of the medium.

Solution: We assume that the high frequency limit is acceptable, i.e., that ray theory can be used instead of wave theory. If $c(\mathbf{x})$ denotes the celerity of the waves at point $\mathbf{x}$, let $m(\mathbf{x})$ be the slowness

$$m(\mathbf{x}) = \frac{1}{c(\mathbf{x})} . \tag{1}$$

The i-th datum is the travel time for the i-th ray:

$$d^i = g^i(\mathbf{m}) = \int_{R^i(\mathbf{m})} ds^i \; m(\mathbf{x}^i) , \tag{2}$$

where $R^i(\mathbf{m})$ denotes the i-th ray path. As this ray path depends on $\mathbf{m}$, d^i is a nonlinear functional of m. Given a medium $\mathbf{m}$, the actual ray path is obtained using Fermat's theorem (or, equivalently, the eikonal equation) and some numerical method.

First we wish to obtain the derivative of the nonlinear operator g^i at a point $\mathbf{m}_n$. We have

$$g^i(\mathbf{m}_n+\delta\mathbf{m}) = \int_{R^i(\mathbf{m}_n+\delta\mathbf{m})} ds^i \; (m_n(\mathbf{x}^i) + \delta m(\mathbf{x}^i)) . \tag{3}$$

The travel time being stationary along the actual ray path (Fermat's theorem),

$$\int_{R^i(\mathbf{m}_n+\delta m)} ds^i \; (m(\mathbf{x}^i)_n + \delta m(\mathbf{x}^i)) =$$

$$= \int_{R^i(\mathbf{m}_n)} ds^i \ (m(\mathbf{x}^i)_n + \delta m(\mathbf{x}^i)) \ + \ O^i(\| \delta \mathbf{m} \|^2) \ . \tag{4}$$

This gives

$$g^i(\mathbf{m}_n + \delta \mathbf{m}) \ = \ g^i(\mathbf{m}_n) \ + \int_{R^i(\mathbf{m}_n)} ds^i \ \delta m(\mathbf{x}^i) \ + \ O^i(\| \delta \mathbf{m} \|^2) \ . \tag{5}$$

Comparison with the definition of the derivative of a nonlinear operator (see text) gives

$$(\mathbf{G}_n \ \delta \mathbf{m} \ )^i \ = \int_{R^i(\mathbf{m}_n)} ds^i \ \delta m(\mathbf{x}^i) \ . \tag{6}$$

The derivative of $\mathbf{g}$ at the point $\mathbf{m}_n$ associates the travel time perturbations (6) with the model perturbation $\delta \mathbf{m}$.

Equation (6) is well adapted to all numerical computations involving the derivative operator $\mathbf{G}_n$, but, for analytic developments it is sometimes useful to introduce the kernel of $\mathbf{G}_n$. Introducing a "delta-like" function by

$$\int_{R^i(\mathbf{m}_n)} ds^i \ \delta m(\mathbf{x}^i) \ = \int_{\mathbf{V}} dV(\mathbf{x}) \ \Delta^i(\mathbf{x})_n \ \delta m(\mathbf{x}) \ , \tag{7}$$

and from the integral representation

$$(\mathbf{G}_n \ \delta \mathbf{m} \ )^i \ = \int_{\mathbf{V}} dV(\mathbf{x}) \ G^i(\mathbf{x})_n \ \delta m(\mathbf{x}) \ , \tag{8}$$

we directly obtain

$$(\mathbf{G}_n \ \delta \mathbf{m} \ )^i \ = \int_{\mathbf{V}} d(\mathbf{x}) \ \Delta^i(\mathbf{x})_n \ \delta m(\mathbf{x}) \ , \tag{9}$$

and

$$G^i(\mathbf{x})_n \ = \ \Delta^i(\mathbf{x})_n \ . \tag{10}$$

Now, we wish here to obtain the transpose of $\mathbf{G}_n$. The definition of transpose operator (see text)

$$\langle \ \mathbf{G}_n{}^t \ \delta\hat{\mathbf{d}} \ , \ \delta \mathrm{m} \ \rangle \quad = \quad \langle \ \delta\hat{\mathrm{d}} \ , \ \mathbf{G}_n \ \delta\mathbf{m} \ \rangle \quad , \tag{11}$$

is written, explicitly,

$$\int_{\mathbf{V}} \mathrm{dV}(\mathrm{x}) \quad (\ \mathbf{G}_n{}^t \ \delta\hat{\mathrm{d}} \ )(\mathrm{x}) \quad \delta\mathrm{m}(\mathbf{x}) \quad = \quad \sum_i \delta\hat{\mathrm{d}}^i \quad (\mathbf{G}_n \ \delta\mathbf{m} \ )^i \ , \tag{12}$$

and, using (6),

$$\int_{\mathbf{V}} \mathrm{dV}(\mathrm{x}) \quad (\ \mathbf{G}_n{}^t \ \delta\hat{\mathrm{d}} \ )(\mathrm{x}) \quad \delta\mathrm{m}(\mathbf{x}) \quad = \quad \sum_i \delta\hat{\mathrm{d}}^i \int_{\mathrm{R}^i(\mathbf{m}_n)} \mathrm{ds}^i \quad \delta\mathrm{m}(\mathbf{x}) \ . \tag{13}$$

We will see that this expression characterizes the transpose operator sufficiently well for practical computations in inversion. But let us try here to obtain a more compact representation. Using the definition of the delta-like function $\Delta^i(\mathbf{x})_n$

$$\int_{\mathrm{R}^i(\mathbf{m}_n)} \mathrm{ds}^i \quad \delta\mathrm{m}(\mathrm{x}) \quad = \quad \int_{\mathrm{V}} \mathrm{dV}(\mathrm{x}) \quad \Delta^i(\mathrm{x})_n \quad \delta\mathrm{m}(\mathbf{x}) \ , \tag{14}$$

equation (13) can be rewritten

$$\int_{\mathbf{V}} \mathrm{dV}(\mathrm{x}) \quad \delta\mathrm{m}(\mathbf{x}) \left[(\ \mathbf{G}_n{}^t \ \delta\hat{\mathrm{d}} \ )(\mathrm{x}) \quad - \quad \sum_i \delta\hat{\mathrm{d}}^i \ \Delta^i(\mathbf{x})_n \right] \quad = \quad 0 \ , \tag{15}$$

and, this being satisfied for any $\delta\mathbf{m}$, it follows that

$$(\ \mathbf{G}_n{}^t \ \delta\hat{\mathrm{d}} \ )(\mathrm{x}) \quad = \quad \sum_i \delta\hat{\mathrm{d}}^i \ \Delta^i(\mathrm{x})_n \ . \tag{16}$$

Notice that this last result can also be obtained directly from knowledge of the kernel of $\mathbf{G}_n$:

$$\mathrm{G}^i(\mathbf{x})_n \quad = \quad \Delta^i(\mathbf{x})_n \ , \tag{10 again}$$

and the rule that an operator and its transpose have same kernels (but they act on transpose variables).

Let d_{obs} be the observed travel times, C_D the covariance operator describing experimental uncertainties, $\mathbf{m}_{prior}$ the a priori model, and C_M the covariance operator describing the confidence we have on $\mathbf{m}_{prior}$. The best model in the least-squares sense minimizes

$$S(\mathbf{m}) = \frac{1}{2}\left[(\mathbf{g}(\mathbf{m})-\mathbf{d}_{obs})^t\, \mathbf{C}_D^{-1}(\mathbf{g}(\mathbf{m})-\mathbf{d}_{obs}) + (\mathbf{m}-\mathbf{m}_{prior})^t\, \mathbf{C}_M^{-1}(\mathbf{m}-\mathbf{m}_{prior})\right]. \quad (17)$$

As the equation solving the forward problem is nonlinear, the functional $S(\mathbf{m})$ is not quadratic. Using, for instance, a Newton method of resolution (see Chapter 4) gives the iterative algorithm

$$\mathbf{m}_{n+1} = \mathbf{m}_{prior} - \mathbf{C}_M\, \mathbf{G}_n{}^t \times$$

$$(\mathbf{C}_D + \mathbf{G}_n\, \mathbf{C}_M\, \mathbf{G}_n{}^t)^{-1}\, (\mathbf{g}(\mathbf{m}_n) - \mathbf{d}_{obs} - \mathbf{G}_n\, (\, \mathbf{m}_n - \mathbf{m}_{prior}\,))\, . \quad (18)$$

In order to interpret this equation, I introduce the following partial steps:

$$\delta\mathbf{v}_n = \mathbf{g}(\mathbf{m}_n) - \mathbf{d}_{obs} - \mathbf{G}_n\, (\mathbf{m}_n - \mathbf{m}_{prior})\, , \quad (19)$$

$$\mathbf{S}_n = \mathbf{C}_D + \mathbf{G}_n\, \mathbf{C}_M\, \mathbf{G}_n{}^t\, , \quad (20)$$

$$\delta\hat{\mathbf{w}}_n = \mathbf{S}_n{}^{-1}\, \delta\mathbf{v}_n\, , \quad (21)$$

$$\delta\hat{\mathbf{m}}_n = \mathbf{G}^t\, \delta\hat{\mathbf{w}}_n\, , \quad (22)$$

and

$$\mathbf{m}_{n+1} = \mathbf{m}_{prior} - \mathbf{C}_M\, \delta\hat{\mathbf{m}}_n\, . \quad (23)$$

Introducing the particular kernels corresponding to this problem gives, in explicit notation,

$$\delta v_n{}^i = \int_{R^i(\mathbf{m}_n)} ds^i\, m(\mathbf{x}^i)_n - d_{obs}{}^i - \int_{R^i(\mathbf{m}_n)} ds^i\, (m(\mathbf{x}^i)_n - m(\mathbf{x}^i)_{prior})$$

$$= \int_{R^i(m_n)} ds^i\, m(\mathbf{x}^i)_{prior} - d_{obs}{}^i\, , \quad (24)$$

$$S_n{}^{ij} = C_D{}^{ij} + \int_{R^i(m_n)} ds^i \int_{R^j(m_n)} ds^j\, C_M(\, \mathbf{x}(s^i)\, ,\, \mathbf{x}(s^j)\,)\, , \quad (25)$$

$$\delta \hat{w}_n{}^i = \sum_j (S_n{}^{-1})^{ij} \ \delta v_n{}^j , \tag{26}$$

and

$$m(\mathbf{x})_{n+1} = m(\mathbf{x})_{prior} - \sum_i \delta \hat{w}_n{}^i \ \Psi^i(x)_n , \tag{27}$$

where

$$\Psi^i(\mathbf{x})_n = \int_{R^i(\mathbf{m}_n)} ds^i \ C_M(x, x(s^i)) \ . \tag{28}$$

The interpretation of these formulas is as in section 7.5.2.

Problem 7.2: Inversion of acoustic waveforms. An acoustic medium can be described using the density $\rho(x)$ and the bulk modulus $K(x)$. For simplicity, assume that the density is known. The problem is then to evaluate $K(\mathbf{x})$. At some shot points $\mathbf{x}_s$ $(s=1,2,...,NS)$, we generate acoustic waves, which are recorded at some receiver positions $\mathbf{x}_r$ $(r=1,2,...,NR)$. Let t be a time variable reset to zero at each new shot. The pressure at the receiver location $\mathbf{x}_r$, at time t , for a shot at point $\mathbf{x}_s$ is denoted $p(\mathbf{x}_r,t;\mathbf{x}_s)$. Let $p(\mathbf{x}_r,t;\mathbf{x}_s)_{obs}$ denote the particular measured (observed) values. For a particular model $K(\mathbf{x})$, $p(\mathbf{x}_r,t;\mathbf{x}_s)_{cal}$ denotes the predicted values. Formulate the inverse problem of evaluating the bulk modulus $K(\mathbf{x})$ from the measurements $p(\mathbf{x}_r,t;\mathbf{x}_s)$.

Solution: We assume the source of acoustic waves to be exactly known. For compactness, a bulk modulus model is denoted $\mathbf{K}$, and a data set is denoted $\mathbf{p}$. The computation of the waveforms corresponding to the model $\mathbf{K}$ is written

$$\mathbf{p} = \mathbf{g}(\mathbf{K}) , \tag{1}$$

where the operator $\mathbf{g}$ is of course nonlinear.

Let $\mathbf{p}_{obs}$ represent the observed data set and $\mathbf{C}_p$ the covariance operator describing experimental uncertainties. In what follows, the kernel of the covariance operator is assumed diagonal:

$$C_p(\mathbf{x}_r,t;\mathbf{x}_s;\mathbf{x}_r{}',t';\mathbf{x}_s{}') = \sigma^2(\mathbf{x}_r,t;\mathbf{x}_s) \ \delta^{rr'} \ \delta(t-t') \ \delta^{ss'} \tag{2}$$

so that the expression

$$\delta \mathbf{p} = C_p \ \delta \hat{\mathbf{p}} \tag{3}$$

is written, explicitly,

$$\delta p(\mathbf{x}_r,t;\mathbf{x}_s) = \sum_{s'} \int_0^T dt' \sum_{r'} C_p(\mathbf{x}_r,t;\mathbf{x}_s;\mathbf{x}_r{}',t';\mathbf{x}_s{}') \ \delta\hat{p}(\mathbf{x}_r{}',t';\mathbf{x}_s{}') ,$$

$$= \sigma^2(\mathbf{x}_r,t;\mathbf{x}_s) \ \delta\hat{p}(\mathbf{x}_r,t;\mathbf{x}_s) . \tag{4}$$

Then, the reciprocal relation

$$\delta \hat{\mathbf{p}} = C_p^{-1} \ \delta \mathbf{p} \tag{5}$$

simply gives

$$\delta\hat{p}(\mathbf{x}_r,t;\mathbf{x}_s) = \frac{\delta p(\mathbf{x}_r,t;\mathbf{x}_s)}{\sigma^2(\mathbf{x}_r,t;\mathbf{x}_s)} . \tag{6}$$

The best model (in the least-squares sense) is defined by the minimization of the squared norm

$$S(\mathbf{K}) = \frac{1}{2} \ \| \ g(\mathbf{K}) - \mathbf{d}_{obs} \ \|^2 \tag{7}$$

where, if we use the notations

$$\langle \ \delta\hat{\mathbf{p}}_1 \ , \delta\mathbf{p}_2 \ \rangle = \delta\hat{\mathbf{p}}_1{}^t \ \delta\mathbf{p}_2 = \delta\mathbf{p}_2{}^t \ \delta\hat{\mathbf{p}}_1$$

$$= \sum_{s} \int_0^T dt \sum_{r} \delta\hat{p}(\mathbf{x}_r.t;\mathbf{x}_s)_1 \ \delta p(\mathbf{x}_r,t;\mathbf{x}_s)_2 , \tag{8}$$

the norm $\| \ \delta\mathbf{p} \ \|$ is defined by

$$\| \ \delta\mathbf{p} \ \|^2 = \langle \ C_p^{-1} \ \delta\mathbf{p} \ , \delta\mathbf{p} \ \rangle = \delta\mathbf{p}^t \ C_p^{-1} \ \delta\mathbf{p} . \tag{9}$$

One then usually writes

$$S(\mathbf{K}) = (\ g(\mathbf{K}) - \mathbf{d}_{obs} \ )^t \ C_p^{-1} \ (\ g(\mathbf{K}) - \mathbf{d}_{obs} \ ) . \tag{10}$$

I do not explicitly introduce the a priori information in the model space: I will use gradient methods, which are naturally robust. If the minimum of (10) is a subspace rather than a single point, they converge to the point which is the closest to the starting point.

The *Fréchet derivative* of the nonlinear operator g at a point $\mathbf{K}_n$ of the model space is the linear operator $\mathbf{G}_n$ that associates any model perturbation $\delta\mathbf{K}$ with the data perturbation $\mathbf{G}_n\ \delta\mathbf{K}$ defined by the first-order development

$$g(\mathbf{K}_n+\delta\mathbf{K}) = g(\mathbf{K}_n) + \mathbf{G}_n\ \delta\mathbf{K} + \text{higher order terms}\ . \tag{11}$$

Introducing notations equivalent to those given by (8) in the model space:

$$\langle\ \delta\hat{\mathrm{K}}_1\ ,\ \delta\mathrm{K}_2\ \rangle = \delta\hat{\mathrm{K}}_1{}^t\ \ \delta\mathrm{K}_2 = \delta\mathrm{K}_2{}^t\ \delta\hat{\mathrm{K}}_1 = \int_{\mathbf{V}} \mathrm{d}V(x)\ \delta\hat{\mathrm{K}}(x)_1\ \delta\mathrm{K}(\mathbf{x})_2\ , \tag{12}$$

and given a linear operator $\mathbf{G}_n$, the *transpose operator* $\mathbf{G}_n{}^t$ is defined by the identity (see text)

$$\langle\ \delta\hat{\mathrm{p}}\ ,\ \mathbf{G}_n\ \delta\mathrm{K}\ \rangle = \langle\ \mathbf{G}_n{}^t\ \ \delta\hat{\mathrm{p}}\ ,\ \delta\mathbf{K}\ \rangle \qquad \text{for any } \delta\hat{\mathbf{p}} \text{ and } \delta\mathbf{K}\ . \tag{13}$$

The least-squares minimization problem can then be solved using, for instance, a preconditioned steepest descent algorithm (see Chapter 4). This gives

$$\mathbf{K}_{n+1} = \mathbf{K}_n - \mu_n\ \hat{\mathbf{S}}_0\ \mathbf{G}_n{}^t\ \mathbf{C}_D{}^{-1}\ (g(\mathbf{K}_n) - \mathbf{d}_{obs}) \tag{14}$$

where $\hat{\mathbf{S}}_0$ is an arbitrary positive definite operator, called the "preconditioning operator" and which is suitably chosen to accelerate the convergence (see below).

Let us now turn to the computation of the Fréchet derivatives corresponding to this problem. The solution of the forward problem is defined by the differential system

$$\frac{1}{K(x)}\ \frac{\partial^2 p}{\partial t^2}(\mathbf{x},t;\mathbf{x}_s) - \mathrm{div}\left[\ \frac{1}{\rho(x)}\ \mathbf{grad}\ p(\mathbf{x},t;\mathbf{x}_s)\ \right] = S(\mathbf{x},t;\mathbf{x}_s) \tag{17a}$$

$$p(\mathbf{x},t;\mathbf{x}_s) = 0 \qquad (\text{ for } \ \mathbf{x} \in \mathbf{S}\ ) \tag{17b}$$

$$p(\mathbf{x},0;\mathbf{x}_s) = 0 \tag{17c}$$

$$\dot{p}(\mathbf{x},0;\mathbf{x}_s) = 0 \ . \tag{17d}$$

Here $S(\mathbf{x},t;\mathbf{x}_s)$ is the function describing the source, and $\mathbf{S}$ denotes the surface of the medium. Practically, this differential system can, for instance, be solved using a finite-difference algorithm (see, for instance, Alterman and Karal, 1968).

The Green function $\Gamma(\mathbf{x},t;\mathbf{x}_s,t')$ is defined by

$$\frac{1}{K(\mathbf{x})}\frac{\partial^2\Gamma}{\partial t^2}(\mathbf{x},t;\mathbf{x}',t') - \operatorname{div}\left(\frac{1}{\rho(\mathbf{x})}\ \mathrm{grad}\ \Gamma(\mathbf{x},t;\mathbf{x}',t')\right) = \delta(\mathbf{x}-\mathbf{x}_s)\ \delta(t-t') \tag{18a}$$

$$\Gamma(\mathbf{x},t;\mathbf{x}',t') = 0 \qquad (\text{ for } \ \mathbf{x} \in \mathbf{S}\ ) \tag{18b}$$

$$\Gamma(\mathbf{x},t;\mathbf{x}',t') = 0 \qquad (\text{ for } \ t < t') \tag{18c}$$

$$\dot{\Gamma}(\mathbf{x},t;\mathbf{x}',t') = 0 \ (\text{ for } t < t')\ , \tag{18d}$$

and we have the integral representation (see, for instance, Morse and Feshbach, 1953)

$$p(\mathbf{x},t;\mathbf{x}_s) = \int_{\mathbf{V}} dV(\mathbf{x}')\ \Gamma(\mathbf{x},t;\mathbf{x}',0) * S(\mathbf{x}',t;\mathbf{x}_s)\ . \tag{19}$$

In order to obtain the Fréchet derivative of the displacements with respect to the bulk modulus (as defined by equation 11), let us introduce the wavefield $p(\mathbf{x},t;\mathbf{x}_s)_n$ propagating in the medium $K(\mathbf{x})_n$:

$$\frac{1}{K(\mathbf{x})_n}\ \frac{\partial^2 p(\mathbf{x},t;\mathbf{x}_s)_n}{\partial t^2} - \operatorname{div}\left(\frac{1}{\rho(\mathbf{x})}\ \mathrm{grad}\ p(\mathbf{x},t;\mathbf{x}_s)_n\right) = S(\mathbf{x},t;\mathbf{x}_s) \tag{20a}$$

$$p(\mathbf{x},t;\mathbf{x}_s)_n = 0 \qquad (\text{ for } \ \mathbf{x} \in \mathbf{S}\ ) \tag{20b}$$

$$p(\mathbf{x},0;\mathbf{x}_s)_n = 0 \tag{20c}$$

$$\dot{p}(\mathbf{x},0;\mathbf{x}_s)_n = 0\ , \tag{20d}$$

and the corresponding Green function

$$\frac{1}{K(\mathbf{x})_n}\ \frac{\partial^2\Gamma(\mathbf{x},t;\mathbf{x}',t')_n}{\partial t^2} - \operatorname{div}\left(\frac{1}{\rho(\mathbf{x})}\ \mathrm{grad}\ \Gamma(\mathbf{x},t;\mathbf{x}',t')_n\right) = \delta(\mathbf{x}-\mathbf{x}_s)\ \delta(t-t') \tag{21a}$$

$$\Gamma(\mathbf{x},t;\mathbf{x}',t')_n = 0 \qquad (\text{ for } \mathbf{x} \in S) \tag{21b}$$

$$\Gamma(\mathbf{x},t;\mathbf{x}',t')_n = 0 \qquad (\text{ for } t < t') \tag{21c}$$

$$\dot{\Gamma}(\mathbf{x},t;\mathbf{x}',t')_n = 0 \qquad (\text{ for } t < t') \,. \tag{21d}$$

A perturbation of bulk modulus $K(x)_n \rightarrow K(\mathbf{x})_n+\delta K(\mathbf{x})$ will produce a field $p(x,t;\mathbf{x}_s)_n+\delta p(\mathbf{x},t;\mathbf{x}_s)$ defined by

$$\frac{1}{K(\mathbf{x})_n+\delta K(\mathbf{x})} \frac{\partial^2 (p(\mathbf{x},t;\mathbf{x}_s)_n+\delta p(\mathbf{x},t;\mathbf{x}_s))}{\partial t^2}$$
$$- \operatorname{div}\left[\frac{1}{\rho(\mathbf{x})} \mathbf{grad}\ (p(\mathbf{x},t;\mathbf{x}_s)_n+\delta p(x,t;\mathbf{x}_s))\right] = S(\mathbf{x},t;\mathbf{x}_s) \tag{22a}$$

$$p(\mathbf{x},t;\mathbf{x}_s)_n+\delta p(\mathbf{x},t;\mathbf{x}_s) = 0 \qquad (\text{ for } x \in S) \tag{22b}$$

$$p(\mathbf{x},0;x_s)_n+\delta p(\mathbf{x},0;\mathbf{x}_s) = 0 \tag{22c}$$

$$\dot{p}(\mathbf{x},0;\mathbf{x}_s)_n+\delta\dot{p}(\mathbf{x}0;\mathbf{x}_s) = 0 \,. \tag{22d}$$

This gives

$$\frac{1}{K(\mathbf{x})_n} \frac{\partial^2 \delta p(\mathbf{x},t;\mathbf{x}_s)}{\partial t^2} - \operatorname{div}\left(\frac{1}{\rho(x)} \mathbf{grad}\ \delta p(\mathbf{x},t;\mathbf{x}_s) \right]$$
$$= \frac{\partial^2 p(\mathbf{x},t;\mathbf{x}_s)_n}{\partial t^2} \frac{\delta K(\mathbf{x})}{K(\mathbf{x})_n{}^2} + O(\| \delta K \|^2) \tag{23a}$$

$$\delta p(\mathbf{x},t;\mathbf{x}_s) = 0 \qquad (\text{ for } \mathbf{x} \in \mathbf{S}) \tag{23b}$$

$$\delta p(\mathbf{x},0;\mathbf{x}_s) = 0 \tag{23c}$$

$$\delta\dot{p}(x,0;\mathbf{x}_s) = 0 \,, \tag{23d}$$

and, using theorem (19),

$$\delta p(\mathbf{x}_r,t;\mathbf{x}_s) = \int_{\mathbf{V}} dV(\mathbf{x})\ \Gamma(\mathbf{x}_r,t;x,0)_n * \frac{\partial^2 p}{\partial t^2}(x,t;\mathbf{x}_s)_n \frac{\delta K(\mathbf{x})}{K(\mathbf{x})_n{}^2} + O(\| \delta \mathbf{K} \|^2). \tag{24}$$

The Fréchet derivative operator $\mathbf{G}_n$ introduced in (11) is then defined by

$$(\mathbf{G}_n \, \delta\mathbf{K})(\mathbf{x}_r,t;\mathbf{x}_s) = \int_{\mathbf{V}} dV(\mathbf{x}) \ \Gamma(\mathbf{x}_r,t;\mathbf{x},0)_n * \frac{\partial^2 p}{\partial t^2}(\mathbf{x},t;\mathbf{x}_s)_n \ \frac{\delta K(\mathbf{x})}{K(\mathbf{x})_n{}^2} , \tag{25}$$

where $\Gamma(\mathbf{x},t;\mathbf{x}',t')_n$ is the Green function corresponding to the medium $K(\mathbf{x})_n$ (defined by equation 21) , and $p(\mathbf{x},t;\mathbf{x}_s)_n$ is the wavefield also corresponding to $K(x)_n$ (defined by equation 20).

We now turn to the characterization of the transpose operator. We have just defined the Fréchet derivative operator G_n . Its transpose $\mathbf{G}_n{}^t$ was defined by equation (13):

$$\langle \ \delta\hat{p} \ , \mathbf{G}_n \, \delta\mathbf{K} \ \rangle = \langle \ \mathbf{G}_n{}^t \ \delta\hat{p} \ , \delta K \ \rangle \qquad \text{for any } \delta\hat{\mathbf{p}} \text{ and } \delta\mathbf{K} . \tag{13 again}$$

To solve the inverse problem, we need to be able to compute $\mathbf{G}_n{}^t \ \delta\hat{p}$ for arbitrary $\delta\hat{p}$. Using the notations introduced in (8) and (12) , equation (13) is written

$$\sum_s \int_0^T dt \sum_r \delta\hat{p}(\mathbf{x}_r,t;\mathbf{x}_s) \ (\mathbf{G}_n \ \delta\mathbf{K})(\mathbf{x}_r,t;\mathbf{x}_s)$$

$$= \int_V dV(\mathbf{x}) \ (\mathbf{G}_n{}^t \ \delta\hat{\mathbf{p}})(\mathbf{x}) \ \delta K(\mathbf{x}) , \tag{26}$$

and, using (25),

$$\sum_s \int_0^T dt \sum_r \delta\hat{p}(\mathbf{x}_r,t;\mathbf{x}_s) \int_{\mathbf{V}} dv(\mathbf{x}) \ \Gamma(\mathbf{x}_r,t;\mathbf{x},0)_n * \frac{\partial^2 p}{\partial t^2}(\mathbf{x},t;\mathbf{x}_s)_n \ \frac{\delta K(\mathbf{x})}{K(\mathbf{x})_n{}^2}$$

$$= \int_V dV(\mathbf{x}) \ (\mathbf{G}_n{}^t \ \delta\hat{\mathbf{p}})(\mathbf{x}) \ \delta K(\mathbf{x}) , \tag{27}$$

i.e.,

$$\int_{\mathbf{V}} dV(\mathbf{x}) \ \delta K(\mathbf{x}) \{(\mathbf{G}_n{}^t \ \delta\hat{\mathbf{p}})(\mathbf{x})$$

$$- \frac{1}{K(\mathbf{x})_n{}^2} \sum_s \int_0^T dt \sum_r \Gamma(\mathbf{x}_r,t;\mathbf{x},0)_n * \frac{\partial^2 p}{\partial t^2}(\mathbf{x},t;\mathbf{x}_s)_n \ \delta\hat{p}(\mathbf{x}_r,t;\mathbf{x}_s)\} = 0 . \tag{28}$$

As this has to be valid for any $\delta K(\mathbf{x})$, we obtain

$$(G_n{}^t\ \delta\hat{p})(\mathbf{x}) =$$

$$= \frac{1}{K(\mathbf{x})_n{}^2} \sum_s \int_0^T dt \sum_r \Gamma(\mathbf{x}_r,t;\mathbf{x},0)_n * \frac{\partial^2 p}{\partial t^2}(\mathbf{x},t;\mathbf{x}_s)_n\ \delta\hat{p}(\mathbf{x}_r,t;\mathbf{x}_s)\ . \tag{29}$$

I introduce a field $\Psi(x,t;x_s)_n$ defined by the differential system

$$\frac{1}{K(\mathbf{x})_n} \frac{\partial^2\Psi(\mathbf{x},t;\mathbf{x}_s)_n}{\partial t^2} - \mathrm{div}\left(\frac{1}{\rho(x)}\ \mathbf{grad}\ \Psi(\mathbf{x},t;\mathbf{x}_s)_n \right) = \Phi(\mathbf{x},t;\mathbf{x}_s) \tag{30a}$$

$$\Psi(x,t;\mathbf{x}_s)_n = 0 \qquad (\text{ for } \mathbf{x} \in \mathbf{S}\ ) \tag{30b}$$

$$\Psi(x,T;\mathbf{x}_s)_n = 0 \tag{30c}$$

$$\dot{\Psi}(x,T;\mathbf{x}_s)_n = 0\ , \tag{30d}$$

where

$$\Phi(\mathbf{x},t;\mathbf{x}_s) = \sum_r \delta(x-\mathbf{x}_r)\ \delta\hat{p}(\mathbf{x}_r,t;\mathbf{x}_s)\ . \tag{31}$$

It should we noticed that the field Ψ satifies *final* (instead of initial) conditions, using the property

$$\Gamma(\mathbf{x},t;\mathbf{x}_r,t')_n = \Gamma(\mathbf{x},t+\tau;\mathbf{x}_r,t'+\tau)_n\ , \tag{32}$$

and reversing time in theorem (19), one obtains

$$\Psi(x,t;\mathbf{x}_s)_n = \sum_r \int_0^T dt'\ \Gamma(\mathbf{x},0;x_r,t-t')_n\ \ \delta\hat{p}(x_r,t';x_s)\ . \tag{33}$$

We have

$$\dot{\Psi}(\mathbf{x},t;\mathbf{x}_s) = \sum_r \int_0^T dt' \frac{\partial}{\partial t} \Gamma(\mathbf{x},0;x_r,t-t')\ \ \delta\hat{p}(\mathbf{x}_r,t';\mathbf{x}_s)$$

$$= \sum_r \int_0^T dt' \frac{\partial}{\partial t} \Gamma(\mathbf{x},t'-t;\mathbf{x}_r,0)\ \ \delta\hat{p}(\mathbf{x}_r,t';\mathbf{x}_s)$$

$$= - \sum_r \int_0^T dt' \; \dot{\Gamma}(\mathbf{x},t'-t;\mathbf{x}_r,0) \quad \delta\hat{p}(\mathbf{x}_r,t';\mathbf{x}_s) \; , \tag{34}$$

where

$$\dot{\Gamma}(\mathbf{x},t;\mathbf{x}',t') \;=\; \frac{\partial}{\partial t}\,\Gamma(\mathbf{x},t;\mathbf{x}',t') \; . \tag{35}$$

Using an integration by parts, we have

$$\begin{aligned}
\Gamma(\mathbf{x}_r,t;\mathbf{x},0) * \frac{\partial^2 p}{\partial t^2}(\mathbf{x},t;\mathbf{x}_s) \;&=\; \int_0^T dt' \;\; \Gamma(\mathbf{x}_r,t-t';\mathbf{x},0) \;\; \frac{\partial^2 p}{\partial t^2}(\mathbf{x},t';\mathbf{x}_s) \; , \\
&=\; \Gamma(\mathbf{x}_r,t-T;\mathbf{x},0)\; \dot{p}(\mathbf{x},T;\mathbf{x}_s) - \Gamma(\mathbf{x}_r,t;\mathbf{x},0)\; \dot{p}(\mathbf{x},0;\mathbf{x}_s) \\
&\quad - \int_0^T dt' \;\; \dot{\Gamma}(\mathbf{x}_r,t-t';\mathbf{x},0)\; \dot{p}(\mathbf{x},t';\mathbf{x}_s) \; ,
\end{aligned}$$

and using the initial conditions (20d) and (21c),

$$\Gamma(\mathbf{x}_r,t;\mathbf{x},0) * \frac{\partial^2 p}{\partial t^2}(\mathbf{x},t;\mathbf{x}_s) \;=\; - \;\; \dot{\Gamma}(\mathbf{x}_r,t;\mathbf{x},0) * \dot{p}(\mathbf{x},t;\mathbf{x}_s) \; . \tag{36}$$

Using the last equation gives

$$\begin{aligned}
&\sum_s \int_0^T dt \sum_r \;\; \Gamma(\mathbf{x}_r,t;\mathbf{x},0)_n \; * \; \frac{\partial^2 p}{\partial t^2}(\mathbf{x},t;\mathbf{x}_s) \;\; \delta\hat{p}(\mathbf{x}_r,t;\mathbf{x}_s) \;\; = \\
&\quad = - \sum_s \int_0^T dt \sum_r \;\; \dot{\Gamma}(\mathbf{x}_r,t;\mathbf{x},0) * \dot{p}(\mathbf{x},t;\mathbf{x}_s) \;\; \delta\hat{p}(\mathbf{x}_r,t;\mathbf{x}_s) \\
&\quad = - \sum_s \int_0^T dt \int_0^T dt' \sum_r \;\; \dot{\Gamma}(\mathbf{x}_r,t-t';\mathbf{x},0)\; \dot{p}(\mathbf{x},t';\mathbf{x}_s) \;\; \delta\hat{p}(\mathbf{x}_r,t;\mathbf{x}_s) \; ,
\end{aligned}$$

whence, using (34), we obtain

$$\sum_s \int_0^T dt \sum_r \;\; \Gamma(\mathbf{x}_r,t;\mathbf{x},0)_n \; * \; \frac{\partial^2 p}{\partial t^2}(\mathbf{x},t;\mathbf{x}_s) \;\; \delta\hat{p}(\mathbf{x}_r,t;\mathbf{x}_s) \;\; =$$

$$= \sum_s \int_0^T dt \ \dot{p}(\mathbf{x},t;\mathbf{x}_s) \ \dot{\Psi}(\mathbf{x},t;\mathbf{x}_s) \ . \tag{37}$$

From equation (29) we then finally obtain the result characterizing the transpose operator:

$$(\mathbf{G}_n{}^t \ \delta\hat{\mathbf{p}})(\mathbf{x}) = \frac{1}{K(\mathbf{x})_n{}^2} \sum_s \int_0^T dt \ \dot{p}(\mathbf{x},t;\mathbf{x}_s)_n \ \dot{\Psi}(\mathbf{x},t;\mathbf{x}_s)_n \ . \tag{38}$$

The preconditioned steepest descent algorithm (14) is written, step by step,

$$\delta\mathbf{p}_n = \mathbf{g}(\mathbf{K}_n) - \mathbf{d}_{obs} \ , \tag{39}$$

$$\delta\hat{\mathbf{p}}_n = \mathbf{C}_p{}^{-1} \ \delta\mathbf{p}_n \ , \tag{40}$$

$$\hat{\gamma}_n = \mathbf{G}_n{}^t \ \delta\hat{\mathbf{p}}_n \ , \tag{41}$$

$$\mathbf{d}_n = \hat{\mathbf{S}}_0 \ \hat{\gamma}_n \ , \tag{42}$$

and

$$\mathbf{K}_{n+1} = \mathbf{K}_n - \mu_n \ \mathbf{d}_n \ . \tag{43}$$

To compute the residuals (38) we need to compute $\mathbf{g}(\mathbf{K}_n)$, i.e., we need to solve the forward problem (using any numerical method). The weighted residuals (40) are easily obtained using (6). $\hat{\gamma}_n$ is computed using (38), where the field $\Psi(\mathbf{x},t;\mathbf{x}_s)_n$ is obtained by solving the system (30) (using, for instance, the same method used to solve the forward problem). To obtain $\mathbf{d}_n$ we have to apply the preconditioning operator $\hat{\mathbf{S}}_0$ to $\hat{\gamma}_n$. It may simply consist in some ad-hoc geometrical correction (see, for instance, Gauthier et al., (1986)). To end one iteration, we have to estimate μ_n in (43). A linearized estimation of μ_n can be obtained as follows.

For given $\mathbf{K}_n$, we have

$$S(\mathbf{K}_n-\mu_n \ \mathbf{d}_n)$$
$$= \frac{1}{2} \Big[(\ \mathbf{g}(\mathbf{K}_n-\mu_n \ \mathbf{d}_n) - \mathbf{p}_{obs} \ )^t \ \mathbf{C}_p{}^{-1} \ (\ \mathbf{g}(\mathbf{K}_n-\mu_n \ \mathbf{d}_n) - \mathbf{p}_{obs} \ ) \Big] \ . \tag{44}$$

If μ_n is small enough, using the definition of Fréchet derivatives, we have

$$\mathbf{g}(\mathbf{K}_n-\mu_n \ \mathbf{d}_n) \simeq \mathbf{g}(\mathbf{K}_n) - \mu_n \ \mathbf{G}_n \ \mathbf{d}_n \ , \tag{45}$$

which gives

$$S(\mathbf{K}_n-\mu_n\,\mathbf{d}_n) \simeq S(\mathbf{K}_n) - \mu_n\,(\,\mathbf{G}_n\,\mathbf{d}_n\,)^t\,\mathbf{C}_p^{-1}\,(\,\mathbf{g}(\mathbf{K}_n) - \mathbf{p}_{obs}\,)$$

$$+\ \frac{1}{2}\,\mu_n^2\,(\,\mathbf{G}_n\,\mathbf{d}_n\,)^t\ \mathbf{C}_p^{-1}\,(\,\mathbf{G}_n\,\mathbf{d}_n\,)\ . \tag{46}$$

The condition $\partial S/\partial\mu_n = 0$ gives

$$\mu_n \simeq \frac{(\,\mathbf{G}_n\,\mathbf{d}_n\,)^t\,\mathbf{C}_p^{-1}\,(\,\mathbf{g}(\mathbf{K}_n) - \mathbf{p}_{obs}\,)}{(\,\mathbf{G}_n\,\mathbf{d}_n\,)^t\ \mathbf{C}_p^{-1}\,(\,\mathbf{G}_n\,\mathbf{d}_n\,)}\ , \tag{47}$$

and using the definition of the transpose operator (equation 13), we finally obtain

$$\mu_n \simeq \frac{\mathbf{d}_n{}^t\,\mathbf{G}_n{}^t\,\mathbf{C}_p^{-1}\,(\,\mathbf{g}(\mathbf{K}_n) - \mathbf{p}_{obs}\,)}{(\,\mathbf{G}_n\,\mathbf{d}_n\,)^t\ \mathbf{C}_p^{-1}\,(\,\mathbf{G}_n\,\mathbf{d}_n\,)}$$

$$= \frac{\mathbf{d}_n{}^t\,\hat{\boldsymbol{\gamma}}_n}{(\,\mathbf{G}_n\,\mathbf{d}_n\,)^t\ \mathbf{C}_p^{-1}\,(\,\mathbf{G}_n\,\mathbf{d}_n\,)}\ . \tag{48}$$

To compute $\mathbf{G}_n\,\mathbf{d}_n$ we could use the result (25), but it is more practical to use directly the definition of derivative operator and a finite-difference approximation. We then may use

$$\mathbf{G}_n\,\mathbf{d}_n \simeq \frac{1}{\epsilon}\,(\,\mathbf{g}(\mathbf{K}_n+\epsilon\mathbf{d}_n) - \mathbf{g}(\mathbf{K}_n)\,) \tag{49}$$

with a sufficiently small value of ϵ .

The physical interpretation of the obtained results is as in Section 7.6. Figures 7.14-7.17 show some of the results obtained by Gauthier et al. (1986) using this method.

Problem 7.3: At some points $\mathbf{x}_s$ (s=1,2,...) on the Earth's surface, we shot seismic sources producing elastic waves. For each shot point, we place receivers at some other points $\mathbf{x}_r$ (r=1,2,...) which record the displacement $u^i(\mathbf{x}_r,t;\mathbf{x}_s)$ of the surface. Use the observed displacements to infer the structure of the Earth.

Solution:

Choice of parameters

From a seismological point of view, the Earth may be described using the density $\rho(\mathbf{x})$, the elastic coefficients $c^{ijkl}(\mathbf{x})$, and some parameters describing attenuation of waves. At first order, the Earth is isotropic (although heterogeneous) and non-attenuating. Only three parameters are then needed. For instance, in addition to the density $\rho(\mathbf{x})$, usual choices are:

the bulk modulus $K(\mathbf{x})$ and the the shear modulus $\mu(\mathbf{x})$,

the Lamé parameters $\lambda(\mathbf{x})$ and $\mu(\mathbf{x})$,

the velocity of compressional (longitudinal, or P) waves, $\alpha(\mathbf{x})$, and the velocity of shear (transverse, or S) waves, $\beta(\mathbf{x})$.

I consider here a small-scale experiment, as in petroleum exploration. Using a seismic reflection data set (sources and receivers at the surface of the Earth), all three parameters are not well resolved. To avoid an intrinsic ill-posedness of the problem, it is important to identify independent parameters (i.e., parameters for which correlation between posterior errors will be small). Using physical arguments, Tarantola (1986) suggests using density $\rho(\mathbf{x})$, longitudinal-wave impedance

$$IP(\mathbf{x}) = \rho(\mathbf{x})\,\alpha(\mathbf{x}) = \sqrt{\rho(\mathbf{x})\,(\,\lambda(\mathbf{x}) + 2\,\mu(\mathbf{x})\,)} \tag{2a}$$

and transverse-wave impedance

$$IS(\mathbf{x}) = \rho(\mathbf{x})\,\beta(\mathbf{x}) = \sqrt{\rho(\mathbf{x})\,\mu(\mathbf{x})}\ . \tag{2b}$$

The least-squares criterion of goodness of fit

I define the best Earth's model ($IP(\mathbf{x})$, $IS(\mathbf{x})$, $\rho(\mathbf{x})$) as the model that minimizes the least-squares expression

$$S(\mathbf{IP},\mathbf{IS},\rho) =$$

$$= \frac{1}{2}\left(\| \mathbf{u}_{cal}-\mathbf{u}_{obs} \|^2 + \| \mathbf{IP}-\mathbf{IP}_{prior} \|^2 + \| \mathbf{IS}-\mathbf{IS}_{prior} \|^2 + \| \boldsymbol{\rho}-\boldsymbol{\rho}_{prior} \|^2 \right) \quad (3)$$

where

$$\| \mathbf{u}_{cal}-\mathbf{u}_{obs} \|^2 = \sum_s \sum_r \int_0^T dt \int_0^T dt' \left(u^i(\mathbf{x}_r,t;\mathbf{x}_s)_{cal} - u^i(\mathbf{x}_r,t;\mathbf{x}_s)_{obs} \right)$$

$$W^{ij}(t,t',\mathbf{x}_r,\mathbf{x}_s) \left(u^j(\mathbf{x}_r,t';\mathbf{x}_s)_{cal} - u^j(\mathbf{x}_r,t';\mathbf{x}_s)_{obs} \right) \quad (4a)$$

$$\| \mathbf{IP}-\mathbf{IP}_{prior} \|^2 = \int_V dV(\mathbf{x}) \int_V dV(\mathbf{x}') \left(IP(\mathbf{x})-IP(\mathbf{x})_{prior} \right) W_p(\mathbf{x},\mathbf{x}')$$

$$\left(IP(\mathbf{x}')-IP(\mathbf{x}')_{prior} \right) \quad (4b)$$

$$\| \mathbf{IS}-\mathbf{IS}_{prior} \|^2 = \int_V dV(\mathbf{x}) \int_V dV(\mathbf{x}') \left(IS(\mathbf{x})-IS(\mathbf{x})_{prior} \right) W_s(\mathbf{x},\mathbf{x}')$$

$$\left(IS(\mathbf{x}')-IS(\mathbf{x}')_{prior} \right) \quad (4c)$$

and

$$\| \boldsymbol{\rho}-\boldsymbol{\rho}_{prior} \|^2 = \int_V dV(\mathbf{x}) \int_V dV(\mathbf{x}') \left(\rho(\mathbf{x})-\rho(\mathbf{x})_{prior} \right) W_\rho(\mathbf{x},\mathbf{x}')$$

$$\left(\rho(\mathbf{x}')-\rho(\mathbf{x}')_{prior} \right) . \quad (4d)$$

Here $u^i(\mathbf{x}_r,t;\mathbf{x}_s)_{cal}$ represents the data predicted by the model ($IP(\mathbf{x})$, $IS(\mathbf{x})$, $\rho(\mathbf{x})$) , and $W^{ij}(t,t',\mathbf{x}_r,\mathbf{x}_s)$, $W_p(\mathbf{x},\mathbf{x}')$, $W_s(\mathbf{x},\mathbf{x}')$, and $W_\rho(\mathbf{x},\mathbf{x}')$ represent weighting functions to be discussed below. A priori uncertainties on density and impedance are assumed uncorrelated.

The physical interpretation of the criterion (14) is simple: we require a model ($IP(\mathbf{x})$, $IS(\mathbf{x})$, $\rho(\mathbf{x})$) not too far from the a priori model ($IP(\mathbf{x})_{prior}$, $IS(\mathbf{x})_{prior}$, $\rho(\mathbf{x})_{prior}$) and such that the predicted data are not too far from the observed data. The constraint that the final model has to be not too far from some a priori model is necessary to avoid ill-posedness of the problem: if very different models give approximately the same seismograms, we prefer the model which is the closest to some simple model.

When so defined, the problem is fully nonlinear (the best model is defined without invoking any linear approximation of the basic equations). In particular, I do not use Born's approximation. It should be noticed that, as the computed seismograms are *nonlinear* functionals of the model parameters, the functional (4) is a *nonquadratic* function of the parameters.

Choice of the weighting functions

In the context of least squares, a weighting function is the integral kernel of the inverse of a covariance operator. For instance, if $C(\mathbf{x},\mathbf{x}')$ is a covariance function, the associated weighting function verifies

$$\int_{\mathbf{V}} dV(\mathbf{x}') \;\; C(\mathbf{x},\mathbf{x}') \;\; W(\mathbf{x}',\mathbf{x}'') \;=\; \delta(\mathbf{x}-\mathbf{x}'') \, . \tag{5}$$

Taking the covariance function

$$C(\mathbf{x},\mathbf{x}') \;=\; C(x,y,z,x',y',z') \;=\; K\ \delta(x-x')\ \delta(y-y')\ \mathrm{Min}(z,z') \tag{6}$$

gives (see problem 7.8)

$$\|\ \phi-\phi_{\mathrm{prior}}\ \|^2 = \frac{1}{K} \int_{\mathbf{V}} dV(\mathbf{x}) \left(\frac{\partial\phi}{\partial z}(x,y,z) - \frac{\partial\phi_{\mathrm{prior}}}{\partial z}(x,y,z) \right)^2 , \tag{7}$$

which is an adequate norm to impose on impedances or density in the Earth: we do not wish our final model to be close to the initial model, but we wish the vertical gradient of the model to be close to the vertical gradient of the a priori model. Taking, for instance, a homogeneous a priori model in impedances and density, the norm (7) will impose that the final model has small vertical gradients. So, for each of the model parameters, I choose

$$C_p(x,y,z,x',y',z') \;=\; K_p\ \delta(x-x')\ \delta(y-y')\ \mathrm{Min}(z,z') \ , \tag{8a}$$

$$C_s(x,y,z,x',y',z') \;=\; K_s\ \delta(x-x')\ \delta(y-y')\ \mathrm{Min}(z,z') \ , \tag{8a}$$

and

$$C_\rho(x,y,z,x',y',z') \;=\; K_\rho\ \delta(x-x')\ \delta(y-y')\ \mathrm{Min}(z,z') \ , \tag{8a}$$

which gives respectively

$$\| \mathbf{IP}\text{-}\mathbf{IP}_{\text{prior}} \|^2 = \frac{1}{K_p} \int_{\mathbf{V}} dV(\mathbf{x}) \left[\frac{\partial IP}{\partial z}(x,y,z) - \frac{\partial IP_{\text{prior}}}{\partial z}(x,y,z) \right]^2 , \tag{9a}$$

$$\| \mathbf{IS}\text{-}\mathbf{IS}_{\text{prior}} \|^2 = \frac{1}{K_s} \int_{V} dV(\mathbf{x}) \left[\frac{\partial IS}{\partial z}(x,y,z) - \frac{\partial IS_{\text{prior}}}{\partial z}(x,y,z) \right]^2 , \tag{9b}$$

and

$$\| \rho\text{-}\rho_{\text{prior}} \|^2 = \frac{1}{K_\rho} \int_{V} dV(\mathbf{x}) \left[\frac{\partial \rho}{\partial z}(x,y,z) - \frac{\partial \rho_{\text{prior}}}{\partial z}(x,y,z) \right]^2 . \tag{9c}$$

Thus, final models of impedance and density will have small vertical gradients (more precisely, small vertical gradient differences from the a priori model).

If there are uncorrelated errors in the data set, depending on time or source and receiver position, then

$$C^{ij}(t,t',\mathbf{x}_r,\mathbf{x}_s) = \sigma^2(t,\mathbf{x}_r,\mathbf{x}_s)\, \delta^{ij}\, \delta(t-t') \tag{10}$$

and

$$\| \mathbf{u}_{\text{cal}}\text{-}\mathbf{u}_{\text{obs}} \|^2 = \sum_s \sum_r \int_0^T dt \sum_{i=1}^{i=3} \frac{(u^i(\mathbf{x}_r,t;\mathbf{x}_s)_{\text{cal}} - u^i(\mathbf{x}_r,t;\mathbf{x}_s)_{\text{obs}})^2}{(\sigma^2(\mathbf{r}_r,t;\mathbf{x}_s))^2} . \tag{11}$$

Usually, only the vertical component u^3 is recorded. The sum over i then disappears from (11).

I now turn to the description of the forward problem.

The elastodynamic wave equation

Let us consider an isotropic elastic Earth. In what follows, $\mathbf{x}$ represents a point inside (or at the surface of) the Earth, and t is the time variable, running in an interval $0 \le t \le T$. If $f^i(\mathbf{x},t;\mathbf{x}_s)$ is the volume density of internal forces for the s-th shot, $T^i(\mathbf{x},t;\mathbf{x}_s)$ the stress vector (traction) at the Earth's surface $\mathbf{S}$, and $n^i(\mathbf{x})$ the unit normal at the surface, then the displacement $u^i(\mathbf{x},t;\mathbf{x}_s)$ corresponding to that s-th shot is uniquely defined by the differential equations

$$\rho(\mathbf{x})\, \ddot{u}^i(\mathbf{x},t;\mathbf{x}_s) - \frac{\partial}{\partial x^i} \left(\lambda(\mathbf{x})\, u^{kk}(\mathbf{x},t;\mathbf{x}_s) \right) - 2 \frac{\partial}{\partial x^j} \left(\mu(\mathbf{x})\, u^{ij}(\mathbf{x},t;\mathbf{x}_s) \right) =$$

$$= f^i(\mathbf{x},t;\mathbf{x}_s) \tag{12a}$$

$$\lambda(\mathbf{x})\ u^{kk}(\mathbf{x},t;\mathbf{x}_s)\ n^i(\mathbf{x}) + 2\ \mu(\mathbf{x})\ u^{ij}(\mathbf{x},t;\mathbf{x}_s)\ n^j(\mathbf{x}) = T^i(\mathbf{x},t;\mathbf{x}_s) \qquad \mathbf{x} \in S \tag{12b}$$

$$u^i(\mathbf{x},0;\mathbf{x}_s) = 0 \tag{12c}$$

$$\dot{u}^i(\mathbf{x},0;\mathbf{x}_s) = 0\ , \tag{12d}$$

where $u^{ij}(\mathbf{x},t;\mathbf{x}_s)$ represents the strain tensor

$$u^{ij}(\mathbf{x},t;\mathbf{x}_s) = \frac{1}{2}\left[\frac{\partial u^i}{\partial x^j}(\mathbf{x},t;\mathbf{x}_s) + \frac{\partial u^j}{\partial x^i}(\mathbf{x},t;\mathbf{x}_s)\right]\ , \tag{13}$$

and where an implicit sum over repeated indexes is assumed.

For a given Earth model $\rho(\mathbf{x})$, $\lambda(\mathbf{x})$, $\mu(\mathbf{x})$, and given volume and surface sources $f^i(\mathbf{x},t;\mathbf{x}_s)$, $T^i(\mathbf{x},t;\mathbf{x}_s)$, the displacement field $u^i(\mathbf{x},t;\mathbf{x}_s)$ can be evaluated directly from equations (2) using numerical methods such as finite-differences (Altermann and Karal, 1968). The sources of seismic waves can be either tractions at the surface, described by $T^i(\mathbf{x},t;\mathbf{x}_s)$, or internal sources (such as borehole explosions), described by $f^i(\mathbf{x},t;\mathbf{x}_s)$.

Green's function

Green's function (or the impulse response) of the problem is denoted $\Gamma^{ij}(\mathbf{x},t;\mathbf{x}',t')$ and is defined by

$$\rho(\mathbf{x})\ \frac{\partial^2\Gamma^{ij}}{\partial t^2}(\mathbf{x},t;\mathbf{x}',t') - \frac{\partial}{\partial x^i}\left(\lambda(\mathbf{x})\ \Gamma^{kkj}(\mathbf{x},t;\mathbf{x}',t')\right) - 2\ \frac{\partial}{\partial x^k}\left(\mu(\mathbf{x})\ \Gamma^{ikj}(\mathbf{x},t;\mathbf{x}',t')\right) =$$
$$= \delta^{ij}\ \delta(\mathbf{x}-\mathbf{x}')\ \delta(t-t') \tag{14a}$$

$$\lambda(\mathbf{x})\ \Gamma^{kkj}(\mathbf{x},t;\mathbf{x}',t')\ n^i(\mathbf{x}) + 2\ \mu(\mathbf{x})\ \Gamma^{ikj}(\mathbf{x},t;\mathbf{x}',t')\ n^k(\mathbf{x}) = 0 \qquad \mathbf{x} \in S \tag{14b}$$

$$\Gamma^{ij}(\mathbf{x},t;\mathbf{x}',t') = 0 \qquad \text{for} \quad t < t' \tag{14c}$$

$$\dot{\Gamma}^{ij}(\mathbf{x},t;\mathbf{x}',t') = 0 \qquad \text{for} \quad t < t'\ , \tag{14d}$$

where $\Gamma^{ijk}(\mathbf{x},t;\mathbf{x}',t')$ is the strain associated with $\Gamma^{ik}(\mathbf{x},t;\mathbf{x}',t')$:

$$\Gamma^{ijk}(\mathbf{x},t;\mathbf{x}',t') = \frac{1}{2}\left[\frac{\partial\Gamma^{ik}}{\partial x^j}(\mathbf{x},t;\mathbf{x}',t') + \frac{\partial\Gamma^{jk}}{\partial x^i}(\mathbf{x},t;\mathbf{x}',t')\right]\ , \tag{15}$$

and where δ^{ij} , $\delta(\mathbf{x})$, and $\delta(t)$ respectively represent Kronecker's symbol and Dirac's delta functions in space and time. It is usual to say that

$\Gamma^{ij}(\mathbf{x},t;\mathbf{x}',t')$ represents "the i-th component of the displacement at point $\mathbf{x}$ and time t corresponding to a unit impulse in the j-th coordinate direction at point $\mathbf{x}'$ and time t' (for homogeneous boundary and initial conditions)".

As the values of the elastic parameters of the medium are assumed not to depend on time, Green's function is "invariant by time translation" :

$$\Gamma^{ij}(\mathbf{x},t;\mathbf{x}',t') = \Gamma^{ij}(\mathbf{x},t-t';\mathbf{x}',0) . \tag{16}$$

The introduction of Green's function allows the following representation of the general solution of (14) (see for instance Aki and Richards, 1980):

$$u^i(\mathbf{x},t;\mathbf{x}_s) = \int_V dV(\mathbf{x}') \; \Gamma^{ij}(\mathbf{x},t;\mathbf{x}',0) * f^j(\mathbf{x}',t;\mathbf{x}_s)$$
$$+ \int_S dS(\mathbf{x}') \; \Gamma^{ij}(\mathbf{x},t;\mathbf{x}',0) * T^j(\mathbf{x}',t;\mathbf{x}_s) , \tag{17}$$

where " * " denotes time convolution, and V and S are the Earth's volume and surface respectively. The Green function introduced here is not arbitrary, and it is also not the Green function corresponding to an unbounded space, but the actual solution of equations (14). As Green's function has infinite bandwidth, it is not possible to evaluate it using standard numerical methods. Although its introduction is useful for analytical developments, it will not appear in the final computations, so that the problem of numerical evaluation will not arise.

Another important property of Green's function is *reciprocity* :

$$\Gamma^{ij}(\mathbf{x},t;\mathbf{x}',0) = \Gamma^{ji}(\mathbf{x}',t;\mathbf{x},0) . \tag{18}$$

This means that the response at point $\mathbf{x}$ in the i-th direction due to a source at point $\mathbf{x}'$ in the j-th direction is identical to the response at point $\mathbf{x}'$ in the j-th direction due to a source at point $\mathbf{x}$. For a theoretical demonstration of this property, see for instance Aki and Richards (1980).

The linearized forward problem.

Let $\rho(\mathbf{x})_n$, $\lambda(\mathbf{x})_n$, $\mu(\mathbf{x})_n$ represent an arbitrary ("current" or "unperturbed") medium and let $u^i(\mathbf{x},t;\mathbf{x}_s)_n$ represent the displacement field which propagates in this reference medium for given surface and volume sources. A model perturbation

$$\rho(\mathbf{x})_n \;\rightarrow\; \rho(\mathbf{x})_n + \delta\rho(\mathbf{x}) \tag{19a}$$

$$\lambda(\mathbf{x})_n \;\rightarrow\; \lambda(\mathbf{x})_n + \delta\lambda(\mathbf{x}) \tag{19b}$$

$$\mu(\mathbf{x})_n \;\rightarrow\; \mu(\mathbf{x})_n + \delta\mu(\mathbf{x}) \tag{19c}$$

will produce a perturbation of the displacement field

$$u^i(\mathbf{x},t;\mathbf{x}_s)_n \;\rightarrow\; u^i(\mathbf{x},t;\mathbf{x}_s)_n + \delta u^i(\mathbf{x},t;\mathbf{x}_s) \;. \tag{20}$$

I wish here to obtain the first order approximation to $\delta u^i(\mathbf{x},t;\mathbf{x}_s)$.

By definition, $u^i(\mathbf{x},t;\mathbf{x}_s)_n$ is the field propagating in the unperturbed medium. It then satifies

$$\rho(\mathbf{x})_n \; \ddot{u}^i(\mathbf{x},t;\mathbf{x}_s)_n - \frac{\partial}{\partial x^i}\left(\lambda(\mathbf{x})_n \; u^{kk}(\mathbf{x},t;\mathbf{x}_s)_n\right) - 2\,\frac{\partial}{\partial x^j}\left(\mu(\mathbf{x})_n \; u^{ij}(\mathbf{x},t;\mathbf{x}_s)_n\right) = f^i(\mathbf{x},t;\mathbf{x}_s) \tag{21a}$$

$$\lambda(\mathbf{x})_n \; u^{kk}(\mathbf{x},t;\mathbf{x}_s)_n \; n^i(\mathbf{x}) + 2\,\mu(\mathbf{x})_n \; u^{ij}(\mathbf{x},t;\mathbf{x}_s)_n \; n^j(\mathbf{x}) = T^i(\mathbf{x},t;\mathbf{x}_s) \qquad \mathbf{x}\in S \tag{21b}$$

$$u^i(\mathbf{x},0;\mathbf{x}_s)_n = 0 \tag{21c}$$

$$\dot{u}^i(\mathbf{x},0;\mathbf{x}_s)_n = 0 \;. \tag{21d}$$

Then, the field $u^i(\mathbf{x},t;\mathbf{x}_s)_n + \delta u^i(\mathbf{x},t;\mathbf{x}_s)$ verifies

$$(\rho_n+\delta\rho)(\mathbf{x})\;(\ddot{u}_n{}^i+\delta\ddot{u}^i)(\mathbf{x},t;\mathbf{x}_s) - \frac{\partial}{\partial x^i}\Big[(\lambda_n+\delta\lambda)(\mathbf{x})\;(u_n{}^{kk}+\delta u^{kk})(\mathbf{x},t;\mathbf{x}_s)\Big]$$
$$- 2\,\frac{\partial}{\partial x^j}\Big[(\mu_n+\delta\mu)(\mathbf{x})\;(u_n{}^{ij}+\delta u^{ij})(\mathbf{x},t;\mathbf{x}_s)\Big] = f^i(\mathbf{x},t;\mathbf{x}_s) \tag{22a}$$

$$(\lambda_n+\delta\lambda)(\mathbf{x})\;(u_n{}^{kk}+\delta u^{kk})(\mathbf{x},t;\mathbf{x}_s)\;n^i(\mathbf{x}) + 2\,(\mu_n+\delta\mu)(\mathbf{x})\;(u_n{}^{ij}+\delta u^{ij})(\mathbf{x},t;\mathbf{x}_s)\;n^j(\mathbf{x}) =$$
$$= T^i(\mathbf{x},t;\mathbf{x}_s) \qquad \mathbf{x}\in S \tag{22b}$$

$$(u_n{}^i+\delta u^i)(\mathbf{x},0;\mathbf{x}_s) = 0 \tag{22c}$$

$$(\dot{u}_n{}^i+\delta\dot{u}^i)(\mathbf{x},0;\mathbf{x}_s) = 0 \;. \tag{22d}$$

Using (21), equations (22) simplify to

$$\rho(\mathbf{x})_n\ \delta\ddot{u}^i(\mathbf{x},t;\mathbf{x}_s) - \frac{\partial}{\partial x^i}(\lambda(\mathbf{x})_n\ \delta u^{kk}(\mathbf{x},t;\mathbf{x}_s)) - 2\ \frac{\partial}{\partial x^j}(\mu(\mathbf{x})_n\ \delta u^{ij}(\mathbf{x},t;\mathbf{x}_s)) = \delta f^i(\mathbf{x},t;\mathbf{x}_s) \tag{23a}$$

$$\lambda(\mathbf{x})_n\ \delta u^{kk}(\mathbf{x},t;\mathbf{x}_s)\ n^i(\mathbf{x}) + 2\ \mu(\mathbf{x})_n\ \delta u^{ij}(\mathbf{x},t;\mathbf{x}_s)\ n^j(\mathbf{x}) = \delta T^i(\mathbf{x},t;\mathbf{x}_s) \qquad \mathbf{x} \in S \tag{23b}$$

$$\delta u^i(\mathbf{x},0;\mathbf{x}_s) = 0 \tag{23c}$$

$$\delta\dot{u}^i(\mathbf{x},0;\mathbf{x}_s) = 0\ , \tag{23d}$$

where

$$\delta f^i(\mathbf{x},t;\mathbf{x}_s) = -\ \ddot{u}^i(\mathbf{x},t;\mathbf{x}_s)_n\ \delta\rho(\mathbf{x}) + \frac{\partial}{\partial x^i}(\ u^{kk}(\mathbf{x},t;\mathbf{x}_s)_n\ \delta\lambda(\mathbf{x})\ ) + 2\ \frac{\partial}{\partial x^j}(\ u^{ij}(\mathbf{x},t;\mathbf{x}_s)_n\ \delta\mu(\mathbf{x})\ ) - \delta\ddot{u}^i(\mathbf{x},t;\mathbf{x}_s)\ \delta\rho(\mathbf{x}) + \frac{\partial}{\partial x^i}(\delta u^{kk}(\mathbf{x},t;\mathbf{x}_s)\ \delta\lambda(\mathbf{x})\ ) + 2\ \frac{\partial}{\partial x^j}(\ \delta u^{ij}(\mathbf{x},t;\mathbf{x}_s)\ \delta\mu(\mathbf{x})\ ) \tag{24a}$$

$$\delta T^i(\mathbf{x},t;\mathbf{x}_s) = -\ u^{kk}(\mathbf{x},t;\mathbf{x}_s)_n\ \delta\lambda(\mathbf{x})\ n^i(\mathbf{x}) - 2\ u^{ij}(\mathbf{x},t;\mathbf{x}_s)_n\ \delta\mu(\mathbf{x})\ n^j(\mathbf{x}) - \delta u^{kk}(\mathbf{x},t;\mathbf{x}_s)\ \delta\lambda(\mathbf{x})\ n^i(\mathbf{x}) - 2\ \delta u^{ij}(\mathbf{x},t;\mathbf{x}_s)\ \delta\mu(\mathbf{x})\ n^j(\mathbf{x})\ . \tag{24b}$$

As I am seeking the first-order approximation to $\delta u^i(\mathbf{x},t;\mathbf{x}_s)$, I can drop second-order terms in (24). Then, up to the first order,

$$\delta f^i(\mathbf{x},t;\mathbf{x}_s) = -\ \ddot{u}^i(\mathbf{x},t;\mathbf{x}_s)_n\ \delta\rho(\mathbf{x}) + \frac{\partial}{\partial x^i}(\ u^{kk}(\mathbf{x},t;\mathbf{x}_s)_n\ \delta\lambda(\mathbf{x})\ ) + 2\ \frac{\partial}{\partial x^j}(\ u^{ij}(\mathbf{x},t;\mathbf{x}_s)_n\ \delta\mu(\mathbf{x})\ ) \tag{25a}$$

$$\delta T^i(\mathbf{x},t;\mathbf{x}_s) = -\ u^{kk}(\mathbf{x},t;\mathbf{x}_s)_n\ \delta\lambda(\mathbf{x})\ n^i(\mathbf{x}) - 2\ u^{ij}(\mathbf{x},t;\mathbf{x}_s)_n\ \delta\mu(\mathbf{x})\ n^j(\mathbf{x})\ . \tag{25b}$$

Equations (23) and (25) are interpreted as follows: up to the first order, the perturbation $\delta u^i(\mathbf{x},t;\mathbf{x}_s)$ of the displacement field due to perturbations $\delta\rho(\mathbf{x})$, $\delta\lambda(\mathbf{x})$, and $\delta\mu(\mathbf{x})$ can be interpreted as the field propagating in the medium $\rho(\mathbf{x})_n$, $\lambda(\mathbf{x})_n$, and $\mu(\mathbf{x})_n$ (equations (23)) and created by the "secondary sources" (25).

Using Green's function $\Gamma^{ij}(\mathbf{x},t;\mathbf{x}',0)_n$ corresponding to the reference medium $\rho(\mathbf{x})_n$, $\lambda(\mathbf{x})_n$, and $\mu(\mathbf{x})_n$, the solution of (23) at the receiver locations is given by

$$\delta u^i(\mathbf{x}_r,t;\mathbf{x}_s) = \int_{\mathbf{V}} dV(\mathbf{x}) \ \Gamma^{ij}(\mathbf{x}_r,t;\mathbf{x},0)_n * \delta f^j(\mathbf{x},t;\mathbf{x}_s)$$

$$+ \int_{\mathbf{S}} dS(\mathbf{x}) \ \Gamma^{ij}(\mathbf{x}_r,t;\mathbf{x},0)_n * \delta T^j(\mathbf{x},t;\mathbf{x}_s) \ , \tag{26}$$

i.e.,

$$\delta u^i(\mathbf{x}_r,t;\mathbf{x}_s) = - \int_{\mathbf{V}} dV(\mathbf{x}) \ \Gamma^{ij}(\mathbf{x}_r,t;\mathbf{x},0)_n * \ddot{u}^j(\mathbf{x}',t;\mathbf{x}_s)_n \ \delta\rho(\mathbf{x})$$

$$+ \int_{\mathbf{V}} dV(\mathbf{x}) \ \Gamma^{ij}(\mathbf{x}_r,t;\mathbf{x},0)_n * \frac{\partial}{\partial x^j} \left(u^{kk}(\mathbf{x},t;\mathbf{x}_s)_n \ \delta\lambda(\mathbf{x}) \right)$$

$$+ 2 \int_{\mathbf{V}} dV(\mathbf{x}) \ \Gamma^{ij}(\mathbf{x}_r,t;\mathbf{x},0)_n * \frac{\partial}{\partial x^k} \left(u^{jk}(\mathbf{x},t;\mathbf{x}_s)_n \ \delta\mu(\mathbf{x}) \right)$$

$$- \int_{\mathbf{S}} dS(\mathbf{x}) \ \Gamma^{ij}(\mathbf{x}_r,t;\mathbf{x},0)_n * u^{kk}(\mathbf{x},t;\mathbf{x}_s)_r \ n^j(\mathbf{x}) \ \delta\lambda(\mathbf{x})$$

$$- 2 \int_{\mathbf{S}} dS(\mathbf{x}) \ \Gamma^{ij}(\mathbf{x}_r,t;\mathbf{x},0)_n * u^{jk}(\mathbf{x},t;\mathbf{x}_s)_n \ n^k(\mathbf{x}) \ \delta\mu(\mathbf{x}) \ . \tag{27}$$

Using

$$\Gamma^{ij}(\mathbf{x},t;\mathbf{x}',0) * \frac{\partial}{\partial x'^j} \left(u^{kk}(\mathbf{x}',t;\mathbf{x}_s) \ \delta\lambda(\mathbf{x}') \right)$$

$$= - \frac{\partial \Gamma^{ij}}{\partial x'^j}(\mathbf{x},t;\mathbf{x}',0) * u^{kk}(\mathbf{x}',t;\mathbf{x}_s) \ \delta\lambda(\mathbf{x}')$$

$$+ \frac{\partial}{\partial x'^j} \left(\Gamma^{ij}(\mathbf{x},t;\mathbf{x}',0) * u^{kk}(\mathbf{x}',t;\mathbf{x}_s) \ \delta\lambda(\mathbf{x}') \right) \ , \tag{28}$$

$$\Gamma^{ij}(\mathbf{x},t;\mathbf{x}',0) * \frac{\partial}{\partial x'^k} \left(u^{jk}(\mathbf{x}',t;\mathbf{x}_s) \ \delta\mu(\mathbf{x}') \right)$$

$$= - \frac{\partial \Gamma^{ij}}{\partial x'^k}(\mathbf{x},t;\mathbf{x}',0) * u^{jk}(\mathbf{x}',t;\mathbf{x}_s) \ \delta\mu(\mathbf{x}')$$

$$+ \frac{\partial}{\partial x'^k} \left(\Gamma^{ij}(\mathbf{x},t;\mathbf{x}',0) * u^{jk}(\mathbf{x}',t;\mathbf{x}_s) \, \delta\mu(\mathbf{x}') \right) , \tag{29}$$

and Green's theorem

$$\int_{\mathbf{V}} dV(\mathbf{x}) \; \frac{\partial Q}{\partial x^k}(\mathbf{x}) \;=\; \int_{\mathbf{S}} dS(\mathbf{x}) \; n^k(\mathbf{x}) \, Q(\mathbf{x}) , \tag{30}$$

equations (27) simplify to

$$\begin{aligned} \delta u^i(\mathbf{x}_r,t;\mathbf{x}_s) \;=\; & - \int_{\mathbf{V}} dV(\mathbf{x}) \; \Gamma^{ij}(\mathbf{x}_r,t;\mathbf{x},0)_n * \ddot{u}^j(\mathbf{x},t;\mathbf{x}_s)_n \;\; \delta\rho(\mathbf{x}) \\ & - \int_{\mathbf{V}} dV(\mathbf{x}) \; \frac{\partial \Gamma^{ij}}{\partial x'^j}(\mathbf{x}_r,t;\mathbf{x},0)_n * u^{mm}(\mathbf{x},t;\mathbf{x}_s)_n \;\; \delta\lambda(\mathbf{x}) \\ & - 2 \int_{\mathbf{V}} dV(\mathbf{x}) \; \frac{\partial \Gamma^{ip}}{\partial x'^m}(\mathbf{x}_r,t;\mathbf{x},0)_n * u^{pm}(\mathbf{x},t;\mathbf{x}_s)_n \;\; \delta\mu(\mathbf{x}) . \end{aligned} \tag{31}$$

If the perturbations $\delta\rho(\mathbf{x})$, $\delta\lambda(\mathbf{x})$, $\delta\mu(\mathbf{x})$ are sufficiently small (in a sense to be discussed below), this first order approximation can be used for computing the displacement field in the medium defined by ($\rho(\mathbf{x})_n+\delta\rho(\mathbf{x})$, $\lambda(\mathbf{x})_n+\delta\lambda(\mathbf{x})$, $\mu(\mathbf{x})_n+\delta\mu(\mathbf{x})$) . It is called the (first) *Born approximation.* I will *not* use such an approximation. But we need Fréchet's derivative of the displacement field with respect to the model parameters, and, clearly, these Fréchet derivatives are easily obtained from the first-order development (31).

Instead of parametring the Earth using Lamé's parameters $\lambda(\mathbf{x})$ and $\mu(\mathbf{x})$, I can use the P-wave impedance $IP(\mathbf{x})$ and the S-wave impedances $IS(\mathbf{x})$ (see above). I then have

$$\lambda(\mathbf{x}) \;=\; \frac{1}{\rho(\mathbf{x})} \left(IP^2(\mathbf{x}) - 2\, IS^2(\mathbf{x}) \right) , \tag{32a}$$

$$\mu(\mathbf{x}) \;=\; \frac{1}{\rho(\mathbf{x})} \; IS^2(\mathbf{x}) , \tag{32b}$$

which gives

$$\delta\lambda(\mathbf{x}) \;=\; - \left(\alpha^2(\mathbf{x}) - 2\,\beta^2(\mathbf{x})\right) \delta\rho(\mathbf{x}) + 2\,\alpha(\mathbf{x})\,\delta IP(\mathbf{x}) - 4\,\beta(\mathbf{x})\,\delta IS(\mathbf{x}) \tag{33a}$$

$$\delta\mu(\mathbf{x}) \;=\; - \beta^2(\mathbf{x})\,\delta\rho(\mathbf{x}) \;+\; 2\,\beta(\mathbf{x})\,\delta IS(\mathbf{x}) . \tag{33b}$$

Equation (31) then becomes

$$
\begin{aligned}
\delta u^i(\mathbf{x}_r,t;\mathbf{x}_s) = & - \int_V dV(x) \Bigg\{ \Gamma^{ij}(\mathbf{x}_r,t;\mathbf{x},0)_n * \ddot{u}^j(\mathbf{x},t;\mathbf{x}_s)_n \\
& - (\alpha^2(x)_n - 2\,\beta^2(x)_n) \frac{\partial \Gamma^{ij}}{\partial x^j}(\mathbf{x}_r,t;\mathbf{x},0)_n * u^{mm}(\mathbf{x},t;\mathbf{x}_s)_n \\
& - 2\,\beta^2(x)_n \frac{\partial \Gamma^{ij}}{\partial x^m}(\mathbf{x}_r,t;\mathbf{x},0)_n * u^{jm}(\mathbf{x},t;\mathbf{x}_s)_n \Bigg\} \delta\rho(\mathbf{x}) \\
& - 2 \int_V dV(\mathbf{x})\, \alpha(\mathbf{x})_n \frac{\partial \Gamma^{ij}}{\partial x^j}(\mathbf{x}_r,t;\mathbf{x},0) * u^{mm}(\mathbf{x},t;\mathbf{x}_s)_n\ \delta IP(\mathbf{x}) \qquad (34) \\
& - 4 \int_V dV(\mathbf{x})\, \beta(\mathbf{x})_n \Bigg\{ \frac{\partial \Gamma^{ij}}{\partial x^m}(\mathbf{x}_r,t;\mathbf{x},0)_n * u^{jm}(\mathbf{x},t;\mathbf{x}_s)_n \\
& - \frac{\partial \Gamma^{ij}}{\partial x^j}(\mathbf{x}_r,t;\mathbf{x},0)_n * u^{mm}(\mathbf{x},t;\mathbf{x}_s)_n \Bigg\} \delta IS(\mathbf{x}) .
\end{aligned}
$$

The validity of Born's approximation

As previously stated, Born's approximation consists in using the first-order approximation (34) for estimating the displacement field $u_0{}^i(x,t;\mathbf{x}_s) + \delta u^i(\mathbf{x},t;\mathbf{x}_s)$ corresponding to the medium $(\rho_0(\mathbf{x})+\delta\rho(x) , IP_0(\mathbf{x})+\delta IP(\mathbf{x}) , IS_0(x)+\delta IS(x))$. Although it is possible to obtain *rigorous* conditions for the validity of such an approximation (see, for instance, Hudson and Heritage, 1981), it is not so easy to obtain *useful* conditions. Common physical sense suggests that a necessary condition for Born's approximation to be adequate is that *travel times* in the perturbed medium are adequately modeled by the travel times in the unperturbed medium, i.e., that the unperturbed medium *contains the low spatial frequency part of the P-wave and S-wave velocities.*

The Fréchet derivatives of the displacements

Using equation (34), we see that the *Fréchet derivative* (at the point $\rho(\mathbf{x})_n$, $IP(\mathbf{x})_n$, $IS(\mathbf{x})_n$) *of the displacements* $u^i(\mathbf{x},t;\mathbf{x}_s)$ with respect to the P-wave impedance $IP(x)$ is the linear operator that associates an arbitrary perturbation $\delta IP(\mathbf{x})$ with the displacement perturbation corresponding to the first order-development

$$\delta u^i(\mathbf{x}_r,t;\mathbf{x}_s) = -2 \int_{\mathbf{V}} dV(\mathbf{x})\ \alpha(\mathbf{x})_n\ \frac{\partial \Gamma^{ij}}{\partial x^j}(\mathbf{x}_r,t;\mathbf{x},0)_n * u^{mm}(\mathbf{x},t;\mathbf{x}_s)_n\ \delta IP(\mathbf{x}) \tag{35a}$$

Introducing the kernel $A^i(\mathbf{x}_r,t,\mathbf{x})_n$ of this linear operator by

$$\delta u^i(\mathbf{x}_r,t;\mathbf{x}_s) = \int_{\mathbf{V}} dV(\mathbf{x})\ A^i(\mathbf{x}_r,t,\mathbf{x})_n\ \delta IP(\mathbf{x}') \tag{35b}$$

gives

$$A^i(\mathbf{x}_r,t,\mathbf{x})_n = -2\ \alpha(\mathbf{x})_n\ \frac{\partial \Gamma^{ij}}{\partial x^j}(\mathbf{x}_r,t;\mathbf{x},0)_n * u^{mm}(\mathbf{x},t;\mathbf{x}_{s_n})\ . \tag{36}$$

Similarly, introducing the kernels $B^i(\mathbf{x}_r,t,\mathbf{x})_n$ and $C^i(\mathbf{x}_r,t,x)_n$ corresponding respectively to the Fréchet derivatives of the displacements with respect to the S-wave impedance and to the density,

$$\delta u^i(\mathbf{x}_r,t;\mathbf{x}_s) = \int_{\mathbf{V}} dV(\mathbf{x})\ B^i(\mathbf{x}_r,t,x)_n\ \delta IS(\mathbf{x}) \tag{37}$$

$$\delta u^i(\mathbf{x}_r,t;\mathbf{x}_s) = \int_{\mathbf{V}} dV(\mathbf{x})\ C^i(\mathbf{x}_r,t,\mathbf{x})_n\ \delta\rho(\mathbf{x})\ , \tag{37}$$

gives respectively

$$B^i(x_r,t,\mathbf{x})_n = -4\ \beta(x)_n \left\{ \frac{\partial \Gamma^{ij}}{\partial x^m}(\mathbf{x}_r,t;\mathbf{x},0)_n * u^{jm}(\mathbf{x},t;\mathbf{x}_s)_n - \frac{\partial \Gamma^{ij}}{\partial x^j}(\mathbf{x}_r,t;\mathbf{x},0)_n * u^{mm}(\mathbf{x},t;\mathbf{x}_s)_n \right\} \tag{38}$$

and

$$\begin{aligned} C^i(\mathbf{x}_r,t,\mathbf{x})_n = & - \Gamma^{ij}(\mathbf{x}_r,t;\mathbf{x},0)_n * \ddot{u}^j(\mathbf{x},t;\mathbf{x}_s)_n \\ & + (\alpha^2(\mathbf{x})_n - 2\beta^2(\mathbf{x})_n)\ \frac{\partial \Gamma^{ij}}{\partial x^j}(\mathbf{x}_r,t;\mathbf{x},0)_n * u^{mm}(\mathbf{x},t;\mathbf{x}_s)_n \\ & + 2\ \beta^2(\mathbf{x})_n\ \frac{\partial \Gamma^{ij}}{\partial x^m}(\mathbf{x}_r,t;\mathbf{x},0)_n * u^{jm}(\mathbf{x},t;\mathbf{x}_s)_n\ . \end{aligned} \tag{39}$$

Transpose operators

a) P-wave impedance. The Fréchet derivative (at the point ρ_n , $\mathbf{IP}_n$, $\mathbf{IS}_n$) of displacements with respect to the P-wave impedance is the linear operator $\mathbf{A}_n$ that associates an arbitrary perturbation $\delta\mathbf{IP}$ with the displacement

$$\delta\mathbf{u} = \mathbf{A}_n \ \delta\mathbf{IP} \ , \tag{40}$$

i.e.

$$\delta u^i(\mathbf{x}_r,t;\mathbf{x}_s) = \int_{\mathbf{V}} dV(\mathbf{x}) \ A^i(\mathbf{x}_{r,t},\mathbf{x})_n \ \delta IP(\mathbf{x}) \ , \qquad \text{(35 again)}$$

where $A^i(\mathbf{x}_r,t,\mathbf{x})_n$ is given by equation (36). By definition of the transpose of an operator (see text, the operator $\mathbf{A}_n{}^t$ associates any $\delta\hat{u}^i(\mathbf{x}_r,t;\mathbf{x}_s)$ with $\delta\hat{IP}(\mathbf{x})$ given by

$$\delta\hat{\mathbf{IP}} = \mathbf{A}_n{}^t \ \delta\hat{\mathbf{u}} \ , \tag{41}$$

i.e.,

$$\delta\hat{IP}(\mathbf{x}) = \sum_r \int_0^T dt \ A^i(\mathbf{x}_r,t,\mathbf{x})_n \ \delta\hat{u}^i(\mathbf{x}_r,t;\mathbf{x}_s) \tag{42}$$

(remember that implicit sum is assumed over repeated indexes). This gives

$$\delta\hat{IP}(\mathbf{x}) = -\,2\ \alpha(\mathbf{x})_n \sum_r \int_0^T dt \ \frac{\partial\Gamma^{ij}}{\partial x^j}(\mathbf{x}_r,t;\mathbf{x},0)_n * u^{mm}(\mathbf{x},t;\mathbf{x}_s)_n \ \delta\hat{u}^i(\mathbf{x}_r,t;\mathbf{x}_s). \tag{43}$$

Defining

$$\delta\Psi^i(\mathbf{x},t;\mathbf{x}_s) = \sum_r \Gamma^{ij}(\mathbf{x},0;\mathbf{x}_r,t;\mathbf{x}_s)_n * \delta\hat{u}^j(\mathbf{x}_r,t;\mathbf{x}_s) \tag{44}$$

and

$$\delta\Psi^{ij}(\mathbf{x},t;\mathbf{x}_s) = \frac{1}{2}\left[\frac{\partial\delta\Psi^i}{\partial x^j}(\mathbf{x},t;\mathbf{x}_s) + \frac{\partial\delta\Psi^j}{\partial x^i}(\mathbf{x},t;\mathbf{x}_s) \right] , \tag{45}$$

and using Identity 1 of problem 7.4 this gives

$$\delta\hat{IP}(\mathbf{x}) = -\,2\ \alpha(\mathbf{x})_n \int_0^T dt \ u^{mm}(\mathbf{x},t;\mathbf{x}_s)_n \ \delta\Psi^{kk}(\mathbf{x},t;\mathbf{x}_s) \ . \tag{46}$$

Given the vector $\delta\hat{\mathbf{u}}$, to compute $\mathbf{A}_n{}^t \ \delta\hat{\mathbf{u}}$, we can then use (46), where $\delta\mathbf{\Psi}$ is defined by (44). But equation (44) should not be used as it stands. I will now show that the field $\delta\mathbf{\Psi}$ satisfies the equations

$$\rho(\mathbf{x})_n \ \frac{\partial^2 \delta\Psi^i}{\partial t^2}(\mathbf{x},t;\mathbf{x}_s) - \frac{\partial}{\partial x^i}(\lambda(\mathbf{x})_n \ \delta\Psi^{kk}(\mathbf{x},t;\mathbf{x}_s))$$

$$- 2 \ \frac{\partial}{\partial x^j}(\mu(\mathbf{x})_n \ \delta\Psi^{ij}(\mathbf{x},t;\mathbf{x}_s)) = 0 \ , \qquad (47a)$$

$$\lambda(\mathbf{x})_n \ \delta\Psi^{kk}(\mathbf{x},t;\mathbf{x}_s) \ n^i(\mathbf{x}) + 2 \ \mu(\mathbf{x})_n \ \delta\Psi^{ij}(\mathbf{x},t;\mathbf{x}_s) \ n^j(\mathbf{x}) =$$

$$= \sum_r \delta(\mathbf{x}-\mathbf{x}_r) \ \delta\hat{u}^i(\mathbf{x}_r,t;\mathbf{x}_s) \qquad \mathbf{x} \in \mathbf{S} \ , \qquad (47b)$$

$$\delta\Psi^i(\mathbf{x},T;\mathbf{x}_s) = 0 \ , \qquad (47c)$$

$$\delta\dot{\Psi}^i(\mathbf{x},T;\mathbf{x}_s) = 0 \ , \qquad (47d)$$

where it should be noticed that $\delta\Psi^i(\mathbf{x},t;\mathbf{x}_s)$ satisfies homogeneous *final* conditions (47c-d), instead of initial conditions. The "source" of the field $\delta\Psi^i(\mathbf{x},t,\mathbf{x}_s)$ is $\delta\hat{u}^i(\mathbf{x}_r,t;\mathbf{x}_s)$, acting as if it were a traction (47b). Using the representation theorem (17), with reversed time, shows directly that $\delta\Psi^i(\mathbf{x},t;\mathbf{x}_s)$ satisfies equation (44).

The field $\delta\Psi^i(\mathbf{x},t;\mathbf{x}_s)$ can, for instance, be obtained numerically using a finite-difference code, with the time running backwards from $t = T$ to $t = 0$, and where, for a given shot point $\mathbf{x}_s$, we consider virtual sources, one at each receiver, radiating the weighted residuals backwards in time. See Gauthier et al. (1987) for a numerical implementation in an acoustic example.

b) S-wave impedance. Analogously, the Fréchet derivative of displacements with respect to the S-wave impedance is the linear operator $\mathbf{B}$ that associates an arbitrary perturbation δIS with the displacement

$$\delta\mathbf{u} = \mathbf{B} \ \delta IS \ , \qquad (48)$$

i.e.,

$$\delta u^i(\mathbf{x}_r,t;\mathbf{x}_s) = \int_{\mathbf{V}} dV(\mathbf{x}) \ B_0{}^i(\mathbf{x}_r,t,\mathbf{x}) \ \delta IS(\mathbf{x}) \ , \qquad (37 \text{ again})$$

where $B^i(\mathbf{x}_r,t,\mathbf{x})_n$ is given by equation (38). To any $\delta\hat{u}$ the operator $\mathbf{B}_n{}^t$ associates $\delta\hat{\mathbf{IS}}$ given by

$$\delta\hat{\mathbf{IS}} = \mathbf{B}^t\ \delta\hat{\mathbf{u}}\ , \tag{49}$$

i.e.,

$$\delta\hat{IS}(\mathbf{x}) = \sum_r \int_0^T dt\ B^i(\mathbf{x}_r,t,\mathbf{x})_n\ \ \delta\hat{u}^i(\mathbf{x}_r,t;\mathbf{x}_s)\ . \tag{50}$$

This gives

$$\delta\hat{IS}(\mathbf{x}) = -\,4\ \ \beta(\mathbf{x})_n \sum_r \int_0^T dt \left\{ \frac{\partial\Gamma^{ij}}{\partial x^m}(\mathbf{x}_{r,t;\mathbf{x},0})_n\ *\ u^{jm}(\mathbf{x},t;\mathbf{x}_s)_n \right.$$

$$\left. -\ \frac{\partial\Gamma^{ij}}{\partial x^j}(\mathbf{x}_r,t;\mathbf{x},0)_n\ *\ u_0{}^{mm}(\mathbf{x},t;\mathbf{x}_s)_n \right\}\ \delta\hat{u}^i(\mathbf{x}_r,t;\mathbf{x}_s)\ . \tag{51}$$

Using identities 1 and 2 of problem 7.4, this gives

$$\delta\hat{IS}(\mathbf{x}) = -4\ \beta(\mathbf{x})_n \int_0^T dt \left(u^{km}(\mathbf{x},t;\mathbf{x}_s)_n\ \ \delta\Psi^{km}(\mathbf{x},t;\mathbf{x}_s)\ -\ u^{mm}(\mathbf{x})_n\ \ \delta\Psi^{kk}(\mathbf{x},t;\mathbf{x}_s) \right), \tag{52}$$

where the field $\delta\boldsymbol{\Psi}$ has been defined by equations (46).

c) Density. Finally, the Fréchet derivative of displacements with respect to the density is the linear operator $\mathbf{C}$ that associates an arbitrary perturbation $\delta\rho$ with the displacement

$$\delta\mathbf{u} = \mathbf{C}\ \delta\rho\ , \tag{53}$$

i.e.,

$$\delta u^i(\mathbf{x}_r,t;\mathbf{x}_s) = \int_{\mathbf{V}} dV(\mathbf{x})\ \ C_0{}^i(\mathbf{x}_r,t,\mathbf{x})\ \ \delta\rho(\mathbf{x})\ , \tag{37 again}$$

where $C_0{}^i(\mathbf{x}_r,t,\mathbf{x})$ is given by equation (39). The operator $\mathbf{C}_n{}^t$ associates $\delta\hat{\mathbf{u}}$ with $\delta\hat{\rho}$ given by

$$\delta\hat{\rho} = \mathbf{C}^t\ \delta\hat{u}\ , \tag{54}$$

i.e.,

$$\delta\hat{\rho}(\mathbf{x}) \;=\; \sum_{r} \int_0^T dt \;\; C^i(\mathbf{x}_r,t,\mathbf{x})_n \;\; \delta\hat{u}^i(\mathbf{x}_r,t;\mathbf{x}_s) \; . \tag{55}$$

This gives

$$\delta\hat{\rho}(\mathbf{x}) \;=\; \sum_{r} \int_0^T dt \; \Bigg\{ - \;\; \Gamma^{ij}(\mathbf{x}_r,t;\mathbf{x},0)_n \; * \; \ddot{u}^j(\mathbf{x},t;\mathbf{x}_s)_n$$

$$+ \; (\; \alpha^2(\mathbf{x})_n \; - \; 2 \; \beta^2(\mathbf{x})_n \;) \; \frac{\partial\Gamma^{ij}}{\partial x^j}(\mathbf{x}_r,t;\mathbf{x},0)_n \; * \; u^{mm}(\mathbf{x},t;\mathbf{x}_s)_n$$

$$+ \; 2 \; \beta^2(\mathbf{x})_n \;\; \frac{\partial\Gamma^{ij}}{\partial x^m}(\mathbf{x}_r,t;\mathbf{x},0)_n \; * \; u^{jm}(\mathbf{x},t;\mathbf{x}_s)_n \Bigg\} \; \delta\hat{u}^i(\mathbf{x}_r,t;\mathbf{x}_s) \; . \tag{56}$$

Using identities 1, 2, and 3 of problem 7.4, this gives

$$\delta\hat{\rho}(\mathbf{x}) \;=\; \int_0^T dt \; \Bigg\{ \dot{u}^i(\mathbf{x},t;\mathbf{x}_s)_n \;\; \delta\dot{\Psi}^i(\mathbf{x})$$

$$+ \;\; (\; \alpha^2(\mathbf{x})_n \; - \; 2 \; \beta^2(\mathbf{x})_n \;) \; u^{mm}(\mathbf{x},t;\mathbf{x}_s)_n \;\; \delta\Psi^{kk}(\mathbf{x},t;\mathbf{x}_s)$$

$$+ \;\; 2 \; \beta^2(\mathbf{x})_n \;\; u^{km}(\mathbf{x},t;\mathbf{x}_s)_n \;\; \delta\Psi^{km}(\mathbf{x},t;\mathbf{x}_s) \Bigg\} \; , \tag{57}$$

where the field $\delta\boldsymbol{\Psi}$ has been defined by equations (46).

Methods of resolution

We have seen throughout this book that there exist two classes of methods for minimizing a functional.

The first contains the methods implying extensive exploration of the model space. The exploration can be systematic or random (Monte-Carlo), as explained in Chapter 3. As we seek for high-resolution Earth models, we typically need, say, millions or billions of parameters. Thus, our problem has too many degrees of freedom for these methods to be useful.

The second class contains iterative methods which, using the local properties (at a given point of the model space) of the functional to be minimized, define a "descent direction" along which a new and better point in the model space will be obtained. The more efficient methods are gradient methods (in the wide sense).

My personal experience with the present problem suggests the following strategy. First, as for all nonlinear problems, it is important to start iterating at a point as close as possible to the final solution, so that the nonlinearity of the problem may be as small as possible. In the present context, this means starting from a model for which the data residuals can be explained as well as possible using Born's approximation, i.e., using a model for which the low spatial frequencies of the P-wave and S-wave velocities are as good as possible.

The three parameters $IP(\mathbf{x})$, $IS(z)$, and $\rho(\mathbf{x})$ have been chosen to be as independent as possible. Furthermore, the importance of these parameters is very different. Most of the data features can be explained with P-waves alone. This suggests starting iterating using a gradient method for the P-wave impedance alone (i.e., maintaining fixed S-wave impedance and density). This requires a reasonably good model of the low spatial frequency part of the P velocity.

Once a good model $IP(\mathbf{x})$ has been obtained, the remaining data residuals will, in particular, contain S waves. If a reasonably good model for the long spatial wavelengths of the S velocity can then be obtained, some gradient iterations should be performed to obtain a good model of S-wave impedance. If the remaining residuals contain any information, it will concern the density. Some gradient iterations for the density will end the process.

As the problem is nonlinear, the entire process should, in principle, be iterated until convergence. As the chosen parameters are acceptably independent, I hope that the first model obtained after a single loop will be good enough (if the long spatial wavelengths of the P-wave and S-wave velocity in the starting model are right).

The starting models of the low spatial frequency part of the P velocity and S velocity have to be obtained using an independent method (not discussed here). It is not clear at present to what extent the gradient iterations for $IP(\mathbf{x})$, $IS(\mathbf{x})$, and $\rho(\mathbf{x})$ will be able to correct for the imperfections of these velocity models. I am not very optimistic about this, because preliminary results on nonlinear inversion for one-dimensional models suggest that the number of iterations needed to modify the low spatial frequencies of the model using gradient methods may be enormous.

In what follows, it is assumed that an Earth model ($\rho(\mathbf{x})_n$, $IP(\mathbf{x})_n$, $IS(\mathbf{x})_n$) that contains the long spatial wavelength component of the P-wave and S-wave velocities is given, and I will discuss how to ameliorate this model, i.e., how to obtain a model with lower value of the functional (3). From the previous discussion, it follows that I can discuss separately the problem of ameliorating the P-wave impedance model, the S-wave impedance model, and the density model.

Optimization of the P-wave impedance

Denote by $\rho(\mathbf{x})_n$, $IP(\mathbf{x})_n$, and $IS(\mathbf{x})_n$ the model already obtained, which we wish to optimize further for the impedance $IP(\mathbf{x})$. I will use a gradient iterative method which will give models $IP(\mathbf{x})_{n+1}$, $IP(\mathbf{x})_{n+2}$, ...

Using the algorithm (4.93) in the text, and the results already obtained for the transpose operators, we obtain the equations corresponding to an iteration of the steepest descent method for the P-wave impedance:

$$\delta\hat{u}^i(\mathbf{x}_r,t;\mathbf{x}_s)_n = \frac{u^i(\mathbf{x}_r,t;\mathbf{x}_s)_n - u^i(\mathbf{x}_r,t;\mathbf{x}_s)_{obs}}{\sigma^2(\mathbf{x}_r,t;\mathbf{x}_s)} , \tag{58a}$$

$$\delta\Psi^i(\mathbf{x},t;\mathbf{x}_s)_n = \sum_r \Gamma^{ij}(\mathbf{x},0;\mathbf{x}_r,t;\mathbf{x}_s)_n * \delta\hat{u}^j(\mathbf{x}_r,t;\mathbf{x}_s)_n , \tag{58b}$$

$$\delta\hat{IP}(\mathbf{x})_n = -\,2\,\alpha(\mathbf{x})_n \sum_s \int_0^T dt\ u^{ii}(\mathbf{x},t;\mathbf{x}_s)_n\ \delta\Psi^{jj}(\mathbf{x},t;\mathbf{x}_s)_n , \tag{58c}$$

$$\delta IP(x,y,z)_n = K_p \int_0^{Z_{max}} dz'\ Min(z,z')\ \delta\hat{IP}(x,y,z')_n , \tag{58d}$$

$$\Delta IP(\mathbf{x})_n = \mathbf{Precond}(\ \delta IP(\mathbf{x})_n\ ) , \tag{58e}$$

$$IP(\mathbf{x})_{n+1} = IP(\mathbf{x})_n - a_n\ (\ \Delta IP(\mathbf{x})_n + IP(\mathbf{x})_n - IP(\mathbf{x})_{prior}\ ) , \tag{58f}$$

where a_n is the real constant which renders $S(\ IP_{n+1}\ ,\ IS_n\ ,\ \rho_n\ )$ minimum (and which can be estimated analytically (see Chapter 4, or problem 7.1) or simply obtained by trial and error).

Let me now turn to the physical interpretation of this result.

Equation (58a): $u^i(\mathbf{x}_r,t;\mathbf{x}_s)_n$ are the data predicted for the model $\mathbf{IP}_n$, $\mathbf{IS}_n$, ρ_n . Its effective computation requires a numerical resolution of the system (12). $s^2(\mathbf{x}_r,t;\mathbf{x}_s)$ represents the (squared) estimated error at time t in the i-th component of the dispacement measured at $\mathbf{x}_r$, for the source at $\mathbf{x}_s$. Then $\delta\hat{u}^i(\mathbf{x}_r,t;\mathbf{x}_s)$ clearly represent the weighted residuals.

Equation (58b): $\Gamma^{ij}(\mathbf{x},t;\mathbf{x}_r,0)_n$ is Green's function for the model $\mathbf{IP}_n$, $\mathbf{IS}_n$, ρ_n . We have already seen that equation (59a) should not be used as it stands. Instead, the field $\delta\boldsymbol{\Psi}_n$ has to be obtained solving numerically the system (47) (backwards in time). As $\delta\Psi^i(\mathbf{x},t;\mathbf{x}_s)_n$ is obtained when propagating the wheighted residuals $\delta\hat{u}^i(\mathbf{x}_r,t;\mathbf{x}_s)$ *backwards in time*, it can be intuitively interpreted as a "current missing field".

Equation (58c): This is the most important of the equations, because here inversion is being performed. After some corrections (58d-e-f), $\delta\hat{IP}(\mathbf{x})_n$ will essentially be the correction to be applied to $\mathbf{IP}_n$ to obtain $\mathbf{IP}_{n+1}$ (as shown by (58f)). Equation (58c) shows that this correction at a given point $\mathbf{x}$, for given shot $\mathbf{x}_s$, equals the time correlation of the dilatation $u^{ii}(\mathbf{x},t;\mathbf{x}_s)_n$ of the current predicted field with the dilatation $\delta\Psi^{jj}(\mathbf{x},t;\mathbf{x}_s)_n$ of the current missing field. The physical interpretation is as follows: if for a given source point $\mathbf{x}_s$, and at a given point $\mathbf{x}$, the dilatation of the current predicted field is time correlated with the dilatation of the missing field, we should create this missing field by adding a P impedance diffractor at point $\mathbf{x}$. This interpretation is strikingly similar to the imaging principle of Claerbout (1971), but is here in an elastic context and resulting from a very general optimization criterion.

Equation (58d): The "migrated" field $\delta\hat{IP}(\mathbf{x})_n$ is here operated with the covariance operator incorporating a priori information. Here I have chosen the kernel Min(z,z′) corresponding to the hypothesis that real impedance sequences look like random walks. The sum here essentially corresponds to taking twice the primitive of $\delta\hat{IP}(x,y,z)_n$ with respect to z . The parameter K_p will control the trade-off between the importance of the a priori information and the information obtained from the data set.

Equation (58e): As usual in all gradient methods, some "preconditioning" may greatly speed convergence. At least, a preconditioning operator has to simulate the action of the inverse Hessian appearing in the Newton-like methods of optimization. Using physical intuition, operators can be defined that may be better than the inverse Hessian. The simplest operator in this example corresponds to a correction for spherical divergence of waves (Gauthier et al., 1986), i.e., to a multiplication by z^n , where $n \simeq 1$ for a 2D problem, and $n \simeq 2$ for a 3D problem.

Equation (58f): The new model $IP(\mathbf{x})_n$ is obtained here. An optimum value of α_n (for which the costs function is minimum) is obtained by trial and error.

Each iteration corresponds to a sort of generalized elastic "prestack" migration. Hopefully, a few iterations will suffice (if we do not wish to improve the long spatial wavelengths). Readers not interested in inversion, but only in prestack migration, may consider these equations as a serious candidate for replacing acoustic migration equations.

Optimization of the S-wave impedance

Turning now to S-wave impedance, denote by $\rho(\mathbf{x})_n$, $IP(\mathbf{x})_n$, and $IS(\mathbf{x})_n$ the model already obtained, which we wish to optimize further for the impedance $IS(\mathbf{x})$. The gradient iterative method will give models $IS(\mathbf{x})_{n+1}$, $IS(\mathbf{x})_{n+2}$, ...

Using the algorithm (4.93) in the text, and the results already obtained for the transpose operators, we obtain the equations corresponding to an iteration of the steepest descent method for the S-wave impedance:

$$\delta\hat{u}^i(\mathbf{x}_r,t;\mathbf{x}_s)_n = \frac{u^i(\mathbf{x}_r,t;\mathbf{x}_s)_n - u^i(\mathbf{x}_r,t;\mathbf{x}_s)_{obs}}{\sigma^2(\mathbf{x}_r,t;\mathbf{x}_s)} , \tag{58a}$$

$$\Psi^i(\mathbf{x},t;\mathbf{x}_s)_n = \sum_r \Gamma^{ij}(\mathbf{x},0;\mathbf{x}_r,t;\mathbf{x}_s)_n * \delta\hat{u}^j(\mathbf{x}_r,t;\mathbf{x}_s)_n , \tag{59b}$$

$$\delta\hat{IS}(\mathbf{x})_n = - 4\,\beta(\mathbf{x})_n$$

$$\times \sum_s \int_0^T dt \left(u^{km}(\mathbf{x},t;\mathbf{x}_s)_n \;\; \Psi^{km}(\mathbf{x},t;\mathbf{x}_s)_n - u^{ii}(\mathbf{x},t;\mathbf{x}_s)_n \;\; \Psi^{jj}(\mathbf{x},t;\mathbf{x}_s)_n \right) , \tag{59c}$$

$$\delta IS(x,y,z)_n = K_s \int_0^{Z_{max}} dz' \, Min(z,z') \, \delta\hat{IS}(x,y,z')_n , \tag{59d}$$

$$\Delta IS(\mathbf{x})_n = \mathbf{Precond}(\, \delta IS(\mathbf{x})_n \,) , \tag{59e}$$

$$IS(\mathbf{x})_{n+1} = IS(\mathbf{x})_n - a_n \left(\Delta IS(\mathbf{x})_n + IS(\mathbf{x})_n - IS(\mathbf{x})_{prior} \right) , \tag{59f}$$

where a_n is the real constant which renders $S(\mathbf{IP}_n , \mathbf{IS}_{n+1} , \boldsymbol{\rho}_n)$ minimum. The physical interpretation is as for the P-wave impedance.

Optimization of the density

Finally, turning to the density, denote by $\rho(\mathbf{x})_n$, $IP(\mathbf{x})_n$, and $IS(\mathbf{x})_n$ the model already obtained, which we wish to optimize further for the density $\rho(\mathbf{x})$. The gradient iterative method will give models $\rho(\mathbf{x})_{n+1}$, $\rho(\mathbf{x})_{n+2}$, ...

Using the algorithm (4.93) in the text, and the results already obtained for the transpose operators, we obtain the equations corresponding to an iteration of the steepest descent method for the density:

$$\delta\hat{u}^i(\mathbf{x}_r,t;\mathbf{x}_s)_n = \frac{u^i(\mathbf{x}_r,t;\mathbf{x}_s)_n - u^i(\mathbf{x}_r,t;\mathbf{x}_s)_{obs}}{\sigma^2(\mathbf{x}_r,t;\mathbf{x}_s)} , \tag{60a}$$

$$\Psi^i(\mathbf{x},t;x_s)_n = \sum_r \Gamma^{ij}(\mathbf{x},0;x_r,t;\mathbf{x}_s)_n * \delta\hat{u}^j(\mathbf{x}_r,t;x_s)_n \; , \tag{60b}$$

$$\begin{aligned}\delta\hat{\rho}(\mathbf{x})_n = \sum_s \int_0^T dt \Big\{ & \dot{u}^i(\mathbf{x},t;\mathbf{x}_s)_n \; \dot{\Psi}^i(\mathbf{x},t;\mathbf{x}_s)_n \\ & + (\alpha^2(\mathbf{x})_n - 2\,\beta^2(\mathbf{x})_n) \; u^{ii}(\mathbf{x},t;\mathbf{x}_s)_n \;\; \Psi^{jj}(\mathbf{x},t;x_s)_n \\ & + 2\,\beta^2(\mathbf{x})_n \; u^{km}(x,t;\mathbf{x}_s)_n \;\; \Psi^{km}(\mathbf{x},t;\mathbf{x}_s)_n \Big\} \; ,\end{aligned} \tag{60c}$$

$$\delta\rho(x,y,z)_n = K_\rho \int_0^{Z_{max}} dz' \; \mathrm{Min}(z,z') \; \delta\hat{\rho}(x,y,z')_n \; , \tag{60d}$$

$$\Delta\rho(\mathbf{x})_n = \mathbf{Precond}(\; \delta\rho(x)_n \;) \; , \tag{60e}$$

$$\rho(\mathbf{x})_{n+1} = \rho(x)_n - a_n \; (\; \Delta\rho(x)_n \; + \; \rho(x)_n \; - \; \rho(\mathbf{x})_{prior} \;) \; , \tag{60f}$$

where a_n is the real constant which remders $S(\mathbf{IP}_n \; , \; \mathbf{IS}_n \; , \; \rho_{n+1})$ minimum. The physical interpretation is as for the P-wave impedance.

Problem 7.4: Demonstrate the following identities (used in the previous problem):

$$\sum_r \int_0^T dt \; \frac{\partial\Gamma_0{}^{ij}}{\partial x^j}(\mathbf{x}_r,t;\mathbf{x},0) * u_0{}^{mm}(\mathbf{x},t) \;\; \delta\hat{u}^i(\mathbf{x}_r,t)$$

$$= \int_0^T dt \;\; u_0{}^{mm}(\mathbf{x},t) \;\; \Psi_0{}^{kk}(\mathbf{x}) \; , \tag{1}$$

$$\sum_r \int_0^T dt \; \frac{\partial\Gamma_0{}^{il}}{\partial x^m}(\mathbf{x}_r,t;\mathbf{x},0) * u_0{}^{lm}(\mathbf{x},t) \;\; \delta\hat{u}^i(\mathbf{x}_r,t)$$

$$= \int_0^T dt \;\; u_0{}^{lm}(x,t) \;\; \Psi_0{}^{lm}(\mathbf{x},t) \; , \tag{2}$$

and

$$\sum_r \int_0^T dt \quad \Gamma_0^{ij}(\mathbf{x}_r,t;x,0) * \ddot{u}_0^j(x,t) \quad \delta\hat{u}^i(\mathbf{x}_r,t)$$

$$= - \int_0^T dt \quad \dot{u}_0^i(\mathbf{x},t) \quad \dot{\Psi}_0^i(\mathbf{x},t) \ , \tag{3}$$

where $\Psi_0^i(\mathbf{x},t)$ is defined by

$$\Psi_0^i(\mathbf{x},t) = \sum_r \Gamma_0^{ij}(\mathbf{x},0;\mathbf{x}_r,t) * \delta\hat{u}^j(\mathbf{x}_r,t) \ , \tag{4}$$

and where $\Psi_0^{ij}(\mathbf{x},t)$ is the associated strain :

$$\Psi_0^{ij}(\mathbf{x},t) = \frac{1}{2} \left[\frac{\partial \Psi_0^i}{\partial x^j}(\mathbf{x},t) + \frac{\partial \Psi_0^j}{\partial x^i}(\mathbf{x},t) \right] . \tag{5}$$

Solution: From equations (4)-(5) we have

$$\Psi_0^{ik}(\mathbf{x},t) = \sum_r \frac{1}{2} \left[\frac{\partial \Gamma_0^{ij}}{\partial x^k}(x,0;\mathbf{x}_r,t) + \frac{\partial \Gamma_0^{kj}}{\partial x^i}(x,0;\mathbf{x}_r,t) \right] * \delta\hat{u}^j(\mathbf{x}_r,t) \ ) \ ,$$

and, using the reciprocity property of Green's function (see previous problem),

$$\Psi_0^{ik}(\mathbf{x},t) = \sum_r \frac{1}{2} \left[\frac{\partial \Gamma_0^{ji}}{\partial x^k}(\mathbf{x}_r,0;\mathbf{x},t) + \frac{\partial \Gamma_0^{jk}}{\partial x^i}(\mathbf{x}_r,0;\mathbf{x},t) \right] * \delta\hat{u}^j(\mathbf{x}_r,t) \ . \tag{6}$$

For the trace $\Psi_0^{kk}(\mathbf{x},t)$ this gives

$$\Psi_0^{kk}(\mathbf{x},t) = \sum_r \frac{\partial \Gamma_0^{jk}}{\partial x^k}(\mathbf{x}_r,0;\mathbf{x},t) * \delta\hat{u}^j(\mathbf{x}_r,t) \ . \tag{7}$$

For the time derivative $\dot{\Psi}_0^i(\mathbf{x},t)$ we have successively

$$\dot{\Psi}_0^i(\mathbf{x},t) = \frac{\partial}{\partial t} \sum_r \Gamma_0^{ij}(\mathbf{x},0;\mathbf{x}_r,t) * \delta\hat{u}^j(\mathbf{x}_r,t)$$

$$= \frac{\partial}{\partial t} \sum_r \int_0^T dt' \ \Gamma_0^{ij}(\mathbf{x},0;\mathbf{x}_r,t-t') * \delta\hat{u}^j(\mathbf{x}_r,t')$$

$$= \frac{\partial}{\partial t} \sum_r \int_0^T dt' \; \Gamma_0^{ij}(\mathbf{x},t'-t;\mathbf{x}_r,0) * \delta\hat{u}^j(\mathbf{x}_r,t')$$

$$= - \sum_r \int_0^T dt' \; \dot{\Gamma}_0^{ij}(\mathbf{x},t'-t;\mathbf{x}_r,0) * \delta\hat{u}^j(\mathbf{x}_r,t')$$

$$= - \sum_r \int_0^T dt' \; \dot{\Gamma}_0^{ij}(\mathbf{x},0;\mathbf{x}_r,t-t') * \delta\hat{u}^j(\mathbf{x}_r,t')$$

$$= - \sum_r \dot{\Gamma}_0^{ij}(x,0;\mathbf{x}_r,t) * \delta\hat{u}^j(\mathbf{x}_r,t) \; , \qquad (8)$$

and, using the reciprocity property,

$$\dot{\Psi}_0^{i}(\mathbf{x},t) = - \sum_r \dot{\Gamma}_0^{ji}(x_r,0;\mathbf{x},t) * \delta\hat{u}^j(\mathbf{x}_r,t) \; . \qquad (9)$$

We have successively

$$\sum_r \int_0^T dt \; \frac{\partial\Gamma_0^{ij}}{\partial x^j}(\mathbf{x}_r,t;\mathbf{x},0) * u_0^{mm}(x,t) \;\; \delta\hat{u}^i(\mathbf{x}_r,t) =$$

$$= \sum_r \int_0^T dt \int_0^T dt' \; \frac{\partial\Gamma_0^{ij}}{\partial x^j}(\mathbf{x}_r,t-t';\mathbf{x},0) \;\; u_0^{mm}(\mathbf{x},t') \;\; \delta\hat{u}^i(\mathbf{x}_r,t)$$

$$= \sum_r \int_0^T dt \int_0^T dt' \; \frac{\partial\Gamma_0^{ij}}{\partial x^j}(\mathbf{x}_r,0;x,t'-t) \;\; u_0^{mm}(\mathbf{x},t') \;\; \delta\hat{u}^i(\mathbf{x}_r,t)$$

$$= \sum_r \int_0^T dt' \; \frac{\partial\Gamma_0^{ij}}{\partial x^j}(\mathbf{x}_r,0;x,t') * \delta\hat{u}^i(\mathbf{x}_r,t') \;\; u_0^{mm}(\mathbf{x},t') \; , \qquad (10)$$

from where, using (7), Identity (1) follows.

We have successively

$$\sum_r \int_0^T dt \; \frac{\partial\Gamma_0^{il}}{\partial x^m}(\mathbf{x}_r,t;\mathbf{x},0) * u_0^{lm}(\mathbf{x},t) \;\; \delta\hat{u}^i(\mathbf{x}_r,t) =$$

$$= \sum_r \int_0^T dt \int_0^T dt' \; \frac{\partial \Gamma_0^{il}}{\partial x^m}(\mathbf{x}_r,t-t';\mathbf{x},0) \;\; u_0^{lm}(\mathbf{x},t') \;\; \delta\hat{u}^i(\mathbf{x}_r,t)$$

$$= \sum_r \int_0^T dt \int_0^T dt' \; \frac{\partial \Gamma_0^{il}}{\partial x^m}(\mathbf{x}_r,0;x,t'-t) \;\; u_0^{lm}(\mathbf{x},t') \;\; \delta\hat{u}^i(\mathbf{x}_r,t)$$

$$= \sum_r \int_0^T dt' \; \frac{\partial \Gamma_0^{il}}{\partial x^m}(\mathbf{x}_r,0;x,t') * \delta\hat{u}^i(\mathbf{x}_r,t) \;\; u_0^{lm}(\mathbf{x},t')$$

$$= \sum_r \int_0^T dt' \frac{1}{2}\left[\frac{\partial \Gamma_0^{il}}{\partial x^m}(\mathbf{x}_r,0;\mathbf{x},t') + \frac{\partial \Gamma_0^{im}}{\partial x^l}(\mathbf{x}_r,0;\mathbf{x},t') \right] * \delta\hat{u}^i(\mathbf{x}_r,t) \; u_0^{lm}(\mathbf{x},t'), \qquad (11)$$

the last equality being due to the symmetry of the strain tensor ($u_0^{lm}(\mathbf{x},t) = u_0^{ml}(\mathbf{x},t)$) . Using (6), Identity (2) follows.

For Identity (3), we need first to obtain an intermediate result. We have

$$\Gamma_0^{ij}(\mathbf{x}_r t,\mathbf{x},0) * \ddot{u}_0^j(\mathbf{x},t) \;=\; \int_0^T dt' \;\; \Gamma_0^{ij}(\mathbf{x}_r t-t',\mathbf{x},0) \;\; \ddot{u}_0^j(\mathbf{x},t') \; , \qquad (12)$$

and, integrating by parts,

$$\Gamma_0^{ij}(\mathbf{x}_r t,\mathbf{x},0) * \ddot{u}^j(x,t) \;=\; \Gamma_0^{ij}(\mathbf{x}_r,t-T;\mathbf{x},0) \; \dot{u}_0^j(\mathbf{x},T) \; - \; \Gamma_0^{ij}(\mathbf{x}_r,t;\mathbf{x},0) \; \dot{u}_0^j(\mathbf{x},0)$$
$$+ \int_0^T dt' \;\; \dot{\Gamma}_0^{ij}(\mathbf{x}_r t-t',\mathbf{x},0) \;\; \dot{u}_0^j(\mathbf{x},t') \; . \qquad (13)$$

As the initial conditions (equations (12c-d) of the previous problem) impose $\Gamma_0^{ij}(\mathbf{x}_r,t;\mathbf{x},t') = 0$ for $t < t'$, and $\dot{u}_0^j(\mathbf{x},0) = 0$, this gives

$$\Gamma_0^{ij}(\mathbf{x}_r t,\mathbf{x},0) * \ddot{u}^j(x,t) \;=\; \dot{\Gamma}_0^{ij}(\mathbf{x}_r t,\mathbf{x},0) * \dot{u}_0^j(\mathbf{x},t) \; . \qquad (14)$$

Now, using this last equality we obtain successively

$$\sum_r \int_0^T dt \;\; \Gamma_0^{ij}(\mathbf{x}_r,t;\mathbf{x},0) * \ddot{u}_0^j(\mathbf{x},t) \; \delta\hat{u}^i(\mathbf{x},t) \;=$$

$$= \sum_r \int_0^T dt \;\; \dot{\Gamma}_0^{ij}(\mathbf{x}_r,t;\mathbf{x},0) * \dot{u}_0^j(x,t) \; \delta\hat{u}^i(x,t)$$

$$= \sum_r \int_0^T dt \int_0^T dt' \; \dot{\Gamma}_0^{ij}(\mathbf{x}_r,t-t';x,0)\; \dot{u}_0^j(x,t')\; \delta\hat{u}^i(\mathbf{x}_r,t)$$

$$= \sum_r \int_0^T dt \int_0^T dt' \; \dot{\Gamma}_0^{ij}(\mathbf{x}_r,0;x,t'-t)\;\; \dot{u}_0^j(\mathbf{x},t')\; \delta\hat{u}^i(\mathbf{x}_r,t)$$

$$= \sum_r \int_0^T dt' \;\; \dot{\Gamma}_0^{ij}(\mathbf{x}_r,0;x,t') * \delta\hat{u}^i(x_{r,t'}) \;\; \dot{u}_0^j(\mathbf{x},t') \; , \tag{15}$$

from where, using (9), Identity (3) follows.

Problem 7.5: Figure 7.21 represents a borehole in which we are able to introduce a sensor. For three different depths z_1, z_2, z_3, we have measured the travel times t_1, t_2, t_3 of acoustic waves from the top of the borehole to the sensor. We obtain the following results

z_1 = 2.3 m $\qquad$ t_1 = 2.0 ms
z_2 = 8.1 m $\qquad$ t_2 = 8.1 ms
z_3 = 10.1 m $\qquad$ t_3 = 10.2 ms

We are interested in obtaining the celerity (velocity) $c(z)$ of the acoustic waves in the medium. Why is it better (for computational purposes) to parameterize the problem using the slowness $m(z) = 1/v(z)$ instead of the velocity? Use the Backus and Gilbert method for estimating the slowness $m(z)$. Represent the resolving kernel $R(z,z')$ for $z = 10$ m and $0 \le z' \le \infty$.

Solution: Let $\mathbf{m}$ represent a model of slowness (i.e., a particular function $z \rightarrow m(z)$), and $\mathbf{d}$ the column matrix of travel times predicted from the model $\mathbf{m}$. Formally

$$\mathbf{d} = \mathbf{G}\,\mathbf{m} \tag{1}$$

where $\mathbf{G}$ is the linear operator defined by

$$d^i = \int_0^{z_i} dz \;\; m(z) \; . \tag{2}$$

The kernels of the operator $\mathbf{G}$ are introduced by

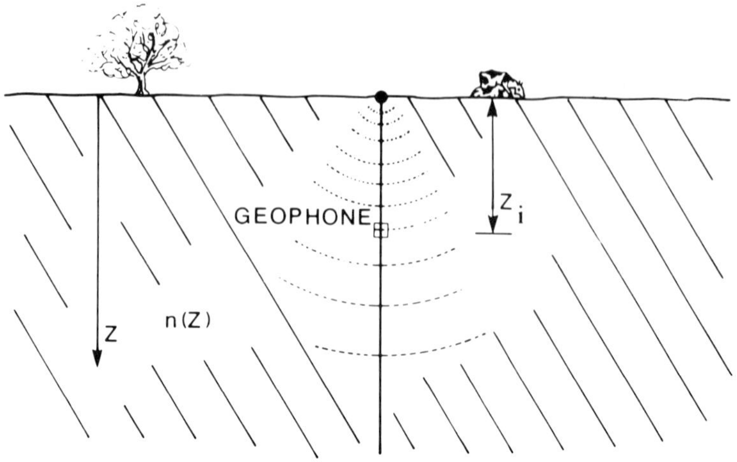

Figure 7.21.

$$d^i = \int_0^{\infty} dz \; G^i(z) \; m(z) \; . \tag{3}$$

This gives

$$G^1(z) = \begin{cases} 1 \text{ for } 0 \le z < z_1 \\ 0 \text{ for } z_1 \le z < \infty \; , \end{cases} \tag{4a}$$

$$G^2(z) = \begin{cases} 1 \text{ for } 0 \le z < z_2 \\ 0 \text{ for } z_2 \le z < \infty \; , \end{cases} \tag{4b}$$

$$G^3(z) = \begin{cases} 1 \text{ for } 0 \le z < z_3 \\ 0 \text{ for } z_3 \le z < \infty \; , \end{cases} \tag{4c}$$

Let $\mathbf{d}_{obs}$ be the column matrix of observed travel times

$$\mathbf{d}_{obs} = \begin{bmatrix} 2.0 \text{ ms} \\ 8.1 \text{ ms} \\ 10.2 \text{ ms} \end{bmatrix} . \tag{5}$$

The Backus and Gilbert solution of the problem of estimating the model is (equation (11′) of Box 7.2)

$$\mathbf{m} = \mathbf{G}^t \, (\mathbf{G} \, \mathbf{G}^t)^{-1} \, \mathbf{d}_{obs} = \mathbf{G}^t \, (\mathbf{S})^{-1} \, \mathbf{d}_{obs} . \tag{6}$$

where

$$\mathbf{S} = \mathbf{G} \, \mathbf{G}^t . \tag{7}$$

Explicitly,

$$m(z) = \sum_i G^i(z) \;\; \Psi^i , \tag{8}$$

where

$$\Psi^i = \sum_j (\mathbf{S}^{-1})^{ij} \;\; d_{obs}{}^j , \tag{9}$$

and

$$S^{ij} = \int_0^\infty dz \;\; G^i(z) \;\; G^j(z) . \tag{10}$$

Using (4) we obtain

$$\mathbf{S} = \begin{bmatrix} 2.3 & 2.3 & 2.3 \\ 2.3 & 8.1 & 8.1 \\ 2.3 & 8.1 & 10.1 \end{bmatrix} . \tag{11}$$

This gives

$$\mathbf{S}^{-1} = \frac{1}{26.68} \begin{bmatrix} 16.2 & -4.6 & 0 \\ -4.6 & 17.94 & -13.34 \\ 0 & -13.34 & 13.34 \end{bmatrix} . \tag{12}$$

From (5) and (9) we then obtain

$$\boldsymbol{\Psi} = \mathbf{S}^{-1} \, \mathbf{d}_{obs} = \begin{bmatrix} -0.182 \\ +0.002 \\ +1.050 \end{bmatrix} . \tag{13}$$

Equation (8) finally gives the Backus and Gilbert estimate:

$$m(z) = -0.182\ G^1(z) + 0.002\ G^2(z) + 1.050\ G^3(z)\ , \tag{14}$$

which is represented in Figure 7.22.

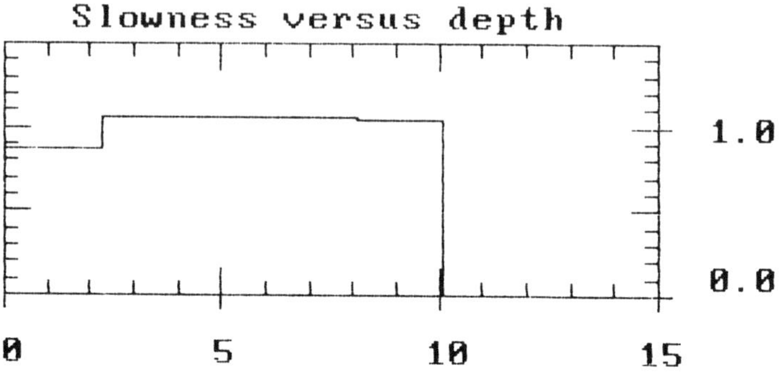

Figure 7.22.

The resolving operator is given by (equation (13′) of Box 7.2):

$$R = G^t\ (G\ G^t)^{-1}\ G = G^t\ S^{-1}\ G\ . \tag{15}$$

Explicitly, the resolving kernel is given by

$$R(z,z') = \sum_i \sum_j G^i(z)\ (S^{-1})^{ij}\ G^j(z')\ . \tag{16}$$

The terms of the problem ask for the values of $R(z,z')$ for $z = 10$ m . We have

for $0.0 \le z' < 2.3$: $R(11.0\ m\ ,\ z') = (S^{-1})^{31}+(S^{-1})^{32}+(S^{-1})^{33} = 0.0$

for $2.3 \le z' < 8.1$: $R(11.0\ m\ ,\ z') = (S^{-1})^{32}+(S^{-1})^{33} = 0.0$

for $8.1 \le z' < 10.1$: $R(11.0\ m\ ,\ z') = (S^{-1})^{33} = 0.5$

for $10.1 \le z' < \infty$: $R(11.0\ m\ ,\ z') = 0.0$.

The corresponding result is represented in Figure 7.23.

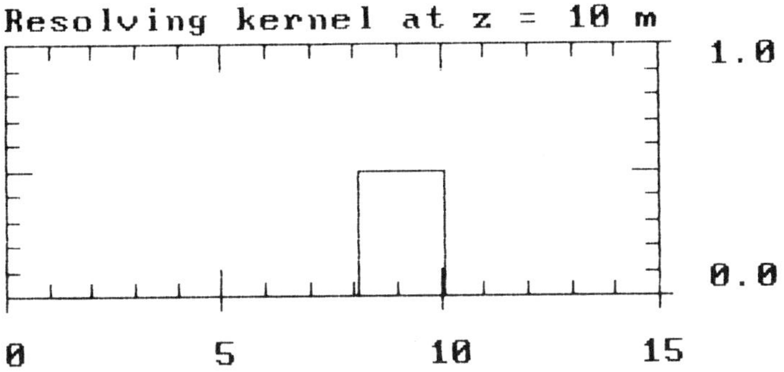

Figure 7.23.

Discussion: The solution shown in Figure 7.22 is "the simplest" solution predicting the observed travel times exactly.

The kernel shown in Figure 7.23 says that the value of slowness estimated at $z = 10$ m is the mean of the true value for $8.1 \text{ m} < z < 10.1 \text{ m}$. This can be physically understood: the values of the slowness for $z \leq 8.1$ m are fixed by the first and second observed travel times; it is the third travel time which gives information for $8.1 \text{ m} < z < 10.1 \text{ m}$, and it only gives information on the integrated slowness, i.e., on the mean value between $z = 8.1$ m and $z = 10.1$ m .

The method gives a null value for the slowness for $z > 10.1$ m . This is of course unphysical, but this value is totally unresolved. In fact, the essence of the Backus & Gilbert method is better obtained when the unknown is a "correction" to some current model: where there are no data, there is no correction.

Problem 7.6: In the Backus and Gilbert method the problem arises of obtaining the coefficients $Q^i(r_0)$ which minimize the expression

$$J(r_0) = \int dt \left[R(r_0,r) - \delta(r_0-r) \right]^2 , \tag{1}$$

where

$$R(r_0,r) = \sum_i Q^i(r_0) \, G^i(r) , \tag{2}$$

and where the $G^i(r)$ are given functions. Show that the coefficients $Q^i(r_0)$ are given by

$$Q^i(r_0) = \sum_j (S^{-1})^{ij} \, G^j(r_0) \, , \tag{3}$$

where

$$S^{ij} = \int dr \;\; G^i(r) \, G^j(r) \, . \tag{4}$$

Solution: Using the summation convention over repeated indexes, and the notation Q^i for $Q^i(r_0)$ gives

$$J(r_0) = \int dr \left[Q^i \, G^i(r) - \delta(r_0-r) \right]^2$$

$$= \int dr \;\; [\, Q^i \, G^i(r) \, Q^j \, G^j(r) - 2 \, Q^i \, G^i(r) \, \delta(r_0-r) + \delta^2(r_0-r) \,]$$

$$= Q^i \, S^{ij} \, Q^j - 2 \, Q^i \, G^i(r_0) + \delta(0) \, , \tag{5}$$

where

$$S^{ij} = \int dr \;\; G^i(r) \, G^j(r) \, , \tag{6}$$

and where the infinite value $\delta(0)$ can formally be handled as a constant. At the minimum of $J(r_0)$,

$$\frac{\partial J(r_0)}{\partial Q^i} = 0 \quad \Rightarrow \quad S^{ij} \, Q^j = G^i(r_0) \, , \tag{7}$$

whence the result follows (the matrix S^{ij} is regular if the $G^i(r)$ are linearly independent functions).

Problem 7.7: Let $C(t,t')$ be the covariance function considered in examples 7.4 and 7.11:

$$C(t,t') = \sigma^2 \; \exp\left(- \frac{|t-t'|}{T} \right) . \tag{1}$$

The covariance operator **C** corresponding to the integral kernel (1), associates any function $\hat{e}(t)$ with the function

$$e(t) = \int_{t_1}^{t_2} dt' \;\; C(t,t') \;\; \hat{e}(t') \qquad\qquad t \in [t_1,t_2] \,. \tag{2}$$

Obtain the inverse operator and the associated scalar product and norm.

Solution: Noticing that if

$$\phi(t) = \sigma^2 \;\exp\left(- \frac{|\,t\,|}{T} \right), \tag{3}$$

then

$$\frac{\partial \phi}{\partial t}(t) = - \frac{1}{T} \;\; sg(t) \;\; \phi(t) \tag{4}$$

and

$$\frac{\partial^2 \phi}{\partial t^2}(t) = \frac{1}{T^2} \;\phi(t) \; - \; \frac{2\sigma^2}{T} \;\; \delta(t) \;, \tag{5}$$

we easily obtain

$$\frac{\partial e}{\partial t}(t) = - \frac{1}{T} \int_{t_1}^{t_2} dt' \; sg(t-t') \;\; C(t,t') \;\; \hat{e}(t') \;, \tag{6}$$

and

$$\frac{\partial^2 e}{\partial t^2}(t) = \frac{1}{T^2} \; e(t) \; - \; \frac{2\sigma^2}{T} \; \hat{e}(t) \;. \tag{7}$$

Using (2), equation (6) shows that the values at $t = t_1$ and $t = t_2$ of $e(t)$ and $\partial e/\partial t \;(t)$ are not independent:

$$\frac{\partial e}{\partial t}(t_1) = \frac{1}{T} \; e(t_1) \qquad\qquad \frac{\partial e}{\partial t}(t_2) = - \frac{1}{T} \; e(t_2) \;. \tag{8}$$

Equation (7) then gives

$$\hat{e}(t) = \frac{1}{2\sigma^2 T} \; e(t) \; - \; \frac{T}{2\sigma^2} \; \frac{\partial^2 e}{\partial t^2}(t) \qquad\qquad t \in [t_1,t_2] \,. \tag{9}$$

We see thus that the domain of definition of the operator C^{-1} is the set of functions verifying the conditions (8). To any such function, the operator C^{-1} associates the function given by (9).

Using for C^{-1} the integral representation

$$\hat{e}(t) = \int_{t_1}^{t_2} dt' \; C^{-1}(t,t') \; e(t') \tag{10}$$

gives

$$C^{-1}(t,t') = \frac{1}{2\sigma^2} \left[\frac{1}{T} \delta(t-t') - T \, \delta^2(t-t') \right] , \tag{11}$$

where I have used the definition of the derivative of Dirac's Delta distribution:

$$\int dt' \; \delta^{(n)}(t-t') \; \mu(t') = (-1)^n \frac{d^n}{dt^n} \mu(t) \; . \tag{12}$$

The scalar product of two functions $\mathbf{e}_1$ and $\mathbf{e}_2$ may be defined by

$$(\mathbf{e}_1 , \mathbf{e}_2) = \mathbf{e}_1{}^t \, \mathbf{C}^{-1} \, \mathbf{e}_2 = \int_{t_1}^{t_2} dt \; e_1(t) \left(\frac{1}{T} e_2(t) - T \frac{\partial^2 e_2}{\partial t^2}(t) \right) . \tag{13}$$

Integrating by parts gives

$$(\mathbf{e}_1 , \mathbf{e}_2) = \frac{1}{T} \int_{t_1}^{t_2} dt \; e_1(t) \, e_2(t) + T \int_{t_1}^{t_2} dt \; \frac{\partial e_1}{\partial t}(t) \, \frac{\partial e_2}{\partial t}(t)$$

$$+ T \left[\frac{\partial e_1}{\partial t}(t) \, e_2(t) \right] \Bigg|_{t_1}^{t_2} . \tag{14}$$

Similarly,

$$(\mathbf{e}_2 , \mathbf{e}_1) = \frac{1}{T} \int_{t_1}^{t_2} dt \; e_2(t) \, e_1(t) + T \int_{t_1}^{t_2} dt \; \frac{\partial e_2}{\partial t}(t) \, \frac{\partial e_1}{\partial t}(t)$$

$$+ \; T \left[\frac{\partial e_2}{\partial t}(t) \; e_1(t) \right] \Bigg|_{t_1}^{t_2} . \tag{15}$$

and we see that the scalar product is symmetric

$$(e_1 \, , \, e_2) \;=\; (e_2 \, , \, e_1) \tag{16}$$

only if the functions e_1 and e_2 satisfy the dual boundary conditions.

$$\left[\frac{\partial e_1}{\partial t}(t) \; e_2(t) \; - \; e_1(t) \; \frac{\partial e_2}{\partial t}(t) \right] \Bigg|_{t_1}^{t_2} \;=\; 0 \; . \tag{17}$$

The norm of an element $e \in E$ can be computed by

$$\| \, e \, \|^2 \;=\; (e \, , \, e) \;=\; e^t \, C^{-1} \, e \;=\; \int_{t_1}^{t_2} dt \;\; \hat{e}(t) \;\; e(t) \; , \tag{18}$$

where

$$\hat{e} \;=\; C^{-1} \, e \; . \tag{19}$$

We have

$$\| \, e \, \|^2 \;=\; \int_{t_1}^{t_2} dt \; \frac{1}{2\sigma^2} \left(\frac{1}{T} \, e(t) \, - \, T \, \frac{\partial^2 e}{\partial t^2}(t) \right) e(t)$$

$$= \; \frac{1}{2\sigma^2} \left(\frac{1}{T} \int_{t_1}^{t_2} dt \;\; e(t)^2 \; - \; T \int_{t_1}^{t_2} dt \;\; \frac{\partial e^2}{\partial t^2}(t) \; e(t) \right) , \tag{20}$$

and, integrating by parts,

$$\| \, e \, \|^2 \;= \tag{21}$$

$$= \frac{1}{2\sigma^2} \left(\frac{1}{T} \int_{t_1}^{t_2} dt \; \left[e(t) \right]^2 + T \int_{t_1}^{t_2} dt \; \left[\frac{\partial e}{\partial t^2}(t) \right]^2 + T \; \frac{\partial e}{\partial t}(t_1) \; e(t_1) \; - \; T \; \frac{\partial e}{\partial t}(t_2) \; e(t_2) \right) ,$$

which, using equations (8), can be rewritten in either of the two following forms:

$$\| \mathbf{e} \|^2 = \tag{22}$$

$$= \frac{1}{2\sigma^2} \left(\frac{1}{T} \int_{t_1}^{t_2} dt \left[e(t) \right]^2 + T \int_{t_1}^{t_2} dt \left[\frac{\partial e}{\partial t}(t) \right]^2 + \left[e(t_1) \right]^2 + \left[e(t_2) \right]^2 \right) ,$$

$$\| \mathbf{e} \|^2 = \tag{23}$$

$$= \frac{1}{2\sigma^2} \left(\frac{1}{T} \int_{t_1}^{t_2} dt \left[e(t) \right]^2 + T \int_{t_1}^{t_2} dt \left[\frac{\partial e}{\partial t}(t) \right]^2 + T^2 \left[\frac{\partial e}{\partial t}(t_1) \right]^2 + T^2 \left[\frac{\partial e}{\partial t}(t_2) \right]^2 \right) .$$

As these expressions are sums of squares, and vanish only for a null function, we verify a posteriori that the covariance operator defined by the covariance function $C(t,t')$ is a positive definite operator.

In many applications, the first two terms in (22) and (23) are approximately proportional to $t_2 - t_1$, and, as usually $t_2 - t_1 >> T$, the last two terms can be dropped, thus giving

$$\| \mathbf{e} \|^2 \simeq \frac{1}{2\sigma^2} \left(\frac{1}{T} \int_{t_1}^{t_2} dt \left[e(t) \right]^2 + T \int_{t_1}^{t_2} dt \left[\frac{\partial e}{\partial t}(t) \right]^2 \right) . \tag{24}$$

This corresponds to the usual norm in the Sobolev space H^1 (see Appendix 7.2), which is the sum of the usual L_2 norm of the function and of the L_2 norm of its derivative.

Problem 7.8: For $0 \leq t \leq T$, let $C(t,t')$ be the covariance function

$$C(t,t') = \beta \, \mathrm{Min}(t,t') . \tag{1}$$

Notice that the variance at the point t is $\sigma^2 = C(t,t) = \beta \, t$. With any function $\hat{e}(t)$, the covariance operator $\mathbf{C}$, whose kernel is the covariance function (1), associates the function

$$e(t) = \int_0^T dt' \; C(t,t') \; \hat{e}(t') . \tag{2}$$

Obtain the inverse operator and the associated norm.

Solution: We have

$$e(t) = \beta \left[\int_0^t dt' \; t' \; \hat{e}(t') + t \int_t^T dt' \; \hat{e}(t') \right] . \tag{3}$$

Using

$$\frac{\partial}{\partial t} \int_t^T dt' \; f(t') = - f(t) \tag{4}$$

gives

$$\frac{\partial e}{\partial t}(t) = \beta \int_t^T dt' \; \hat{e}(t') , \tag{5}$$

and

$$\frac{\partial^2 e}{\partial t^2}(t) = - \beta \; \hat{e}(t) . \tag{6}$$

Equation (3) gives the condition

$$e(0) = 0 , \tag{7a}$$

while (5) gives the condition

$$\frac{\partial e}{\partial t}(T) = 0 . \tag{7b}$$

Equation (6) gives

$$\hat{e}(t) = - \frac{1}{\beta} \frac{\partial^2 e}{\partial t^2}(t) . \tag{8}$$

The domain of definition of the operator $\mathbf{C}^{-1}$ is the set of functions verifying the conditions (7). Any such a function is associated by the operator $\mathbf{C}^{-1}$ with the function given by (8).

A formal introduction of the integral kernel of $\mathbf{C}^{-1}$ gives

$$\hat{e}(t) = \int_0^T dt' \; C^{-1}(t,t') \; e(t') , \tag{9}$$

and, by comparison with (8), we directly obtain

$$C^{-1}(t,t') = -\frac{1}{\beta}\,\delta^2(t-t') \,. \tag{10}$$

The norm of an element can be computed by

$$\|\, \mathbf{e} \,\|^2 = \mathbf{e}^t\, \mathbf{C}^{-1}\, \mathbf{e} = \langle\, \hat{\mathbf{e}} \,,\, \mathbf{e} \,\rangle = \int_0^T dt\ \hat{e}(t)\ e(t) \,, \tag{11}$$

where

$$\hat{\mathbf{e}} = \mathbf{C}^{-1}\, \mathbf{e} \,. \tag{12}$$

We have

$$\|\, \mathbf{e} \,\|^2 = -\frac{1}{\beta} \int_0^T dt\ e(t)\ \frac{\partial^2 e}{\partial t^2}(t) \,, \tag{13}$$

and, integrating by parts,

$$\|\, \mathbf{e} \,\|^2 = \frac{1}{\beta} \left(\int_0^T dt \left[\frac{\partial e}{\partial t}(t) \right]^2 - e(t)\ \frac{\partial e}{\partial t}(t) \Bigg|_0^T \right) . \tag{14}$$

Using conditions (7) gives the final result:

$$\|\, \mathbf{e} \,\|^2 = \frac{1}{\beta} \int_0^T dt \left[\frac{\partial e}{\partial t}(t) \right]^2 . \tag{15}$$

In particular, this demonstrates that $C(t,t')$ is a positive definite function (because $\|\, \mathbf{e} \,\|$ is nonnegative for any $\mathbf{e}$, if $\|\, \mathbf{e} \,\|$ is null, then $e(t)$ must be constant (almost everywhere), and then, it follows from (7) that the constant is necessarily zero).

The result (15) is interesting, because we see that a criterion of least norm associated with the covariance function (1) imposes that the *derivative* of the function is small (and not the function itself).

Notice that the condition (7a) (i.e., that the function will vanish for $t = 0$) could be predicted directly from the fact that the variance at $t = 0$ is null.

Problem 7.9: The exponential covariance function in 3D. Let $\mathbf{x}$ denote a point in the euclidean three-dimensional space , and let $C(\mathbf{x},\mathbf{x}')$ be the covariance function

$$C(\mathbf{x},\mathbf{x}') = \sigma^2 \exp\left(- \frac{\|\mathbf{x}-\mathbf{x}'\|}{L} \right) , \tag{1}$$

where $\|\mathbf{x}-\mathbf{x}'\|$ denotes the euclidean distance between points $\mathbf{x}$ and $\mathbf{x}'$. The corresponding covariance operator associates any function $\hat{\phi}(\mathbf{x})$ with the function

$$\phi(\mathbf{x}) = \int_V dV(\mathbf{x}') \; C(\mathbf{x},\mathbf{x}') \; \hat{\phi}(\mathbf{x}') , \tag{2}$$

Demonstrate that the inverse operator gives

$$\hat{\phi}(\mathbf{x}) \simeq \frac{1}{8\pi\sigma^2} \left(\frac{1}{L^3} \phi(\mathbf{x}) - \frac{2}{L} \Delta\phi(\mathbf{x}) + L \; \Delta\Delta\phi(\mathbf{x}) \right) . \tag{3}$$

The least-squares norm associated with the covariance function can be defined by

$$\| \phi \|^2 = \int_V dV(\mathbf{x}) \; \phi(\mathbf{x}) \, \hat{\phi}(\mathbf{x}) . \tag{4}$$

Demonstrate that this gives

$$\| \phi \|^2 \simeq \tag{5}$$

$$\simeq \frac{1}{8\pi\sigma^2} \left\{ \frac{1}{L^3} \int_V dV(\mathbf{x}) \; \left[\phi(\mathbf{x})\right]^2 + \frac{2}{L} \int_V dV(\mathbf{x}) \; \left[\mathbf{grad} \; \phi(\mathbf{x})\right]^2 \right.$$

$$\left. + L \int_V dV(\mathbf{x}) \; \left[\Delta\phi(\mathbf{x})\right]^2 \right\} .$$

Solution: (from G. Jobert, personal communication). First, we have to solve the following equation for $\hat{\phi}(\mathbf{x})$

$$\phi(\mathbf{x}) = \sigma^2 \int_V dV(\mathbf{x}') \exp\left(-\frac{\|\mathbf{x}-\mathbf{x}'\|}{L}\right) \hat{\phi}(\mathbf{x}') . \tag{6}$$

Let

$$g(\mathbf{x}) = \sigma^2 \exp\left(-\frac{\|\mathbf{x}\|}{L}\right) \tag{7}$$

and let $\Phi(\mathbf{k})$, $\hat{\Phi}(\mathbf{k})$, and $G(\mathbf{k})$ be the Fourier transforms of $\phi(\mathbf{x})$, $\hat{\phi}(\mathbf{x})$, and $g(\mathbf{x})$ respectively. As equation (6) is clearly a spatial convolution

$$\phi(\mathbf{x}) = g(x) * \hat{\phi}(x) , \tag{8}$$

it becomes, in the Fourier domain,

$$\Phi(\mathbf{k}) = G(k) \hat{\Phi}(k) . \tag{9}$$

This gives

$$\hat{\Phi}(k) = H(\mathbf{k}) \Phi(k) , \tag{10}$$

where

$$H(\mathbf{k}) = \frac{1}{G(k)} . \tag{11}$$

Letting now $h(x)$ be the inverse Fourier transform of $H(k)$ gives

$$\hat{\phi}(\mathbf{x}) = h(\mathbf{x}) * \phi(\mathbf{x}) , \tag{12}$$

which formally solves the problem. The task now is to compute $G(\mathbf{k})$ and $h(\mathbf{x})$.

We have

$$g(\mathbf{k}) = \sigma^2 \int_V dV(\mathbf{x}) \; e^{-\|\mathbf{x}\|/L \, + \, i\mathbf{k}\cdot\mathbf{x}} , \tag{13}$$

and taking spherical coordinates with the polar axis colinear with $\mathbf{k}$,

$$G(k) = \sigma^2 \int_0^{\pi} d\theta \int_0^{2\pi} d\phi \int_0^{\infty} dr \; r^2 \sin\theta \; e^{-r/L \, + \, i \, \|k\| \, r \cos\theta}$$

$$= 2\pi\sigma^2 \int_{-1}^{+1} du \int_0^{\infty} dr\ r^2\ e^{-r/L\ +\ i\ ||\mathbf{k}||\ r\ u} \ , \tag{14}$$

where $u = \cos\theta$. As

$$\int_{-1}^{+1} du\ e^{2i\pi\ ||k||\ r\ u} = 2\ \frac{\sin\ r\ ||\mathbf{k}||}{r\ ||k||} \ , \tag{15}$$

we have

$$G(k) = 4\pi\sigma^2 \int_0^{\infty} dr\ r^2\ e^{-r/L}\ \frac{\sin(2\pi\ r\ ||\mathbf{k}||)}{r\ ||\mathbf{k}||}$$

$$= \frac{4\pi\sigma^2}{||k||} \int_0^{\infty} dr\ \ r\ \ \sin(r\ ||k||)\ \ e^{-r/L} \tag{16}$$

Using

$$I(p) = \int_0^{\infty} dt\ t\ \sin t\ e^{-pt} = \frac{2p}{(1+p^2)^2} \ , \tag{17}$$

with

$$p = \frac{1}{L\ ||\mathbf{k}||} \ , \tag{18}$$

and

$$t = r\ ||\mathbf{k}|| \ , \tag{19}$$

gives

$$G(k) = \frac{8\pi\sigma^2 L^3}{(1+L^2\ ||k||^2)^2} \ . \tag{20}$$

Then,

$$H(\mathbf{k}) = \frac{1}{8\pi\sigma^2} \left(\frac{1}{L^3} + \frac{2}{L}\ ||\mathbf{k}||^2 + L\ ||\mathbf{k}||^4 \right) . \tag{21}$$

As the Fourier transform of $\delta(\mathbf{x})$ is 1 , that of $\Delta\delta(\mathbf{x})$ is $-||\mathbf{k}||^2$, and that of $\Delta\Delta\delta(\mathbf{x})$ is $||\mathbf{k}||^4$, we directly obtain

$$h(\mathbf{x}) = \frac{1}{8\pi\sigma^2} \left(\frac{1}{L^3}\,\delta(x) - \frac{2}{L}\,\Delta\delta(x) + L\,\Delta\Delta\delta(\mathbf{x}) \right). \tag{22}$$

This gives

$$\hat{\phi}(\mathbf{x}) = \int_{\mathbf{V}} dV(x)\ h(\mathbf{x}-\mathbf{x}')\ \phi(\mathbf{x}')$$

$$= \frac{1}{8\pi\sigma^2}\left\{\frac{1}{L^3}\int_{\mathbf{V}} dV(x')\ \delta(x-x')\ \phi(\mathbf{x}') - \frac{2}{L}\int_{\mathbf{V}} dV(x')\ \Delta\delta(\mathbf{x}-\mathbf{x}')\ \phi(x')\right.$$

$$\left. + L\int_{\mathbf{V}} dV(x')\ \Delta\Delta\delta(x-x')\ \phi(\mathbf{x}')\right\}$$

$$= \frac{1}{8\pi\sigma^2}\left(\frac{1}{L^3}\,\phi(x) - \frac{2}{L}\,\Delta\phi(x) + L\,\Delta\Delta\phi(x) \right), \tag{23}$$

which demonstrates equation (3).

The (squared) norm of a function is then given by

$$\|\,\phi\,\|^2 = \int_{\mathbf{V}} dV(x)\ \phi(x)\ \hat{\phi}(x) \tag{24}$$

$$= \frac{1}{8\pi\sigma^2}\left\{\frac{1}{L^3}\int_{\mathbf{V}} dV(\mathbf{x})\ [\,\phi(\mathbf{x})\,]^2 - \frac{2}{L}\int_{\mathbf{V}} dV(x)\ \phi(\mathbf{x})\ \Delta\phi(\mathbf{x})\right.$$

$$\left. + L\int_{\mathbf{V}} dV(\mathbf{x})\ \phi(\mathbf{x})\ \Delta\Delta\phi(\mathbf{x})\right\}.$$

Using

$$-\,\phi\,\Delta\phi = \left[\,\mathbf{grad}\ \phi\,\right]^2 - \mathrm{div}(\,\phi\ \mathbf{grad}\ \phi\,)\,, \tag{25}$$

gives

$$-\int_{\mathbf{V}} dV(\mathbf{x})\ \phi(\mathbf{x})\ \Delta\phi(x) = \int_{\mathbf{V}} dV(x)\ \left[\mathbf{grad}\ \phi(\mathbf{x})\right]^2 - \int_{\mathbf{V}} dV(\mathbf{x})\ \mathrm{div}(\phi(\mathbf{x})\ \mathbf{grad}\ \phi(x))$$

$$= \int_{\mathbf{V}} dV(\mathbf{x}) \ \left[\mathbf{grad} \ \phi(\mathbf{x})\right]^2 - \int_{\partial\mathbf{V}} dS(\mathbf{x}) \ \phi(\mathbf{x}) \ \mathbf{n}(\mathbf{x}) \cdot \mathbf{grad}$$

$\phi(\mathbf{x})$,

where $\partial\mathbf{V}$ denotes the boundary of $\mathbf{V}$, $\mathbf{n}(\mathbf{x})$ the unit normal to the boundary, and $dS(\mathbf{x})$ the element of area on the boundary. Using

$$\phi \ \Delta\Delta\phi \ = \ - \ \mathbf{grad} \ \phi \cdot \mathbf{grad} \ \Delta\phi + \mathrm{div}(\ \phi \ \mathbf{grad} \ \Delta\phi \ )$$

$$= \ (\ \Delta\phi \ )^2 - \mathrm{div}(\ \Delta\phi \ \mathbf{grad} \ \phi \ ) + \mathrm{div}(\ \phi \ \mathbf{grad} \ \Delta\phi \ ) \ , \tag{27}$$

gives

$$\int_{\mathbf{V}} dV(\mathbf{x}) \ \phi(\mathbf{x}) \ \Delta\Delta\phi(\mathbf{x}) \ = \ \int_{\mathbf{V}} dV(\mathbf{x}) \ \left[\Delta\phi(\mathbf{x})\right]^2$$

$$+ \int_{\mathbf{V}} dV(\mathbf{x}) \ \mathrm{div}(\phi(\mathbf{x}) \ \mathbf{grad} \ \Delta\phi(\mathbf{x}))$$

$$- \int_{\mathbf{V}} dV(\mathbf{x}) \ \mathrm{div}(\Delta\phi(\mathbf{x}) \ \mathbf{grad} \ \phi(\mathbf{x}))$$

$$= \int_{\mathbf{V}} dV(\mathbf{x}) \ \left[\Delta\phi(\mathbf{x})\right]^2$$

$$+ \int_{\partial\mathbf{S}} dS(\mathbf{x}) \ \phi(\mathbf{x}) \ \ \mathbf{n}(\mathbf{x}) \cdot \mathbf{grad} \ \Delta\phi(\mathbf{x})$$

$$- \int_{\partial\mathbf{S}} dS(\mathbf{x}) \ \Delta\phi(\mathbf{x}) \ \mathbf{n}(\mathbf{x}) \cdot \mathbf{grad} \ \phi(\mathbf{x}) \ . \tag{28}$$

In most practical applications, the boundary terms can be dropped. This gives

$$\| \ \phi \ \|^2 \ \simeq \ \frac{1}{8\pi\sigma^2} \ \{ \frac{1}{L^3} \int_{\mathbf{V}} dV(\mathbf{x}) \ \left[\phi(\mathbf{x})\right]^2 + \frac{2}{L} \int_{\mathbf{V}} dV(\mathbf{x}) \ \left[\mathbf{grad} \ \phi(\mathbf{x})\right]^2$$

$$+ L \int_{\mathbf{V}} dV(\mathbf{x}) \left[\Delta\phi(\mathbf{x})\right]^2 \} , \tag{29}$$

which demonstrates equation (5).

REFERENCES AND REFERENCES FOR GENERAL READING

Abdelmadek, N., 1977. Computing the strict Chebyshev solution of overdetermined linear equations, Math. Comp., **31**, 974-983.

Abramowitz, M., and Stegun, I.A. (editors), 1970. Handbook of mathematical functions. Dover Publications Inc., New York.

Afifi, A.A., and Azen, S.P., 1979. Statistical Analysis: A computer oriented approach, Academic Press.

Aki, K., Chistofferson, A., and Husebye, E.S., 1977. Determination of the three-dimensional seismic structure of the lithosphere, J. Geophys. Res., **82**, 277-296.

Aki, K., and Lee, W.H.K., 1976. Determination of three-dimensional velocity anomalies under a seismic array using first P arrival times from local earthquakes. 1. A homogeneous initial model, J. Geophys. Res., **81**, 4381-4399.

Aki, K., and Richards, P.G., 1980. Quantitative seismology (2 volumes), Freeman and Co.

Alterman, Z.S., and Karal, F.C., Jr., 1968. Propagation of elastic waves in layered media by finite difference methods, Bull. Seismological Soc. of America, **58**, 367-398.

Anderssen, R.S., and Seneta, E., 1971. A simple statistical estimation procedure for Monte-Carlo inversion in Geophysics, Pageoph, **91**, 5-13.

Anderssen, R.S., and Seneta, E., 1972. A simple statistical estimation procedure for Monte-Carlo inversion in Geophysics. II: Efficiency and Hempel's paradox, Pageoph, **96**, 5-14.

Angelier, J., Tarantola, A., Valette, B., and Manoussis, S., 1982. Inversion of field data in fault tectonics to obtain the regional stress, Geophys. J. R. astr. Soc., 69, 607-621.

Armstrong, R.D., and Golfrey, J.P., 1979. Two linear programming algorithms for the linear L_1 norm problem, Math. Comp., **33**, 145, 289-300.

Avriel, M., 1976. Non linear programming: Analysis and methods, Prentice-Hall series in automatic computation, London.

Azlarov, T.A., and Volodin, N.A., 1982. Kharaterizatsionnye Zadachi, Sviazannye S Pokazatel'nym Raspredeleniem, Taschkent, Izdatel'stvo Fan Uzbekckoi CCR. English translation: Characterization problems associated with the exponential distribution, Springer-Verlag, 1986.

Backus, G., 1970. Inference from inadequate and inaccurate data: I, Proceedings of the National Academy of Sciences, **65**, 1, 1-105.

Backus, G., 1970. Inference from inadequate and inaccurate data: II, Proceedings of the National Academy of Sciences, **65**, 2, 281-287.

Backus, G., 1970. Inference from inadequate and inaccurate data: III,

Proceedings of the National Academy of Sciences, **67**, 1, 282-289.

Backus, G., and Gilbert, F., 1967. Numerical applications of a formalism for geophysical inverse problems, Geophys. J. R. astron. Soc., **13**, 247-276.

Backus, G., and Gilbert, F., 1968. The resolving power of gross Earth data, Geophys. J. R. astron. Soc., **16**, 169-205.

Backus, G., and Gilbert, F., 1970. Uniqueness in the inversion of inaccurate gross Earth data, Philos. Trans. R. Soc. London, **266**, 123-192.

Backus, G., 1971. Inference from inadequate and inaccurate data, Mathematical problems in the Geophysical Sciences: Lecture in applied mathematics, **14**, American Mathematical Society, Providence, Rhode Island.

Bakhvalov, N.S., 1977. Numerical Methods, Mir publishers, Moscow.

Balakrishnan, A.V., 1976. Applied functional analysis, Springer-Verlag.

Bamberger, A., Chavent, G, Hemon, Ch., and Lailly, P., 1982. Inversion of normal incidence seismograms, Geophysics, **47**, 757-770.

Barrodale I., 1970. On computing best L_1 approximations, in: Approximation theory, edited by A. Talbot, Academic Press.

Barrodale, I., and Phillips, C., 1975a. An improved algorithm for discrete Chebyshev linear approximation, in: Proceedings of the Fourth Manitoba Conference on Numerical Mathematics, edited by B.L. Hartnell and H.C. Williams, Utilitas Math. Pub. Co.

Barrodale, I., and Phillips, C., 1975b. Algorithm 495: Solution of an overdetermined system of linear equations in the Chebyshev norm, A.C.M. Trans. Math. Software, **1**, 264-270.

Barrodale, I., and Roberts, F.D.K., 1973. An improved algorithm for discrete L_1 linear approximation, SIAM J. Numer. Anal., **10**, 839-848.

Barrodale, I., and Roberts, F.D.K., 1974. Algorithm 478: Solution of an overdetermined system of equations in the ℓ_1 norm, CACM, 14, 319-320.

Barrodale, I. and Young, A., 1966. Algorithms for best L_1 and L_∞ linear approximations on a discrete set, Num. Math., **8**, 295-306.

Bartels, R.H., 1971. A stabilisation of the simplex method, Num. Math., **16**, 414-434.

Bartels, R.H., 1980. A penalty linear programming method using reduced gradient basis exchanges techniques, Lin. Alg. and its Appl., **29**, 17-32.

Bartels, R.H., and Conn, A.R., 1981. An approach to nonlinear ℓ_1 data fitting, Lecture Notes in Mathematics no. 909, in: Numerical Analysis, J.P. Hennart (ed.), Cocoyoc.

Bartels, R.H., Conn, A.R., and Sinclair, J.W., 1978. Minimisation techniques for piecewise differentiable functions: the ℓ_1 solution to an overdetermined linear system, SIAM J. Numer. Anal., **15**, 224-241.

Bartels, R.H., Conn, A.R., and Charalambous, C., 1978. On Cline's direct method for solving overdetermined linear systems in the ℓ_∞ sense, SIAM J. Num. Anal., **15**, 255-270.

Bartels, R.H., and Golub, G.H., 1969. The simplex method of linear programming using LU decomposition, Com. ACM, **12**, 206-268.

Bartels, R.H., Stoer, J., and Zenger, C.H., 1971. A realization of the simplex method based on triangular decomposition, in: Handbook for automatic computation, J.H. Wilkinson and C. Reinsch (eds.), Springer Verlag.

Bayer, R., and Cuer, M., 1981. Inversion tri-dimensionnelle des données aéromagnétiques sur le massif volcanique du Mont-Dore: Implications structurales et géothermiques, Ann. Géophys., *t.* 37, fasc. 2, 347-365.

Bayes, Thomas (Reverend), (1702-1761), 1958. Essay towards solving a problem in the doctrine of chances, 1763, republished in Biometrika, Vol. 45, 298-315.

Ben-Menahem, A., and Singh, S.J., 1981. Seismic waves and sources, Springer-Verlag.

Bender, C.M., and Orszag, S.A., 1978. Advanced mathematical methods for scientists and engineers, McGraw-Hill.

Berkhout, A.J., 1980. Seismic migration, imaging of acoustic energy by wave field extrapolation, Elsevier.

Beydoun, W., 1985. Asymptotic wave methods in heterogeneous media, Ph.D Thesis, Massachusetts Institute of Technology.

Bierman, G.J., 1977. Factorization methods for discrete sequential estimation, Academic Press.

Binder, K. (ed.), 1979. Monte Carlo methods in statistical physics, Springer-Verlag.

Binder, K., 1984. Applications of the Monte Carlo method in statistical physics, Springer-Verlag.

Bland, R.G., 1976. New finite pivoting rules for the simplex method, CORE-research paper 7612.

Bleistein, N., 1984. Mathematical methods for wave phenomena, Academic Press.

Bloomfield, P., and Steiger, W.L., 1980. Least absolute deviations curve-fitting, SIAM J. Scientific and Statistical Computing, 1, 290-301.

Bloomfield, P., and Steiger, W.L., 1983. Least absolute deviations, theory, applications, and algorithms, Birkhäuser, Boston.

Boothby, W.M., 1975. An introduction to differentiable manifolds and Riemannian Geometry, Academic Press.

Box, G.E.P., Leonard, T., and Chien-Fu Wu, (eds.), 1983. Scientific inference, data analysis, and robustness, Academic Press.

Box, G.E.P., and Tiao, G.C., 1973. Bayesian inference in statistical analysis, Addison-Wesley.

Bradley, S.P., Hax, A.C., and Magnanti, T. L., 1977. Applied mathematical programming, Addison-Wesley.

Brecis, H., 1983. Analyse fonctionnelle, théorie et applications, Masson, Paris.

Brillouin, L., 1962. Science and information theory, Academic Press.

Broyden, C.G., 1967. Quasi-Newton methods and their application to function minimization, Maths. Comput., **21**, 368-381.

Bunch, J.R., and Rose D.J. (eds.), 1976. Sparse matrix computations, Academic Press.

Campbell, S.L., Meyer, C.D., and Rosi, N.J., 1981. Results of an informal survey on the use of linear algebra in industry and government, Lin. Alg. and its Appl., **38**, 289-294.

Carroll, Lewis, (Rev. Charles Lutwidge Dodgson) (1832-1898), 1871. Alice's adventures in Wonderland, Macmillan. I recommend to the reader "The annotated Alice", by Martin Gardner, Penguin Books, 1970.

Cartan, H., 1967. Cours de calcul différentiel, Hermann, Paris.

Céa, J., 1971. Optimisation, théorie et algorithmes, Dunod, Paris.

Censor, Y., and Elfving, T., 1982. New methods for linear inequalities, Lin. Alg. and Appl., **42**, 199-211.

Censor, Y., and Herman, G.T., 1979. Row-generation methods for feasibility and optimization problems involving sparse matrices and their applications, in: Sparse matrix proceedings 1978, I.S. Duff and G.W. Stewart, eds., SIAM, Philadelphia, Pennsylvania.

Chapman, N.R., and Barrodale, I., 1983. Deconvolution of marine seismic data using L_1 norm, Geophys. J. Royal astr. Soc., **72**, 93-100.

Chernov, L.A., 1960. Wave propagation in a random medium, Dover Publications Inc., New York.

Ciarlet, P.G., 1982. Introduction à l'analyse numérique matrcielle et à l'optimisation, Masson, Paris.

Ciarlet, P.G., and Thomas, J.M., 1982. Exercices d'analyse numérique matricielle et d'optimisation, Masson, Paris.

Cimino, G., 1938. Calcolo approximato per le solusioni del sistemi di equazioni lineari, La Ricerca Scientifica, Roma XVI, Ser. II, Anno IX, Vol. 1, 326-333.

Claerbout, J.F., 1971. Toward a unified theory of reflector mapping, Geophysics, **36**, 467-481.

Claerbout, J.F., 1976. Fundamentals of Geophysical data processing, McGraw-Hill.

Claerbout, J.F., 1985. Imaging the Earth's interior, Blackwell.

Claerbout, J.F., and Muir, F., 1973. Robust modelling with erratic data, Geophysics, **38**, 5, 826-844.

Cottle, R.W., and Dantzig, G.B., 1968. Complementary pivot theory of mathematical programming, Lin. Anal. and its appl., **1**, 103-125.

Courant, R., and Hilbert, D., 1966. Methods of mathematical physics, Interscience Publishers, INC., New York.

Cuer, M., and Bayer, R., 1980a. A package of routines for linear inverse problems, Cahiers Mathématiques, Université des Sciences et Techniques du Languedoc, UER de Mathématiques, 34060 Montpellier, France.

Cuer, M., and Bayer, R., 1980b. Fortran routines for linear inverse problems, Geophysics, **45**, 1706-1779.

Cuer, M., 1984. Des questions bien posées dans des problèmes inverses de gravimétrie et géomagnétisme. Une nouvelle application de la programmation linéaire, Thesis, Université des Sciences et Techniques du Languedoc, UER de Mathématiques, 34060 Montpellier, France.

Dantzig, G.B., 1963. Linear programming and extensions, Princeton University Press, New York.

Dautray, R., and Lions, J.L., 1984 and 1985. Analyse mathématique et calcul numérique pour les sciences et les techniques (3 volumes), Masson, Paris.

Davidon, W.C., 1959. Variable metric method for minimization, AEC Res. and Dev., Report ANL-5990 (revised).

Davidon, W.C., 1968. Variance algorithms for minimization, Computer J., **10**, 406-410.

de Ghellinck, G., and Vial, J.Ph., 1985. An extension of Karmarkar's algorithm for solving a system of linear homogeneous equations on the simplex (preprint), ISSN, 0771-3894.

de Ghellinck, G., and Vial, J.Ph., 1986. A polynomial Newton method for linear programming (preprint), CORE, Université de Louvain.

Denel, J., Fiorot, J.C., and Huard, P., 1981. The steepest ascent method for linear programming problems, RAIRO Analyse Numérique, **15**, 3, 195-200.

Devaney, A.J., 1984. Geophysical diffraction tomography, IEEE trans. Geos. remote sensing, Vol. GE-22, No. 1.

Dickson, J.C., and Frederick, F.P., 1960. A decision rule for improved efficiency in solving linear problems with the simplex algorithm, Comm. ACM, **3**, 509-512.

Dixmier, J., 1969. Cours de mathématiques du premier cycle (2 volumes), Gauthier-Villars, Paris.

Dixon, L.C.W. (ed.), 1976. Optimization in action, Academic Press, London.

Draper, N, and Smith, H., 1981. Applied regression analysis, second edition, Wiley.

Dubes, R.C., 1968. The theory of applied probability, Prentice-Hall.

Duijndam, A.J.W., 1987. Detailed inversion of seismic data, Ph.D. Thesis, Delft University of Technology, 1987.

Ecker, J.C., and Kupferschmid, 1985. A computational comparison of the ellipsoid algorithm with several nonlinear programming algorithms, SIAM J. Control and Opt., Vol. 23, No. 5, 657-674.

Edgeworth, F.Y., 1887. A new method of reducing observations relating to several quatities, Phil. Mag. (5th. Ser.), **24**, 222-223.

Edgeworth, F.Y., 1888. On a new method of reducing observations relating to several quatities, Phil. Mag. (5th. Ser.), **25**, 184-191.

Ekblom, H., 1973. Calculation of linear best L_p approximations, BIT, **13**, 292-300.

Feynman, R.P., Leighton, R.B., and Sands, M., 1963. The Feynman lectures on Physics, Addison-Wesley, Reading, Mass.

Fiacco, A.V., and McCormick, G.P., 1968. Nonlinear programming, Wiley.

Fletcher, R., 1980. Practical methods of optimization, Volume 1: Unconstrained optimization, Wiley.

Fletcher, R., 1981. Practical methods of optimization, Volume 2: Constrained optimization, Wiley.

Fletcher, R., and Powell, M.J.D., 1963. A rapidly convergent descent method for minimization, Computer J., **6**, 163-168.

Fletcher R. and Reeves, C.M., 1964. Function minimization by conjugate gradients, The Computer Journal, **7**, 149-154.

Fox, L., 1964. An introduction to numerical linear algebra, Clarendon Press, Oxford.

Franklin, J.N., 1970. Well posed stochastic extensions of ill posed linear problems, J. Math. Anal. Applic., **31**, 682-716.

Gacs, P., and Lovasz, L., 1981. Khachiyan's algorithm for linear programming, in: Mathematical programming study, H. Konig, B. Korte, and K. Ritter (eds.), North-Holland.

Gale, D., 1960. The theory of linear economic models, McGraw-Hill.

Gass, S. I., 1975. Linear programming, methods and applications, fourth edition, McGraw-Hill.

Gauthier, O., Virieux, J., and Tarantola, A., 1986. Two-dimensional inversion of seismic waveforms: numerical results, Geophysics, **51**, 1387-1403.

Gauss, Carl Friedrich (1777-1855), 1809. Theoria Motus Corporum Coelestium.

Gel'fand I.M., and Levitan B.M., 1955. On the determination of a differential equation by its spectral function, Amer. Math. Soc. Transl., **1**, Ser.2, 253-304 (translated from the russian paper of 1951).

Geman, S., and Geman, D., 1984. Stochastic relaxation, Gibbs distributions, and the Bayesian restoration of images, IEEE Transactions on Pattern Analysis and Machine Intelligence, **PAMI-6**, 721-741.

Genet, J., 1976. Mesure et intégration, théorie élémentaire, Vuibert, Paris.

Gill, P.E., and Murray, W., 1973. A numerical stable form of the simplex algorithm, Lin. Alg. and its appl., **7**, 99-138.

Gill, P.E., Murray, W., Saunders, M.A., and Wright, M.H., 1984. Trends in nonlinear programming software, European journal of operational research, **17**, 141-149.

Gill, P.E., Murray, W., and Wright, M.H., 1981. Practical optimization, Academic Press.

Glashoff, K., and Gustafson, S.Å, 1983. Linear optimization and approximation, An introduction to the theoretical analysis and numerical treatment of semi-infinite programs, Springer-Verlag.

Goldfarb, D., 1976. Using the steepest edge simplex algorithm to solve linear programs, in: Sparse matrix computations, J.R. Bunch, and D.J. Rose (eds.), Academic Press.

Grasso, J.R., Cuer, M., and Pascal, G., 1983. Use of two inverse techniques. Application to a local structure in the New Hebrides island arc, Geophys. J. Royal astr. Soc., **75**, 437-472.

Gutenberg, B., and Richter, C.F., 1939. On seismic waves, G. Beitr., Vol. 54, 94-136.

Guyon, R., 1963. Calcul tensoriel, Vuibert, Paris.

Hacijan, L.G., 1979. A polynomial algorithm in linear programming, Soviet Math. Dokl., **20**, 1, 191-194.

Hadamard, J., 1902. Sur les problèmes aux dérivées partielles et leur signification physique, Bull. Univ. Princeton, **13**.

Hadamard, J., 1932. Le problème de Cauchy et les équations aux dérivées partielles linéaires hyperboliques, Hermann, Paris.

Hadley, G., 1965. Linear programming, Addison-Wesley.

Hampel, F.R., 1978. Modern trends in the theory of robustness, Math. Operationsforsch Statist., Ser. Statistics, **9**, 3, 425-442.

Hammersley, J.M., and Handscomb, D.C., 1964. Monte-Carlo methods, Chapman and Hall.

Hanson, R.J., and Wisnieski, J.A., 1979. A mathematical programming updating method using modified Givens transformations and applied to LP problems, Comm. ACM, **22**, 245-250.

Harris, P.M.J., 1975. Pivot selection methods of the Devex LP code, in: Mathematical Programming Study 4, M.L. Balinski and E. Hellerman (eds.), North-Holland.

Herman, G.T., 1980. Image reconstruction from projections, the fundamentals of computerized tomography, Academic Press.

Hirahara, K., 1986. Calculation of resolution matrix by an iterative tomographic inversion method: ARTB, Geophys. J. R. astr. Soc., (in press).

Hirn, A., Jobert, G., Wittlinger, G., Xu Zhong-Xin, and Gao En-Yuan, 1984. Main features of the upper lithosphere in the unit between the high Himalayas and the Yarlung Zangbo Jiang suture, Annales Geophysicæ, 2, 2, 113-118.

Ho, J.K., 1978. Princig for sparsity in the revised simplex method, RAIRO, 12, 3, 285-290.

Hoel, P.G., 1947. Introduction to mathematical statistics, Wiley.

Hofstadter, D., 1979. Gödel, Escher, Bach: an Eternal Golden Braid, Basic Books, Inc, Publishers, New York.

Huard, P., 1979. La méthode simplex sans inverse explicite, Bull. Dir. Etud. Etud. Rech. EDF, Série C, 79-98.

Huard, P., 1980. Compléments concernant la méthode des paramètres, Bull. Dir. Etud. Rech. EDF, Série Math. et Info., **2**, 63-67.

Huber, P.J., 1977. Robust statistical procedures, CBNS-NSF Regional Conference series in applied mathematics, SIAM, Philadelphia.

Huber, P.J., 1981. Robust statistics, Wiley.

Ikelle, L.T., Diet, J.P., and Tarantola, A., 1986. Linearized inversion of multi offset seismic reflection data in the ω-k domain, Geophysics, **51**, 1266-1276.

Jackson, D.D., 1972. Interpretation of inaccurate, insufficient and inconsistent data, Geophys. J. R. astr. Soc., **28**, 97-110.

Jackson, D.D., 1979. The use of a priori data to resolve nonuniqueness in linear inversion, Geophys. J. R. astr. Soc., **57**, 137-157.

Jackson, D.D., and Matsu'ura, M., 1985. A Bayesian approach to nonlinear inversion, Journal of Geophysical Research, Vol. 90, No. B1, 581-591.

Jaynes, E.T., 1968. Prior probabilities, I.E.E.E. Transactions on systems, science, and cybernetics, Vol. SSC-4, No. 3, 227-241.

Jeffreys, H., 1968. Theory of probability, Clarendon Press, Oxford, (third edition in 1983).

Jeffreys, H., 1957. Scientific Inference, Cambridge University Press, London.

Jeroslov, R.G., 1973. The simplex algorithm with the pivot rule of maximizing criterion improvement, Discrete Mathematics, **4**, 367-378.

Johnson, G.R., and Olhoeft, G.R., 1984. Density of Rocks and Minerals, in: CRC handbook of physical properties of rocks, edited by Carmichael, R.S., CRC, Boca Ratón, Florida.

Journel, A.G., and Huijbregts, Ch.J., 1978. Mining geostatistics, Academic Press.

Kantorovitch, L., and Akilov, G., 1977. Functional analysis, Vol. 1 and Vol. 2 (in russian), Nauka, Moscow.

Kagan, A.M., Linnink, Yu.,V., and Rao, C.R., 1973. Characterization problems in mathematical statistics, Wiley.

Karmarkar, N., 1984. A new polynomial-time algorithm for linear programming, Combinatorica, Vol. 4, No. 4, 373-395.

Kauffman, A., 1977. Introduction à la théorie des sous-ensembles flous, Masson, Paris.

Kennett, B.L.N., 1978. Some aspects of non-linearity in inversion, Geophys. J. R. astr. Soc., **55**, 373-391.

Keilis-Borok, V.J., 1971. The inverse Problem of Seismology, Proceedings of the International School of Physics Enrico Fermi, Course L, Mantle and Core in Planetary Physics, J. Coulomb and M. Caputo (editors), Academic Press, New York.

Keilis-Borok, V.J., Levshin, A., and Valus, V., 1968. S-wave velocities in the upper mantle in Europe, report on the IV International Symposium on Geophysical Theory and Computers 1967, Dokladi Akademii

Nauk SSSR 185, No.3, *p.* 564.
Keilis-Borok, V.J., and Yanovskaya, T.B., 1967. Inverse Problems of Seismology (Structural Review), Geophys. J. R. astr. Soc., **13**, 223-234.
Keller, J.B., 1978. Rays, waves, and asymptotics, Bull. AMS, **54**, 5, 727-750.
Kennett, B., and Nolet, G., 1978. Resolution analysis for discrete systems, Geophys. J. R. astr. Soc., **53**, 413-425.
Kirkpatrick, S., Gelatt, C.D., Jr., and Vecchi, M.P., 1983. Optimization by simulated annealing, Science, **220**, 671-680.
Klee V., and Minty, G.J., 1972. How good is the simplex algorithm?, in: Inequalities III, O. Shisha (ed.), Academic Press, New York.
Kolmogoroff, A.N., 1956. Foundations of the theory of probability, Chelsea, New York.
Kônig, H., and Pallaschke, D., 1981. On Krachian's algorithm and minimal ellipsoids, Numer. Math., **36**, 211-223.
Kuhn, H.W., and Quandt, R.E., 1962. An experimantal study of the simplex method, Proceed. 15th. symp. in Applied Mathematics of the American Mathematical Society, 107-124.
Kullback, S., 1959. Information theory and statistics, Wiley.
Kullback, S., 1967. The two concepts of information, J. Amer. Statist. Assoc., 62, 685-686.
Lanczos C., 1957. Applied Analysis, Prentice Hall.
Lang, S., 1962. Introduction to differentiable manifolds, Interscience Publishers, New York.
Laplace, Pierre Simon (Marquis de), (1749-1827), 1799. Mécanique céleste, Tome III, No. 39.
Lavrent'ev, M.M., Romanov, V.G., and Sisatskij, S.P., 1980. Nekorrektnye zadachi matematicheskoi fisiki i analisa (Ill posed problems in mathematical physics and analysis) (in russian), Nauka, Moscow. Translated in italian: Pubblicazioni dell'Istituto di Analisi Globale e Applicazioni, Firenze, 1983.
Lawson, Ch.L., and Hanson, R.J., 1974. Solving least squares problems, Prentice-Hall.
Levenberg, K., 1944. A method for the solution of certain nonlinear problems in least squares, Quart. Appl. Math., 2, 164-168.
Levitan, B.M., 1962. Generalized translation operators and their applications (in russian), Fiz. Mat. Gosudarstv. Izdat., Moscow.
Licknerowicz, A., 1960. Eléments de Calcul Tensoriel, Armand Collin, Paris.
Lions, J.L., 1968. Contrôle optimal de systèmes gouvernés par des équations aux dérivées partielles, Dunod, Paris. English translation: Optimal control of systems governed by partial differential equations, Springer, 1971.
Lions, J.L., 1974. Sur la Théorie du Contrôle, Actes du Congrès International des Mathématiciens, Vancouver, 2, 139-154.
Luenberger, D.G., 1969. Optimization methods by vector space methods,

Wiley.
Luenberger, D.G., 1973. Introduction to Linear and Nonlinear Programming, Addison-Wesley.
Magnanti, T.L., 1976. Optimization for sparse systems, in: Sparse matrix computations, J.R. Bunch, and D.J. Rose (eds.), Academic Press.
Mandelbaum, A., 1984. Linear estimators and measurable linear transformations on a Hilbert space, Z. Wahrscheinlichkeitstheorie verw. Gebiete, **65**, 385-397.
Mandelbrot, B.B., 1977a. Fractals: Form, Chance, and Dimension, Freeman.
Mandelbrot, B.B., 1977b. The Fractal geometry of Nature, Freeman.
Mangasarian, O.L., 1981. Iterative solution of linear programs, Siam J. Numer. Anal., **18**, 4, 606-614.
Mangasarian, O.L., 1983. Least norm linear programming solution as an unconstrained minimization problem, J. Math. Anal. and Appl., **92**, 240-251.
Marcenko, V.A., 1978. Sturm-Liouville operators and their applications (in russian), Naukova Dumka, Kiev.
Marquardt, D.W., 1963. An algorithm for least squares estimation of nonlinear parameters, SIAM J., **11**, 431-441.
Marquardt, D.W., 1970. Generalized inverses, ridge regression, biased linear estimation and non-linear estimation, Technometrics, **12**, 591-612.
Martz, H.F., and Waller, R.A., 1982. Bayesian reliability analysis, Wiley, New York.
Matsu'ura, M., and Hirata, N., 1982. Generalized least-squares solutions to quasi-linear inverse problems with a priori information, J. Phys. Earth, **30**, 451-468.
McCall E.H., 1982. Performance results of the simplex algorithm for a set of real world linear programming models, Comm. ACM, **25**, 3, 207-212.
McNutt, M.K., and Royden, L., 1986. Extremal bounds on geotherms in eroding mountain belts from metamorphic pressure-temperature conditions, Geophys. J. R. astr. Soc., (in press).
McNutt, M.K., 1986. Temperature beneath midplate swells: the inverse problem, submitted to Seamounts, Islands, and Atolls, Amer. Geophys. Union monograph.
Meissl, P., 1976. Hilbert spaces and their applications to geodetic least-squares problems, Boll. Geod. Sci. Affini, **35**, 49-80.
Menke, W., 1984. Geophysical data analysis: discrete inverse theory, Academic Press.
Meschkowsky, H., 1962. Hilbertsche räume mit kernfunction, Springer, Berlin.
Metropolis, N., Rosenbluth, A., Rosenbluth, M., Teller, A., and Teller, E., 1953. Equation of state calculations by fast computing machines, Journal of Chemical Physics, **21**, 1081-1092.
Miller, D. (ed.), 1985. Popper Selections, Princeton University Press.

Milne, R.D., 1980. Applied functional Analysis, Pitman Advanced Publishing Program, Boston.

Minster, J.B., and Jordan, T.M., 1978. Present-day plate motions, J. Geophys. Res., **83**, 5331-5354.

Minster, J.B., Jordan, T.H., Molnar, P., and Haines, E., 1974. Numerical modelling of instantaneous plate tectonics, Geophys. J. R. astr. Soc., 36, 541-576.

Misner, Ch.W., Thorne, K.S., and Wheeler, J.A., 1973. Gravitation, Freeman.

Morgan, B.W., 1968. An introduction to Bayesian statistical decision processes, Prentice-Hall.

Moritz, H., 1980. Advanced physical geodesy, Herbert Wichmann Verlag, Karlsruhe, Abacus Press, Tunbridge Wells, Kent.

Moritz, H., and Sünkel, H., 1978. Approximation methods in geodesy, H. Wichmann, Karlsruhe.

Morse, P.M., and Feshbach, H., 1953. Methods of theoretical physics, McGraw-Hill.

Müller, G., and Kind, R., 1976. Observed and computed seismogram sections for the whole Earth, Geophys. J. R. astr. Soc., **44**, 699-716.

Murtagh, B.A., and Sergent, R.W.H., 1969. A constrained minimization method with quadratic convergence, in: Optimization, edited by R. Fletcher, Academic Press.

Murty, K., 1976. Linear and combinational programming, Wiley.

Narasimhan, R., 1968. Analysis on real and complex manifolds, Masson, Paris, and North-Holland Publishing Company, Amsterdam.

Nazareth, L., 1984. Numerical behaviour of LP algorithms based upon the decomposition principle, Lin. Alg. and its Appl., **57**, 181-189.

Nercessian, Al., Hirn, Al., and Tarantola, Al., 1984. Three-dimensional seismic transmission prospecting of the Mont-Dore volcano, France, Geophys. J.R. astr. Soc., **76**, 307-315.

Nering E.D., 1970. Linear algebra and matrix theory, second edition, Wiley.

Nolet, G., 1981. Linearized inversion of (teleseismic) data, in: The solution of the inverse problem in Geophysical interpretation, R. Cassinis (ed.), Plenum Press, 9-37.

Nolet, G., 1985. Solving or resolving inadequate and noisy tomographic systems, J. Comp. Phys., **61**, 463-482.

Oppenheim, A.V., (ed.) 1978. Applications of digital signal processing, Prentice-Hall.

Oran Brigham, E., 1974. The fast Fourier transform, Prentice-Hall.

Parker, R.L., 1975. The theory of ideal bodies for gravity interpretation, Geophys. J. R. astron. Soc., 42, 315-334.

Parker, R.L., 1977. Understanding inverse theory, Ann. Rev. Earth Plan. Sci., 5, 35-64.

Parzen, E., 1959. Statistical inference on time series by Hilbert space methods, I, reprinted in Time Series Analysis Papers, Holden-Day, San

Francisco, 251-282.
Plackett, R.L., 1972. Studies in the history of probability and statistics; chapter 29, The discovery of the method of least squares, Biometrika, 59, 239-251.
Poincaré, H., 1929. La Science et l'Hypothèse, Flammarion, Paris.
Polack, E. et Ribière, G., 1969. Note sur la convergence de méthodes de directions conjuguées, Revue Fr. Inf. Rech. Oper., 16-R1, 35-43.
Powell, M.J.D., 1981. Approximation theory and methods, Cambridge University Press.
Press, F., 1968. Earth models obtained by Monte-Carlo inversion, J. Geophys. Res., Vol. 73, No.16, 5223-5234.
Press, F., 1971. An introduction to Earth structure and seismotectonics, Proceedings of the International School of Physics Enrico Fermi, Course L, Mantle and Core in Planetary Physics, J. Coulomb and M. Caputo (editors), Academic Press.
Press, W.H., Flannery, B.P., Teutolsky, S.A., and Vetterling, W.T., 1986. Numerical recipes: the art of scientific computing, Cambridge University Press. (See also Vetterling et al., 1986).
Price, W.L., 1977. A controlled random search procedure for global optimization, The computer Journal, Vol. 20, 367-370.
Pugachev, V.S., 1965. Theory of random functions and its application to control problems, Pergamon Press, Oxford, (translation of the original russian "Teoriya sluchainykh funktsii", 2nd edition, Fizmatgiz, Moscow, 1962).
Rand Corporation, 1955. A million random digits with 100,000 normal deviates, The Free Press, Glencoe, Ill.
Rao, C.R., 1973. Linear statistical inference and its applications, Wiley.
Ray Smith, C., and Grandy, W.T., Jr., (eds.), 1985. Maximum-entropy and Bayesian methods in Inverse Problems, Reidel.
Rebbi, C., 1984. Monte Carlo calculations in lattice gauge theories, in: Applications of the Monte Carlo method, K. Binder (ed.), Springer-Verlag, 277-298.
Reid, J.K., 1977. Sparse matrix, in: The state of the art in numerical analysis, D. Jacobs (ed.), Academic Press.
Richard, V., Bayer, R., and Cuer, M., 1984. An attempt to formulate well-posed questions in gravity: Application of linear inverse techniques to mining exploration, Geophysics, **49**, 1781-1793.
Richter, C.F., 1958. Elementary seismology, Freeman.
Rietsch, E., 1977. The maximum entropy approach to inverse problems, J. Geophys., 42, 489-506.
Roach, G.F., 1982. Green's functions, Cambridge University Press.
Robinson, E.A., 1981. Least squares regression analysis in terms of linear algebra, Goose Pond Press, Houston, Texas.
Rockafellar, R.T., 1974. Augmented Lagrange multiplier functions and dual-

ity in non convex programming, SIAM J. Control, **12**, 2, 268-285.

Robinson, E.A., and Treitel, S., 1980. Geophysical signal analysis, Prentice-Hall.

Rodgers, C.D., 1976. Retrieval of atmospheric temperature and composition from remote measurements of thermal radiation, Reviews of Geophysics and Space Physics, Vol. 14, No. 4, 609-624.

Rothman, D.H., 1985. Large near-surface anomalies, seismic reflection data, and simulated annealing (Ph. D. Thesis), Stanford University.

Rothman, D.H., 1985. Nonlinear inversion, statistical mechanics, and residual statics estimation, Geophysics, **50**, 2797-2807.

Rothman, D.H., 1986. Automatic estimation of large residual statics corrections, Geophysics, **51**, 332-346.

Ruffié, J., 1982. Traité du vivant, Fayard, Paris. (Now and then, Luck brings confusion in the biological order established by selection. It periodically shifts its too restrictive barriers, and allows natural evolution to change its course. Luck is anti-conservative.)

Sabatier, P.C., 1977a. On geophysical inverse problems and constraints, J. Geophys., **43**, 115-137.

Sabatier, P.C., 1977b. Positivity constraints in linear inverse problems: I) General theory, Geophys. J. R. astr. Soc., **48**, 415-441.

Sabatier, P.C., 1977c. Positivity constraints in linear inverse problems: II) Applications, Geophys. J. R. astr. Soc., **48**, 443-459.

Safon, C., Vasseur, G., and Cuer, M., 1977. Some applications of linear programming to the inverse gravity problem, Geophysics, **42**, 1215-1229.

Savage, L.J., 1954. The foundations of statistics, Wiley.

Savage, L.J., 1962. The foundations of statistical inference, Methuen, London.

Scales, L. E., 1985. Introduction to non-linear optimization, Macmillan.

Schmitt, S.A., 1969. Measuring uncertainty: an elementary introduction to Bayesian statistics, Addison-Wesley.

Schwartz, L., 1965. Méthodes mathématiques pour les sciences physiques, Hermann, Paris.

Schwartz, L., 1966. Théorie des distributions, Hermann, Paris.

Schwartz, L., 1970. Analyse (topologie générale et analyse fontionelle), Hermann, Paris.

Shannon, C.E., 1948. A mathematical theory of communication, Bell System Tech. J., 27, 379-423.

Snay, R.A., 1978. Applicability of array algebra, Reviews of Geophysics and Space Physics, Vol. 16, No. 3.

Sobczyk, K., 1985. Stochastic wave propagation, Elsevier, Amsterdam.

Spyropoulos, K., Kiountouzis, E., and Young, A., 1973. Discrete approximation in the L_1 norm, Comp. J., **16**, 180-186.

Tanimoto, A., 1985. The Backus-Gilbert approach to the three-dimensional structure in the upper mantle. - I. Lateral variation of surface wave phase velocity with its error and resolution., Geophys. J. R. astr. Soc., **82**, 105-123.

Tarantola, A., 1981. Essai d'une approche générale du problème inverse, Thèse de doctorat d'Etat, Université de Paris VI.

Tarantola, A., 1984. Linearized inversion of seismic reflection data, Geophysical Prospecting, 32, 998-1015.

Tarantola, A., 1984. Inversion of seismic reflection data in the acoustic approximation, Geophysics, **49**, 1259-1266.

Tarantola, A., 1984. The seismic reflection inverse problem, in: Inverse Problems of Acoustic and Elastic Waves, edited by: F. Santosa, Y.-H. Pao, W. Symes, and Ch. Holland, SIAM, Philadelphia.

Tarantola, A., 1986. A strategy for nonlinear elastic inversion of seismic reflection data, Geophysics, **51**, No. 10, 000-000 (in press).

Tarantola, A., 1987. Inversion of travel time and seismic waveforms, in: Seismic tomography, edited by G. Nolet, Reidel.

Tarantola, A., Jobert, G., Trézéguet, D., and Denelle, E., 1987. The use of depth extrapolation for the nonlinear inversion of seismic data, Submitted to Geophysics.

Tarantola, A. and Nercessian, A., 1984. Three-dimensional inversion without blocks, Geophys. J. R. astr. Soc., **76**, 299-306.

Tarantola, A., Ruegg, J.C., and Lépine, J.C., 1979. Geodetic evidence for rifting in Afar: A brittle-elastic model of the behaviour of the lithosphere, Earth and Planet. Sci. Letters, **45**, 435-444.

Tarantola, A., Ruegg, J.C., and Lépine, J.C., 1980. Geodetic evidence for rifting in Afar. 2: Vertical displacements, Earth and Planet. Sci. Letters, **48**, 363-370.

Tarantola, A., Trygvasson, E., and Nercessian, A., 1985. Volcanic or seismic prediction as an inverse problem, Annales Geophysicæ, 1, 6, 443-450.

Tarantola, A., and Valette, B., 1982a. Inverse Problems = Quest for Information, J. Geophys., **50**, 159-170.

Tarantola, A., and Valette, B., 1982b. Generalized nonlinear inverse problems solved using the least-squares criterion, Rev. Geophys. Space Phys., 20, No. 2, 219-232.

Taylor, A.E., and Lay, D.C., 1980. Introduction to functional analysis, Wiley.

Taylor, S.J., 1966. Introduction to measure and integration, Cambridge University Press.

Teo, K.L., and Wu, Z.S., 1984. Computational methods for optimizing distributed systems, Academic Press.

Tikhonov, A.N., 1963. Resolution of ill-posed problems and the regularization method (in russian), Dokl. Akad. Nauk SSSR, **151**, 501-504.

Tikhonov, A.N., and Arsenine, V., 1974. Methods of resolution of Ill-posed problems (in russian), Nauka, Moscow. French translation: Méthodes de résolution de problèmes mal posés, Mir, Moscow, 1976.

Tolla, P., 1984. Amélioration de la stabilité numérique d'algorithmes de résolution de programmes linéaires à matrices de contraintes clairsemées, RAIRO Recherche Opérationelle, **18**, 1, 19-42.

Tscherning, C.C., 1978. Introduction to functional analysis with a view to its applications in approximation theory, in: Approximation methods in geodesy, edited by H. Moritz and H. Sünkel, H. Wichmann, Karlsruhe.

Tukey, J.W., 1960. A survey of sampling from contaminated distributions, in: Contributions to Probability and Statistics I, Olkin (ed.), Stanford University Press, Stanford.

Tukey, J.W., 1962. Ann. Math. Stat., Vol. 33, page 13.

Tukey, J.W., 1965. Data analysis and the frontiers of geophysics, Science, Vol. 148, No. 3675, 1283-1289.

Twomey, S., 1977. Introduction to the mathematics of inversion in remote sensing and indirect measurements, Developments in Geomathematics 3, Elsevier Scientific Publishing Company, Amsterdam.

Van Campenhout, J.M., and Cover, T.M., 1981. Maximum entropy and conditional probability, IEEE transactions on information theory, Vol. IT-27, No. 4.

Vetterling, W.T., Teutolsky, S.A., Press, W.H., and Flannery, B.P., 1986. Numerical recipes: example book, Cambridge University Press. (See also Press et al., 1986).

Von Dam, W.B., and Tilanus, C.B., 1984. Mathematical programming in the Netherlands, Eur. J. of Oper. Res., **18**, 315-321.

Von Newmann, J., and Morgenstern, O., 1947. Theory of games and economic behaviour. 2nd. ed., Princeton University Press.

Walsh, G.R., 1975. Methods of optimization, Wiley.

Watson, G.A., 1980. Approximation theory and numerical methods, Wiley.

Wiggins, R.A., 1972. The general inverse problem: implication of surface waves and free oscillations for Earth structure, Rev. Geophys. Space Phys., **10**, 251-285.

Williamson, J.H., 1968. Least squares fitting of a straight line, Canadian J. of Physics, **46**, 1845-1848.

Winkler, R.L., 1972. Introduction to Bayesian inference & decision, Holt, Rinehart & Winstons, New York.

Wold, H., 1948. Random normal deviates, Tracts for computers 25, Cambridge University Press.

Wolfe, J.M., 1979. On the convergence of an algorithm for discrete L_p approximation, Num. Math., **32**, 439-459.

Woodhouse, J. H., and Dziewonski, A.M., 1984. Mapping the upper mantle: Three-dimensional modeling of Earth structure by inversion of seismic waveforms, Journal of Geophysical Research, Vol. 89, No. B7, 5953-5986.

York, 1969. Least squares fitting of a straight line with correlated errors,

Earth and Planetary Science Letters, **5**, 320-324.

INDEX

D

E

F

I

J

K

L

M

T